The Neurobiology of Australian Marsupials

Australian marsupials represent a parallel adaptive radiation to that seen among placental mammals. This great natural experiment has produced a striking array of mammals with structural and behavioural features echoing those seen among primates, rodents, carnivores, edentates and ungulates elsewhere in the world. Many of these adaptations involve profound evolutionary changes in the nervous system, and occurred in isolation from those unfolding among placental mammals. Ashwell provides the first comprehensive review of the scientific literature on the structure and function of the nervous system of Australian marsupials. The book also includes the first comprehensive delineated atlases of brain structure in a representative diprotodont marsupial (the tammar wallaby) and a representative polyprotodont marsupial (the stripe-faced dunnart). For those interested in brain development, the book also provides the first comprehensive delineated atlas of brain development in a diprotodont marsupial (the tammar wallaby) during the critical first four weeks of pouch life.

Ken W. S. Ashwell is Professor of Anatomy at the University of New South Wales, Australia, and has nearly 30 years' experience in teaching medical anatomy, neuroscience, comparative anatomy and anthropology. He has published extensively on the comparative neuroscience of living and recently extinct Australasian mammals and birds, with a particular focus on monotreme and marsupial neuroanatomy.

The Neurobiology of Australian Marsupials

Brain Evolution in the Other Mammalian Radiation

Editor and Principal Author

Ken W. S. Ashwell

School of Medical Sciences,
The University of New South Wales,
Australia

CAMBRIDGE UNIVERSITY PRESS

Cambridge, New York, Melbourne, Madrid, Cape Town, Singapore,
São Paulo, Delhi, Dubai, Tokyo, Mexico City

Cambridge University Press
The Edinburgh Building, Cambridge CB2 8RU, UK

Published in the United States of America by Cambridge University Press, New York

www.cambridge.org
Information on this title: www.cambridge.org/9780521519458

© Cambridge University Press 2010

First published 2010

Printed in the United Kingdom at the University Press, Cambridge

A catalogue record for this publication is available from the British Library

Library of Congress Cataloguing in Publication data
The neurobiology of Australian marsupials / [edited by] Ken Ashwell.
    p. ; cm.
  Includes bibliographical references and index.
  ISBN 978-0-521-51945-8 (hardback)
  1. Marsupials.   2. Neurobiology.   I. Ashwell, Ken W. S.   II. Title.
  [DNLM: 1. Marsupialia–anatomy & histology.   2. Anatomy, Comparative.
  3. Brain.   4. Evolution.   5. Marsupialia–physiology.   6. Nervous System. QL 737.M3]
  QL737.M3N48 2010
  573.8′192–dc22   2010027382

ISBN 978-0-521-51945-8 Hardback

Additional resources for this publication at www.cambridge.org/9780521519458

Dedicated to
Richard F. Mark
(1934–2003)
&
John R. Haight
(1938–)

Earlier in the evening I had been lying on a sunny bank and reflecting on the strange character of the Animals of this country as compared to the rest of the World. A Disbeliever in everything beyond his own reason, might exclaim, 'Surely two distinct Creators must have been (at) work; their object however has been the same and certainly in each case the end is complete.'

Charles Darwin, quoted in Frame (2009), p 57.

Marsupials represent the great alternative case of mammalian adaptive radiation, and when the same result happens in two such separate phylogenetic lines, we can begin to identify the determining factors in brain evolution.

J. I. Johnson (1977).

Contents

Contributors

Editor and principal author

Ken W. S. Ashwell,
Dept of Anatomy, School of Medical Sciences,
The University of New South Wales, NSW 2052, Australia.

Chapter contributors (in alphabetical order)

Lindsay M. Aitkin,
Bionic Ear Institute, 384–388 Albert Street, East Melbourne,
VIC 3002, Australia.

Catherine Arrese,
School of Animal Biology, University of Western Australia,
Nedland, WA 6009, Australia.

Lynda D. Beazley,
School of Animal Biology, University of Western Australia,
Nedland, WA 6009, Australia.

David M. Hunt,
UCL Institute of Ophthalmology, 11–43 Bath Street,
London, EC1V 9EL, UK.
and
School of Animal Biology, University of Western Australia,
35 Stirling Highway, Crawley, Perth, WA 6009, Australia.

Catherine A. Leamey,
Department of Physiology and Bosch Institute for Medical
Research, University of Sydney, NSW 2006, Australia.

Jürgen K. Mai,
Department of Neuroanatomy, Heinrich Heine University
of Düsseldorf, Düsseldorf, D40001, Germany.

Lauren R. Marotte,
Visual Sciences Group, Research School of Biology,
Australian National University, ACT 0200, Australia.

Bronwyn M. McAllan,
School of Medical Sciences, Discipline of Physiology
and Bosch Institute, Anderson Stuart Building (F13), The

University of Sydney, NSW 2006, Australia.

Samantha J. Richardson,
School of Medical Sciences, Building 201, RMIT University,
PO Box 71, Bundoora, VIC 3083, Australia.

Robert K. Shepherd,
Bionic Ear Institute, 384–388 Albert Street, East Melbourne,
VIC 3002, Australia.

Phil M. E. Waite,
Department of Anatomy, School of Medical Sciences, The
University of New South Wales, NSW 2052, Australia.

Preface

Marsupials in general, and Australian marsupials in particular, have much to offer in broadening our understanding of brain function. Both Charles Darwin and J. I. Johnson recognised that the native Australian mammals represent a 'second creation' or, to couch this in more modern terms, a parallel adaptive radiation. Australian marsupials have made a unique journey through the Tertiary, evolving nervous systems to deal with the special features of the changing Australian environment. For much (but not all) of that time, Australian marsupials were (apparently) completely isolated from placental mammals and many of the megafaunal species of the Pleistocene and recent times evolved large and complex brains. This raises many questions. Have marsupials used the same neural solutions to meet the demands of their environment as their distant placental cousins, or have they explored novel adaptations? How does the dramatically different developmental timetable of the marsupial lifestyle influence brain development and adult structure? Does the marsupial brain mediate sexual behaviour in the same way as the placental brain? These are some of the abiding questions of marsupial neurobiology.

The reader should note that, for the purposes of this work, the term Australian is used to include all those parts of Greater Australia to the east and south of the Wallace line (i.e. including New Guinea and its associated islands, mainland Australia and Tasmania). This corresponds to the continental land mass of Greater Australia (Sahul) as it was at the height of the last glaciation of the Pleistocene, when sea-levels were at their lowest and Australian marsupials could move relatively freely between the Australian mainland and surrounding islands.

The textual component of this book has been organised into four broad parts. The first part (Chapters 1 to 3) deals with general aspects of marsupial classification, evolution, brain organisation and brain development. The second

(Chapters 4 to 8) deals with the parts of the marsupial brain from a regional approach. The third (Chapters 9 to 13) includes chapters on important systems, while the fourth part (Chapters 14 and 15) is concerned with marsupials as research models. The book also includes three atlases. The first of these (Chapter 16) is a non-stereotaxic atlas of a representative polyprotodont marsupial (*Sminthopsis macroura*), since this species breeds well in captivity and may provide a useful model for studies of brain ageing. The second (Chapter 17) is a stereotaxic atlas of the brain of a representative diprotodont marsupial (the tammar wallaby, *Macropus eugenii*). Since the tammar is also used widely for developmental studies, the third atlas (Chapter 18) is a developmental series of five ages of the tammar pouch young. The atlases are provided in the hope that they will stimulate interest in Australian marsupial neurobiology and facilitate their study.

In these times of fierce competition for government grant money, marsupials are often seen as convenient models for solving human health problems and scientists interested in marsupial biology must often pitch their applications for research money with this in mind. There is no doubt that the special reproductive features and life cycles of Australian marsupials make them excellent models for studying problems relevant to human development and ageing, and, with this in mind, the book includes chapters on those aspects of marsupial neurobiology. On the other hand, I am certain that the neurobiology of Australian marsupials is worthy of study in its own right and will continue to fascinate scientists.

I have prepared this work with two broad audiences in mind. The first group would be students and researchers in zoology who are fully acquainted with the physiology, reproductive biology, evolution and palaeontology of marsupials, but who are interested in knowing more about the nervous systems of marsupials. The second would be neuroscientists who usually study placental mammals, but want to know more about what makes marsupials different. I have endeavoured to place the neurobiology of Australian marsupials in the context of their evolution, general and reproductive physiology. Naturally, as a neuroanatomist my perspectives

on these may occasionally suffer from naïvety and over-generalisation, and I beg the forgiveness of my zoologist colleagues should this be the case. I welcome any feedback from colleagues to improve future editions.

While I was preparing this book, a colleague asked me if I could have found a more obscure topic to write a book about. The natural answer to this is: 'Yes, monotreme neurobiology!' Unfortunately, marsupials (and monotremes) still suffer from prejudice and misinformation. In some quarters they continue to be seen as 'not quite right', 'primitive', 'evolutionary dead-end' mammals, whose only scientific value is to inform our thinking on how the 'advanced' mammals evolved. The quotations at the front of this book highlight another approach. As Darwin noted concerning the northern hemisphere and Gondwanan radiations, 'certainly in each case the end is complete.' The Gondwanan mammalian radiation is in no sense incomplete, inferior or deficient, simply intriguingly different.

To paraphrase David Attenborough's lines in the television documentary *Life on Earth*, I see Australian mammals as alternative solutions to the problem of staying alive. As such, study of the unique and convergent features of their nervous system anatomy and physiology has much to teach us about neurological solutions to organism survival. It is my enduring hope that this work will make some contribution to dispelling misconceptions about marsupials and stimulate further research into the nervous systems of these fascinating and beautiful creatures.

This field of study owes much to the work of many outstanding scientists in Australia and internationally. I am fortunate that some of these talented people have contributed to this work, but there are many who are no longer active in science for various reasons. Many of these scientific pioneers have been given due recognition in the body of this book, and I have made every effort to acknowledge their contribution in the text. In my opinion two names in recent times stand out for their foresight, scientific leadership and imagination. These are: Richard F. Mark and John R. Haight, to whom this book is dedicated. Australian science is much richer for their work.

Acknowledgements

Much of the atlas work within this book was done with the financial support of the Alexander von Humboldt Foundation, which has generously supported my collaborative work with Professor Jürgen Mai over more than a decade. I am very grateful to the Foundation for its vision and commitment to support the study of Australian marsupials. This continues a long tradition of involvement by German science in the study of marsupials and monotremes that dates back to Ziehen's pioneering work in the late nineteenth century.

Many esteemed colleagues, friends and family have helped make this work possible. Several talented histology technicians made the sections and immunohistochemical preparations illustrated in this book. In particular, I am very grateful to Mr Gavin McKenzie of the University of New South Wales and Frau Marieta Kazimirek, Frau Sabina Lensing-Höhn and Frau Ulla Lammersen of Heinrich Heine University, Düsseldorf.

I am indebted to Dr Sandy Ingleby for access to the marsupial collection of the Australian Museum, Sydney, and to Dr Wayne Longmore for use of the material at the Museum of Victoria.

I am particularly grateful to Professor Novotny of the Department of Anatomy in the Heinrich Heine University, Düsseldorf for permission to use the *.slide* photomicrographic system to photograph the sections of dunnart brain and wallaby pouch young. I am also grateful to Professors George Paxinos and Charles Watson for permission to use their abbreviation system for the atlases of adult dunnart and wallaby brain and pouch-young wallabies, albeit with some modifications. George also kindly allowed me to use the *SPOT* photomicrographic system and *Olympus Provis* microscope in his laboratory for photography of the adult wallaby brain sections. Bronwyn McAllan and Charles Watson provided comments on the diencephalon and cerebellum chapters, respectively.

Finally, I would like to thank my wife, Jennifer, who patiently tolerated her husband camping in the lounge room with a laptop and piles of books and papers while this book was being written.

Part I

Introduction and overview

Classification, evolution and behavioural ecology of Australian marsupials

K. W. S. Ashwell

Summary

Mammals are traditionally divided into the monotremes, marsupials and placentals, although this terminology is not entirely appropriate. Marsupial and placental mammals are usually considered to be more closely related to each other (as therians) than either is to the monotremes, but this has been questioned by the results of DNA hybridisation studies.

Marsupials have characteristic features of their physiology and anatomy, some of which are of particular significance for their neurobiology. These characteristics include: dentition; reproductive system embryology and anatomy; small size of the newborn and prolonged pouch dependency; control of reproductive organ differentiation; and metabolism.

There are seven living orders of marsupials and at least four extinct orders. Three of the seven modern orders live in the Americas (Didelphimorphia, Paucituberculata and Microbiotheria) and at least one of the Australian orders is extremely diverse (Diprotodontia). Three of the extinct orders are from South America, while the fourth is Australian.

Molecular data calibrated against the fossil record suggest that the diversification of the extant marsupial orders occurred between 65 and 50 million years ago (mya), before the fragmentation of Gondwana isolated the Australian and American marsupials. Diversity of Australian marsupials probably peaked in the early Miocene (about 23 to 16 mya) when conditions were warm and wet. Subsequent decline in rainfall has reduced the coverage of Australian forests and encouraged the spread of grasslands, thereby favouring the diversification of macropods, bandicoots and small dasyurids. New Guinea remained warm and wet and consequently is the current refuge of many marsupial species adapted to life in warm, moist and complex forest environments.

Recent Australian marsupials occupy niches ranging from cursorial predator (the thylacine), through the subterranean (marsupial mole), arboreal folivores and nectar feeders to grazers. In fact, when considered together with the monotremes, Australian marsupials reside in all the terrestrial niches occupied by placental mammals in the northern hemisphere. Some of the possums show striking specialisations, which are suggestive of convergence with primates (binocular vision, clawless expanded digit tips, specialised elongated digits like the aye-aye, prehensile tails like New World monkeys). The Macropodoidea are probably the most successful marsupial family of recent times and may be the only marsupial superfamily to have diversified exclusively in Australia. They are now represented by 19 species in New Guinea and 49 species in Australia (making up one-third of all marsupial species in Australia).

Introduction: what's in a name?

Marsupials were originally defined by de Blainville in 1816 on the basis of the anatomy of the reproductive tract (Tyndale-Biscoe, 2005). Female marsupials have two vaginae, two uteri and two oviducts, whereas there is only a single vagina, single cervix and single uterus in the reproductive tracts of placental or eutherian mammals. Only the uterine tubes and ovaries are paired in placental females. On the basis of these divisions, de Blainville named the marsupials *Di/delphia*, meaning two uteri, whereas all other mammals were called the *Mono/delphia*, meaning single uterus. Unfortunately, the platypus and echidna did not fit into this classification, since they have a reproductive tract superficially more like that of birds and reptiles, sharing a common opening with the digestive tract and urinary bladder. For these mammals, de Blainville coined the term *Ornitho/delphia* or bird uterus, to recognise these similarities. The broad division of mammals into three groups remains to the present day, even though the terms used by de Blainville are no longer suitable descriptors.

The Ornithodelphia of de Blainville are now known as the Monotremata, an allusion to the presence of a single opening (the cloaca) for reproductive, gastrointestinal and urinary systems in these mammals. The Didelphia are now known as the Marsupialia, a reference to the presence of a pouch, pocket or depression for prolonged postnatal support of dependent offspring. Finally, the Monodelphia are now referred to as the Placentalia, to acknowledge that the young of these mammals are supported for a significant proportion of gestation by the placenta – a specialised organ for gas, nutrient and waste exchange, and hormonal regulation. Of course, these terms are still unsatisfactory because the defining criteria are not exclusive (Tyndale-Biscoe, 2005). Several exceptions need to be recognised: (i) Monotremes and marsupials share a single opening for discharges from the reproductive and urinary tracts. In monotremes this is a true cloaca because it is a common opening for the gastrointestinal tract as well, whereas in marsupials the common opening is more correctly known as a urogenital sinus. (ii) The echidna may develop a pouch to protect and maintain hydration of the egg and during lactation after hatching. (iii) Not all female marsupials possess a pouch. (iv) All marsupials have a yolk sac placenta for at least some period during development (see Chapter 15) and one group, the bandicoots, have a complex allantoic placenta during development.

An alternative terminology is that proposed by Huxley (1880), who coined terms for modern mammals based on their supposed evolutionary sequence or successional stages. Therefore, the supposedly most primitive mammals, the monotremes, were referred to as the *Proto/theria*, or first mammals. The supposed next stage of mammalian evolution, the marsupials, were referred to as the *Meta/theria*, meaning middle or halfway mammals, and the allegedly most advanced mammals, the placentals, were named the *Eu/theria*, meaning good, or true, mammals. Once again, this classification is flawed, since the concept of evolutionary succession upon which it was based is no longer considered valid. Indeed it is likely that all three mammalian groups have evolutionary histories as old as each other and none of them can, or should be, considered primitive or ancestral to the other.

Nevertheless, despite the limitations of the two systems of nomenclature considered above, one needs to be able to use agreed terms for the living mammals when comparing and contrasting their nervous systems. In modern palaeontology, the term Metatheria is used to unite marsupials and their presumed extinct relatives (Deltatheroidea and Asiadelphia) (Rose, 2006). Similarly, Eutheria includes Placentalia and additional stem taxa that are closer to placentals than to marsupials (Rose, 2006). For the purposes of the present book, the modern echidnas and platypus will be referred to as the monotremes; Australian and American marsupials will be referred to as simply marsupials, or occasionally, metatheria; and all other mammals will be referred to as placentals, or eutheria. These terms are used while still recognising their limitations and, where Huxley's terminology is occasionally used, no phylogenetic sequence is implied.

The relationship between monotremes, marsupials and placentals

Before turning to the characteristics and classification of Australian marsupials, it is useful to consider the question of whether marsupials are more closely related to monotremes or placental mammals. This question is important when considering the structure and function of the nervous system and the relevance of brain traits to phylogeny. Usually the marsupial and placental mammals have been considered to be more closely related to each other than either is to the monotremes, because live-bearing of young is regarded as a shared, derived feature. For this reason, placentals and marsupials are usually collectively referred to as therians. Recent DNA hybridisation studies have suggested that there is a special marsupial–monotreme relationship, which would lead to grouping these mammal groups together as the Marsupionta (see discussion by Archer and Kirsch, 2006). This would require that viviparity evolved independently in marsupials and placentals and that the similarities in the nervous (and other) systems of marsupial and placental mammals are the result of convergence. Since the fossil history of monotremes dates to the early Cretaceous (115 to 100 mya), somewhat older than the earliest marsupials, a common ancestry would imply that marsupials evolved from monotremes (Archer and Kirsch, 2006). The highly specialised and unique nervous systems of all modern monotremes, compared to the generalised neuroanatomy of many modern marsupials, would tend to argue against this. Furthermore, other authorities consider that metatherians are more closely related to eutherians than any other mammals (Rose, 2006). While this question remains open, the term therian will continue to be used in this book, and marsupials will be considered to be more closely related to placental mammals than either is to the monotremes.

Anatomical and physiological characteristics of marsupials

Although the main focus of the current work is on the nervous system of Australian marsupials, it is useful to consider those general biological features, which may be broadly, or specifically, characteristic of modern marsupials. As seen above, some of these features may overlap with other mammals, but as a suite of features they are biologically useful for developing a concept of what a marsupial is.

Dentition and skull. A general characteristic of marsupials is that only the third premolars among the postcanine teeth are replaced during maturation. There are also more upper than lower incisors (Rose, 2006). Other marsupial dental features include metacones being equal to or larger than the paracones on the upper molars and hypoconulids being closer to the entoconids than the hypoconids on the lower molars (Archer and Kirsch, 2006). Marsupials have an auditory bulla composed primarily of the alisphenoid bone, large openings in the palate and an inflected angular process of the dentary (Rose, 2006).

Postcranial skeleton. Marsupials have epipubic bones, but these are also present in monotremes, multituberculates and basal eutherians (Rose, 2006).

Reproductive tract anatomy. As noted above, marsupials have paired vaginae, cervices, uteri and uterine tubes, whereas placental mammals have a single vagina, cervix and uterus. These anatomical differences arise from developmental dissimilarity in the migration of the ureters to meet the urinary bladder: outside and below the genital ducts in placentals; inside and above the genital ducts in marsupials (Tyndale-Biscoe, 2005). At birth, young marsupials pass through a newly formed canal in the tissues between the ureters (pseudovaginal or birth canal), direct from the lateral vaginae to the urogenital sinus (Tyndale-Biscoe, 2005).

Small birth size and pouch dependency. The 'unusual' anatomy of the marsupial reproductive tract considered above may be the reason for the birth of marsupial young at a relatively immature stage, since large newborns may not readily be passed down the pseudovaginal canal. Newborn marsupials range in body weight from 4 mg for the honey possum to 830 mg for large macropods. The newborn kangaroo is 0.003% of the maternal weight, compared to about 5% for a newborn rodent or primate. It is not true that the small size of the newborn marsupial is a consequence of a short gestation, since the gestational periods of several marsupials are longer than comparably sized placentals (Tyndale-Biscoe, 2005). Regardless of whatever the operative factor in determining the small size of marsupial newborns is, the bulk of marsupial growth occurs in a pouch, whereas placental young are nourished internally to larger birth size by an allantoic placenta. Placental females therefore make their major investment in their young during gestation, whereas marsupials are adapted to provide the major investment in their young during postnatal life. The small size and immature state of newborn marsupials has major implications for the timetabling of neural development (see Chapter 3).

Control of sexual differentiation. In placental mammals, sexual differentiation of the internal organs of the fetus takes place during gestation, when the external genitalia of both sexes are morphologically similar. The external genitalia of males subsequently develop under the influence of testosterone secreted by the testes. Both placental and marsupial males have a Y chromosome, with a sex-determining region (SRY gene), but the developing testes of marsupial males do not influence the differentiation of mammary glands, pouch or scrotum. In other words, scrotal bulges develop in genetic males and mammary glands and the pouch develop in genetic females many days before the gonads have differentiated. Furthermore, the maintenance of scrotum, mammary glands and pouch are not affected by levels of sex hormones in later life (Tyndale-Biscoe, 2005). Therefore, sexual differentiation of external reproductive organs in male marsupials is probably controlled by the intrinsic sex chromosome constitution of the cells themselves.

Metabolic rate. Marsupials maintain their body temperature at a relatively constant level, but their set point is below that normally seen for placental mammals. The average normal body temperature for marsupials is 35.5 °C, approximately 2.5 °C below that for most placentals (Hume, 1999; Tyndale-Biscoe, 2005). Any homeothermic vertebrate maintaining a core body temperature above the external environment must expend significant amounts of energy and that expenditure will be higher the greater the difference between core and external temperature, so a lower set point is less metabolically expensive. The basal or standard metabolic rate (BMR or SMR) is the energy required to maintain essential functions of the body at a constant body temperature under laboratory conditions and is usually determined by O_2 consumed or CO_2 produced under these circumstances. The average SMR for 56 marsupials is 204 kJ/kg0.75 per day, whereas SMR for a group of 272 placental mammals averages 289 kJ/kg0.75 per day (Tyndale-Biscoe, 2005). In other words, the average SMR for marsupials is 70.6% of the average value for placental mammals. The cellular basis for this may lie with the proportion of polyunsaturated and mono-unsaturated lipids in mitochondrial membranes, and the physiological consequences affect diverse aspects of metabolism from nerve conduction through respiration to growth rates (Tyndale-Biscoe, 2005). It should be noted that with regard to field metabolic rate (FMR), which is a measure of the actual energy expenditure in the natural environment, the difference between marsupial and placental mammals varies with body weight. In other words, small marsupials (10 to 20 g body weight) have FMR levels comparable to placental mammals of the same body weight, whereas large-bodied marsupials (10 to 40 kg) have values of FMR much less than placental mammals of similar body weight.

The above represents only a brief overview of the major physiological and structural differences between marsupial and placental mammals. Interested readers who wish to explore the above points in more detail are referred to several excellent books on general aspects of marsupial biology (Hume, 1999; Tyndale-Biscoe, 2005; Armati *et al.*, 2006).

Even the brief consideration of marsupial characteristics raises several important questions concerning the marsupial nervous system:

- How does birth at an early stage of development affect the time-course of maturation of neural systems? Are some systems advanced relative to others in the newborn marsupial? How does this influence adult neuroanatomy?
- How does the newborn marsupial navigate its climb from the urogenital sinus to the pouch and teat? How are the rhythmic movements of this climb controlled while the nervous system is in an immature state?
- How does the prolonged period of pouch life influence the maturation of neural systems? Are there disadvantages or advantages in forebrain maturation occurring after birth? Has the timetable of neural system maturation in marsupials been tailored to the timing of pouch life and seasonal cycles.
- How does the mode of development of marsupials influence the sexual differentiation of the brain? Are levels of circulating gonadal steroid more important than cellular chromosomal constitution in determining how sexual dimorphism develops in the brain?
- How are the reproductive cycles of marsupials controlled by the central nervous system? Are the same nuclei, transmitters and hormones controlling reproductive cycles in placental and marsupial mammals?
- How is the mating behaviour of marsupials controlled by neural systems? Are the same systems at play in placentals and marsupials? Are there sexually dimorphic structures in the marsupial brain concerned with control of mating behaviour?
- How do the profound metabolic differences between marsupials and placentals influence brain/body weight relationships? Marsupials exhibit a wider range in ratios of field to basal metabolic rate (i.e. FMR/BMR) than placentals (Koteja, 1991). Do differences in field metabolic rate and encephalisation influence competitive success in the natural environment?

Classification of marsupials

Marsupials include seven living (Figure 1.1) and at least four extinct orders, but some of these orders include a much broader range of lifestyles than would be found for placental orders. In terms of the number of species, modern marsupials are an order of magnitude less diverse than modern placentals, but marsupials encompass almost the same range of lifestyles as seen in placental mammals, with the exception of bats and cetaceans (Archer and Kirsch, 2006).

Four of the seven living marsupial orders live in Australia and New Guinea (Dasyuromorphia, Notoryctemorphia, Peramelemorphia, Diprotodontia) and three in the Americas (Paucituberculata, Didelphimorphia and Microbiotheria). Extinct orders include the carnivorous Sparassodontia, rodent-like Groeberidans and possibly hopping Argyrolagidians from South America; and the Yalkaparidontia (known colloquially as 'Thingodonta') from Australia (Archer and Kirsch, 2006).

The carnivorous species of Australian marsupials, the bandicoots and all the living American marsupials have four molars, three premolars, one prominent canine, four or five incisors in the upper jaw and three incisors in the lower jaw (Tyndale-Biscoe, 2005). On this basis they are grouped together as the Polyprotodontia, which means 'many front teeth'. On the other hand, herbivorous Australian marsupials have fewer premolars, small or absent canines, only one to three incisors in the upper jaw and a single large pair of incisors in the lower jaw, giving rise to the term Diprotodontia (two front teeth) for these marsupials (Tyndale-Biscoe, 2005).

The Dasyuromorphia includes three families of recent members (Dasyuridae, Myrmecobiidae and Thylacinidae). The Dasyurids are the most diverse and range in size from the tiny 3 g planigales to the 9 kg Tasmanian devil. Myrmicobiidae is represented by a single modern species, the numbat, which weighs about 500 g and feeds on termites in a relatively unspecialised way. The dog-like carnivorous Thylacinidae became extinct recently (1936) with the tragic and shameful hunting to extinction of the Tasmanian thylacine. Thylacines have been found in deposits as old as the late Oligocene and are represented by several extinct genera (Long *et al.*, 2002).

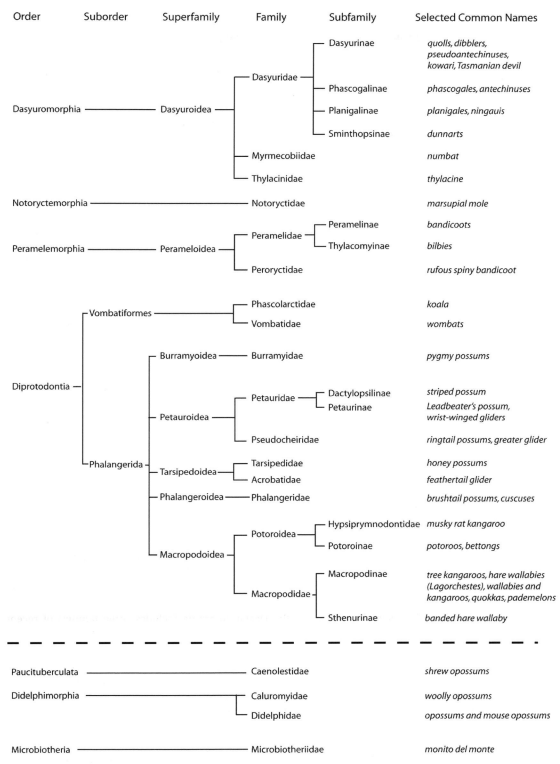

Figure 1.1 Diagram of the classification of the seven living orders of modern marsupial and their constituent suborders, superfamilies, families and subfamilies. The dashed line separates Australian orders (above) from American orders (below), but note that this subdivision may not be a phylogenetically significant division: Microbiotheria have been considered as members of a superorder within Australidelphia (Rose, 2006).

The Peramelemorphia (marsupial bandicoots) are small to rabbit-sized burrowing or terrestrial insectivores, omnivores and herbivores from Australia and New Guinea. There are two families of Peramelemorphia (Peramelidae – Australian bandicoots and bilbies; and the Peroryctidae – rainforest bandicoots from New Guinea and far northern Australia), although some authors consider the bilbies and Australian bandicoots as distinct families. There are also at least two extinct families and the group dates back to the Paleocene (family Yaralidae, Long *et al.*, 2002). On the other hand, recent analysis of nuclear genes has suggested that the living bandicoots are a relatively recent radiation originating in the later Oligocene or early Miocene with subsequent radiations in the late Miocene to early Pliocene (Meredith *et al.*, 2008a).

The order Notoryctemorphia contains a single genus (and species) of living insectivorous marsupial mole, which is morphologically convergent with placental golden moles. The relationship of this strange creature with other marsupials remains uncertain, but it is likely to be most closely related to dasyuromorphians. Fossil members of the order may date to the Miocene.

The Diprotodontia is the most diverse marsupial order, at least as assessed by the number of species. The order includes 10 living families and over 115 living species and is particularly disparate in terms of the many anatomically and ecologically different families, both living and extinct. Living members include the wombats and koalas, brush-tailed possums, honey possums, gliders, cuscuses, rat kangaroos, tree kangaroos and large macropods. The range of modern possums is particularly large, most of which are omnivorous or herbivorous. Extinct browsing macropods include many large-bodied members of the genus *Protemnodon*, *Procoptodon* and *Sthenurus* (Helgen *et al.*, 2006) and modern kangaroos exhibit a variety of anatomical adaptations to hopping (Chen *et al.*, 2005). Other important extinct members of the Diprotodontia include the Diprotodontidae, a family of widespread medium- to large-sized browsing quadrupeds from the late Miocene to Pleistocene (Long *et al.*, 2002; Price, 2008), which measured up to 3 m in length and 2 m in height; the Palorchestidae, a family of unusual marsupials with large robust forelimbs and small hindlimbs from the middle Miocene to Pleistocene (Long *et al.*, 2002), which may have had trunks; and the Thylacoleonidae, lion-sized carnivores with strikingly large blade-like third premolars, from the middle Miocene to the Pleistocene (Long *et al.*, 2002).

The Paucituberculata are a South American order represented in the modern world by the caenolestids. Extinct members are known from South America and Antarctica (Archer and Kirsch, 2006).

The Didelphimorphia, or opossums, are carnivorous, omnivorous or insectivorous and range in body weight from 20 g to several kg. Didelphimorphians are usually considered to be most like the earliest marsupials and two species (the North American or 'Virginia' opossum – *Didelphis virginiana* and the grey short-tailed opossum – *Monodelphis domestica*) have become standard laboratory species in the neurosciences. The problems with this are that: (1) both of these species are relatively generalised and therefore give no idea of the range of diversification and specialisation possible on the marsupial plan; (2) it is certain that modern Didelphimorphians were preceded by several forms (Archer and Kirsch, 2006) and therefore are not quite the original 'primitive' therian; and (3) the most studied (*Didelphis virginiana*) is probably the most poorly encephalised of any marsupial, contributing to the condescending perception of marsupials by some neuroscientists during the twentieth century.

The order Microbiotheria is represented by a single living species (*Dromiciops gliroides* – monito del monte), which has been considered an evolutionary link between South American and some of the Australian marsupials (Archer and Kirsch, 2006). The living *Dromiciops* is a small, semiarboreal marsupial found only in the wet forests of southern Chile and Argentina, but fossil members of the Microbiotheria have been found in Australia and Antarctica (Archer and Kirsch, 2006). Assessment of marsupial relationships based on a nuclear multigene molecular data set groups *Dromiciops* within Australidelphia and the sister taxon to a monophyletic Australasian clade (Meredith *et al.*, 2008b).

Origins of Australian marsupials

Molecular studies calibrated against fossil findings suggest that none of the living marsupial orders can be older than the end of the Cretaceous (about 65 mya) and probably all the diversification of marsupial orders was completed by 50 to 55 mya (Archer and Kirsch, 2006; Meredith *et al.*, 2008b). This appears to be more recent than the radiation of placental orders, which probably originated more than 90 mya, as assessed by their molecular trees (Archer and Kirsch, 2006). This might suggest that very few marsupials survived the end-Cretaceous extinction and that the group diversified within Gondwana before the break-up of that supercontinent. An overview of marsupial evolution supported by Archer and Kirsch (2006) is that marsupials first evolved in the northern continents before dispersing into South America towards the end of the Cretaceous. Several groups dispersed to Western Antarctica and some didelphimorphians and the microbiotheriids reached Australia between 70 and 55 mya. Exactly where and when the now characteristically Australian marsupial orders appeared remains unknown. Molecular data based on nuclear genes suggest that all intraordinal divergence within Australian marsupials occurred in the mid to late Cenozoic (Meredith *et al.*, 2008b).

The earliest South American marsupials come from early Paleocene (61 to 63 mya) deposits in Argentina and Bolivia, whereas the oldest known marsupials from Australia are found in the Tingamarra assemblage from south-eastern Queensland, dated to at least 55 mya (Long *et al.*, 2002) and the oldest known marsupials from Antarctica are dated to the middle Eocene (45 mya). Very little is known about marsupial evolution in Australia between the late Paleocene and late Oligocene (a period of 55 to 24 mya), but, by the early Miocene, Australian marsupials had reached extraordinary levels of disparity and diversity (Long *et al.*, 2002; Archer and Kirsch, 2006). Declines in rainfall from that time and the arrival of humans have contributed to an ongoing decline in marsupial diversity, although some groups (dasyuromorphians, peramelemorphians, macropods and wombats) have benefited from the reduction in forests and spread of grasslands (Archer and Kirsch, 2006).

The continuity of marsupial populations across Gondwana during the early Tertiary and the close relationship between some South American marsupials (e.g. *Dromiciops*) and modern Australian marsupials means that it is not phylogenetically valid to consider Australian and South American marsupials as separate groups. Nevertheless, the implication of marsupial phylogeny for comparative study of marsupial neurobiology is that modern Australian diprotodontia are a relatively advanced and special group. Furthermore, from the point of view of the neuroscience literature, study of marsupial neurobiology is either focussed on generalised didelphid species (e.g. *Didelphis virginiana* or *Monodelphis domestica*) or Australian species (often diprotodontia). For this reason, the terms Ameridelphian and Australidelphian are useful handles for summarising the comparative neuroscience literature, even if not phylogenetically valid.

Marsupials and the history of Australia and New Guinea

The evolution of Australian marsupials reflects the changing conditions on the Australian continent during the last 65 million years (Figure 1.2). As noted above, conditions during the early Tertiary were warmer and wetter than currently. During the Eocene much of the Australian continent, even the now arid centre, was covered by dense rainforests dominated by flowering plants. Species of *Nothofagus* were particularly common, but are now found only in isolated pockets of Greater Australia, in environments that are continuously warm and wet. So, when marsupials first entered Australia they began their radiation in dense, warm and wet rainforests. It is not surprising then that the initial diversification of marsupials was of forms adapted to complex forests and these are represented in the faunal assemblage of the Tingamurra site near Murgon in south-eastern Queensland, dated to 55 mya.

Unfortunately, after Tingamurra there is a 30 million year gap in the fossil record of Australian mammals (Archer *et al.*, 1999), and fossils are not available again until the mid Oligocene. Nevertheless, diversity of marsupials in those deposits is high and continues to rise into the early Miocene, when there are about 50% more marsupial families than in the Pleistocene (Archer *et al.*, 1999). Conditions began to cool and dry from the early Eocene, eventually giving rise to aridity in the interior of Australia. The *Nothofagus* forests were largely lost by the middle of the Miocene and began to be replaced by vegetation more suited to arid conditions (open and dry forest and woodland, shrublands and grasslands). The predominantly dry fruited Myrtaceae (including the modern genus *Eucalyptus*) became more widespread. These changed conditions were largely the result of the emergence of a circum-Antarctic current and the loss of heat exchange between the tropics and the Antarctic region. Antarctica began to freeze over and conditions in southern Australia became cooler. More importantly, because heat exchange with the tropics was reduced, a greater latitudinal variation in temperature developed and there was more differentiation of climate zones across the continent. These drier conditions were accompanied by more frequent burning of the vegetation, as indicated by increased charcoal levels in deposits.

The drying trend caused the extinction of the specialised rainforest marsupial groups with the result that marsupial diversity at the family level declined from the mid to late Miocene and never fully recovered (Archer *et al.*, 1999; Johnson, 2006). On the other hand, the drier conditions favoured the expansion of dry-country species. In the dasyuromorph marsupials, there was a surge of diversification that coincided with the expansion of dry vegetation and insect species across the continent. This led to the current abundance of small insectivorous and carnivorous species like dunnarts, planigales and phascogales.

The drier environments also led to opportunities for grazing forms like the kangaroos. The main trend in the diversification of kangaroos, diprotodonts and wombats during the Pliocene and Pleistocene was the accumulation of adaptations for feeding on tough abrasive vegetation of low

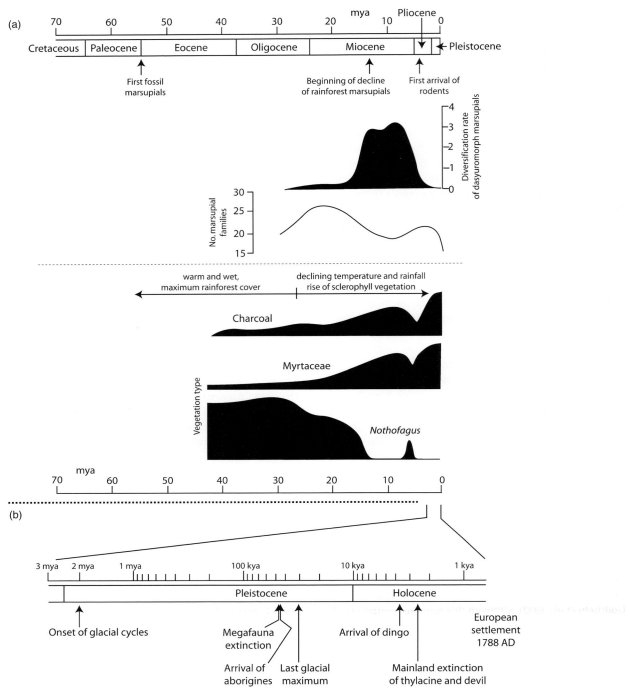

Figure 1.2 The history of Australia, its marsupial and placental species and the vegetation they rely on from the late Cretaceous to the present. This diagram draws on data cited by Johnson (2006) and Archer *et al.* (1999). The upper part of the diagram (a) is a linear timescale representing key events in mammal history in Australia correlated with changes in charcoal level and flora. The lower part of the diagram (b) is a non-linear scale representation of key recent events that have influenced the evolution and diversity of modern Australian marsupials. mya – millions of years ago; kya – thousands of years ago.

nutritional value. The increase in body size and species richness of these herbivores was matched by the evolution of large marsupial carnivores (thylacinids and thylacoleonids).

During the Pleistocene (from about 2.6 mya), global temperatures began to fluctuate in concert with the cycle of advance and retreat of the northern hemisphere ice sheets. In Australia, the consequences of these changes were oscillations between cool and dry conditions (during glacial periods) and warm and wet conditions (during the brief interglacials). These fluctuations in temperature were accompanied by changes in sea level (around 130 metres below current sea level), facilitating movement of marsupials between mainland Australia and Tasmania during the glaciations.

The history of New Guinea is also important to understanding modern marsupials, because some modern marsupial families (tree kangaroos and cuscuses) have their greatest diversity in that island. During the early Tertiary, New Guinea was underwater and there have been more recent periods of immersion, but mountain building in the last 15 million years has progressively elevated New Guinea and its associated islands. Significantly, New Guinea became an island of immense topographic diversity and retained its humid climate while the rest of Australia dried out. For this reason, New Guinea remains the refuge of marsupial species adapted to life in complex wet forests (Johnson, 2006). New Guinea was joined to continental Australia and Tasmania as part of Greater Australia (Sahul) during the periods of low sea level during the Pleistocene, but differences in habitat between New Guinea and Australia by that time meant that there was little exchange of species during that period.

A final point to be made concerns the other mammals present on the Australian continent, Tasmania and New Guinea (apart from marine mammals on the coast). When the first marsupials arrived in Australia during the early Tertiary, the continent was already occupied by monotremes, which were probably by then specialised for the niches they occupy today. There may also have been early placental mammals present in Australia at 55 mya (*Tingamara porterorum*, Godthelp *et al.*, 1992), although this remains controversial. Nevertheless, it is certain that as the Australian continent moved north during the mid to late Tertiary, a further opportunity for invasion by placental species was available. Bats were already present in Australia by the early Miocene and rodents arrived during the Pliocene. Therefore modern Australian marsupials clearly did not evolve in splendid isolation from placental competitors.

More significantly, two highly encephalised placental predators entered the Australian continent during the late Pleistocene and Holocene and their effects on the marsupial fauna of Australia have been profound. Humans crossed the Wallace line into New Guinea and thence Australia and Tasmania at least 40 000 years ago, before the last glacial maximum. Their arrival coincided with the extinction of the marsupial megafauna and has been considered to have been the major contributing factor, although argument continues on this (Johnson, 2006). Aboriginal contact with Asian peoples probably also led to the introduction of the dingo or warrigal, a descendant of the Asiatic wild dog, sometime around 3800 years ago. This may well have been the major contributing factor in the mainland extinction of the thylacine and devil.

Finally, recent history has seen the catastrophic effects of European settlement from 1788. The last 200 years have witnessed the rise of populations of introduced feral placental mammals (cat, fox, rabbit), which have been disastrous for Australia in general and Australian marsupials in particular.

Behavioural ecology of Australian marsupials

Australian marsupials show much greater diversity and disparity than their American cousins and this has profound implications for understanding the unique neurobiology of Australian marsupials.

Modern Australian marsupials occupy niches ranging from the subterranean (marsupial mole), through arboreal gliders to grazers. In fact, together with the monotremes, Australian marsupials inhabit all the terrestrial niches occupied by placental mammals in the northern hemisphere. Each of these niches requires particular adaptations and behaviours, which have implications for the neuroanatomy and neurophysiology of the species under consideration.

Some of these marsupials show intriguing specialisations, which suggest some convergence with primates (e.g. the elongated digits of the dactylopsilines – striped possums; the clawless expanded digit tips and prehensile tail of the honey possum; and the binocular vision of the feathertail glider).

In this part of the chapter we will be examining the varied niches occupied by Australian marsupials and considering how these niches might influence brain structure and function.

Marsupial carnivores: small, large and strange

Marsupial carnivores (Figure 1.3) range from tiny ningauis and planigales (Figure 1.3a, a´) to the recently extinct thylacine (Figure 1.3e, e´, f), and this broad group encompasses families with unusual reproductive cycles and dietary habits. Modern marsupial carnivores occupy a range of arid to

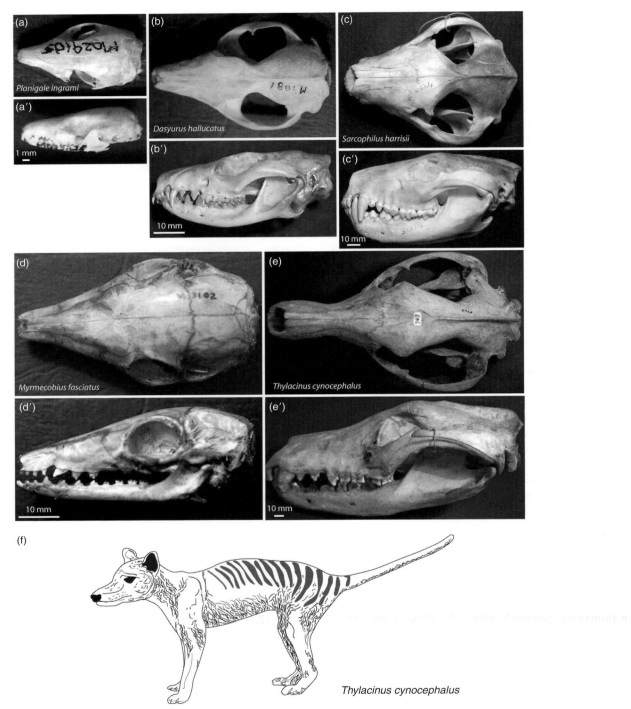

Figure 1.3 Dorsal and left lateral views of the skulls of Australian marsupial carnivores ranging from the tiny long-tailed planigale (4.2 g body weight; a, a′), through the northern quoll (400 to 900 g body weight; b, b′), Tasmanian devil (7 to 9 kg body weight; c, c′), the ant-eating numbat (300 to 700 g body weight; d, d′) to the recently extinct thylacine (Tasmanian 'tiger' or 'wolf'; 15 to 35 kg body weight; e, e′). Panel f is a line drawing of a young thylacine demonstrating the striping of the back and rump and semirigid tail. All skull specimens are from the collection of the Australian Museum, Sydney.

tropical habitats: most Australian marsupial carnivores are members of the family Dasyuridae, which has 47 species in mainland Australia and Tasmania and 17 species in New Guinea and associated islands (Tyndale-Biscoe, 2005). The three remaining recent species of this group occupy families of their own: the recently extinct thylacine (*Thylacinus cynocephalus*), the endangered ant-eating numbat (*Myrmecobius fasciatus*, Figure 1.3d, d´) and the bizarre marsupial mole (*Notoryctes typhlops*).

The small dasyurids, up to 160 g of body weight, eat mainly insects and other small invertebrates. They hunt during the night and appear to rely on a combination of olfaction and vision in finding their prey. Laboratory studies indicate that the visual acuity of fat-tailed dunnarts (*Sminthopsis crassicaudata*) is 2.3 cycles per degree compared to visual acuities of 0.5 and 1.2 cycles per degree for laboratory mice and rats, respectively (Arrese *et al.*, 1999; Bonney and Wynne, 2002, 2004). Furthermore, given their small absolute brain size, dunnarts show remarkable behavioural flexibility and exhibit rapid acquisition of learning set and reversal set, as well as efficient delayed spatial reversal learning (Bonney and Wynne, 2002). The mode of feeding of these small dasyurids is to use their paws to orient and subdue their arthropod prey before biting through the soft intersegmental areas. The prey is then decapitated before being dismembered by tearing the carapace and removing the internal parts. The manipulation of the prey before dismemberment implies some fine independent control of the forelimbs, comparable to that seen in many placental mammal orders.

Larger dasyurids like quolls (Figure 1.3b, b´) weighing up to 1 kg eat a diet ranging from invertebrates to small vertebrates (reptiles, birds and small mammals). The two largest species of dasyurid (the spotted tailed quoll, *Dasyurus maculatus*; and the Tasmanian devil, *Sarcophilus harrisii*) catch and eat small possums and pademelons (Tyndale-Biscoe, 2005). In addition, the Tasmanian devil (Figure 1.3c, c´) is a scavenger on the carcasses of larger mammals such as kangaroos, wallabies and the common wombat. Modern devils in Tasmania forage singly, with each individual occupying a home range of 8 to 20 square kilometres, although several individuals may feed simultaneously on a large carcass (van Dyck and Strahan, 2008).

The limbs of dasyurids are usually unspecialised in that the forefeet have five clawed toes radiating from a circular palm. In the climbing dasyurids, the forefoot provides a simple grip by flexion of all toes into the palm. Many species lack a pouch (e.g. phascogales) and others simply develop folds on either side of the abdomen during lactation (e.g. antechinus). This means that the young of many species have little protection and dangle from the mother's belly.

The numbat (Figure 1.3d, d´)(*Myrmecobius fasciatus*) is an unusual marsupial in that its diet consists exclusively of ants and termites and it is most active during the day. When feeding, the numbat searches by scent for underground termite galleries in the floor of open woodland. It turns over sticks and small branches and makes small excavations in the soil to expose termites, before extracting them with a long slender tongue (van Dyck and Strahan, 2008). Numbats are essentially solitary and live in arid regions of Western Australia with home ranges varying in size from 25 to 50 hectares.

With the extinction of the 15 to 35 kg thylacine (Figure 1.3e. e´, f) in 1936, the niche of top marsupial predator was left vacant. Thylacines probably hunted either singly or in small family groups and were generalist predators of small-to medium-sized species (Paddle, 2000). Captive thylacines were said to have been mute, although wild animals were reported to communicate by yapping while in pursuit of prey (Paddle, 2000; van Dyck and Strahan, 2008).

The strangest marsupial carnivore is undoubtedly the marsupial mole (*Notoryctes typhlops*, Figure 1.4a, b, b´), which has a remarkable external resemblance to the placental golden mole of Africa (van Dyck and Strahan, 2008). The marsupial mole spends most of its life underground and is blind, with the eyes reduced to vestigial subcutaneous lenses (Sweet, 1906). The snout is protected by a horny shield and the external ear consists of only tiny holes surrounded by dense hair. The marsupial mole is probably solitary, although little is known about its social organisation. Above ground the mole moves in a sinuous manner, dragging its legs to leave parallel furrows on either side of a central groove made by the body and tail (van Dyck and Strahan, 2008). Its adoption of subterranean life has led to some remarkable anatomical changes in its cerebellum and olfactory system (see Chapters 5 and 12).

Some species of small dasyurid (members of the family Dasyurinae) are of special scientific interest for understanding hormonal control of mating and aggressive behaviour because of their unusual reproduction. These species include all species of antechinus from Tasmania to northern Australia, both species of phascogale, the little red kaluta (*Dasykaluta rosamondae*), the dibbler (*Parantechinus apicalis*) and the northern quoll (*Dasyrurus hallucatus*) (Tyndale-Biscoe, 2005). The males of these species show a striking phenomenon of a single cycle of spermatogenesis per lifetime, a highly synchronised and brief mating period, followed by death at 11 months of age of the entire male population (male semelparity). The death of all the males appears to be due to the action of a concert of behavioural, immunological and hormonal factors. Testosterone levels peak immediately prior to mating and testosterone appears

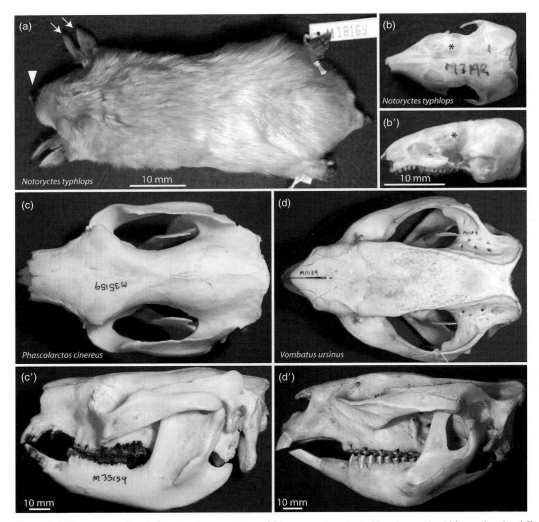

Figure 1.4 The marsupial mole (postmortem specimen: (a) has a snout protected by a horny shield (arrowhead in (a)) and the tail is reduced to a leathery stub. Each forelimb is tipped by two flattened claws (paired arrows in (a)). Dentition is reduced and the skull shows a bulbous expansion where the enlarged olfactory bulbs are located (asterisks in b, b´). Skulls of the koala (c, c´) and common wombat (d, d´). The koala has three pairs of upper incisors, whereas the wombats have only two. Cheek teeth in both the koala and wombat have broad grinding surfaces. All teeth in the wombat have open roots and grow continuously. All specimens are from the collection of the Australian Museum, Sydney.

to be the main driver of the phenomenon. Cortisol levels are also high and predominantly unbound in the blood. The high free cortisol level contributes to mobilisation of glucose stores for frenetic mating, but also has adverse effects on the males' immune systems. During the mating period, males cease to hunt and devote all their efforts to fighting other males and finding females. Without the intake of food, males must mobilise glucose from their structural protein and go into a negative nitrogen balance, losing as much as half their body weight by the end of the breeding season. Because their immune system is impaired due to the high

levels of cortisol, the males develop a heavy burden of blood and intestinal parasites and die. Control of the onset of mating appears to be under the influence of day length, implying circuitry connecting visual input with the pineal gland and hypothalamus (see Chapter 6).

Bandicoots and bilbies

The members of order Peramelemorphia show very little somatic morphological variation among themselves, and this is reflected in the similarities in brain size and encephalisation within the order. Bandicoots are opportunists, who exploit the environment by taking a wide variety of foods. They have large litters, grow rapidly and have short life-spans (Tyndale-Biscoe, 2005). All members are strictly terrestrial and have long pointed heads and compact bodies (van Dyck and Strahan, 2008). Bandicoots dig conical holes with their forelimbs and use olfaction to explore for insects, other arthropods, and small vertebrates. Fruit and soft tubers may also supplement the diet. The hindlimbs of bandicoots and bilbies are large, with powerful thighs and elongated feet. This results in a gait that is a gallop at high speeds and a bunny hop at slow speeds (van Dyck and Strahan, 2008). Bandicoots are also distinguished by the presence of a complex allantoic placenta, which invades and fuses with the wall of the maternal uterus much as the placenta of eutherian mammals. Nevertheless, the young are born as immature as any marsupial and the anatomy of the female reproductive tract is undoubtedly marsupial.

Pygmy possums

The family Burramyidae includes the smallest possums. Pygmy possums are mainly insectivorous, but the eastern pygmy possum consumes nectar, the little pygmy possum can eat small lizards and the mountain pygmy possum may incorporate hard seeds in its diet (van Dyck and Strahan, 2008). All pygmy possums build nests, either in a tree hollow or on the ground and all are able to become torpid during winter months.

Wrist-winged gliders and striped possums

The family Petauridae includes the subfamilies Petaurinae and Dactylopsilinae. Petaurines feed mainly on insects, but can consume the sap and gum of eucalypts and acacias. The wrist-winged gliders have a gliding membrane attached to the tip of the fifth finger of the forelimb and the first toe of the hindlimb. Sugar gliders (*Petaurus breviceps*) can glide for at least 50 metres by leaping from a tree and spreading the membrane. Steering and stability are achieved by varying the tension and curvature of the membrane and, when the target tree is approached, the glider brings its hindlimbs up to land with all four feet. Sugar gliders live in social groups of up to seven adults and scent-marking glands are well developed.

The Dactylopsilinae (four species of striped possum from New Guinea and Cape York in north-eastern Australia) are of particular interest neurologically because the fourth finger is extremely long and terminates in a powerful curved claw. Dactylopsilines are insectivorous and use the long fourth finger much like the primate aye-aye, i.e. to remove adult and larval insects that burrow into trees.

Ringtail possums and the greater glider

Most ringtail possums have a strongly prehensile tail that is about the same length as the body and the strength of the tail is sufficient to support the body without any assistance from the limbs. On the other hand, the related greater glider has a tail that is more adapted to gliding than grasping. The molar teeth of these folivorous marsupials are adapted to grinding leaves to a fine paste and all employ hindgut fermentation.

The greater glider is the largest of the gliding possums and appears to employ gliding as a means to reduce energy demands. Glides may cover up to 100 metres and involve changes of direction up to 90°. Territories are marked by scent from anal glands in males.

The honey possum

The honey possum (*Tarsipes rostratus*) is specialised for a diet of nectar and has a long, pointed snout and an elongated, brush-tipped tongue, which can be protruded to probe flowers for nectar and pollen. Females are only 10 to 15 g in body weight and males weigh 7 to 9 g. The standard metabolic rate of the honey possum is the highest of any marsupial and 158% of the average placental rate (Withers *et al.*, 1990). The honey possum does not climb with claws, but has nail-like structures on the upper surfaces of the expanded tips of the fingers and toes. In this respect the feet of the honey possum have a superficial resemblance to those of the primate tarsier (hence the genus name *Tarsi/pes*) and the first clawless digit of the hindfoot is opposable to the other four. The long slender prehensile tail is also able to support the body. Vision is an important sense for the honey possum and this mammal achieves a visual spatial resolution of 0.63 cycles per degree in daylight and 0.60 cycles per degree in moonlight and starlight (Arrese *et al.*, 2002). The honey possum gives birth to the smallest of all mammalian young (weighing less than 5 mg), but the species is also notable for the longest mammalian spermataozoon (approximately 0.3 mm).

Feathertail glider

The feathertail glider (*Acrobates pygmaeus*) is the world's smallest gliding mammal and its common name derives from the somewhat prehensile tail, which has very short fur on the upper and lower surfaces and a conspicuous fringe of long, stiff hairs on either side. It has large forward-directed eyes, which provide nocturnal binocular vision for judging distances during leaps from tree to tree. The feathertail appears to be an adaptation for flight, and there is a thick gliding membrane extending between the elbows and knees. Its other specialisations include several distinctive features of the auditory system. These are a small bony disc in front of the tympanic membrane and an unusual alignment of bones around the ear, which may contribute to selective sensitivity to very high or very low frequencies. This mammal also has hypertrophied cochlear nuclei, which extend laterally from the medulla to produce an enlarged lateral lobe with connections with the auditory nerve and cerebellum (Aitkin and Nelson, 1989; see Chapters 5 and 11). Group nesting has also been observed with this species and captive animals breed best when a complex environment rich in climbing surfaces is provided.

Koalas and wombats

The koala (Figure 1.4c, c′) is an arboreal folivore, which feeds almost exclusively on the leaves of eucalyptus trees. It is solitary and occupies an extended, but fragmented, range down the eastern coast of Australia. Eucalyptus foliage is poor-quality food and contains high concentrations of indigestible lignin, as well as toxic terpenes and phenols. An adaptation to surviving on such poor nourishment is to adopt a low energy lifestyle and koalas are usually inactive 20 hours of each day (van Dyck and Strahan, 2008). Perhaps as an adaptation to the low-energy diet, the brain is astonishingly small and fills only a little over 60% of the interior of the skull (Haight and Nelson, 1987). Stranger still that such a dopey and somnolent creature should be so beloved by international tourists visiting Australian wildlife parks and zoos.

Wombats are primarily grazers, although their other major adaptation is a burrowing habit. The teeth have open roots (Figure 1.4d, d′) and grow continuously, unlike other marsupials (van Dyck and Strahan, 2008). Wombats have powerful fore- and hindlimbs, capable of burrowing into firm soil, and the head and hindquarters can be pressed against the wall of the burrow to compact the earth. *Vombatus ursinus* (the common wombat) lives in forests and grazes in clearings, whereas the two species of hairy-nosed wombat live in grassland or savannah. Wombats build extensive burrow systems and use olfactory communication to mark territory with urine, faeces and scratchings.

Macropods

The superfamily Macropodoidea includes 19 species in New Guinea and 49 species in Australia and comprises approximately one-third of all marsupials in Australia. Members of this superfamily range in body weight from 500 g to 80 kg

and live in a wide variety of habitats from rainforest to spinifex desert and rocky cliffs. Some species of kangaroo are adapted to travel long distances in search of food, whereas others live in trees or caves. Much of the multiplicity of kangaroos is relatively recent, since the first macropod fossil is known from only the late Oligocene (26 mya) and much of the diversification of macropods occurred in only the last five million years (Tyndale-Biscoe, 2005). The suite of adaptations, which have allowed the success of the macropods, includes foregut fermentation, bipedal hopping and the embryonic diapause. The last two of these have particular significance for the embryology, brain and behaviour of the kangaroos.

Bipedal hopping is a highly efficient form of locomotion, especially in a desert environment, where large distances need to be covered to find enough nourishment. Locomotion at low speeds is an ungainly walk in macropods, but this changes dramatically once a speed is reached where hopping becomes practical. Studies with red kangaroos (Dawson and Taylor, 1973) have shown that energy expenditure with speed of locomotion increases steeply with walking until 10 km per hour. At this point, hopping begins and oxygen consumption does not increase until a speed of 35 km per hour is reached. By contrast, energy consumption in quadrupedal placental mammals increases linearly with speed. This means that a dingo chasing a kangaroo at 35 km per hour consumes almost twice the amount of energy required by the kangaroo.

Vision and hearing are the dominant senses of the macropods. As might be expected by the niche that they occupy, the large kangaroos and wallabies have visual capacities similar to rodents and ungulates with retinal ganglion distributions in the retina concentrated into a horizontal visual streak, but with an additional area of concentration into an area centralis for acuity vision (Tancred, 1981; Wimborne et al., 1999). The eyes of the large macropods are situated high up on the skull, so that they have a wide field of vision (324° in the tammar wallaby, *Macropus eugenii*), while still maintaining a significant binocular part of the visual field (50° in *Macropus eugenii*). This combination allows the wallaby to keep an eye out for predators, while still being able to apply close vision to objects held in the forepaws (Wimborne et al., 1999).

Embryonic diapause is the ability of some marsupials (and placentals) to hold the blastocyst stage of the embryo in a state of dormancy, during which cell division and growth either cease or continue at a slow pace until a change in the hormonal milieu signals the embryo to resume development. Embryonic diapause is found in almost all kangaroos, wallabies and rat kangaroos and is an adaptation to allow these macropods to delay development of offspring until habitat conditions are suitable. Embryonic diapause is also found in pygmy possums, the feathertail glider and the honey possum.

Concluding remarks

Australian marsupials have a rich and unique history throughout the Tertiary, although significant gaps in the fossil record leave many questions unanswered. Modern marsupials occupy a remarkable range of niches and exhibit an extraordinary range of adaptations to those roles. Many of these adaptations are reminiscent of those seen in placental mammals (carnivores, primates and ungulates). To date, the exploration of how these adaptations and behaviour influence the nervous systems of Australian marsupials has been only superficial. Australian marsupials continue to be a fascinating group in which to explore how the evolving nervous system responds to selection pressures of the Australian environment.

2

Overview of marsupial brain organisation and evolution

K. W. S. Ashwell

Summary

The central nervous system consists of the brain and spinal cord, while the peripheral nervous system is defined as the nerves and nerve cell bodies outside the central nervous system. Subdivisions of the mammalian brain include the brainstem, including medulla oblongata, pons and midbrain; the cerebellum attached to the pons; the diencephalon (including pretectum, thalamus, epithalamus, prethalamus and hypothalamus); and the telencephalon (including olfactory regions, a laminated cortex or pallium, a septal region, striatum and pallidum, and amygdala).

The external appearance of the brains of Australian marsupials falls into four types. These are: type I – most polyprotodonts (Dasyuromorphia and Peramelemorphia); type II – non-macropodoid diprotodonts (e.g. possums and gliders); type III – macropodids (kangaroos and wallabies); and type IV – notoryctids (the modern marsupial mole).

Smaller members of Dasyuromorphia and Diprotodontia have similar brain size to placental mammals of comparable body weight, whereas larger-bodied marsupial carnivores and herbivores have brains consistently smaller than placentals of similar body weight. There is no evidence to indicate that relative brain size among Australian marsupials is associated with either geographical distribution, aridity of habitats or risk of extinction. Play behaviour appears to be best correlated with absolute brain size and body weight, probably as a reflection of longevity and prolonged dependency. There is also a significant positive correlation between relative brain size and both field metabolic activity and metabolic scope (FMR/BMR ratio). The diversity of extant marsupials means that some species within the Burramyoidea and Petauridae show endocranial volume comparable with supposedly 'advanced' placentals such as prosimians.

There are several characteristics of marsupial brains that are sufficiently distinctive as to be useful for constructing phylogeny. These include end arteries in the brain leading into segregated arteriovenous loops, distinctive arrangements of forebrain commissures (absence of corpus callosum and presence of the fasciculus aberrans in diprotodonts) and developmentally significant differences in brainstem nerves and nuclei compared to placentals (i.e. the course of the facial nerve and the position of the ventral accessory nucleus of the inferior olivary complex).

Expansion of the cerebral mantle and the acquisition of gyrencephalic (folded) cortex probably occurred independently in marsupial and placental lineages, since all the available mammalian cranial endocasts from the Cretaceous are lissencephalic (smooth-surfaced isocortex). Since marsupial and placental mammals probably diverged some time in the early Cretaceous and all available placental endocasts from the Cretaceous have exposed superior colliculi, it is likely that the caudal expansion of the cerebral cortex to cover the colliculi occurred independently in placentals and marsupials.

Analysis of cranial endocasts of Australian marsupials from the Miocene suggests that the distinctive brain shapes and cortical topography of phalangerid and macropod marsupials were present as early as that time. The distinctive flexion of the cerebrum in macropod brains (type III) is probably a reflection of adaptations in dentition and the masticatory apparatus to grazing and browsing.

The central and peripheral nervous systems

The central nervous system of all vertebrates consists of the brain and spinal cord, whereas the peripheral nervous system consists of the nerves and aggregated nerve cells (located in sensory or autonomic ganglia) outside the central nervous system. Nerves of the peripheral nervous system may serve motor or sensory functions as part of the somatic nervous system, or be involved in processing sensory information from the viscera and controlling the glands, smooth and cardiac muscle of the body's internal organs (autonomic nervous system).

The junction of the brain and spinal cord lies at approximately the level of the foramen magnum of the skull. Bones of the cranial skeleton protect the brain, whereas the spinal cord is enclosed within the neural canal of the vertebral column. It should be noted that some sensory and motor elements and their outflow (e.g. trigeminal sensory nuclei, spinal accessory nucleus) cross the supposed border between the brain and spinal cord, suggesting that the distinction between the two is more due to gross anatomical convenience rather than being indicative of a true functional division. Furthermore, there is a similarity in function between nuclei of the caudal brainstem and spinal cord in that both are concerned with control of somatic (body wall and limb) muscle and automatic functions.

Spinal cord structure

The spinal cord is involved in the initial processing of somatosensory information (i.e. simple and discriminative touch, vibration, joint position sense, pain and temperature) from the body surface, joints, muscles and bones; and in mediating local reflexes to control muscle tension and minimise damage to the body from noxious and damaging objects (i.e. sharp, abrasive or excessively hot or cold objects). The spinal cord also contains conduits (white matter columns or tracts) for transmitting information up and down the spinal cord. Most of these continue without interruption into the caudal medulla or as far as the thalamus.

The spinal cord gives off spinal nerves (Figure 2.1a) and is subdivided segmentally according to the vertebral level through which these nerves exit. In all mammals the cervical (neck region) spinal cord has eight spinal nerves associated with it and these exit either between the occiput of the skull and the first cervical vertebra (spinal nerve C1) or below the seven cervical vertebrae (spinal nerves C2 to C8). The remaining subdivisions are the thoracic (chest), lumbar (postcostal or abdominal), sacral (pelvic) and coccygeal (tail) segments. The number of spinal nerves associated with each of these may vary from one mammalian species to another, but usually there are around 12 thoracic spinal nerves, 3–6 lumbar spinal nerves, 4–5 sacral nerves and a highly variable number of coccygeal nerves.

The internal structure of the spinal cord consists of a grey matter core of nerve cell bodies (Figure 2.1b) and their shorter input processes (dendrites), surrounded on all sides by white matter tracts (funiculi or columns; lfu, dfu, vfu – Figure 2.1b, c) containing the myelin-coated axons of ascending and descending pathways.

The dorsal part of the spinal cord (dorsal horn) is concerned with processing sensory information from primary sensory afferents (the sensory axons from receptors in the skin, joints, muscles and bones). These fibres (the central processes of dorsal root or spinal ganglion cells) are collectively called general somatic afferents to signify that they come from the limbs, body surface or wall and that they are concerned with general sensation (i.e. touch, pain and temperature, as opposed to special senses like hearing and vision). The dorsal horn at thoracolumbar and sacral segments also receives information from visceral organs (general visceral afferents). The ventral part of the spinal cord (ventral horn) contains motorneurons concerned with control of somatic (skeletal) muscle. These nerve cells and their outgoing axons are known collectively as general somatic efferents. The dorsal and ventral horns are separated by a region called the intermediate grey zone. This region contains some tract neurons contributing axonal pathways destined for the brain and other levels of the spinal cord, neuronal circuitry concerned with processing sensory information from internal organs (intermediomedial cell column) and (in the thoracic and upper lumbar segments) an intermediolateral cell column for control of internal organs.

In the thoracic and upper lumbar regions of the spinal cord, effector neurons of the sympathetic division of the autonomic nervous system are located in the lateral horn of the intermediate grey matter (Figure 2.1c). A further group of effector neurons of the parasympathetic division of the autonomic nervous system are located in the sacral segments. These effector nerve cells for internal organs are collectively known as general visceral efferent neurons. Sensory information from visceral organs is conveyed to sensory nuclei in the intermediate grey matter medial to the lateral horn (intermediomedial cell column, IMM).

The brainstem

The brainstem contains nuclei concerned with the control of muscles and glands within the head and neck. These include muscles surrounding and moving the eye, the muscles of facial expression and of chewing (mastication), and the muscles of the pharynx, the soft palate and the larynx. Glands and smooth muscle controlled by the brainstem include the

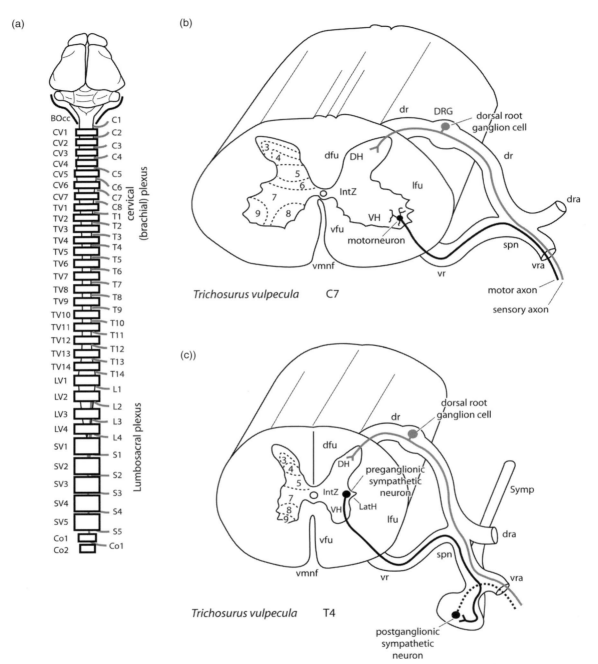

(a)

BOcc
CV1
CV2
CV3
CV4
CV5
CV6
CV7
TV1
TV2
TV3
TV4
TV5
TV6
TV7
TV8
TV9
TV10
TV11
TV12
TV13
TV14
LV1
LV2
LV3
LV4
SV1
SV2
SV3
SV4
SV5
Co1
Co2

C1
C2
C3
C4
C5
C6
C7
C8
T1
T2
T3
T4
T5
T6
T7
T8
T9
T10
T11
T12
T13
T14
L1
L2
L3
L4
S1
S2
S3
S4
S5
Co1

cervical (brachial) plexus

Lumbosacral plexus

(b)

Trichosurus vulpecula   C7

(c)

Trichosurus vulpecula   T4

Figure 2.1 Diagrammatic representation of the spinal cord in relation to the vertebral column (a) showing the emergence of spinal nerves (e.g. C1, C2 etc.) in relation to the vertebrae (e.g. CV1, CV2 etc.). Selected cross-sections through the spinal cord at cervical (b) and thoracic (c) levels show the organisation of the grey and white matter into functional columns and connections with nerves of the peripheral nervous system. 3 to 9 – Rexed's laminae; BOcc – basioccipital bone; dfu – dorsal funiculus of white matter; DH – dorsal horn; dr – dorsal root; dra – dorsal ramus of spinal nerve; DRG – dorsal root ganglion; IntZ – intermediate zone of spinal cord; LatH – lateral horn of spinal cord; lfu – lateral funiculus of white matter; spn – spinal nerve; Symp – sympathetic trunk; vfu – ventral funiculus of white matter; VH – ventral horn; vmnf – ventral median fissure; vr – ventral roots; vra – ventral ramus of spinal nerve.

muscles within the eye which control focussing of the lens and pupil diameter, salivary glands, lacrimal glands (producing tear fluid), and glands and smooth muscle tissue of the thoracic and upper abdominal organs (by the vagus cranial nerve, 10n). The brainstem is also concerned with the control of lung ventilation and cardiovascular function (heart rate, force and speed of heart muscle contraction). Sensory information from internal organs (blood vessel dilation, lung distension, gut expansion, taste) is also processed initially in the brainstem.

The brainstem is classically divided into the midbrain (including SC, IC, PAG and MeTg; Figure 2.2), pons (Pn and PnRt; Figure 2.2) and medulla oblongata (Md; Figure 2.2); but, at least in the case of the pons, the boundaries are inadequately defined and both developmentally and phylogenetically invalid (see below and Chapter 3 on development). The medulla oblongata (or just medulla) contains a number of vital centres concerned with the control of ventilation, cardiovascular function and gastrointestinal activity. Although the basic structure of the brainstem is highly conserved across vertebrates, the size of particular sensory or motor nuclei in the brainstem tends to correlate with the importance of the relevant sensory or motor function for the animal's behavioural repertoire. Apart from the motor and sensory nuclei, the medulla contains scattered populations of nerve cells collectively known as the reticular formation. The nuclei of the reticular formation are concerned with integrating information from a number of sensory modalities and their component neurons have large dendritic trees to facilitate this input. The output from the reticular nuclei influences the spinal cord and brain.

The term pons is useful in human neuroanatomy, but inadequate for comparative purposes. The pons in human anatomy is usually defined as a region on the basis of the rostrocaudal extent of the pontine nuclei, which are involved in circuits controlling fine motor co-ordination through the cerebellum, but the size of these nuclei will naturally vary depending on the number of pontine neurons in a particular species. Therefore the 'pons' may have a large rostrocaudal extent in mammals with extensive pontocerebellar systems for control of fine motor activity in the forelimb (e.g. primates), or be quite short rostrocaudally in mammals with poorly developed pontocerebellar circuitry (e.g. the marsupial mole or the Virginia opossum). Strictly speaking, most of the so-called pontine tegmentum, i.e. that region dorsal to the pontine nuclei, is more properly considered to be an extension of the medulla and has a similar suite of functions.

The midbrain or mesencephalon is divided into a dorsally located tectum and an underlying tegmentum. The tectum contains components concerned with vision, i.e. the optic tectum or superior colliculus (SC), and audition, i.e. the torus semicircularis or inferior colliculus (IC). Although its most obvious input is visual, the superior colliculus has a broader integrative function in that its deeper layers bring together information from other sensory modalities to form a three-dimensional construct of sensory stimuli in space. The inferior colliculus is an important relay centre conveying auditory information to the thalamus and thence the cerebral cortex. The midbrain or mesencephalic tegmentum (MeTg) contains some important groups of nerve cells concerned with control of motor function (e.g. the substantia nigra and red nucleus) and the activation of eye movement (oculomotor and trochlear nuclei), whereas the tissue surrounding the cerebral aqueduct through the midbrain (PAG) is concerned with implementation of physiological and behavioural responses to stressful stimuli.

The brainstem has cranial nerves entering and leaving it. These cranial nerves may be sensory or motor or mixed (i.e. both sensory *and* motor). General sensory nerves attached to the brainstem are concerned with somatosensation (i.e. touch, pain, temperature) from the face (trigeminal nerve; 5n), soft palate, pharynx and larynx (glossopharyngeal and vagus nerves; 9n and 10n, respectively) or viscerosensation (organ distension, peritoneal traction) from thoracic and upper abdominal organs (vagus nerve; 10n). Special sensory nerves (special somatic afferents) attached to the brainstem are concerned with hearing, balance and linear or angular acceleration (vestibulocochlear nerve, 8n), or visceral sensation like taste from the tongue and epiglottis (special visceral afferents; facial glossopharyngeal and vagus nerves; 7n, 9n and 10n, respectively). Motor control of the eye muscles (oculomotor, trochlear and abducens nerves; 3n, 4n, 6n respectively) and tongue (hypoglossal nerve; 12n) is provided by motor nuclei along the midbrain of the brainstem (somatic efferents), whereas nerves mediating motor control of the muscles of facial expression (facial nerve, 7n), masticatory muscles (trigeminal nerve, 5n), muscles of the palate, pharynx and larynx (glossopharyngeal and vagus nerves, 9n and 10n) and upper trapezius and sternocleidomastoid muscle (spinal accessory nerve, 11n) are grouped together as branchiomeric (sometimes called special visceral) efferents.

The cerebellum

The cerebellum (Figure 2.2i, ii and e) is a cortical or laminated structure attached to the brainstem and is derived developmentally as an extension of the roof of the rostral hindbrain (rhombic lip). It is classically considered as a

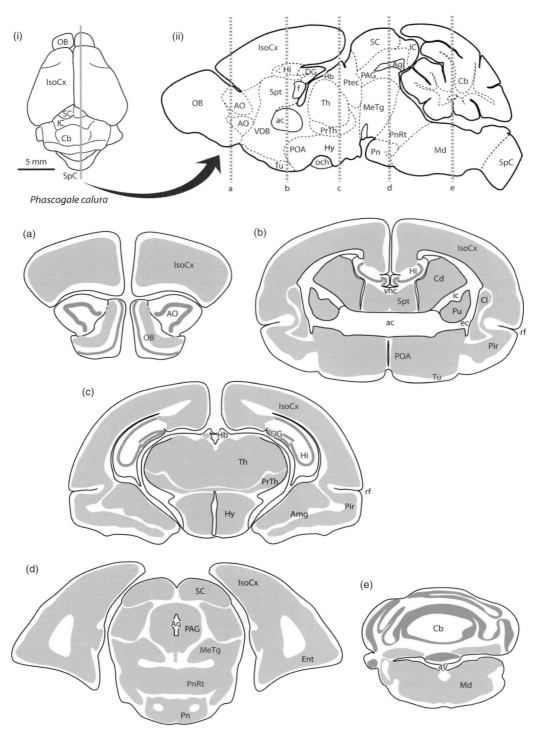

Figure 2.2 Summary diagram of brain structure as seen in a sagittal section and multiple coronal (frontal) sections through the brain of a representative dasyurid (red-tailed phascogale, *Phascogale calura*). (i) dorsal view of the brain showing main regions; (ii) sagittal section showing the main parts of the brain from rostral to the left to caudal to the right. Lines on (ii) marked (a) to (e) indicate the rostrocaudal positions from which the respective coronal (frontal) diagrams have been taken. 4V – fourth ventricle; ac – anterior commissure; Amg – amygdala; AO – anterior olfactory nuclei; Aq – cerebral aqueduct; Cb – cerebellum; Cd – caudate nu.; Cl – claustrum; DG – dentate gyrus; ec – external capsule; Ent – entorhinal cortex; f – fornix; Hb – habenular nuclei; Hi – hippocampus; Hy – hypothalamus; IC – inferior colliculus; ic – internal capsule; IsoCx – isocortex; Md – medulla; MeTg – mesencephalic tegmentum; OB – olfactory bulb; och – optic chiasm; PAG – periaqueductal grey; Pir – piriform cortex; Pn – pontine nuclei; PnRt – pontine reticular formation; POA – preoptic area; PrTh – prethalamus; Ptec – pretectum; Pu – putamen; rf – rhinal fissure; SC – superior colliculus; SpC – spinal cord; Spt – septum; Th – thalamus; Tu – olfactory tubercle; VDB – nuclei of vertical limb of diagonal band; vhc – ventral hippocampal commissure.

motor centre involved in balance, co-ordination and the smooth execution of rapid movements, but it clearly serves a broader range of functions (e.g. analysis of electrosensory information in some vertebrates).

The diencephalon

The diencephalon lies rostral to the midbrain and is traditionally said to be composed of five main areas in the mammalian brain (Figure 2.2ii, c). In a rough caudorostral order they are: pretectum (Ptec), epithalamus (including the habenula – Hb), dorsal thalamus (or just 'thalamus' – Th), ventral thalamus (or prethalamus – PrTh), hypothalamus (Hy) and preoptic area (POA).

The mammalian pretectum contains at least five nuclei, which receive visual input and participate in visuomotor reflexes. The epithalamus is the most dorsal part of the diencephalon and consists of the habenular nuclei (Hb) and the pineal gland. The dorsal thalamus (or simply 'thalamus') receives inputs from many sensory modalities (visual, auditory, somatosensory pathways), participates in motor feedback pathways and projects mainly to the telencephalon (see below). The ventral thalamus (or more properly called the 'prethalamus' because it develops rostral to the thalamus in the neuraxis) is relatively small in mammals and contains the thalamic reticular nucleus, the subthalamic nucleus and the pregeniculate (formerly ventral lateral geniculate) nucleus. The hypothalamus is concerned with homeostatic control (feeding, fluid balance, thermoregulation, hormonal control of the pituitary) as well as reproductive behaviour. Immediately rostral to the decussation of the optic nerve (optic chiasm) is the preoptic area, which is usually considered part of the hypothalamus, although some authors have placed it with the telencephalon.

The telencephalon

The developmental and phylogenetic elaboration of the telencephalon is a major feature of mammals (Figure 2.2). The most rostral part of the telencephalon is the olfactory bulb (OB, Figure 2.2a) – including main and accessory parts, which receive input from the olfactory epithelium of the dorsal nasal cavity and vomeronasal gland (Jacobson's organ), respectively. The more caudal parts of the telencephalon of mammals (Figure 2.2b, c) include a laminated pallium (more properly called cortex in mammals), septum, globus pallidus, striatum and basal amygdala.

The pallium includes the three layered hippocampal formation (hippocampus – Hi, and dentate gyrus – DG), six layered isocortex (IsoCx), three to five layered olfactory (piriform) cortex (Pir), claustrum (Cl) and endopiriform

nuclei (DEn, VEn) and cortical parts of the amygdala (Amg) (Figure 2.2b, c). The cerebral hemispheres of mammals are largely made up of isocortical pallial tissue and this may be folded into gyri (elevated regions) separated by fissures or sulci (intervening grooves). The hippocampal formation (sometimes called archicortex) is concerned with memory and the olfactory (piriform and associated) cortex is naturally concerned with processing sensory information from the olfactory nasal cavity and vomeronasal organ. The roles of the claustrum and endopiriform nuclei remain obscure, although the widespread connections of the claustrum with many pallial regions suggest some integrative role. The septal nuclei (Spt, Figure 2.2b) have been implicated in a wide range of motor (autonomic and neuroendocrine), behavioural and cognitive functions. In truth, although topographically distinct, the septum probably does not form a discrete functional unit by itself, but plays a key role within looped neural circuits involving the hypothalamus and hippocampus. The globus pallidus and striatum play important roles in motor function. In broad terms, the basal ganglia as a group (mainly the globus pallidus and striatum, the latter including caudate and putamen, Cd and Pu, respectively) connect the cerebral cortex with neural systems that influence behaviour. Many cortical regions project to the basal ganglia, which in turn influence the activity of thalamic nuclei that project to isocortical regions involved in the planning and execution of movements. The amygdala receives olfactory input and plays a pivotal role in anger, fear and anxiety.

External morphology of marsupial brains

Figure 2.3 shows the external features of a variety of Australian marsupial brains. These views facilitate comparison of the relative sizes of the olfactory bulbs and cortical surface in the different marsupial families. As a proportion of total brain size, the olfactory structures are largest in dasyuridae, notoryctidae and peramelidae, and decrease in relative size with increased overall brain volume in the diprotodonts. It is also clear that the degree of folding of the cortical surface varies substantially between species. Some species, such as small dasyurids (e.g. *Antechinus swainsonii*), have quite smooth (lissencephalic) cortical surfaces, whereas other species, like the eastern grey kangaroo (*Macropus giganteus*), have highly folded (gyrencephalic) cerebral cortex. As in placental mammals, gyrification (the folding of the cerebral cortex) is more extensive in those species with larger brains and higher body weight.

Haight and Murray (1981) have argued that, with respect to gross morphology, the brains of Australian marsupials fall into three types. The first type (type I of Figure 2.3) is characteristic of Australian polyprotodonts (Dasyuromorphia and Peramelemorphia). The brains of these marsupials have

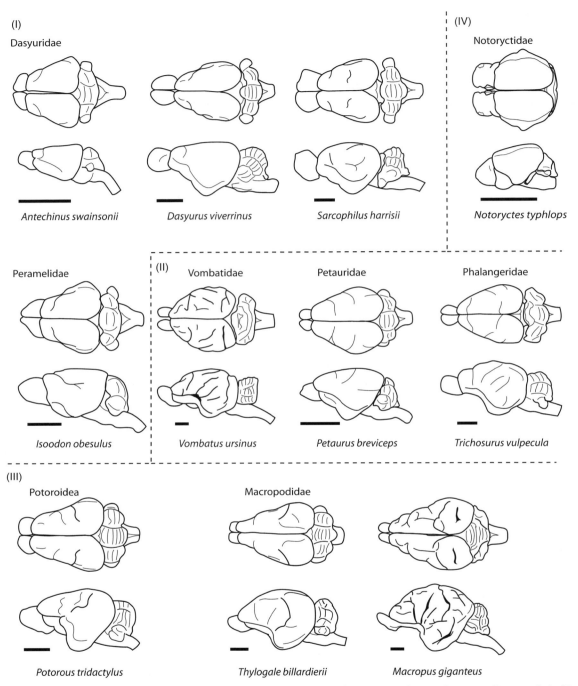

Figure 2.3 Dorsal and left lateral views of the brains of representative Australian marsupials. Note the range of lissencephalic (flat surfaced cortex) and gyrencephalic (folded cortex surface) forms. Four basic morphological types can be identified; three of them as described by Haight and Murray (1981), the fourth being the olfactory specialist marsupial mole. Scale bars indicate 10 mm.

a trapezoid outline when viewed from above, with large olfactory bulbs, a linear profile when viewed from the side, and little or no dorsoventral flexion. The degree of foliation or folding of the cerebellar cortex may vary, being extensive in dasyurids and less so in bandicoots. The second type of brain is found in non-macropod members of Diprotodontia. When these brains are viewed from above they exhibit a more ovoid shape than the brains of polyprotodonts. This is due to the lateral expansion of the parietofrontal cortex in type II brains. The olfactory bulbs are relatively small compared to type I brains and the olfactory cortical regions are correspondingly less extensive. Type II brains have a more rounded dorsal profile because of a slight dorsoventral flexion and the volumetric proportion of isocortex is greater. The third group of brains (type III) is characteristic of macropods (e.g. *Potorous tridactylus*, *Thylogale billardierii*, *Macropus giganteus* in Figure 2.3). The hemispheres of these brains are triangular in outline when viewed from above, although the potoroos may have a more trapezoidal shape because of blunting of the frontal pole. When viewed from the side, the cerebral hemispheres of all macropod brains exhibit a pronounced flexion due to the elevation of the rostral parts of the skull. The conspicuous flexion in dolichocephalic grazing kangaroos like *Macropus giganteus* may be related to non-neural factors such as more active molar progression, perpendicular cropping angle of the incisors and perhaps higher masticatory strength in the molar region of the maxilla (Kear, 2003). Haight and Murray (1981) did not include the bizarre brain of the marsupial mole in their consideration and this would undoubtedly constitute a fourth type, with exaggeration of the olfactory structures (see Chapter 12, especially Figure 12.2), no dorsoventral flexion of the forebrain and a poorly differentiated cerebellum.

Cranial endocasts of modern Australian marsupials

Figures 2.4 and 2.5 illustrate the skulls and cranial endocasts of selected Australian marsupials. For most marsupials, the brain occupies the bulk of the endocranial cavity. In fact, Haight and Murray found that the difference between endocranial volume (ECV) and brain weight never exceeded 5% for a range of poly- and diprotodont marsupials (Haight and Murray, 1981). Often this leads to clear impressions of fine features of the cerebral cortex (e.g. cortical sulci) being left on the inside of the skull.

The notable exception to this is the koala (*Phascolarctos cinereus*), which has an oddly small and lissencephalic brain occupying only 61% of the ECV (Haight and Nelson, 1987; Tyndale-Biscoe 2005). This is probably an adaptation

peculiar to the koala, which allows it to conserve energy in response to the poor nutritional quality of its terpene- and phenol-rich eucalyptus-leaf diet (Flannery, 1994; Tyndale-Biscoe, 2005).

Brain size of marsupials: graphical plots

Encephalisation is a measure of the relative brain size of a species for its body weight. The relationship between brain weight (or its proxy, endocranial volume – ECV) and body weight may be represented graphically, as shown in Figure 2.6. Such graphical presentations usually employ logarithmic scales for both axes to accommodate the large range in body and brain sizes. They allow the broad comparison of large-scale differences in brain sizes between groups of similar body weight. For example, it is evident from examination of Figure 2.6a that the marsupial mole has a smaller ECV for its body weight than do most Phascogalinae and Dasyurinae. Similarly it can be seen in Figure 2.6b that the Peramelemorphia as a group have smaller ECVs than most Diprotodontia of a similar body weight (e.g. Phalangeroidea, Petauroidea, Macropodoidea).

Graphical comparisons of brain weight against body weight for monotremes, and marsupial and placental mammals, have usually concluded that, for a range of body weights, marsupials have smaller brains than placental mammals (Striedter, 2005), but there are several problems with these comparisons. Firstly, these analyses are often made on the basis of only a few or even a single specimen of each species. Secondly, and more importantly, these comparisons are biased by the preponderance of large-bodied Australian or American marsupial species in the data set: smaller-bodied dasyuromorph and diprotodont species are often missing from these comparisons.

When a broad representation of Australian marsupial species is used for the analysis (Ashwell, 2008a), a more complex picture emerges. Larger marsupials (both Australian and American) do indeed have a smaller ECV than most placental mammals of the same body weight, but many smaller-bodied Australian marsupials have ECVs comparable to placental mammals of the same body weight. In fact, there is considerable overlap between marsupial and placental ECVs in the 10–200 g body weight range (Ashwell, 2008a). The rapid learning and retention of learning strategies by small marsupial carnivores (e.g. *Sminthopsis crassicaudata*; Bonney and Wynne, 2002, 2004) should therefore be of no surprise. Observations that small marsupial carnivores may actually exceed larger-bodied diprotodonts in learning and problem-solving (Bonney and Wynne, 2004) may also be partially explained by the higher relative brain size

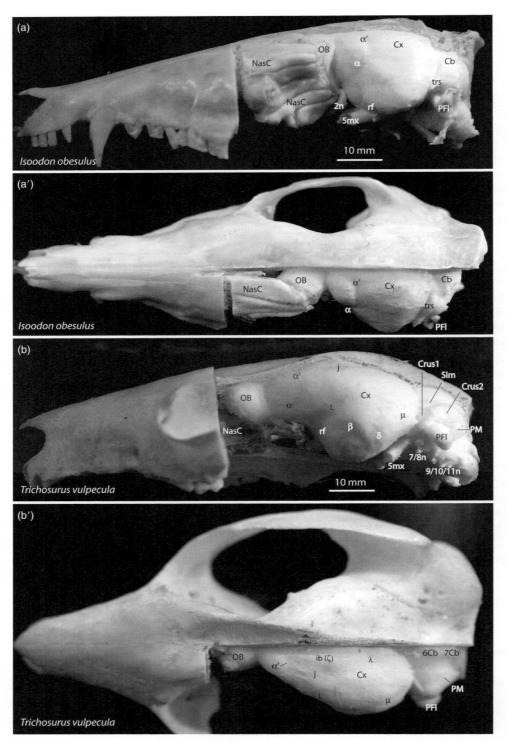

Figure 2.4 Photographs of dorsal and lateral views of skulls and cranial endocasts of the brown bandicoot (*Isoodon obesulus*, a, a′) and brush-tailed possum (*Trichosurus vulpecula*, b, b′). Features of the brain as a whole and cortical surface in particular (sulci α, α′, β, ib, j, L, λ, μ of the cerebral cortex) can be seen. Note that labels for cranial nerves (2n, 5mx, 7/8n, 9/10/11n) refer to endocasts of the relevant foramina transmitting those nerves. 2n – optic nerve; 5mx – maxillary division of trigeminal nerve; 6Cb – lobule 6 of the Cb vermis; 7/8n – 7,8 cranial nerves; 7Cb – lobule 7 of the Cb vermis; 9/10/11n – 9,10,11 cranial nerves; Cb – cerebellum; Crus1 – crus 1 of the ansiform lobule; Crus2 – crus 2 of the ansiform lobule; Cx – cerebral cortex; NasC – nasal cavity; OB – olfactory bulb; PFl – paraflocculus; PM – paramedian lobule; rf – rhinal fissure; Sim – simplex lobule of Cb; trs – transverse dural venous sinus.

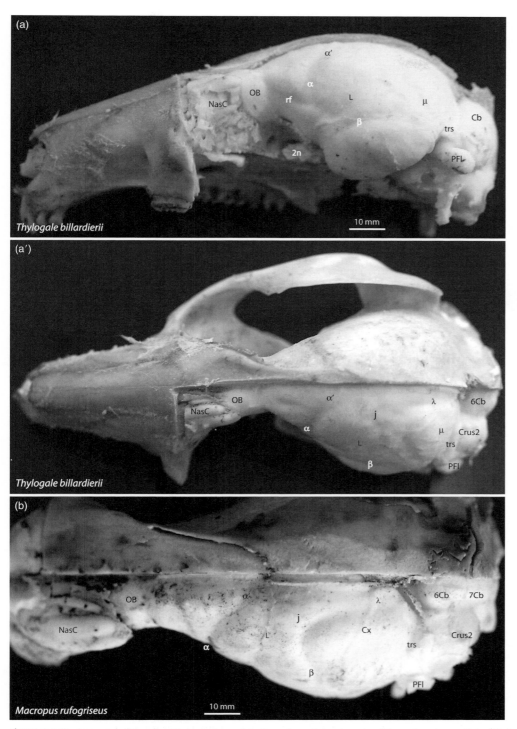

Figure 2.5 Photographs of dorsal and lateral views of skulls and cranial endocasts of the pademelon wallaby (*Thylogale billardierii*, a, a′) and a dorsal view of the skull and endocast of the red-necked wallaby (*Macropus rufogriseus*, b). Labels of cortical sulci as for Figure 2.4. 2n – optic nerve; 6Cb – lobule 6 of the Cb vermis; 7Cb – lobule 7 of the Cb vermis; Cb – cerebellum; Crus2 – crus 2 of the ansiform lobule; Cx – cerebral cortex; NasC – nasal cavity; OB – olfactory bulb; PFl – paraflocculus; rf – rhinal fissure; trs – transverse dural venous sinus.

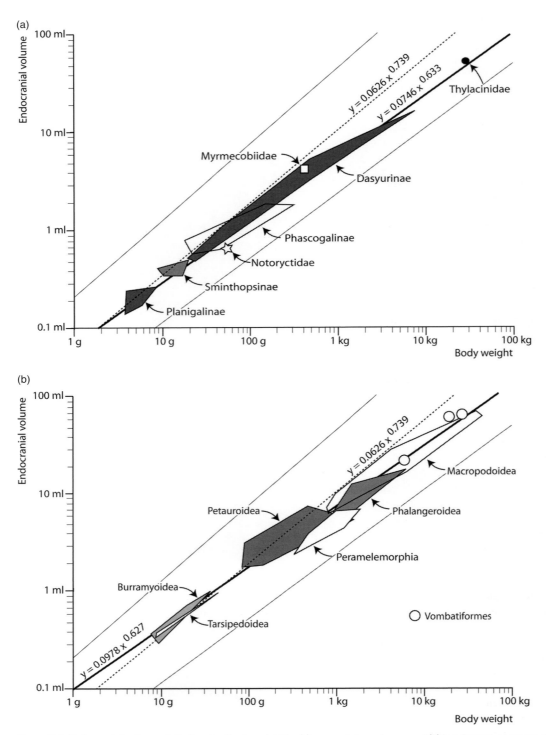

Figure 2.6 Endocranial volume plotted against body weight for (a) marsupial carnivores and (b) bandicoots, possums, gliders, koalas, wombats and macropods. Plots are based on data from Ashwell (2008a). Minimum side convex polygons have been drawn around the values for representatives of most families, but Vombatiformes, Thylacinidae, Myrmecobiidae and Notoryctidae have been represented as single points for each species. The dashed line indicates the regression line for all mammals derived from van Dongen's sample (1998). Light unbroken lines indicate the boundaries for all mammals from van Dongen's sample. The thick unbroken line in (a) indicates the regression line for marsupial carnivores, while the thick line in (b) indicates the regression line for the Diprotodontia.

Table 2.1 Allometric parameters relating ECV (in ml)[a] or brain weight (in g)[b] to body weight (in g)

Group	α	k	r
Dasyuromorphia[a]	+ 0.633	0.0746	+ 0.990
Peramelemorphia[a]	+ 0.507	0.1421	+ 0.884
Diprotodontia[a]	+ 0.627	0.0978	+ 0.982
American marsupials as a group[a]	+ 0.492	0.1169	+ 0.972
Mammals as a group[b]	+ 0.739	0.0626	+ 0.980
Reptiles as a group[b]	+ 0.530	0.0179	+ 0.954

Equation 2.1: $\log_{10}$ (ECV or brain weight[c]) = α (body weight) + $\log_{10}k$.

[a] based on ECV (Ashwell, 2008a)

[b] data from van Dongen (1998)

[c] ECV used as alternative measure to brain weight, assuming that brain has a density of 1 g/ml of ECV

of smaller marsupials. When comparing marsupial and placental brain size, it should also be remembered that many petaurids have values for ECV which plot to regions of the brain/body weight graph occupied by prosimians (Ashwell, 2008a).

Encephalisation of marsupials: numerical comparisons

Statistical comparison of relative brain size for species of widely divergent body weight requires some numerical measure that controls for those differences in brain weight that are due to variation in body weight. These comparisons imply that the allometric expansion of brain size with body weight is in large part due to extra brain mass being required to handle increased body size (e.g. more motorneurons to control increased muscle mass; more sensory neurons to process afferent information from a larger body surface). Since brain tissue is expensive to grow and maintain, boosting of brain size beyond that expected for a standard mammal of that body weight might indicate the addition of neural tissue to perform more complex, behaviourally important functions that enhance Darwinian fitness.

Relative numerical measures for comparing brain size across divergent body sizes include the encephalisation quotient and residuals from regression analysis. The encephalisation quotient is defined as the ratio of actual brain weight to that expected for a 'standard mammal' of that body weight (Jerison, 1973). Herein lies one of several problems with such evaluations. To be able to calculate these quotients, one must have a broad, unbiased sample of mammalian species to generate a best-fit line to make comparisons against, and it is well recognised that best-fit lines vary considerably depending on which data are included in the

analysis (Striedter, 2005). For example, Ashwell (2008a) has shown that all three orders of Australidelphians, as well as the Ameridelphians, exhibit slightly different relationships between ECV and body weight and each has a lower slope (α) than the relationship derived by van Dongen (1998) for mammals as a whole (Table 2.1). Nevertheless, some method for comparing relative brain size across widely varying body size is useful and desirable, provided that one appreciates the limitations and caveats associated with using encephalisation quotients. These are that: (i) the allometric parameters of the comparison group must be clearly stated and (ii) undue significance should not be placed on small differences in relative brain size. The reader should also bear in mind that, whereas ECV can be accurately measured by very simple techniques, estimations of what constitutes the true body weight of a specimen can be very difficult, particularly when archived museum material is used. How, for example, does one compare the encephalisation of captivity-bred (possibly overweight) specimens with that of emaciated wild-caught specimens? Furthermore, the larger gut mass of herbivores would tend to yield a lower measure of relative brain size compared to carnivores of similar body mass, without any behavioural implications.

Nevertheless, some consideration of EQ is useful in our discussion of relative brain size and Tables 2.2 to 2.4 summarise some data for key marsupial species. The reader is referred to Ashwell (2008a) for the complete data set. Note that the values for EQ stated in Tables 2.2, 2.3 and 2.4 have been calculated using the allometric parameters derived by van Dongen (1998) for all mammals (slope $\alpha = 0.739$; y-intercept $k = 0.626$). As noted above, EQ for any given species depends critically on the values of α and k used in the calculation and there are several commonly quoted values used by different authors. Therefore interpretation of the values for EQ must be considered in the context of the referent sample.

Table 2.2 Encephalisation and allometric parameters of selected Australian and New Guinean marsupial carnivores

Species	No. specimens	ECV ± SD (ml)	Body weight ± SD (g)	EQ ± SD	α	k	r
Antechinus agilis	18	0.56 ± 0.05	22.1 ± 3.2	0.925 ± 0.114	+0.159	0.344	+0.262
Antechinus stuartii	21	0.76 ± 0.09	29.8 ± 12.3	1.074 ± 0.284	+0.189	0.404	+0.621
Antechinus swainsonii	13	1.09 ± 0.12	63.5 ± 48.1	0.977 ± 0.302	+0.116	0.683	+0.615
Dasyuroides byrnei	13	1.50 ± 0.19	98.4 ± 25.1	0.837 ± 0.146	+0.270	0.436	+0.645
Dasyurus albopunctatus[NG]	14	3.79 ± 0.57	403 ± 147	0.769 ± 0.173	+0.298	0.639	+0.784
Dasyurus hallucatus	14	3.38 ± 0.43	513 ± 271	0.585 ± 0.126	+0.213	0.904	+0.684
Dasyurus maculatus	55	9.41 ± 0.92	2 599 ± 905	0.478 ± 0.108	+0.203	1.927	+0.797
Dasyurus viverrinus	26	5.33 ± 0.88	813 ± 278	0.633 ± 0.118	+0.362	0.476	+0.783
Myrmecobius fasciatus	2	4.25 ± 0.21	405.4 ± 38.4	0.804 ± 0.016	–	–	–
Notoryctes typhlops	2	0.78 ± 0.01	55	0.649 ± 0.010	–	–	–
Phascogale tapoatafa	20	1.81 ± 0.19	150.0 ± 50.9	0.759 ± 0.189	+0.187	0.710	+0.538
Planigale maculata	10	0.20 ± 0.04	6.7 ± 2.1	0.808 ± 0.125	+0.468	0.082	+0.859
Sarcophilus harrisii	34	16.09 ± 1.40	7 430 ± 1436	0.360 ± 0.040	+0.249	1.755	+0.541
Sminthopsis crassicaudata	20	0.36 ± 0.04	13.3 ± 3.6	0.889 ± 0.202	+0.099	0.277	+0.275
Sminthopsis murina	15	0.44 ± 0.06	15.9 ± 3.8	0.934 ± 0.155	+0.439	0.128	+0.569
Thylacinus cynocephalus	18	51.2 ± 6.9	28 500 ± 11 650	0.454 ± 0.107	+0.243	4.299	+0.798

NG – from New Guinea and associated islands

Table 2.3 Encephalisation and allometric parameters for ECV of selected Peramelemorphia

Species	No. specimens	ECV ± SD (ml)	Body weight ± SD(g)	EQ ± SD	α	k	r
Echymipera kalubu[NG]	18	3.67 ± 0.80	697 ± 373	0.525 ± 0.165	+0.327	0.443	+0.902
Echymipera rufescens[NG]	12	3.73 ± 0.77	798 ± 576	0.492 ± 0.119	+0.306	0.498	+0.877
Isoodon macrourus	34	4.42 ± 1.14	822 ± 319	0.513 ± 0.062	+0.556	0.107	+0.921
Isoodon obesulus	11	3.89 ± 0.85	691 ± 215	0.523 ± 0.110	+0.448	0.209	+0.844
Macrotis lagotis	11	6.57 ± 1.63	1 859 ± 663	0.415 ± 0.067	+0.529	0.122	+0.836
Microperoryctes longicauda[NG]	14	3.81 ± 0.64	479 ± 155	0.658 ± 0.087	+0.439	0.255	+0.902
Perameles nasuta	16	4.96 ± 0.92	1 273 ± 403	0.421 ± 0.073	+0.431	0.231	+0.921

NG – from New Guinea and associated islands

Values for EQ for marsupial carnivores range from 0.360 for the Tasmanian devil (*Sarcophilus harrisii*) to over 1.000 for some smaller dasyuromorphs (e.g. *Antechinus stuartii* – 1.074, *Antechinus wilhelmina* – 1.458; *Ningaui ridei* – 1.378). The values of the slope α for individual marsupial carnivores (range of +0.116 to +0.468; Ashwell, 2008a) are always below that for Dasyuromorphia as a whole (+0.633) (Table 2.2). Many of these species overlap in body weight with other species, so ECV at the top of the body-weight range for one species is always below that for members at the lower end of the body-weight range of the next heaviest species. In general the EQ of larger-bodied polyprotodont carnivores is lower than diprotodonts of similar body size (see below), but this stands in contrast with the relatively high EQ (around

0.70) attained by the extinct South American marsupial carnivore (*Thylacosmilus atrox*, Quiroga and Dozo, 1988). The relatively larger size of the brain of *Thylacosmilus atrox* may have been selected for by the demands of predation of ungulates (Quiroga and Dozo, 1988).

The Peramelemorphia (bilbies and bandicoots) are all relatively similar in body shape and lifestyle, although their habitats may vary from moist tropical to arid zones. This is reflected in the similar absolute and relative ECV of the group (Table 2.3). The allometric functions for this order are also more alike than for the Dasyuromorphia, with α values ranging only from +0.306 to +0.556 (compared to an α for the order as a whole of +0.507) (Ashwell, 2008a).

Table 2.4 Encephalisation and allometric parameters for ECV of selected Diprotodontia

Species	No. specimens	ECV ± SD (ml)	Body weight ± SD (g)	EQ ± SD	α	k	r
Acrobates pygmaeus	14	0.35 ± 0.05	8.9 ± 2.1	1.117 ± 0.138	+0.376	0.152	+0.645
Cercartetus caudatus	11	0.71 ± 0.12	19.3 ± 5.6	1.298 ± 0.202	+0.421	0.203	+0.740
Dactylopsila trivirgata	12	5.83 ± 0.41	506 ± 154	1.002 ± 0.314	−0.083	9.716	−0.387
Macropus agilis	10	33.3 ± 4.3	14 066 ± 4 241	0.470 ± 0.059	+0.361	1.062	+0.800
Macropus rufogriseus	14	36.0 ± 4.7	12 090 ± 3 902	0.577 ± 0.091	+0.341	1.469	+0.341
Petauroides volans	20	4.69 ± 0.32	900 ± 329	0.540 ± 0.167	−0.012	5.080	−0.068
Petaurus breviceps	20	2.09 ± 0.38	89.1 ± 21.9	1.226 ± 0.155	+0.489	0.232	+0.674
Petaurus norfolcensis	14	3.02 ± 0.31	191 ± 45	1.016 ± 0.131	+0.317	0.573	+0.819
Petrogale penicillata	10	23.9 ± 1.4	5 990 ± 630	0.618 ± 0.034	+0.381	0.868	+0.708
Phalanger mimicus[NG]	12	8.90 ± 0.76	1 473 ± 317	0.664 ± 0.109	+0.061	5.676	+0.150
Phalanger sericeus[NG]	10	9.96 ± 1.17	1 896 ± 504	0.635 ± 0.153	+0.257	1.444	+0.797
Phalanger vestitus[NG]	10	9.54 ± 0.75	1 672 ± 212	0.639 ± 0.083	−0.132	25.25	−0.208
Phascolarctos cinereus[a]	20	20.2 ± 2.3	5 888 ± 600	0.528 ± 0.051	+0.655	0.068	+0.576
Potorous tridactylus	12	9.53 ± 0.72	1 283 ± 309	0.791 ± 0.130	+0.166	2.915	+0.547
Pseudocheirus peregrinus	15	4.52 ± 0.77	526 ± 211	0.774 ± 0.235	+0.271	0.839	+0.777
Setonix brachyurus	11	13.9 ± 1.3	1 931 ± 605	0.870 ± 0.169	+0.065	8.527	+0.210
Trichosurus vulpecula	16	11.0 ± 1.3	1 902 ± 549	0.696 ± 0.155	+0.205	2.348	+0.571
Vombatus ursinus	18	59 ± 6.7	27 192 ± 7 230	0.520 ± 0.094	+0.297	2.872	+0.829

[a] note that the koala has a brain occupying only 61% of the ECV (Haight and Nelson, 1987) and encephalisation calculated on the basis of ECV (as here) provides an over-estimation.

NG – from New Guinea and associated islands

Body weights for New Guinea marsupials have been checked against data from Flannery (1995).

The Diprotodontia are the most diverse order of living marsupials, with members occupying a strikingly wide range of niches (arboreal nectar feeders to grazers). This is reflected in the diversity of values for absolute and relative ECV (Table 2.4). Endocranial volume for this order range from 0.30 ml for *Cercartetus concinnus* (the western pygmy possum) to 70 ml for *Macropus giganteus* (the eastern grey kangaroo). Three superfamilies of Diprotodontia (Burramyoidea, Petauroidea and, to a limited extent, Tarsipedoidea) have representatives with ECV and EQ above the average for mammals as a whole. In the case of the Petauroidea, EQ of many members lie within the region of modern prosimians (Ashwell, 2008a). These highly encephalised species included Australian species such as the threatened Leadbeater's possum (*Gymnobelideus leadbeateri*) and yellow-bellied glider (*Petaurus australis*) and New Guinean species such as *Dactylonax palpator* and *Dactylopsila tatei*. Allometric parameters for the Diprotodontia exhibit a much broader range of regression slopes and intercepts than that seen for either the Dasyuromorphia or Peramelemorphia, with α ranging from −0.132 for *Phalanger vestitus* to + 0.655 for the koala (*Phascolarctos cinereus*) and k ranging from 0.068 to 25.25 (Ashwell, 2008a).

The significance of marsupial encephalisation

What exactly does encephalisation mean in functional terms? Jerison proposed that relative brain size correlates with 'biological intelligence' (Jerison, 1976; Striedter, 2005), but it is recognised that 'biological intelligence' or the alternative term 'cognitive ability' is difficult to define for a range of vertebrates (see review in Striedter, 2005). Some authors have proposed that all vertebrates except humans possess the same level of general intelligence (Macphail, 1982), but there is some support for the idea that social intelligence correlates at least loosely with relative brain size (Striedter, 2005).

Many authors have noted the correlation between encephalisation and indicators of metabolism (see discussion in Striedter, 2005). With respect to Australian marsupials, an analysis of the relationship between encephalisation and metabolic indicators (Ashwell, 2008a) found a moderate positive correlation between relative brain size as calculated by EQ and both field metabolic rate – FMR ($r = + 0.538$, $p = 0.004$, df = 24) and metabolic scope (the ratio of FMR to basal metabolic rate, indicating the ability of a mammal to raise aerobic metabolism above basal levels) ($r = + 0.554$, $p = 0.011$, df = 18). Leadbeater's possum (*Gymnobelideus leadbeateri*) was found to have both a high EQ (1.631) and high FMR (2.99 ml CO_2 g/h) and FMR/BMR ratio (6.2), consistent with the high levels of activity and the well-developed territorial social system noted for this species (Hume, 1999). Although the data set with respect to metabolic scope for Australian marsupials is limited, the findings suggest that those mammals capable of maintaining high levels of free-range activity tend to be more encephalised. Nevertheless, three points should be noted: (i) the correlation coefficient is not high, i.e. other factors must also be significant in determining EQ; (ii) even if a high FMR/BMR ratio is causally linked to high EQ, the question remains open as to whether high EQ is the consequence of (or permitted by) high metabolic scope or vice versa and (iii) Koteja (1991) has noted that there is a poor correlation between basal and field metabolic rates in marsupials.

One final intriguing point may be made concerning the potential metabolic significance of encephalisation in Australian marsupials. This concerns the relationship between BMR, FMR and body weight in placental and marsupial mammals. Both marsupials and placentals have similar allometric scaling slopes (approximately +0.75; Dawson and Hulbert, 1970) for the relationship between BMR and body weight, whereas the relationship between FMR and body weight has a lower slope for marsupials (approximately +0.58) (Nagy, 2005) than for placentals (around +0.75) (Hume, 1999; Tyndale-Biscoe, 2005; Nagy, 2005). In fact, marsupials appear to be unusual among amniotes in having a lower allometric scaling slope for FMR than for BMR (Nagy, 2005). This means that the regression line relating FMR to body weight for marsupials intersects with that for placentals at about 10 g body weight (see Tyndale-Biscoe, 2005), much as brain weight for the two groups is similar in that body weight range. Whether this observation is purely a coincidence or reflects some underlying metabolic determinant of brain weight remains unknown.

There is no clear correlation between relative brain size and the risk of extinction for Australian marsupials (Ashwell, 2008a). The coincidence of the arrival of the placental dingo and the mainland extinction of the thylacine and devil naturally prompts the questions of whether there was a causal relationship between the two events and whether this might be due to differences in encephalisation. Mean EQ of recent Tasmanian thylacines and devils are 0.454 ± 0.107 and 0.360 ± 0.040, respectively, compared to 1.004 ± 0.126 for the dingo (Ashwell, 2008a). Did low encephalisation place the thylacine and devil at a competitive disadvantage relative to the dingo? The social structures of the three predators is and was profoundly different. Many modern dingos belong to socially integrated packs, which may be very stable and number from 3 to 12 individuals in areas where dingos are undisturbed by humans (Corbett, 1995). By contrast, modern devils in Tasmania scavenge alone, with home ranges of 8 to 20 square kilometres, although several individuals may feed together on a large carcass (van Dyck and Strahan, 2008). The thylacine is thought to have hunted at night, locating its prey by scent, and usually hunting either singly or in pairs (van Dyck and Strahan, 2008), although family groups of up to three adults and first-generation young were sometimes observed (Paddle, 2000). Wild animals may have communicated by yapping while pursuing prey, but thylacines studied in captivity were reported to have been mute (Paddle, 2000; van Dyck and Strahan, 2008). It is therefore possible that the lower relative brain size of both the thylacine and the devil may have been correlated with less effective hunting strategies and thereby contributed to their extinction on mainland Australia (Ashwell, 2008a).

On the other hand, it is clear that differences in brain size do not play a significant role in extinction across the broader range of Australidelphian carnivores, herbivores and omnivores (Dasyuromorphia, Diprotodontia and Peramelemorphia), because there are no significant differences in either EQ or residuals between extinct/threatened and common/abundant species for any of those three orders (Ashwell, 2008a). In fact, some of the most threatened modern marsupials have high relative brain size (e.g. Leadbeater's possum, *Gymnobelideus leadbeateri*, EQ of 1.631). By contrast, some of the most successful marsupials, which have adapted so well to life in the proximity of humans that they have become pests (e.g. *Didelphis virginiana* in North America and *Trichosurus vulpecula* in urban Australia and throughout New Zealand), have relatively low absolute brain weight and EQ (Ashwell, 2008a).

Arid-zone marsupials have 35% lower FMR than marsupials from non-arid habitats (Nagy and Bradshaw, 2000) raising the possibility that small brain size might be an adaptation for energy conservation, but the small relative brain size of larger-bodied Australian marsupials does not seem to be linked with survival in the nutrient-poor or arid habitats common in mainland Australia. Specifically, there is no clear correlation for any marsupial order between relative brain size and either aridity or type of vegetation of habitat (Ashwell, 2008a).

Absolute brain size vs. encephalisation

Striedter (2005) and Marino (2006) have noted that absolute brain size tends to be ignored in comparative neuroscience. Larger animals tend to have larger brains and even allowing for the increased number of nerve cells necessary for running a large body, the rise in brain size with body size must allow additional neurons for more 'computing power'. In support of this proposition, Deaner and colleagues (Deaner *et al.*, 2007) found that overall brain size was a superior predictive factor of cognitive ability among primates, with no indication that even cerebral-cortex-based measures were superior to whole brain size in predicting general cognitive ability. With respect to this, the range of brain size for Australian marsupials (0.14 ml for the narrow-nosed planigale *Planigale tenuirostris* to about 70 ml for the eastern grey kangaroo, *Macropus giganteus*) lies within that for placental mammals, but the largest marsupial brain is several orders of magnitude smaller than the largest placental brain (Striedter, 2005).

The type and frequency of play behaviour shown by Australian marsupials (Byers, 1999) correlates better with absolute brain size and body size than with encephalisation or regression residuals (Ashwell, 2008a). By contrast, Iwaniuk *et al.* (2001) found a significant inter-order effect of relative brain size on play, but not at the level of species within orders. On the other hand, it should also be remembered that brain size is positively correlated with body size, which in turn is positively correlated with longevity. Animals with short life-spans, no matter how encephalised (e.g. Phascogalinae and Planigalinae), have little time within their short lives to devote to play and its putative positive effects on brain development, whereas longer-lived and larger-brained species (e.g. larger Dasyuridae and Macropodidae) can gain from prolonged juvenile dependence and the benefits of play activity for brain maturation (Ashwell, 2008a).

What major features of marsupial brains make them different?

This entire book is broadly concerned with what makes Australian marsupial brains different and we will explore these in turn, but there are several major structural differences that are worth summarising at this early stage. These features, or characters, of central nervous system organisation have been used by Johnson and colleagues in reconstructing mammalian phylogeny (Johnson *et al.*, 1982a, 1982b, 1994; Kirsch *et al.*, 1983). Some of the more useful characters for separating marsupial orders from each other and from monotremes and placental mammals have been summarised in Table 2.5.

The first of these concerns the microcirculation of the brain. In monotremes and placental mammals, the cerebral microvasculature is characterised by freely anastomosing arteriovenous nets which are considered the primitive condition (0), whereas for those marsupial orders studied so far,

end arteries in the brain lead into segregated arteriovenous loops (Gillilan, 1972), which are regarded as the derived condition (1) (Johnson *et al.*, 1982a, 1982b). This feature of the microcirculation gives rise to the distinctive appearance of paired vessels surrounded by a single basement membrane in histological sections of marsupial brain and may be an adaptive counter-current exchange system that maximises oxygen transfer from capillaries (Snyder *et al.*, 1989) or facilitates heat transfer from incoming warm arterial blood to outgoing veins (Maloney *et al.*, 2008). Interestingly, the characteristic vascular loops of marsupials form even when marsupial vessels grow into implants of placental brain tissue (Johnson, 1977), indicating that this may be an inherent feature of marsupial cerebral vasculature. Marsupials do not possess a carotid rete at the base of the brain, such as has been claimed to contribute to selective brain cooling in bovids, but the counter-current arteriovenous loops probably contribute to the cerebral thermal inertia seen in large macropods as an adaptation to survival in arid-zone Australia (Maloney *et al.*, 2008). Note that this aspect of the microvasculature has not been specifically investigated in some marsupial orders (see '?' in Table 2.5), but is likely to be widely present.

The next character in Table 2.5 concerns the course of the dorsal lateral olfactory tract. In monotremes and most marsupials these fibres pass under the accessory olfactory formation, or a few fibres pass through it in some members of Dasyuromorphia and Diprotodontia. In many orders of placental mammals all fibres of the dorsal lateral olfactory tract pass through the accessory olfactory formation, although in some cases the less-derived condition is present (e.g. some carnivores) or the accessory olfactory formation is absent altogether (e.g. condition '4' seen in Chiroptera).

The presence of jugal (paired or twin) cones, both with oil droplets (1), in members of dasyuromorph and diprotodont marsupials may be a derived feature that distinguishes Australian from American marsupials. In monotremes and didelphids, paired cones are present (but without oil droplets) and placental mammals have only unpaired cones (all the underived condition '0').

Two important neural features pertain to the commissural pathways of the forebrain (Table 2.5). Both monotremes and marsupials rely mainly on the axons of the anterior commissure for transmission of information between the two cerebral hemispheres (0), whereas placental mammals have an additional isocortical commissure, the corpus callosum (1) that penetrates the hippocampal formation, and only a few fibres from temporal isocortex use the anterior commissural route. In monotremes and most marsupial orders, fibres approaching the anterior commissure do so by passing through the external capsule

Table 2.5 Characters of marsupial vs. monotreme and placental brains[a]

Character Order	Arteriovenous loops	Course of olfactory tract past accessory olfactory formation	Presence of jugal cones with oil droplets	Hemispheres joined by corpus callosum	Presence of fasciculus aberrans	Somatosensory cortex with barrels	Emergence of 7n over or beneath 5n	Medial or lateral ventral nuclei of inferior olive
Monotremata	0	0	0	0	0	0	0	0
Didelphimorphia	1	0	0	0	0	0	0	0
Paucituberculata	?	0	?	0	0	0	0	0
Microbiotheria	?	0	?	0	0	0	0	0
Peramelemorphia	?	0	?	0	0	0	0	0
Dasyuromorphia	1	0 to 1	1	0	0	0 to 1?	0	0
Diprotodontia	1	0 to 1	1	0	1	0 to 1	0	0
Rodentia	0	3	0	1	0	0 to 1	3	3
Primates	0	3 to 4	0	1	0	0	3	3
Carnivora	0	0 to 2	0	1	0	0	3	3
Chiroptera	0	3 to 4	0	1	0	0	3	1 to 3

[a] Adapted from Johnson *et al.* (1982b).

outside the striatum. A derived condition seen in diprotodont marsupials is the presence of a fibre bundle (the fasciculus aberrans) running from the internal capsule to join the anterior commissure. The large anterior commissure of marsupials and monotremes is nevertheless just as effective a commissural pathway as the corpus callosum of placental brains. The efficacy of inter-hemispheric transfer of visual information in the brush-tailed possum is comparable to that described for primate and carnivores and superior to that described for rodents and lagomorphs (Robinson, 1982).

The aggregation of neurons into so-called 'barrel' structures in layer 4Cx of the primary somatosensory cortex as a somatotopic reflection of vibrissae on the face is a derived condition seen unequivocally at some stage of development in some rodents and diprotodonts (e.g. brush-tailed possum), but is not seen in the somatosensory cortex of monotremes, other marsupials and most placental mammals. The presence of this character in diprotodonts and rodents is likely to be a case of convergence.

The last two characters to be considered probably arise as a consequence of differences in the timetabling of developmental events in the brainstem, in turn the result of the birth of monotremes and marsupials in a relatively immature state compared to placental mammals. In both monotremes and marsupials, the seventh cranial nerve (the facial nerve, 7n) emerges from the brainstem dorsal to the trigeminal sensory nuclei (primitive condition '0'), whereas in placental mammals this nerve passes either through the trigeminal spinal nuclei (1 or 2) or completely ventral

to them (3). The second of these developmentally derived brainstem characters concerns the inferior olivary nuclear complex of the medulla, which projects to the cerebellum in a topographically ordered fashion (see Chapter 5). The inferior olivary nuclear complex (IO) has an accessory nucleus (IOV) which lies predominantly ventral and lateral in the monotremes and marsupials (0), but mainly in a medial position in placental mammals (3), although some placental mammals (e.g. members of Chiroptera) have an intermediate position.

Both the course of the facial nerve and the positioning of the ventral accessory nucleus of the IO are undoubtedly reflections of the relative timing and mechanisms of migration of neuronal groups during development and will be discussed further in the relevant chapters (Chapters 4 and 5, respectively).

Evolution of therian brains, with special attention to marsupials

What do we know about the early evolution of the brains of marsupials and how they compared structurally with other contemporaneous mammals? Unfortunately, direct evidence on this question is rare and we must make deductions based on the broad range of mammalian fossil material. The first Mesozoic marsupials probably diverged from eutherian mammals during the early Cretaceous (Rose, 2006). There are no suitable cranial endocasts of Mesozoic marsupials to throw light on the early brain evolution in this group, but a consideration of the endocasts of other Cretaceous mammals gives useful insights into the likely brain structure of early marsupials.

A selection of the endocasts of Cretaceous multituberculates, stem therians and eutherian mammals is shown in Figure 2.7, along with the endocast of a modern Ameridelphian and the brain of a modern dasyuromorph. Multituberculates were the longest-lived order of mammals (except for the monotremes) being found in deposits from the Upper Jurassic to the Upper Eocene (Rose, 2006). Although their relationship to stem therians remains controversial, some authors have considered multituberculates to be more closely related to therians than eutriconodonts (Luo and Wibble, 2005). Multituberculates provide a perspective on the broad range of endocranial morphology of Cretaceous mammals and how this compares to modern mammals. The endocasts of both multituberculates and eutriconodonts differ from those of primitive therian mammals in that there is no apparent impression of the tectum (superior colliculus) on the inside of the cranium (Kielan-Jaworowska and Lancaster, 2004; Figure 2.7a). Both multituberculates and eutriconodonts also have a triangular bulge in the midline between the caudal cerebral hemispheres. Kielan-Jaworowska and Lancaster (2004) interpreted this bulge (stimulated by a personal communication from Harry Jerison) as a large superior cistern of the subarachnoid space covering the superior colliculus and vermis, rather than an impression of the superior vermis of the cerebellum alone. Coincidentally, a similarly large superior cistern has been observed in a modern Australian marsupial, the koala (Kielan-Jaworowska and Lancaster, 2004).

Multituberculates like *Kryptobaatar* are also of interest because of their high EQ compared to other Cretaceous mammals. This has been estimated at 0.71 by Kielan-Jaworowska and Lancaster (2004), based on the allometric parameters of Jerison ($\alpha = 0.74$; $k = 0.055$), or 0.629 (present author) based on the allometric parameters of van Dongen (1998)($\alpha = 0.739$; $k = 0.0626$). The range of EQ for multituberculates was 0.37 to 0.71, larger than that for many contemporaneous mammals, but smaller than that for modern mammals. Note that low encephalisation cannot be invoked as an explanation of multituberculate extinction, just as there is no correlation between encephalisation and risk of extinction for Australian marsupials.

An important fossil for the study of therian brain evolution is *Vincelestes neuquenianus*, from the Early Cretaceous of Argentina. This specimen has been considered a stem therian near crown group Theria (marsupials, placentals and all descendants of their common ancestor). The cerebral hemispheres of the cranial endocast of *Vincelestes*

(Figure 2.7b, b´) are ovoid and lissencephalic, with no clearly discernable rhinal fissure (Macrini *et al.*, 2007a). Casts of accessory olfactory bulbs are visible dorsolateral to the main olfactory bulb, as seen in modern therians. The cerebellar hemispheres and vermis are clearly visible on the endocast, unlike the endocasts of multituberculates, which lack cerebellar hemispheres (Kielan-Jaworowska and Lancaster, 2004). Encephalisation quotients for this species are quite low, and have been estimated as ranging from 0.27 to 0.37 depending on equations used (Macrini *et al.*, 2007a).

Examples of the cranial endocasts of Late Cretaceous eutherians are illustrated in Figure 2.7c, d, e. Note that these eutherian mammals are from an epoch which may be after the date of divergence of eutherian and metatherian lineages (placed at about 110 to 120 mya by some authors, Rose, 2006). These show very large olfactory bulbs, cerebral hemispheres separated widely posteriorly, large dorsal midbrain exposure with superior colliculi impressions and a short wide cerebellum with distinct cerebellar hemispheres and a clearly demarcated paraflocculus (Kielan-Jaworowska, 1984). Encephalisation quotients for Late Cretaceous eutherians range from 0.36 to 0.70 (Kielan-Jaworowska, 1984).

Comparison of the endocasts of these Cretaceous mammals with the brains of modern marsupials (Figure 2.7f, f´, g, g´) raises several points. Firstly, the encephalisation of Cretaceous theriiformes, e.g. *Vincelestes neuquenianus*, was lower than that for some multituberculates. This indicates that the range of encephalisation quotients of the diverse groups of Cretaceous mammals was wide and bears no clear relationship with the subsequent survival of those groups and their descendants. Secondly, all the available mammalian cranial endocasts from the Cretaceous exhibit impressions for either a large superior cistern over the superior colliculus and superior cerebellar vermis or distinct impressions for the superior colliculus and cerebellar vermis. This contrasts with the coverage of the superior colliculus in many modern Ameridelphians and Australidelphians (Fig. 2.7f, f´, g, g´) and placentals, suggesting that the caudal extension of the cortical mantle in both marsupials and placentals occurred independently after the end of the Cretaceous. Thirdly, all Cretaceous mammals (including *Vincelestes neuquenianus*, the theriiforme near crown group Theria) appear to have had lissencephalic brains, and this is likely to be true for Cretaceous marsupials. In other words, eutherians did not develop gyrencephalic cortical surfaces until after the beginning of the Cenozoic, well after divergence from metatherians, indicating that the gyrencephalic elaboration of the isocortex occurred independently in marsupials and

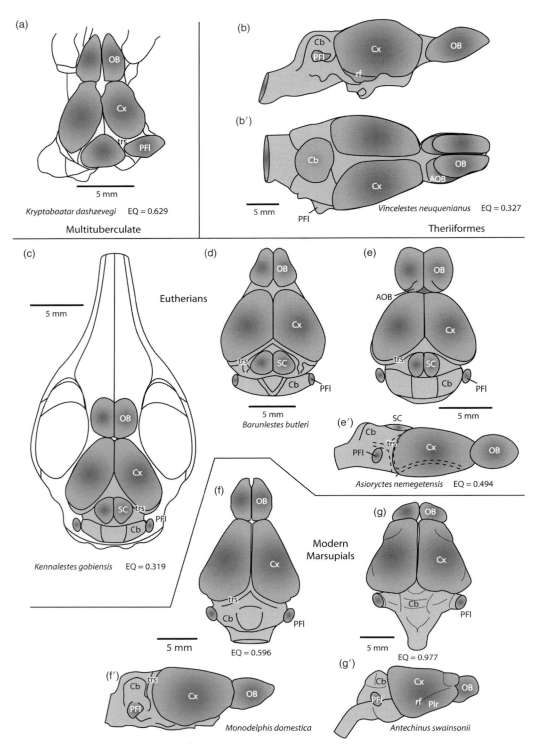

Figure 2.7 Drawings of cranial endocasts or brains of some important mammalian fossils and modern marsupials. The dorsal view of the endocast of the Late Cretaceous multituberculate *Kryptobaatar dashzevegi* has been drawn from photographs in Kielan-Jaworowska and Lancaster (2004). The dorsal (b) and lateral (b′) views of the endocast of the stem therian *Vincelestes neuquenianus* from the Early Cretaceous of Argentina has been redrawn and adapted from Macrini *et al.* (2007a). The illustrations of the dorsal (c, d, e) and lateral (e′) views of the endocasts of eutherian mammals from the Late Cretaceous of Mongolia have been based on illustrations in Kielan-Jaworowska (1984) and Kielan-Jaworowska and Trofimov (1986). The drawings of the dorsal and lateral views (f, f′, respectively) of the modern Ameridelphian, *Monodelphis domestica*, have been drawn from photographs in Macrini *et al.* (2007b), while the dorsal and lateral views of the brain of the Dasyuromorphian *Antechinus swainsonii* has been drawn from photographs in Haight and Murray (1981). Note that the values for EQ given for some mammals have been calculated using values for ECV and estimated body weight in the respective source publications, but with the allometric parameters of van Dongen (1998). Therefore, the stated values for EQ may differ slightly from those quoted in the source publications, but may be compared with other EQ values in the present work. AOB – accessory olfactory bulb; Cb – cerebellum; Cx – cerebral cortex; OB – olfactory bulb; PFl – paraflocculus; Pir – piriform cortex; rf – rhinal fissure; SC – superior colliculus; trs – transverse dural venous sinus.

placentals. Analyses of cranial endocasts of early Tertiary metatherians (Macrini *et al.*, 2007c; Sánchez-Villagra *et al.*, 2007) also suggest that the brains of American marsupials of this age were broadly similar in morphology to those of modern didelphids like *Monodelphis domestica* (Macrini *et al.*, 2007b).

Cranial endocasts of Australian fossil marsupials

A critically important fossil for understanding brain evolution of Australian marsupials is the early Miocene *Wynyardia bassiana*. This specimen was discovered in northern Tasmania at the beginning of the twentieth century and has been dated to 21.4 mya. The nearly intact neurocranium of this fossil has allowed an analysis of the external brain morphology and comparison with modern forms (Haight and Murray, 1981). The ECV of *Wynyardia* has been estimated at 18 ml and body weight is thought to have been between 4000 and 4500 g (Haight and Murray, 1981). Using the allometric parameters for mammals applied in Tables 2.1 to 2.4, this would indicate an EQ of 0.599, comparable to that of many extant petaurid and phalangerid species. Key morphological features of the *Wynyardia* endocast (Figure 2.8a, a′) include: (i) moderately sized olfactory bulbs and large olfactory cortex separated from the isocortex by a prominent rhinal fissure; (ii) a relatively smooth isocortical surface, but with some major sulci visible (e.g. α, β, δ, μ, ν'); (iii) flexion of the neuraxis of 152°; (compared to an average flexion angle of 166° for type I brains, 156° for type II brains and 142° for type III brains, Haight and Murray, 1981) and (iv) shallowly foliated cerebellum with a not especially prominent paraflocculus. Haight and Murray (1981) concluded that, based on morphology of the endocast, *Wynyardia* should be placed firmly with the non-macropod diprotodonts (type II) although they felt that the animal was not as gyrencephalic as modern phalangers and *Wynyardia* probably had a higher proportion of olfactory cortex to isocortex than do modern phalangers. As far as the deduced functional organisation of the isocortex is concerned, *Wynyardia* probably was similar to the extant and relatively well-studied brush-tailed possum, *Trichosurus vulpecula* (Haight and Neylon, 1978b). Based on similarities in the sulcal pattern of the isocortex, they were able to allocate putative primary somatosensory cortex with somatotopic organisation (i.e. separate regions of isocortex concerned with different parts of the body in an orderly mapping) as well as primary visual and auditory cortex (Figure 2.8b, c). The conclusion that *Wynyardia* was behaviourally and morphologically similar to modern phalangerids like the cuscus (*Phalanger*

maculatus) appears to be well supported by both the postcranial skeleton and the neuroanatomy (Haight and Murray, 1981).

Another important study in this area is an analysis of two partial macropodoid cranial endocasts from the early Miocene of the Riversleigh site in Queensland (Kear, 2003). Both specimens are natural endocasts formed by infilling with limestone, so the internal features are not as well preserved as for *Wynyardia bassiana*, and the olfactory bulbs and rostral isocortical endocast are missing. Nevertheless, isocortical sulcal patterns and flexion of the brain can be assessed from the endocasts with sufficient accuracy. Specimen QM F40188 has been attributed to *Balbaroo* sp., while specimen QM F40187 ('type 2') was found as an isolated endocast and is attributed as a macropodoid on the basis of endocast morphology. In lateral view, both endocasts show the characteristic macropodoid flexion of the cerebral hemisphere (type III brains of Haight and Murray, 1981). In dorsal view, although both endocasts are incomplete rostrally, the weakly flared posterior margins of the cortex suggest that the typically macropodoid triangular outline is also present. Sulcal patterns are better developed on QM F40187, where jugular, α, β, δ and μ sulci can all be distinguished. By extrapolation from modern macropods (e.g. the tammar wallaby, Ashwell *et al.*, 2005a) the major sensory areas can be assigned to the surface of this endocast (see Fig. 2.8e′). Kear (2003) also suggested that the sulcal pattern of QM F40187 indicates expansion of the mandibular and maxillary regions of the somatosensory cortex, consistent with browsing and grazing; and large visual and auditory cortical regions, consistent with progression towards crepuscular to diurnal habits. Both endocasts are therefore morphologically similar to the brains of plesiomorphic macropodoids and the findings suggest that the general morphology of the macropod brain was well established by the early Miocene.

What remains unknown?

There are intriguing suggestions of a relationship between encephalisation on the one hand and both field metabolic rate and metabolic scope (FMR/BMR ratio) for Australian marsupials on the other. Nevertheless, the data set, particularly with regard to FMR/BMR, is limited and collection of further metabolic data is necessary to properly test for a relationship between field metabolism and encephalisation.

The importance of relative vs. absolute brain size in determining/limiting/shaping behaviour remains an enduring

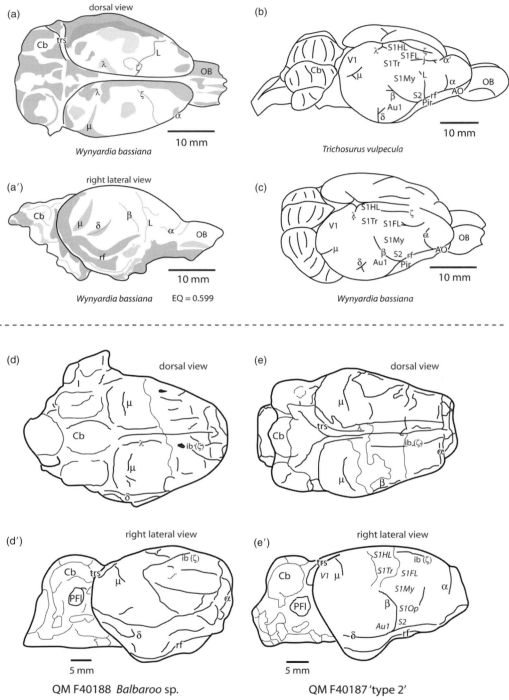

Figure 2.8 Illustrations of the cranial endocast of the phalangerid *Wynyardia bassiana* from the early Miocene of Tasmania (a, a′), and dorsolateral views of the isocortical sensory regions of *Trichosurus vulpecula* (b) and *Wynyardia bassiana* (c). Figures a, a′ and c have been redrawn from Haight and Murray (1981) with kind permission of S Karger AG, Basel. The lower half of the figure shows drawings of two macropodoid endocasts (d, d′, e, e′) from the early Miocene of north-western Queensland, based on photographs in Kear (2003). For both groups of illustrations, presumed isocortical sulci (α, β, δ, λ, μ, j – jugular sulcus, ib – interbrachial sulcus, rf – rhinal fissure) and putative sensory regions (i.e. subregions of primary somatosensory cortex – S1FL, S1HL, S1My, S1Op, S1Tr; primary auditory cortex – Au1; primary visual cortex – V1) have been indicated. AO – anterior olfactory nuclei; Cb – cerebellum; L – sulcus L; OB – olfactory bulb; PFl – paraflocculus; Pir – piriform cortex; rf – rhinal fissure; S2 – somatosensory area 2; trs – transverse dural venous sinus.

puzzle for neuroscientists. An additional question concerns the relative importance of metabolic rate and cerebral blood flow in modulating brain function.

There is a yawning gulf in the evolutionary history of the brains of Australian marsupials before the Miocene, simply because of the paucity of fossils from prior to that time. Analysis of the cranial endocasts of *Wynyardia bassiana* and the Riversleigh macropodoids indicate that the distinctive brain morphology of phalangerid and macropodoid marsupials had emerged by that time, naturally raising the question of when these features first appeared. It seems likely that the flexion of type III macropodoid brains emerged as a result of changes in the masticatory apparatus of early kangaroos as they adapted to grassland grazing, but fossils from the Oligocene would be necessary to confirm this.

3

Development and sexual dimorphism

K. W. S. Ashwell

Summary

In all vertebrates, the developing brain is divided into neuromeres, which segment the neural tube transversely. The developing hindbrain or rhombencephalon is derived from six or seven rhombomeres and gives rise to the pons, medulla and cerebellum of the adult. The mesencephalon or midbrain gives rise to the superior and inferior colliculi and midbrain tegmentum. The prosencephalon or forebrain develops from three prosomeres and an unsegmented proneuromere: prosomeres 1 to 3 give rise to the pretectum, thalamus/habenula and prethalamus, respectively, whereas the proneuromere rostral to this gives rise to the hypothalamus, preoptic area, septal nuclei, cerebral cortex, striatum and pallidum.

Studies of pouch young tammar wallabies have shown that there are two phases in growth of the brain: an initial rapid phase proceeding from around the time of birth to about P140 to P180, followed by a period of slower growth. The transition point at five to six months corresponds with physiologically important changes in maternal milk and thermoregulation, preparing the young for forays outside the pouch. The cerebellum has the most rapid postnatal growth of all brain regions, whereas the brainstem has the lowest.

Forebrain tract development in the tammar is almost exclusively postnatal, although the medial forebrain bundle and stria medullaris thalami are visible at the time of birth. All the major diencephalic tracts are in place within the first week of postnatal life, but descending projections from the developing cortex do not appear until late in the second postnatal week. The relative timing of forebrain tract development is broadly similar in all therians, although development of the hippocampal commissure appears to be delayed in diprotodont marsupials relative to the other commissural pathways. The tempo of forebrain tract development in mammals appears to be related to reproductive strategy regardless of phylogeny, in that r-selection strategists (e.g. rodents, polyprotodont marsupials) have a period of forebrain tract development that occupies a large proportion of the developmental period, whereas K-selection strategists (e.g. macropods, primates) have a prolonged period of forebrain development after tract outgrowth has been completed.

Sexual dimorphism in the basal forebrain of Australian marsupials appears to be less than in placental mammals, with profound significance for the neural circuitry controlling sexual behaviour.

Introduction

In all vertebrates, the central nervous system develops from a flat neural plate, which folds to form a neural tube (neurulation). Neurons of the central nervous system are predominantly derived from cellular proliferation in the walls of the developing neural tube and its derivatives. In marsupials, prenatal development involves neurulation and the establishment of neuromeric (segmental) organisation, formation of the cephalic and hindbrain flexures, and generation of some neuronal groups of the brainstem and caudal diencephalon. Postnatal brain development in marsupials sees the elaboration of the pallium into the cerebral cortex, generation of striatal and rostral diencephalic cell groups, and formation of the cerebellum. Fibre tract development occurs almost exclusively after birth in marsupials.

Prenatal development and neuromeric organisation

Neurons of the nervous system are generated either from proliferative cell populations in the wall of the neural tube (central nervous system neurons and many glia) or from neural crest cells given off ventrolaterally from the neural folds as the tube closes (peripheral nervous system ganglia and Schwann cells). The early neural tube has roof (rfp) and floor plates (fp), dorsally and ventrally, respectively; and the lateral wall of the neural tube is divided into basal and alar primary longitudinal zones, separated by the sulcus limitans (Figure 3.1a). At the rostral end of the neural tube, the alar and basal plates converge at the midline in the space between the sites of the developing neurohypophysis (posterior pituitary) and future anterior commissure (Puelles *et al.*, 2004).

At the level of the developing spinal cord, the alar and basal plates give rise to the dorsal and ventral horn neurons, respectively (Figure 3.1a). The floor plate generates a gradient of chemical signals that establishes progenitor domains in the neuroepithelial ventricular zone, which in turn give rise to distinct neuronal subtypes in the basal plate mantle zone (Price and Briscoe, 2004; Zhuang and Sockanathan, 2006). Dorsal root (spinal) ganglion cells are derived from neural crest cells, which migrate from the rim of the closing neural plate to lie alongside the developing spinal cord. The dorsal root ganglion cells extend central processes into the spinal cord and peripheral processes to receptors in the skin, tendons, muscles or peritoneum. The marginal zone surrounding the developing alar and basal plates will give rise to the white matter of the adult spinal cord, when ascending and descending fibre pathways develop during the first few weeks of postnatal life.

Like the spinal cord, the developing medulla is divided into a dorsal sensory alar plate region and a ventral motor basal plate region (Figure 3.1b), although the opening of the rostral medulla to form the fourth ventricle shifts the alar plate and its derivatives (e.g. trigeminal sensory nuclei, cochlear and vestibular nuclei) laterally, whereas the basal plate derivatives (e.g. abducens and hypoglossal nucleus) eventually lie close to the midline. Derivatives of both the alar and basal plate contribute to many cranial nerves with mixed sensory and motor function (Figure 3.1b).

The rostral end of the neural tube becomes elaborated to form transverse domains (previously called primary and secondary brain vesicles). The primary brain vesicles are the fore-, mid- and hindbrain vesicles (respectively, prosencephalon, mesencephalon, rhombencephalon; Figure 3.1c), which in the traditional description give rise to five secondary brain vesicles (Figure. 3.1d). The lumen of the neural tube is retained into adult life as the ventricular system – lateral, third and fourth ventricles, which are sites of cerebrospinal fluid production.

Modern analyses of brain embryogenesis emphasise the segmental nature of the developing brain (Figure 3.2). The most caudal region is the hindbrain, segmented as rhombomeres 1 to 7 (r1 to r7; Figure 3.2a), giving rise to the adult hindbrain (medulla, pons and cerebellum of the adult brain; Figure 3.2b). The cranial nerve nuclei develop from particular rhombomeres (Figure 3.2c), but may migrate through the brainstem during postnatal life to their final resting positions. Further rostrally are the isthmic (is) and midbrain or mesencephalic (mes) domains, giving rise to the adult midbrain nuclei (Figure 3.2a), including the oculomotor and trochlear nuclei and superior and inferior colliculi (3N, 4N, SC, IC, respectively; Figure 3.2c). In the forebrain, prosomeres 1 to 3 (p1, p2, p3) give rise to the pretectum, thalamus and prethalamus, respectively (Figure 3.2b, c); whereas the secondary prosencephalon at the most rostral end of the neural tube gives rise to the pallium and subpallial regions (Figure 3.2b, c). Note that the hypothalamus, a region traditionally considered part of the diencephalon (Figure 3.1d), is actually derived mainly from the unsegmented proneuromere rostral to the prethalamus (Figure 3.2b), although some contribution from p3 to the caudal hypothalamus is likely. The neural retina is also traditionally considered a derivative of the embryonic diencephalon (Figure 3.1d), because it develops from lateral outgrowths of the diencephalic part of the brain during late embryonic life. The point of origin of the eyecup from the side of the forebrain proneuromere is marked by the position of the optic stalk (os; Figure 3.2a, b), which connects the eyecup and forebrain and will provide a guide for growth of optic nerve axons during postnatal life.

Gene expression analysis reveals the existence of developmental zones comprising more than one anatomical segment. The term 'tagma' has been used to refer to a series of adjacent segments, which share special regional characteristics (see review in Puelles *et al.*, 2004). Gene expression and tagmas have not been specifically studied in marsupials or monotremes, but it is likely that the major boundaries and expression patterns are similar in all mammals. The caudal epichordal diencephalic segments (prosomeres 1 and 2) in combination with the midbrain may build a caudal forebrain tagma, based on the common early expression of *Irx3* gene. Prosomere 3 and the secondary prosencephalon can be grouped into a rostral forebrain tagma, which includes all forebrain structures that lie rostral to the zona limitans interthalamica (i.e. between p2 and p3; Figure 3.2b) on the basis of alar expression of *Dlx* and *Arx* genes and basal expression of *Nkx2.1* (Puelles *et al.*, 2004).

The region of the forebrain rostral to p3 will also become greatly elaborated during late prenatal and early postnatal life to give rise to the embryonic cortex, olfactory bulb, septal region, striatum and pallidum (Figure 3.2b). The telencephalic vesicle can be divided into subpallial and pallial domains, both of which may be derived from the dorsal

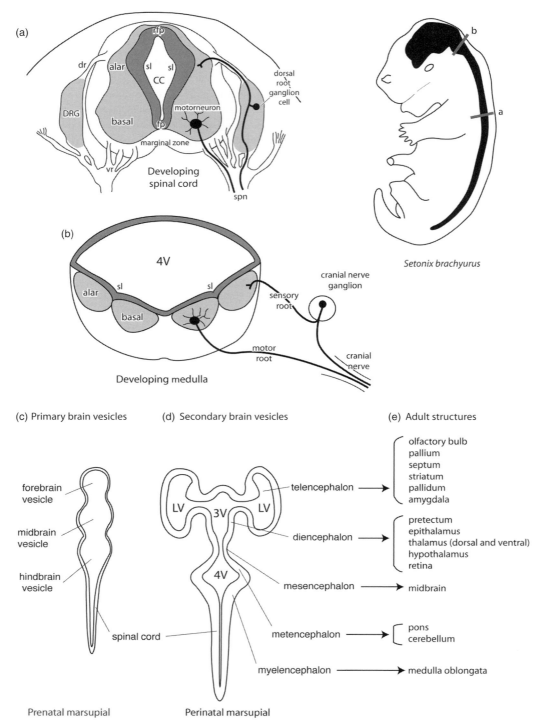

Figure 3.1 Diagrammatic representation of the generalised structure of the developing spinal cord (a) of a late embryonic stage quokka (*Setonix brachyurus*), showing the derivation of the dorsal and ventral horns from the alar and basal plates, respectively. The marginal zone gives rise to the white matter of the adult spinal cord. Although the opening of the fourth ventricle shifts the alar plate derivatives laterally (b), the medulla is developed from the same longitudinal organisation. Schematic diagram showing the traditional view of the brain as derived from three primary brain vesicles (c) during embryonic development, which give rise to five secondary brain vesicles (d) present immediately before birth in marsupials. These in turn give rise to the adult brain structures as listed (e). 3V – third ventricle; 4V – fourth ventricle; alar – alar plate; basal – basal plate; CC – central canal; dr – dorsal root; DRG – dorsal root ganglion; fp – floor plate; LV – lateral ventricle; rfp – roof plate; sl – sulcus limitans; spn – spinal nerve; vr – ventral roots.

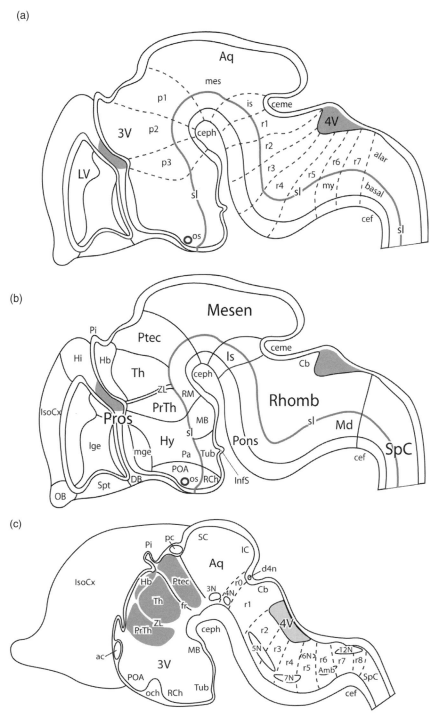

Figure 3.2 Gene mapping studies suggest that the origins of the diencephalon and telencephalon are more complex than the schema outlined in Figure 3.1c, d, e. Diagrams (a) and (b) depict the neuromeres (a) and corresponding brain regions (b) of an embryonic day (E)12.5 mouse embryo seen in sagittal view (based on Bulfone *et al.*, 1993; Puelles and Rubenstein, 1993; Puelles *et al.*, 2004). In all vertebrates, the hindbrain is derived from eight consecutive rhombomeres (a, c). The diencephalon consists of three prosomeres, with the region rostral to p3 (giving rise to the hypothalamus and preoptic area) regarded as an unsegmented proneuromere. The entire telencephalon is also

alar derivatives of the secondary prosencephalon (Puelles *et al.*, 2004). The medial ganglionic eminence (mge), which will give rise to the pallidum, appears before the lateral ganglionic eminence (lge), which will form the striatum. The pallium (future cortex) can be divided into medial, dorsal, lateral and ventral portions. The medial pallium gives rise to the hippocampal and parahippocampal cortex and the dorsal pallium gives rise to the isocortex, whereas the lateral and ventral pallium separately contribute to piriform cortex and the underlying claustrum, endopiriform and basolateral amygdaloid nuclei (see review in Puelles *et al.*, 2004).

Compartmentation within the developing brain is provided by glial boundaries, and these can be seen in immature therian brains with the aid of immunohistochemistry to the carbohydrate antigen CD15 (Mai *et al.*, 1998; Ashwell *et al.*, 2004). Compartmental boundaries in the developing diencephalon of the tammar wallaby are best seen during the period from P5 to P25, when there is clear demarcation of epithalamus, thalamus and prethalamus (Ashwell *et al.*, 2004). One difference that has been observed in developing marsupial brain compared to placental is the pattern of CD15 immunoreactivity in the brainstem. In marsupials (i.e. tammar wallaby), CD15 is intense in the raphe or ventromedial wedge region of the medulla (from P5 to P25), but less strong at the sulcus limitans, where CD15 immunoreactivity is much stronger in rodents (Ashwell and Mai, 1997). The functional significance of these differences remains to be seen.

The pace of systems development in prenatal marsupials

The extremely altricial state of newborn marsupials and the requirements of the climb to the pouch have profound effects on the pace of development of particular skeletal elements and neural systems (Smith, 2001). In all marsupials, the facial skeleton (dentary, maxillary, premaxillary and exoccipital) and cranial musculature are accelerated in development relative to other elements of the skeleton, probably facilitated by the early development of the neural crest in the facial region (Smith, 2001). In marsupials, the neural crest of the face begins differentiation and migration much earlier than the comparable site in placental mammals, and the rostrocaudal temporal gradient of neural crest development is much stronger than in placentals. For example, first pharyngeal arch neural crest migration begins before any somites appear in marsupials, but at the four somite stage in rodents (Smith, 2001). The first pharyngeal arch (with the associated trigeminal nerve fibres) is extremely large in early marsupial embryos and the development of the olfactory placode is also advanced relative to the forebrain, although the subsequent development of functional olfactory connections with the developing forebrain will be delayed relative to placentals (see below and Chapter 12). In turn, the forebrain is delayed in development relative to the hindbrain and facial region. This is reflected in the relatively delayed closure of the rostral end of the neural tube and the delayed evagination of the telencephalic vesicle in marsupials relative to the brainstem developmental stage (Smith, 2001).

It should also be noted that not all marsupials are morphologically the same at birth. In diprotodont marsupials (e.g. wallabies and possums), the newborns have prominent external ear and eye primordia, retinal pigmentation, digits on the hindlimb and pronounced differentiation of the mandible (Smith, 2001). By contrast, the newborn dasyurids are ultra-atricial (Figure 3.3; see also Figure 14.1), with barely visible ear and eye primordia and a head that is almost entirely nose and mouth (Smith, 2001). These differences may reflect the varied requirements of dissimilar modes of transfer of the young from the urogenital sinus to the pouch and teat (see below).

Caption for Figure 3.2 (Cont.)

derived from the rostral proneuromere region of the neural tube. Figure 3.2c shows the relationship between rhombomeres and the major cranial nerve efferent nuclei and the positions of the major tracts in the developing diencephalon of a representative therian. The position of efferent nuclei relative to rhombomeres is based on Gilland and Baker (2005). 3N – oculomotor nu.; 3V – third ventricle; 4N – trochlear nu.; 4V – fourth ventricle; 5N – motor trigeminal nu.; 6N – abducens nu.; 7N – facial nu.; 12N – hypoglossal nu.; ac – anterior commissure; alar – alar plate; Amb – ambiguus nu.; Aq – cerebral aqueduct; basal – basal plate; Cb – cerebellum; cef – cervical flexure; ceme – cerebellomesencephalic fissure; ceph – cephalic flexure; d4n – decussation of 4n; DB – diagonal band nuclei; fr – fasciculus retroflexus; Hb – habenular nuclei; Hi – hippocampus; Hy – hypothalamus; IC – inferior colliculus; InfS – infundibular stem; Is – isthmus; is – isthmic neuromere; IsoCx – isocortex; lge – lateral ganglionic eminence; LV – lateral ventricle; MB – mamillary body; Md – medulla; mes – mesencephalic neuromere; Mesen – mesencephalon; mge – medial ganglionic eminence; my – myelomeres; OB – olfactory bulb; och – optic chiasm; os – optic stalk; p1 to p3 – prosomeres 1 to 3; pc – posterior commissure; Pi – pineal gland; POA – preoptic area; Pros – prosencephalon; PrTh – prethalamus; Ptec – pretectum; Pa – paraventricular hypothalamic nuclei; r0 to r8 – rhombomeres 0 to 8; RCh – retrochiasmatic area; Rhomb – rhombencephalon; RM – retromamillary nu.; SC – superior colliculus; sl – sulcus limitans; SpC – spinal cord; Spt – septum; Th – thalamus; Tub – tuberal hypothalamus; ZL – zona limitans.

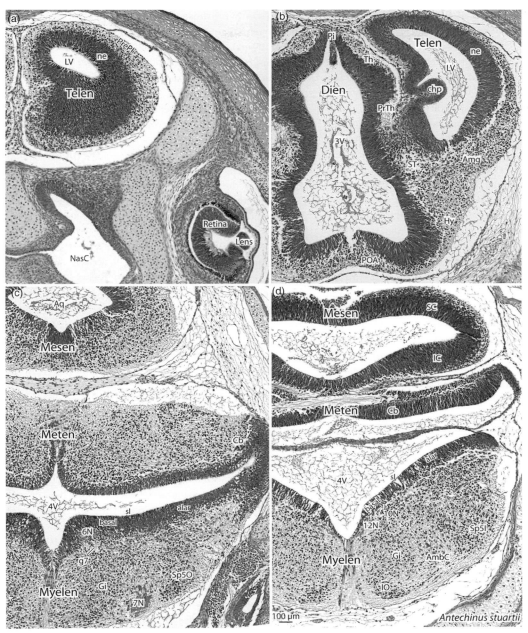

Figure 3.3 Photomicrographs of coronal sections through the brain of an early pouch young *Antechinus stuartii* (crown–rump length – 6.7 mm, body weight – 0.68 g), illustrating the relative immaturity of rostral telencephalic as compared to caudal rhombencephalic parts of the brain at this stage. Note that the telencephalon and retina are rudimentary, with only a preplate region present around the telencephalic neuroepithelium. By contrast, the maturation of many nuclei of the myelencephalon (medulla oblongata) is well advanced (e.g. facial motor nu., inferior olivary nu., nuclei of the trigeminal spinal tract). 3V – third ventricle; 4V – fourth ventricle; 6N – abducens nu.; 7N – facial nu.; 10N – vagus nerve nu.; 12N – hypoglossal nu.; alar – alar plate; AmbC – ambiguus nu., compact part; Amg – amygdala; Aq – cerebral aqueduct; basal – basal plate; ST – bed nu. stria terminalis; Cb – cerebellum; chp – choroid plexus; Dien – diencephalon; g7 – genu of the facial nerve; Gi – gigantocellular reticular nu.; Hy – hypothalamus; IC – inferior colliculus; IO – inferior olivary nu.; LV – lateral ventricle; Mesen – mesencephalon; Meten – metencephalon; Myelen – myelencephalon; NasC – nasal cavity; ne – neuroepithelium; Pi – pineal gland; POA – preoptic area; PrePl – preplate of cortex; PrTh – prethalamus; SC – superior colliculus; sl – sulcus limitans; sol – solitary tract; Sp5I – spinal trigeminal nu., interpolaris; Sp5O – spinal trigeminal nu., oralis; Telen – telencephalon; Th – thalamus.

Neural systems available at birth for the transfer to the pouch

For all the subsequent discussion of development in this and other chapters, days of postnatal life are indicated as P followed by the number of days.

There are essentially three ways for the newborn marsupial to be transferred from the uterus to the pouch (Gemmell *et al.*, 1999; Veitch *et al.*, 2000; Nelson *et al.*, 2003). The first is seen in marsupials with a forward facing pouch (i.e. diprotodonts such as wallabies and possums) and involves the mother adopting a birth position whereby the urogenital sinus is positioned below the pouch and the newborn climbs the intervening distance against gravity. The second is seen in bandicoots with backward facing pouches, where the mother lies on her side with the urogenital sinus above the pouch and the young slither down. The third is observed in dasyurids, where the mother stands on four legs with the urogenital sinus above the pouch depression. Naturally the first is the most demanding for the newborn and has excited the most scientific interest with respect to guidance and locomotion of the newborn marsupial.

Many studies of the developing nervous system of diprotodont marsupials have focussed on which neural systems are available to contribute to the newborn's climb from the urogenital sinus and the finding of the pouch and teat (Renfree *et al.*, 1989). Several sensory systems have been put forward as potential contributors to either the guidance of the initial climb against gravity (e.g. vestibular system) or the location of pouch entrance and nipple (olfactory and trigeminal systems) (Krause, 1991; Veitch *et al.*, 2000; Tyndale-Biscoe, 2005). The newborn must also be able to engage in rhythmic climbing movements using the forelimbs, with appropriate timing of forelimb extension, retraction and grasping manoeuvres (see Chapter 13). Remarkably, the climb from the urogenital sinus to the pouch rim may take only two to three minutes and the attachment to the teat occurs within five to ten minutes after birth (Veitch *et al.*, 2000). There are three main mechanisms, which may contribute to the climb of the newborn against gravity. The details of these systems will be discussed in the development sections of the appropriate chapters, but an overview will be presented in summary here.

The first of these might be called the 'plumb-bob mechanism'. At birth, the hindlimbs and caudal trunk of the developing marsupial are extremely immature and are essentially dead weight, whereas the forelimbs are able to actively reach and grasp. The pendulous hindlimbs therefore may act as a stabilising weight, keeping the forequarters and upper trunk in the proper vertical orientation (Larsell *et al.*, 1935). Clearly this mechanism would be effective only for those marsupials where the opening of the urogenital sinus is held directly below the pouch (e.g. diprotodonts). The simplicity of this essentially passive guidance mechanism is attractive, in that no sensory neuronal pathways need be present, at least for the initial stages of the climb.

The second possible mechanism for guidance against gravity was proposed by Cannon and co-workers (Cannon *et al.*, 1976), who observed a transient active righting reflex in newborn quokkas (*Setonix brachyurus*). They suggested that this reflex could be mediated by muscle stretch receptors in the neck, but it is not clear how muscle stretch receptors could directly detect orientation against gravity. Furthermore, studies in other marsupials have concluded that righting reflexes develop well after birth (Pellis *et al.*, 1992). Nevertheless, GAP43 immunoreactivity in the dorsal column pathways of the spinal cord is strong in newborn tammar wallabies (Hassiotis *et al.*, 2002), suggesting that the dorsal column pathways, which would be required to carry such information to the brainstem, are actively growing at birth and could convey sensory information as far as the medulla.

The third mechanism involves detection of gravity by static receptors in the inner ear and the use of this information to control spinal-cord pattern generators to turn the newborn. Some authors have argued that the newborn inner ear is too immature to mediate this mechanism (Cannon *et al.*, 1976), whereas others have claimed that the vestibular apparatus is sufficiently differentiated. Ultrastructural examination of the vestibular systems of various marsupials (both poly- and diprotodont) have indicated that at least one sensory region, the macula of the utricle, is more mature at birth than the remainder of the inner ear, and may be sufficiently developed to enable the newborn to detect and respond to gravity (Gemmell and Nelson, 1989, 1992; Gemmell and Rose, 1989; Krause, 1991; Gemmell and Selwood, 1994). Of course, a functional peripheral vestibular apparatus is of no benefit unless the central vestibular connections are sufficiently developed to process this input and influence locomotor centres. Pathways between the vestibule, vestibular nuclei and the spinal cord have been identified in the Ameridelphian *Monodelphis domestica* at birth (Pflieger and Cabana, 1996). A study of the development of central vestibular connections in the newborn Australidelphian *Macropus eugenii* found that central and peripheral connections of the vestibular ganglion were also present (McCluskey *et al.*, 2008), but that there was no direct projection from the vestibular nuclei to the cervical spinal cord until P5. Nevertheless, the possibility remains that an indirect projection between the vestibular nuclei and the brainstem motor centres is present by birth and can mediate control of the climb. It is conceivable that a combination of two of the above processes may control the climb against gravity. The geometric properties of the newborn suspended from its forelimbs intuitively best equip the young for vertical motion, but fine-tuning of the negative geotropism may be further aided by a primitive and perhaps transient vestibular righting reflex.

Some authors have argued that the manner in which the newborn brush-tailed possum or tammar wallaby redirects its movements when it reaches the edge of the pouch and enters it indicates a response to either the odour or the texture of the moist hairless skin of the pouch (Veitch et al., 2000; Tyndale-Biscoe, 2005), implying that olfactory and rostral somatosensory pathways are critical. Certainly, olfactory knobs have been identified in the nasal epithelium of the newborns of several species of diprotodont and polyprotodont marsupials (Gemmell and Nelson, 1988a; Gemmell and Rose, 1989; Gemmell and Selwood, 1994) and well-developed nerve fibres have been shown to connect the olfactory epithelium to the relatively large olfactory bulbs at birth in the native cat and the tammar (Hill and Hill, 1955; Hassiotis et al., 2002). On the other hand, neuronal activity in the olfactory bulb of the tammar wallaby develops much later than this (Ellendorff et al., 1988). Analysis of the developing central olfactory pathways confirms that the connections necessary for the olfactory system to influence brainstem or spinal cord centres are not present until well after birth (Ashwell et al., 2008a; see Chapter 12). Therefore, the olfactory system cannot be contributing to location of the pouch or nipple.

The trigeminal somatosensory pathways make up the other candidate sensory system, which would logically be considered for the role of detecting the pouch entrance and/ or teat (Gemmell and Nelson, 1988a; Hughes et al., 1989). This would require that the newborn be able to detect with its facial skin changes in the texture of the maternal skin and/or warm humid air emanating from the pouch opening. Veitch et al. (2000) noticed that the area surrounding the teat at birth in brush-tailed possums is studded with prominent glandular projections, which are not present around the non-pregnant teat. These may provide a tactile 'aiming mark' for the newborn seeking the teat. At birth, the developing vibrissal follicles on the snout consist of only solid epidermal plugs surrounded by dermal condensations, but even at this early age these structures are innervated by afferents of the maxillary division of the trigeminal nerve (Waite et al., 1994). Central processes of the trigeminal nerve have already entered the trigeminal tract at birth, reaching as far rostrally as the developing main (or principal) trigeminal sensory nucleus of the pons (Pr5) and as far caudally as the caudalis division of the spinal trigeminal nucleus (Sp5C) in the upper cervical cord (Waite et al., 1994). Since locomotion is likely to be controlled from the brainstem reticular formation, these central trigeminal afferents are sufficiently close and well developed enough to influence activity in the medullary and pontine reticular formation, which in turn may project to pattern generators (Ho, 1997) in the cervical enlargement of the spinal cord. Central projections of the trigeminal nerve pathways therefore appear to be sufficiently developed at birth to allow for somatosensory information from the snout to contribute to control of brainstem locomotor activity. Note that this function occurs well before any ascending projections of the trigeminal system reach the dorsal thalamus (P10 to P15; Leamey et al., 1996) and that thalamocortical axons do not reach the somatosensory cortex until P15 (Marotte et al., 1997). Therefore these connections are purely reflexive at the brainstem level.

The final element required for the newborn marsupial to climb to the pouch is a motor system to regulate rhythmic reaching, grasping and retraction movements of the forelimb. This involves gigantocellular and magnocellular reticular nuclei in the medial medulla, descending pathways to the cervical spinal cord, pattern generators in the cervical cord grey matter and peripheral nerves to the relevant muscle groups. All of these elements appear to be present and functional at birth (Pflieger and Cabana, 1996; Ho, 1997; McCluskey et al., 2008) and are discussed in detail in Chapters 4 and 13.

Pre- vs. postnatal neurogenesis

At the time of birth in marsupials, some neural systems in the brain and spinal cord have reached at least some functional capacity, but many are little advanced from the embryonic stage. In general, the newborn hindbrain is greatly advanced relative to the midbrain and forebrain.

Those components of the tammar wallaby hindbrain, which appear to be generated prior to birth, (and have therefore begun to progress to cytoarchitectural maturation) include the reticular nuclei (particularly the medial large-celled groups), somatic efferent cranial nerve nuclei (e.g. trigeminal motor nucleus, hypoglossal nucleus) and the trigeminal sensory nuclear complex (main or principal trigeminal and spinal trigeminal nuclei). Notably the cochlear nuclei, cerebellum and precerebellar nuclei are absent or poorly differentiated at birth. They emerge during the first to second month, in the case of cochlear and precerebellar nuclei, and the first three months, in the case of cerebellar granule cells.

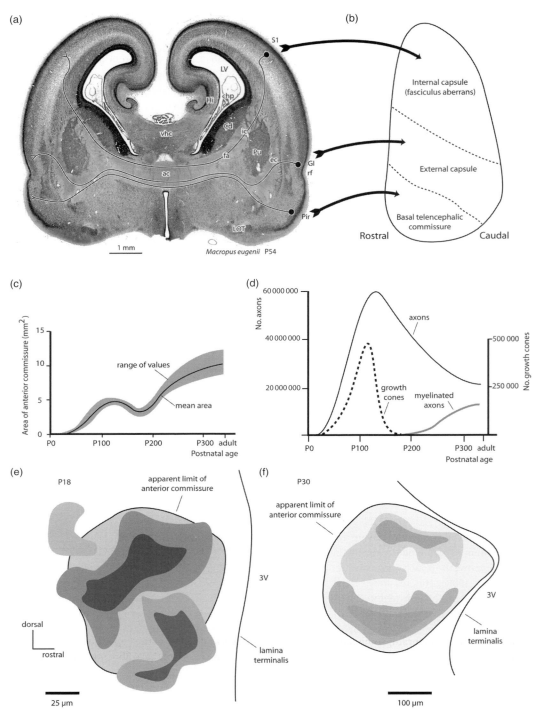

Figure 3.4 Structure and the time-course of development of the anterior commissure in a macropod (tammar wallaby; *Macropus eugenii*). Panel (a) shows a sagittal section through the anterior commissure of a developing tammar wallaby with the three zones of fibre origin indicated in (b). Panels (c) and (d) show the change in cross-sectional area (c) and total number of axons, axonal growth cones and myelinated axons (d). Panels (e) and (f) show schematic representations of growth cone density in sagittal sections through the developing anterior commissure at P18 and P30, respectively. Deeper shading indicates higher growth-cone density. Note the two 'hotspots' of growth-cone density in the dorsal and ventral parts of the anterior commissure. Illustrations are based on data from Ashwell *et al.* (1996b). 3V – third ventricle; ac – anterior commissure; Cd – caudate nu.; chp – choroid plexus; ec – external capsule; fa – fasciculus aberrans; GI – granular insular cortex; Hi – hippocampus; ic – internal capsule; LOT – nu. of lateral olfactory tract; LV – lateral ventricle; Pir – piriform cortex; Pu – putamen; rf – rhinal fissure; S1 – primary somatosensory cortex; vhc – ventral hippocampal commissure.

Those forebrain regions which have at least some neurons generated prior to birth, include the caudatoputamen, septum, bed nuclei of the stria terminalis and lateral hypothalamus. Nevertheless, the bulk of neurogenesis of the diencephalon and telencephalon occurs postnatally. Similarly, the pallium is at an undifferentiated two-layered stage at birth (Warner, 1980) and does not develop a cortical plate until P10 to P15 (Reynolds *et al.*, 1985).

Quantitative assessment of postnatal brain growth

Of all Australian marsupials, gross development of the brain has been best studied quantitatively in the tammar wallaby (Renfree *et al.*, 1982). General findings concerning macroscopic brain development in this species are certainly applicable to other macropods and probably also applicable to diprotodonts in general.

Postnatal growth of the tammar brain as a whole follows a biphasic pattern. There is an initial rapid growth phase that extends from around the time of birth until a body weight of about 100 to 130 g is reached (roughly equivalent in postnatal age to P140 to P180) followed by a phase of much slower growth into adulthood (Renfree *et al.*, 1982). During the initial phase, brain weight increases from 90 mg at P14 to 12 g at P180, representing a change from 4.5% of adult brain weight to 60% of adult brain weight.

The transition point from rapid to sedate phases of brain growth (at about P140 to P180) is physiologically significant for the young tammar for several reasons. This is the time at which thermoregulatory ability develops and is preceded slightly by the development of thyroid function (Setchell, 1974; Renfree *et al.*, 1982). There are also significant changes in maternal milk composition and metabolic rates of major visceral organs (see discussion in Renfree *et al.*, 1982).

As might be expected, given the differential demands of postnatal life on the various parts of the brain, some parts of the brain grow at different rates from others. The brainstem, for example, has an essentially linear growth between P30 and P180 and grows more slowly than other parts of the brain during the rapid growth phase. Presumably this is because most of the neural systems of the brainstem have already connected by the first postnatal month. Brainstem growth during the later mature phase of growth is at a similar pace to that of other brain parts. The cerebellum is the most rapidly growing part of the brain during early postnatal life, increasing in mass by a factor of 240 between P30 and P180, compared to only 25 for the brainstem and 40 for the forebrain (Renfree *et al.*, 1982).

Development of major fibre pathways

Timing of tract development in the tammar wallaby

Although the telencephalon is extremely immature at birth in the tammar wallaby, two adult forebrain tracts are present in the newborn. These are the medial forebrain bundle, which is a longitudinal fibre tract running the length of the diencephalon through that region which will later become the lateral hypothalamus, and the stria medullaris thalami, which connects the septal region with the habenular nuclei (Ashwell *et al.*, 1996a). The origin and nature of fibres in the early medial forebrain bundle are not known for certain, but studies in Ameridelphians suggest that at least some of these early medial forebrain bundle fibres are serotonergic axons arising from the superior central and dorsal raphe nuclei of the brainstem and projecting rostrally to the early forebrain structures (Martin *et al.*, 1987). There is also an intriguing fibre tract (possibly thalamostriate or striatothalamic fibres) in the lateral telencephalon of the newborn tammar (lateral telencephalic tract; see lt in atlas plate WP0/3). These fibres must be confined to the forebrain because they have never been labelled in carbocyanine dye studies of projections to the brainstem or spinal cord (Cheng *et al.*, 2002), and their function remains unknown. This tract disappears by the end of the first postnatal week, possibly by incorporation into the developing internal capsule.

Fibre pathways appearing early in postnatal life in the tammar wallaby (P2) include: (i) the posterior commissure, which provides for communication between the two sides of the brain at the level of the pretectum and mediates bilateral visual reflex responses; ii) the fasciculus retroflexus (habenulo-interpeduncular tract), which connects the habenular nuclei with the interpeduncular nuclei and is a central element in the forebrain control of brainstem autonomic centres; and iii) the mamillothalamic tract, which connects the early developing mamillary nuclei with the thalamic anlage, and is part of the Papez circuit controlling emotions and memory. All of these fibre bundles serve communication within the diencephalon and midbrain.

Descending projections from the pallium (isocortex and hippocampal allocortex) do not appear until the second week of postnatal life, consistent with the delayed maturation of pallial regions in marsupials. Fibres of the fornix, which connect the hippocampus with septal nuclei, first appear at P8. These fibres appear to be only the corticoseptal precommissural part of the fornix; the postcommissural fibres to the mamillary nuclei and other regions of the hypothalamus don't develop until several days later. Descending axons from the early-generated neurons of the isocortical pallium begin to form the internal and external capsules from P10.

Significance and development of the marsupial anterior commissure

The major forebrain commissure in marsupials is the anterior commissure, serving interhemispheric transfer of information for both isocortical and olfactory allocortical parts of the pallium (Putnam *et al.*, 1968; Heath and Jones, 1971; Robinson, 1982). In diprotodont marsupials, the dorsal part of this pathway receives axons from the isocortex by a special bundle arising from the internal capsule (the fasciculus aberrans; Smith, 1902; Ariëns-Kappers *et al.*, 1960; Friant, 1961; Figure 3.4a). In the adult wallaby, the anterior commissure has three subdivisions (Figure 3.4b): (i) a rostroventral basal telencephalic commissure transmitting commissural fibres from olfactory structures; (ii) an intermediate component containing axons which have approached the anterior commissure by the external capsule and (iii) a dorsocaudal component carrying commissural axons from the internal capsule by the fasciculus aberrans (Ashwell *et al.*, 1996b).

Development of the anterior commissure in the tammar wallaby (*Macropus eugenii*) is exclusively postnatal, with the first pioneer fibres crossing in the lamina terminalis (rostral limit of the third ventricle) at about P14, and the fasciculus aberrans appearing at about P18 (Ashwell *et al.*, 1996a, 1996b). These initial fibres appear to grow through a relatively cell sparse region, but whether the increased extracellular space facilitates the early growth of the commissure remains to be seen. Certainly, there do not appear to be any unusual concentrations of macrophages or specialised glial cells that would contribute to clearance of the growth zone or guidance of pioneer axons. The early postnatal growth (at P5 to P7) of the anterior commissure in *Monodelphis domestica* also occurs through a region of large extracellular space and may involve interaction with glial fibrillary acidic protein immunoreactive glial processes to provide guidance (Cummings *et al.*, 1997). The subsequent growth in cross-sectional area of the wallaby anterior commissure proceeds in three phases (Figure 3.4c): (i) an initial period of growth from P14 to about P120, followed by (ii) a period of steady state or small decline in cross-sectional area from about P120 to P170 and (iii) a period of more gradual growth to adult size from about P170. The number of axons in the anterior commissure (Figure 3.4d) rises from about 100 000 to 150 000 at P18 to a peak of almost 63 million at P148 (Ashwell *et al.*, 1996b), representing the addition of almost half a million axons per day. As seen for fibre tracts and nerves studied in many

therian mammals, there is a subsequent period of axonal loss, occurring between P148 and P216 and resulting in the elimination of approximately 41 million axons or 65% of the maximum axon number at P148. This axonal loss probably arises from elimination of the collaterals of descending isocortical pathways.

The subdivisions of the anterior commissure into basal telencephalic, external and internal capsular components become apparent at about P216, but the pattern of growth cones at much earlier ages (e.g. P18, P30; Figure 3.4e, f) indicates that there are two zones of high growth-cone density, which probably correspond to iso- and olfactory allocortical commissure growth. This indicates that segregation of these major components of the telencephalic commissure is present from the earliest stages.

Timing of tract development compared to placental mammals

Although the broad sequence of forebrain tract development in marsupials is similar to that in placental mammals, there are some important differences (Ashwell *et al.*, 1996a). Firstly, forebrain tract development is almost exclusively postnatal in marsupials (with the possible exception of the medial forebrain bundle and the stria medullaris thalami mentioned earlier), whereas the development of these tracts is exclusively prenatal in placental mammals. Secondly, development of the hippocampal commissure relative to the isocortical commissures appears to be delayed in marsupials. In the tammar wallaby for example, the hippocampal commissure develops from P35, compared to P14 for the fasciculus aberrans of the anterior commissure, whereas the hippocampal commissure appears at embryonic day (E)17 in the laboratory rat, compared to E18.5 for the corpus callosum (the placental isocortical commissure, see discussion in Ashwell *et al.*, 1996a). This difference in relative timing may arise from the different morphogenic pathways followed by the developing rostral forebrain in marsupial and placental mammals. Formation of the corpus callosum in placental mammals depends on an initial fusion of the forebrain hemispheres during fetal life, followed by growth of pioneer axons of the corpus callosum between the hemispheres at the pallial/septal border (Silver *et al.*, 1982; Valentino and Jones, 1982a). It is likely that this prenatal fusion of the two hemispheres facilitates early growth of the dorsal hippocampal commissure in developing placental mammals.

As noted above, the sequence of forebrain tract development is broadly similar across all therians (Ashwell *et al.*, 1996a). Where differences occur between mammals in the tempo of forebrain tract development, these reflect reproductive strategies rather than phylogeny. For example, those mammals with r-selection reproductive patterns (i.e. large litter sizes, many litters per reproductive lifetime, rapid development of offspring, e.g. rodents and polyprotodont marsupials) have forebrain tract development occupying a relatively large proportion of the period from conception to the attainment of behavioural autonomy (Figure 3.5a). On the other hand, those animals with K-selection reproductive patterns (i.e. one or two offspring per litter, few offspring per reproductive lifetime, relatively prolonged development, e.g. macropod diprotodonts, primates) have the period of forebrain tract development occupying a relatively short period of the total developmental time-span. These differences are irrespective of whether the mammals in question are marsupial or placental and probably reflect the longer period of time necessary for the offspring of large-brained K-selection strategists to complete the later phases of cortical development (e.g. elaboration of cortical circuitry, dendritic tree growth, synapse overproduction and elimination; Ashwell *et al.*, 1996a).

Developmental changes in GAP43 immunoreactivity in marsupial brain

The calmodulin binding phosphoprotein GAP43, which also serves as a substrate for protein kinase C, plays a critical role in axonal growth and GAP43 is enriched in growth cones of axons during the period of axonal elongation in all therians (see discussion in Hassiotis *et al.*, 2002). In addition, elevated levels of GAP43 in regions of the mature brain are thought to be associated with retained capacity to reorganise synaptic connections in response to physiological activity.

Fibre tracts of the developing wallaby brain can be divided into two groups on the basis of the pattern of developmental changes in GAP43 immunoreactivity (Figure 3.5b). One group of tracts (including the stria medullaris thalami, optic nerve and tract, spinal trigeminal tract and dorsal columns) have high levels of GAP43 immunoreactivity at birth. Immunoreactivity drops steadily until there is either no GAP43 immunoreactivity or a low baseline level, which is maintained into adulthood. The second group of tracts are pathways associated with the pallium or cortex (e.g. internal and external capsules, anterior commissure, fornix), which have no GAP43 immunoreactivity at birth, show a burst of intense immunoreactivity at around P10 to P50 and low or no immunoreactivity into adulthood. This contrast highlights the temporal differences between pallial and subpallial development in marsupials.

Neurotrophins in marsupial nervous system

Neurotrophins are a group of proteins that function as survival factors during development of the peripheral nervous system. In the central nervous system, neurotrophins may play important roles in neuronal differentiation and plasticity. The overlapping patterns of neurotrophin and neurotrophin-receptor mRNA expression in mammals and birds suggest that the major functions of these proteins are similar in all amniotes (Kullander *et al.*, 1997). Four neurotrophin proteins have been identified in mammals: NGF (nerve growth factor); BDNF (brain-derived neurotrophic factor); NT-3 (neurotrophin-3) and NT-4 (neurotrophin-4). The amino-acid sequences for BDNF and NT-3 are essentially the same when comparing rodents, monotremes and marsupials, whereas NGF appears to be more divergent and has accumulated more changes in its amino-acid sequence during evolution. Sequencing studies of the neurotrophins in marsupials clearly separate the four marsupial orders, but also emphasise the similarities of the neurotrophins in all therians (Kullander *et al.*, 1997).

Development of the blood–brain and blood–cerebrospinal fluid barriers

The blood–brain and blood–cerebrospinal fluid (CSF) barriers prevent or control the transfer of proteins and other large molecules from the blood into the relatively privileged environment of the brain and its surrounding fluid. Anatomically, the blood–brain barrier consists of the linkage of vascular endothelial cells by tight junctions, thereby preventing passage of large molecules from the blood to the central extracellular space. The barriers also serve to protect the brain tissue from pathogenic micro-organisms. Note that there are some regions of the brain and associated glands (e.g. pineal gland, some hypothalamic nuclei, the area postrema of the brainstem) where circulating hormones modulate neural activity, and these regions have fenestrated capillaries allowing freer movement of large molecules into the extracellular space of the brain. In placental mammals, the blood–brain barrier forms around the end of embryonic life (about week 8 to 12 of human postconceptional life). Ultrastructural studies of marsupial cerebral blood vessels indicate that the structural elements of the blood–brain and blood–CSF barrier (tight junctions between endothelial and choroid plexus epithelial cells, respectively) are present even at birth (Dziegielewska *et al.*, 1988), although the blood–CSF barrier may be bypassed by a transcellular route for passage of plasma proteins across the choroid plexus.

(a)

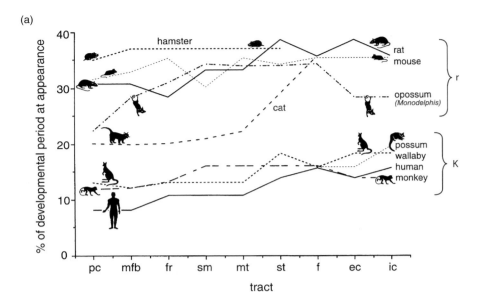

(b)

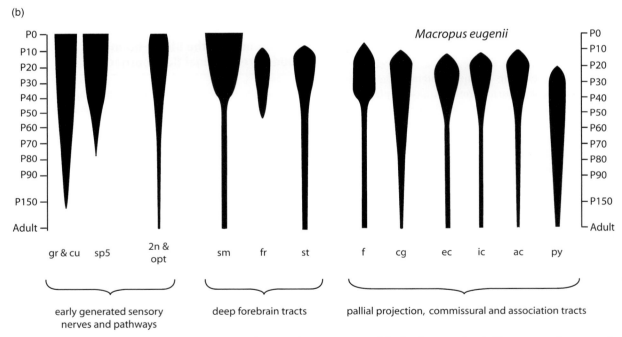

Figure 3.5 Relative timing of major tract development (expressed as a percentage of the developmental period from conception to weaning) in a variety of mammals (a). Note that in r-selection strategists like rodents and polyprotodont marsupials, the time to development of the major forebrain tracts occupies a high proportion of the developmental period, i.e. very little developmental time follows the establishment of major tracts, whereas K-selection strategists, such as diprotodont marsupials and primates, have a large proportion of the developmental period occurring after the major tracts have developed. This illustration has been reproduced from Ashwell *et al.* (1996a) with kind permission of S Karger AG, Basel. Panel (b) shows the relative strength of GAP43 immunoreactivity in major tracts within the developing tammar wallaby brain. This figure is based on data from Hassiotis *et al.* (2002). 2n – optic nerve; ac – anterior commissure; cg – cingulum; cu – cuneate fasciculus; ec – external capsule; f – fornix; fr – fasciculus retroflexus; gr – gracile fasciculus; ic – internal capsule; mfb – medial forebrain bundle; mt – mamillothalamic tract; opt – optic tract; pc – posterior commissure; py – pyramidal tract; sm – stria medullaris thalami; sp5 – spinal trigeminal tract; st – stria terminalis.

Comparison of developmental event timing between marsupials and placentals: the visual system perspective

Given the striking differences between marsupials and placentals in modes of development, one might expect comparable differences in the timing of developmental events in the central nervous system, but the currently available data are incomplete. In this regard, the part of the nervous system, which has been best studied in the widest range of mammals, is the visual system. It is clear that all the significant developmental events observed in the visual systems of placental mammals also take place in marsupials (Dreher and Robinson, 1988; Robinson and Dreher, 1990). When the relative timing of visual system development is calculated as a proportion of the relevant developmental time-span (i.e. percentage of the caecal or 'blind' period – the time from conception to eye opening), some general observations may be made. Developmental events in the early part of the caecal period (e.g. generation of neurons of visual centres in the brain such as the dorsal lateral geniculate thalamic nucleus, superior colliculus and visual cortex) appear to occur 10 to 15% earlier in poly- and diprotodont marsupials (opossum and wallabies) than in placental mammals (represented by rodents, carnivores and primates). Growth of retinal ganglion cell axons also occurs relatively earlier in marsupials than in placentals (24 to 31% of the caecal period in marsupials like the tammar wallaby and Virginia opossum, compared to 38 to 43% for placental mammals) (Robinson and Dreher, 1990). On the other hand, visual system developmental events in the later half of the caecal period occur at approximately the same relative time in marsupial and placental mammals (Robinson and Dreher, 1990). These later events include: the timing of the peak in retinal ganglion cell populations (at about 52 to 62% of the caecal period in marsupials and 53 to 58% in placental mammals); the timing of segregation of ipsi- and contralateral retinal ganglion cell terminals in the visual system centres of the thalamus and midbrain – the dorsal lateral geniculate nucleus and superior colliculus (at about 60 to 82% of the caecal period in marsupials compared to 73 to 85% in placental mammals); and the arrival of visual cortex terminals in the superior colliculus (at about 65% of the caecal period in the tammar wallaby, compared to a range of 68 to 70% in three placental mammals) (Robinson and Dreher, 1990).

The proportionally longer time-span between early and late developmental events in the visual system probably reflects the slower pace of pouch development in marsupials compared to intrauterine development of placental mammals.

Correlation of structural and behavioural development

The tammar wallaby is probably the best-studied Australian marsupial in terms of the depth and breadth of analyses of its development. This allows correlation of structural and functional events both within the range of body systems and with observed behaviour.

The newborn tammar weighs about 350 mg and will remain continuously attached to the maternal teat until P100, by which age its body weight has reached a little over 100 g. This period corresponds to the phase 2a period of lactation (Tyndale-Biscoe, 2005) and includes the bulk of neurogenesis (nerve-cell production) of the central nervous system (Figure 3.6). Within the CNS, the generation of cervical spinal-cord neurons, brainstem reticular formation and cranial-nerve nuclei neurons is an early event (prenatally to end of first week), reflecting the role of these in vital locomotor and cardiorespiratory functions around the time of birth. Macroneurons (i.e. neurons with cell-body diameters over 15 μm) of the cerebellum (deep cerebellar nuclei, Purkinje neurons) are also generated in the first week of postnatal life, although it is unclear at present whether these are functionally important in the life of the early pouch young marsupial. The next cohorts of neurons to be generated are in the diencephalon and caudalmost spinal cord, within a relatively narrow time-frame from around birth to P25. The former reflects the establishment of relay centres for, and growth of ascending connections from, the brainstem sensory nuclei, whereas the latter allows the generation of neurons for the mediation of ascending sensory information from the hindlimb and the establishment of lumbosacral spinal-cord reflex centres. Neurons destined for telencephalic structures and precerebellar nuclei are generated and mature over a protracted period (into the second or third postnatal month). Major ascending fibre tracts from the brainstem are established in a caudal-to-rostral (i.e. tail-to-head) sequence following the neurogenetic gradients, although growth of olfactory projections proceeds in the opposite direction (i.e. from nasal cavity towards caudal olfactory cortex).

Major behavioural changes occur during the second hundred days of postnatal life and by the end of this period the young tammar is physiologically independent of its mother (Tyndale-Biscoe, 2005). During the period from P100 to P200, the young tammar is intermittently attached to the teat (phase 2b of lactation) and the major neurological feature of this period is the growth of fine axonal and dendritic processes, the formation of corticocortical (association and commissural) connections and the maturation of function, although some neurogenesis is still in process. For example,

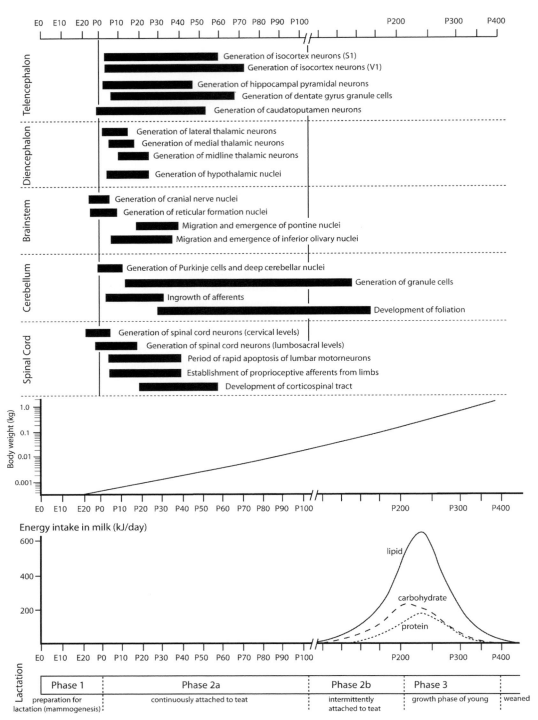

Figure 3.6 Time-course of significant developmental events in major parts of the brain of the tammar wallaby (*Macropus eugenii*) related to changes in body weight, nutrient consumption and lactation phases. Note the change in time scale at P100. Data for body weight, nutrient consumption in milk and lactation phases are drawn from Tyndale-Biscoe (2005). Nutrient consumption in milk is minimal until after P100, when the bulk of somatic growth occurs. Note the precocious development of cervical spinal-cord and brainstem nuclei in contrast to the extended development of other brain regions.

small neurons of the cerebellar cortex do not complete their neurogenesis until the later part of this period (i.e. fifth postnatal month) long after all other neuronal populations (with the possible exception of small neurons of the olfactory bulb) have been generated.

Phase 2b sees the completion of functional maturation of all levels of the major sensory systems. As an indicator of this, eye opening occurs at about P140 and the first auditory evoked response happens at P114. Functional maturation of the hypothalamus and its descending pathways to brainstem autonomic centres is indicated by the emergence of functional indicators. The ability to shiver develops in the tammar at P150, indicating the onset of thermoregulation in the face of low ambient temperature. The ability to pant and lick fur appears from about P200 in young tammars, suggesting that the somewhat more behaviourally complex thermoregulation in response to elevated temperatures lags behind cold resistance. Locomotor development proceeds through this period so that the young tammar can stand at about P200 and hop at P210, but little is known about those structural and physiological changes in the motor pathways of this species, which allow these changes. Certainly rubrospinal and reticulospinal pathways mediating descending control of the hindlimbs are probably anatomically mature long before the end of phase 2a, so structural maturation is unlikely to be the limiting factor determining locomotor maturation.

Most morphological changes in the brain during phase 2b involve growth of dendritic processes and axon collaterals, with concomitant deepening of cerebral fissures. Since foliation of the cerebellum is largely driven by expansion of the cerebellar granule cell population, foliation is completed shortly after granule-cell neurogenesis (i.e. at the sixth postnatal month).

It is significant that all structural and functional maturation of the central nervous system occurs well in advance of the major somatic growth period of the young tammar (i.e. phase 3) and before energy and nutrient intake peaks.

Sexual dimorphism of the nervous system and its development

Sexual dimorphism in the brains of placental mammals

In placental mammals, differences between the two sexes in reproductive and defensive behaviour appear to be based on morphological differences in specific neural circuits. These gender differences (sexual dimorphisms) in neural organisation have naturally been studied most extensively in rodents and are particularly striking in the hypothalamus, the bed nuclei of the stria terminalis (ST) and the basal forebrain (Shah et al., 2004).

One of the most striking sexual dimorphisms in the rodent brain concerns the medial preoptic nucleus (MPO), which has been shown to have an intensely staining component (sexually dimorphic nucleus of the medial preoptic area – SDN-POA), which is 2.4 to 5.0 times larger in males than in females (Gorski et al., 1978, 1980; Jacobson et al., 1980). The cytoarchitectural sexual dimorphism of the SDN-POA in the laboratory rat is established during the first 10 days of postnatal life under the influence of gonadal steroids and depends on the presence of aromatase activity to catalyse the conversion of androgens like testosterone to oestrogen (Jacobson et al., 1980, 1985).

Central and posteromedial subnuclei of the ST have also been shown to be larger in male compared to female rats, whereas lateral and medial anterior ST and the anteroventral periventricular nucleus of the hypothalamus have been found to be bigger in females (Chung et al., 2000). Homologous regions of the ST are also larger in men than women (Allen and Gorski, 1990). Experiments in rodents have shown that sexual dimorphism in these structure develops around the time of birth under the influence of testosterone (Chung et al., 2000).

More recent studies have shown the existence of an additional sexually dimorphic area in the basal forebrain of rodents, identified on the basis of gender differences in the distribution of androgen receptor (AR) expressing neurons. This area consists of AR-positive clusters of neurons in the region of the olfactory tubercle, ventral pallidum and islands of Calleja (granule-cell clusters) (Shah et al., 2004).

Most of these studies have been undertaken in laboratory rats with attempts to extend the findings to humans because of the clinical significance of sexual dimorphism. Some studies have used a comparative approach looking for a relationship between the magnitude of sexual dimorphism and mating behaviours in a variety of species of placental mammals (Commins and Yahr, 1984). For example, Shapiro and colleagues found that polygamous voles display greater levels of sexual dimorphism in the medial preoptic area and anterior hypothalamus than monogamous vole species (Shapiro et al., 1991), indicating that mating behaviour in quite closely related species is correlated with sexual dimorphism of neural structures.

Sexual dimorphism in the brains of marsupials

To date, only a few sexually dimorphic regions have been found in the brains and spinal cords of marsupials (lateral septal nucleus, spinal nucleus of the bulbocavernosus muscle) and all of these studies have been confined to Ameridelphians (for review see Gilmore, 2002). No apparent sex differences in oestrogen- and androgen-receptor immunoreactive structures have been found in American marsupial brains (Gilmore, 2002), but aromatase activity is present in the juvenile marsupial central nervous system (Callard *et al.*, 1982). This raises the question of whether structural sexual dimorphism of the brain and spinal cord is less striking in marsupials compared to placental mammals, perhaps because of the very different developmental environment. This hypothesis could be falsified by examining sexual dimorphism in a wider range of marsupials (e.g. diprotodont and polyprotodont marsupials from Australia), but remains essentially untested.

On the other hand, there are clear similarities between the brains of marsupials and placentals in the distribution of sex-hormone receptors. Localisation of androgen receptors in the brains of Ameridelphian marsupials is comparable to that in rodents; i.e. concentrations in lateral septal nuclei; medial divisions of the bed nuclei of the stria terminalis; medial preoptic area and constituent nuclei; nucleus of the lateral olfactory tubercle; central, anterior cortical and posterior amygdaloid nuclei; subiculum of the hippocampal formation; and various hypothalamic nuclei (ventromedial, arcuate and ventral premamillary) (Iqbal *et al.*, 1995b). Oestrogen binding is undetectable in newborn *Monodelphis domestica*, but emerges at P4 and peaks at about P16, roughly corresponding in developmental stage with the temporal changes of oestrogen binding found in rodents (Etgen and Fadem, 1989; Fox *et al.*, 1991b). The binding sites for oestrogen in the adult *Monodelphis* brain (medial preoptic area, ventral septal nuclei, medial division of the bed nuclei of the stria terminalis, lateral part of the ventromedial hypothalamic nucleus, premamillary and arcuate hypothalamic nuclei, posterior amygdaloid nucleus) are broadly similar to those seen for androgen binding and are also very similar to that for the laboratory rat (Fox *et al.*, 1991a).

Sexual dimorphism in the localisation of cholecystokinin immunoreactive structures and the distribution of arginine vasopressin immunoreactive fibres has been reported in Ameridelphians (Gilmore, 2002). In the Brazilian opossum, males have more cholecystokinin immunoreactive cells and fibres in the medial and periventricular preoptic area of the hypothalamus than females (Fox *et al.*, 1990, 1991c) and this difference emerges at about P25 (Fox *et al.*, 1991c) coincident with emergence of cholecystokinin binding (Kuehl-Kovarik *et al.*, 1993). Cholecystokinin in the brains of Australian marsupials (macropods and a quoll) is structurally similar to that in many placental mammals (Fan *et al.*, 1988; Johnsen and Shulkes, 1993), but no information is currently available concerning whether cholecystokinin has a sexually dimorphic distribution in the brains of Australian marsupials.

No oestrogen or aromatisation activity has been found in adult diprotodont brains (i.e. *Macropus eugenii* and *Trichosurus vulpecula*; Callard *et al.*, 1982), but high levels of aromatase activity are present in the behaviourally important 'limbic areas' (hypothalamus, preoptic area, amygdala, septal nuclei) of P10 to P25 male tammar wallabies and progressively decline with age. This follows a similar time-course to the changes in aromatase activity seen in an Ameridelphian with demonstrated sexual dimorphism (Fadem *et al.*, 1993).

The fine-scale effects of testosterone on the central nervous system of Australian marsupials has not been studied in detail, but castration, whether at P25, P150 or adulthood, has profound effects on male sexual behaviour (Rudd *et al.*, 1996). In male and female tammar wallabies, the expression of male-type sexual behaviour appears to be completely dependent on the adult steroid hormone environment. For example, testosterone-implanted female tammars exhibit just as much masculine sexual behaviour as normal male tammars. Conversely, the adoption of female behaviour (e.g. the taking up of the characteristic tammar birth posture) can be induced with equal ease in both castrated males and females. Rudd and colleagues argue that this indicates that brain mechanisms regulating masculine sexual behaviour in the tammar are not structurally differentiated sexually, in contrast to the obvious neural sexual dimorphism seen in placental mammals (Rudd *et al.*, 1996; discussion above). These observations are consistent with the lack of any obvious sexual dimorphism in the preoptic area of the tammar hypothalamus (Cheng *et al.*, 2003).

What remains unknown?

While it is likely that the broad patterns of gene expression in the embryonic forebrain are similar across all mammals (and indeed all tetrapods), the details of how these develop both topographically and temporally in the prenatal marsupial brain have never been explored. It is likely that the differential timing of forebrain development in marsupials vs. placentals is reflected in functionally significant differences in morphogenic gene expression, but this remains to be explored.

Sexual dimorphism in the CNS of Australian marsupials is an unexplored territory, and one which demands attention. It may be that macropod brains do not have major structural differences between the genders; but this makes the contrast with placental brains all the more fascinating and provides an intriguing model to explore the circuitry controlling sexual behaviour. Just as fascinating is the question of how the remarkable life cycle of small dasyurids like *Antechinus stuartii* influences, or arises from, gender differences in brain circuitry. The range of gender-specific behaviours shown by small marsupial carnivores provides an excellent opportunity to examine the correlation between sexual dimorphism in the preoptic area/hypothalamus, the bed nuclei of the stria terminalis and the basal forebrain, and behaviour. Furthermore, the marsupials in general and Australian marsupials in particular present unique developmental models in which to study the factors associated with the development of sexual dimorphism in the brain and gender-specific behaviour, because they spend much of their early developmental life accessible to experimenters in the maternal pouch.

Part II

Regional neurobiology

4

Ventral hindbrain and midbrain

K. W. S. Ashwell

Summary

The adult hindbrain consists of the pons and medulla oblongata (ventral hindbrain), and the cerebellum, but, as noted in a previous chapter, the pons is not a consistently defined brain segment. Key functional elements within the ventral hindbrain are the cranial nerve nuclei (both sensory or afferent, and motor or efferent), somatosensory relay nuclei, precerebellar nuclei, reticular formation and cardiorespiratory control centres.

Organisation and structure of the ventral hindbrain structures are highly conserved across all mammals. Some particular examples of specialisation in the hindbrain of Australian marsupials may be seen in the enlarged lateral lobe (presumably expanded cochlear nuclei) of *Acrobates pygmaeus* and the trigeminal nuclei of some diprotodonts (see Chapter 10). Ventral hindbrain centres (in particular cardiorespiratory and reticulospinal locomotor units) are critically important in the behaviour of early pouch young marsupials and this is reflected in their precocious development relative to forebrain centres.

The midbrain, or mesencephalon, lies between the forebrain and hindbrain and is an important thoroughfare for pathways connecting the two. Broadly speaking, the midbrain consists of three regions: the tectum, the tegmentum and the isthmus. The tectum is the most dorsal part and contains the superior and inferior colliculi. The deepest part of the tectum includes the periaqueductal grey matter surrounding the cerebral aqueduct. Nuclei within the midbrain tegmentum include the red nucleus and substantia nigra, which are both involved in motor pathways, although in differing ways. The isthmus is the most caudal part of the midbrain, located at the transition to the hindbrain.

The superior and inferior colliculi are laminated structures, which receive input predominantly from the visual and auditory pathways, respectively. Lamination of the superior colliculus is similar across all mammals, with superficial grey and optic layers receiving visual input, whereas deeper layers receive somatosensory input. The inferior colliculus is divided into a central nucleus, a dorsal cortical part and a lateral external cortical part. Lamination of the external part is similar in marsupials to that in rodents.

The periaqueductal grey is concerned with co-ordinating the diverse responses that mammals show to physically and emotionally stressful situations. Although no functional studies have been done of the periaqueductal region in Australian marsupials, the similar chemoarchitectural features in all mammals suggest that this longitudinal functional organisation is common to monotreme, marsupial and placental mammals.

The midbrain/isthmus also contains two nuclei concerned with eye muscle control (i.e. the oculomotor and trochlear nuclei) located ventral to the periaqueductal grey. Additional functionally important nuclei include the terminal nuclei of the optic tract and the cholinergic interpeduncular nucleus.

Serotonergic neurons in the brainstem of a representative diprotodont (the tammar wallaby) are broadly similar in both morphology and distribution to those in American polyprotodont marsupials, monotremes and placental mammals. The degree of lateral spread of serotonergic neurons of the rostral brainstem is associated with the emergence during evolution of more robust ascending serotonergic projections and the extent of lateral spread of serotonergic neurons in diprotodonts is comparable to that seen in rodents and carnivores.

Introduction to the hindbrain

The hindbrain (rhombencephalon) is usually said to consist of the pons, medulla and cerebellum; but, as noted earlier,

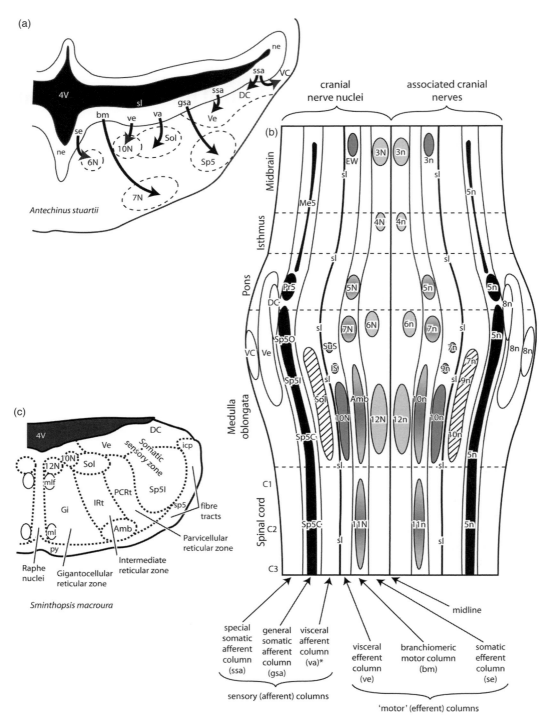

Figure 4.1 Developmental origin of the cranial nerve nuclei and zonal organisation of the brainstem. Figure (a) shows a coronal section through the rostral medulla of a representative early pouch young polyprotodont marsupial, illustrating the origin of the cranial nerve nuclei from discrete regions of the neuroepithelium on the floor of the fourth ventricle. 'Motor' nuclei (somatic efferent – se; branchiomeric – bm; visceral efferent – ve) arise from the neuroepithelium (germinal zone) medial to the sulcus limitans (sl), whereas sensory nuclei (visceral afferent – va; general somatic afferent – gsa; special somatic afferent – ssa) arise from the neuroepithelium lateral to the sl. Note that branchiomeric and general somatic afferent neurons subsequently migrate towards the medial and lateral parts of the ventral medullary surface, respectively. Figure (b) shows a schematic dorsal view representation of the cranial nerve nuclear columns in

the pons is not a valid developmental entity and not consistent in extent in adult mammals. The cerebellum and precerebellar nuclei will be considered in Chapter 5, so the current section will focus on the cranial nerve and reticular nuclei of the ventral hindbrain.

The cranial nerve nuclei of the rostral spinal cord, hind- and midbrain (Figure 4.1a, b) can be divided into sensory components (general somatic afferent – trigeminal sensory nuclei complex; special somatic afferent – cochlear and vestibular nuclei; visceral afferent – nucleus of the solitary tract) and motor (efferent) components (somatic efferent neurons such as those in the hypoglossal, abducens, trochlear and oculomotor nuclei; branchiomeric – neurons, such as those in the spinal accessory nucleus, nucleus ambiguus, facial and trigeminal motor nuclei; and visceral efferent neurons (such as those in the superior and inferior salivatory, lacrimal nuclei, and in the dorsal motor nucleus of the vagus). Note that somatic efferent nuclei (oculomotor and trochlear) and visceral efferent nuclei (Edinger–Westphal nucleus) of the midbrain/isthmus will be considered in the following section of this chapter.

The reticular formation receives input from a wide variety of sensory modalities and is concerned with an assortment of automatic motor functions, in particular those involving rhythmic motor patterns (e.g. respiration, walking, hopping and chewing) and the regulation of motor tone. The reticular formation influences the activity of the brainstem and spinal cord by reticulobulbar and reticulospinal projections, respectively. In addition, the reticular formation includes nuclei that provide ascending projections to the forebrain.

The general topography of the pons and medulla is organised according to wedge-shaped segments radiating from the fourth ventricle (Figures 4.1c, 4.2a to c). The following discussion will deal with the major sensory nuclei of the pons and medulla, followed by motor nuclei, reticular formation nuclei and assorted catecholaminergic nuclei.

Sensory nuclei of the hindbrain

Nucleus of the solitary tract

The nucleus of the solitary tract (Sol) receives input from gustatory and somatosensory axons from the oral cavity (by cranial nerves 7 and 9; Figure 4.1b) and general visceral afferents from thoracic and abdominal organs (by cranial nerve 10). Strictly speaking, the Sol is really two nuclei (Butler and Hodos, 2005): a rostral gustatory nucleus, serving taste from ecto- and endodermally derived receptors (i.e. in rostral foregut – oral cavity, tongue and pharynx), and hence a mixture of somatic and visceral afferents; and a caudal general visceral afferent component. As in placental mammals, Sol of the marsupial brainstem shows cytoarchitectural and chemoarchitectural subdivisions (Figure 4.2d to g). These subnuclei are arranged adjacent (mostly medial and dorsal) to the heavily myelinated solitary tract (sol) that extends from the level of the facial nucleus to as far caudally as the spinomedullary junction. The sol contains fibres from the facial nerve rostrally, the glossopharyngeal nerve in its intermediate levels and the vagus nerve in its caudal parts. The Sol is the initial relay for input from baroreceptor (i.e. arterial blood pressure), chemoreceptor (O_2 levels in arterial blood), cardiac afferents (right atrial stretch receptors) and other vagal afferents from airways and the gastrointestinal tract.

Dorsal column nuclei

The somatosensory system (dorsal column and trigeminal sensory nuclei) will be dealt with in detail in Chapter 10, but a brief consideration focussed on nuclear topography will be given here.

The medullary nuclei that receive somatosensory afferents from the body are the gracile, cuneate and external

---

Caption for Figure 4.1 (Cont.)

the midbrain/isthmus, pons, medulla and cervical spinal cord (left half of the diagram) and the associated cranial nerves (right half of the diagram). *Note that the rostral parts of the Sol receive input from ectodermally derived taste receptors on the rostral tongue, so the rostral Sol is more properly considered a somatic afferent input region (Butler and Hodos, 2005). Figure (c) illustrates the organisation of the brainstem reticular formation into wedge-shaped regions radiating from the fourth ventricle floor of the medulla of a representative polyprotodont. From medial to lateral these are the raphe nuclei, giganto-/magnocellular reticular, intermediate reticular and parvicellular reticular zones. Fibre tracts are arranged predominantly around the ventrolateral margin of the medulla (icp, sp5, py), or alongside the raphe nuclei (mlf, ml). 3N – oculomotor nu.; 3n – oculomotor nerve; 4N – trochlear nu.; 4n – trochlear nerve; 4V – fourth ventricle; 5N – motor trigeminal nu.; 5n – trigeminal nerve; 6N – abducens nu.; 6n – abducens nerve; 7N – facial nu.; 7n – facial nerve; 8n – vestibulocochlear nerve; 9n – glossopharyngeal nerve; 10N – vagus nerve nu.; 10n – vagus nerve; 11N – accessory nerve nu.; 11n – accessory nerve; 12N – hypoglossal nu.; 12n – hypoglossal nerve; Amb – ambiguus nu.; DC – dorsal cochlear nu.; EW – Edinger–Westphal nu.; Gi – gigantocellular reticular nu.; icp – inferior cerebellar peduncle; IRt – intermediate reticular nu.; IS – inferior salivatory nu.; Me5 – mesencephalic trigeminal nu.; ml – medial lemniscus; mlf – medial longitudinal fasciculus; ne – neuroepithelium; PCRt – parvicellular reticular nu.; Pr5 – principal sensory trigeminal nu.; py – pyramidal tract; sl – sulcus limitans; Sol – solitary nu.; Sp5 – spinal trigeminal nu.; sp5 – spinal trigeminal tract; Sp5C – spinal trigeminal nu., caudalis; Sp5I – spinal trigeminal nu., interpolaris; Sp5O – spinal trigeminal nu., oralis; SuS – superior salivatory nu.; VC – ventral cochlear nu.; Ve – vestibular nuclei.

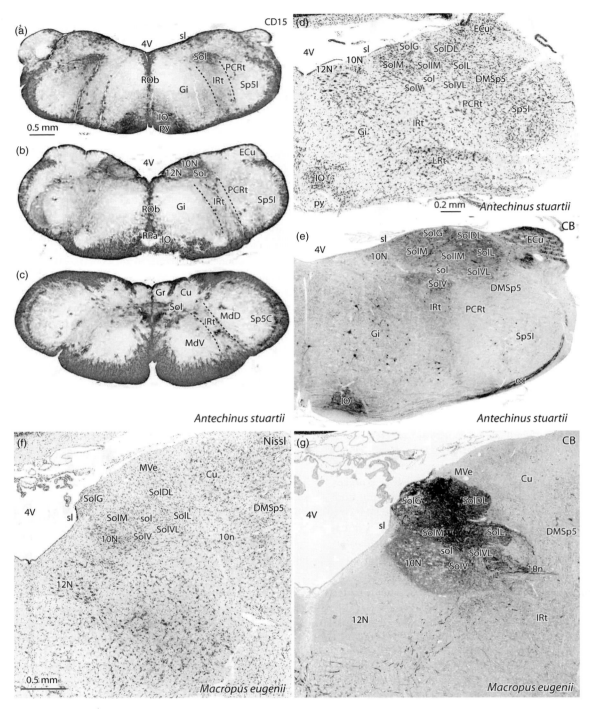

Figure 4.2 Topography of the brainstem as seen in coronal sections through the medulla of representative poly- and diprotodont marsupials. Immunoreactivity for CD15 in the brown antechinus reveals distinct glial populations in the wedge-shaped subdivisions of the medulla (rostral to caudal medulla, (a) to (c)). From medial to lateral these are: (i) the raphe nuclei in the midline (e.g. ROb, RPa); flanked by (ii) large-celled reticular formation nuclei (e.g. Gi, rostrally and MdV caudally); (iii) the intermediate reticular zone (IRt) in line with the vagal sensorimotor complex (10N, Sol); (iv) parvicellular reticular formation nuclei (e.g. PCRt rostrally and MdD caudally) and (v) most laterally, the somatic sensory nuclei, e.g. spinal trigeminal nuclei (Sp5I and Sp5C) and external cuneate nuclei. The sensory nu. of the vagal complex can be subdivided into a variety of subnuclei on the basis of cytoarchitecture and the relationship to the solitary tract. These

cuneate nuclei (Figure 4.3) and nucleus Z, all of which can be identified in marsupial brainstem. In addition, a median accessory nucleus of the medulla is present near the midline for the tail representation of diprotodonts (Johnson, 1977). The dorsal column fibre tracts are organised somatotopically, with input from the most caudal spinal segments in the most medial parts of the gracile fasciculus and fibres from progressively more rostral segments being added to the lateral side of the dorsal columns. The gracile and cuneate nuclei are also somatotopically organised in all mammals (Figure 4.3a inset), and inputs from muscles and joints are distributed to the external cuneate and nucleus Z, whereas cutaneous inputs are distributed to the gracile and cuneate nuclei. Unfortunately, only limited information is available as to how these aspects of dorsal column organisation are arranged in Australian marsupials. In the brush-tailed possum (Culberson, 1987), somatotopic organisation of the dorsal columns and their termination in the dorsal column nuclei is broadly similar to that seen in placental mammals. Cervical dorsal root fibres from rostral segmental levels enter more lateral parts of the cuneate nucleus, but terminal projection fields are large and extensively overlap. The dorsal part (or 'shell') of the cuneate nucleus has the more topographically precise projection. In placental mammals, the input to this more dorsal part of the cuneate is derived from cutaneous mechanoreceptors, which are most numerous in those cervical roots innervating the distal forelimb; and this region is also likely to carry distal forepaw representation in the brush-tailed possum (Culberson, 1987).

The spatial arrangement of termination of cervical dorsal root fibres in the external cuneate nucleus is more complex than that in the cuneate nucleus, with even more extensive overlap. Caudal cervical root fibres are distributed to the dorsal and medial parts of the external cuneate, whereas the terminals from rostral cervical root fibres are spread to ventral and ventrolateral parts of the external cuneate nucleus (Culberson, 1987).

The cuneate system of the brush-tailed possum is better-developed than the gracile component of the dorsal column system, consistent with the manipulative capabilities of possums. Nevertheless, neither the cytoarchitecture of the cuneate nuclei, nor the cervical afferent fibres appear to be as precisely arranged as in those placental mammals (e.g. primates) with well-developed digital dexterity (Culberson, 1987). In particular, the region of the cuneate nucleus likely to receive cutaneous forepaw afferent input in the brush-tailed possum does not exhibit the cytoarchitectural specialisations (cell 'clusters', 'bricks' or 'columns') seen in these nuclei in primates (Culberson, 1987).

The dorsal column nuclei project to the ventral posterolateral thalamus and their axons follow an initially ventral course (ia – internal arcuate fibres, Figure 4.3a) before crossing the midline and ascending as the medial lemniscus. Descending projections from the limb area of the somatosensory/motor cortex are extensive to the gracile nucleus in the brush-tailed possum (Martin et al., 1975b), consistent with the large hindlimb representation in possum somatosensory cortex, although the functional significance of these corticobulbar connections remains unexplored.

Trigeminal sensory nuclei

The trigeminal pathways are concerned with sensory inputs from the head and face. The primary sensory neurons of the pathway are located primarily in the trigeminal ganglion and the central processes of the primary neurons are distributed in the spinal trigeminal tract (Figure 4.3). In all mammals, the trigeminal sensory nuclear complex consists of the mesencephalic trigeminal nucleus (Me5), the main (or principal) trigeminal nucleus (Pr5) and the spinal trigeminal nucleus, which is further subdivided into oralis, interpolaris and caudalis components (Figure 4.1b). There is also a supratrigeminal nucleus concerned with masticatory reflexes and parasolitary nuclei, which receive input from the teeth pulp. In the brush-tailed possum, descending cortical projections to the trigeminal sensory nuclei are mainly derived from the somatosensory cortex located in ventral parietal cortex (Martin et al., 1971).

The mesencephalic trigeminal nucleus contains the primary neurons concerned with information from ipsilateral masticatory muscle spindles and periodontal receptors. Neurons in the main trigeminal nucleus and spinal

---

Caption for Figure 4.2 (Cont.)

subdivisions are equally evident in the small shrew-like polyprotodont *Antechinus stuartii* (d, e) as in the large macropod *Macropus eugenii* (f, g) in Nissl-stained (d, f) and calbindin-immunoreacted (CB) material (e, g). 4V – fourth ventricle; 10N – vagus nerve nu.; 10n – vagus nerve; 12N hypoglossal nu.; Cu – cuneate nu.; DMSp5 – dorsomedial spinal trigeminal nu.; ECu – external cuneate nu.; Gi – gigantocellular reticular nu.; Gr – gracile nu.; IO – inferior olivary nu.; IRt – intermediate reticular nu.; LRt – lateral reticular nu.; MdD – medullary reticular nu., dorsal; MdV – medullary reticular nu., ventral; MVe – medial vestibular nu.; oc – olivocerebellar tract; PCRt – parvicellular reticular nu.; py – pyramidal tract; ROb – raphe obscurus nu.; RPa – raphe pallidus nu.; sl – sulcus limitans; sol – solitary tract; Sol – solitary nu.; SolDL – dorsolateral part; SolG – gelatinous part; SolIM – intermediate part; SolL – lateral part; SolM – medial part; SolV – ventral part; SolVL – ventrolateral part; Sp5C – spinal trigeminal nu., caudal; Sp5I – spinal trigeminal nu., interpolaris.

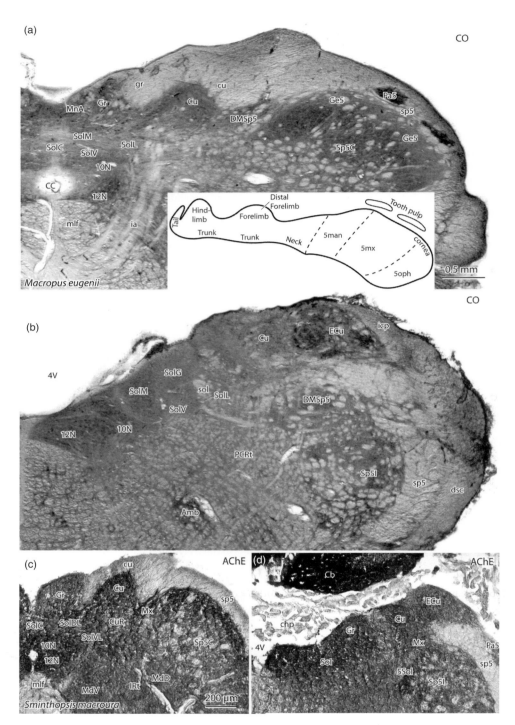

Figure 4.3 Topography of somatosensory nuclei in coronal sections through the medulla of the tammar wallaby (caudal medulla – (a), midstrocaudal medulla – (b)) and stripe-faced dunnart (caudal medulla – (c), midstrocaudal medulla – (d)). Sections through the wallaby medulla are stained for cytochrome oxidase (CO) while those through the dunnart have been stained for acetylcholinesterase (AChE). The line diagram in (a) illustrates the putative somatotopic organisation of these somatosensory nuclei (median accessory for the tail, gracile for hindlimb, cuneate for forelimb, spinal trigeminal and paratrigeminal nuclei for head and neck and tooth pulp). Note the presence of matrix (Mx) and trigeminosolitary (5Sol) transitional regions between the trigeminal nuclear complex and the solitary sensory

trigeminal nuclei receive input from the ipsilateral guard hairs, facial skin and teeth.

Auditory system nuclei

Although at least one Australian marsupial (*Acrobates pygmaeus*) has a highly specialised auditory pathway, most Australian marsupials have auditory nuclei that deviate very little in size from the mammalian norm (Glendenning and Masterton, 1998). The cochlear nuclei are the first relay centres for the ascending auditory pathways (see Chapter 11) and in all mammals are located laterally and/or superficially in the brainstem. As in rodents, marsupials have dorsal and ventral cochlear nuclei (Aitkin, 1996). In many placentals, the cochlear division of the vestibulocochlear nerve often subdivides the ventral cochlear nucleus into anteroventral (VCA) and posteroventral (VCP) components. These subdivisions are also visible in marsupials, but the line of separation is not always marked by the cochlear division of the vestibulocochlear nerve. In many marsupials, the cochlear nuclear complex lies medial to the restiform body of the inferior cerebellar peduncle (Aitkin *et al.*, 1986a; Aitkin *et al.*, 1991), but in macropods there is some migration of ventral cochlear nucleus neurons around the side of the inferior cerebellar peduncle (Cowley, 1973). Cochlear nerve afferents terminate within the cochlear nuclei in a tonotopic fashion, i.e. different frequencies are mapped onto the physical structure of the cochlear nuclei in a spatially ordered fashion. The dorsal cochlear nucleus of marsupials also shows a laminated structure (Cowley, 1973; Figure 4.4a to h) in contrast to the more homogeneous structure of primates and this may reflect the greater extent of processing of auditory input at the level of entry of the cochlear nerve, rather than at higher forebrain levels of the auditory pathway. Ascending connections from the cochlear nuclei to the superior olivary nuclei, nuclei of the lateral lemniscus and central nucleus of the inferior colliculus are tonotopically organised in placental mammals (Aitkin *et al.*, 1986b; see Chapter 11).

The superior olivary complex is involved in sound localisation and amplification, and Aitken and colleagues have identified four nuclei consistently observed in marsupials (Aitkin *et al.*, 1986a, 1991; Figure 4.4i, j). These are: (i) the triangular-shaped lateral superior olive (LSO), which is usually the largest component in arboreal marsupials (Aitkin, 1996); (ii) a column of fusiform cells, the medial superior olive (MSO); (iii) an ovoid mass of multipolar cells, the superior paraolivary nucleus (SPO) and (iv) the medial nucleus of the trapezoid body (MTz), which forms lines of neurons connecting the superior olivary complex to the midline. These subdivisions are particularly distinct in parvalbumin and calbindin immunoreacted sections (Figure 4.4i and j) and are surrounded by diffuse cellular regions, which are collectively referred to as the periolivary areas. A ventral periolivary nucleus with medial and lateral subdivisions can be identified in macropods and small polyprotodonts (see atlas plates D24, D25, W38 to W40; Figure 4.4i and j).

The nuclei of the lateral lemniscus lie at the boundary between the pons and midbrain. In marsupials, as in rodents, three nuclei can be identified, i.e. the ventral, intermediate and dorsal nuclei of the lateral lemniscus (see atlas plates D21, D22, W36, W37). In many placental mammals, the ventral nucleus is part of a monaural system (i.e. receiving input from one ear), whereas the dorsal nucleus is part of a binaural system.

The inferior colliculus will be considered with the midbrain/isthmic structures.

Cranial nerve motor nuclei of the hindbrain

Oromotor nuclei

The oromotor nuclei (trigeminal motor, facial and hypoglossal nuclei) are more than simple clusters of efferent motorneurons; all of these nuclei include interneurons and may even contribute projections to other areas of the brainstem. Another common feature of oromotor nuclei is myotopic organisation, i.e. particular muscles or muscle groups are driven by motorneurons in distinct parts of the nucleus.

Caption for Figure 4.3 (Cont.)

complex. These regions probably serve somatosensory input from rostral foregut structures. 4V – fourth ventricle; 5man – mandibular division; 5mx – maxillary division; 5oph – ophthalmic division; 10N – vagus nerve nu.; 12N – hypoglossal nu.; Amb – ambiguus nu.; Cb – cerebellum; CC – central canal; chp – choroid plexus; Cu – cuneate nu.; cu – cuneate fasciculus; CuR – cuneate nu., rotundus part; DMSp5 – dorsomedial spinal trigeminal nu.; dsc – dorsal spinocerebellar tract; ECu – external cuneate nu.; Ge5 – gelatinous layer of caudal Sp5; Gr – gracile nu.; gr – gracile fasciculus; ia – internal arcuate fibres; icp – inferior cerebellar peduncle; IRt – intermediate reticular nu.; MdD – medullary reticular nu., dorsal; MdV – medullary reticular nu., ventral; mlf – medial longitudinal fasciculus; MnA – median accessory nu. medulla; Pa5 – paratrigeminal nu.; PCRt – parvicellular reticular nu.; sol – solitary tract; Sol – solitary nu.; SolC – commissural part; SolDL – dorsolateral part; SolG – gelatinous part; SolL – lateral part; SolM – medial part; SolV – ventral part; SolVL – ventrolateral part; sp5 – spinal trigeminal tract; Sp5C – spinal trigeminal nu., caudal; Sp5I – spinal trigeminal nu., interpolaris.

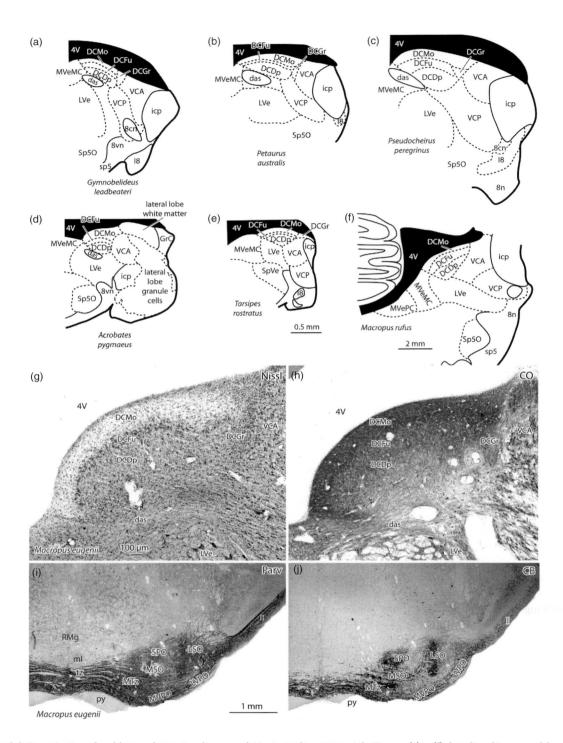

Figure 4.4 Organisation of cochlear and superior olivary nuclei in Australian marsupials. Figures (a) to (f) show line diagrams of dorsal (DC) and ventral cochlear (VCA, VCP) nuclei in three petaurids (a to c), the acrobatid feathertail glider (d), the tarsipedid honey possum (e) and a large macropod (f). Note the presence of the large lateral lobe in the feathertail glider. This strange structure appears to contain large numbers of granule cells and may be derived developmentally from one or more of the cochlear nuclei. Although undoubtedly part of the auditory system, it appears to also have connections with the cerebellum and is believed to play a key role in coordinating and guiding gliding behaviour (Aitkin and Nelson, 1989). The VCP is visible in more caudal sections through the feathertail glider brainstem.

The motor trigeminal nucleus of rodents is usually divided into a large dorsomedial division and a smaller ventromedial division. Some suggestion of subdivision is apparent in the main Mo5 of small polyprotodont marsupials (e.g. *Sminthopsis macroura*), but the division is more into a larger-celled rostral part and a smaller-celled caudal region. Nevertheless, a small accessory trigeminal motor nucleus (Acs5; atlas plates D23 to D25) is clearly present in marsupials, which may correspond to the ventromedial Mo5 of rodent brainstem. If this is correct, then Mo5 of marsupials probably contains motorneurons controlling jaw-closing muscles (masseter, temporalis, medial and lateral pterygoid muscles), whereas Acs5 probably contains motorneurons driving jaw-opening movements, i.e. the anterior belly of the digastric and the mylohyoid muscle.

Motorneurons within the facial motor nucleus are grouped into clusters or subnuclei concerned with driving particular muscles or muscle groups, separated by intervening white matter (see atlas plates D26, D27, W42). Precise data on myotopic organisation are only available for the brush-tailed possum (Provis, 1977) of all Australian marsupials, but even these limited data suggest that the myotopic organisation of the facial motor nucleus is similar in all therians. Motorneurons innervating the vibrissal or proboscal muscles of the possum are located laterally within the nucleus, but the lateral subnucleus (7L) is relatively small in the brush-tailed possum. Those driving the posterior auricular muscle are located in a large medial nucleus (which may be divided into dorsal and ventral components; 7DM, 7VM), whereas the anterior auricular (and probably orbicularis oculi) motorneurons are located in the dorsal intermediate subnucleus (7DI). Motorneurons innervating muscles moving the malar vibrissae are situated within the centre of the intermediate region of the nucleus and mentalis motorneurons are found in the most ventral intermediate subnucleus (7VI). In rodents (and probably marsupials), motorneurons driving deep facial muscles (posterior belly of the digastric, stylohyoid and stapedius) are located outside the main body of the facial motor nucleus, in the accessory facial nucleus (Acs7; see atlas plates D26, W41 and W42). In all therians, these latter motorneurons are found dorsal to the main nucleus, along the exiting root of the facial nerve, but their innervation of the deep facial muscles has not yet been confirmed in poly- or diprotodont marsupials.

The hypoglossal nucleus contains motorneurons driving the intrinsic and extrinsic muscles of the tongue. The hypoglossal nucleus has a clear myotopic organisation in rodents, but there are no data concerning this type of organisation in any Australian marsupial.

Abducens nucleus

The abducens nucleus lies relatively close to the floor of the fourth ventricle adjacent to the medial longitudinal fasciculus, a fibre tract interconnecting the eye muscle nuclei and accessory oculomotor nuclei. It also has a close relationship to the internal genu of the facial nerve (D26, W40), which bends around the abducens nucleus before exiting dorsal to the spinal trigeminal tract nuclei in marsupials. The abducens nucleus supplies the lateral rectus muscle. In macropods, it consists of a cytoarchitecturally homogeneous central region (6N) surrounded by a sparsely populated para abducens region (see atlas plate W40). The latter probably contains neurons concerned with internuclear connections co-ordinating eye movements.

Reticular formation of the hindbrain

The reticular formation derives its name from the net-like arrangement of constituent neurons and their processes interspersed between the specific sensory and motor nuclei of the brainstem. The reticular formation is arranged in longitudinal columns through the brainstem (Figure 4.1c). The

---

Caption for Figure 4.4 (Cont.)

Photomicrographs (g) and (h) show the dorsal cochlear nu. and its lamination in *M. eugenii* in Nissl-stained and cytochromes-oxidase-reacted coronal sections. Note the strong CO activity in the superficial input zone (DCMo) of the dorsal cochlear nu.. Photomicrographs (i) and (j) show parvalbumin and calbindin immunoreacted coronal sections through the superior olivary nuclear complex of the tammar wallaby. Drawings of cochlear nuclei are based on photomicrographs in Aitkin (1996) (a to e) and drawings in Cowley (1973) (f). Scale bar in (e) applies to (a) to (e). 4V – fourth ventricle; 8cn – cochlear root of eighth nerve; 8n – vestibulocochlear nerve; 8vn – vestibular root of eighth nerve; das – dorsal acoustic stria; DC – dorsal cochlear nu.; DCDp – deep layer; DCFu – fusiform part; DCGr – granular layer; DCMo – molecular layer; GrC – granule cell layer, cochlear nu.; I8 – interstitial nu. of eighth nerve; icp – inferior cerebellar peduncle; ll – lateral lemniscus; LSO – lateral superior olive; LVe – lateral vestibular nu.; LVPO – lateroventral periolivary nu.; ml – medial lemniscus; MSO – medial superior olive; MTz – medial trapezoid nu.; MVeMC – medial vestibular nu., magnocellular; MVePC – medial vestibular nu., parvicellular; MVPO – medioventral periolivary nu.; py – pyramidal tract; RMg – raphe magnus nu.; sp5 – spinal trigeminal tract; Sp5O – spinal trigeminal nu., oralis; SPO – superior paraolivary nu.; SpVe – spinal vestibular nu.; tz – trapezoid body; VCA – ventral cochlear nu., anterior; VCP – ventral cochlear nu., posterior.

most medial of these is a column of raphe nuclei close to the midline. These are flanked by magnocellular (large-celled) and/or gigantocellular columns, which are, in turn, flanked by the intermediate reticular zone. Lateral to the intermediate reticular zone is the parvicellular (small-celled) neuron column.

In all vertebrates, the parvicellular column nuclei tend to be the sensory or afferent zone of the reticular formation and receive input from the nearby cranial nerve sensory nuclei (Sol, Sp5) in the dorsal and lateral side of the brainstem. The magno- and gigantocellular columns are the efferent or output divisions of the reticular formation, with projections to the spinal cord or higher levels of the central nervous system. The intervening intermediate reticular zone is slightly more reactive for acetylcholinesterase than surrounding reticular nuclei and developmentally occupies the regions between the derivatives of the alar and basal plates of the prenatal brainstem. The intermediate zone contains many catecholaminergic neurons and engages in projections to the parabrachial nuclei, nucleus of the solitary tract and phrenic motorneurons of the cervical spinal cord. Descending cortical influence on the midline and medial magnocellular components of the reticular formation in the brush-tailed possum appear to be mainly derived from the cerebral cortex rostral to the α sulcus, whereas cortical influences on the lateral parvicellular components of the reticular formation come mainly from the cerebral cortex caudal to the α sulcus (Martin et al., 1971).

The raphe nuclei of the brainstem are the main site of serotonergic neurons in the brain, but note that not all raphe nuclei are serotonergic. The boundaries of serotonergic cell groups are often poorly defined, so these cell groups are commonly classified using an alphanumeric system, numbering them from caudal to rostral. Groups B1 to B4 develop from a caudal group in the medulla during late embryonic life in rodents (or early postnatal life in marsupials), whereas groups B5 to B9 are generated in a rostral group at the same stages.

Development of the marsupial ventral hindbrain

Cytoarchitectural and connectional development

As noted in the general-development chapter, several nuclear elements of the marsupial hindbrain (e.g. magno- and gigantocellular reticular nuclei, somatic efferent cranial nerve nuclei, trigeminal sensory nuclei) exhibit a more advanced level of maturation at birth compared to others (e.g. small cells of the cerebellum, precerebellar nuclei, auditory nuclei), reflecting the special demands of early postnatal life.

The constituent neurons of most somatic efferent nuclei (e.g. 6N, 5N, 12N) are all in position by the time of birth in poly- and diprotodont marsupials (Swanson et al., 1999; see atlas plates WP0/6 to WP0/11), with the possible exception of the facial motor nucleus, for which at least some constituent neurons appear to be still in the process of migration at birth in diprotodonts (WP0/8). In at least one polyprotodont, facial nucleus motorneurons are still situated near the ventricular floor at birth and take up to five days to reach their final settling site (Swanson et al., 1999). Nevertheless, the robust internal genu, efferent intramedullary limb and extracranial course of the facial nerve at birth in the tammar wallaby and Brazilian opossum indicate that motor fibres of the facial nerve are already distributed to at least the caudal facial musculature in early pouch young marsupials (Swanson et al., 1999). Migration of facial motorneurons to their final resting site appears to occur in close association with radial glial fibres (Swanson et al., 1999). The characteristic marsupial course of the intramedullary facial nerve fibres (i.e. dorsal to the spinal trigeminal nuclear complex, see Chapter 2) probably arises because trigeminal sensory nuclei migrate ventrally before the efferent facial motor fibres reach the lateral hindbrain, although this has never been directly studied because the key events would occur prior to birth.

The detailed development of the trigeminal sensory nuclei will be considered in the relevant systems chapter, but a brief summary of cytoarchitectural development in the context of other hindbrain nuclei is relevant here. All components of the trigeminal sensory nuclear complex are visible at birth in marsupials and (at least in the case of the tammar wallaby) receive afferent fibres, but internal features in the spinal trigeminal nucleus do not emerge clearly until about P20 in the tammar (Waite et al., 1994). Rostrocaudal subdivisions of the spinal nucleus emerge by P30, coincident with the first appearance of cytochrome oxidase reactive patches in the principal nucleus corresponding to afferents from the rows of facial whiskers. Although the functional and chemoarchitectural development of the trigeminal sensory nuclear complex is protracted in marsupials, there is no doubt that the basic neuronal machinery for mediating simple reflex responses is present at the time of birth.

The solitary tract is present by birth in marsupials, as would be expected, given the early development of the cranial nerve ganglia (7Gn, 9Gn, 10Gn) contributing central processes to it. Nevertheless, cytoarchitectural maturation of the nucleus of the solitary tract in marsupials is delayed and prolonged relative to placental mammals, which may have significance for the control of ventilation (see below). In the tammar wallaby, cytoarchitectural maturation of the nucleus proceeds over at least the first two months of post-natal life.

Surprisingly, at least some neurons of the raphe nuclei appear quite early in the marsupial hindbrain (see atlas plates WP0/8 and 9). Some neurons of the presumptive raphe magnus, pallidus and obscurus can be seen even at the time of birth in the tammar wallaby. Nevertheless, nothing is known about the functional and connectional development of these neuronal groups in any Australian marsupial. Studies in Ameridelphians suggest that raphe nuclei engage in descending projections to the spinal cord within a few days of birth, if not earlier (Martin *et al.*, 1988, 1991). Ascending projections, presumably from the superior central and dorsal raphe nuclei, also appear to reach the diencephalon by birth in the opossum (Martin *et al.*, 1987). It is therefore likely that serotonergic projections to the brainstem are present from birth in other marsupials, but it is not clear at present whether these projections are functionally significant. Most raphespinal pathways in rodents are concerned with modulation of pain transmission and are distributed to the dorsal horn, which is immature in most segmental levels of the spinal cord in newborn marsupials.

Large neurons in the hindbrain auditory nuclei (e.g. large pyramidal cochlear nucleus neurons and superior olivary nucleus) are generated around the time of birth in the brush-tailed possum and have largely completed neurogenesis by the end of the first postnatal week (Sanderson and Aitkin, 1990). The subdivisions of the superior olivary nucleus in the tammar wallaby also begin to be recognisable by P5 (see atlas plate WP5/7), suggesting that neurogenesis and migration of these neurons is completed in the early postnatal period. A similar time-frame has been found for *Dasyurus hallucatus* (neurogenesis of MTz and LSO from P5 to P9, with a peak at P7; Aitkin *et al.*, 1991). Similarly, the nuclei of the lateral lemniscus emerge during the first postnatal week in the tammar. By contrast, cochlear granule cells are mainly produced during the period from P9 to P21 in the brush-tailed possum, less frequently between P28 and P46 and only occasionally after P50. In both the brush-tailed possum and the tammar wallaby (Sanderson and Aitkin, 1990; WP25/9), the dorsal cochlear nucleus does not develop lamination until the fourth postnatal week.

Some elements of the brainstem reticular formation are quite precocious in development in marsupials. In particular, the maturation of the magnocellular medial medullary reticular formation nuclei (gigantocellular and ventral gigantocellular nuclei) appears to be advanced at birth in both poly- and diprotodont marsupials (Martin *et al.*, 1988; McCluskey *et al.*, 2008). These magnocellular nuclei are probably critically important for providing descending projections to the spinal cord locomotor neuronal groups in newborn marsupials. A few neurons in the ventral intermediate reticular formation can also be labelled by tracer injections into the spinal cord around the time of birth in the opossum (Martin *et al.*, 1988) and tammar wallaby (unpublished observations) and it is likely that these include projection neurons from the ventral respiratory group and Bötzinger complex of the medulla to the spinal cord phrenic nerve motorneurons for control of respiratory rhythm.

Very few (average of 1.2%, range of 0.2 to 2.4%) of the immature brainstem reticular formation neurons in the pouch young Virginia opossum appear to engage in collateral projections to multiple targets, suggesting that transient projections from the medullary reticular neurons are extremely rare and that the early medullary projections are relatively stereotyped. This is in contrast to the early development of forebrain systems, where transient projections mediated by collateral outgrowth are a common feature and facilitate competitive establishment of appropriate connections.

Development of respiratory function

A key function of the hindbrain is the generation and modulation of respiratory rhythm. The early maturation of this essential function is clearly important in marsupials and apparently occurs despite the immaturity of some of the hindbrain nuclei. Functional maturation of respiratory centres has not been studied in diprotodonts, but has received some attention in Ameridelphians.

In several respects, the respiratory-rhythm-generating centres in newborn and mature *Monodelphis domestica* and *Didelphis virginiana* resemble those in developing and mature placental mammals. Firstly, respiratory neurons are located adjacent to the ventral surface of the medulla around the nucleus ambiguus (extending from about 150 μm rostral to the obex to about 500 μm caudal to the obex in newborn opossums; Zou, 1994; Farber, 1995). Secondly, neurons in the respiratory region of the caudal medulla fall into several classes (early inspiratory, inspiratory, postinspiratory and expiratory; Zou, 1994), which are suppressed by microinjections of the neurotransmitter GABA (γ-aminobutyric acid) and show increased discharge in response to application of the $GABA_A$ receptor antagonist, bicuculline (Farber, 1993a, 1995). Thirdly, rate and amplitude of ventilation can be varied independently (Eugenín and Nicholls, 2000). Finally, development of afferent input from the lungs and airways plays an important role in maturation of respiratory rhythm (Farber *et al.*, 1984).

Nevertheless, there are some important functional differences between newborn marsupials and mature marsupial and placental mammals with respect to ventilation. Most very early ventilatory regulation in marsupials is due to the activity of inspiratory neurons in the caudal medulla and these neurons increase in activity with age, coincidentally with increasing myelination of their axons (Farber, 1988). In *Didelphis virginiana*, expiratory neurons do not appear to be active during the early period of postnatal life, indicating that active expiratory commands play no part in regulating ventilation in very early postnatal life (Farber, 1989). In adult placental mammals (and presumably adult marsupials) the ventilatory rate is primarily influenced by vagal input, which influences the expiratory, rather than the inspiratory, phase. Maturation of vagal afferents in marsupials may play an important part in the development of a more mature expiratory neuron activity pattern (Farber, 1983; 1985). Respiratory neurons of newborn *Didelphis* also exhibit low discharge rates compared to mature neurons, but this is not due to any intrinsic limitation of the neurons (Farber, 1993b).

An unusual feature of ventilation in isolated preparations of *Monodelphis* brainstem maintained *in vitro* is that the frequency of ventilation depends primarily on changes in duration of the expiratory, rather than inspiratory, events. This means that the functional compartment controlling the timing of the ventilatory cycle in newborn *Monodelphis* represents an 'expiratory off switch', rather than an 'inspiratory off switch' (Eugenín and Nicholls, 2000). It would appear that the principal influence on ventilation in the newborn *Monodelphis* (like that in *Didelphis virginiana*) is by direct chemical influences on the brainstem surface itself, rather than vagal inputs. This functional difference may be the consequence of the relative immaturity of the vagus nerve and/or solitary vagal complex in marsupials, but it remains to be seen when vagal influence begins to assume the major importance. Certainly the vagus nerve is quite immature morphologically in both didelphids, taking until about P50 to reach a comparable state of myelination to that seen in newborn kittens and rabbits (Krous *et al.*, 1985)

A further distinctive feature of isolated newborn *Monodelphis* brainstem is that the amplitude and frequency of ventilation appear to be coupled; increases in amplitude produced by chemical and cholinergic stimulation are associated with a shortened duration of expiration (Eugenín and Nicholls, 2000). This is in contrast to the ventilatory mechanisms in *Didelphis virginiana* at a slightly older age (P7 to P20), in which increased ventilation in response to chemical stimulation occurs mainly by increases in depth of ventilation. How much this disparity depends on species differences and/or methodology of preparation remains to be determined.

Introduction to the midbrain

The midbrain develops from the mesencephalic or midbrain vesicle of the embryonic brain. It is formed from the extensive mesencephalic neuromere, rostrally, and the rostrocaudally short isthmic region, caudally. In the adult mammal, the midbrain can be divided into three regions: tegmentum, tectum and isthmus (Butler and Hodos, 2005).

As for the rest of the brainstem, the ventral part of the midbrain is called the tegmentum. This region contains ascending and descending pathways of the auditory, somatosensory and motor systems, the oculomotor nucleus (for controlling eye movement) and reticular formation nuclei.

The most dorsal part of the midbrain is the tectum (roof), which contains the laminated superior and inferior colliculi. The superior colliculus receives topographically organised projections, mainly from the visual pathways, whereas the inferior colliculus mainly receives auditory input. The tectum as a whole is responsible for forming maps of the sensory space around the animal, by integrating information from visual, tactile and auditory pathways. In marsupials (and several other mammalian orders) the size of the superior colliculus and the number of neurons contributing axons to the tectospinal tract to the spinal cord are positively correlated with predatory behaviour (Barton and Dean, 1993).

The isthmus abuts the rostral end of the hindbrain. It includes some important cell groups (i.e. locus coeruleus, dorsal raphe nuclei), that project to diverse areas of the brain, the trochlear nucleus (for controlling eye movement) and reticular formation nuclei. The raphe nuclei lie near the midline and contain serotonin. Pathways from the serotonergic neurons of the isthmus and hindbrain project to the spinal cord, the tectal regions and diverse areas of the forebrain. Serotonin has a wide variety of functions, including regulation of sleep and mood. The locus coeruleus contains pigmented neurons that contain the catecholamine noradrenaline (norepinephrine) and project to the diverse areas of the forebrain and brainstem.

Locus coeruleus and parabrachial nuclei

Locus coeruleus

In Nissl-stained sections, the locus coeruleus is visible as a densely packed cluster of darkly stained neurons in the rostral tegmentum of the isthmus and rostral hindbrain (Figure 4.5). The locus coeruleus (A6) itself is a dense cluster of noradrenergic neurons, which are immunoreactive for tyrosine hydroxylase (Figure 4.5b). Other catecholaminergic neuronal groups nearby include the subcoeruleus, the

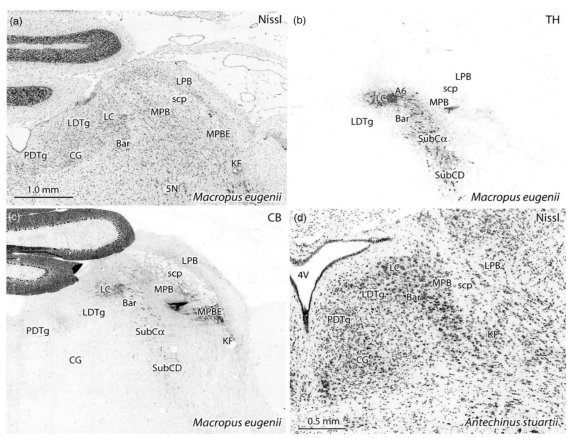

Figure 4.5 Topography of the isthmic tegmentum as seen in coronal sections through the tammar wallaby (a, b, c) and brown antechinus (d). Photomicrographs (a) to (c) show matching Nissl-stained, tyrosine-hydoxylase- and calbindin-immunoreacted sections illustrating the relationship between the superior cerebellar peduncle, parabrachial nuclei, locus coeruleus, subcoeruleus nu., Barrington's nu. (Bar) and tegmental nuclei. Scale bar in (a) applies to (a) to (c). Photomicrograph (d) shows the same nuclei in a Nissl-stained section through the isthmic tegmentum of the brain of the brown antechinus. 4V – fourth ventricle; 5N – motor trigeminal nu.; A6 – A6 noradrenergic cell group; CG – central grey; KF – Kölliker-Fuse nu.; LC – locus coeruleus; LDTg – laterodorsal tegmental nu.; LPB – lateral parabrachial nu.; MPB – medial parabrachial nu.; MPBE – medial parabrachial nu., external; PDTg – posterodorsal tegmental nu.; scp – superior cerebellar peduncle; SubCα – subcoeruleus nu., alpha; SubCD – subcoeruleus nu., dorsal.

ventral extension of the locus coeruleus and the A4 group extending dorsally around the aqueduct. The locus coeruleus projects to a wide variety of structures throughout the forebrain (olfactory bulb, isocortex, hippocampus, anterior thalamic nuclei, visual and somatosensory thalamic nuclei, bed nuclei of the stria terminalis, amygdala and hypothalamus), hindbrain (cerebellum, sensory nuclei of the pons and medulla) and the spinal cord.

Parabrachial nuclei

The parabrachial nuclei are important structures in autonomic control because they act as the interface between the medullary reflex control centres and the forebrain centres integrating behaviour and autonomic control. The most obvious division of the parabrachial nuclei is into medial and lateral components separated by the superior cerebellar peduncle, and this subdivision is clear in all marsupials. In rodents, where the subdivisions and connections of the parabrachial nuclei have been studied in detail, as many as 13 distinct nuclei and regions have been identified (Saper, 2004b). In diprotodont marsupials, the lateral part of the parabrachial complex appears to be less complex (atlas plate W37) than in rodents or polyprotodonts (see atlas plate D24 for comparison), whereas the medial part of the complex is similar in cytoarchitecture to that in rodents and polyprotodonts. Whether these cytoarchitectural differences reflect functional divergence remains to be seen, because no functional studies have been undertaken of these nuclei in marsupials.

The parabrachial nuclei receive ascending projections from the nucleus of the solitary tract with distinct terminal fields from functionally diverse components of the Sol. In rodents, the visceral projections from the parabrachial nuclei are sorted along chemically coded lines and the organisation of the parabrachial nuclei as a whole reflects their role in particular regulatory functions throughout the body (e.g. control of body fluids, energy metabolism, respiratory function). In this respect, the parabrachial nuclei differ from the Sol, which is organised by specific organs or autonomic reflexes, and the forebrain, which is organised according to behavioural contexts (Saper, 2004b). Nothing is currently known about the functional organisation of the parabrachial nuclei in marsupials, but given the adaptations of some marsupials to life in the arid zone, study of this region could prove a rich source of information on the relationship between structure and function in this nuclear complex.

In placental mammals, Barrington's nucleus is part of a pontine micturition centre and has direct excitatory connections with the parasympathetic neurons controlling urinary bladder muscle.

Midbrain dopaminergic neurons

Midbrain dopaminergic neurons (located in the substantia nigra pars compacta, ventral tegmental area, retrorubral area) engage in reciprocal connections with the basal ganglia and cortical circuits and modulate a broad range of behaviours from learning and working memory to motor control.

Substantia nigra

The substantia nigra lies in the ventral midbrain and is a key element of the basal ganglia pathways. In all mammals, the substantia nigra contains two main neuron types: one that contains dopamine and the other using GABA (γ-amino butyric acid) as a neurotransmitter. The dorsal part of the substantia nigra is known as the pars compacta and contains most of the dopaminergic neurons (A9 catecholaminergic group of Dahlström and Fuxe, 1964a, 1964b). Dopaminergic neurons are also distributed into the nearby retrorubral area (A8) and adjacent ventral tegmental area (A10). Some dopaminergic neurons are distributed loosely in the more neuron-sparse ventral part of the substantia nigra (pars reticulata), whereas neurons containing GABA are largely localised in the pars reticulata. All of these areas are identifiable in sections through the brains of poly- and diprotodont marsupials, with medial and lateral compact regions identifiable within the substantia nigra nuclear complex (Figure 4.6a to d).

In placental mammals, dopaminergic neurons of the substantia nigra pars compacta project to the striatum, whereas the GABAergic neurons of the pars reticulata project to the thalamus, superior colliculus and pedunculopontine nucleus. The organisation of these regions appears to be similar in all therians, but the projections of the pars compacta and reticulata have not been specifically studied in Australian marsupials.

Ventral tegmental area

Dopaminergic neurons are also located in the ventral tegmental area (VTA), medial to the substantia nigra. In fact, the dopaminergic neurons of the ventral tegmental area are mediolaterally continuous with the pars compacta (Figure 4.6b). Neurons of both the VTA and pars compacta also share immunoreactivity for the calcium-binding protein calbindin. Dopaminergic neurons of the VTA engage in a closed (i.e. reciprocal) connection with the shell of the nucleus accumbens in the forebrain, suggesting a role in the mediation of reward and motivation.

Red nucleus

The red nucleus was originally given its name because of its pinkish coloration as seen in sections through human midbrain. The pink coloration is, in turn, the result of the increased vascularisation of the nucleus relative to the surrounding midbrain tegmentum. The red nucleus is divided into large- and small-celled parts (i.e. magno- and parvicellular) and the relative sizes of each component vary between mammals. Both magnocellular and parvicellular components are clearly distinguishable in poly- and diprotodont marsupials (Figure 4.6d). Afferents to the red nucleus in birds and mammals arise from the somatomotor parts of the cortex or pallium and the deep cerebellar nuclei (Butler and Hodos, 2005). Efferents of the red nucleus include a major projection to contralateral spinal cord (mainly from the magnocellular red nucleus) and the inferior olivary complex (from the parvicellular part). The red nucleus is also considered elsewhere in this book (Chapters 6 and 13) because of its central role in descending motor pathways (rubrospinal tract) and cerebellar outflow.

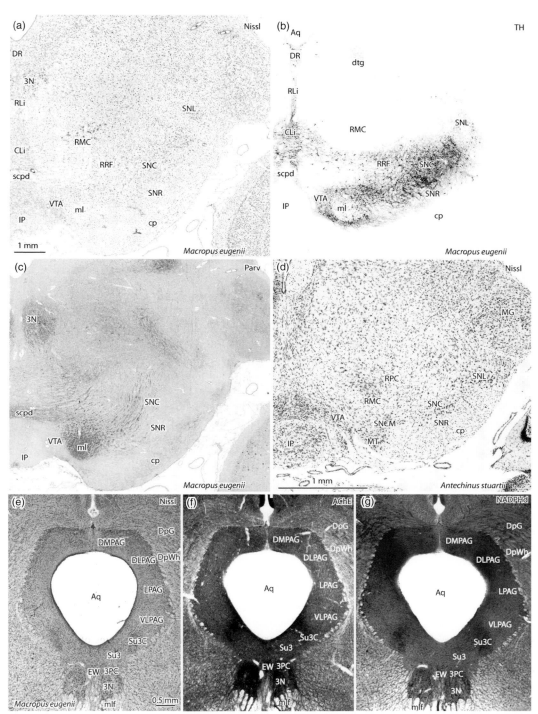

Figure 4.6 Cyto- and chemoarchitecture of the substantia nigra and red nu. in representative diprotodont (a to c) and polyprotodont (d) Australian marsupials. Note the continuity between the dopaminergic cell groups of the SNC, VTA and RRF (TH immunoreactive cell groups in (b)). Figures (e) to (g) show cyto- and chemoarchitecture of the periaqueductal grey and associated structures in the tammar wallaby. Adjacent sections have been stained for Nissl substance (e), acetylcholinesterase (f) and NADPH diaphorase (g). 3N – oculomotor nu.; 3PC – oculomotor nu., parvicellular part; Aq – cerebral aqueduct; CLi – caudal linear nu. of the raphe; cp – cerebral peduncle; DLPAG dorsolateral periaqueductal grey; DMPAG – dorsomedial periaqueductal grey; DpG – deep grey superior colliculus; DpWh – deep white superior colliculus; DR – dorsal raphe nu.; dtg – dorsal tegmental tract; EW – Edinger–Westphal nu.; IP – interpeduncular nu.; LPAG – lateral periaqueductal grey; MG – medial geniculate nu.; ml – medial lemniscus; mlf – medial longitudinal fasciculus; MT – medial terminal nu.; RLi – rostral linear nu. (midbrain); RMC – red nu., magnocellular part; RPC – red nu., parvicellular part; RRF – retrorubral field; scpd – decussation superior cerebellar peduncle; SNC – substantia nigra, compact part; SNCM – substantia nigra, pars compacta, medial; SNL – substantia nigra, lateral part; SNR – substantia nigra, reticular part; Su3 – supraoculomotor PAG; Su3C – supraoculomotor cap; VLPAG – ventrolateral periaqueductal grey; VTA – ventral tegmental area.

Periaqueductal grey

The periaqueductal grey (PAG) is a cell-dense region that surrounds the part of the ventricular system known as the midbrain, or cerebral, aqueduct. The PAG extends from the level of the posterior commissure, where it is continuous with the periventricular grey matter of the diencephalon, to the expansion of the cerebral aqueduct to become the fourth ventricle. Electrical stimulation of the PAG in rodents evokes a variety of behavioural and sensory changes (aversive or defensive reactions, elevated blood pressure, sexual behaviour and analgesia).

It is now appreciated that the mammalian PAG is principally organised into rostrocaudally elongated functional columns (see Keay and Bandler, 2004, for review). These longitudinal columns have been revealed as a result of studies using excitatory amino-acid microinjections into the rodent PAG, but the basic organisational plan is likely to be similar in marsupials. Injections into the dorsal PAG evoke active defensive reactions, with injections in the rostral dorsal PAG eliciting confrontational responses, whereas injections in the caudal dorsal PAG elicit escape or flight responses. Injections of excitatory amino acids into the ventrolateral part of the caudal PAG evoke a passive coping reaction or disengagement from the environment. This more passive response is characterised by less reactivity, slowing of the heart rate and reduced blood pressure and immobility. Stimulation of both the dorsolateral and ventrolateral regions produces analgesia, but the passive coping reaction arising from stimulation of the VLPAG is associated with an opioid-mediated analgesia, whereas the active defence reactions from stimulation of the dorsal PAG are associated with non-opioid-mediated analgesia (see Keay and Bandler, 2004, for review).

Although the PAG is cytoarchitecturally homogeneous, there is a good basis for subdividing the dorsal PAG on chemoarchitectural grounds (Figure 4.6e to g). Within the dorsal PAG there is a wedge-shaped dorsolateral zone (DLPAG), which stains strongly for NADPH diaphorase (Figure 4.6g) and nitric oxide synthase. Conversely, flanking the dorsolateral zone, there are dorsomedial and lateral zones, which stain strongly for cytochrome oxidase and weakly for NADPH diaphorase. These chemoarchitectural patterns have led to the suggestion that the dorsal PAG should be divided into three longitudinal columns: dorsomedial, dorsolateral and lateral PAG. The different PAG columns appear to play important roles in co-ordinating the distinct behavioural strategies for dealing with physically and emotionally stressful situations. It has been proposed (for review see Keay and Bandler, 2004) that: (i) the lateral PAG is preferentially activated by so-called 'escapable' physical stressors to which an active defensive reaction is appropriate and useful; (ii) the dorsolateral PAG is involved in dealing with 'escapable' psychological stressors for which the primary response is an active defensive reaction and (iii) the ventrolateral PAG is engaged in 'inescapable' physical or psychological stressors for which passive coping is appropriate, or as a delayed response to promote recovery and healing following any stressor.

No studies have been made of the PAG or its functional organisation in any Australian marsupial, but similarities in the chemoarchitectural organisation of the PAG to that described for placental mammals suggests that a similar longitudinal zonation is present in all mammals.

Interpeduncular nucleus

The interpeduncular nucleus of the midbrain (see atlas plates D18 to D20, W31 to W33) is a key element in the brainstem centres for expression of emotions and receives a large fibre bundle, the fasciculus retroflexus (habenulointerpeduncular tract), from the habenular nuclei. The major outflow tract from the interpeduncular nucleus is the interpedunculotegmental tract, which arises from the rostral third of the central interpeduncular nucleus and projects to the dorsal tegmental nuclei of the hindbrain. The interpeduncular nucleus plays a key role in an important pathway for conveying information from the striatal and limbic areas of the forebrain to the brainstem in a wide range of vertebrates. The fasciculus retroflexus, the interpeduncular nucleus and the interpedunculotegmental tract are all distinguished in a range of therian mammals by intense enzyme reactivity for acetylcholinesterase (Wilson and Watson, 1980). Concomitant with this, the interpeduncular nucleus has the highest levels of acetylcholine of any part of the brain. The interpeduncular nucleus probably also receives an afferent projection from the ventral tegmental nucleus (Wilson and Watson, 1980).

Superior colliculus and accessory optic system

As noted above, the superior colliculus receives topographically ordered visual inputs from the retina and plays an important role in visuomotor behaviour. In the tammar wallaby, the surface of each superior colliculus has a representation of the visual field extending from 25° ipsilaterally to 120° contralaterally (Flett et al., 1988; Mark et al., 1993). The horizon is represented by a line that runs rostrocaudally across the SC; whereas lines of azimuth (i.e. vertical lines in the visual field) are represented by lines running mediolaterally. Consistent with the behavioural importance of detecting potential predators in the horizon area, there is an enlarged representation on the SC of the visual field 5° above and below the horizon.

The superior colliculus is essential for the generation of an orienting response to an object of visual and auditory interest. The superior colliculus is laminated and is divided into three zones (superficial, intermediate and deep or periventricular). The three zones can in turn be subdivided to give seven layers for the superior colliculus proper (i.e. excluding the PAG). From the pial surface to the PAG these are: zona layer, superficial grey layer, optic layer, intermediate grey layer, intermediate white layer, deep grey layer and deep white layer. The superficial grey and upper optic layers are innervated by retinal axons and process visual information, before projecting to intermediate and deep layers, which also receive inputs from other sensory modalities (auditory and somatosensory). The superior colliculus is said to be best developed in those mammals inhabiting arboreal or partly arboreal habitats (Butler and Hodos, 2005) and the tectospinal tract, which arises from the intermediate and deep layers of the superior colliculus, is best developed in predatory mammals (Barton and Dean, 1993).

Fibres from retinal ganglion cells also terminate in nuclei of an accessory optic system (Hayhow, 1966). These fibres decussate in the optic chiasm before coursing across the cerebral peduncle and terminating in medial, lateral and dorsal terminal nuclei (see atlas plates D17, D18, W32) of the midbrain tegmentum. The arrangement of these nuclei is similar in placental and marsupial mammals (Hayhow, 1966), although the organisation in marsupials appears to be more similar to that in rodents and lagomorphs, as compared to carnivores, artiodactyls and primates (Johnson, 1977).

Inferior colliculus

The inferior colliculus is the auditory midbrain roof of mammals and receives ascending auditory projections from the cochlear and other auditory nuclei of the brainstem (e.g. superior olivary nucleus, nucleus of the lateral lemniscus). The inferior colliculus in therians is classically divided into an ovoid central nucleus, a laterally and rostrally placed external cortex, and a dorsal cortex (Figure 4.7). Further subdivisions can be recognised in many mammals, although how these correspond between species is uncertain. Even in small polyprotodont Australidelphians, like *Sminthopsis macroura*, the external cortex of the inferior colliculus can be divided into three layers on cellular grounds (atlas plate D23), much like that seen in the

laboratory rat, but the dorsal cortex and central nucleus are more cytoarchitecturally homogeneous (Figure 4.7c, d). Neurons in the dorsal and external cortex use nitric oxide and are reactive for NADPH diaphorase, whereas the central nucleus and deeper layers of the dorsal and external cortex are strongly immunoreactive for parvalbumin (Figure 4.7a, d).

In rodents, where connections of the inferior colliculus have been studied in detail, the central nucleus receives its main input from more caudal auditory centres (i.e. cochlear nuclei, nucleus of the lateral lemniscus and superior olivary nuclei) by a fibre bundle known as the lateral lemniscus, whereas the external and dorsal cortex of the inferior colliculus receive mainly descending projections from auditory and non-auditory cerebral cortex. The output of the inferior colliculus is chiefly ascending projections to the medial geniculate nucleus of the thalamus (by a distinct fibre bundle known as the brachium of the inferior colliculus) and descending projections to the superior olivary nuclear complex and the cochlear nuclei. There is also an extensive commissural projection (commissure of the inferior colliculus) visible even in Nissl-stained material (see atlas plates D22, D23, W36).

Oculomotor and trochlear nuclei

The oculomotor nucleus contains motorneurons controlling most of the extraocular muscles (the ipsilateral medial and inferior rectus muscles, the ipsilateral inferior oblique muscle and the contralateral superior rectus muscle). Motorneuron subgroups are usually organised in the oculomotor nucleus in a topographic map, but this organisation has not been adequately investigated in Australian marsupials. There does appear to be a dorsal parvicellular component of the oculomotor complex in both poly- and diprotodont marsupials (Figure 4.6e to g; see also atlas plates D19, D20, W32, W33), which may be homologous to the parvicellular oculomotor neuronal groups of placental mammals (nucleus of Perlia, interoculomotor nucleus). These small motorneurons probably supply slow non-twitch fibres of the medial rectus, inferior rectus, inferior oblique and inferior rectus extraocular muscles.

The adjacent Edinger–Westphal nucleus contains parasympathetic preganglionic neurons, which drive the ipsilateral constrictor pupillae and ciliary muscles of the eye. The trochlear nucleus of the isthmus lies in the caudal midbrain ventral to the cerebral aqueduct. It contains motorneurons controlling the contralateral superior oblique muscle.

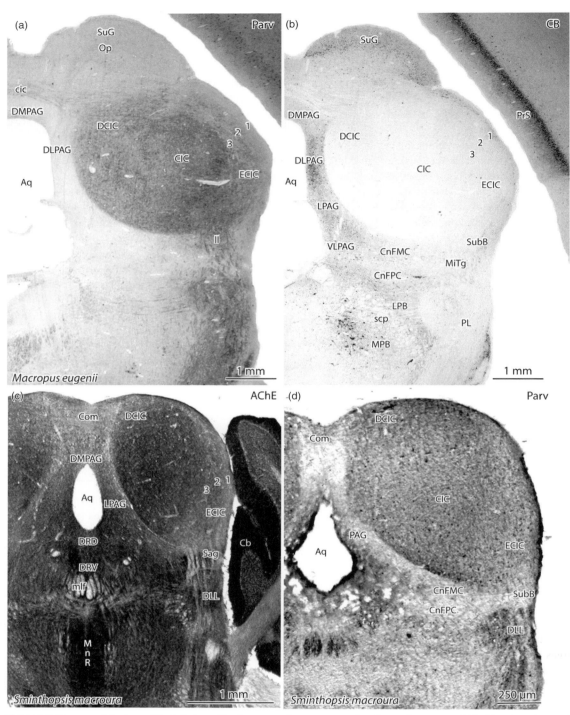

Figure 4.7 Chemoarchitecture of the inferior colliculus as seen in a representative diprotodont marsupial (a, b) and a polyprotodont marsupial (c, d). Note the complementary pattern of immunoreactivity to parvalbumin and calbindin in the central nu. of the inferior colliculus of the tammar wallaby (a, b). The subdivisions and topography of the components of the inferior colliculus are similar in both poly- and diprotodont marsupials to that seen in rodents and carnivores. Aq – cerebral aqueduct; Cb – cerebellum; CIC – central nu. of the inferior colliculus; cic – commissure of the inf colliculus; CnFMC – cuneiform nu., magnocellular part; CnFPC – cuneiform nu., parvicellular part; Com – commissural nu. of inferior colliculus; DCIC – dorsal cortex of inferior colliculus; DLL – dorsal nu. of lateral lemniscus; DLPAG – dorsolateral periaqueductal grey; DMPAG – dorsomedial periaqueductal grey; DRD – dorsal raphe nu., dorsal; DRV – dorsal raphe nu., ventral; ECIC – external cortex of inferior colliculus; ll – lateral lemniscus; LPAG – lateral periaqueductal grey; LPB – lateral parabrachial nu.; MiTg – microcellular tegmental nu.; mlf – medial longitudinal fasciculus; MnR – median raphe nu.; MPB – medial parabrachial nu.; Op – optic nerve layer superior colliculus; PAG – periaqueductal grey; PL – paralemniscal nu.; PrS – presubiculum; Sag – sagulum nu.; scp – superior cerebellar peduncle; SubB – sub-brachial nu.; SuG – superficial grey of superior colliculus; VLPAG – ventrolateral periaqueductal grey.

Cortical projections to the midbrain

Several cerebral cortical regions in diprotodont marsupials project to the midbrain structures considered above. Cortex of the frontal pole region in the brush-tailed possum projects to the ventral tegmental area, nucleus linearis, periaqueductal grey matter and the deep midbrain tegmentum (Martin and Megirian, 1972). Martin and Megirian (1972) noted that the preorbital cortex of the Virginia opossum (*Didelphis virginiana*) projected specifically to the ventral part of the periaqueductal grey, but that this pathway did not appear to be present in the brush-tailed possum. The forelimb region of the motor cortex also projects to the periaqueductal grey, nucleus of Darkschewitsch and the nucleus of the medial longitudinal fasciculus. Why the forelimb region should project to midbrain nuclei usually considered part of the accessory optic system (nucleus of Darkschewitsch, nucleus of the medial longitudinal fasciculus) is uncertain. It may be that Martin and Megirian were actually seeing the effects of damage to fibres of passage from the marsupial equivalent of the frontal eye field, a cortical region that controls eye movements. Pathways from the prefrontal cortex to the periaqueductal grey are topographically organised in placental mammals, but this has not been studied in Australian marsupials. Nevertheless, the presence of a projection from the prefrontal cortex to the PAG in marsupials suggests a similar functional role for the PAG in dealing with psychological stressors in marsupials to those studied in placental mammals (see above).

The forelimb and hindlimb motor/sensory cortical areas of the possum have a robust projection to the red nucleus and nearby dorsolateral midbrain tegmentum (Martin and Megirian, 1972). The corticospinal tract in diprotodonts extends only to the upper to mid-thoracic levels, so the corticorubral and rubrospinal pathways probably provide the major (indirect) route for the hindlimb motor/sensory cortex to influence the caudal spinal cord segmental levels and therefore hindlimb function (see Chapter 13).

Visual areas of the cortex of the brush-tailed possum project to the superficial grey and optic layers of the ipsilateral superior colliculus (Martin and Megirian, 1972). This corticotectal projection is similar to that reported for *Didelphis virginiana* (Martin, 1968) and a variety of placental mammals and is involved in cortical control of visually guided behaviour. The corticotectal projection appears to be visuotopically organised in all therians (Martin and Megirian, 1972), with convergence of cortical and retinal input in the superficial layers of the superior colliculus.

The inferior colliculus of the brush-tailed possum also receives a projection from the temporal cortex providing a cortical feedback on auditory relay nuclei (Martin and Megirian, 1972).

Serotonergic neurons

The distribution and morphology of serotonergic neurons has been studied in an American polyprotodont marsupial (*Didelphis virginiana*; Martin *et al.*, 1985) and a macropod (*Macropus eugenii*; Ferguson *et al.*, 1999). Brainstem serotonergic neurons are broadly similar in both distribution and morphology in all mammals studied (Martin *et al.*, 1985; Ferguson *et al.*, 1999; Manger *et al.*, 2002), indicating that these cell groups are highly conserved across all living mammals. Caudal and rostral groups are clearly present in all therians and monotremes, but some differences exist in the relative size of some serotonergic nuclei and the separation between nuclei. For example, the dorsal raphe nucleus of the wallaby makes up a smaller proportion of all brainstem serotonergic neurons (32%) than does the same structure in rodents (56%) and carnivores (40%) (Ferguson *et al.*, 1999). Nevertheless, the internal organisation of the dorsal raphe nucleus of the wallaby is at least as complex as that seen in rodents and carnivores, although not as complex as that in primates. On the other hand, the median raphe nucleus of the wallaby is not as clearly divisible into subnuclei as has been reported for the laboratory cat. Developmental spread of serotonergic neurons away from the midline of the rostral brainstem has been linked with the evolution of more complex brains and is probably associated with differences in ascending serotonergic projections. Interestingly, the proportion of laterally placed serotonergic neurons in the wallaby (27%) (Ferguson *et al.*, 1999) is greater than in the laboratory rat (12.5%) and comparable to a representative carnivore (the domestic cat, 22.5%).

Development of the midbrain

Neurogenesis and immunoreactivity

At the time of birth in all marsupials, the midbrain consists of a dorsally placed mesencephalic vesicle with a thick germinal zone (please refer back to Figure 3.3d), much like that seen in the telencephalon at this stage. By contrast, the ventral part of the marsupial mesencephalon at birth is better developed than the basal telencephalon; postmitotic neuronal groups have already emerged to provide an outline of the mature structure of the midbrain (see atlas plates WP0/7 and 8). Those parts of the midbrain already showing significant neuronal populations at birth in the tammar wallaby include the ventral tegmental area, red nucleus, substantia nigra, retrorubral fields, oculomotor and trochlear nuclei, and mesencephalic reticular formation. The cytoarchitecture of the tammar midbrain is largely mature by the end of the first postnatal month.

Although development of immunoreactivity to tyrosine hydroxylase is quite a late event in the locus coeruleus, some evidence from Ameridelphians suggests that the ascending projections from the locus coeruleus reach targets in the diencephalon and midbrain at least as early as the first postnatal week (i.e. about P5; Martin *et al.*, 1987). Ascending projections from the parabrachial nuclei to the diencephalon also appear to develop at around this time (Martin *et al.*, 1987). Tyrosine hydroxylase immunoreactivity develops in the substantia nigra and ventral tegmental area of *Didelphis virginiana* from around the time of birth, but most dopaminergic fibres do not reach the striatum until the second week of postnatal life (Martin *et al.*, 1989).

Red nucleus neurons appear in the midbrain tegmentum as early as birth in diprotodont marsupials, but differentiation into parvi- and magnocellular components does not occur until late in the second week of postnatal life. Similarly, although some of the neurons of the substantia nigra have already settled into the ventral tegmentum by the time of birth in the tammar wallaby, emergence of distinct compact and reticular subdivisions does not occur until about P12 (see WP12/10). Immunoreactivity for tyrosine hydroxylase emerges in the compact part of the wallaby substantia nigra around the end of the third postnatal week (unpublished observations) suggesting that dopamine production is probably not significant until the second month of postnatal life. This would correspond approximately with completion of neurogenesis in the major forebrain target of the pars compacta outflow (the striatum, see Chapter 7). Judging from observations in Ameridelphians (see above), the relatively late development of dopamine in pars compact neurons of the tammar wallaby may not preclude the earlier establishment of connections between those neurons and the striatum.

The regional organisation of the mature periaqueductal grey is matched by postnatal neurogenetic gradients, as shown by studies in the brush-tailed possum (Sanderson and Aitkin, 1990). The earliest generated parts of the periaqueductal grey are those situated laterally, i.e. LPAG and VLPAG, the neurons of which leave the mitotic cycle mainly during the first postnatal week (and perhaps earlier) and complete neurogenesis by about P9. By contrast, neurons of the more dorsal parts of the PAG of the brush-tailed possum (i.e. DMPAG and DLPAG) appear to be generated quite late in postnatal life (maximally at P21 to P32; Sanderson and Aitkin, 1990).

The superior colliculus develops according to an inside-out neurogenetic gradient during a relatively short postnatal time-frame for diprotodont marsupials (Sanderson and Aitkin, 1990). In the brush-tailed possum, neurons of the future deep layers of the superior colliculus (deep grey and white layers) leave the mitotic cycle around the time of birth to P5, the bulk of neurons in the intermediate and superficial layers are generated around P12, and neurogenesis is largely complete by the end of the third postnatal week. A similar time-course of neurogenesis is evident in the SC of the quokka (birth to P18; Harman, 1991).

By contrast, the inferior colliculus of the brush-tailed possum develops according to an outside-to-inside neurogenetic gradient (Sanderson and Aitkin, 1990); the earliest generated neurons appear to be those of the external cortex of the inferior colliculus, ventrolateral central nucleus, and the adjacent cuneiform and precuneiform nuclei (around P5). The bulk of the neurons of the central nucleus and the dorsal cortex of the inferior colliculus are generated around P18. By contrast, neurogenesis of the neurons of the inferior colliculus in the northern quoll occurs slightly earlier in postnatal life (Aitkin *et al.*, 1994). The bulk of neurons of the central nucleus in the quoll can be labelled following ^{3}H thymidine injection on P7 to P9, but it is not until P45 to P50 that a cytoarchitectural distinction between external cortex, dorsal cortex and central nucleus emerges.

Development of the retinocollicular pathway

The development of the projection from the retina to the SC has received some attention in macropods because of its potential for testing hypotheses of the control of visual pathway development and topographic map formation. Experiments involving monocular rotation at the time of eye opening indicate that discordance between the visual field representations of the two eyes does not alter the topography of the retinocollicular projection (James *et al.*, 1993), but do not address the mechanisms of development.

Models of how the topographic map develops during the early stages of ingrowth of the retinocollicular axons emphasise the importance of chemical topographic gradients in both the axon source (retina) and target (SC) for production of an organised landscape of connections. A topographic gradient of brain-derived neurotrophic factor (BDNF) is present in glial endfeet in a caudal (high) to rostral (low) gradient across the SC of the developing tammar at the same time that the BDNF receptor (trkB) is present in the retinal ganglion cells (Marotte *et al.*, 2004), although there is no topographic gradient for trkB in the retina. The differential topography of BDNF concentration may provide a gradient of attractant or branching signal in the SC. The Ephrin A3 receptor and its ligands are probably better candidates for helping to regulate the development of the topographic projection (Vidovic *et al.*, 1999; Stubbs *et al.*, 2000), although they cannot be the only factors responsible. Ephrin B receptors and their ligands have also been considered for topographic patterning in this system, but gradients are not strong at the critical developmental stages (Vidovic and Marotte, 2003).

What remains unknown?

Most studies of the connectional and functional development of hindbrain nuclei in marsupials have focussed on Ameridelphians and there is a pressing need to extend these studies to Australian marsupials. Nothing is currently known about the development of respiratory centres in any Australian marsupial, and many questions remain about whether the observed dissimilarity in respiratory function development between *Monodelphis* and *Didelphis* reflects species or methodological differences. As noted elsewhere, not all marsupials are at a similar stage of morphological development at birth, and this may be reflected in important functional differences in locomotor and cardiorespiratory hindbrain centres.

The broad similarities in cyto- and chemoarchitecture of midbrain structures between marsupials and placentals suggest that the functions of these regions are also likely to be similar. There are two areas of midbrain function, which may be fruitful for future studies in Australian marsupials. The first of these concerns the connections and functions of the midbrain dopaminergic cell groups and the interaction between cortical areas, nucleus accumbens, and striatum and midbrain dopaminergic groups. These pathways are behaviourally significant because of their role in learned behaviour, reward and motivation and may have important differences in marsupials, as compared to placentals. The second concerns the periaqueductal grey and its role in modulating the ascending pain sensory system and descending limbic emotional motor system. The contribution of the PAG to a wide range of behavioural reactions in placental mammals (e.g. pain modulation, defensive reactions, cardiovascular regulation, vocalisation and reproductive behaviour) make this region and its connections a potentially fruitful place to search for divergence in the evolution of emotional control mechanisms in marsupials and placentals.

5

Cerebellum, vestibular and precerebellar nuclei

K. W. S. Ashwell

Summary

Australian marsupials exhibit a wide variety in the morphology of the cerebellum, commensurate with the broad range of niches that they occupy. The simplest cerebellum of any mammal is that of the marsupial mole, whereas large macropods show a complexity of cerebellar foliation comparable to ungulates. Some Australian marsupials may have specialised circuitry mediating behaviourally important integration of auditory and cerebellar function (e.g. gliding marsupials).

Although only a few studies of cerebellar afferents and efferents have been made in Australian marsupials, chemoarchitectural studies suggest that the sagittal zonation, which underlies cerebellar organisation, is a common feature of all mammals. Spinocerebellar afferents appear to be distributed in parasagittal zones in the possum cerebellar vermis similar to those seen in placental mammals, and the overall features of spinocerebellar projections in the opossum have been reported to be similar to eutherians.

Development of the cerebellum is almost entirely postnatal in Australian marsupials. In both the brush-tailed possum and tammar wallaby, macroneuron production (i.e. generation of Purkinje cells and deep cerebellar nuclei) occurs around the time of birth or the first postnatal week, whereas migration of the external granule-cell layer does not occur until the second postnatal week and generation of the granule cells extends beyond P95 and for the bulk of pouch life (i.e. perhaps as long as P150).

Although the inferior olivary complex of marsupials is as complex as that in placental mammals, there is a significant difference in the arrangement of cell groups homologous to the medial accessory olivary nucleus of placentals. In marsupials, this cell group has many constituent subnuclei lateral to the principal nucleus, an arrangement which may arise from differences between marsupials and placental mammals in the mode and relative timing of the migration of inferior olivary neurons during early postnatal development. Nevertheless, the topographic organisation of the olivocerebellar projection of at least one Australian marsupial (*Trichosurus vulpecula*) shows many similarities with that seen in placental carnivores.

The vestibular apparatus and vestibular nuclei are similar across all mammals, but connectivity around the time of birth may have some peculiarities in marsupials. In both *Monodelphis domestica* and *Macropus eugenii*, there is a substantial bilateral efferent projection at birth from the presumptive paragenual region (future nucleus of origin of vestibular efferent fibres, EVe) to the vestibular ganglia. This projection is significantly reduced by a few weeks after birth. The functional activity of this developmental pathway remains to be studied, but its temporal association with the climb to the pouch raises the intriguing possibility that it may contribute to that function.

Introduction

Early studies of the mammalian cerebellum emphasised its role in motor co-ordination, but it is now recognised that, across vertebrates, it also plays a role in analysis of lateral line system information and electrical-field detection (where present). The mammalian cerebellum receives information from a wide variety of sensory systems (vestibular, somatosensory, visual and auditory) and processes sensory information in order to control the activation of a wide variety of eye, axial and appendicular musculature (Butler and Hodos, 2005). The histological structure of the cerebellum is deceptively simple, with the cerebellar cortex consisting of a relatively cell-free molecular layer, a layer of large Purkinje cells with characteristic planar dendritic trees and an underlying densely packed granule-cell layer, but this relatively simple and stereotypical histological structure underlies a complex regional functional organisation.

In all mammals, afferents to the cerebellum come from primary and secondary nuclei of the relevant sensory nuclei and spinal cord, as well as specialised precerebellar nuclei, principally the inferior olivary nuclear complex and the pontine nuclei. The vestibular component of the vestibulocochlear nerve also terminates directly in the cerebellum. The output from the cerebellar cortex is principally to the deep cerebellar nuclei, which are located in the central cerebellar white matter. Outputs from the deep cerebellar nuclei are to the red nucleus, dorsal thalamus, vestibular nuclei, reticular formation of the brainstem and selected motor nuclei of the brainstem, principally those concerned with eye muscle control.

The mammalian cerebellum is usually considered to consist of three functional subdivisions: the vestibulocerebellum (anatomically – flocculonodular lobe and fastigial deep cerebellar nucleus), receiving and processing information concerned with vestibular input and control of balance; the spinocerebellum (anatomically – anterior lobe, vermal and paravermal zones and interposed deep cerebellar nuclei), concerned with processing information from the spinal cord and co-ordination of axial musculature; and the pontocerebellum (anatomically – cerebellar hemispheres and lateral deep cerebellar nucleus), concerned with descending control from the cerebral cortex (received via the pontine nuclei) and the co-ordination of distal musculature. Nevertheless, the functional areas of cerebellum are not as clearly segregated as this nomenclature would imply.

External features of the cerebellum

The external features of the marsupial cerebellum were first described by Ziehen (1897) and Smith (1903a, 1903b), and later by Riley (1929). Larsell (1970) provided a more recent review and detailed comparative analysis of external morphology of the marsupial cerebellum.

The complexity of the external features of the marsupial cerebellum varies markedly with the body size and lifestyle of the animal (Figure 5.1). As noted by G. E. Smith (1903a), the simplest cerebellum of any mammal is that of the marsupial mole (*Notoryctes typhlops*, Figure 5.1a), whereas large species of kangaroo (genus *Macropus*) exhibit complex foliation of the cerebellar vermis comparable to ungulates (Figure 5.1e). The basic pattern of the marsupial cerebellum is identical to that of placentals and the relative complexity of folding is correlated, as is the case in eutherian cerebellum, with the body size and relative functional importance of cerebellar and precerebellar systems (Larsell, 1970).

The simplicity of the cerebellum of the marsupial mole is probably due to phylogenetic factors. Although *Notoryctes* has a body weight greater than shrews and small chiroptera, it has a cerebellum which is less folded than that of the smallest bats and shrews (Larsell, 1970) and this simplicity is undoubtedly related to the specialised locomotion of the marsupial. The blind marsupial mole has a head and body length of up to 160 mm, weighs 40 to 70 g, and spends most of its life underground. A horny shield protects the snout and the mole digs with the aid of two flattened claws on each forelimb. Above ground it moves in a sinuous fashion, dragging its legs to leave parallel furrows flanking a central depression made by the body (van Dyck and Strahan, 2008). Although simple, the cerebellum of the marsupial mole does show division into a relatively poorly differentiated anterior lobe (lobules 1Cb to 5Cb) rostral to the primary fissure (prf) and a more differentiated posterior lobe (lobules 6Cb to 9Cb) with a secondary fissure and a posterolateral fissure demarcating the flocculonodular lobe. The poor development of the anterior lobe in the marsupial mole compared to placental moles is thought to reflect the poor functional capacity of the hindlimbs and therefore lack of development of caudal spinocerebellar systems (Larsell, 1970).

The cerebellum of small marsupial carnivores such as the Ameridelphians *Caenolestes obscurus* and *Orolestes inca* (Figure 5.1f, g) and the Australian carnivore *Phascogale calura* (Figure 5.1c) are similar in complexity and were considered by Larsell (1970) to be intermediate in pattern between the simple cerebellum of *Notoryctes* and that of the Peramelidae (Figure 5.1b). All these species show clear preculminate, primary, secondary and posterolateral fissures, but only *Perameles* and *Phascogale* have a distinct precentral fissure (pcn) separating lobules 2Cb and 3Cb.

The cerebellum of the kangaroos (Figure 5.1d, e) is much larger and its fissures and subdivisions are more prominent than in any Ameridelphian (compare *Macropus* and *Didelphis*, Figure 5.1e and h) (Dillon, 1963; Larsell, 1970). Features of the macropod cerebellum (Figure 5.1d) include: (i) vermal and hemispheric zones distinctly separated by a paravermal sulcus; (ii) cerebellar hemispheres expanded rostrally and caudally with subdivision into lobus simplex, crus 1 and crus 2 and paramedian regions and (iii) a prominent group of folia forming a paraflocculus. Most of these features can also be seen in other diprotodont marsupials (e.g. *Trichosurus vulpecula*). Larsell (1970) also noted that both the anterior and posterior lobes of the *Macropus* cerebellum (separated by the primary fissure in Figure 5.1e) are much larger and more complex than in *Didelphis virginiana*, with a more vertically oriented primary fissure in *Macropus*.

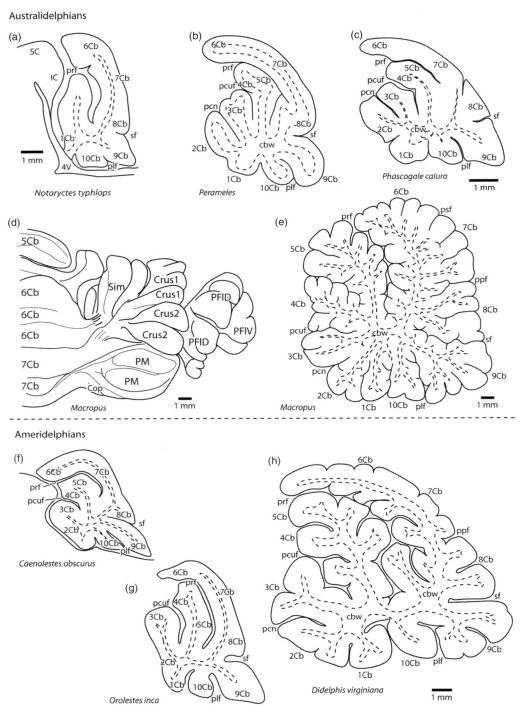

Figure 5.1 The variety of gross morphology of the cerebellum as seen in Australian marsupials (a to e) and Ameridelphians (f to h). Most images are of midline sections through the vermis, whereas (d) shows the dorsal view of the right cerebellar hemisphere and paraflocculus. Note the striking contrast between the simple cerebellum of the marsupial mole *Notoryctes typhlops* and that of large kangaroos (e). All images except that for *Phascogale calura* are based on findings presented in Larsell (1970). 1Cb to 10Cb – lobules 1 to 10 of the Cb vermis; 4V – fourth ventricle; cbw – cerebellar white matter; Cop – copula pyramis; Crus1 – crus 1 of the ansiform lobule; Crus2 – crus 2 of the ansiform lobule; IC – inferior colliculus; pcn – precentral fissure; pcuf – preculminate fissure; PFlD – paraflocculus, dorsal; PFlV – paraflocculus, ventral; plf – posterolateral fissure; PM – paramedian lobule; ppf – prepyramidal fissure; prf – primary fissure; psf – posterior superior fissure; SC – superior colliculus; sf – secondary fissure; Sim – simplex lobule.

Anatomical and functional subdivisions of the cerebellum

The mammalian cerebellar cortex is divided into longitudinal or parasagittal zonal subdivisions based on the cerebellar corticonuclear and olivocerebellar projections. In placental mammals, three corticonuclear projection zones can be distinguished in the vermis (A, X, B) (Voogd, 2004). The most medial of these (A) extends rostrocaudally over the entire vermis and projects to the middle and caudomedial subdivisions of the fastigial deep cerebellar nucleus. The X zone is found in the anterior lobe (and perhaps lobule 6aCb and lateral lobules 9Cb, 10Cb) and projects to the interstitial cell groups, while the B zone occupies the lateral vermis of the anterior lobe and lobule 6aCb and projects to the lateral vestibular nucleus. In the cerebellar hemisphere of rodents and carnivores, the A_2 zone, three C zones and three D zones have been distinguished (Voogd, 2004). The A_2 zone projects to the dorsolateral protuberance of the fastigial nucleus. In carnivores, the C_1 and C_3 zones project to the anterior interposed nucleus, whereas the C_2 zone projects to the posterior interposed nucleus. The three D zones project to the lateral deep cerebellar nucleus. These zonal patterns are matched by chemoarchitectural features, such as the zebrin pattern discovered by Hawkes and co-workers (Hawkes et al., 1985), enzyme histochemistry with 5′-nucleotidase and nerve growth factor receptor protein (see Voogd, 2004, for review). Zebrin II immunoreactivity has been studied in the cerebellum of an Ameridelphian (Monodelphis domestica – Doré et al., 1990), but never in any Australian marsupial.

In the marsupial cerebellum, the longitudinal zonation of the vermis can usually be clearly seen with the aid of immunohistochemistry to calcium-binding proteins (Figure 5.2a), but longitudinal zones beyond (i.e. lateral to) C_1 are difficult to distinguish with these techniques. Nevertheless, these sagittal zones are probably present in all mammals, since they can also be identified by differential calcium-binding immunoreactivity in monotremes (Tachyglossus aculeatus – short-beaked echidna; Ashwell et al., 2007a).

As noted above, there are a number of deep cerebellar nuclei embedded within the cerebellar white matter, which may be clearly seen in sections immunoreacted for calcium-binding proteins. These deep cerebellar nuclei are morphologically similar in both diprotodont and polyprotodont Australian marsupials (Figure 5.2b, c, d; and Figure 5.2e, f, g, respectively) to the corresponding nuclei in placental mammals.

In eutherians, regions dominated by projections from the spinal cord, dorsal column nuclei and trigeminal nuclei (i.e. cerebellar regions concerned with using somatosensory input for motor co-ordination) correspond anatomically to the anterior lobe, lobus simplex, lobule 8, crus II,

the paramedian lobule and the copula pyramidis (Voogd, 2004). Terman and co-workers (Terman et al., 1998) found that spinocerebellar axons in Didelphis virginiana are generally comparable in origin, course and laterality to the same fibres in placental mammals. In the brush-tailed possum, dorsal and ventral spinocerebellar tracts enter the cerebellum through the inferior and superior cerebellar peduncle, respectively (Watson et al., 1976), as is characteristic of placental mammals. The dorsal spinocerebellar tract terminates principally in the ipsilateral anterior lobe, but also sends a major bundle to the cortex of the cerebellar pyramis. The ventral spinocerebellar tract terminates in both sides of the anterior lobe, with most fibres ending on the side contralateral to the side of origin. Collectively, both spinocerebellar tracts terminate in five sagittal zones mainly in the preculmen (lobules 2Cb and 3Cb) and lingula (1Cb), and only a small number of fibres reach the paraflocculus. This pattern is similar to the five sagittal zones identified in rodents for projections from the thoracic and lumbar spinal cord (Voogd, 2004). The regions lying between the sagittal zones of spinocerebellar tract termination in the possum are presumably occupied by cuneocerebellar fibres, as is seen in rodents, but this projection has never been studied in any Australian marsupial.

Output from the cerebellum in rodents and carnivores is ordered according to the deep cerebellar nuclei (Voogd, 2004). The medial deep cerebellar or fastigial nucleus projects via the inferior cerebellar peduncle to the vestibular nuclei, brainstem reticular formation, parafascicular, ventromedial and ventrolateral thalamic nuclei. In the domestic cat, the medial third of the superior cerebellar peduncle contains fibres from the medial cerebellar and posterior interposed nuclei, whereas the lateral two-thirds of the superior cerebellar peduncle contains efferents from the anterior interposed and lateral cerebellar nuclei. The posterior interposed nuclei project to the red nucleus, the central grey of the midbrain, the deep layers of the superior colliculus, the subparafascicular nucleus and the zona incerta of the diencephalon. The anterior interposed nucleus projects to the red nucleus, ventrolateral thalamic nucleus, basilar pons and reticulotegmental nucleus, and the dorsal accessory nucleus of the inferior olivary complex. The lateral deep cerebellar nucleus projects to the contralateral red nucleus, contralateral ventrolateral nucleus of the thalamus, reticulotegmental nucleus, pontine nuclei and the principal nucleus of the inferior olivary complex.

In experiments in which the deep cerebellar nuclei were ablated or the superior cerebellar peduncle cut in Trichosurus vulpecula, projections were identified to two main regions of the contralateral thalamus (Rockel et al., 1972). One focus of terminal degeneration was in the intralaminar nuclei (parafascicular, central lateral and paracentral nuclei), whereas

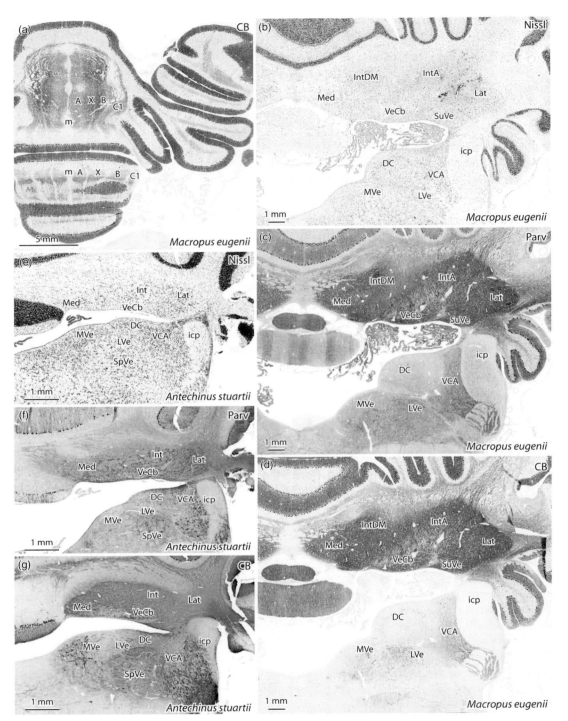

Figure 5.2 Chemoarchitecture of the cerebellar cortex, deep cerebellar nuclei and vestibular nuclei in representative diprotodont (*Macropus eugenii* – a to d) and polyprotodont (*Antechinus stuartii* – e to g) Australian marsupials. Immunoreactivity to calbindin reveals parallel zonation in the cerebellar cortex (a), whereas immunoreactivity to both parvalbumin and calbindin reveal the extent of the deep cerebellar nuclei in both these Australian marsupials. DC – dorsal cochlear nu.; icp – inferior cerebellar peduncle; Int – interposed deep cerebellar nu.; IntA – interposed cerebellar nu., anterior; IntDM – interposed nu., dorsomedial crest; Lat – lateral deep cerebellar nu.; LVe – lateral vestibular nu.; Med – medial deep cerebellar nu.; MVe – medial vestibular nu.; SpVe – spinal vestibular nu.; SuVe – superior vestibular nu.; VCA – ventral cochlear nu., anterior; VeCb – vestibulocerebellar nu.

the other was in the ventrolateral nucleus of the dorsal thalamus. The ventral anterior nucleus was free of terminal degeneration. These patterns of termination are broadly consistent with those observed in placental mammals and suggest a similar organisation of cerebellar nuclear efferents across therians.

As noted in the introduction, the cerebellum receives input from a variety of sensory modalities. Unfortunately, very little consideration has been given to the specific role that the cerebellum may play in Australian marsupials with highly specialised sensory systems. One tantalising hint in this direction is the presence in the feathertail glider (*Acrobates pygmaeus*) of a massive lateral lobe on the medulla with a projection into the cerebellum (Aitkin and Nelson, 1989). This hypertrophied region of the medulla and its cerebellar projection may allow auditory input to influence the motor co-ordination required for gliding behaviour and predator avoidance, although the circuitry that underlies this supposed function remains to be studied.

Vestibular system

The vestibular system is concerned with providing information that enables an animal to maintain its body's equilibrium in the gravitational field and detect and respond to rotation of the head in space. The vestibular apparatus includes five sensory organs: three cristae ampullares (associated with the horizontal, anterior and posterior semicircular ducts) and two otolith organs (the utricular and saccular maculae). In rodents, the utricular and saccular maculae are opposed to each other at approximately 90°, with the utricle lying in a horizontal position (Vidal and Sans, 2004). The cristae ampullares are sensitive to angular acceleration, whereas the otolith organs specifically transduce linear acceleration induced by head movements and gravity. The first-order neurons in the vestibular pathway are bipolar cells in the vestibular (Scarpa's) ganglion and the central processes of these neurons are distributed to the vestibular nuclei and vestibulocerebellum (flocculus and nodule, fastigial nucleus). Not all parts of the vestibular nuclear complex necessarily receive vestibular nerve axons and there is also a significant input to these nuclei from spinovestibular projections and the cerebellum. The vestibular nuclear complex is usually divided into superior, lateral, medial and spinal (inferior) vestibular nuclei. Projections from the vestibular nuclei include pathways to influence the eye muscle nuclei (concerned with control of eye movement and mainly arising from the superior vestibular nucleus), spinal cord (concerned with control of posture and arising mainly from the lateral vestibular nucleus) and the nodulus, uvula and flocculus of the cerebellum. The medial vestibular nucleus is mainly concerned with stabilising gaze and posture in the horizontal plane, while the spinal vestibular nucleus is mainly implicated in the integration of vestibular, spinal and cerebellar messages in the control of body posture.

All the vestibular nuclei familiar in placental mammals can be identified in both poly- and diprotodont Australian marsupials (Figure 5.2b, c, d; and Figure 5.2e, f, g, respectively). Indeed, these nuclei are similar cyto- and chemoarchitecturally in all mammals, ranging from the monotreme *Tachyglossus aculeatus* (short-beaked echidna, Ashwell et al., 2007b), through polyprotodont marsupials (Henkel and Martin, 1977; see atlas plates D24 to D29; Figure 5.2e, f, g) and diprotodont marsupials (see atlas plates W39 to W45; Cowley, 1976; McCluskey et al., 2008) to rodents (Vidal and Sans, 2004). Cowley (1976) thought that all four vestibular nuclei receive input from the vestibular apparatus in the red kangaroo, but this observation was based on tracing myelinated and/or silver-stained fibres from the vestibular nerve root and needs to be confirmed with modern tracing techniques. He also observed contributions of fibres from the medial, lateral, superior and rostral inferior vestibular nuclei to the medial longitudinal fasciculus, presumably mediating vestibulo-ocular reflexes. While the cytoarchitecture of the vestibular nuclei in the red kangaroo is much like that in placental mammals, Cowley (1976) was impressed by the large size of the lateral vestibular nucleus and the (ventro)lateral vestibulospinal tract in macropods. He considered this to be due to the importance of precise and rapid changes in the application of hindlimb force in response to vestibular information during hopping locomotion. Unfortunately, this hypothesis has never been put to the test in any macropod and the precise role of the lateral vestibulospinal tract in macropod locomotion remains speculative.

Precerebellar nuclei

Inferior olivary nucleus and olivocerebellar projection

Figure 5.3 illustrates the cyto- and chemoarchitecture of the inferior olivary complex and lateral reticular nucleus in the tammar wallaby and brown antechinus. The most detailed consideration of the organisation of the inferior olivary nuclear complex in marsupials and the homologies of its constituent nuclei with that in eutherians was undertaken by Watson and Herron (1977). Those authors concluded that the principal and dorsal accessory nuclei (as seen in representatives of eight marsupial families) are

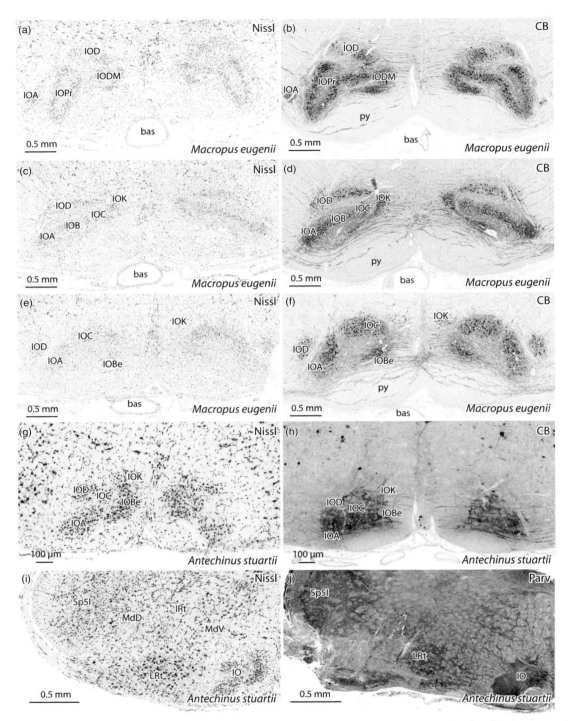

Figure 5.3 Cyto- and chemoarchitecture of the inferior olivary complex in representative diprotodont (a to f) and polyprotodont (g, h) Australian marsupials. Cyto- and chemoarchitecture of the lateral reticular nu. in *Antechinus stuartii* is illustrated in photomicrographs (i) and (j). bas – basilar artery; IO – inferior olivary nu.; IOA – IO, subnu. A; IOB – IO, subnu. B; IOBe – IO, beta subnu.; IOC – IO, subnu. C; IOD – IO, dorsal nu.; IODM – IO, dorsomedial cell group; IOK – IO, cap of Kooy; IOPr – IO, principal nu.; IRt – intermediate reticular nu.; LRt – lateral reticular nu.; MdD – medullary reticular nu., dorsal; MdV – medullary reticular nu., ventral; py – pyramidal tract; Sp5I – spinal trigeminal nu., interpolaris.

directly comparable to similarly named nuclei in eutherians. On the other hand, the probable homologue of the placental medial accessory olivary nucleus has certain features unique to marsupials. Most of the neurons of the marsupial 'medial' accessory nucleus (subnucleus a or IOA, Figures 5.3a and b, 5.4b and c) actually lie *lateral* to the principal nucleus. In Ameridelphians (*Didelphis virginiana*, *Philander*, *Marmosa*, *Lestoros* and *Caenolestes*), those few cells of IOM which lie medial to the IOPr (subnucleus a_1) are connected to the main lateral portion by a sheet of neurons that lies ventral to the IOPr. Some Australian marsupials (*Petaurus breviceps* and *Macropus eugenii*) have broken strands of neurons connecting subnucleus a_1 and a, but no Australian species has a continuous linking strand of neurons like the Ameridelphians. On the basis of these findings, Watson and Herron (1977) proposed that the medial accessory nucleus homologue should be named the ventral accessory nucleus (IOV) in marsupials. They also argued that the lateral position of subnucleus a probably arises from the failure of most neurons of the IOV to pass ventral to the IOPr during early postnatal development.

Watson and Herron (1977) also noted variation in the morphology of IOPr among marsupials. In particular, they pointed out that the IOPr of large diprotodont marsupials such as kangaroos and wallabies showed both absolute and relative increase in size (compared to other subnuclei of the inferior olivary complex). In the wombat, the dorsal lamella of the IOPr is greatly expanded to form a ball-like structure, whereas the ventral lamella of the IOPr is greatly enlarged and folded. Enlargement of the IOD was also seen in the eastern grey kangaroo, with the lateral portion greatly elongated in coronal sections. These distinctive features of inferior olivary organisation in kangaroos may reflect the highly specialised locomotion of these marsupials, but the precise functional significance of these features remains to be investigated.

Martin and colleagues (Martin *et al.*, 1975a) reported that the cerebello-olivary projection of *Didelphis virginiana* is broadly similar to that in placental mammals, with only a few minor differences related to the arrangement of subnuclei within the marsupial olivary complex. This particular projection has not been studied in detail in Australian marsupials.

Data concerning the organisation of the olivocerebellar projection of Australian marsupials are also very sparse. Holst (1986) showed that the topographical organisation of the olivocerebellar projection in the phalangerid, the brush-tailed possum *Trichosurus vulpecula* (Figure 5.4) is broadly similar to that reported for the cat and *Didelphis virginiana* (Martin *et al.*, 1975a; Linauts and Martin, 1978).

However, she also argued that the brush-tailed possum exhibited greater input to all regions of the cerebellum from the minor nuclei of the inferior olive, as well as a projection from subnucleus c to the hemispheres and a contribution from the ventrolateral portion of IOPr to zone C1 of the cerebellum, in contrast to the opossum. The details of the olivocerebellar projection found by Holst for *Trichosurus vulpecula* support the contention by Watson and Herron (1977) that the only major difference between the marsupial and eutherian inferior olivary complex is the lateral position of part of the IOV.

Olivocerebellar axons are known as climbing fibres when they are seen in the cerebellar cortex. Quantitative analysis of neuronal numbers in the inferior olivary complex of *Trichosurus vulpecula* indicates that a single inferior olivary neuron supplies, on average, 15 Purkinje cells (Furber and Watson, 1979) and the total number of Purkinje cells in this species is of the order of 881 000. The ratio of inferior olivary complex neurons to Purkinje cells in the possum is similar to that reported for the echidna (1:10), many placentals (1:7 to 1:15) and the chick (1:19) (see Holst, 1986, for review). Clusters of inferior olivary complex neurons in *Trichosurus vulpecula* probably project to rows of from 120–300 Purkinje cells and the total number of such projection microzones within the cerebellum of the possum is likely to be between 3000 and 7000. Similar analysis in the laboratory cat indicate that each neuron of the inferior olivary complex projects to about eight Purkinje cells (on average) and that the estimated number of microzones in the cat cerebellar cortex is of the order of 6000 to 17 000 (Mlonyeni, 1973), although Ito (1984) obtained a much lower figure of 5000 for the number of subunits in human cerebellar cortex. Clearly, the numerical matching of cerebellar processing units to inferior olivary neurons is broadly similar across all mammals.

Lateral reticular nucleus

The lateral reticular nucleus is situated along the ventral medulla, roughly co-extensive rostrocaudally with the inferior olivary complex. The lateral reticular nucleus is a major source of mossy fibres to the cerebellar cortex in placental mammals (Ruigrok, 2004) and is believed to play an important role in the control of motor activity and co-ordination, particularly when related to the execution of fine and target-directed movements of the forelimb (Ruigrok, 2004). Although the lateral reticular nucleus is clearly recognisable in the monotreme and Australian marsupial brainstem (Ashwell *et al.*, 2007b; Figure 5.4i; see atlas plates D30, W46), nothing is known about its function in these groups.

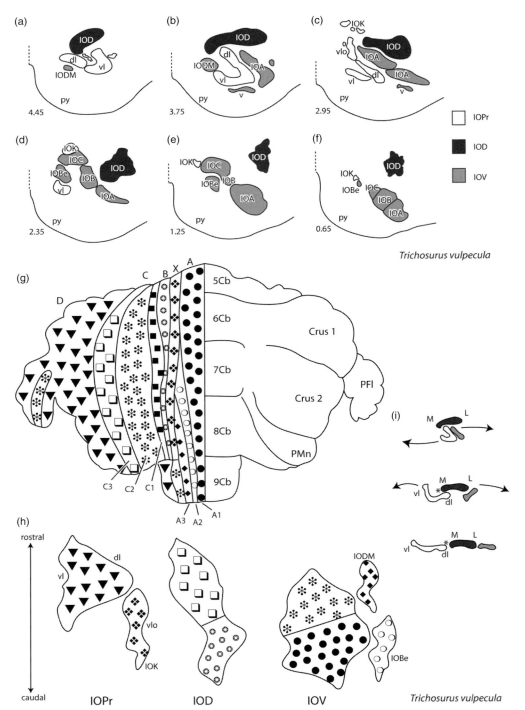

Figure 5.4 Diagrammatic representation of the organisation of the inferior olivary nuclear complex in the brush-tailed possum, based on data and illustrations from Holst (1986). The diagrams are represented from rostral (a) to caudal (f) with accompanying values for distance in mm from the caudal extremity of the complex. The midline is to the left (dashed line) and the key indicates the subdivision into dorsal (IOD), principal (IOPr) and ventral (IOV; homologous to the medial nu. of the placental inferior olivary nuclear complex) components. Diagrams g) and h) illustrate the pattern of the olivocerebellar projection in this species. Diagram g) is a dorsal view of the cerebellum of the brush-tailed possum with rostral uppermost, showing the parasagittal zonation alluded to earlier (see Figure 5.2). Note that diagram h) shows an unfolded view of the inferior olivary complex as used by Holst (1986) and diagram i) shows how the unfolding was performed. The asterisk marks the pivotal point for the unfolding process. 5Cb to 9Cb – lobules 5 to 9 of the Cb vermis; Crus1, Crus 2 – crus 1, crus 2 of ansiform lobule; dl – dorsal lamina of IOPr; IOA – IO, subnu. A; IOB – IO, subnu. B; IOBe – IO, beta subnu.; IOC – IO, subnu. C; IOD – IO, dorsal nu.; IODM – IO, dorsomedial cell group; IOK – IO, cap of Kooy; IOPr – inferior olive, principal nu.; IOV – IO, ventral nu.; PFl – paraflocculus; PMn – paramedian reticular nu.; py – pyramidal tract; v – neurons associated with VLO; vl – ventral lamina of IOPr; v/o – ventrolateral outgrowth of IOPr.

Pontine nuclei

The pontine nuclei are a large cluster of small- to medium-sized neurons located around the ventral surface of the rostral hindbrain, although the rostrocaudal extent of the pontine nuclei will vary between species, depending on the neuronal population size. In therians (and probably all mammals), the pontine nuclei receive a major input from the cerebral cortex and some subcortical structures (deep cerebellar nuclei, superior colliculus, mamillary nuclei of the hypothalamus, spinal cord) and project to the cerebellum through the middle cerebellar peduncle to terminate in the cerebellar cortex as mossy fibres. The pontine reticulotegmental nucleus, which lies dorsal to the pontine nuclei, is usually considered with the pontine nuclei, because it also extends mossy fibres to the cerebellar cortex, although some regard it as a specialised nucleus of the pontine reticular formation (Ruigrok, 2004).

The pontine nuclei and their connections have not been extensively studied in any Australian marsupial, although they received some attention in Ameridelphians (King *et al.*, 1968; Mihailoff and King, 1975; Mihailoff *et al.*, 1980). Nevertheless, the cytoarchitecture of these nuclei in marsupials (see plates D19 to D21 and W33 to W37) appears to be broadly similar to those in eutherians. Indeed, pontine nuclei are cytoarchitecturally similar across all mammals, i.e. from eutherians to monotremes (Ashwell *et al.*, 2007b). A brief report using degeneration techniques found that corticopontine projections in two diprotodonts (*Trichosurus vulpecula*, *Setonix brachyurus*) were bilateral in distribution, in contrast to the predominantly ipsilateral projection reported for placental mammals (Way, 1970). The pontine nuclei of the brush-tailed possum receive projections from a broad area of neocortex, including areas rostral to sulcus α and as far caudally as the occipital cortex (Martin *et al.*, 1971).

Development of the cerebellum

The mammalian cerebellum is derived from neurons produced in two generative zones of the developing hindbrain. The macroneurons of the cerebellum (e.g. deep cerebellar nuclei neurons and Purkinje cells) are generated relatively early (i.e. early fetal life in placental mammals and around the time of birth in marsupials) from the ventricular lining of the rhombic lip, whereas the microneurons of the cerebellum (e.g. granule cells) are generated from the external granule-cell layer relatively late (i.e. around the time of birth to early postnatal life in placental mammals and exclusively postnatally in marsupials). As a general principle, therians

that have their eyes open at birth tend to have the most mature cerebelli at birth, and the most advanced stage of cerebellar development tends to coincide with the opening of the eyes (Sánchez-Villagra and Sultan, 2002).

Very few studies have been made of cerebellar development in Australian marsupials and those available are confined to diprotodonts (a ³H thymidine study of the brush-tailed possum, and a histological and immunochemical study of the tammar wallaby). The cerebellum of the brush-tailed possum (*Trichosurus vulpecula*) is only rudimentary at birth and the lobular pattern develops gradually from P10 to P47, although the paraflocculus development continues until the fourth month of postnatal life (Sanderson and Weller, 1990b). As with rodents, the output neurons of the deep cerebellar nuclei are the first to develop, with most deep cerebellar neurons generated before P5. Similarly, Purkinje cell generation in the possum begins before P5 and concludes around P9, and there are neurogenetic gradients for Purkinje cell generation, with those from the lingula part of the rostral vermis generated last. Macroneurons of the granule-cell layer (e.g. Golgi cells) are generated from birth to the beginning of the third postnatal week. Microneurons like the abundant granule cells are generated later than the macroneurons considered above; generation of granule cells in the possum follows an extended time-course, with the first few cells arising at P12 and granule-cell birth extending beyond P95 (perhaps to P150). The peak granule-cell generation appears to be around P82 (Sanderson and Weller, 1990). The protracted period of development of granule cells is much longer than in eutherians (e.g. mouse – E17 to P15; rat – P0 to P21; see Sanderson and Weller, 1990, for discussion), but is still within the period of pouch dependence of young brush-tailed possums (approximately 5.5 months, Tyndale-Biscoe, 2005). This protracted postnatal development is similar to that in an Ameridelphian (*Didelphis virginiana*), where the external granular layer proliferative zone does not grow over the cerebellar anlage until P5, begins to generate granule cells from about P30 (O'Donoghue *et al.*, 1987) and persists until after P105 (Laxson and King, 1983).

The development of the cerebellum in the tammar wallaby (Hassiotis *et al.*, 2002) follows a similarly protracted time-course to that described above for the possum. As seen above for the brush-tailed possum, the cerebellum of the newborn tammar wallaby is extremely rudimentary. At birth and during the first postnatal week in the wallaby (WP0/10, WP5/10), the post-mitotic components of the cerebellum are confined to cortical and nuclear transitory zone macroneuron populations and there is no external granular layer. The external granular layer grows over the cerebellar anlage during the second postnatal week and a distinct Purkinje cell layer does not emerge until P12 (atlas plate

WP12/13). Generation of granule cells from the external granular layer probably begins during the third postnatal week, because the (internal) granule-cell layer of the cerebellar cortex begins to emerge from P19. Ingrowth of afferents is probably most active during the first four postnatal weeks, because the intensity of GAP-43 immunoreactivity of the cerebellar white matter declines after P25 (Hassiotis *et al.*, 2002). Cerebellar folding or foliation, which develops as a consequence of expansion of the internal granule and Purkinje cell layers, begins at about the end of the first postnatal month (WP25/9). A study of gross brain growth in the tammar wallaby (Renfree *et al.*, 1982) has found that the cerebellum grows rapidly to P180 (increasing in mass from 5 mg at P30 to 1.2 g at P180, a factor of 240), followed by a much slower pace of growth to adult size. The cerebellum appears to be the fastest growing brain structure during postnatal life in the wallaby and, judging from Sanderson and Weller's findings in the possum (Sanderson and Weller, 1990), it is likely that this extraordinary growth is largely due to generation of the granule-cell populations over this period.

Although the development of afferent and efferent cerebellar connections has not been studied in Australian marsupials, it is likely that efferent connections are similarly late to develop, since efferents from the cerebellum to the ventrolateral thalamus do not develop until after P12 in the North American opossum (Martin *et al.*, 1987).

Development of the vestibular system

Several studies have focussed on the development of the vestibular apparatus in newborn marsupials (Larsell *et al.*, 1935; Krause, 1991) because of its perceived central role in control of the climb to the pouch. Early studies in didelphids noted that the vestibular division of the vestibulocochlear nerve myelinated significantly earlier than the acoustic division (see discussion in Johnson, 1977), but only recently has the development of central pathways of this system been explored in Australian marsupials.

Gemmell and Nelson (1989, 1992) identified a sensory epithelium (with kinocilia and stereocilia) and overlying otoliths in the utricle of the newborn northern native cat (*Dasyurus hallucatus*), but the saccule and three cristae ampullares of the semicircular canals did not develop until later. Gemmell and Nelson (1989) concluded that the rudimentary vestibular receptors might aid the newborn in its climb to the pouch, but behavioural studies suggest that functional maturation of the vestibular system in this species is delayed for many weeks after birth. Pellis and colleagues (Pellis *et al.*, 1992) did not observe tactile-induced righting behaviour in juvenile *Dasyurus hallucatus* until

P40 and vestibular righting reflexes were not seen until even later. Vestibular righting triggered by falling supine in the air did not develop until P80.

Gemmell and co-workers also reported that the sensory region of the utricle is present at birth in the stripe-faced dunnart (*Sminthopsis macroura*) and that the semicircular canals and cochlea of the dunnart are at a similar stage at birth as the newborn *Dasyurus hallucatus* (Gemmell and Selwood, 1994). Similarly, Gemmell and Rose (1989) claimed that the newborn brush-tailed possum (*Trichosurus vulpecula*) and rat kangaroos (*Potorous tridactylus*, *Bettongia gaimardi*) have sensory utricles similar in structure to that of *Dasyurus hallucatus*. Nevertheless, these purely ultrastructural studies ignore the maturation of central vestibular pathways, which are essential if the vestibular apparatus of the inner ear is to influence behaviour of the newborn marsupial.

A more recent study of the development of the vestibular apparatus and its central connections in the tammar wallaby (McCluskey *et al.*, 2008) found that the otocyst of newborn wallabies has distinct utricle, saccule and semicircular canals with immature sensory regions receiving innervation by GAP-43 immunoreactive fibres. At birth, vestibular nerve fibres can be traced using carbocyanine dyes into the brainstem to the developing vestibular nuclei, although these nuclei are not yet cytoarchitectonically distinct and are unlikely to be fully functional. On the other hand, the vestibular nuclei do not contribute direct projections to the lower cervical spinal cord at birth in the wallaby and most bulbospinal projections in the newborn appear to be derived bilaterally from the gigantocellular, lateral paragigantocellular reticular and ventral medullary nuclei. These findings cast doubt on the proposal that the vestibular system directly influences the brachial motor system at birth, although indirect projections through the brainstem reticular formation remain a possibility.

An interesting finding in newborn poly- and diprotodont marsupials concerns the vestibular efferent projection. A substantial bilateral projection to the vestibular ganglion and apparatus from the region of the gigantocellular and lateral paragigantocellular nuclei was seen at birth in the tammar wallaby, but not at subsequent ages (McCluskey *et al.*, 2008). This is similar to the findings of Pflieger and Cabana (1996), who also reported a transient commissural projection to the vestibular ganglion in *Monodelphis domestica*, although they mistakenly assigned these efferent neurons to the superior olivary nucleus. The position of the labelled cells and the course of their efferent axons along the path of the facial nerve fibres are reminiscent of the paragenual vestibular efferent system, as seen in many non-mammals (see Lysakowski and Goldberg, 2004, for review). On the other hand, the extent of the retrogradely labelled cells in the immature wallaby brainstem is substantially greater than that seen in adult mammals (Lysakowski and Goldberg, 2004). Given that this extensive field of vestibular efferent neurons is present for only a short time after birth in newborn marsupials (*Monodelphis domestica* and *Macropus eugenii*), it seems likely that it is of some developmental significance, possibly in co-ordinating the newborn's climb from urogenital sinus to pouch, although further physiological studies would be needed to investigate this possibility.

Despite the early maturation of some elements (McCluskey *et al.*, 2008), most of the development of the vestibular system and its projections to the spinal cord occurs well after birth in the tammar wallaby. From P5, the vestibular apparatus has extensive projections to all vestibular nuclei and it is at this time that the first neurons projecting to the spinal cord via the lateral vestibulospinal tract may be seen in the lateral vestibular nucleus. Cytoarchitectonic differentiation of the vestibular nuclei continues for the first postnatal month with emergence of discrete parvicellular and magnocellular components of the medial vestibular nucleus by P19. GAP-43 immunoreactivity was found to remain high in the lateral vestibulospinal tract for several months after birth, suggesting that development of this tract follows a prolonged timecourse throughout the dependency of pouch life.

Development of the inferior olivary and pontine nuclei in Australian marsupials

The inferior olivary and pontine nuclei arise from the rhombic lip neurogenerative zone. In placentals (and probably all mammals), these neuronal populations migrate around the circumference of the brainstem to come to rest in the caudal medulla and pons, respectively. As might be expected from the delayed and prolonged time-course of cerebellar development in marsupials, the major precerebellar nuclei are also generated relatively late. In the North American opossum, the inferior olivary complex is not present at birth, but appears at P3 (Maley and King, 1980). The inferior olive is produced by migration of neurons via two streams: the marginal and submarginal migratory streams. The marginal migratory stream disappears by P5 to P7 in the opossum, but the submarginal stream persists for several days. Major subdivisions of the inferior olivary complex begin to emerge by P10 to P14 and projections to spinal cord, midbrain and cerebellum are present by P21 (Bauer-Moffat and King, 1981).

At the time of birth in the tammar wallaby, there are no inferior olivary neurons present in the ventral medulla (WP0/11), but by P5, an inferior olivary migratory stream is visible on the ventrolateral surface of the medulla and postmigratory neurons begin to accumulate in the ventromedial medulla (WP5/10). Initially the developing inferior olivary complex does not show any internal subdivisions, but by P12 an internal division into medial and dorsal compartments begins to appear (WP12/14). The differentiation of the inferior olivary complex proceeds smoothly from P19 to P37 and all the subnuclei of the mature inferior olive emerge at about the end of the first month of postnatal life of the wallaby (WP25/10).

In the North American opossum, basilar pontine nuclei begin to collect at P7 and dendritic growth of pontine neurons occurs from P25 to P80 (King *et al.*, 1987). Growth of pontine origin mossy fibres into the cerebellum proceeds from P17 and corticopontine fibres reach the pontine nucleus at about P25 and expand their terminal fields by P29. The time of arrival of afferents from the cerebral cortex and deep cerebellar nuclei is closely correlated with the initiation of dendritic outgrowth by pontine neurons and the outgrowth of pontocerebellar axons.

The pontine nuclei of the tammar wallaby appear to follow a similar development timetable to that found for *Didelphis virginiana*. At the time of birth, and at least until the end of the second postnatal week, there are no pontine nuclei visible in the wallaby brainstem, although the developing superior olivary nuclei and ventral nucleus of the lateral lemniscus have already appeared on the ventral edge of the pons. It is not until the second week of postnatal development (WP12/10) that a pontine migratory stream can be identified along the rostroventral surface of the pons and postmigratory neurons begin to accumulate in the developing pontine nuclear complex. By P25 (WP25/7), the pontine nuclei have expanded across the ventral surface of the pons and a middle cerebellar peduncle can be recognised passing laterally and caudally towards the developing cerebellum.

What remains unknown?

What little is known about the chemoarchitecture and connectivity of the cerebellum in Australian marsupials suggests that functional zonation of the cerebellum is similar in all therians. Nevertheless, it remains to be seen whether zebrin immunoreactivity and afferent and efferent connectivity in Australian marsupials correspond with those seen in placental mammals and Ameridelphians. This would require immunohistochemical studies of zebrin II expression correlated with tracing studies of cerebellar corticonuclear, spinocerebellar and olivocerebellar pathways.

An unexplored area of cerebellar function in Australian marsupials concerns possible specialisation of the cerebellar and associated nuclei for the unusual locomotion and behaviour of some marsupials. A fertile group in which to ascertain this relationship would be the gliding marsupials, like the acrobatid *Acrobates pygmaeus*. Exploration of the interaction between auditory function and cerebellar activity is likely to reveal anatomical and physiological specialisations in this mammal. At the other end of the scale is the marsupial mole (*Notoryctes typhlops*), which would represent an excellent model in which to study the loss of cerebellar circuitry and function in a mammal adapting to subterranean life. The phylogeny of this mammal remains uncertain, but it is probable that it is derived from a surface-dwelling insectivore similar to modern small dasyurids. Presumably the ancestor of the marsupial mole possessed a cerebellum comparable to modern marsupial insectivores with 10 cerebellar lobules and substantial spinocerebellar projections from lumbar segments. The change in locomotion in this mammal presumably led to a decline in the significance of hindlimb co-ordination with reorganisation of spinocerebellar afferents in the cerebellar cortex.

Watson and Herron (1977) hypothesised that the division of the marsupial IOV into components both medial and lateral to the principal nucleus arose due to failure of some neurons of the marginal migratory stream to pass under the principal nucleus during early postnatal development. While an eminently reasonable explanation of the 'unusual' organisation of the inferior olivary nuclei in marsupials, this hypothesis has never been tested. In fact it may be quite technically difficult to test this proposition, because some form of molecular tagging of IOV neurons (quite independent of the topographic patterning of their olivocerebellar projection) would be required to identify the neurons during early stages of migration. At present such molecular markers are not available.

It is most likely that such 'special' features of the marsupial brainstem arise from the relatively early birth of marsupials compared to placental mammals, with consequent demands on the early functional maturation of locomotor and ventilatory circuitry in the medulla. It is unknown as to whether the development of mossy and climbing fibre types of cerebellar afferents is affected by the developmental constraints of early delivery and prolonged pouch life, since no study has ever examined the development of afferents to the developing cerebellum in any Australian marsupial.

Studies in Ameridelphians and Australian marsupials suggest that there may be vestibular efferent circuitry in the brainstem of these newborn marsupials which are: (i) lost by a few weeks after birth and (ii) much more extensive than in developing placental mammals. It is tempting to speculate that this circuitry plays a role in the unique behaviour of newborn marsupials, but there have been no studies of the function of this circuitry during early postnatal life. Since vestibular circuitry is obliged to be more active in young marsupials than in placentals, this is likely to be a fertile area for exploring evolutionary divergence in the maturational timetables and circuitry of the vestibular system.

6

Diencephalon and associated structures: prethalamus, thalamus, hypothalamus, pituitary gland, epithalamus and pretectal area

K. W. S. Ashwell

Summary

The pretectum, thalamus and prethalamus are derived from the first, second and third prosomeres of the embryonic forebrain (i.e. p1, p2, p3), respectively. The pretectum is mainly concerned with visual reflexes and has five nuclei (anterior pretectal, medial pretectal, olivary pretectal, tectal grey or posterior pretectal, and nucleus of the optic tract) consistently observed in mammals. The prethalamus (also called the ventral thalamus) consists of the zona incerta, subthalamic nucleus, reticular nucleus and ventral lateral geniculate nucleus (pregeniculate nucleus). The thalamus (also called the dorsal thalamus) is divided anatomically into dorsomedial and ventrolateral parts and is functionally separated into specific and non-specific relay nuclei. The organisation of the thalamic nuclei in poly- and diprotodont Australian marsupials has been reported to be more like each other than American didelphids. In dasyurids, the midline/intralaminar nuclei are more complex than in diprotodonts, but conversely the somatosensory and visual thalamic relay nuclei are much more complex in diprotodonts compared to Australian polyprotodonts.

The hypothalamus is derived from the proneuromere of the embryonic brain, with some possible contribution from the p3 prosomere. In all mammals it may be divided into three longitudinal zones (periventricular, medial and lateral) and four rostrocaudal regions (preoptic, anterior, tuberal and mamillary). Although there are some minor differences in the organisation of the hypothalamus in marsupials as compared to placental mammals (e.g. supraoptic nucleus, posterior mamillary nucleus), the same broad cytoarchitectural, chemoarchitectural and functional organisation applies. Furthermore, many of the major hypothalamic nuclei in marsupials (e.g. paraventricular and ventromedial hypothalamic nuclei) show similar subdivisions to those identified in placental mammals as underlying functionally important segregation of pathways, although functional analysis of hypothalamic pathways in marsupials lags behind that in placental mammals by many decades. Development of the hypothalamus in all therians follows a lateral to medial gradient in terms of neurogenesis, cytoarchitectural development and functional maturation. As in Ameridelphians, the development of the hypothalamus of diprotodont marsupials like the tammar wallaby is almost exclusively postnatal. Most of the connections of the hypothalamus are also developed after birth in the wallaby, although a projection from the lateral hypothalamus to the brainstem is present at birth and may contribute to early control of suckling and lung ventilation.

The pituitary gland lies below the hypothalamus and is influenced by regulatory factors and hormones released from the hypothalamic nuclei into the hypothalamo-hypophyseal portal system and tract. Ultrastructure of both the anterior and posterior hypothalamus is similar in marsupials to that in placental mammals, but there are important differences in the posterior pituitary hormones (e.g. mesotocin vs. oxytocin) and their control of milk ejection. The time-course of maturation of anterior pituitary cell types correlates closely with the physiological requirements of the pouch young.

The epithalamus (a p2 prosomere derivative) includes the habenular nuclei and pineal gland. The epithalamus is highly conserved in structure across mammals and the pineal gland and its projections play an important role in controlling seasonality of reproductive behaviour.

Introduction: subdivisions of the diencephalon

The diencephalon is defined as the region of the embryonic brain derived from the three prosomeres (Puelles *et al.*, 2007). Prosomere 1 gives rise to the pretectum, prosomere 2 forms the thalamus (and epithalamus) and prosomere 3 gives rise to the prethalamus.

The prethalamus (formerly called the ventral thalamus, but more properly considered on developmental grounds to be the 'rostral' thalamus) intervenes between the thalamus proper and the rostral hypothalamus and includes the zona incerta, reticular nucleus (formerly called the thalamic reticular nucleus), pregeniculate nucleus (formerly called the ventral lateral geniculate nucleus), subgeniculate nucleus and subthalamus. The thalamus proper was formerly called the dorsal thalamus and has close functional and connectional relationships with the cerebral cortex, basal ganglia and amygdala. It is more properly considered as the 'caudal' thalamus, given that it is positioned and develops caudal to the prethalamus in young marsupials and fetal placental mammals. The pretectum consists of a number of distinct nuclear groups involved in visual reflexes.

Prethalamus

The prethalamus (formerly known as the ventral thalamus) is involved with a mixture of functions including some sensory processing and motor control. Component nuclei include the reticular components, subthalamus and zona incerta.

The reticular components of the prethalamus (reticular, pregeniculate and subgeniculate nuclei) are characterised by projections to the dorsal thalamus and a lack of projections to the cerebral cortex. The reticular or reticular thalamic nucleus forms a thin sheet of neurons around the rostral, ventral and lateral margins of the dorsal thalamus. The pregeniculate and subgeniculate nuclei form the dorsolateral and caudal extensions of the reticular thalamic nucleus and can be considered the reticular components of the visual thalamus. All the afferent and efferent projections of the dorsal thalamus pass through the reticular elements of the prethalamus and most projections from the dorsal thalamus to the cerebral cortex give off reciprocal connections to the reticular nucleus. This means that the reticular prethalamic nuclei like the reticular, subgeniculate and pregeniculate nuclei are strategically placed to modulate all incoming and outgoing information of the thalamus proper. The reticular nucleus also receives descending projections from the cortex (Haight et al., 1980). Reticular neurons are all GABAergic and exhibit parvalbumin immunoreactivity in all mammals (Ashwell and Paxinos, 2005).

The subthalamus receives GABAergic inhibitory projections from the lateral part of the globus pallidus and excitatory projections from the cortex and in turn provides an excitatory output to the lateral globus pallidus,

entopeduncular nucleus (medial globus pallidus) and substantia nigra (pars reticulata). These pathways form the indirect component of the basal ganglia circuitry.

The zona incerta lies dorsal and rostral to the subthalamic nucleus. The zona incerta has diverse connections with motor cortex, association cortex, superior colliculus, pretectum and the brainstem.

Thalamus

The common characteristic of the nuclei of the thalamus is that they provide reciprocal connections with various parts of the cortex. The component nuclei provide relays between extrinsic (i.e. sensory) and intrinsic (i.e. motor relay) information and the cerebral cortex. Thalamic nuclei are usually divided into specific and non-specific on the basis of the nature of the ascending connections. Specific nuclei are concerned with the special senses (e.g. vision, touch and hearing), motor relay and associational relays; whereas non-specific nuclei are concerned with arousal and attentional mechanisms. The overall function of the thalamus is to oversee and modulate the streaming of information to the cortex from various sensory modalities (much as the Hausverwaltung or concierge of a European apartment block keeps an eye on the comings and goings of the block's occupants).

General comments on thalamic organisation

The delineation of the thalamic nuclei in all mammals is based on a combination of cytoarchitectonic, chemoarchitectonic and connectional features. A fibre bundle known as the external medullary lamina, which incorporates one of the prethalamic nuclei (the reticular nucleus) surrounds the thalamus proper. The third and lateral ventricles bound the thalamus medially and dorsally, but in all marsupials the thalamic masses meet to a large extent across the midline, thereby shifting most of the third ventricle ventrally. In therian mammals, an internal medullary lamina divides the thalamus into dorsomedial and ventrolateral groups of nuclei (Figure 6.1) and some nuclei are embedded within the internal medullary lamina (intralaminar nuclei). Some therians have commissural bundles joining the thalamic nuclei of each side. These commissural fibres appear to be more extensive in polyprotodont (e.g. opossum) than in diprotodont (e.g. brushtailed possum) marsupials (Goldby, 1941), a difference that may reflect the shift of commissural information transfer to the telencephalic commissures in diprotodont marsupials.

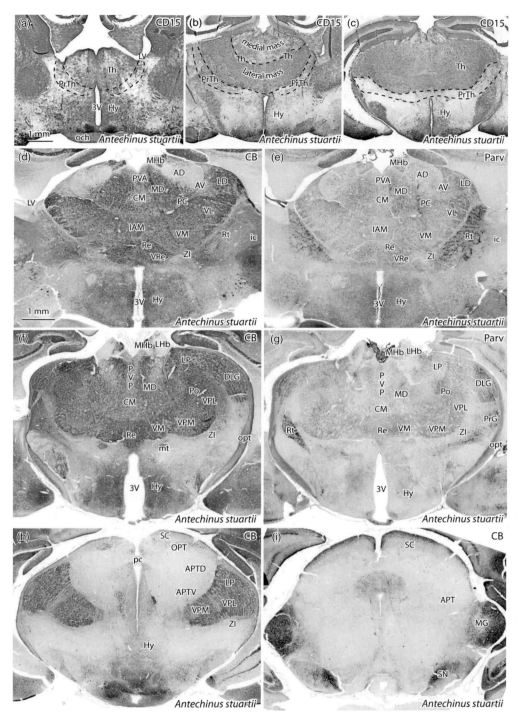

Figure 6.1 Chemoarchitecture of the thalamus and prethalamus in a representative polyprotodont Australian marsupial (*Antechinus stuartii*). Panels (a) to (c) show coronal (frontal) sections at rostral to caudal levels of the adult diencephalon stained for CD15 (Lewis x). Differential immunoreactivity to CD15 allows distinction of the medial mass and midline nuclei from the lateral mass nuclei (b) and demarcates the dorsal thalamus (Th) from the prethalamus (PrTh) at rostral (a) to caudal levels (c). Panels (d), (f), (h) and (i) show calbindin-immunoreacted coronal sections at successive rostrocaudal levels of the thalamus with nuclei indicated. Panels (e) and (g) show a parvalbumin immunoreacted section at a similar level to (f) illustrating the complementary immunoreactivity in some nuclei

The classical functional subdivision of the thalamus is usually made on the basis of the type of information transmitted to the cerebral cortex. The first group are known as the principal- or specific-relay-type nuclei which receive sensory, motor and associational information through ascending or descending projections and transmit this information to precise areas and layers of the cortex. This group includes the anterior, mediodorsal, ventral anterior, ventrolateral, ventromedial, lateral dorsal, lateral posterior, ventral posterior, posterior, medial geniculate and dorsal lateral geniculate nuclei. The second group includes the intralaminar and midline nuclei, which are also known as the non-specific thalamic nuclei. These nuclei receive more general information from arousal and visceral regions of the brainstem and forebrain and project to the cerebral cortex in a less precise manner than the specific thalamic nuclei, as well as to subcortical areas like the basal ganglia and amygdala.

Thalamic cytoarchitecture in general has been studied in three Australian marsupials, the polyprotodont eastern quoll, *Dasyurus viverrinus* (Haight and Neylon, 1981a, 1981b) and the diprotodonts, the brush-tailed possum, *Trichosurus vulpecula* (Goldby, 1941; Haight and Neylon, 1978) and the tammar wallaby, *Macropus eugenii* (Mayner, 1989c). In general, the organisation and connections of the nuclei within the thalamus of various species of Australian marsupials are more similar to each other (whether poly- or diprotodont) than each is to a representative Ameridelphian like the Virginia opossum. Haight and Neylon (1981a) felt that the major cytoarchitectural differences between the thalamic nuclei of *Dasyurus viverrinus* and *Trichosurus vulpecula* lie in three regions: the midline/intralaminar nuclei, the ventral posterior nuclei and the dorsal lateral geniculate nuclei. In dasyurids, the midline/intralaminar region appears to be larger and more cytoarchitecturally complex than in either the brush-tailed possum or the tammar wallaby (Haight and Neylon, 1981a; present work), but the functional or behavioural significance of this (if any) is unknown. On the other hand, the major somatosensory and visual relay nuclei of the dorsal thalamus are much more complex in diprotodont than in polyprotodont species (see *Sminthopsis macroura*, this work; Haight and Neylon, 1981a). The latter certainly reflects the greater specialisation of somatosensory and visual pathways in the so-called advanced Australian marsupials. Haight and Neylon (1981a) also reported that the submedius nucleus is present in *Dasyurus viverrinus* and *Didelphis virginiana*, but is not apparent in the thalamus of *Trichosurus vulpecula*. This stands in contrast to the identification of the submedius nucleus in brush-tailed possum brain by Rockel and colleagues (Rockel *et al.*, 1972).

Association thalamic nuclei

This group of thalamic nuclei includes the mediodorsal, submedius, anterior and lateral thalamic nuclei. The anterior thalamic nuclear group can be further subdivided into anterodorsal, anteroventral and anteromedial nuclei. The lateral dorsal thalamic nucleus is also sometimes included among the anterior nuclei, because of similarities in connections.

The mediodorsal thalamic nucleus is situated medial and dorsal to the internal medullary lamina and is surrounded by the midline and intralaminar nuclei. In rodents the mediodorsal nucleus is subdivided into medial, central and lateral segments on the basis of cytoarchitecture, connections and calcium-binding protein immunoreactivity, and these subdivisions are also visible (at least cytoarchitectonically) in poly- and diprotodont thalamus (see atlas plates D12, W26). In diprotodont marsupials, the lateral segment may also be divided into dorsolateral and ventrolateral parts. Goldby noted that the differentiation of the mediodorsal nucleus of diprotodont marsupials resembles the corresponding nucleus in carnivores, rodents and *Tupaia glis* (Goldby, 1941). In placental mammals, the mediodorsal thalamic nucleus is reciprocally connected with the frontal cortex rostral to the motor and premotor cortex and the cortical extent of this projection is often used to define the prefrontal cortex. This corticothalamic relationship is similar in the brush-tailed possum in that the mediodorsal nucleus engages in reciprocal connections with the orbital and

---

Caption for Figure 6.1 (Cont.)

(e.g. Rt). Scale bar in (a) applies to (a) to (c). Scale bar in (d) applies to (d) to (i). 3V – third ventricle; AD – anterodorsal thalamic nu.; APT – anterior pretectal nu.; APTD – APT, dorsal; APTV – APT, ventral; AV – anteroventral thalamic nu.; CM – central medial thalamic nu.; DLG dorsal lateral geniculate nu.; Hy – hypothalamus; IAM – interanteromedial thalamic nu.; ic – internal capsule; LD – laterodorsal thalamic nu.; LHb – lateral habenular nu.; LP – lateral posterior thalamic nu.; LV – lateral ventricle; MD – mediodorsal thalamic nu.; MG – medial geniculate nu.; MHb – medial habenular nu.; mt – mamillothalamic tract; och – optic chiasm; OPT – olivary pretectal nu.; opt – optic tract; PC – paracentral thalamic nu.; pc – posterior commissure; Po – post thalamic nuclear group; PrG – pregeniculate nu.; PrTh – prethalamus; PVA – paraventricular thalamic nu., anterior; PVP – paraventricular thalamic nu., posterior; Re – reuniens thalamic nu.; Rt – reticular thalamic nu. (prethalamus); SC – superior colliculus; SN – substantia nigra; Th – thalamus; VL – ventrolateral thalamic nu.; VM – ventromedial thalamic nu.; VPL – ventral posterolateral thalamic nu.; VPM – ventral posteromedial thalamic nu.; VRe – ventral reuniens thalamic nu.; ZI – zona incerta.

medial surfaces of the rostral isocortical pole (Broomhead, 1974), but neither the mediodorsal thalamic nucleus nor the presumptive prefrontal cortex is large in either poly- or diprotodonts. Functionally, the mediodorsal thalamic nucleus is concerned with learning and 'working' memory, behavioural control and visceral functions.

The submedius is a relatively small nucleus in approximately the middle of the rostrocaudal extent of the thalamus, which may be involved in feedback circuits that modulate pain sensation. It is situated between the ventromedial and midline thalamic nuclei (Re) and is easiest to define by the absence of calbindin immunoreactivity (see below). In rodents, the submedius is divided into a rostromedial part, which receives predominantly olfactory input; and a caudo-dorsal part, which receives somatosensory input, but Rockel et al. (1972) did not report termination of ascending spinal cord axons in this nucleus in the brush-tailed possum.

The anterodorsal thalamic nucleus is always easy to identify in a wide range of therians because of the distinctive large, darkly staining neurons that it contains (see atlas plate W27). The anterodorsal thalamic nucleus is also distinguished by the absence of calbindin immunoreactivity (see Figure 6.1d and below). The anterior thalamic nuclei are involved in a circuit involving cingulate cortex, the hippocampus and the mamillary bodies of the hypothalamus. In this sense, the anterior thalamic nuclei can be considered part of a larger or extended hippocampal system. This looped (Papez) circuit is important for attentional processes and various types of learning and memory.

The lateral thalamic complex consists of the rostromedially located lateral dorsal nucleus and the caudally positioned lateral posterior nucleus. As noted above, studies with placental mammals have found that the lateral dorsal nucleus has connections similar to those of the anterior nuclei, with topographically ordered projections from the retrosplenial cortex. By contrast, the lateral posterior thalamic nucleus has its major input from the occipital, parietal and temporal association cortex. In rodents, the lateral dorsal thalamic nucleus is most closely related to the limbic cortex, whereas the lateral posterior thalamic nucleus is most strongly linked with visual association cortex. In diprotodonts, the lateral posterior nucleus is better developed than the lateral dorsal nucleus, which is in keeping with the respective relative sizes of visual and limbic cortex. The lateral posterior nucleus of diprotodont marsupials like the brush-tailed possum appears to be more developed than in the opossum (Goldby, 1941). In fact, Goldby (1941) likened the size and differentiation of the lateral posterior nucleus to that seen in prosimians. In the brush-tailed possum, the lateral posterior nucleus can be subdivided into medial and lateral components on the basis of cytoarchitecture (Goldby,

1941) and afferent source (Rockel et al., 1972). While neither subdivision receives direct projections from the retina, the medial part of the LP (but not the lateral) receives a dense projection from the superior colliculus, whereas the lateral part of the LP (but not the medial) receives a dense projection from the visual cortex (Rockel et al., 1972). Both subdivisions of the LP project to the primary visual (V1) and secondary (V2) areas in diprotodonts (Haight et al., 1980).

Midline and intralaminar thalamic nuclei

The midline nuclei are naturally distributed along the midline of the thalamus from the rostral end to about 70 to 80% of the rostrocaudal extent of the thalamus. This group includes the paraventricular, parataenial, intermediodorsal, reuniens and rhomboid nuclei. Goldby (1941) observed that the reuniens nucleus of the brush-tailed possum is somewhat smaller and less differentiated than in the opossum and likened the reuniens of the brush-tailed possum to that in the macaque.

The intralaminar nuclei are embedded in the internal medullary lamina of the thalamus. The group can be divided into a rostral set of nuclei (central medial, paracentral and central lateral nuclei) and a caudal set (parafascicular centre median complex, subparafascicular nucleus). Very little is known about the connections of these nuclei in Australian marsupials, but some nuclei of the caudal and rostral intralaminar groups in the brush-tailed possum (parafascicular and central lateral; PF and CL, respectively) receive dense termination of afferents from both sides of the cervical spinal cord (Rockel et al., 1972). By contrast, afferents from the lumbar spinal cord are sparse and distributed only to limited parts of the dorsal CL. The rostral intralaminar nuclei (in particular CL, but also the paracentral) also receive extensive terminations from the contralateral deep cerebellar nuclei and from the sensory/motor cortex (Rockel et al., 1972). The intralaminar nuclei also have reciprocal connections with the posterior parietal cortex (Haight et al., 1980).

Dorsal lateral geniculate nucleus

Of all the specific thalamic nuclei, the dorsal lateral geniculate nucleus has received the most attention in Australian marsupials (both polyprotodont species and macropods) by reason of its role as the major retinal ganglion cell target and the existence of species differences in the pattern of retinogeniculate axon terminations.

In all marsupials studied to date, the dorsal lateral geniculate nucleus of the thalamus (DLG) is subdivided into a medial and relatively less-differentiated β segment and a (usually) more complex lateral α segment, although the α

segment is poorly differentiated in Ameridelphians (Kahn and Krubitzer, 2002). The α segment is often further subdivided into laminae on the basis of cytoarchitecture, even in quite small Australian marsupials (Packer, 1941; Johnson et al., 1969; Sanderson and Pearson, 1977; Sanderson et al., 1979; Sanderson et al., 1984, 1987; Haight and Sanderson, 1988). There is considerable variation both between and even within species in the degree of α segment lamination (Sanderson et al., 1987). It should also be noted at the outset that some Ameridelphians show a similar subdivision of the DLG; for example, the small opossum Monodelphis domestica has a DLG with clear divisions into α and β segments (Olkowicz et al., 2008).

Among polyprotodont Australian marsupials, it is perhaps surprising that it is the more derived Tasmanian devil (Sarcophilus harrisii) which has the least differentiated DLG (Sanderson et al., 1979). The devil has only two α segment laminae (α1 and α2) and there is no well-defined transition from the α to the β zones. Perhaps related to this poorly differentiated lamination, projections from the contralateral and ipsilateral eye overlap extensively in the DLG of the devil. By contrast, the much smaller (100 to 120 g) kowari (Dasyuroides byrnei) and 10 to 20 g fat-tailed dunnart (Sminthopsis crassicaudata) have more complex cytoarchitecture within the α segment of the DLG (Haight and Sanderson, 1988). The kowari has four distinct cell laminae within the α segment, plus one cell-sparse zone, whereas the DLG of the fat-tailed dunnart has only three α segment cell laminae and one cell-sparse zone. Both the kowari and fat-tailed dunnart are prey species as well as predators, and therefore good visual ability may be important for avoiding predation as well as catching prey. Like the devil, the DLG of both the kowari and fat-tailed dunnart receives overlapping projections from both eyes. In other small dasyurids (e.g. the 10 to 12 g common planigale Planigale maculata and the 40 to 50 g yellow-footed antechinus Antechinus flavipes) the α segment also has three laminae, with the two lateral laminae being only two or three cells wide (Haight and Sanderson, 1988). The laminar pattern of the contralateral retinogeniculate projection is not dependent on interaction between axons from each eye, because it is retained even in the absence of binocular competition, as shown by eye-removal experiments undertaken in early pouch young dasyurids (Dasyurus hallucatus; Crewther et al., 1988).

Sanderson et al. (1984, 1987) observed considerable species differences in the cytoarchitecture of the dorsal lateral geniculate nucleus (DLG) and retinogeniculate termination in several species of macropod diprotodonts. In two rat kangaroos (the bettong Bettongia gaimardi and the potoroo Potorous tridactylus), the cytoarchitecture of the DLG was found to be relatively simple, although the cytoarchitecture of the DLG was slightly more complex in the potoroo compared to the bettong (Sanderson et al., 1984). In all macropods the DLG can be clearly divided into lateral α and medial β segments (Flett et al., 1988). As seen for the polyprotodonts, the α segment may vary considerably in complexity, whereas the β segment tends to be more homogeneous in all species. In both rat kangaroos, the α segment can be further divided into three laminae (α_0, α_1, α_2), with α_2 exhibiting further subdivisions in the potoroo. The DLG of the pademelon (Thylogale billardierii), a small wallaby, was described by Sanderson and colleagues (1984) as being intermediate in complexity between the bettong and the potoroo, in that sublamination of α_2 is present, but not as complex as that in the potoroo. The cytoarchitecture of the DLG of the tammar wallaby (Macropus eugenii) and eastern grey kangaroo (Macropus giganteus) are very similar and exhibit the same general organisation plan seen in the pademelon, although the kangaroo has slightly more complex lamination and sublamination of the α segment. Retinogeniculate termination in these macropods can be complex, with up to ten terminal bands seen in the eastern grey kangaroo (four from the contralateral eye and six from the ipsilateral eye). The retinotopic organisation of the DLG has been analysed for the tammar, showing that the nasal retina is represented rostrally and the temporal retina is represented caudally; the dorsal retina projects to the ventral DLG, whereas the ventral retina projects to the dorsal DLG (Flett et al., 1988).

Cytoarchitecture of the DLG and the pattern of retinogeniculate projections in other (non-macropod) diprotodont marsupials (e.g. the brush-tailed possum Trichosurus vulpecula, the koala Phascolarctos cinereus, the wombat Vombatus ursinus, the sugar glider Petaurus breviceps, the ring-tailed possum Pseudocheirus peregrinus and the feathertail glider Acrobates pygmaeus) are also more complex than those seen in polyprotodont marsupials (Hayhow, 1967; Johnson et al., 1969; Pearson et al., 1976; Sanderson et al., 1987). Interestingly, the feathertail glider shows considerable overlap in the ipsi- and contralateral retinogeniculate projections; something not seen in other diprotodont marsupials, and this may be a reflection of the binocular convergence of this interesting marsupial.

Studies of thalamocortical projections from the DLG of the brush-tailed possum indicate that there are morphologically distinct types of relay neurons in the two regions: class A neurons in the α segment and class B neurons in the β segment (Robinson and Webster, 1985). The two segments also differ in their projection to visual cortex (Haight et al., 1980) and may represent an anatomical substrate of functional streaming, like the A and C layers of cat DLG or the magnocellular/parvicellular components of primate DLG (Henry and Mark, 1992). Enzyme reactivity for acetylcholinesterase is also differentially distributed in DLG of marsupials, with most reactivity in any given species being found in the medial parts of the α region (Wilson and Astheimer, 1989). Reactivity to AChE is also strongest in those regions of the α segment which have the highest retinal terminal density.

Ventral posterior thalamic nucleus

The details of the organisation of this somatosensory relay nucleus will be dealt with in Chapter 10, but a brief overview will be given here. As in placental mammals, the ventral posterior thalamic nucleus can be divided into a medial division, the ventral posteromedial (VPM) nucleus, and a ventrolateral division, the ventral posterolateral (VPL) nucleus. The VPM receives a topographically ordered projection from the trigeminal sensory nuclear complex, whereas the VPL receives topographically ordered input principally from the dorsal column nuclei of the medulla, but also from the spinal cord. The VPM has cellular aggregations in some diprotodont marsupials corresponding more or less to vibrissae on the snout (Haight and Neylon, 1978a). In the brush-tailed possum these show a distinctive correspondence with individual vibrissae, but in the tammar wallaby the field consists of dorsoventrally directed finger-like cell bands separated by myelin bundles (Leamey et al., 1996). Haight and Neylon (1978a, 1978b) also identified a cytoarchitecturally distinct ventral posterior posterior nucleus (VPP) of unknown input in the brush-tailed possum. The VPL nucleus projects in a precise topographic relationship to those areas of somatosensory cortex concerned with body representation. In the brush-tailed possum, most of these are to the primary somatosensory area (S1), but about 10% project to the secondary somatosensory area (S2). The VPM nucleus projects to head regions of somatosensory cortex (S1Op, S1My, S1Md), while the VPP projects to the rostromedial parts of S1 (Haight and Neylon, 1978b).

Although there is considerable overlap of the somatosensory and motor representations on the cortex of marsupials (see Chapter 8), afferents to the somatosensory and motor thalamic nuclei show a clear segregation between motor and sensory pathways, with very sharp borders in the distribution of thalamic afferents from deep cerebellar nuclei and dorsal column nuclei (Rockel et al., 1972).

Ventromedial thalamic nucleus

The ventromedial thalamic nucleus is an elongated nucleus in the medial part of the ventral thalamic tier. It is usually considered part of the motor thalamic nuclear group due to its input from the entopeduncular nucleus and substantia nigra pars reticulata, but the ventromedial thalamic nucleus also receives substantial nociceptive input from caudal brainstem reticular nuclei (Rockel et al., 1972). This eclectic input has been interpreted as allowing the ventromedial nucleus to prime the cerebral cortex in preparation for coordinated behavioural responses to painful stimuli (see review in Groenewegen and Witter, 2004).

Posterior thalamic nuclei

A posterior thalamic nuclear complex has been identified in diprotodont marsupials on the basis of afferent connections (Rockel et al., 1972). These include a suprageniculate nucleus with a strong input from the deep layers of the superior colliculus and the spinal cord (Rockel et al., 1972), a posterior thalamic nucleus (subsequently further subdivided into medial and lateral parts by Neylon and Haight, 1983) and a parageniculate nucleus (alternatively called the subparafascicular nucleus). Neylon and Haight (1983) subsequently examined the thalamocortical projections of two of the component nuclei of the posterior thalamic complex. They found that the suprageniculate nucleus in the brush-tailed possum projects extensively to the auditory cortex, as well as that part of the posterior parietal cortex immediately adjacent to the primary auditory cortex. This pronounced auditory function is in contrast to the pattern of projections of the suprageniculate nucleus of placental mammals and the Virginia opossum, where it appears to play a role in transferring multimodal sensory information from deep layers of the superior colliculus to areas of cortex adjacent to, or overlapping, the somatosensory, visual and auditory representations. On the other hand, the posterior thalamic nucleus of the brush-tailed possum also projects widely to motor/somatosensory and posterior parietal cortex, much like many other mammals (with the exception of New and Old World monkeys). Neylon and Haight (1983) also identified some somatotopic organisation within the posterior thalamic nucleus projection, in that the lateral Po projects to the postcranial representation on the somatosensory cortex, whereas the medial Po projects to the head region of the somatosensory cortex. A similar projection from the posterior thalamic nucleus to somatosensory cortex has been found for Dasyurus viverrinus (Haight and Neylon, 1981b).

Ventral anterior and ventral lateral nucleus

The ventral anterior and ventral lateral nuclei are the motor thalamic nuclei and occupy an extensive area of the ventral tier of the lateral mass of the thalamus. These two motor nuclei are in contact with the intralaminar and anterior nuclei rostrally and medially, and the posterior, ventral posterior and lateral dorsal nuclei further caudally.

In mammals, the ventral anterior and ventral lateral thalamic nuclei receive projections from the deep cerebellar nuclei and basal ganglia (medial globus pallidus or entopeduncular nucleus, pars reticulata of the substantia nigra) and project, in turn, to motor areas of the cortex. In the brush-tailed possum, dense terminal fields to the ventral lateral thalamus arise from the deep cerebellar nuclei (Rockel et al., 1972). These motor thalamic nuclei also receive descending projections from somatosensory/motor cortex in therian mammals (Rockel et al., 1972).

Haight and Neylon (1979) found that the VL of the brush-tailed possum can be subdivided into a small rostromedial division (VLi) and a larger caudolateral division (VLe) on the basis of differing cortical projections. The projection field of VLe corresponds with the area of cerebral cortex from which body movements can be elicited by electrical stimulation (i.e. motor cortex), but is confined to the anteromedial part of frontoparietal cortex and does not project to the face region of S1. On the other hand, VLi does make some discontinuous projections to somatosensory cortex (in particular the barrel field of the mystacial somatosensory cortex).

Chemoarchitecture of thalamic nuclei in marsupials

Immunoreactivity to calcium-binding proteins is differentially distributed in the thalamus of Australian marsupials (Figures 6.1 and 6.2) and facilitates discrimination of nuclei in a similar fashion to that seen in placental mammals. In rodent thalamus, neuropil immunoreactivity for calbindin (CB) is found mainly in midline thalamic nuclei, and anteroventral, centrolateral, dorsal lateral geniculate and ventromedial thalamic nuclei (Celio, 1990; Arai et al., 1994). Somata immunoreactive for CB have been reported in midline thalamic nuclei (reuniens, paraventricular, central medial), and ventromedial, dorsal lateral geniculate, laterodorsal and posterior thalamic nuclei (Celio, 1990; Frassoni et al., 1991; Arai et al., 1994). In rodent thalamus, immunoreactivity for parvalbumin (Parv) is also quite widespread, but with a largely complementary distribution to that for CB. Rodents show strong neuropil immunoreactivity in VPM, VPL, laterodorsal and reticular thalamic nucleus (Celio, 1990; Covenas et al., 1991; Frassoni et al., 1991; Arai et al., 1994; Amadeo et al., 2001; Contreras-Rodriguez et al., 2002). Parvalbumin immunoreactive somata are largely confined to the reticular nucleus and dorsal zona incerta in the rodent diencephalon (Celio, 1990; Covenas et al., 1991; Frassoni et al., 1991; Arai et al., 1994). On the other hand, parvalbumin immunoreactive somata have been found in other thalamic nuclei of nonrodent placentals and didelphids (e.g. macaque VP – Rausell

et al., 1992; VP of the galago and *Tupaia* – Diamond et al., 1993; AV, LP and PrG nucleus of the guinea pig – De Biasi et al., 1994; VP of the raccoon – Herron et al., 1997; AD, VPM of *Monodelphis domestica* – Olkowicz et al., 2008).

In both poly- and diprotodont marsupials, medial thalamic nuclei have large numbers of CB immunoreactive somata and the anterodorsal thalamic nuclei have low or no CB immunoreactivity in both somata and neuropil. This pattern is similar to that seen in laboratory rats and stands in contrast to the pattern seen in monotremes (Ashwell and Paxinos, 2005).

On the other hand, other thalamic nuclei show distinct differences in chemoarchitecture, both between poly- and diprotodonts, and between polyprotodonts and placental mammals. For example, calbindin immunoreactivity is poor in the neuropil of many lateral mass nuclei of both diprotodont marsupials and rodents, but relatively strong in the lateral mass neuropil of polyprotodont marsupials (compare LD in *Antechinus stuartii*, Figure 6.1d, with *Macropus eugenii*, Figure 6.2b). Other features of thalamic chemoarchitecture in diprotodont marsupials like the tammar show similarities with rodents, as exemplified by the pattern of calbindin immunoreactivity in LP and Po thalamic nuclei (Figure 6.2 c, d).

Immunoreactivity to calcium-binding proteins also provides a striking contrast between α and β segments of the DLG in diprotodont marsupials (compare Figure 6.2d and e), revealing a complex subdivision to this nucleus comparable to that seen in carnivores and primates.

Development of the prethalamus and thalamus in marsupials

The thalamus and epithalamus (habenular nuclei) are derived from prosomere 2 (p2) of the embryonic brain, whereas the prethalamus is derived from p3. In the developing brain, p2 and p3 are separated by the cell-free zona limitans interthalamica. In developing placental and marsupial mammals, the zona limitans is marked by stronger immunoreactivity to CD15 than either the thalamic or prethalamic primordium (Ashwell et al., 2004). The thalamus is demarcated from the more caudal pretectum by a fibre bundle, the fasciculus retroflexus, which arises from the habenular nuclei and terminates in the interpeduncular nuclei.

Like the cortex or pallium, the dorsal thalamus and prethalamus are very rudimentary at the time of birth in marsupials, but are clearly separated by a pronounced sulcus medius on the ventricular wall (Warner, 1969). The epithalamic rudiment is also separated from the dorsal thalamus

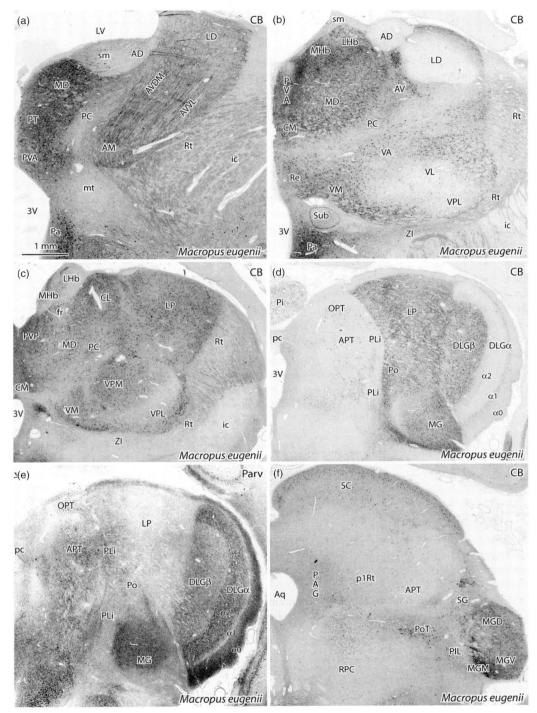

Figure 6.2 Chemoarchitecture of the thalamus and prethalamus in a representative diprotodont Australian marsupial (*Macropus eugenii*). Panels (a), (b), (c), (d) and (f) show coronal sections immunoreacted for calbindin (CB) and evenly spaced in a rostrocaudal sequence. Panel (e) shows a coronal section immunoreacted for parvalbumin (Parv), which is adjacent to that shown in (d). The midline is to the left in all photomicrographs. Note the complementary immunoreactivity for Parv and CB in the α segment of the DLG. The scale bar in (a) is applicable to all the photomicrographs. 3V – third ventricle; AD – anterodorsal thalamic nu.; AM – anteromedial thalamic nu.; APT – anterior pretectal nu.; Aq – cerebral aqueduct; AV – anteroventral thalamic nu.; AVDM – AV, dorsomedial part; AVVL – AV, ventrolateral part; CL – centrolateral

by a distinct sulcus dorsalis (Warner, 1969). The prethalamus is slightly more advanced than the dorsal thalamus in the newborn tammar wallaby, as judged by the size of the postmitotic pool of neurons migrated to the side of the diencephalon. Both regions have extremely thick ventricular germinal zones at the time of birth (see atlas plates WP0/4 and 5), perhaps the thickest of all proliferative zones at any stage of development in this marsupial. The large size of the ventricular proliferative pool at birth probably reflects the rapid production of thalamic neurons, which will occur in the first two weeks of postnatal life in diprotodonts. For example, neurons of the somatosensory (VP) and motor (VL) thalamic nuclei of the brush-tailed possum are all generated by P12 (Sanderson and Weller, 1990b). A similar short and intense period of thalamic neurogenesis occurs in small Ameridelphians (e.g. P3 to P16 in *Monodelphis domestica*, Glendenning, 2006).

Neurogenesis in the thalamus proceeds in a lateral-to-medial temporospatial gradient in all therians (Altman and Bayer, 1979a, 1979b; Sanderson and Aitkin, 1990; Sanderson and Weller, 1990b; Glendenning, 2006). In the brush-tailed possum, neurons from the lateral thalamic regions can be labelled by ^{3}H-thymidine injection on P5 to P12, whereas neurons in medial thalamic nuclei can be labelled following injection on P9 to P18. Neurons of the midline thalamic nuclei are generated even later (Sanderson and Weller, 1990). Consistent with observations in the brush-tailed possum, neurogenesis in the DLG of the quokka proceeds from birth to P10 with a peak at about P3 (Harman, 1991).

As might be expected, given the relatively short time-course of thalamic neurogenesis (compared to cortex), the cytoarchitecture of the thalamus emerges relatively rapidly in the first postnatal month (Warner, 1969; see also the atlas of the developing tammar wallaby brain – compare, for example WP0/5 with WP19/6). Thalamic structure is still quite rudimentary by the end of the first postnatal week, but the major components of both the pre- and (dorsal) thalamus can be distinguished by the end of the second postnatal week. This process is more advanced in the prethalamus, in that distinct pregeniculate, reticular thalamic and zona incerta components can be seen by P12 in the tammar wallaby. The major sensory nuclei of the dorsal thalamus are all homogeneous fields at P12, although the boundaries of the medial geniculate nucleus are distinct at that age. Cytoarchitectural subdivisions within the major sensory nuclei emerge during the third and fourth postnatal weeks and this probably coincides with a period of developmental cell death, as seen in the DLG of the quokka (peaking at P85, Harman, 1991). Refinement of the topography of the retinal projections to the DLG is completed by about P110, shortly before the visual pathways become functional at about 20 weeks of postnatal life (Marotte, 1990). In the case of the VPM, the first hints of segmentation of the nucleus appear around P50 and neuronal aggregations emerge over the subsequent three to four weeks (Leamey *et al.*, 1996).

The details of the connectional and functional development of visual, somatosensory and auditory thalamic nuclei will be discussed in the relevant systems Chapters (9, 10 and 11, respectively).

Hypothalamus

Function and organisation of the mammalian hypothalamus

The hypothalamus plays a key role in maintaining homeostasis by integrating endocrine, visceral and somatomotor systems. To this end, the mammalian hypothalamus regulates the cardiovascular system, body temperature, visceral function, cellular metabolic activities, defensive/aggressive behaviour and ingestion of water and nutrients. Furthermore, the hypothalamus ensures species survival by controlling reproductive and maternal behaviours, and regulating endocrine secretions associated with reproduction and lactation. The hypothalamus has vascular and neural connections with the anterior and posterior pituitary, respectively, and has robust neural connections with all the major parts of the brain and spinal cord.

The hypothalamus occupies the ventral half of the diencephalon on either side of the third ventricle (Figure 6.3).

Caption for Figure 6.2 (Cont.)

thalamic nu.; CM – central medial thalamic nu.; DLGα,β – dorsal lateral geniculate nu., alpha, beta segments; fr – fasciculus retroflexus; ic – internal capsule; LD – laterodorsal thalamic nu.; LHb – lateral habenular nu.; LP – lateral posterior thalamic nu.; LV – lateral ventricle; MD – mediodorsal thalamic nu.; MG – medial geniculate nu.; MGD, MGM, MGV – MG, dorsal, medial, ventral subnuclei; MHb – medial habenular nu.; mt – mamillothalamic tract; OPT – olivary pretectal nu.; p1Rt – p1 reticular formation; Pa – paraventricular hypothalamic nu.; PAG – periaqueductal grey; PC – paracentral thalamic nu.; pc – posterior commissure; Pi – pineal gland; PIL – posterior intralaminar thalamic nu.; PLi – posterior limitans thalamic nu.; Po – post thalamic nuclear group; PoT – posterior thalamic nu., triangular; PT – paratenial thalamic nu.; PVA – paraventricular thalamic nu., ant; PVP – paraventricular thalamic nu., post; Re – reuniens thalamic nu.; RPC – red nu., parvicellular part; Rt – reticular thalamic nu. (prethalamus); SC – superior colliculus; SG – suprageniculate thalamic nu.; sm – stria medullaris; Sub – submedius thalamic nu.; VA – ventral anterior thalamic nu.; VL – ventrolateral thalamic nu.; VM – ventromedial thalamic nu.; VPL – ventral posterolateral thalamic nu.; VPM – ventral posteromedial thalamic nu.; ZI – zona incerta.

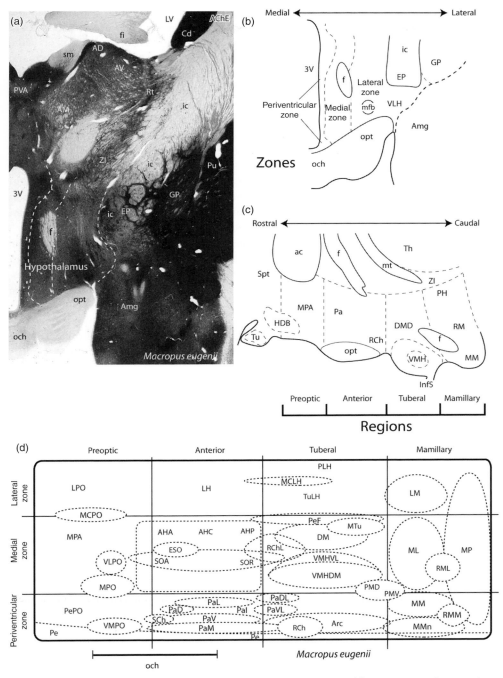

Figure 6.3 Basic structure of the hypothalamus in the tammar wallaby. Panel (a) shows a coronal section through the optic chiasm stained for acetylcholinesterase (AChE), illustrating the topography of the hypothalamus relative to other diencephalic and telencephalic structures. Panel (b) is a schematic line diagram illustrating the subdivision of the hypothalamus into three mediolateral zones. Panel (c) is a line diagram drawn from a sagittal section through the wallaby hypothalamus illustrating the four rostrocaudal regions in relation to major fibre tracts of the forebrain. The last panel (d) is a schematic diagram of the major nuclei in the hypothalamus of the tammar wallaby (based on data from Cheng *et al.*, 2003). 3V – third ventricle; ac – anterior commissure; AD – anterodorsal thalamic nu.; AHA – anterior hypothalamic area, anterior; AHC – anterior hypothalamic area, central; AHP – anterior hypothalamic area, posterior; AM – anteromedial thalamic nu.; Amg – amygdala; Arc – arcuate hypothalamic nu.; AV – anteroventral thalamic nu.; Cd – caudate nu.; DM – dorsomedial

It is bounded laterally by the fibres of the internal capsule and dorsally by a component of the prethalamus (the zona incerta; Figure 6.3a, b). The ventral boundary of the hypothalamus throughout its rostrocaudal extent is the pial surface of the brain, although this curves dorsally at the caudal end of the hypothalamus (mamillary and retromamillary regions). The hypothalamus is penetrated by dorsoventrally running fibre bundles like the fornix, mamillothalamic and mamillotegmental tracts and is traversed rostrocaudally by a diffuse longitudinal fibre tract, the medial forebrain bundle.

The hypothalamus of placental mammals is divided into three mediolateral zones and four rostrocaudal regions (Figure 6.3b to d) and these subdivisions are just as applicable to monotremes and marsupials (Cheng *et al.*, 2003). The three mediolateral longitudinal zones are, from medial to lateral: the periventricular, medial and lateral zones. The periventricular zone contains many of the hypothalamo-neurohypophyseal neurons (which project to the posterior pituitary) and nuclei secreting releasing hormones for control of the anterior pituitary gland. The medial zone of the hypothalamus contains nuclei which receive their major inputs from the limbic components of the telencephalon, such as the amygdala and septal nuclei. These nuclei play key roles in motivated behaviours like copulation, aggression and appetitive behaviours. The lateral zone consists of nuclei distributed among the diffuse fibres of the medial forebrain bundle.

The four rostrocaudal regions of the mammalian hypothalamus are the preoptic, anterior, tuberal and mamillary (Figure 6.3d). The most rostral region of the hypothalamus is the preoptic area, although some authors treat this as a discrete region. In all mammals, this region is bounded rostrally by the diagonal band nuclei (components of the septal region) and dorsally by the anterior commissure. The large size of the anterior commissure in marsupials compresses the preoptic area in the dorsoventral direction, but the nuclei of the marsupial preoptic area are similar to those identified in placental mammals. Important elements of the preoptic region include the median preoptic (MnPO) and anteroventral periventricular (AVPe) nucleus of the periventricular zone, the medial preoptic nucleus (MPO) of the medial zone, and the magnocellular preoptic (MCPO) and lateral preoptic (LPO) nuclei of the lateral zone. It is known from studies of rodents that the MnPO plays a critical role in cardiovascular responses and fluid homeostasis, the MPO is important for reproductive and maternal behaviour and has a sexually dimorphic subnucleus (the cell-dense central part) in rodents and the MCPO is a major source of cholinergic projections to the cortex and olfactory bulb (for a review see Simerly, 2004).

The anterior region of the hypothalamus is characterised by the presence of the paraventricular nuclear complex (discussed below) in the periventricular zone, the anterior hypothalamic area in the medial zone and the continuation of the lateral hypothalamus from the preoptic region. The supraoptic nucleus extends from the caudal preoptic region, through the anterior region and into the rostral part of the tuberal region. In both poly- and diprotodont marsupials (Cheng *et al.*, 2003), the supraoptic nucleus has an extensive retrochiasmatic part (SOR) medial to the optic tract, caudal to the optic chiasm. The SOR is continuous with the anterior supraoptic nucleus (SOA) via a scattering of cells over the rostral optic tract.

The tuberal region of the hypothalamus contains the arcuate nucleus in the periventricular zone, the large ventromedial and dorsomedial hypothalamic nuclei in the medial zone and the continuation of the lateral hypothalamic zone from more rostral levels. The most caudal region of the hypothalamus is largely taken up by the mamillary and retromamillary nuclei. The caudal pole of the mamillary region in the tammar wallaby is occupied by an

Caption for Figure 6.3 (Cont.)

hypothal. nu.; DMD – dorsomedial hypothal. nu., dorsal; EP – entopeduncular nu.; ESO – episupraoptic nu.; f – fornix; fi – fimbria of the hippocampus; GP – globus pallidus; HDB – nu. horizontal limb diagonal band; ic – internal capsule; InfS – infundibular stem; LH – lateral hypothalamic area; LM – lateral mamillary nu.; LPO – lateral preoptic area; LV – lateral ventricle; MCLH – magnocellular nu. lateral hypothal; MCPO – magnocellular preoptic nu.; mfb – medial forebrain bundle; ML – medial mamillary nu., lateral; MM – medial mamillary nu., medial; MMn – medial mamillary nu., median; MP – posterior mamillary nu.; MPA – medial preoptic area; MPO – medial preoptic nu.; mt – mamillothalamic tract; MTu – medial tuberal nu.; och – optic chiasm; opt – optic tract; Pa – paraventricular hypothalamic nu.; PaD – Pa, dorsal; PaDL – Pa, dorsolateral; PaI – Pa, intermediate; PaL – Pa, lateral; PaM – Pa, medial magnocellular; PaV – Pa, ventral; PaVL – Pa, ventrolateral; Pe – periventricular hypothal. nu.; PeF – perifornical nu.; PePO – Pe, preoptic; PH – posterior hypothal. nu.; PLH – peduncular part of lateral hypothal.; PMD – premamillary nu., dorsal; PMV – premamillary nu., ventral; Pu – putamen; PVA – paraventricular thal. nu., ant; RCh – retrochiasmatic area; RChL – retrochiasmatic area, lateral; RM – retromamillary nu.; RML – retromamillary nu., lateral; RMM – retromamillary nu., medial; Rt – reticular thal. nu. (prethalamus); SCh – suprachiasmatic nu.; sm – stria medullaris; SOA – supraoptic nu., anterior; SOR – supraoptic nu., retrochiasmatic; Spt – septum; Th – thalamus; Tu – olfactory tubercle; TuLH – tuberal region of lateral hypothal.; VLH – ventrolateral hypothal. nu.; VLPO – ventrolateral preoptic nu., VMPO – ventromedial preoptic nu.; VMH – ventromedial hypothalamic nu.; VMHDM – VMH, dorsomedial; VMHVL – VMH, ventrolateral; ZI – zona incerta.

extensive and homogeneous posterior mamillary nucleus. While most of the mamillary region nuclei of the wallaby have clear counterparts in the hypothalamus of placental mammals, it is not clear how the large posterior mamillary nucleus of the wallaby corresponds to mamillary nuclei in rodents.

Specific nuclei within the marsupial hypothalamus

The paraventricular nucleus (Pa) of the hypothalamus (Figure 6.4) plays critical roles in the neuroendocrine and autonomic systems. The functions of its component nuclei include control of the pituitary adrenocortical activity in response to stress, the milk ejection reflex, thyroid hormone secretion and appetitive behaviour (see discussion in Cheng et al., 2002). The general topographic pattern of the Pa in the tammar wallaby is broadly similar to that seen in rodents, with a prominent magnocellular subnucleus (PaM), parvicellular subnuclei (PaD, PaV) and subnuclei of mixed small to large neurons (PaL and PaI). The PaM is probably homologous to the medial magnocellular subnucleus of rodents, PaD and PaV are possibly homologous to the dorsal and anterior parvicellular subnuclei of the rat and the PaI may be homologous to the medial and lateral parvicellular subnuclei of rodents.

In rodents, most of the supraoptic nucleus abuts the lateral border of the optic tract (anterior component of the SO – SOA), with a small retrochiasmatic continuation (SOR) caudal to the optic chiasm and medial to the optic tract. A similar topography is also seen in humans, where the bulk of the SO neurons are located dorsolateral to the optic tract and only a few neurons spill medial to the optic tract (Saper, 2004a). In the wallaby, the SO extends from the caudal preoptic region (atlas plates D10 and D11, W23 and W24) to the rostral tuberal region and differs in distribution from eutheria in that a large cohort of SO neurons are in the retrochiasmatic position (Cheng et al., 2003; see atlas plates D12, W25, W26). Nevertheless, chemoarchitectural features of the wallaby SO are quite similar to those in rodents, although tyrosine hydroxylase immunoreactive neurons are less common in the wallaby SO than they are in rodents.

The ventromedial nucleus of the hypothalamus (VMH) is one of the largest and most prominent in the region in all therian mammals (Cheng et al., 2002, 2003). In rodents, the VMH has been subdivided into dorsomedial (VMHDM) and ventrolateral (VMHVL) parts with an intervening cell-sparse core, and a similar division is discernable in macropods (see atlas plate W26). These subdivisions have different connections and serve dissimilar functions. For example, the VMHVL projects to parts of the forebrain and brainstem that are involved in reproductive behaviour, whereas the VMHDM has robust connections with other parts of the hypothalamus concerned with appetitive behaviour (e.g. the paraventricular and dorsomedial nuclei), as well as the anterior hypothalamic area. Although the function and connections of the VMH and its individual constituents have never been directly studied in marsupials, the similarities in the cyto- and chemoarchitecture of this region in the wallaby (Cheng et al., 2003) to that in placental mammals strongly suggests a similar functional subdivision in all therians.

Functional organisation of the marsupial hypothalamus

The hypothalamus is concerned with diverse functions, including providing releasing and neurohypophyseal hormones for the pituitary, controlling water and food intake, regulating respiratory and cardiovascular function, and thermoregulation.

Releasing hormones from the hypothalamus are carried to the anterior pituitary (pars distalis) via the hypothalamo-hypophyseal portal system. The distribution of releasing hormones has not been specifically studied in Australian marsupials, but studies in an Ameridelphian (Monodelphis domestica) suggest that there may be significant differences in the distribution of luteinizing hormone releasing hormone (LHRH) between marsupials and placentals. LHRH immunoreactive cells are much rarer in the preoptic area and hypothalamus of Monodelphis than in placental mammals (Schwanzel-Fukuda et al., 1988), although the median eminence of the grey short-tailed opossum is well-supplied with LHRH immunoreactive fibres from neurons in the medial septal nucleus, the nuclei and tract of the diagonal band and the olfactory tubercle. It remains to be seen whether this paucity of LHRH-immunoreactive neurons in the hypothalamus also applies to Australian marsupials.

Most oxytocic- and vasopressinergic neurons of the hypothalamus project to the posterior pituitary (neurohypophysis), from which site their hormones are released to the circulation, thereby controlling milk ejection and fluid retention, respectively. Some of these neurons project within the brain and may even play significant roles in controlling development. In all therians, oxytocin- and vasopressinergic neurons are located primarily in the supraoptic and paraventricular nuclei and the lateral hypothalamus (Iqbal and Jacobson, 1995a, 1995b). Note that diprotodonts release mesotocin from their posterior pituitary, but nevertheless their hypothalamus exhibits oxytocin-like immunoreactivity. While the overall distribution of neurons showing oxytocin- and vasopressinergic-like immunoreactivity is similar in diprotodonts to that in other mammals, there are some important differences (Cheng et al., 2003).

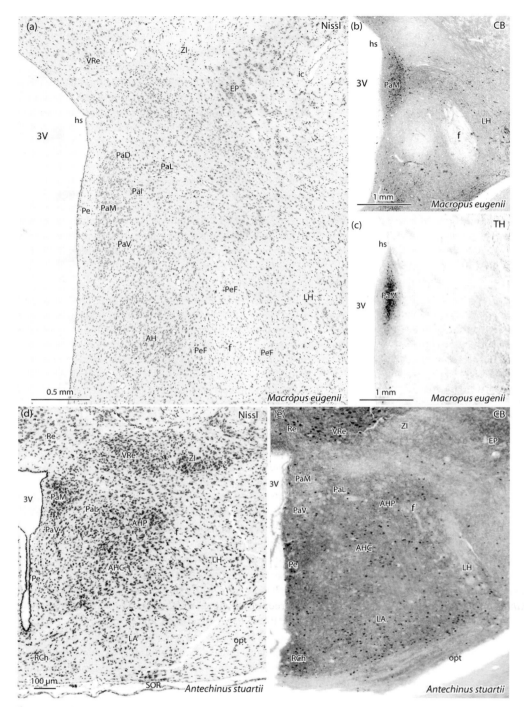

Figure 6.4 The paraventricular nu. of the hypothalamus and its subdivisions in representative diprotodont (a to c) and polyprotodont (d, e) Australian marsupials. The PaM of the wallaby is also strongly immunoreactive for calbindin and tyrosine hydroxylase (b and c, respectively), but the PaM of *A. stuartii* does not contain calbindin-immunoreactive neurons (e). 3V – third ventricle; AH – anterior hypothalamic area; AHC – AH, central; AHP – AH, posterior; EP – entopeduncular nu.; f – fornix; hs – hypothalamic sulcus; ic – internal capsule; LA – lateroanterior hypothal. nu.; LH – lateral hypothal. area; opt – optic tract; PaD – paraventricular hypothal. nu., dorsal; PaI – Pa, intermediate; PaL – Pa, lateral; PaM – Pa, medial magnocellular; PaV – Pa, ventral; Pe – periventricular hypothal. nu.; PeF – perifornical nu.; RCh – retrochiasmatic area; Re – reuniens thalamic nu.; SOR – supraoptic nu., retrochiasmatic; VRe – ventral reuniens thalamic nu.; ZI – zona incerta.

Firstly, there appears to be less segregation of oxytocic(-like) and vasopressinergic neurons in the tammar wallaby hypothalamus than has been reported for placental mammals. For example oxytocic(-like) and vasopressinergic neurons appear to mingle in the wallaby PaM, whereas oxytocic(-like) neurons are situated medial to vasopressinergic neurons in the magnocellular paraventricular nucleus of the cat (Caverson et al., 1987). Secondly, there is no clear aggregation of oxytocic(-like)- or vasopressinergic neurons into a nucleus circularis in the wallaby hypothalamus, in contrast to the presence of this nucleus in some placental mammals (e.g. rodents and felids), but a nucleus circularis is also not seen in laboratory rabbits (see Cheng et al., 2003, for discussion). Finally, neurons immunoreactive for vasopressin are found in the suprachiasmatic nucleus of most theria, including Ameridelphians (Iqbal and Jacobson, 1995a), but these neurons are quite rare in this site in the tammar wallaby (Cheng et al., 2003).

The neuropeptides orexin-A and orexin-B are involved in regulation of feeding behaviour, sleep–wake cycles and autonomic function. In adult macropods, immunoreactivity for orexin-A and orexin-B is found in neurons in the dorsomedial hypothalamus, perifornical nucleus, and lateral and posterior hypothalamic areas (Yamamoto et al., 2006). This pattern of distribution is very similar across all therian mammals (Nambu et al., 1999), indicating that these systems are highly conserved.

Regulation of feeding has also been studied in a small dasyurid (the fat-tailed dunnart Sminthopsis crassicaudata; Vozzo et al., 1999; Hope et al., 2000). These studies confirm that nitric oxide (NO) plays a significant role as a messenger in the regulation of food intake in all therians (Vozzo et al., 1999). Hope and colleagues (Hope et al., 2000) have also shown that corticotrophin releasing factor (CRF) has a potent effect of decreasing food intake in Sminthopsis, much like occurs in rodents. These effects appear to be mediated by the CRF2 receptor, possibly within the ventromedial and paraventricular hypothalamic nuclei, but the sites of action of CRF in Sminthopsis brain have not been specifically studied.

Development of the hypothalamus in diprotodont marsupials

The hypothalamus develops from an unsegmented proneuromere rostral and ventral to p3 (from rostral to caudal: anterior, tuberal, tuberomamillary and mamillary regions), with the preoptic area derived from the ventricle wall immediately rostral to the optic stalk. The posterior hypothalamus may be derived in part from p3.

Of all Australian marsupials, development of the hypothalamus has been most extensively studied in the tammar wallaby (Cheng et al., 2002), but the major findings in the tammar are probably applicable to a broad range of diprotodont marsupials and the time-course of development in the tammar is similar to that in Ameridelphians (Iqbal et al., 1995a). Several lines of evidence indicate that the hypothalamus is very immature structurally for the first two weeks of postnatal life in macropods and is unlikely to serve any significant functional roles during that time.

Until about 12 days after birth, the tammar hypothalamic anlage consists of only the proliferative cell populations of the ventricular germinal zone, plus a thin marginal zone with a few postmitotic neurons. Nevertheless, a period of rapid cytoarchitectural development follows for the next 10 to 12 days, such that most of the major hypothalamic nuclei appear well differentiated by P25. Of the major fibre bundles associated with the hypothalamus, the medial forebrain bundle is evident even at the time of birth, whereas the fornix does not appear until almost the end of the second week of postnatal life. Analysis of neuronal birthdates with ^{3}H-thymidine autoradiography reveals neurogenetic gradients in the wallaby and brush-tailed possum hypothalamus (Sanderson and Wilson, 1997; Cheng et al., 2002), much as have been described for the rodent hypothalamus. In general, laterally placed neurons (i.e. those in the lateral hypothalamic zone) are generated before those in the medial and periventricular zones, and, within any given zone, neurons in dorsally placed nuclei are generated before those situated more ventrally. The possible exception to this is the suprachiasmatic nucleus, which is generated over the first one to two weeks of postnatal life in the brush-tailed possum (Sanderson and Aitkin, 1990). There is also a suggestion from the ^{3}H-thymidine data that neurons of the rostral preoptic area are generated after those in more caudal regions of the hypothalamus. Neurons of the lateral hypothalamus appear to be all generated before P12, whereas some parts of the medial, periventricular and ventral preoptic area (e.g. PePO, MPO, VMPO) are generated as late as P24 to P32 (Cheng et al., 2002).

As suggested by the timing and gradients of cytoarchitectural development, establishment of connections of the wallaby hypothalamus with major targets (pituitary, brainstem and spinal cord) occurs predominantly after birth and with a lateral to medial gradient. Projections to the posterior pituitary from the supraoptic and paraventricular nuclei are not evident until P5 and P10, respectively (Cheng et al., 2002), indicating that the posterior pituitary cannot be functional until after that time. The early generated lateral hypothalamic area has a few neurons projecting to the medulla at birth, but more medial nuclei, such as those of the paraventricular nuclear complex, do not project to the medulla until P10. Projections from the hypothalamus to the lateral horn of the spinal cord (i.e. to the intermediolateral cell column for control of sympathetic outflow) are not found until P20 and the cells of origin at this early stage are confined to the first generated lateral hypothalamic zone (Cheng *et al.*, 2002). The exact age at which oxytocin-like immunoreactivity is first detected in the wallaby hypothalamus is not known, but vasopressin expression has been found in tammar wallaby pituitary as early as P7 (Wilkes and Janssens, 1988), contemporaneous with the formation of the hypothalamo-neurohypophyseal tract (Cheng *et al.*, 2002).

Fully independent endothermy of pouch young wallabies develops at a relatively late stage (about P140, Setchell, 1974), but even newborn wallabies begin to regulate their metabolism in response to the ambient environment (Hulbert, 1988). Some of these early thermoregulatory mechanisms may depend on the early maturing lateral hypothalamus and its projection to the medulla and/or spinal cord (Cheng *et al.*, 2002), but the functional anatomy of this clearly needs further study.

The early development of the projection from the lateral hypothalamus to the medulla may also play a critical role in the feeding behaviour of the young marsupial. The lateral hypothalamus is known in placental mammals as a feeding regulatory centre (Saper, 2004a) and it is likely that it also serves this function in marsupials. Therefore the early projection from the lateral hypothalamus to the medulla may be important in control of suckling if the early pathway terminates on vagal sensorimotor complex nuclei, but more detailed analysis of these early projections in newborn marsupials is necessary for confirmation.

Observations on the structural development of the hypothalamus in the wallaby are consistent with findings on the maturation of respiration in this species. Although there is lung ventilation in newborn wallabies, control of respiratory function in response to O_2 and CO_2 levels is not established until P7 and the regulatory function remains immature until P70 (Baudinette *et al.*, 1988). Carbocyanine dye-tracing studies (Cheng *et al.*, 2002) indicate that a projection from the paraventricular nuclei to the brainstem is not present at birth, but emerges by P10, consistent with a role of this early projection in establishing respiratory function responsiveness to gas levels.

Development of orexin-A and orexin-B systems in the hypothalamus is a late event in marsupial brain. In the preweaning period of development of the eastern grey kangaroo (up to about six months postnatal life), very few orexin-A and orexin-B immunoreactive neurons are present in the hypothalamus. Those present at this stage are only weakly stained and confined to the lateral hypothalamus, dorsomedial hypothalamic nucleus, perifornical nucleus and posterior hypothalamic area. Very few orexin immunoreactive fibres are present at this stage, suggesting that those few orexin-containing neurons present have very few connections. Peak numbers of orexin-A- and orexin-B-immunoreactive neurons can be found after weaning, i.e. from about 10 months postnatal age, when there is a marked increase in the number and staining intensity of orexin-immunoreactive neurons and fibres. This time-course is consistent with observations in placentals, in that orexin-A and orexin-B also develop quite late in rodent hypothalamus (i.e. after weaning begins at P15; Yamamoto *et al.*, 2000). In rodents, the increase in orexin-A and orexin-B coincides with profound changes in a number of functions (decline in REM sleep, regulation of body temperature, opening of eyes and ears; Yamamoto *et al.*, 2000). The reported peak in orexin immunoreactivity in the eastern grey kangaroo (after nine months; Yamamoto *et al.*, 2006) is somewhat later than eye-opening (around P150) and probably after development of endothermy (around P200), but this may reflect inaccuracies in estimating the age of the kangaroos studied by Yamamoto and colleagues.

Pituitary

Adult structure and function

Although part of the endocrine rather than the nervous system, the pituitary (or hypophysis) will be considered here because of its close functional and anatomical relationship to the hypothalamus. The pituitary consists of an anterior lobe (pars distalis or adenohypophysis), intermediate lobe (pars intermedia) and posterior lobe (pars nervosa or neurohypophysis). In brush-tailed possums, the pars intermedia encircles the pars nervosa (Leatherland and Renfree, 1983a, 1983b). The pituitary is attached to the hypothalamus by a short hypophyseal stalk, which carries the fine vessels of the hypothalamo-neurohypophyseal portal system between the hypothalamus and pituitary and axons of the hypothalamo-neurohypophyseal tract.

The pars distalis receives releasing hormones from the neurosecretory centres in the hypothalamus by a system of fine vessels (hypothalamo-hypophyseal portal system) and contains a variety of secretory cells, which respond to hypothalamic releasing hormones. In many mammals, including diprotodonts (Leatherland and Renfree, 1983a, 1983b), the pars distalis can be divided into three distinct regions on the basis of the types of secretory cells. In the brush-tailed possum, cells secreting growth hormone (somatotropes) predominate in the posterior pars distalis, cells secreting adrenocorticotrophic hormone (ACTH)(adrenotropes) and thyroid-stimulating hormone (thyrotropes) predominate in the anterior pars distalis, whereas prolactin- and gonadotropin-secreting cells (mammotropes and gonadotropes, respectively) are spread throughout the pars distalis (Leatherland and Renfree, 1983a, 1983b). The ultrastructural morphology of the respective secretory cells in marsupials is similar to that in placental mammals and, where this has been studied, there appears to be a high degree of conservation in the sequences of releasing hormones and pars distalis secretions across all therians (Farmer et al., 1981; McFarlane et al., 1997; Curlewis et al., 1998; Fidler et al., 1998; Saunders et al., 1998; King et al., 2000), although greater diversity is apparent in some releasing-hormone receptors (Cheung and Hearn, 2002). In some marsupials, alternate forms of releasing hormones may co-exist, even some with similarities to avian forms (King et al., 1989).

The pars intermedia of the pituitary contains cells secreting melanocyte-stimulating hormone and has not been specifically studied in Australian marsupials.

The pars nervosa receives axons of the hypothalamo-neurohypophyseal tract, which arises in the supraoptic, paraventricular and other assorted hypothalamic nuclei. The enlarged terminals of these axons secrete hormones concerned with milk ejection and uterine contraction (oxytocin and mesotocin), or fluid balance/pressor function (vasopressin). Mesotocin ([Ile^8]-oxytocin) has been identified in all non-mammalian tetrapods and lungfish, whereas oxytocin is found in the pituitary of monotremes, Ameridelphians (Chauvet et al., 1984) and placental mammals. In Australian marsupials, mesotocin is the exclusive uterotrophic and milk-ejection hormone (e.g. dasyurids and diprotodonts; Chauvet et al., 1981, 1983a, 1983b, 1983c, 1987; Curlewis et al., 1988) or co-exists in neurohypophyseal secretions with oxytocin (e.g. bandicoots; Rouillé et al., 1988; Bathgate et al., 1992a, 1995). Interestingly, although mesotocin is the only oxytocic hormone secreted from the pituitary of diprotodonts, both oxytocin and mesotocin are present in the hypothalamus of the brush-tailed possum (Gemmell and Sernia, 1989). Presumably the oxytocin present in diprotodont hypothalamus is the neurotransmitter concerned with mediating transmission of information along intracerebral oxytocic pathways. It would appear that Australian marsupials have retained mesotocin from a non-mammalian ancestor and reclaimed its use for milk ejection, whereas Ameridelphians, monotremes and placental mammals have converted to the use of oxytocin for this function.

Some observations of milk ejection in marsupials suggest important differences from placental mammals. Lincoln and Renfree (1981) noted that the mammary glands of tammar wallabies are highly sensitive to circulating oxytocic hormones and the ejection pressures developed in glands supplying small pouch young are five times higher than those seen in laboratory rats. This may facilitate the forcing of milk into the mouths of early pouch young with even limited suckling. They also noted that milk ejection in wallabies occurs in response to electrical stimulation of the hypothalamo-neurohypophyseal tract at much lower frequencies (5 Hz) than that seen to be effective in rabbits and rats (20 Hz or more). They suggested that this low threshold level of stimulation is comparable to the tactile stimulus produced by the limited movements of early marsupial pouch young. Finally, there is a decline in the sensitivity of mammary glands to oxytocic hormones as lactation progresses, which may help mediate differential patterns of milk ejection in response to changing patterns of nipple stimulation by pouch young of different ages.

The two vasopressor neurohypophyseal hormones identified in macropod marsupials (Macropus eugenii, Macropus giganteus, Macropus rufus, Setonix brachyurus) are lysine vasopressin and phenypressin (Phe^2-Arg^8-vasopressin; Chauvet et al., 1980, 1983a, 1983b, 1983c; Hurpet et al., 1980, 1982). By contrast, bandicoots and Ameridelphians have both lysine and arginine vasopressin in their neurohypophyseal secretions (Chauvet et al., 1984; Rouillé et al., 1988) and dasyurids, phalangerids and phascolarctids have arginine vasopressin (Chauvet et al., 1987; Bathgate et al., 1992b). Lysine vasopressin is found in the neurohypophyseal secretions of pigs and the hippopotamus, but is relatively rare in placental mammals, which usually use arginine vasopressin. Chauvet and colleagues (Chauvet et al., 1984) have proposed that the dual vasopressin types of marsupials arose by gene duplication with divergent neutral drifts in Ameridelphians and various Australidelphians.

Development of the pituitary

The pituitary develops from dual sources during embryonic life. The pars distalis or anterior lobe of the pituitary is derived (mainly) from Rathke's pouch, which provides ectodermal cells from the rostral embryonic foregut, whereas the pars nervosa or posterior lobe is derived from a descending process of the hypothalamus.

The pituitary is at a relatively early stage of morphological development in most newborn marsupials, but the cytological appearance differs between groups. In the northern brown bandicoot (Isoodon macrourus), Rathke's pouch first appears at E10.5 as an ectodermal thickening of the roof of the early oral cavity and becomes a large vesicle with upper, middle and lower chambers by E11.5 (Hall and Hughes, 1985). A more caudal pouch (Seessel's pouch) derived from the endodermal part of the gut tube may contribute to the anterior pituitary in the northern brown bandicoot, koala and wombat, even though this structure is not believed to contribute to the pituitary in placental mammals. At the time of birth in the northern brown bandicoot, the anterior pituitary consists of only a multichambered vesicle connected to the oral cavity by a thin cord of tissue. The anterior pituitary at birth in this species is also avascular, suggesting that the hypothalamo-hypophyseal portal system is yet to develop (Hall and Hughes, 1985). The anterior pituitary of the newborn tammar wallaby is also poorly vascularised (Leatherland and Renfree, 1983b).

The timing of development of cell types in the anterior pituitary of Australian marsupials coincides with the functional requirements of pouch young life. At birth in both the northern brown bandicoot and tammar wallaby, pars distalis cells have only a small amount of cytoplasm with few organelles. Most cells have a scattering of electron-dense granules at birth, but the anterior pituitary of newborn marsupials is unlikely to contribute to parturition. The commonest anterior pituitary cell type during early pouch life of the northern brown bandicoot is the somatotrope (growth-hormone-secreting cell, about 70% of total cell cohort at birth; Leatherland and Renfree, 1983b) consistent with the rapid postnatal growth of this period. Presumptive prolactin-, gonadotropin- and thyrotropin-secreting cells increase in number steadily between P1 and P50, whereas presumptive ACTH-secreting cells appear to peak at about P25 to P30. The rise in thyrotropin (thyroid-stimulating hormone)-producing cells coincides with the development of intraluminal colloid in thyroid follicles and the development of thermoregulation (Leatherland and Renfree, 1983b). Similarly, the increase in ACTH-producing cells at around P25 to P30 coincides with a rapid increase in adrenal weight and adrenocortical secretion. Virilisation of male tammars begins about three weeks after the onset of testosterone synthesis (i.e. about 3.5 weeks after the time of birth, Butler et al., 1998), but puberty in male tammar wallabies does not begin until about 19 months of age (Williamson et al., 1990). Androgen secretion was found to occur from the male testes of the tammar wallaby from P2, when the testis cords have differentiated (Renfree et al., 1992, 1996). In tammar wallabies, the concentration of testosterone in the testis starts to rise at P2 to plateau for the period from P10 to P40, after which it drops to reach basal levels at P70 (Renfree et al., 1992, 1996). Leydig cells in the testes of the grey short-tailed opossum (*Monodelphis domestica*) are evident by day three post partum, but testosterone secretion is low until four weeks after birth, although 3β-hydroxysteroid activity is found in adrenals and gonads of both male and female grey short-tailed opossums from birth (Xie et al., 1998; Russell et al., 2003). In tammar wallabies (*Macropus eugenii*), the

differentiation of the mammary primordia, gubernaculum and processus vaginalis occurs early in fetal development under genetic control, whereas later development of the vas deferentia, epididymides, prostate, male urethra and phallus is under the control of androgens (Renfree et al., 1992; Butler et al., 1998; Shaw et al., 2000).

By contrast, all secretory cells types have been reported to be present in the anterior pituitary of the newborn *Didelphis virginiana*, with growth-hormone-, ACTH- and prolactin-secreting cell types present even during late embryonic life (Sherman and Krause, 1990). Similarly, the anterior pituitary of newborn *Monodelphis domestica* exhibits moderate or strong immunoreactivity for ACTH, growth hormone and melanocyte-stimulating hormone (Gasse and Meyer, 1995b). It remains to be seen whether these findings represent a real difference in the timing of anterior pituitary development between American and Australian marsupials.

As noted earlier under the section on hypothalamic development, projections to the posterior pituitary from the supraoptic and paraventricular nuclei in the tammar wallaby are not evident until P5 and P10, respectively. This stands in contrast with the finding that the head of newborn brush-tailed possums contains 0.26 and 0.34 ng of mesotocin and vasopressin, respectively (Gemmell et al., 1993), even though the posterior pituitary at this stage is avascular and poorly differentiated. Curiously, Gemmell and colleagues identified presumptive axons with secretory granules in newborn possum posterior pituitary even though the hypothalamic nuclei, which should be projecting those axons, could not be found. Perhaps the granulated axons are derived from another source in the forebrain. Clearly the question of whether the posterior pituitary secretions contribute to parturition or fluid balance in the newborn possum remains open. Hypothalamo-neurohypophyseal development in Ameridelphians also appears to be insufficiently advanced at birth to provide effective secretory output, although arginine vasopressin binding does appear early in the posterior pituitary (Gasse and Meyer, 1995a; Kuehl-Kovarik et al., 1997).

Epithalamus

Organisation and function of the adult epithalamus

The epithalamus can be divided into the habenular nuclei and associated nuclei (anterior and posterior paraventricular nuclei) on either side of the dorsal third ventricle and the epiphysis (largely equivalent to the pineal gland in mammals), which forms the roof of the third ventricle between the habenula and the posterior commissure. The epithalamus is highly conserved in its structure and is broadly similar in configuration in all therians, although variations within marsupials have been noted in the size and presence of a pineal recess of the third ventricle (Kenny and Scheelings, 1979). In macropods, the pineal gland or body is a conspicuous structure, whereas it is intermediate in size in the koala and small in most dasyurids. The ventricular surface of the pineal recess in brush-tailed possums has three distinct zones: central, paracentral and peripheral (Tulsi, 1979a), like those seen in *Didelphis virginiana* (Cosenza *et al.*, 1990). The ventricular lining of the paracentral zone is covered by microvilli, supraependymal cells and CSF-contacting nerve processes, whereas the central zone is free of both cilia and microvilli (Samarasinghe and Delahunt, 1980) and the peripheral zone is heavily ciliated. The surface features of the pineal recess of diprotodonts may facilitate sweeping of CSF with dissolved pineal secretions towards the median eminence to influence hypothalamic activities (Tulsi, 1979a), particularly with regard to seasonal control of sexual behaviour and reproductive function. Nevertheless, no season- or gender-related differences have been found in the morphology or ultrastructure of the pineal recess (Samarasinghe and Delahunt, 1980).

Projections from the pineal gland are more restricted in mammals than in many other vertebrates, but terminate in the medial habenular nucleus and the pretectum. The habenular nuclei and posterior commissure also project to the pineal gland in monotremes and marsupials (Kenny and Scheelings, 1979; Butler and Hodos, 2005). The major function of the epithalamus is the regulation of cyclical behaviour, i.e. circadian rhythms and reproductive cycles. In placental mammals, the pineal gland and its hormone melatonin play an important role in the timing of seasonal reproduction, and information about day length appears to be transmitted to the central nervous system by the pattern of secretion of melatonin (Goldman, 2001; see below).

In diprotodont marsupials, sympathetic denervation of the pineal gland by bilateral removal of the superior cervical ganglia was initially reported to eliminate seasonal embryonic diapause, but had no effect on embryonic diapause due to concurrent lactation (Renfree *et al.*, 1981). Bilateral sympathectomy was found to abolish the natural nocturnal rise in melatonin, prompting the attribution of the elimination of seasonal diapause to loss of melatonin. On the face of it, the changes in both amplitude and duration of the melatonin cycle appear to be important in signalling seasonal change for macropods (McConnell, 1986; McConnell *et al.*, 1986), but the mechanism may not be as simple as it first seems. Whereas pinealectomy itself abolishes the nocturnal rise in melatonin (McConnell and Hinds, 1985), it does not abolish the seasonal reproductive

quiescence of the female wallabies. Plasma progesterone profiles of both pinealectomised and sham-operated wallabies indicate that the corpora lutea remain quiescent until activation at the normal season, despite the loss of the pineal gland. Tulsi and Kennaway (1979) have also reported that the pineal of the brush-tailed possum has a low capacity to store and secrete melatonin and have questioned the importance of the pineal in controlling seasonal breeding in possums. Many of the studies on diprotodont pineal function, via melatonin administration, have been equivocal, with studies reporting either no effect, significant effects or confused results, and this may relate to the marsupial response to temperate-zone photoperiods (see McAllan, 2003). Clearly the functional role of the pineal in diprotodont marsupials warrants further investigation.

Dasyurids have also attracted attention with regard to control of seasonal behaviour and physiology. In particular, there are profound seasonal changes in reproductive and non-reproductive anatomy and physiology and reproductive behaviour in those small dasyurids exhibiting male semelparity (McAllan *et al.*, 1996, 1997, 1998). This is best studied in *Antechinus stuartii*, which has a short and highly synchronised mating period lasting for only about one to two weeks. The environmental cue, which synchronises the reproductive life history of both males and females, is a specific rate of change of the photoperiod (McAllan *et al.*, 1991). This can be confirmed by experimentally inducing a phase delay in the photocycle, which leads to a corresponding change in the reproductive cycle timing (McAllan *et al.*, 1991). Administration of melatonin in drinking water to *A. stuartii* of both sexes changes the normal response to an increasing rate of change in photoperiod (McAllan *et al.*, 2002). In the case of females, melatonin administration from the winter solstice shifts the induction of oestrus two weeks earlier; in the case of males, postmating decline and death is accelerated. Daily activity patterns of animals administered melatonin are also disturbed (McAllan *et al.* 2008a). These results indicate that melatonin is important in the synchronisation of the dasyurid reproductive cycle, but the neural pathways mediating this natural synchronisation remain unexplored. Presumably the key anatomical elements include the suprachiasmatic nucleus, which receives retinal afferents, the pineal gland, producing the melatonin, secretory regions in the hypothalamus, to mediate regulatory hormone production, and a variety of forebrain nuclei (septal nuclei, amygdala, bed nuclei of the stria terminalis and limbic cortex) and hindbrain output (PAG, nucleus retroambiguus) to control and mediate the behavioural manifestations.

The subcommissural organ is situated ventral to the posterior commissure and has a broadly similar morphology throughout all mammals (Kenny and Scheelings, 1979; Collins, 1983), although Johnson noted its large size in kangaroos (Johnson, 1977). On the other hand, caudal to the subcommissural organ of both the brush-tailed possum and red-necked wallaby are paired ciliated paramedian folds, which have not been reported in placental brain and may be a distinctive feature of marsupials (Tulsi, 1979b; Collins, 1983). Large neurons have been observed in the subcommissural organ of Australian marsupials (both poly- and diprotodont), but not in monotremes. Kenny and Scheelings (1979) felt that these were simply the result of rostral and medial spread of neurons from the midbrain tectum, rather than functional elements of the subcommissural organ itself. The apical regions of cells of the subcommissural organ show evidence of three methods of releasing secretions into the CSF: microapocrine secretion, terminal microvillus eletion and apical membrane rupture (Tulsi and Kennaway, 1979). Secretory function of the subcommissural organ does not appear to be related to gender or season and the precise function of the organ in marsupials is not known. Certainly the subcommissural organ is not a significant source of melatonin in the brush-tailed possum, but the subcommissural organ is responsible for secretion of Reissner's fibre, which contributes to tethering of the spinal cord (Tulsi, 1982; see also the Chapter 13 section on the spinal cord).

The habenular nuclei of each side may be asymmetric in size in some vertebrates (Butler and Hodos, 2005), but do not appear to be so in marsupials. The habenula contains medial and lateral nuclei: the medial habenular nucleus is interconnected with the limbic system, receiving input from the septal nuclei (including the nuclei of the diagonal band), lateral preoptic area and lateral hypothalamus, whereas the lateral habenular nucleus receives inputs mainly from the striatum and limbic system. The habenular nuclei have two major fibre tracts associated with them. They receive input (largely from the septal nuclei) by the stria medullaris and in turn give rise to the habenulointerpeduncular tract (fasciculus retroflexus). The fasciculus retroflexus is usually said to project to the highly cholinergic interpeduncular nuclei (see Chapter 4), but Way and Kaelber (1969) reported that, in didelphids, the tract extends to other regions: midbrain tegmentum, raphe nuclei, the superior colliculus and even rostral telencephalic structures such as the preoptic area and olfactory tubercle. The projection from the habenular nuclei is somewhat more restricted in the quokka (Way, 1975) with most projections being to the interpeduncular nuclei, midbrain tegmentum and central grey of the isthmus.

Development of the epithalamus

The habenular nuclei show a similar lateral-to-medial neurogenetic gradient to that seen for the adjacent dorsal thalamus. Lateral habenular neurons begin to leave the mitotic cycle at P5 and the last neuron of the medial habenular nucleus completes mitosis at around P32 (Sanderson and Weller, 1990b). Consequently, all the neuronal elements of the habenular nuclei are in place long before cortical development is significantly underway.

The fasciculus retroflexus appears during the second week of postnatal life in the developing tammar wallaby (Hassiotis et al., 2002). GAP-43 immunoreactivity peaks in the fasciculus retroflexus at about P20 and declines during P30 to P50. Unlike many other forebrain tracts, the fasciculus retroflexus does not retain GAP-43 immunoreactivity into adult life, suggesting that the axons of this tract do not retain a capacity for plasticity.

Unfortunately, nothing is known concerning the development of function in the marsupial epithalamus and its circuits. This is disappointing considering the likely significance of the epithalamus for controlling photoperiod-regulated behaviour in marsupials.

Pretectal area

Organisation and function of the pretectal area in adult marsupials

The pretectal area lies between the caudal thalamus and the midbrain. The pretectal nuclei receive afferents predominantly from the retina and optic tectum and are important for modifying motor behaviour in response to visual information (Sefton et al., 2004; Butler and Hodos, 2005). In nonmammalian amniotes, ten or more pretectal nuclei have been recognised (Butler and Hodos, 2005; Puelles et al., 2007). In all mammals, there are also distinct nuclei within the pretectum, although the number recognised is usually around five. These include the nucleus of the optic tract (OT), the olivary pretectal nucleus (OPT), the anterior pretectal nucleus (APT), posterior pretectal nucleus (now called tectal grey – TG) and the medial pretectal nucleus (MPT). The anterior pretectal nucleus is often further subdivided into two regions (APTD and APTV). All of these nuclei can be identified in both monotreme and marsupial brains and the presence of a differentiated pretectum in both of the living monotremes (Ashwell and Paxinos, 2007), with some similar chemoarchitectonic features to at least one group of placental mammals, suggests that a differentiated pretectum may have been present in the stem mammal at the divergence of the therian and prototherian lineages. It is not surprising then that all five pretectal nuclei can be identified readily in all Australian marsupials studied so far (Figure 6.5).

Pretectal nuclei in all therian and prototherian mammals exhibit consistent and distinctive cytoarchitecture (Goldby, 1941) and chemoarchitectural signatures (Ashwell and Paxinos, 2007). For example, the APT has strong reactivity for acetylcholinesterase in a wide variety of mammals (lagomorph, rodent, echidna; Ashwell and Paxinos, 2007; Caballero-Bleda *et al.*, 1992; Paxinos and Watson, 2007), although the APT is not strongly reactive for AChE in the platypus (Ashwell and Paxinos, 2007). Similarly, both the MPT and TG have patches of strong AChE reactivity in a wide variety of mammals (see Figure 6.5). In diprotodont marsupials like the tammar wallaby, the nuclei of the pretectal area have consistent regions of NADPH diaphorase (Figure 6.5).

In well-studied placental mammals like the laboratory rat, the OPT, OT and TG receive projections from the retina. In the tammar, retinal ganglion cells project almost exclusively to the contralateral pretectum with terminations of fibres in the OT and TG (Ibbotson *et al.*, 2002). Several pretectal nuclei receive projections from the primary and association areas of visual cortex in rodents and diprotodonts (OPT, OT and TG; Haight *et al.*, 1980; Sefton *et al.*, 2004), but Ibbotson *et al.* (2002) did not find any cortical projection to the OT in the tammar wallaby. Other sources of afferents include the superior colliculus (projecting to the APT and TG), pregeniculate nucleus of the thalamus (formerly called the ventral lateral geniculate; projecting to the APT, OPT, OT) and nuclei of the accessory optic system (Sefton *et al.*, 2004). Efferents of the pretectum in laboratory rodents include ascending and descending projections to visual cortex, DLG, pregeniculate nucleus, PAG and nuclei of the oculomotor system (Berman, 1977; Sefton *et al.*, 2004). The pretectum is thought to play a central role in oculomotor reflexes, including the pupillary light reflex and stabilisation of visual images on the retina during head or image movement (Simpson *et al.*, 1988). In the tammar wallaby, the OT contains direction selective and non-directional cells which contribute to the mediation of ocular following movements (Ibbotson and Mark, 1994; Ibbotson *et al.*, 1994). Studies in *Didelphis virginiana* indicate that there is a significant commissural connection between the OT/dorsal terminal nucleus of each side mediating responses to horizontal optokinetic stimulation and that this commissural pathway is probably excitatory to GABAergic interneurons (Vargas *et al.*, 1997).

Development of the pretectal area

The development of the pretectal area has been studied with routine histological methods in the brush-tailed possum and bandicoot (*Perameles nasuta*; Warner, 1970) and

has been illustrated in the accompanying development series of the tammar wallaby. Both the early pouch young bandicoot and brush-tailed possum (10 and 15.5 mm crown rump length – CRL, respectively) have a very immature pretectal area, with few or no postmitotic neurons present, but component neurons appear very soon (by 16 mm CRL in *Perameles nasuta* and by 19 mm CRL in *Trichosurus vulpecula*; Warner, 1970). On the other hand, in the tammar wallaby, differentiation of the pretectal nuclei is slightly more advanced than the thalamus or prethalamus, and distinct medial and lateral compartments can be seen in the pretectum even at the time of birth (see atlas plate WP0/6). The cytoarchitectural subdivision into five distinct nuclei emerges at about P19 to P25 in the tammar wallaby (see atlas plate WP25/5).

What remains unknown?

Probably the most important differences between marsupial and placental mammals lie in the area of reproduction. In Chapter 3 we considered the question of sexual dimorphism in the hypothalamus of marsupials and its relevance to control of mating behaviour, but there are several other important questions concerning marsupial reproductive function, which probably involve characteristic functional differences in the diencephalon: (i) What are the pathways involved in control of seasonal breeding cycles? Clearly retinal afferents to the suprachiasmatic nucleus of the hypothalamus and secretion of melatonin from the pineal gland are fundamentally important, but the actual neural pathways mediating these physiological changes are still speculative. (ii) How is the complex lactation pattern of diprotodont marsupials, which allows the supply of milk with different composition to pouch young of different ages, achieved? Does it involve hypothalamic mechanisms or is it mediated solely at the level of the mammary gland?

Detection of photoperiod change is so exquisitely sensitive in the case of Australian marsupials that a change of only one to two minutes per day induces behavioural and physiological effects. Furthermore, the amplitude of the photoperiod change trigger is quite species-specific and related to the latitude and ecology of the species (Dickman and Woodford Ganf, 2007). In the case of small Australian marsupial carnivores, ovulation may be triggered by day length increases of only 77 to 97 seconds for the yellow-footed antechinus (*Antechinus flavipes*), 99 to 111 seconds for the brown antechinus (*Antechinus stuartii*) and 127 to 137 seconds for agile antechinus (*Antechinus* sp.). Based on studies in placental mammals, it is likely that mechanisms for detecting this change lie at the pars tuberalis interface between the

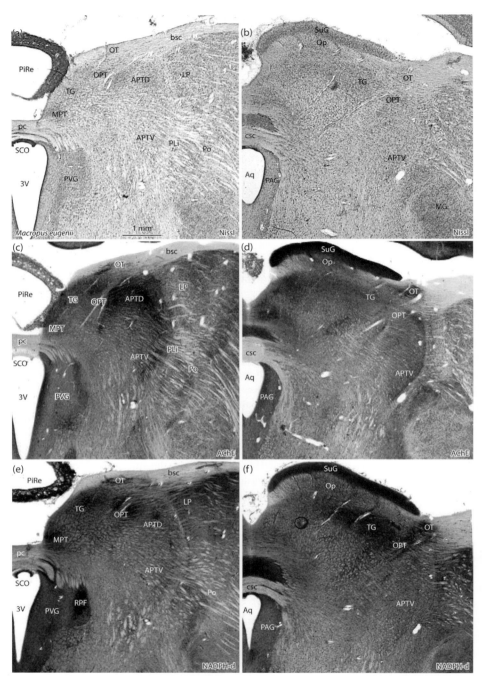

Figure 6.5 Cyto- and chemoarchitecture of the pretectal nuclei in a representative diprotodont marsupial (*Macropus eugenii*). Panels (a) and (b) show Nissl-stained sections through the rostral and caudal parts of the pretectum, respectively. Panels (c), (d) and (e), (f) show acetylcholinesterase and NADPH-diaphorase-reacted sections, respectively, at adjacent rostrocaudal levels. The scale bar in (a) is applicable to all the photomicrographs. 3V – third ventricle; APTD – anterior pretectal nu., dorsal; APTV – anterior pretectal nu., ventral; Aq – cerebral aqueduct; bsc – brachium of superior colliculus; csc – commissure of superior colliculus; LP – lateral posterior thalamic nu.; MG – medial geniculate nu.; MPT – medial pretectal nu.; Op – optic nerve layer of superior colliculus; OPT – olivary pretectal nu.; OT – nu. of the optic tract; PAG – periaqueductal grey; pc – posterior commissure; PiRe – pineal recess of 3V; PLi – posterior limitans thalamic nu.; Po – posterior thalamic nuclear group; PVG – periventricular grey; RPF – retroparafascicular nu.; SCO – subcommissural organ; SuG – superficial grey of superior colliculus; TG – tectal grey.

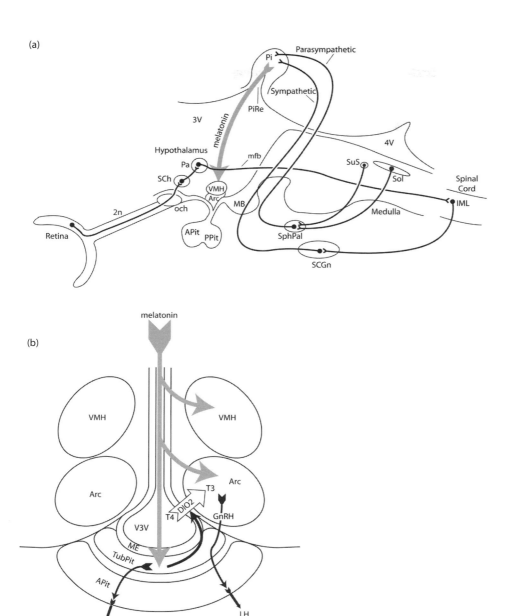

Figure 6.6 Schematic representation of (a) the neural pathways in marsupial brain likely to be involved in the production of melatonin by the pineal gland and (b), the hormonal interactions between melatonin and the hypothalamo-pituitary axis likely to mediate the effects of photoperiod on reproductive behaviour and physiology, and metabolism. Note that both neural and hormonal interactions are hypothesised largely on the basis of studies in placental mammals (for review see Hazlerigg and Loudon, 2008; Morgan and Hazlerigg, 2008; Revel *et al.*, 2009). Although the axonal pathways controlling the pineal gland are unlikely to differ across therians, the involvement of specific hypothalamic nuclei in the hormonal interactions is quite species-specific in placental mammals and is therefore likely to exhibit differences between marsupials and placentals and between marsupial species. The diagrams therefore represent useful working hypotheses to stimulate the investigation of these pathways in Australian marsupials, not definitive conclusions. 2n – optic nerve; 3V – third ventricle; 4V – fourth ventricle; APit – anterior lobe of pituitary gland; Arc – arcuate hypothalamic nu.; IML – intermediolateral cell column; LH – lateral hypothalamic area; MB – mamillary body; ME – median eminence; mfb – medial forebrain bundle; och – optic chiasm; Pa – paraventricular hypothalamic nu.; Pi – pineal gland; PiRe – pineal recess of 3V; PPit – posterior lobe pituitary; PrL – prelimbic cortex; Retina – retina; SCGn – sup cervical ganglion; SCh – suprachiasmatic nu.; Sol – solitary nu.; SphPal – sphenopalatine ganglion; SuS – superior salivatory nu.; TubPit – tuberal pituitary; V3V – ventral third ventricle; VMH – ventromedial hypothalamic nu.

median eminence and anterior pituitary and may depend on the temporal overlap between genes activated at the onset and offset of melatonin secretion (Hazlerigg and Loudon, 2008).

Figure 6.6 summarises the neural and hormonal interactions likely to be involved in detection of changes in photoperiod and the mediation of effects on metabolism and reproductive organs. At this stage, our understanding of the pathways and interactions is based solely on photoperiod-regulated physiology in placental mammals, but at least some neural pathways are likely to be the same in Australian marsupials and candidate nuclei, glands and hormones remain the best starting points for any detailed investigation of photoperiod-mediated physiological changes in Australian marsupials. A circuitous neural pathway involving the retina, suprachiasmatic and paraventricular hypothalamic nuclei, sympathetic preganglionic neurons of the intermediolateral cell column of the spinal cord, and sympathetic innervation of the pineal gland, stimulates melatonin secretion (Revel et al., 2009; Figure 6.6a). Melatonin diffusing through the cerebrospinal fluid of the third ventricle then acts on various glandular and neuroglandular (median eminence, tuberal and distal parts of pituitary) and neural (Arc, VMH) structures in and near the walls of the ventral third ventricle (Figure 6.6b). Data on the localisation of melatonin receptors in marsupial brain are limited, but the receptors appear to have similar sensitivity to those in other vertebrates (Paterson et al., 1992). Regulation of thyroid hormone deiodinase (DIO2, producing more active tri-iodothyronine, T3, from thyroxin, T4) is believed to play a central part in the hypothalamic responses of placental mammals to melatonin (Morgan and Hazlerigg, 2008; Revel et al., 2009). Control of DIO2, and therefore regulation of GnRH from the Arc, is believed to depend on a melatonin-dependent signal relayed through the tuberal pituitary, while a simultaneous melatonin-dependent signal controls prolactin secretion from the anterior pituitary (Morgan and Hazlerigg, 2008).

7

Deep telencephalic structures: striatum, pallidum, amygdala, septum and subfornical organ

K. W. S. Ashwell

Summary

A variety of neuronal groups (striatum, pallidum, amygdala, septal nuclei) lie within the deep parts of the telencephalon, having developed from the medial and ventral parts of the embryonic telencephalic vesicle.

The striatum consists of the caudate, putamen and nucleus accumbens. As the major part of the basal ganglia, the striatum works in concert with the motor areas of the cerebral cortex to orchestrate and execute planned motivated behaviours requiring motor, cognitive and limbic circuits. Neurogenesis of the marsupial striatum (like that in placental mammals) appears to occur in both primary basal ganglia neuroepithelium alongside the lateral ventricle and the surrounding subventricular zone.

The pallidum includes large neurons of the entopeduncular nucleus (also called the internal, or medial, globus pallidus), globus pallidus (also called the external, or lateral, globus pallidus) and substantia innominata. In both placentals and marsupials, the pallidal neurons are generated from the basal ganglia neuroepithelium well before the striatal neurons. In the case of diprotodont marsupials, the pallidal neurons appear to be produced around the time of birth.

The amygdala consists of a grey matter mass in the temporal part of the brain (the temporal amygdala) and extended amygdaloid nuclei distributed in a band towards the anterior commissure and along one of the major fibre bundles of the amygdala, the stria terminalis. The amygdala develops entirely after birth in diprotodont marsupials like the tammar wallaby, but the major subdivisions of the amygdala emerge rapidly during the second and third postnatal week.

The septal nuclei lie medial to the anterior horn of the lateral ventricle rostral and dorsal to the anterior commissure. Four groups of nuclei are recognised in the septal region in both poly- and diprotodont marsupials. The septal nuclei have traditionally been considered part of the limbic system, concerned with emotions and their expression, and most of the septal nuclei have topographical morphofunctional links with the hippocampus and amygdala. The proximity of the septal nuclei to the placental corpus callosum means that the gross morphogenesis of this region is fundamentally different in marsupials compared to placentals, largely due to differences in the proportion of the rostrocaudal extent of the septum that undergoes midline fusion during development.

The subfornical organ is one of the circumventricular organs, sites where blood-borne humoral agents can exert central nervous system effects.

Introduction

Basal telencephalic structures include the striatopallidal components (striatum and pallidum), as well as elements of the limbic system (the amygdala and septal nuclei) and several sets of cholinergic neuronal populations. In all mammals, the dorsal striatopallidal complex is composed of the caudate and putamen (often combined as a single caudatoputamen – the dorsal striatum) and the globus pallidus and entopeduncular nucleus (dorsal pallidum), whereas the ventral striatopallidal complex is composed of the nucleus accumbens and deep olfactory tubercle (together known as the ventral striatum) and the ventral pallidum. The older term 'basal ganglia' is mainly used with reference to the dorsal striatopallidal complex. The cells of the ventral pallidum lie scattered within the rostral part of the ambiguously named substantia innominata in the basal forebrain.

The dorsal striatopallidal complex is concerned with voluntary movements for the body and is part of a looped circuit involving inputs from the cerebral cortex and midbrain dopaminergic neurons, and projections to the motor thalamic nuclei (VA, VL) and thence back to the cerebral cortex. The ventral striatopallidal complex is mainly involved with the limbic system (concerned with emotions, and social and motivated behaviour), receiving projections from the prefrontal and olfactory components of the cerebral cortex and has reciprocal connections with the ventral tegmental area and substantia nigra of the midbrain. Output from the ventral pallidum is to the mediodorsal thalamic nucleus, which in turn projects to the prefrontal cortex, a region involved in control of social behaviour.

The amygdala has temporal and extended components. The nuclei of the amygdala have been divided into four major functional groups (main, accessory, autonomic and frontotemporal). An alternative subdivision considers the amygdala and adjacent claustrum and endopiriform complex as divided into olfactory pallial amygdala, striatal amygdala and claustroamygdalar (pallial) complex (Swanson and Petrovich, 1998; Butler and Hodos, 2005).

The septal nuclei (medial and lateral nuclei, nucleus of the diagonal band of Broca, septofimbrial and triangular septal nucleus) receive most of their input from the limbic cortex, the amygdala and the hypothalamus. Septal nuclei are involved in a wide range of autonomic, neuroendocrine, behavioural and cognitive activities.

Striatum

Adult structure and function

The striatum includes the dorsally situated caudate and putamen (often combined in smaller mammals as the caudatoputamen), and the ventrally positioned nucleus accumbens, below and medial to the anterior commissure. These subdivisions are as clear in marsupials (Figure 7.1; Hamel, 1967) as in eutherians. In placental mammals, the main cell type in the striatum is the medium spiny projection neuron, which is spread uniformly throughout the caudate and putamen and makes up well over 90% of the neuronal population of the striatum. The other neurons in the striatum are interneurons, which do not project out of the striatum and mainly have synapses on the projection neurons. Interneurons of the striatum are more diverse than projection neurons and include cholinergic and GABAergic types.

In placental mammals, the most abundant type of GABAergic striatal interneurons expresses the calcium-binding protein parvalbumin and has distinct neurophysiologic characteristics (fast-spiking interneurons). Parvalbumin interneurons receive projections from the cerebral cortex and provide inputs to medium spiny

projection neurons. In rodents, parvalbumin neurons are more frequent in the dorsolateral caudatoputamen and exhibit a dorsolateral to ventromedial numerical density gradient. Topographic gradients are also seen in the distribution of parvalbumin immunoreactive neurons in the caudatoputamen of representative marsupials and the echidna (Ashwell, 2008b). These topographic gradients probably underlie a subdivision of the dorsal striatum in all mammals into an associative/behavioural part situated dorsally and a sensorimotor component situated ventrally, although the precise positions of these subdivisions along the mediolateral and rostrocaudal axes may vary from species to species (Ashwell, 2008b).

Development of the striatum in Australian marsupials

Based on studies of rodents, it is known that the neuroepithelium of the mammalian ventral telencephalic vesicle produces the neurons of the striatum, nucleus accumbens, olfactory tubercle, pallidum, amygdala, septum and part of the olfactory cortex (for review see Bayer and Altman, 2004). The lateral ('ganglionic') part of the developing telencephalic neuroepithelium is divided into medial and lateral 'ganglionic' (also called ventricular) eminences, both of which can still be seen in newborn marsupials (see atlas plate WP0/3). The earlier-generated medial ganglionic eminence produces the pallidal region, whereas the lateral ganglionic eminence generates the caudate and putamen.

Rodent studies suggest that the medium spiny projection neurons of the striatum are generated in two steps by the basal ganglia neuroepithelium. Some of the earliest projection neurons of the ventrolateral striatum appear to be generated directly from the primary neural stem cells of the lateral ganglionic eminence, whereas other medium spiny projection neurons appear to be derived from secondary neural stem cells which have migrated from the basal ganglia neuroepithelium into the subventricular zone and continued to proliferate (see Bayer and Altman, 2004).

Some observations of marsupial pouch young suggest that a similar biphasic production may be occurring in the developing marsupial striatum. At least a few neurons of the putative caudate and putamen are already present at birth in tammar wallabies (refer to atlas plates WP0/3 and 4), before a subventricular zone has emerged, suggesting that at least some early neurons are generated directly from the lateral eminence neuroepithelium. On the other hand, the caudatoputamen of the tammar continues to expand during the period from P12 to P40, during which stage the subventricular zone is the major proliferative zone of the ventral and lateral telencephalon (see atlas plates WP12/5 and WP19/4).

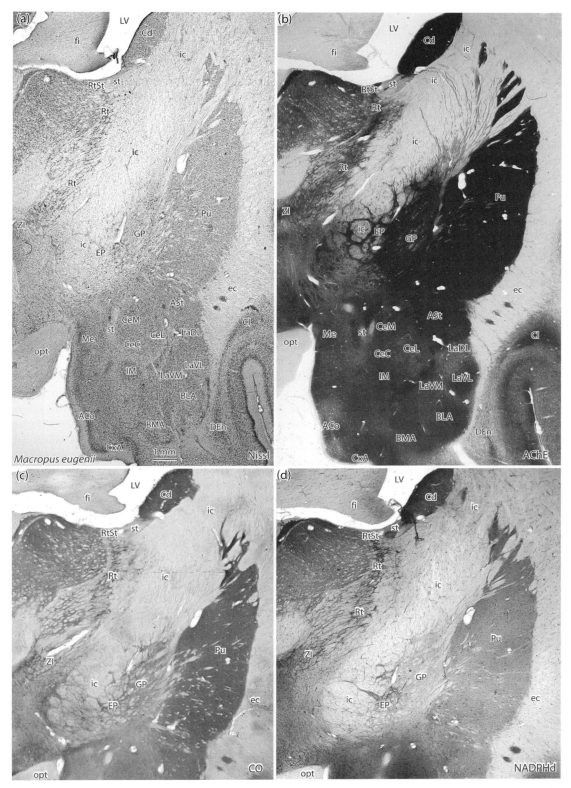

Figure 7.1 Topography and chemoarchitecture of the dorsal striatopallidal complex (caudate – Cd, putamen – Pu, entopeduncular nu. – EP, and globus pallidus – GP) and amygdala in coronal sections through the brain of the tammar wallaby. Scale bar in (a) applies to all photomicrographs. ACo – anterior cortical amygdaloid area; ASt – amygdalostriatal transition area; BLA – basolateral amygdaloid nu., anterior; BMA – basomedial amygdaloid nu., anterior; CeC – central amygdaloid nu., capsular; CeL – central amygdaloid nu., lateral div'n; CeM – central amygdaloid nu., medial div'n; Cl – claustrum; CxA – cortex-amygdala transition zone; DEn – dorsal endopiriform nu.; ec – external capsule; fi – fimbria of the hippocampus; ic – internal capsule; IM – intercalated amygdaloid nu., main; LaDL – lateral amygdaloid nu., dorsolateral; LaVL – lateral amygdaloid nu., ventrolateral; LaVM – lateral amygdaloid nu., ventromedial; LV – lateral ventricle; Me – medial amygdaloid nu.; opt – optic tract; Rt – reticular thalamic nu. (prethalamus); RtSt – reticulostrial nu.; st – stria terminalis; ZI – zona incerta.

Tritiated thymidine studies of striatum development in another diprotodont marsupial (*Trichosurus vulpecula*) are also indicative of similar generative mechanisms for caudatoputamen neurons to those observed in rodents. Sanderson and Weller (1990b) found that neurons of the possum caudatoputamen are generated from P5 to P55, with a ventrolateral to dorsomedial neurogenetic gradient (i.e. neurons in the ventrolateral putamen are generated before those in the dorsomedial caudate). Labelled neurons in the adult caudate are often seen in clusters, particularly in the early-injected animals (P5), consistent with labelled mother cells migrating to a subventricular zone and generating only a few clustered daughter neurons in the adult caudate and putamen. Neurons in the nucleus accumbens are generated from P5 (or earlier) through to P39. There is a striking neurogenetic gradient for the nucleus accumbens in this species, with labelled neuron clusters confined to the lateral accumbens shell after injection at P5, in the ventral accumbens shell and accumbens core after injections on P12 and P21, and confined to the most medial accumbens shell after injection on P39.

The neurogenetic gradients for the dorsal striatum may reflect the previously mentioned functional divisions of the caudate and putamen (i.e. into early generated sensorimotor components vs. later-generated associative elements), but this remains to be tested experimentally.

Pallidum

Organisation of the adult pallidum

The pallidum consists of a diffuse collection of large neurons scattered throughout the basal telencephalon (entopeduncular nucleus or internal globus pallidus, (external) globus pallidus, substantia innominata, ventral pallidum deep to the olfactory tubercle). The magnocellular neurons of the pallidum get their major projection from either the caudate and putamen, olfactory tubercle or the nucleus accumbens. The entopeduncular nucleus is embedded in the caudoventral part of the internal capsule, whereas the globus pallidus is a group of large neurons interposed between the lateral border of the internal capsule and the ventromedial edge of the striatum. Johnson (1977) has commented that the globus pallidus is small in marsupials compared to many placentals, particularly primates, but this has not been analysed quantitatively. The substantia innominata is a region ventral to the striatum and globus pallidus, while large polymorphic neurons of the ventral pallidum deep to the olfactory tubercle make up the most rostral pallidal neurons.

The globus pallidus is the target of direct and indirect striatal output systems. The internal globus pallidus (along with

the reticularis part of the substantia nigra, considered in Chapter 4) make up a single nuclear complex that provides output to the thalamus. The precise nuclei receiving these outputs may vary between species and it is not presently known how these pathways are organised in Australian marsupials.

Development of the pallidum

In rodents, the magnocellular pallidal neurons are among the first generated in the ventral telencephalon, with most neurons leaving the mitotic cycle before E16. Of the pallidal components, the entopeduncular neurons are the first generated, followed by the globus pallidus, substantia innominata and magnocellular neurons of the olfactory tubercle (Bayer and Altman, 2004). Tritiated thymidine studies of the brush-tailed possum show that neurons of the globus pallidus are generated very early (mostly before P5) in contrast to the prolonged neurogenetic timetable for striatal neurons in this species (Sanderson and Weller, 1990b). Cytoarchitectural findings in the developing tammar are consistent with the thymidine labelling studies in the brush-tailed possum. The internal globus pallidus and many neurons of the future external globus pallidus can be identified as early as the first day of pouch life (see atlas plate WP0/3) and the adult topography of the entopeduncular nucleus and globus pallidus relative to medial amygdala and striatum is established by the end of the second postnatal week (see atlas plate WP19/4).

Amygdala

Organisation of the adult amygdala

The amygdala arises from both pallial and subpallial parts of the developing brain and is highly conserved in structure and function across tetrapod vertebrates (Moreno and González, 2007). A major part of the amygdala is an integral component of the main olfactory and vomeronasal systems, while other components are involved in regulation of the autonomic nervous system.

The amygdala proper (temporal amygdala) is a conglomerate of nuclei located in the temporal lobe (Figure 7.1a and b) with contiguity with the dorsal striatum through the amygdalostriatal transition area (Figure 7.1a, b). In well-studied placental mammals like the laboratory rat, the amygdala is known to be involved in a variety of functions (neuroendocrine, visceral effector control, defensive, ingestive, aggressive and reproductive behaviours, as well as memory and learning; see de Olmos *et al.*, 2004, for review). The nuclei of the amygdala are usually divided into superficial (corticomedial) and deep (basal and lateral) groups and this division is as evident in marsupials and monotremes as it is in placental mammals (Figure 7.1a and b; Figure 7.2; see atlas plates of *Sminthopsis macroura* – D12 and D13 – and *Macropus eugenii* – W22 and W23; for the short-beaked echidna see Ashwell *et al.*, 2005b). Johnson (1977) commented that two components of the corticomedial amygdala (nucleus of the lateral olfactory tract and central amygdala) are larger in marsupials compared to placentals, but this does not appear to have been analysed quantitatively.

An important feature of the organisation of the amygdala is the continuity of the amygdala grey matter with nuclei distributed in arcs extending towards the anterior commissure and along the stria terminalis fibre tract (extended amygdala; de Olmos *et al.*, 2004). The extended amygdala can be divided into two divisions: central and medial. The central division is closely affiliated with the main olfactory system and consists of the central nuclei of the amygdala, the lateral division of the bed nucleus of the stria terminalis (e.g. STLD, STLP, STLV) and the interstitial nucleus of the posterior limb of the anterior commissure (IPAC). The medial extended amygdala is largely associated with the vomeronasal part of the olfactory system and includes the medial nucleus of the amygdala, medial division of the bed nucleus of the stria terminalis (e.g. STMA, STMPM, STMPL) and intra-amygdaloidal part of the bed nucleus of the stria terminalis (STIA).

Development of the amygdala in Australian marsupials

The amygdala arises from the basal ganglia neuroepithelium within the telencephalic vesicle (see above). The amygdala cannot be distinguished from other derivatives of the basal ganglia neuroepithelium in the newborn tammar wallaby, but elements of the amygdala and extended amygdala (cortical amygdala and bed nuclei of the stria terminalis) begin to emerge during the first week of postnatal life (see P5 tammar wallaby; WP5/4). In the brush-tailed possum, neurogenesis of the temporal amygdala as a whole proceeds from P5 to P39 (Sanderson and Wilson, 1997). There is a clear mediolateral gradient in neurogenesis, with most nuclei labelled after injection of ^{3}H-thymidine on P5 and the label being progressively confined to more lateral nuclei with injection at subsequent ages. Injections at P32 and P39 labelled only laterally placed nuclei in the magnocellular ventrolateral part of the lateral amygdaloid nucleus.

By contrast, neurogenesis in the bed nuclei of the stria terminalis (ST) of the brush-tailed possum appears to proceed with a lateromedial gradient (Sanderson and Wilson, 1997), in that all nuclei were labelled following injection at P5, but

with neurogenesis concluding first in the lateral ST (around P12), next in the medial ST (around P18) and last in the pre-optic end of the complex (around P21).

The subdivision of the wallaby amygdala into corticomedial and basolateral components becomes clearer towards the end of the second postnatal week (see atlas plate WP12/6) and during the third postnatal week all the major subdivisions (cortical, central, medial, basal, lateral) can be demarcated (WP19/5). The intra-amygdaloidal fibre tracts also become defined during the third postnatal week. The emergence of argyrophilic stria fibres in pouch young and the time-course of development of GAP-43 immunoreactivity within the stria terminalis of the tammar wallaby (Hassiotis *et al.*, 2002) are consistent with the cytoarchitectural changes observed in the developing amygdala. The stria terminalis emerges as a distinct tract during the second postnatal week and reaches peak levels of GAP-43 immunoreactivity during the third postnatal week, coincidentally with the emerging cytoarchitecture. GAP-43 immunoreactivity in the stria terminalis declines sharply between P30 and P40, indicating that most fibres in this tract have completed outgrowth by P40.

Septal nuclei

Arrangement and chemoarchitecture of the septal nuclei

The septal nuclei form the anteromedial wall of the telencephalon, medial to the lateral ventricle. In placental mammals they lie ventral to the corpus callosum and the name of this region refers to the gross appearance of the structure in humans as a membrane-like curtain or 'wall' hanging down from the corpus callosum. In the acallosal marsupial brain, the septal nuclei still form the medial wall of the lateral ventricle, but the dorsal relationships are naturally different from those in placental mammals. In marsupials, the septal nuclei are in contact dorsally with the indusium griseum of the allocortex and the cingulate cortex. Ventrally, the septal region is in contact with the ventral pallidum and medial nucleus accumbens. The septal nuclei are usually considered as a part of the limbic system, concerned with memory and emotions.

In marsupials, as in rodents, the septal region (Figures 7.3 and 7.4) can be divided into a medial group, including the medial septal nucleus and nucleus of the vertical limb of the diagonal band, a posterior group, made up of the triangular septal nucleus and the bed nucleus of the anterior commissure, and a lateral group, comprised of septohippocampal, septofimbrial and lateral septal nuclei. The lateral septal nuclei can be further divided into dorsal, intermediate and

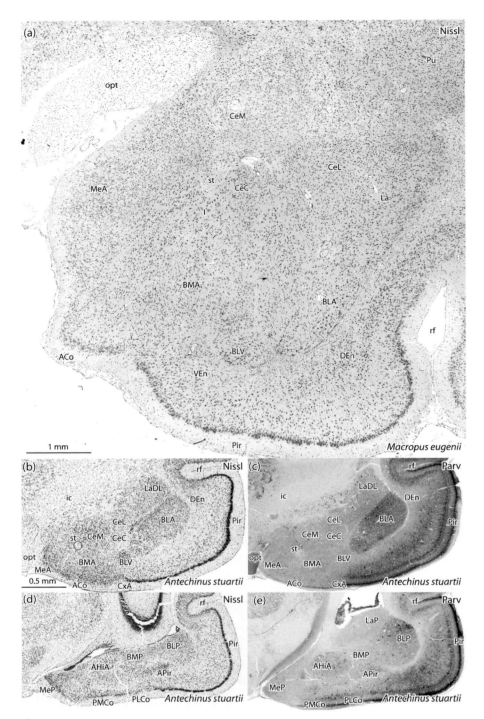

Figure 7.2 Cyto- and chemoarchitecture of the amygdala in representative diprotodont and polyprotodont Australian marsupials. Panel a) shows the cytoarchitecture of the amygdala of a tammar wallaby in a coronal section through the middle of the region's rostrocaudal extent. Panels b) and d) show the cytoarchitecture of the amygdala of *Antechinus stuartii* in coronal sections through the rostral and caudal amygdala, respectively. Panels c) and e) illustrate adjacent sections immunoreacted for parvalbumin. ACo – anterior cortical amygdaloid area; AHiA – amygdalohippocampal area, anterior; APir – amygdalopiriform transition zone; BLA – basolateral amygdaloid nu., ant; BLP – basolateral amygdaloid nu., post; BLV – basolateral amygdaloid nu., ventral; BMA – basomedial amygdaloid nu., ant; BMP – basomedial amygdaloid nu., post; CeC – central amygdaloid nu., capsular div'n; CeL – central amygdaloid nu., lateral div'n; CeM – central amygdaloid nu., medial div'n; CxA – cortex-amygdala transition zone; DEn – dorsal endopiriform nu.; I – intercalated nuclei of the amygdala; ic – internal capsule; La – lateral amygdaloid nu.; LaDL – La, dorsolateral; LaP – La, posterior; MeA – medial amygdaloid nu., ant; MeP – medial amygdaloid nu., post; opt – optic tract; Pir – piriform cortex; PLCo – posterolateral cortical amygdala area; PMCo – posteromedial cortical amygdala area; Pu – putamen; rf – rhinal fissure; st – stria terminalis; VEn – ventral endopiriform nu.

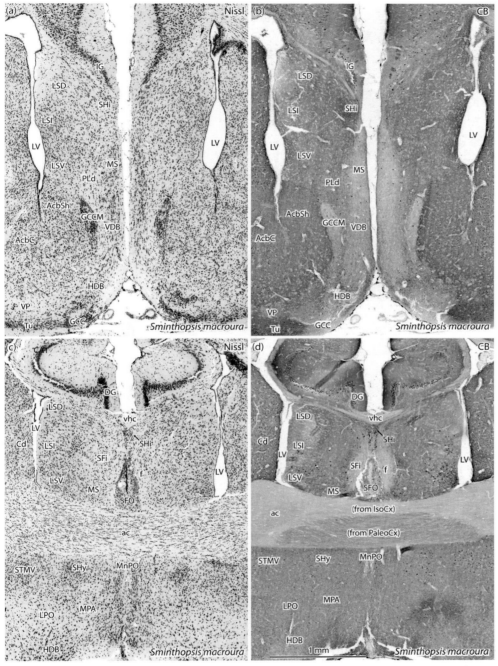

Figure 7.3 Cyto- and chemoarchitecture of the septal nuclei and subfornical organ in a polyprotodont Australian marsupial (*Sminthopsis macroura*). Panels (a) and (c) show Nissl-stained coronal sections through the rostral and caudal levels of the septal nuclei, respectively. Panels (b) and (d) show nearby sections stained for calbindin. Also note the differentiation of isocortical and paleocortical parts of the anterior commissure. ac – anterior commissure; AcbC – accumbens nu., core; AcbSh – accumbens nu., shell; CA1 – field CA1 of hippocampus; Cd – caudate nu.; DG – dentate gyrus; f – fornix; GCC – granule-cell cluster; GCCM – granule-cell cluster magna; HDB – nu. horizontal limb of diagonal band; IG – indusium griseum; IsoCx – isocortex; LPO – lateral preoptic area; LSD – lateral septal nu., dorsal part; LSI – lateral septal nu., intermediate part; LSV – lateral septal nu., ventral part; LV – lateral ventricle; MnPO – median preoptic nu.; MPA – medial preoptic area; MS – medial septal nu.; PLd – paralambdoid septal nu.; SFi – septofimbrial nu.; SFO – subfornical organ; SHi – septohippocampal nu.; SHy – septohypothalamic nu.; STMV – bed nu. of stria terminalis, ventromedial part; Tu – olfactory tubercle; VDB – nu. vertical limb diagonal band; vhc – ventral hippocampal commissure; VP – ventral pallidum.

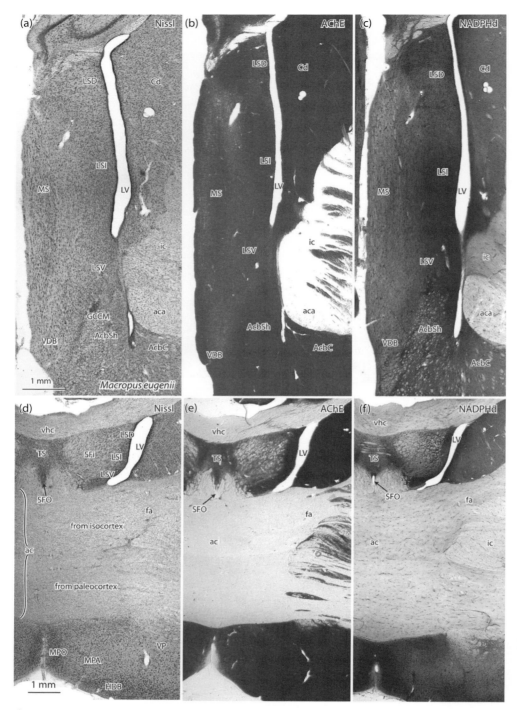

Figure 7.4 Cyto- and chemoarchitecture of the septal nuclei and nu. accumbens in a diprotodont marsupial (*Macropus eugenii*). Photomicrographs (a) and (d) show cytoarchitecture of the septal nuclei at rostral and anterior commissural levels, respectively. Photomicrographs (b) and (e) show adjacent sections reacted for AChE, while photomicrographs (c) and (f) illustrate adjacent sections reacted for NADPH diaphorase. ac – anterior commissure; aca – ac, anterior limb; AcbC – accumbens nu., core; AcbSh – accumbens nu., shell; Cd – caudate nucleus; fa – fasciculus aberrans; GCCM – granule-cell cluster magna; HDB – nu. of horizontal limb of diagonal band; ic – internal capsule; LSD – lateral septal nu., dorsal part; LSI – lateral septal nu., intermediate part; LSV – lateral septal nu., ventral part; LV – lateral ventricle; MPA – medial preoptic area; MPO – medial preoptic nu.; MS – medial septal nu.; SFi – septofimbrial nu.; SFO – subfornical organ; TS – triangular septal nu.; VDB – nu. of vertical limb of diagonal band; vhc – ventral hippocampal commissure; VP – ventral pallidum.

ventral subregions, with distinct cytoarchitecture. The bed nuclei of the stria terminalis may be considered as a ventral component of the septal region, but are also properly regarded on the basis of developmental and immunohistochemical criteria as an extension of the amygdala (see section above on the amygdala and extended amygdala). Most of the septal nuclei develop morphofunctional links with the hippocampus and amygdala.

In placental mammals, the septal nuclei have distinctive immunohistochemical signatures for the calcium-binding proteins parvalbumin, calbindin and calretinin. In rodents (Paxinos *et al.*, 1999a), the medial group of septal nuclei exhibit large numbers of parvalbumin-immunoreactive neurons (in the medial septal nucleus, and nuclei of the vertical and horizontal limbs of the diagonal band), but little neuropil immunoreactivity for parvalbumin, whereas some components of the lateral group (intermediate and ventral parts of the lateral septal nucleus) have abundant calbindin-immunoreactive neurons and moderate to strong neuropil immunoreactivity. The posterior-group nuclei (i.e. triangular septal nucleus and bed nucleus of the anterior commissure) both have intensely calretinin-immunoreactive neurons (Paxinos *et al.*, 1999a; Risold, 2004). By contrast, the demarcation of septal groups by calcium-binding protein immunoreactivity is much less clear in a representative Australian polyprotodont (*Sminthopsis macroura*). The medial septal nuclei show less calbindin immunoreactivity in the neuropil compared to the lateral nuclei, but calbindin-immunoreactive somata are quite rare in the lateral septal nuclei (Figure 7.3).

There are several important fibre tracts associated with the septal region. These include two commissures: (i) the ventral hippocampal commissure, which traverses the triangular nucleus and (ii) the anterior commissure, which passes through the bed nuclei of the stria terminalis. Projections to, and from, the septal region run in four pathways: (i) the fornix, which connects the lateral and medial septum with the hippocampus, (ii) the stria terminalis, which connects the ventral group (bed nuclei of the stria terminalis) with the rest of the amygdala, (iii) the stria medullaris thalami, which contains projections of the posterior group nuclei and (iv) the diffuse medial forebrain bundle, which carries afferents and efferents for all the septal nuclei and runs longitudinally through the lateral hypothalamus.

Development of the septal nuclei in Australian marsupials

In all mammals, the medial and lateral septal nuclei develop from the septal ridge, which forms the ventromedial wall of the rostral telencephalic vesicle. At early developmental stages (E12 in rodents and around the time of birth in marsupials), the septal anlage is situated rostral to the developing striatum and pallidum and ventral to the cortical germinal region. Fusion of the left and right septal primordia into a single midline structure occurs prenatally in placental mammals (about E15 in the laboratory rat) and is associated with the development of the corpus callosum. By contrast, the process of septal fusion occurs late and incompletely in marsupials, e.g. during the third or fourth postnatal week in the tammar wallaby. The fused portion of the septal nuclei extends for about 60% of the rostrocaudal extent of the region in *Sminthopsis macroura* and only about 25% in *Macropus eugenii*, compared to approximately 70 to 75% of the rostrocaudal extent of the septal region for the laboratory rat.

Almost all septal neurons in placental mammals are generated prenatally, even in altricial placental mammals like rodents, but neurogenesis of this region is almost exclusively postnatal in marsupials. In rodents, the neurons of the medial septal nuclei are generated before those of the lateral septal complex. In the brush-tailed possum, there is also a mediolateral neurogenetic gradient, best seen in the lateral septal area (Sanderson and Wilson, 1997). The medial septal area is labelled after injection of ^{3}H-thymidine on P5 to P18, the septal triangular and intermediate lateral septal area after injection on P5 to P12, and the lateral ventral and lateral dorsal septal areas are labelled after injection on P5 to P21. The nuclei of the vertical and horizontal limb of the diagonal band are generated after P12 in the brush-tailed possum.

The stria terminalis and stria medullaris thalami develop very early in placental mammals, about E13, i.e. around the transition from embryonic to fetal stages. The stria medullaris thalami is also present relatively early in wallaby brain development (i.e. by birth) and shows very strong levels of GAP-43 immunoreactivity at that time (Hassiotis *et al.*, 2002). Similarly, a rather diffuse, but nevertheless definable, medial forebrain bundle can be discerned at birth. By contrast, as noted above when amygdala development was being discussed, the stria terminalis develops during the second postnatal week in the tammar (Hassiotis *et al.*, 2002). The first descending fibres of the fornix appear at about E16 to E17 in the laboratory rat, but not until the second postnatal week in the tammar wallaby (Hassiotis *et al.*, 2002).

Subfornical organ

The subfornical organ is one of the circumventricular organs, a group of seven structures located in the walls of the lateral, third and fourth ventricles. The other members of this highly vascularised group of structures are the vascular organ of the lamina terminalis, pineal gland, subcommissural organ, median eminence/neurohypophyseal complex, area postrema and choroid plexus. Where appropriate, these structures have been discussed with the region wherein they are found (see Chapter 6 for diencephalic elements such as the pineal gland, subcommissural organ, neurohypophysis and median eminence).

The subfornical organ is situated in the rostral dorsal wall of the third ventricle immediately ventral to the hippocampal commissure and merges ventrally with the median preoptic nucleus of the hypothalamus (Figures 7.3, 7.4). In marsupials the subfornical organ has a distinctive position. Whereas the subfornical organ of placental mammals is attached to the posteroventral surface of the hippocampal commissure around the midline, in marsupials the subfornical organ is located dorsal and anterior to the anterior commissure (Johnson, 1977).

As for other circumventricular organs, the subfornical organ is a site where blood-borne humoral agents (like angiotensin II, atrial natriuretic peptide, calcitonin, somatostatin and vasopressin) can exert central actions. The renin-angiotensin system is concerned with regulation of blood pressure and fluid balance in all therians. In placental mammals, most angiotensin receptors are found in liver, adrenal glands, blood vessels and the brain (subfornical organ). Although the angiotensin II (AII) receptors have similar functional domains in both marsupial and placental mammals (Sernia *et al.*, 1990) and the effects of AII on mean arterial pressure are similar to those seen in placental mammals, AII receptors do not appear to be present in the brain of the brush-tailed possum (Sernia *et al.*, 1990). In rodents, AII acts on the subfornical organ to stimulate neural pathways serving drinking of water and blood pressure increases. There may be other sites at which AII exerts its effects on the central nervous system of diprotodont marsupials, but this remains to be investigated.

What remains unknown?

Almost nothing is known about the basal ganglia circuitry and function of septal nuclei and amygdala in any marsupial, let alone Australidelphians. Nevertheless, topography and chemoarchitecture of these deep telencephalic structures are similar in the Australian marsupials to rodents and primates, suggesting that structure and function of these regions are probably broadly similar in all therians.

Nevertheless, there may be evolutionarily significant differences between placentals and marsupials that remain to be explored.

For example, the thalamic nuclei involved in the output from the globus pallidus and substantia nigra (pars reticularis) have been noted to show species differences among placental mammals (e.g. rodents vs. primates; see review in Gerfen, 2004). In rodents, output from the internal globus pallidus is to the ventrolateral thalamic nucleus and output from the substantia nigra pars reticulata is primarily to the ventral medial thalamic nucleus and paralaminar medial dorsal thalamus, whereas in primates the principal output targets of the internal globus pallidus are the ventral lateral nucleus pars oralis and ventral anterior pars parvocellularis thalamic nuclei and the projection from the substantia nigra pars reticulata is to the ventral anterior and paralaminar medial dorsal thalamic nuclei. These disparities appear to reflect differences between rodents and primates in the degree of compartmentalisation of the frontal cortex. It remains to be seen whether the nuclei involved in these pathways in marsupials are the same as in some of the more closely studied placental mammals. Since diprotodont marsupials have undergone an expansion of the frontal cortex independently of that occurring in primates and carnivores, evolution of the basal ganglia and thalamic circuitry linked to the frontal cortex may have undergone expansion in novel ways to those seen in primates. Alternatively, perhaps the expansion of basal ganglia circuitry has occurred in a parallel fashion to that in primates. Investigation of these questions may throw light on the evolutionary constraints on forebrain circuitry.

Functional neuroanatomy of the amygdala in marsupials is also an unexplored area. The same broad subdivisions into temporal and extended amygdala are just as applicable to marsupial as placental brains and cyto-, enzyme- and immunohistochemical chemoarchitecture (AChE, NADPH diaphorase, parvalbumin, calbindin) are broadly similar in the constituent nuclei, but nothing is known about the relationship between the amygdala and cortical regions in marsupials.

The overall structure of the septal nuclei in therians is greatly influenced by the degree of fusion accompanying the formation of the corpus callosum (in placental mammals) or enlarged anterior commissure (in monotremes and marsupials), but these differences may be no more than cosmetic. It remains to be seen whether the functional organisation and connections of the medial septal nuclei in marsupials are influenced by the degree of midline fusion.

8

Cerebral cortex and claustrum/endopiriform complex

K. W. S. Ashwell

Summary

The cortical surfaces of the brains of Australian marsupials vary from lissencephalic (smooth) to gyrencephalic (folded), with major consistent sulci separating functionally significant regions. The mammalian cerebral cortex is divided into isocortex, allocortex and transition (periallocortical) regions. Even in small ('mouse-sized') marsupials, a diverse range of iso- and periallocortical regions can be identified on the basis of cyto- and chemoarchitectural differences. Many of these identified areas show similarities to placental cortex in both cellular and chemical markers, such as to strongly suggest homology to topographically similar regions in rodent, carnivore and primate species.

All marsupials share a common plan of sensory cortical organisation (Krubitzer, 1995), which features at least two somatosensory (S1 and S2, plus others), two visual (V1, V2L) and one auditory area (Au1). The two somatosensory areas, V1 and Au1 are present in living monotremes, suggesting that S1, S2, V1 and Au1 arose very early in mammalian evolution, prior to the divergence of prototherian and therian lineages, and have been retained in all mammals. With regard to the overlap between motor and somatosensory cortex, there are two basic types of organisation in the living marsupials: complete sensorimotor overlap of S1 and M in the Didelphimorphia, but partial overlap in more 'advanced' marsupials such as the Diprotodontia.

Australian marsupials exhibit isocortical sensory specialisations consistent with the functional demands of their lifestyle. Marsupial predators have enlarged visual and/ or auditory areas depending on their hunting strategies and prey, marsupials with enhanced manipulative ability have enlarged digital representations, diurnal wallabies exhibit enhanced visual cortical representations for central vision and the sensitive snout vibrissae and forepaw pads of phalangerids are served by barrel fields in S1.

The isocortex of newborn marsupials is extremely immature, with at best only a two-layered structure. Two generative zones are present in the developing mammalian cortex: the ventricular germinal and subventricular zones, but the cortical subventricular zone has been reported to be poorly developed in didelphid marsupials. Cortical plate neurons appear during the first week in diprotodont marsupials and the cortical layers are generated by an inside-to-outside neurogenetic gradient, finishing shortly after the end of the second postnatal month in diprotodonts. The cortical subplate, which is important in placental mammals and has been proposed as a site where thalamocortical axons wait and form synaptic connections before invading the cortical plate, is difficult to distinguish in developing marsupial cortex, and may not be a significant factor in development of marsupial cortex.

Introduction

As in placental mammals, the cerebral cortex (or pallium) of marsupials can be divided into isocortex and allocortex, separated by a transition zone (periallocortex). The isocortex or neocortex is fundamentally a six-layered structure (with some topographic variation) and is usually divided into frontal, parietal, temporal, insular and occipital regions. The allocortex includes the archicortex (hippocampal formation) and paleocortex (olfactory cortex) and has a highly variable laminar structure. An alternative terminology of the cerebral cortex is based around the concept of the pallium, with the telencephalon being covered by a medial pallium (archicortex), dorsal pallium (isocortex or neocortex) and lateral pallium (paleocortex). The claustrum and endopiriform nuclei are also considered to be derivatives of the lateral pallium. In large macropods with gyrencephalic brains, the isocortex occupies approximately 73% of total cortex, compared to 14% for archicortex and 9% for paleocortex (Johnson, 1977). Even in small dasyurids with lissencephalic brains, the isocortex takes up about 50 to 60% of cortical area, compared to 25 to 30% for archicortex and 15 to 20% for paleocortex (Ashwell *et al.*, 2008b).

The isocortex comprises two broad types of cortical areas: one group including primary sensory recipient areas and sensory association areas, the other including areas that are involved in high-level association, executive and analysis functions. The primary visual cortex (V1) and adjacent visual association cortex are located in the occipital lobe. The primary somatosensory cortex (S1) lies within the rostral part of the parietal lobe and is highly interconnected with somatosensory association areas (S2, PV, R, C). The primary auditory cortex (Au1) in the temporal lobe receives ascending auditory projections and is interconnected with the auditory association cortex. The posterior part of the parietal lobe receives both somatosensory and visual information and is concerned with spatial localisation and attention. In humans, the prefrontal cortex is involved in higher conscious cortical function, but the extent of such activity in non-humans in general (and marsupials in particular), remains uncertain. The frontal lobe also includes the primary motor cortex (M), which may be in contact with supplementary and premotor cortex in some placental mammals, although the existence of these in marsupials is not established. The insular cortex usually has representations of visceral organ sensation.

Surface features of the marsupial cortex

As noted in Chapter 2, the cerebral hemispheres of Australian marsupials may range in appearance from smooth-surfaced (lissencephalic) to highly folded (gyrencephalic). The degree of convolution is related to brain and body size (i.e. more folds or gyri in larger brains) rather than any phylogenetic differences. Broadly speaking, grooves or sulci separate major functional areas, presumably arise from topographic variation in cortical growth during development and, for the neuroscientist, provide convenient landmarks for extrapolating functional cortical organisation from one marsupial species to another (Megirian *et al.*, 1972; Haight and Murray, 1981). The pattern of sulci is broadly similar across a wide range of large-brained marsupials, both poly- and diprotodont, and can even be applied to extinct American marsupials (e.g. *Thylacosmilus atrox*, Quiroga and Dozo, 1988). Johnson (1977) was impressed by the consistency of sulcal patterns in marsupials, but Haight and Neylon (1978c) have pointed out the significant intraspecific sulcal variability of the cerebral cortex of brush-tailed possums. Haight and Neylon felt that this sulcal

variability in the brush-tailed possum was even greater than that in both placental mammals and *Didelphis virginiana*, but Karlen and Krubitzer (2006a) recently showed considerable phenotypic diversity in cortical field size within captive short-tailed opossum populations.

Several systems of nomenclature for cortical sulci have been applied to marsupial cortex (Figure 8.1). The earliest of these is based on Greek letters, as advocated by Ziehen (1897) and Johnson (1977). The most consistent groove is the rhinal fissure, which separates the isocortical regions (dorsally and medially) from piriform and entorhinal cortex (ventrally and laterally) at the rostral and caudal ends of the hemisphere, respectively. Within the isocortex, the next most consistently observed sulcus is the α/α' (orbital) sulcus at the rostral end of the hemisphere. This sulcus is found in the larger-bodied members of most marsupial groups (didelphids, dasyurids, diprotodonts), but not in the marsupial mole. Other important sulci include the β sulcus, which often delimits the caudal boundary of somatosensory cortex, and the μ sulcus, which marks the rostrolateral border of the primary visual cortex. Large-brained diprotodont marsupials also have a fairly consistent δ sulcus at the caudolateral border of the primary auditory cortex.

Other authors have used the Roman alphabet to denote additional sulci (Johnson, 1977; Haight and Neylon, 1978b, 1978c) or alternative systems of sulcal nomenclature. These include the interbrachial (ib), jugular (J) and labial (L) sulci. The labial is also (inappropriately) referred to as the pseudosylvian sulcus (pss) by some authors. Mayner (1985, 1989a, 1989b) adopted a system based on lobar position (abbreviated in the current work to a two-Roman-letter system) for nomenclature of sulci in the tammar wallaby. The frontal sulcus (fr) of her system corresponds to sulcus α, the parietal sulcus (pa) largely corresponds to sulci β and ε, prefrontal sulcus (pp) corresponds to the labial (L), and the temporal sulcus (te) corresponds to sulcus δ. In the tammar wallaby brain atlas section of this book, both Greek and Roman systems of sulcal nomenclature have been provided, consistent with the widest use in the literature.

General comments on the cortical organisation of Australian marsupials

Even in the smaller-brained polyprotodont Australian marsupials (Figure 8.2), the degree of iso- and periallocortical differentiation is comparable to less encephalised placentals of similar body size (Ashwell *et al.*, 2008b). Although the diversity of cortical areas in small marsupial carnivores is not quite as great as those reported in laboratory rodent species (Paxinos *et al.*, 1999a; Franklin and Paxinos, 2007;

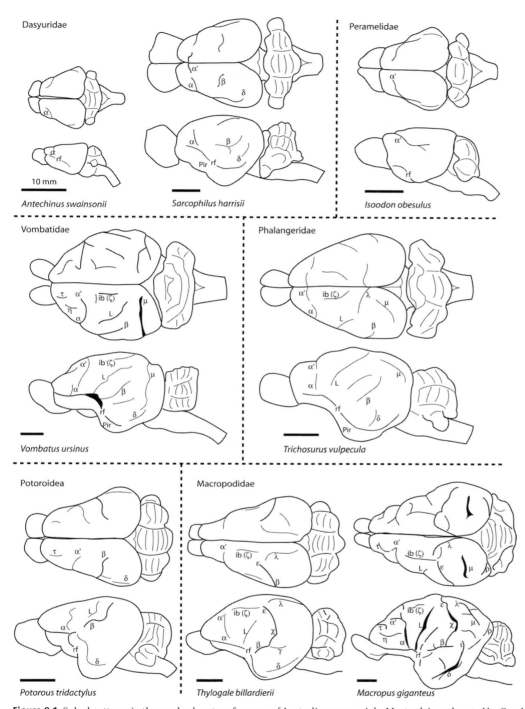

Figure 8.1 Sulcal patterns in the cerebral cortex of a range of Australian marsupials. Most sulci are denoted by Greek letters according to the scheme of Ziehen (1897), as adopted and extended by Johnson (1977), but several additional physiologically important sulci are denoted by Roman letters. ib – interbrachial sulcus; L – labial sulcus; Pir – piriform cortex; rf – rhinal fissure.

Paxinos and Watson, 2007), the diversity of identified cortical regions in these small marsupials calls into question the common use of the word 'primitive' to describe marsupials and their cortical organisation (Frost *et al.*, 2000).

Even in less encephalised marsupials like the didelphids (*Monodelphis domestica*), the number of discrete iso- and periallocortical fields reaches or exceeds 18, but is probably fewer than 25 (Wong and Kaas, 2009).

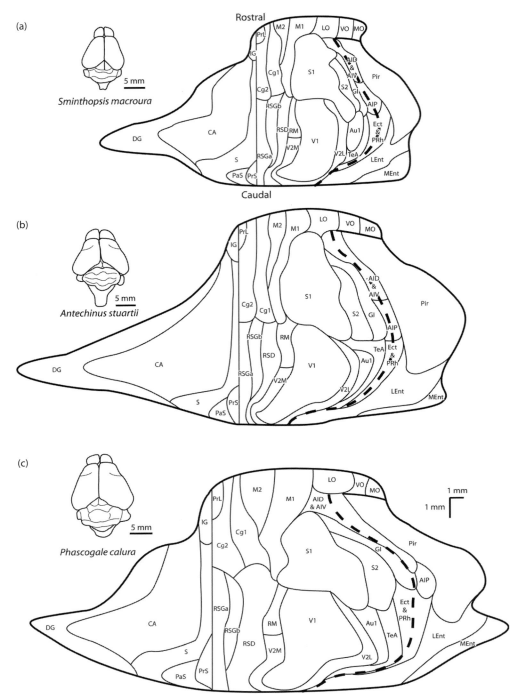

Figure 8.2 Maps of cortical topography in three small polyprotodont marsupial carnivores. The maps have been prepared by measuring the positions of cyto- and chemarchitectural borders around the layer 1Cx/2Cx boundary in isocortical and paleocortical regions or along the molecular/pyramidal or granule-cell boundary in archicortical regions and plotting against rostrocaudal level. The heavy dashed line indicates the position of the rhinal fissure. Note that the scales for circumferential and rostrocaudal position are different. Dimensions are for fixed, undehydrated tissue. These illustrations have been reproduced from Ashwell *et al.* (2008b) with kind permission of S Karger AG, Basel. AID – agranular insular cortex, dorsal; AIP – agranular insular cx, posterior; AIV – agranular insular cx, ventral; Au1 – primary auditory cx; CA – cornu Ammonis; Cg1 – cingulate cx, area 1; Cg2 – cingulate cx, area 2; DG – dentate gyrus; Ect – ectorhinal cx; GI – granular insular cx; IG – indusium griseum; LEnt – lateral entorhinal cx; LO – lateral orbital cx; M1 – primary motor cx; M2 – secondary motor cx; MEnt – medial entorhinal cx; MO – medial orbital cx; PaS – parasubiculum; Pir – piriform cx; PRh – perirhinal cx; PrL – prelimbic cx; PrS – presubiculum; RM – rostromedial visual area; RSD – retrosplenial dysgranular cx; RSGa – retrosplenial granular cx, a; RSGb – retrosplenial granular cx, b; S – subiculum; S1 – primary somatosensory cx; S2 – secondary somatosensory cx; TeA – temporal association cx; V1 – primary visual cx; V2L – secondary visual cx, lateral area; V2M – secondary visual cx, medial area; VO – ventral orbital cx.

Studies of large-brained diprotodont marsupials such as the tammar wallaby and brush-tailed possum suggest that the cerebral cortex of these mammals approaches the complexity of that in rodents. For example, differential topography of neuropil enzyme histochemical activity as well as differential distribution of neurofilament immunoreactive and NADPH-diaphorase-reactive neurons in the tammar wallaby cortex point to the possible existence of as many as eight association areas in addition to the primary motor and sensory cortical regions (Figures 8.3, 8.4, 8.5; Ashwell *et al.*, 2005a). Patterns of distribution of SMI-32 immunoreactive neurons in layer 3Cx of tammar primary somatosensory cortex also suggest the existence of multiple discrete groups of long-distance corticocortical neurons, which are associated with behaviourally significant regions (i.e. in the mystacial representation of the somatosensory cortex and limbic cortical areas).

Neuronal structure and the laminar organisation of isocortex

All mammals have a six-layered isocortex, but this lamination may be substantially modified in specialised cortical regions (Figures 8.3, 8.4). Cortical neurons encountered in marsupial cortex may be divided into three groups: pyramidal, non-pyramidal with spiny dendrites and non-pyramidal with smooth dendrites (Harman *et al.*, 1995). Pyramidal neurons have a pyramid-shaped cell body, a vertically oriented apical dendrite with a variable number of side dendritic branches and basal dendrites ramifying at about the level of the cell body. Non-pyramidal spiny neurons include granule and fusiform neurons. Non-pyramidal smooth neurons include aspinous stellate, bipolar and bitufted neurons.

Layer 1Cx (or the molecular layer) is a cell-sparse fibre layer immediately beneath the pia and consists of mainly incoming axons and the ascending processes of the neurons in deeper layers. Layer 2Cx is also known as the external granular layer, because of its constituent population of small granule cells. Layer 2Cx is often clearly visible as a thin layer of dark cells in Nissl-stained sections (Figure 8.3). Pyramidal neurons predominate in layers 3Cx and 5Cx, which are known as the external and internal pyramidal layers, respectively. Apical processes of pyramidal neurons in layers 3Cx and 5Cx extend through the supervening layers towards the pial surface. Pyramidal neurons of 3Cx engage in ipsilateral corticocortical or commissural connections, whereas layer 5Cx pyramidal neurons project to subcortical structures such as the striatum, the dorsal thalamus, the superior colliculus and the spinal cord. Layer 4Cx, between the two pyramidal layers, is the internal granular layer, contains abundant stellate cells and receives ascending sensory projections from the dorsal thalamus. The deepest multiform layer (6Cx) contains pyramidal neurons that project predominantly to the dorsal thalamus. In some cortical regions of diprotodonts, layer 6Cx may be further subdivided into 6Cxa and 6Cxb regions. Layers 2Cx and 3Cx are often called the supragranular layers, whereas layers 5Cx and 6Cx are collectively known as the infragranular layers.

Sensory cortical organisation of diprotodont Australian marsupials

Somatosensory cortex

Somatosensory cortex of marsupials shows the same broad pattern of somatotopic organisation seen in placental mammals, with ordered representation of the parts of the opposite body side across the cortical surface (Rowe, 1990; Catania *et al.*, 2000; Karlen and Krubitzer, 2007). The somatotopic map in the primary somatosensory cortex (S1) is inverted, in that hindlimbs, trunk, forelimbs and head are represented sequentially from medial to lateral on the cortical surface. Sensorimotor maps for the Virginia and short-tailed opossums, brush-tailed and striped possums, and wallabies all exhibit primary somatosensory cortex (S1) with large face and substantial distal limb representations, but with functionally significant differences between these species (Rowe, 1990; Karlen and Krubitzer, 2007). For example, whereas the maxillary (cheek) representation dominates in the Virginia opossum, both maxillary and mandibular (jaw) fields are large in the tammar wallaby. The relative size of cortex devoted to a given body part follows the principle of proper mass as defined by Jerison (1973): 'The mass of neural tissue controlling a particular function is appropriate to the amount of information processing involved in performing the function'. The area of S1 cortex devoted to a particular body part is more related to the proprioceptive feedback from, and fine control devoted to, the given part, rather than simple muscle mass. For example, although the tail and hindlimb are large structures in the wallaby body, they have quite small representations on wallaby cortex. On the other hand, the tail maps to a large proportion of S1 cortex in both the opossum and brush-tailed possum, consistent with the high peripheral innervation density and prehensile function of the tail in those species (Rowe, 1990).

In rodent and rabbit cortex, the mystacial (bewhiskered maxillary or cheek region) area of S1 has aggregations of small neurons into so-called barrels within layer 4Cx, which are arranged in rows in the cortex corresponding topographically with rows of vibrissal follicles in the whisker pad of the upper lip. These have not been found in Ameridelphians (Woolsey *et al.*, 1975; Wong and Kaas,

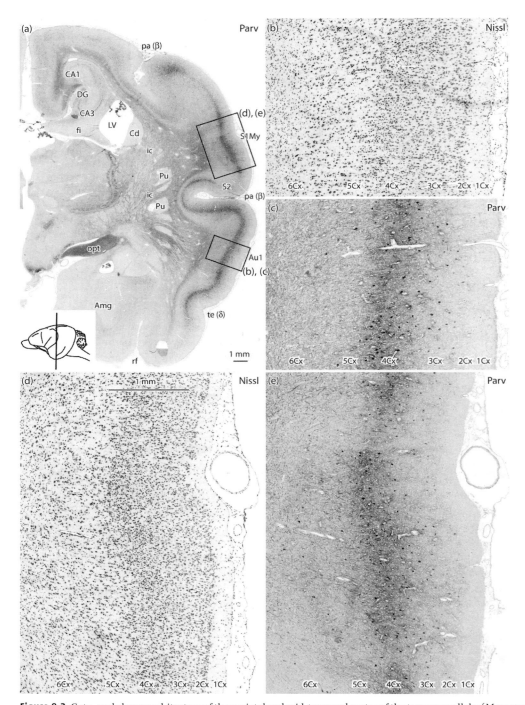

Figure 8.3 Cyto- and chemoarchitecture of the parietal and mid-temporal cortex of the tammar wallaby (*Macropus eugenii*). Insets show higher-power views of the mystacial part of the primary somatosensory cortex (S1My) and rostral primary auditory cortex (Au1) as indicated by the rectangles in the lower-power photomicrographs. 1Cx to 6Cx – layers 1 to 6 of cerebral cortex; Amg – amygdala; CA1 – field CA1 of hippocampus; CA3 – field CA3 of hippocampus; Cd – caudate nu.; DG – dentate gyrus; fi – fimbria of the hippocampus; ic – internal capsule; LV – lateral ventricle; opt – optic tract; Pu – putamen; rf – rhinal fissure; S1My – primary somatosensory cx, mystacial; S2 – secondary somatosensory cx.

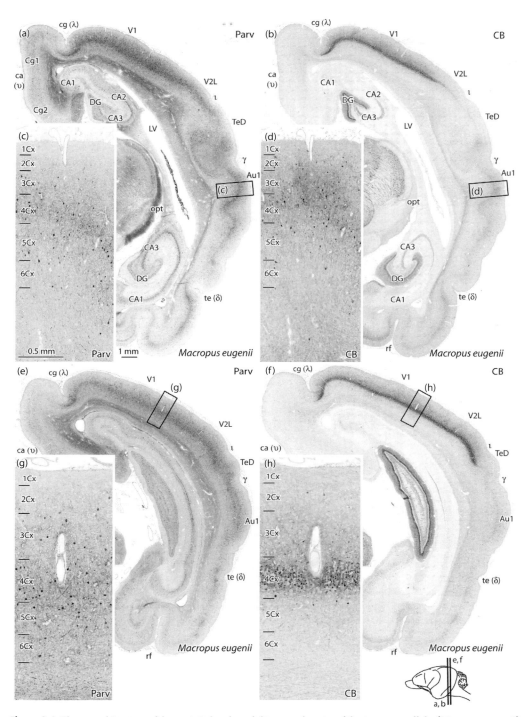

Figure 8.4 Chemoarchitecture of the occipital and caudal temporal cortex of the tammar wallaby (*Macropus eugenii*) with special emphasis on the primary visual (V1) and caudal auditory cortex (Au1). Insets show higher-power views of regions of caudal Au1 (c, d) and V1 (g, h) from positions indicated by the rectangles in the lower-power photomicrographs. Photomicrographs (a), (c), (e) and (g) show sections which have been immunoreacted for parvalbumin and photomicrographs (b), (d), (f) and (h) have been immunoreacted for calbindin. 1Cx to 6Cx – layers 1 to 6 of cerebral cortex; CA1 to CA3 – fields CA1 to CA3 of hippocampus; Cg1 – cingulate cx, area 1; Cg2 – cingulate cx, area 2; DG – dentate gyrus; LV – lateral ventricle; opt – optic tract; rf – rhinal fissure; TeD – dorsal temporal cx; V2L – secondary visual cx, lateral area.

(a)

(b)

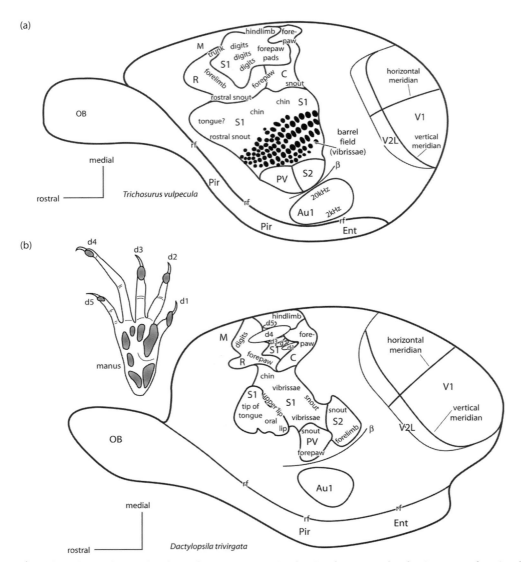

Figure 8.5 Flattened cortical surfaces of two possum species showing the topography of major sensory functional areas. Note the barrel field representation of the facial vibrissae in S1 of the brush-tailed possum (a) and the enlarged representation of digit 4 (d4) and the tongue tip in S1 of the striped possum (b). The palmar aspect of the manus (forepaw) of the striped possum is illustrated to show the enlargement of d4 on the manus of the striped possum, but note that the d4 somatotopic representation on S1 cortex is proportionately even larger. The cortical maps draw on data in Weller and Haight (1973), Gates and Aitkin (1982), Coleman *et al.* (1999), Huffman *et al.* (1999a), Elston and Manger (1999) and Karlen and Krubitzer (2007). Au1 – primary auditory cortex; C – caudal somatosensory cx; d1 to d5 – digits of manus; Ent – entorhinal cx; M – motor cx; OB – olfactory bulb; Pir – piriform cx; PV – parietal ventral somatosensory area; R – rostral somatosensory cx; rf – rhinal fissure; S1 – primary somatosensory cx; S2 – secondary somatosensory cx; V1 – primary visual cx; V2L – secondary visual cx lateral area.

2009), but Weller and colleagues (Weller, 1972, 1993; Weller and Haight, 1973) identified a barrel field in the S1 cortex of the brush-tailed possum (*Trichosurus vulpecula*; Figure 8.5a) and several petaurids (*Petaurus breviceps, Petaurus norfolcensis, Pseudocheirus peregrinus, Petauroides volans*). There is one significant difference between possum and mouse barrels: marsupial barrels have a solid centre

of densely packed neurons surrounded by a rim of loosely packed cells, whereas mouse barrels have a loosely packed centre or core surrounded by a ring of densely packed neurons (Weller and Haight, 1973). Some barrels also serve the forepaw digital pads (see Chapter 10 for more details).

Although barrel fields have been identified in phalangerids, this feature is not present in the mature cortex of other diprotodonts. Unlike the brush-tailed possum, the somatosensory cortex of adult tammar wallaby does not show any cellular aggregations in Nissl-stained sections that can be interpreted as a barrel field (Figure 8.3d; Ashwell *et al.*, 2005a). Nevertheless, barrel-like patches of elevated succinic dehydrogenase activity in clustered thalamocortical axon terminals have been identified in the developing tammar wallaby cortex (at P90; Waite *et al.*, 1991, 1998, 2006), although these are very indistinct in the adult. Another oxidative enzyme, cytochrome oxidase, has been used to identify barrel fields in coronal sections through the tammar wallaby S1 cortex up to the age of P138, but the patches are not striking, even in developing cortex (Mark *et al.*, 2002) and discrete regions of elevated cytochrome oxidase activity are not seen in adult somatosensory cortex (Ashwell *et al.*, 2005a). Loose aggregations of small neurons can also be found in layer 4Cx of the mystacial S1 cortex of some small Australian polyprotodont carnivores, but these do not constitute true barrels (Ashwell *et al.*, 2008).

Although the face representation in S1 is a major site of cortical specialisation in Australian marsupials, other parts of S1 exhibit specialisations in diprotodonts with manipulative and lingual exploratory capabilities. In the striped possum (*Dactylopsila trivirgata*), for example, the entire forepaw takes up 29% of the total area of S1 (Figure 8.5b). Both the behaviourally important digit 4, which is 40% larger than other digits and is used for extracting insects from the bark of trees, and the tongue tip, used for exploring cavities in the bark, have a large proportion of S1 devoted to them (10% and 16%, respectively; Huffman *et al.*, 1999a; Figure 8.5b). These compare to 24% of S1 devoted to the vibrissae of the snout, chin, cheek and lip in this species.

The S1 cortex projects to the other somatosensory cortical regions (S2 and PV, see below) as well as dedicated motor cortex, posterior parietal cortex and perirhinal cortex (Elston and Manger, 1999).

A second somatosensory area (S2) with another (but upright) body representation is found caudal and lateral to S1 in a variety of diprotodont marsupials (Rowe 1990; Coleman *et al.*, 1999; Huffman *et al.*, 1999a). The representations of the face and head in S2 are adjacent to the face representation in the neighbouring S1, and the fore- and hindlimbs are represented progressively caudally within S2. Receptive fields for neurons in S2 are larger than those in S1 and often respond to stimulation on both sides of the body. In most species, S2 can be subdivided into a rostral PV region and a caudal S2 proper (Elston and Manger, 1999; Huffman *et al.*, 1999a; Karlen and Krubitzer, 2007). In the striped possum (Huffman *et al.*, 1999a), PV contains a representation of the head and forelimb surface with the head upright and adjacent to the vibrissal representation in S1.

S2 appears to receive projections directly from the thalamus, indicating the presence of parallel processing in marsupial somatosensory pathways (Coleman *et al.*, 1999), similar to that seen in carnivores, lagomorphs, prosimians and New World monkeys. In the tammar wallaby, S2 occupies the depths of the pa (β) sulcus and extends dorsally towards pp. In AChE-reacted material, S2 is characterised by greater activity in layers 2Cx to 4Cx compared to the adjacent S1 (Ashwell *et al.*, 2005a). Conversely, in corresponding cytochrome oxidase enzyme-reacted sections, activity in the supragranular layers of S2 is slightly lower than in adjacent S1.

Rostral (R) and caudal (C) somatosensory areas have also been reported adjacent to the forelimb representation in S1 of the striped possum (Huffman *et al.*, 1999a). These cortical areas contain neurons that respond to stimulation of deep receptors or high-threshold cutaneous receptors. It has been hypothesised that these areas in marsupials correspond functionally to primate cortical areas 3a and 5, respectively (see discussion in Karlen and Krubitzer, 2007).

We noted in Chapter 2 that the koala has a greatly reduced brain size, with the brain occupying less than two-thirds of the endocranial capacity. This appears to be due in large part to reductions in the extent and thickness of the parieto-frontal cortex, including the somatosensory fields. Whereas other diprotodonts like the brush-tailed possum and wombat have somatosensory cortex 1.8 and 2.1 mm thick, respectively; this region in the koala is only 1.3 mm thick (Haight and Nelson, 1987). The sulci marking the boundaries of motor-somaesthetic cortex are also poorly defined in the koala and the poor development of layer 6Cx in this region suggests greatly reduced output to subcortical forebrain structures (Haight and Nelson, 1987).

Visual cortex

Visual cortical areas in all mammals are visuotopically organised, i.e. parts of the visual world on the opposite side of the body are represented in a point-by-point topographic map across the cortical sheet. The primary visual cortex (V1) is usually identified as receiving a direct projection from the DLG of the thalamus and having high levels of cytochrome oxidase in 4Cx in many marsupials (Martinich *et al.*, 1990), whereas secondary visual areas receive projections from other thalamic nuclei and V1. Discrete visual cortical areas have been recognised in the tammar and parma wallabies (Garrett *et al.*, 1994; Figures 8.4, 8.7). These include the primary visual cortex (V1), defined by the tangential extent of the thalamocortical projection from the main visual thalamic nucleus, i.e. the dorsal lateral geniculate nucleus (Vidyasagar *et al.*, 1992) and a secondary (laterally placed) visual area (V2L), distinguished by reciprocal connections with LP thalamus and V1 (Sheng, 1989; Sheng *et al.*, 1990, 1991) and visual responses qualitatively different from those in V1 (Vidyasagar *et al.*, 1992). In the tammar, V2L appears to be relatively narrow in extent, but further electrophysiological studies would be needed to better define its ventral border. A third region of cortex, which has reciprocal connections with V1 and might also be involved in visual function, has been identified within the dorsal temporal cortex (TeD) (Sheng, 1989; Sheng *et al.*, 1991). This may be homologous to the CT region identified in polyprotodonts (see below).

In some diprotodonts (e.g. the brush-tailed possum – *Trichosurus vulpecula*) transneuronal transport of ^{3}H-fucose from the contralateral retina labels layer 4Cx as far medially as the calcarine sulcus (Sanderson *et al.*, 1980), suggesting an extensive medial visual area. On the other hand, transported label from the contralateral eye does not appear to be so medially extensive in the quokka (*Setonix brachyuris*, Tyler *et al.*, 1998), although another study of the quokka using horseradish-peroxidase labelling of thalamocortical projections from the dorsal lateral geniculate nucleus found that terminals distributed onto the medial cortex at the most caudal levels (Harman *et al.*, 1995). In CO-reacted sections through the tammar wallaby cortex, Ashwell and colleagues (Ashwell *et al.*, 2005a) also found a band of strong reactivity extending around the medial edge of the cortex into retrosplenial regions, suggesting a more medially extensive visual region than hitherto recognised.

Functional properties of neurons in the visual cortex of Australian marsupials are broadly similar to those in placental mammals. In particular, response properties of neurons in V1 of the diurnal tammar wallaby are very similar to those in cats and monkeys in that they are highly orientation selective, directional and tuned to particular spatial frequencies (Ibbotson and Mark, 2003; Ibbotson *et al.*, 2005). Like diurnal placental mammals, the behaviourally important central part of the visual field has an expanded representation in V1.

Auditory cortex

Primary auditory cortex (Au1) in diprotodont marsupials is located in the temporal cortex (Figures 8.3. 8.4, 8.5). In the brush-tailed possum, Au1 is tonotopically organised with high frequencies represented dorsally and medially, and low frequencies represented ventrally and laterally (Gates and Aitkin, 1982; see Figure 8.5a). The primary auditory cortex also receives tonotopically organised projections, mainly from the medial geniculate nucleus, but also from the suprageniculate nucleus of the thalamus (Aitkin and Gates, 1983; Neylon and Haight, 1983).

Ashwell and colleagues (2005a) identified Au1 in the cortex of the tammar wallaby on a low gyrus between the pa and te sulci. Au1 is cytoarchitecturally similar to S1, with a dense layer 4Cx in Nissl-stained tissue. Like other primary and association sensory areas, Au1 of the tammar wallaby has a bilaminar distribution of SMI-32 immunoreactive neurons (i.e. layers 3Cx and 5Cx; Ashwell *et al.*, 2005a), suggesting that it engages in abundant corticocortical connections.

Sensory cortical organisation of small Australian polyprotodonts

With respect to cortical organisation, Australian polyprotodonts can be considered to fall into two broad groups: small insectivores such as dunnarts, antechinuses and phascogales, and larger carnivores such as quolls and the Tasmanian devil. Naturally, electrophysiology is much more difficult in the small cortical sheets of dunnarts and antechinues than in quolls and devils.

Analysis of cortical cyto- and chemoarchitecture of small Australian polyprotodont marsupials (*Sminthopsis macroura*, *Antechinus stuartii*, *Phascogale calura*) shows that primary somatosensory (S1) and visual (V1) areas dominate the isocortex (Ashwell *et al.*, 2008b; Figure 8.2). Primary sensory regions S1 and V1 occupy similar proportions of the total pallial area in all three species (S1: *S. macroura* – 8.4%, *A. stuartii* – 8.7%, *P. calura* – 7.8%; V1: *S. macroura* – 8.7%, *A. stuartii* – 7.8%, *P. calura* – 8.7%), but primary auditory cortex is relatively small, ranging from 1.0% of pallial area (*A. stuartii*) through 1.4% (*S. macroura*) to 1.5% (*P. calura*) (Figure 8.2). The similar proportional sizes of V1 and S1 in the small Australian marsupial carnivores probably reflects the comparable behavioural importance of vision and somatosensation in these marsupials, whereas the relatively small size of V1 compared to S1 in the laboratory mouse (6.9% as compared to 17.4% of pallial surface; Franklin and Paxinos, 2007; Ashwell *et al.*, 2008b) is consistent with greater behavioural dependence of rodents on whiskers and trigeminal somatosensation to explore their environment.

How do the cortical areas of Australian polyprotodonts compare to American didelphids and does this reflect behavioural ecology? Primary somatosensory, visual and auditory cortex occupy 18.9%, 9.1% and 6.9% of the isocortex of the short-tailed opossum (*Monodelphis domestica*), respectively (Karlen and Krubitzer, 2006a), although other studies have emphasised the large size of V1 (Catania *et al.*, 2000; Wong and Kaas, 2009), suggesting that *M. domestica* may be more behaviourally dependent on audition than the small dasyurids (Ashwell *et al.*, 2008b). These differences probably reflect the different environments and diets of the short-tailed opossum as compared to small Australian marsupial carnivores. *Monodelphis domestica* is omnivorous and originates from South American tropical rainforests, whereas Australian dasyurid marsupials are carnivorous and are found in arid zones or open forests of Australia. Larger cortical auditory areas and a reliance on hearing are likely to be essential for survival for *M. domestica*, where concealment by predators (and of *M. domestica*'s prey) is easier in the dense, moist forest (Ashwell *et al.*, 2008b). The visual system is more important for Australian marsupials in their dry, open environment; small Australian marsupial carnivores rapidly learn visual tasks (Bonney and Wynne, 2002, 2004); and the importance of the visual system for detecting predator and prey has been demonstrated behaviourally for *Sminthopsis macroura* (Lippolis *et al.*, 2005).

The question of how many secondary visual areas are present in polyprotodont visual cortex remains open. It is universally accepted that a secondary visual area (V2 or V2L) is located lateral to the primary visual cortex in many marsupial species (Bravo *et al.*, 1990; Kahn *et al.*, 2000; Martinich *et al.*, 2000; Karlen and Krubitzer, 2006a; Ashwell *et al.*, 2008b; Wong and Kaas, 2009), but the topography of the cortical surface medial to V1 makes it technically difficult to ascertain the presence or number of medial secondary visual areas in small polyprotodonts. Transportation of label to the occipital cortex following ³H-proline injection in the contralateral eye of the fat-tailed dunnart (*Sminthopsis crassicaudata*) has identified medially extensive visual areas (Tyler *et al.*, 1998), and at least one medial visual area (possibly one or more limbic visual area(s)) has been delineated in cyto- and chemoarchitectonic studies of didelphids (*Monodelphis domestica*; Wong and Kaas, 2009) and Australian polyprotodonts (Ashwell *et al.*, 2008b). Studies in rodents have identified as many as five medial visual association regions using immunoreactivity to neurofilament protein (van der Gucht *et al.*, 2007), but Ashwell and colleagues (Ashwell *et al.*, 2008b) identified only two chemoarchitecturally distinct areas in small dasyurids: the rostromedial visual (RM) and medial secondary visual (V2M) areas. The RM area of Australian polyprotodonts may correspond to the limbic visual area PS of *Monodelphis domestica* (Wong and Kaas, 2009). Clearly, the question of whether visual association areas exist medial to the primary visual cortex in Australian marsupials (and how these compare functionally to those in rodents) needs further attention.

Studies in Ameridelphians have also reported the existence of a caudal temporal area (CT) in the caudal occipital cortex (Kahn *et al.*, 2000; Karlen and Krubitzer, 2006a, 2009; Wong and Kaas, 2009). This visual area receives projections from both V1 and the thalamus, indicating a probable role in processing visual information beyond the primary visual

cortex (Kahn *et al.*, 2000). It is highly active metabolically as shown by cytochrome oxidase activity and receives densely myelinated afferent axons. This visual region has not been specifically studied in diprotodont marsupials, but may correspond to the medial temporal region of the brush-tailed possum (Crewther *et al.*, 1984; Kahn *et al.*, 2000) or the TeD region of the tammar wallaby (see Figure 8.5). Kahn and colleagues have argued that V1, V2 and CT were present in the brains of ancestral marsupials (Kahn *et al.*, 2000), but given that this region has not been identified in large Australian polyprotodonts (Rosa *et al.*, 1999; see below) more extensive studies of the caudal occipital cortex in a range of Australidelphians are probably needed to establish if this is likely.

Sensory cortical organisation of large Australian polyprotodonts

In the larger-bodied northern quoll (*Dasyurus hallucatus*) a more diverse set of functional somatosensory field can be identified in the cortex (Figure 8.6) than in small dasyurids. Huffman and colleagues (Huffman *et al.*, 1999a) identified S1, S2, PV, R and C somatosensory fields based on somatotopic orientation and receptive field properties. As for diprotodonts, neurons in S1 have small receptive fields and the body representation is inverted, i.e. the caudal-to-rostral sequence of contralateral body parts (from tail to head) is represented in a medial-to-lateral sequence on the cortex. Body parts with magnified representations on S1 of the northern quoll are the snout, chin, cheek and lip vibrissae (35 to 40% of the area of S1) and the forepaw (21% of S1), consistent with the importance of these body parts in predatory behaviour. Rostral (R) and caudal (C) somatosensory areas flank the S1 representation, such that the forelimb digit representation of S1 is contiguous with the forelimb digit areas of R and the hindlimb representation of S1 is contiguous with the hindlimb representation of C. Huffman and colleagues also identified small PV and S2 somatosensory areas at the lateral edge of S1 with upright representations of head, face and limbs.

The northern quoll has at least two distinct visual cortical areas (V1 and V2L). As in all mammals, both areas are visuotopically organised and are in contact along the line of the vertical meridian of the visual field (Rosa *et al.*, 1999; Karlen and Krubitzer, 2007). The horizontal meridian of the visual field passes from V1 to V2 along a line running from caudomedial to rostrolateral.

Primary auditory cortex of the northern quoll is tonotopically organised (Aitkin *et al.*, 1986b), as in placental mammals, with high frequencies (over 32 kHz) represented at the

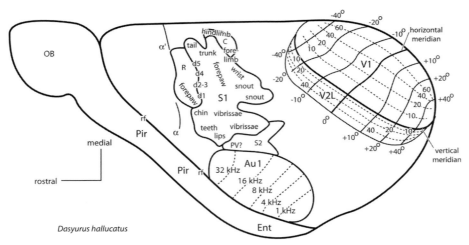

Figure 8.6 Flattened cerebral cortex of the northern quoll illustrating the positions of the major sensory cortical areas and somatotopic, tonotopic and visuotopic organisation of S1, Au1 and V1, respectively. Note the large area of Au1 devoted to high and low frequencies and the enlarged representation on V1 and V2L of visual space up to 20° above and below the horizon. Abbreviations as for Figure 8.5. The figure was drawn using data in Aitkin *et al.* (1986b), Huffman *et al.* (1999a), Rosa *et al.* (1999) and Karlen and Krubitzer (2007).

rostral end of Au1 and low frequencies (below 1 kHz) represented caudally. The areas for representation of frequencies in the range above 10 kHz and below 6 kHz are expanded at the rostral and caudal ends of Au1, respectively. These high and low frequencies are behaviourally significant because vocalisations and other sounds by predators and prey are concentrated in these ranges.

Motor cortex and sensorimotor overlap

The question of the extent of overlap between somatosensory and motor cortex has been a central issue in marsupial neurobiology. Studies in the early twentieth century showed that body movements could be induced by electrical stimulation at the rostromedial pole of the cortex in a range of Australian poly- and diprotodonts (Goldby, 1939; Abbie, 1940). Subsequently, Lende's observations in *Didelphis virginiana*, the short-beaked echidna and placental insectivores led him to propose that mammalian isocortical organisation evolved from a primitive condition in which the somatosensory and motor function were completely superimposed (Lende, 1963). This proposition was strengthened by observations that the projections from motor and somatosensory thalamic nuclei are largely convergent on the same area of cortex in the Virginia opossum (Killackey and Ebner, 1973).

It is certainly true that in Australian marsupials there is at least some overlap between motor and somatosensory representation (Adey and Kerr, 1954; Rees and Hore, 1970),

but this is insufficient to justify the concept of an amalgam for marsupials as a whole (Rowe, 1990). Several observations in Australian marsupials clearly indicate the presence of at least some disjunct motor and somatosensory cortex in diprotodont marsupials. Haight and Neylon (1978b, 1979) showed that the projections from somatosensory and motor thalamic nuclei are not congruent in the brush-tailed possum and that the extent of overlap varies, depending on the body part represented. Overlap is greatest for the hindlimb and trunk representation on medial parietal cortex, but declines progressively laterally as one advances into forelimb and facial representations. The area immediately caudal to the α′ sulcus appears to be exclusively motor, in that it receives a projection from the motor, but not somatosensory, thalamus (Haight and Neylon, 1979; Weller, 1993; Coleman *et al.*, 1999). These detailed findings from *Trichosurus* agree with the findings of a study of thalamocortical interconnections of sensorimotor cortex in the tammar wallaby (Mayner, 1985).

These physiological and hodological observations are consistent with analysis of the cyto- and chemoarchitecture of presumptive motor cortex in diprotodont marsupials. In the tammar wallaby, the presumptive motor area is characterised by a sparsely populated layer 5Cx, with large pyramidal neurons predominating in the upper (i.e. pialward) part (Ashwell *et al.*, 2005a). The lower part of layer 5Cx is relatively cell sparse. Layer 4Cx is present but thin, with densely packed small somata.

Studies of connectivity of the motor cortex in Australian marsupials are limited, but in the brush-tailed possum

(Joschko and Sanderson, 1987), the region receives inputs from neurons in layers 2Cx and 3Cx of the body and head regions of S1, both ipsi- and contralaterally, with evidence for reciprocal connections between the M and S1 cortex. Motor cortex in the brush-tailed possum also appears to engage in reciprocal connections with VL, VM and Po thalamic nuclei, as well as the dorsal claustrum. Other assorted inputs to M in this species are from the raphe nuclei of the midbrain, basal forebrain and hypothalamus. Output from M is to the caudate, the putamen and the globus pallidus.

In the case of small Australian polyprotodont marsupial carnivores, Ashwell *et al.* (2008b) argued for the existence of partially distinct motor and somatosensory areas on the basis of cytoarchitectural differences. They identified a region in the small marsupial carnivores with cytoarchitectural similarities to motor cortex in both diprotodont (Haight and Neylon, 1978b, 1979; Ashwell *et al.*, 2005a), polyprotodont (Haight and Neylon, 1981b) and small-bodied placental species (Franklin and Paxinos, 2007) and argued for the existence of a separate motor cortex in these small marsupial carnivores of comparable size to that seen in rodents of similar body size and neocortical specialisation. In larger Australian polyprotodonts, like the 750 g body-weight northern quoll (*Dasyurus hallucatus*), electrophysiological studies have identified a distinct primary motor cortex with light myelin staining which is unresponsive to sensory stimulation in any modality (Huffman *et al.*, 1999a).

While data support the existence of a complete sensorimotor overlap in Didelphimorphia (*Monodelphis domestica*, *Didelphis virginiana*), more advanced marsupials, such as the diprotodonts and Australian polyprotodonts, appear to have only a partial overlap (Karlen and Krubitzer, 2007). It should also be noted that sensorimotor cortex overlap is a prominent feature of cortical organisation in many placental mammals (e.g. edentates), whereas monotremes like the echidna and platypus have two motor representations, with one being co-extensive with S1 and the other occupying a more rostral position with a mirrored body representation (Lende, 1963), although contention still exists on this delineation (Karlen and Krubitzer, 2007).

In the broader context of cortical evolution, there are two competing hypotheses to explain the observations of sensorimotor cortex in extant mammals (Karlen and Krubitzer, 2007). Lende (1969) proposed that the original ancestral condition for mammals was one of complete sensorimotor overlap and that in some mammals this condition is retained to the present day (i.e. Didelphimorphia). Members of other mammalian lineages progressively developed separation of the somatosensory and motor cortex into either partially overlapping (diprotodonts, edentates) or completely disjunct areas (primates) (Karlen and

Krubitzer, 2007). The alternative hypothesis (Haight and Neylon, 1979; Karlen and Krubitzer, 2007) is that there are three types of sensorimotor representation (Karlen and Krubitzer, 2007): (i) a putative ancestral condition with a dual motor representation, somewhat like modern monotremes (see above), (ii) complete or partial overlap of motor and S1 cortex with the same body orientation – complete overlap in Didelphimorphia and insectivores, less overlap in diprotodonts and other less-specialised placentals and (iii) completely disjunct motor and S1 cortex with mirror-imaged body representations (e.g. primates). Whichever hypothesis provides the best explanation of the evolution of mammalian motor and sensory cortex, it is clear that modern Australian marsupials like the possums have a derived condition for their motor cortex, consistent with selection pressure towards improved cortical control of motor function, and indicative of evolution of somatomotor cortex quite independent from the placental mammals.

Organisation of orbital, transition and cingulate cortex in Australian marsupials

Studies of poly- and diprotodont Australian marsupials have identified several regions within the anteromedial cortex, which show some chemoarchitectural similarities to orbital and prelimbic regions in rodent or primate cortex (Ashwell *et al.*, 2005a, 2008b; Figures 8.2, 8.7). In the tammar wallaby, PrL and MO cortical areas show strong AChE activity deep in layer 3Cx and upper layer 5Cx similar to PrL and MO in rodents (Ashwell *et al.*, 2005a; Paxinos and Watson, 2007). The MO identified in tammars may be homologous to area 14c of the macaque cortex, which also shows a dense AChE-reactive band and is in a similar topographical position (Carmichael and Price, 1994). The less-reactive areas around the ventral circumference of the tammar wallaby cortex (i.e. lateral parts of VO and LO) may be homologous to area 13 of the macaque orbital cortex, which also exhibits lower levels of AChE activity than more medial orbital regions (Carmichael and Price, 1994). In polyprotodont Australian marsupials, orbital cortex can be divided into three zones (MO, VO, LO), with MO and VO distinguished by dense layer 2Cx and sparser layer 3Cx in Nissl-stained section, whereas LO has looser neuronal packing in layer 2Cx. The medial rim of the rostral cortex in Australian marsupial carnivores is divisible into a ventral PrL and the rostral extremity of Cg (see below). These regions are easily distinguished by the relative levels of AChE and parvalbumin immunoreactivity, with patches of stronger AChE in MO and PrL and different patterns of parvalbumin immunoreactivity in the three orbital areas

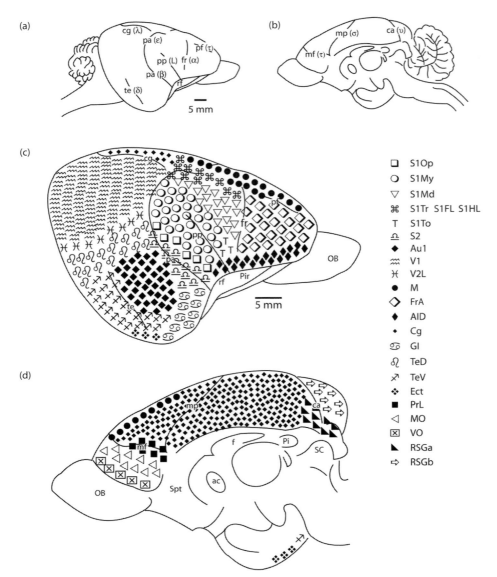

Figure 8.7 Cortical topography of the tammar wallaby, showing major sensory motor, associational and periallocortical areas. Illustrations (a) and (b) show sulci, as seen on the lateral and medial aspects of the right hemisphere, using the nomenclature of Mayner (1985, 1989a) and Ziehen (1897). Illustrations (c) and (d) show cortical regions mapped onto lateral and medial views of the right hemisphere, respectively (adapted from Ashwell *et al.*, 2005a, with kind permission of S Karger AG, Basel). Note that cortical areas deep within the rhinal fissure (e.g. AIV) will not appear on this surface map. ac – anterior commissure; AID – agranular insular cortex, dorsal; Au1 – primary auditory cx; Cg – cingulate cx; Ect – ectorhinal cx; FrA – frontal association cx; GI – granular insular cx; M – motor cx; MO – medial orbital cx; OB – olfactory bulb; Pi – pineal gland; Pir – piriform cx; PrL – prelimbic cx; rf – rhinal fissure; S1 – primary somatosensory cx; S1FL – S1, forelimb; S1HL – S1, hindlimb; S1Md – S1, mandible; S1My – S1, mystacial; S1Op – S1, ophthalmic; S1To – S1, tongue; S1Tr – S1, trunk; S2 – secondary somatosensory cx; SC – superior colliculus; Spt – septum; TeD – dorsal temporal cx; TeV – ventral temporal cx; V1 – primary visual cx; V2L – secondary visual cx, lateral area; VO – ventral orbital cx.

(Ashwell *et al.*, 2008b). Although these cyto- and chemoarchitectural studies point to diversity in frontal cortical organisation in Australian marsupials, there is a clear need for fine-scale physiological or connectional studies of these areas to accurately establish functional or connectional similarities with cortical areas in placental mammals.

Studies of transition cortex in poly- and diprotodont Australian marsupials have also found very similar cyto- and chemoarchitecture of these regions to those in placental mammals (Ashwell *et al.*, 2005a, 2008b; Figures 8.2, 8.7). At rostral levels, the transition to paleocortex around the rhinal fissure includes three cytoarchitecturally distinct regions. From dorsal to ventral, these are a granular insular region (GI), with layer 4Cx present, lying in line with the dorsal claustrum, followed by two cytoarchitecturally distinct agranular insular regions, AID and AIV rostrally, and AIP caudally. Further caudally, transitional cortex around the rhinal fissure includes an agranular dorsal ectorhinal region with clear layers 2Cx, 5Cx and 6Cx, and a perirhinal region with a progressively simpler lamination as the entorhinal cortex is approached.

In poly- and diprotodont Australian marsupials (Ashwell *et al.*, 2008b; atlas plates D5 to D11, W13 to W31), the cingulate cortex is differentiated into two distinct regions: the dorsal Cg1 and the ventral Cg2. In the absence of the corpus callosum, Cg2 merges ventrally with the archicortex (the dorsal tenia tecta, rostrally, and the indusium griseum and cornu Ammonis, caudally). Multiple cingulate regions may be a common feature of marsupial brains, since Wong and Kaas identified three distinct cingulate areas in an Ameridelphian, the short-tailed opossum (Wong and Kaas, 2009). These are dorsal, ventral and rostral cingulate areas, all of which, like homologous areas in diprotodonts, lack a distinct granular layer 4Cx. These regions can be distinguished by differences in patterns of parvalbumin and vesicle glutamate transporter immunoreactivity (Wong and Kaas, 2009).

Further caudally, the cingulate cortex is replaced by a series of 'retrosplenial' regions, with similar cyto- and chemoarchitecture to placental mammals. Of course, the term 'retrosplenial' is a misnomer, given the absence of the corpus callosum (and its splenium) in marsupials, but the term has been retained because of the basic similarities with this region in placental mammals. Even in small Australian polyprotodonts there is a dysgranular dorsal retrosplenial region (RSD, see atlas plates D12 to D17) with an irregular or poorly developed layer 4Cx, flanked ventromedially by a retrosplenial region with a more prominent layer 4Cx (retrosplenial granular – RSG, D13 to D18), which in turn appears to have at least two cytoarchitectonically distinct subregions. Homologous regions may also be seen in the retrosplenial region of Ameridelphians (RSd and RSv of *Monodelphis domestica*, Wong and Kaas, 2009). In diprotodonts like the tammar wallaby, the granular layer (layer 4Cx) appears to be better developed throughout the 'retrosplenial' region, justifying the use of the term RSG in various subdivisions throughout the region (atlas plates W32 to W38).

Chemoarchitecture of iso- and periallocortical regions in Australian marsupials

Topographic and laminar variation in neurofilament protein immunoreactivity

Neurofilament proteins are involved in the stabilisation of the axonal cytoskeleton and their presence in high concentrations has been correlated with large axonal size and high conduction velocity (Morris and Lasek, 1982; Hoffman *et al.*, 1987; Lawson and Waddell, 1991; Nixon *et al.*, 1994; Xu *et al.*, 1996). The proportion of neurofilament protein immunoreactive neurons is low in cortical regions involved in making short corticocortical links, and higher in regions with corpus callosum connections and long association pathways (Campbell and Morrison, 1989; Hof *et al.*, 1995, 1996; Nimchinsky *et al.*, 1996).

Immunoreactivity for non-phosphorylated epitopes of the heavy- and medium-molecular-weight subunits of the neurofilament protein (the SMI-32 antibody) identifies long-distance projection (predominantly layer 5Cx) and corticocortical (layer 2Cx/3Cx) neurons in placental cortex, with distinct topographic differences. Analysis of the distribution of SMI-32 immunoreactive neurons in tammar cortex also reveals variation between cortical regions (Ashwell *et al.*, 2005a), but the pattern is simpler than in primate cortex.

Most of the wallaby cortex has dense bands of SMI-32 immunoreactive neurons within layer 5Cx, but layer 3Cx SMI-32 immunoreactive neurons are present only in selected sensory and association cortex regions, including S1, V1, presumptive Au1, anterodorsal Cg and RSG (Ashwell *et al.*, 2005a). This bilaminar distribution in sensory and limbic regions is consistent with the presence of well-developed corticocortical and commissural connections between sensory and limbic areas of the isocortex and is in a similar laminar pattern to that in placental mammals. On the other hand, a bilaminar distribution of SMI-32 immunoreactive somata in motor cortex and dorsolateral premotor cortex is common in placental mammals (Preuss *et al.*, 1997; Gabernet *et al.*, 1999; Geyer *et al.*, 2000), but is not seen in tammar cortex (Ashwell *et al.*, 2005a). SMI-32-immunoreactive neurons are rare in insular areas of the wallaby cortex (AID, AIV, GI) and in the 'retrosplenial' regions (RSG1) and, in the case of AID and GI, show no particular laminar distribution (Ashwell *et al.*, 2005a). This may indicate that these regions do not engage in significant long-range association or projection connections in this species, but obviously this needs to be confirmed by careful connectional studies. Nevertheless, a bilaminar distribution of SMI-32-immunoreactive neurons in anterodorsal Cg of the tammar is suggestive of a separate limbic functional area in that region, which may be involved in long-distance association connections with other parts of the cortex (Ashwell *et al.*, 2005a).

Topographic and laminar variation in calcium-binding protein immunoreactivity

The morphology and distribution of parvalbumin-immunoreactive local-circuit neurons appears to be broadly similar in all mammals (Hof *et al.*, 1999; Hassiotis *et al.*, 2005). Multipolar parvalbumin immunoreactive neurons are probably the most prevalent type of calcium-binding protein-immunoreactive neuron found in placental cortex. Neurons immunoreactive for parvalbumin make up 24% of neurons in layers 3Cxa and b in primate visual cortex and 9.7% of neurons in the same layer of rodents (Glezer *et al.*, 1993). In cetacean visual cortex (Glezer *et al.*, 1993) parvalbumin immunoreactive neurons comprise 14% of all neurons in layers 3Cxc/5Cx. In marsupial cortex, parvalbumin immunoreactive neurons are common in some association and all major sensory regions (Figures 8.3c, e; 8.4c, g); but, at least in visual cortex, are not as numerous as calbindin-immunoreactive neurons (see below). Parvalbumin-immunoreactive neurons also appear to be more restricted in their laminar distribution in diprotodont marsupials than in either polyprotodont or placental mammals. For example, parvalbumin-immunoreactive neurons in somatosensory cortex of diprotodonts like the tammar wallaby are concentrated in the granular layer (layer 4Cx, Figure 8.3a, e), whereas parvalbumin-immunoreactive neurons are distributed through a wider range of layers in the somatosensory region of rodent cortex (Paxinos *et al.*, 1999a, 1999b).

The morphology and distribution of calbindin-immunoreactive local-circuit neurons in the cerebral cortex of diprotodont marsupials is also similar to that in many polyprotodonts and placental mammals. In particular, the presence of calbindin-immunoreactive bipolar and bitufted neurons in upper layers of the tammar wallaby isocortex (unpublished observations) is comparable to the high density of double-bouquet and bipolar neurons seen in the upper layers (2Cx to 4Cx) of sensory cortex of polyprotodont marsupials, insectivores, microchiroptera, primates and cetaceans (Glezer *et al.*, 1993, 1998; Leuba and Saini, 1997; Letinic and Kostovic, 1998; Hof *et al.*, 1999). On the other hand, diprotodont cortex does not contain any calbindin-immunoreactive pyramidal neurons in any layers, in contrast to reports on rodents (van Brederode *et al.*, 1991) or carnivores and primates (Hof *et al.*, 1999). Dense clusters of strongly calbindin-immunoreactive neurons in layer 4Cx of the primary visual cortex are a prominent feature of diprotodont visual cortex (Figure 8.4f, h).

In the cerebral cortex of placental mammals, calretinin-immunoreactive neurons are most prevalent in supragranular layers 2Cx and 3Cx, although some cells may be found throughout all six cortical layers. In fact, calretinin-immunoreactive neurons make up 11.2% of the total neuronal population and exceed parvalbumin- and calbindin-immunoreactive neurons in frequency in medial prefrontal cortex of the laboratory rat (Gabbott and Bacon, 1996). These calretinin-immunoreactive cortical neurons are commonly bipolar, with long, frequently bifurcated and often varicose processes extending between the pia and the underlying white matter. Calretinin-immunoreactive processes also form a fine meshwork in layer 1Cx of rodent cortex (Résibois and Rogers, 1992). In monkey V1 cortex (Meskenaite, 1997), calretinin-immunoreactive neurons are also predominantly found in supragranular layers. These neurons are polymorphic, including bitufted, multipolar neurons with smooth dendrites and horizontally oriented Cajal–Retzius cells. The majority of calretinin-immunoreactive neurons in placental cortex have been found to be GABAergic. In the superficial layers of placental cortex, calretinin-immunoreactive neurons preferentially form synapses with other GABAergic microneurons, whereas pyramidal neurons are the preferred target of calbindin-immunoreactive neurons in layers 5Cx and 6Cx. Meskenaite (1997) proposed that calretinin-immunoreactive neurons in placental primary visual cortex serve a dual function: to disinhibit superficial layer neurons, while inhibiting pyramidal output neurons in the deeper layers.

Hof *et al.* (1999) reported that calretinin-immunoreactive neurons were also the most prevalent calcium-binding immunoreactive neurons in the cortex of two dasyurids (quoll and dunnart) and one peramelid marsupial (long-nosed bandicoot), with numerous small bipolar neurons being found in layer 2Cx and the upper portion of layer 3Cx. Horizontally oriented Cajal–Retzius and bitufted or bipolar calretinin-immunoreactive neurons are also common in the visual cortex of diprotodont marsupials (tammar wallaby, unpublished observations of the author).

Calretinin-immunoreactive neurons appear to be much more prevalent in therian compared to monotreme cortex, suggesting that these local-circuit neurons emerged as a major cortical neuron type after the divergence of the prototherian and therian lineages. Calretinin immunoreactivity in the platypus cortex is prominent only in the piriform region in small bipolar and round interneurons located in layers Pir2 and Pir3 (Hof *et al.*,1999). Calretinin-immunoreactive neurons are also quite rare in the echidna cortex, never making up more than 1% of all neurons present, even in the regions and layers where they are most prevalent (Hassiotis *et al.*, 2005).

Topographic and laminar distribution of NADPH-diaphorase-reactive neurons

In the cerebral cortex of placental mammals, NADPH-diaphorase (NADPH-d) reactivity is expressed in a particular type of GABAergic non-pyramidal neuron (Valtschanoff *et al.*, 1993; Gabbott and Bacon, 1996) and co-localises with nitric oxide synthase immunoreactivity (Bredt *et al.*, 1991).

Franca *et al.* (2000) reported that strongly NADPH-d reactive (Type I) neurons are densely distributed within functionally significant cortical regions (e.g. snout somatosensory representation) of both placental and polyprotodont marsupials (i.e. *Didelphis* and *Monodelphis*), whereas some cortical regions (e.g. primary visual cortex) have a lower numerical density of Type I neurons. An analysis of the distribution of Type-I-reactive neurons in tammar wallaby cortex found that these neurons are most densely distributed in M, Cg, Pir and AID, with fewer neurons in S1 and V1, and there was no preferential distribution of Type I neurons in lateral cortex as noted in polyprotodont marsupials (Franca *et al.*, 2000). In fact, the tangential distribution of Type I neurons in tammar wallaby cortex (Ashwell *et al.*, 2005a) much more closely resembles that seen in placental mammals like rodents and primates than Ameridelphians. This in turn points to the considerable divergence in cortical organisation that occurred within the marsupial lineage.

Topographic and laminar variation in GAP-43 immunoreactivity

Regional differences in the density of GAP-43 immunoreactivity have been found in the cerebral cortex of many therians, including an Ameridelphian (*Didelphis virginiana*; Zou and Martin, 1995), the tammar wallaby (Hassiotis *et al.*, 2002) and a variety of placental mammals. In light of the association of GAP-43 with the ability of axons to grow (Meiri *et al.*, 1986; Skene *et al.*, 1986; Snipes *et al.*, 1987; Zhu and Julien, 1999), these regional and laminar differences have been interpreted as indicating variation in synaptic plasticity associated with particular cortical regions and their constituent layers (Neve *et al.*, 1988). In the tammar wallaby, isocortical patterning of GAP-43 immunoreactivity develops around the time of weaning by selective loss in particular laminae of discrete isocortical regions, suggesting that some maturing cortical areas progressively lose the ability to support axonal plasticity. For example, GAP-43 immunoreactivity is relatively weak in layer 4Cx of adult primary sensory regions of tammar wallaby cortex (much like that seen in

placental mammals), but this laminar pattern is not identical in all the major sensory areas. Most layers of primary somatosensory cortex in adult tammars (except for the outermost layer 1Cx) exhibit very weak GAP-43 immunoreactivity, whereas the intensity of immunoreactivity is relatively stronger in the presumptive primary auditory cortex (throughout layers 1Cx, 2Cx, 3Cx and upper 6Cx). In primary visual cortex (V1) of the tammar wallaby, GAP-43 immunoreactivity is also strong throughout layers 1Cx, 3Cx, 5Cx and 6Cx, but weak in 4Cx. These findings suggest that the potential for plasticity in the tammar cortex is greater in auditory or visual cortex, relative to somatosensory cortex.

High levels of GAP-43 mRNA have been found in the associative areas (Brodmann's areas 10, 20, 40 and 44) of human isocortex (Neve *et al.*, 1988), particularly in layer 2Cx. The strength of GAP-43 immunoreactivity is also very high in presumptive association cortex of tammar wallaby (Hassiotis *et al.*, 2002) suggesting high retained levels of axonal plasticity. On the other hand, laminar variation in GAP-43 immunoreactivity is more uniform in association cortex of the tammar compared to primary sensory or motor cortex, suggesting that any retained plasticity involves a broad population of neurons and their connections throughout the entire cortical thickness.

Hippocampal region

The hippocampal region (archicortex) includes two sets of cortical structures: the hippocampal formation and the parahippocampal region, which are distinguished by the number of layers and patterns of connectivity. The hippocampal formation has three layers and can be divided into the dentate gyrus, hippocampus proper and the subiculum. The hippocampus proper is further subdivided into three fields (cornu Ammonis or CA1, CA2, CA3), although a fourth field (CA4) between the limbs of the dentate gyrus has been recognised by some authors.

The parahippocampal region includes the entorhinal cortex and adjacent areas. In placental mammals, the entorhinal cortex provides the major (unidirectional) input to the dentate gyrus through the perforant pathway and also engages in reciprocal connections with CA1 and the subiculum. Cortical regions adjacent to the entorhinal cortex (i.e. perirhinal and postrhinal cortex) also engage in reciprocal connections with the hippocampus proper and are often included in the parahippocampal region. Finally, the pre- and parasubiculum are also usually included in the parahippocampal region, because of their laminar structure and reciprocal connections with the subiculum (Witter and Amaral, 2004).

In all marsupials (as in rodents), the hippocampal formation has a complex three-dimensional C-shape extending from the septal nuclei rostrally and medially, around the diencephalon, into the temporal lobe caudally and ventrally. The deep or ventricular surface of the subiculum and hippocampus is covered by a thin sheet of white matter known as the alveus, containing afferent and efferent connections of the hippocampal formation. These fibres collect together as they progress from the temporal to septal end of the formation to form a fibre bundle known as the fimbria, which eventually leaves the hippocampal formation to become the fornix (see atlas plates D8 to D16 and W23 to W29). The fornix bodies extend rostrally, before turning ventrally to become the columns of the fornix. These divide into the precommissural fornix, directed to the septal nuclei and basal forebrain, and postcommissural fornix, directed to the mamillary bodies of the hypothalamus, anterior hypothalamus and anterior thalamic nuclei.

In smaller polyprotodont marsupials, subdivisions within the archicortex are clearly distinguishable with Nissl staining, enzyme histochemistry and immunohistochemistry (Ashwell et al., 2008b) and these patterns of staining largely match those seen in placental mammals (Celio, 1990; Paxinos et al., 1999a).

Neurogenesis in the mature hippocampus has been the focus of much research in placental mammals because of the potential of neural stem cells in this region to allow neuronal regeneration in ageing brains. Neurogenesis also occurs postnatally along the inner margin of the dentate gyrus of the fat-tailed dunnart (Sminthopsis crassicaudata) (Harman et al., 2003a). This adult neurogenesis is susceptible to environmental influences in that the number of dividing cells in the dentate gyrus of stressed, ageing dunnarts is significantly lower than in controls, but the reduction in the case of aged dunnarts may be less severe than that seen in ageing placental mammals (Harman et al., 2003a). This reduced vulnerability of dentate gyrus neurons in ageing dunnart brain might reflect the generally lower metabolic rate of marsupials and lower levels of free-radical damage, but needs to be assessed in a wider range of marsupial species for confirmation.

Olfactory cortex

The adult organisation and development of the cortical components of the olfactory system (anterior olfactory nucleus, superficial olfactory tubercle, piriform cortex) will be discussed in detail in Chapter 12.

Claustrum and endopiriform nucleus

The claustrum is a pallial structure lying in a subcortical position and has been identified in the forebrains of all marsupial and placental mammals (Johnson et al., 1994). It consists of two parts: a dorsal part (the dorsal or insular claustrum), located deep to the insular cortex, and a ventral part (the endopiriform nucleus), situated immediately deep to the dorsal part of the piriform cortex. The claustrum is large in marsupials with extensive isocortical surfaces (e.g. macropods), but difficult to distinguish in small polyprotodonts, like Marmosa and Didelphis (Johnson, 1977).

The insular claustrum has extensive reciprocal connections with many parts of the cortex and is believed to engage in the integration of sensory, motivational and mnemonic information (Dinopoulos et al., 1992). In diprotodont marsupials, the claustrum extends as far rostrally as the olfactory tubercle and as far caudally as the lateral amygdaloid nucleus (Hamel, 1967). The dorsal part of the claustrum is closely associated with the nearby isocortex and in macropod brains contains medium-sized, rounded to angular neurons. The ventral part of the claustrum is made up of slightly darker-staining (in Nissl preparations) angular neurons. The endopiriform nucleus, which is sometimes further subdivided into dorsal, intermediate and ventral components, is connected with the olfactory cortices and is probably involved in information processing in limbic circuits.

Development of the isocortex

General features and time-course of development

As noted previously, the isocortex of newborn marsupials is extremely immature, with at best only a two-layered structure (i.e. germinal zone and surrounding fibre-rich preplate or primordial plexiform layer) present (Warner, 1980; Reynolds et al., 1985; Krause and Saunders, 1994). In the tammar wallaby, neurons of the cortical plate first appear at about P5 to P6 and an intermediate zone does not emerge until P10 to P15. Tritiated thymidine studies show that the neurons settling in the preplate (presumptive Cajal–Retzius cells) are generated before birth, but that the neurons of the cortical plate are generated almost exclusively after birth. The generation of cortical-plate neurons shows the same inside-out pattern of neurogenesis well known from studies of placental mammals.

Not all marsupials have the same degree of cortical development at birth, nor do they follow identical patterns of development (Reynolds and Saunders, 1988; Rowe, 1990).

In particular, dasyurids appear to have the most immature cortex at birth, with *Phascogale calura* (the red-tailed phascogale) showing no development of the fibre-rich marginal zone at this time (Reynolds and Saunders, 1988). Nor is the marginal zone well developed in the newborn kowari (*Dasyuroides byrnei*) or eastern quoll (*Dasyurus viverrinus*). By contrast, didelphids and peramelidae have a well-developed marginal zone at birth. Diprotodont marsupials also have a strikingly laminated loose-packed zone (lpz) of their cortical plate from about the 3rd to 11th postnatal week (see atlas plates WP19–2 to 6 and WP25–2 to 5), with rows of cells aligned parallel with the pial surface.

The emergence of the cortical plate in the tammar wallaby is accompanied by changes in glycoprotein levels that may have functional significance, although this requires further investigation. Immunoreactivity for the glycoprotein fetuin appears in scattered neuronal somata in tammar wallaby pallium from about P5, is present in large numbers of cortical-plate neurons at P15, but disappears from the cortical plate by P20 (Jones *et al.*, 1988, 1991). This sudden disappearance from the cortical plate in the wallaby contrasts with the more prolonged presence of fetuin immunoreactivity in the cortical plate of some placentals (e.g. sheep) and is particularly striking considering the usual prolonged development of forebrain systems in macropods. The timing of fetuin expression is intriguing and suggests some functional significance for cortical development, but the actual role played by this glycoprotein in any therian remains unknown.

Neurogenetic gradients in the isocortex have been mapped in the brush-tailed possum (Sanderson and Aitkin, 1990; Sanderson and Weller, 1990a, b). In the lateral cortex, neuronal generation proceeds in a temporospatial gradient from rostroventrolateral cortex to caudodorsomedial cortex, whereas on the medial surface neurogenesis proceeds from rostroventral to caudodorsal. This means that isocortical neurogenesis is completed first in the transitional cortex around the rhinal fissure (at P32) and not until P68 in caudomedial visual cortex (Sanderson and Aitkin, 1990). On the medial cortical surface, ventral cingulate cortex (i.e. Cg2) also completes neurogenesis before the more caudal RSG cortex (P30 to P40 compared to P68). Neurogenesis in the primary somatosensory cortex of the possum proceeds from P5 to P55 for the head representation and P60 for the body representation. Neurogenesis proceeds from P5 to P68 in both primary visual and auditory cortex.

Molnár and colleagues have argued that there is a fundamental difference in the way that marsupial cortical neurons mature compared to rodents (Molnár *et al.*, 1998). Although all therians have the same inside-out pattern of cortical neurogenesis, the emergence of cortical neurons from the cortical plate is somewhat different in marsupials compared to placental mammals. In a variety of diprotodont marsupials (*Macropus eugenii* – Marotte *et al.*, 1997; Reynolds *et al.*, 1985; *Setonix brachyurus* – Harman *et al.*, 1995) and to a lesser extent in some polyprotodonts (*Monodelphis domestica* – Molnár *et al.*, 1998), cortical neurons mature quickly after completing their migration to the cortical plate, dropping out of the dense cortical plate to form a rapidly expanding layer of loosely packed cells immediately below the plate. This gives rise to characteristic rows of neurons running parallel to the pial surface and interspersed with fibres (the loose-packed zone – lpz – of the cortical plate; atlas plates WP19–2 to 6 and WP25–2 to 5). These loose-packed-zone cortical neurons are destined for the lower (infragranular) layers of the mature cortex. Presumably this laminated lower cortical plate zone arises because of interspersed tangentially oriented fibres and may develop as a consequence of contemporaneous connection formation and rapid differentiation from the cortical plate during the early maturation of diprotodont isocortex.

In some orders of placentals (e.g. carnivores and primates), the subplate region below the cortical plate is thought to serve as a site where thalamocortical axons wait and form synaptic connections before invading the cortical plate (Schatz and Luskin, 1986; Kostovic and Rakic, 1990). On the other hand, a distinct waiting period is not clearly present in rodents, because thalamocortical axons advance at a steady rate into the cortical plate. The subplate is difficult to distinguish in developing marsupial cortex (Harman *et al.*, 1995), perhaps because this region is obscured by development of the loose-packed zone (Molnár *et al.*, 1998). On the other hand, the subplate may not even be a significant player in the development of marsupial cortex, because: (i) ^{3}H thymidine studies in wallabies have not been able to find any population of early generated neurons below the cortical plate and (ii) thalamocortical axons in wallabies grow into the cortical plate without any apparent waiting period in a subplate region (Harman *et al.*, 1995). As noted above, fetuin-immunoreactive neurons disappear from the pallium of early pouch young marsupials very soon after emergence of the cortical plate. Since fetuin-immunoreactive neurons are found in large numbers in the subplate in ungulates, Molnár and colleagues have suggested that subplate neurons disappear within the first few weeks of postnatal life in marsupials.

The pallial subventricular zone and marsupial cortical neurogenesis

In placental mammals, the subventricular zone (SVZ) of the pallium is a secondary proliferative zone that appears between the ventricular germinal zone and the preplate region of the developing cortex in later fetal stages of cortical development. It is believed to give rise to microneurons of the granular and supragranular layers (layers 2Cx to 4Cx) and the emergence of the SVZ is said to coincide with the evolution of a six-layered isocortex, as well as the tangential expansion of the cortical sheet in mammals (Abdel-Mannan *et al.*, 2008). The SVZ is also present deep to subcortical structures and the emergence of subcortical SVZ probably preceded that for cortical SVZ in the mammalian lineage (Charvet *et al.*, 2009).

Early quantitative analysis of neuronal populations led to the contention that the number of neurons in a 30-μm wide standard column of the cerebral cortex is constant across mammals (Rockel *et al.*, 1980), but this notion has since been discredited (Haug, 1987; Herculano-Houzel *et al.*, 2008; Rakic, 2008). More recent studies have claimed that both poly- and diprotodont marsupials have significantly fewer cortical neurons in the 'standard' cortical column (by about 30 to 40%) and that this is the consequence of a deficient pallial subventricular germinal zone during marsupial development (Cheung *et al.*, 2009). The quantitative aspects of this contention with regard to adult marsupial cortex appear to be significantly flawed, in that the sample is heavily biased towards didelphids and, even where diprotodont marsupials are part of the analysis, quite different functional areas are being compared between species (Cheung *et al.*, 2009).

Certainly, studies using immunoreactivity against markers for cell proliferation suggest that the pallial subventricular zone of didelphids like the grey short-tailed opossum is a less active proliferative zone than in placental mammals (Abdel-Mannan *et al.*, 2008), but even didelphids have substantial layers 2Cx to 4Cx of the isocortex, so if the SVZ is deficient, then either other proliferative zones are producing granular and supragranular neurons in this species or we are underestimating the capacity of the SVZ. More significantly for our focus in this book, the pallial SVZ is clearly present in developing diprotodont cortex (see atlas plates WP12–3 to 6 and WP19–4 and 5) and contains mitotic figures for several weeks, suggesting prolonged and active cellular proliferation. These contradictory findings may be explained if one remembers that marsupials in general and diprotodonts in particular have a prolonged pouch development, a phase during which the SVZ should be most active. If the SVZ of marsupials has a more prolonged lifespan than in placental mammals, it may be able to generate cortical neurons with even a low rate of cell division.

Clearly this set of questions can only be cleared up by: (i) accurate quantitative analysis of cortical neuronal populations focussing on homologous cortical areas in a broad range of marsupial species and (ii) developmental studies of pallial SVZ in both poly- and diprotodont marsupials, with quantitative comparisons with placental mammals.

Development of GAP-43 immunoreactivity in marsupial cortex

The Calmodulin-binding phosphoprotein GAP-43 plays a critical role in axonal growth (Meiri *et al.*, 1986; Skene *et al.*, 1986; Snipes *et al.*, 1987; Zhu and Julien, 1999) and is present in high levels within developing nerve fibres during the period of axonal elongation (Jacobson *et al.*, 1986; Karns *et al.*, 1987). The changing distribution and levels of immunoreactivity to GAP-43 therefore have the potential to reveal different stages in development of the marsupial isocortex and its connections (Zou and Martin, 1995; Hassiotis *et al.*, 2002).

At the time of birth in the tammar, GAP-43 immunoreactivity is present in the fibre-rich primordial plexiform layer of the lateral isocortex, at approximately the region of the developing sensorimotor cortex, but the medial parts of the cortex (cingulate, retrosplenial and motor cortex) have only a very thin primordial plexiform layer with very little GAP-43 immunoreactivity. By P12 and P19, the isocortical plate (differentiated into an upper compact cortical zone and a loose-packed zone by P19) of the parietal cortex is visible as an immunonegative region, sandwiched between strongly immunoreactive marginal (developing layer 1Cx) and intermediate zones. The most intense GAP-43 immunoreactivity in the developing sensorimotor isocortex during the second and third months of pouch life is in the deep intermediate zone (immediately adjacent to the subventricular layer), in the marginal zone and in the loose-packed zone of the cortical plate, in descending order of immunoreactivity strength. Strongly immunoreactive fibres in the deeper intermediate layer are due to the addition of long association (Sheng *et al.*, 1990) and commissural cortical connections (Ashwell *et al.*, 1996b; Shang *et al.*, 1997)(see below). The next phase for development of GAP-43 immunoreactivity in sensorimotor isocortex begins around P100, when selective laminar reduction in GAP-43 immunoreactivity in the granular layer (4Cx) commences.

The occipital cortex initially lags behind the parietal region in development of GAP-43, but is largely maturing in step by the end of the third postnatal month. At the time of birth, the marginal zone of the occipital cortex has only a few immunoreactive cells and fibres, and these are confined to the lateral parts. By P12, when the isocortical plate first appears, the pattern of immunoreactivity is similar to that noted for the parietal isocortex. By P115, there is also a selective loss of GAP-43 immunoreactivity in specific regions of visual cortex in a similar fashion to that seen for the somatosensory cortex. At this age, the V1/V2L boundary can be distinguished by the abrupt transition from a GAP-43 immunonegative layer 4Cx of V1 to a GAP-43 immunoreactive layer 4Cx of V2L. The decline in GAP-43 immunoreactivity seen in the main afferent layer of both somatosensory and visual cortex probably indicates restriction in the plasticity of thalamocortical projections to this layer from about P100 and coincides with the change from phase 2a to 2b of lactation (see Chapter 3 and Figure 3.6) and greater behavioural independence of the pouch young.

Development of isocortical connections

Commissural projections from the wallaby isocortex through the anterior commissure are originally (at P15) derived from pyramidal neurons in the compact cell zone of the cortical plate of cortex adjacent to the rhinal fissure at the level of the anterior commissure (Shang *et al.*, 1997), consistent with the relatively early maturation of cortical neurons in this region (see above). Over the subsequent 66 days, pyramidal neurons in surrounding cortical regions (i.e. both rostrally and caudally) and further dorsally and medially contribute axons to the anterior commissure. By about P80, commissural neurons are distributed throughout the entire cerebral hemisphere, but a subsequent period of restriction follows (from about P160 to P290), progressively confining commissural neurons to the more discrete regions where these neurons lie in the adult. This process of progressive refinement of the commissural connection is similar to that seen in placental mammals during the late fetal and/or early postnatal life and presumably involves retraction of inappropriate axons.

The details of development of major sensory thalamocortical and reciprocal projections will be dealt with in the relevant systems chapter, but a short summary will be provided here. Thalamic afferents reach the somatosensory (Marotte *et al.*, 1997) and visual (Sheng *et al.*, 1991; Pearce and Marotte, 2003) cortex shortly after the first cortical plate neurons appear (i.e. by P15), with no sign of a waiting period at a subplate and no significant synapse formation in that zone (Pearce and Marotte, 2003). Indeed, the absence of a waiting period in a subplate for ingrowing afferents is a common feature of cortical development in poly- and diprotodonts (Molnár *et al.*, 1998). During their initial ingrowth, thalamic axons are distributed widely throughout the thickness of the lpz up to the level of the compact zone of the cortical plate, invading the developing layers as they progressively differentiate beneath the compact zone (Sheng *et al.*, 1991; Marotte *et al.*, 1997; Marotte and Sheng, 2000). The first visual thalamocortical synapses can be detected in the loosely packed zone of the cortical plate at about P30 (Pearce and Marotte, 2003). The initial invasion occurs relatively quickly, but the subsequent maturation of terminal fields of visual and somatosensory thalamocortical axons and the corticothalamic projections occurs over a protracted period

in diprotodont marsupials. For example, in the mystacial (whisker pad) somatosensory cortex (Marotte *et al.*, 1997), corticothalamic projections are not established until P60, and thalamocortical terminals in layer 4Cx do not cluster until P76. This is about the time at which the first recordings of an immature cortical evoked potential can be elicited (at P85) and coincides with the earliest signs of succinic-dehydrogenase-positive patches in layer 4Cx of the whisker field of the somatosensory cortex (Waite *et al.*, 1991). On the other hand, the time-course of development in the visual cortex is somewhat later, as might be expected given the slightly delayed neurogenesis of occipital cortex. In the primary visual cortex (Sheng *et al.*, 1990, 1991), descending pathways develop coincidentally with protracted maturation of the afferents: corticothalamic axons reach the lateral geniculate and lateral posterior thalamic nuclei at P48, whereas corticocollicular axons reach the superior colliculus at P71. Maturation of the laminar distribution of visual thalamocortical axon terminal fields occurs by around P118 (Sheng *et al.*, 1991). This is shortly before eye opening, which occurs at around P140 in *Macropus eugenii* (Janssens and Messer, 1988).

Epigenetic factors in the development of cortical functional topography

The special features of marsupial telencephalic development, i.e. that pouch young are accessible to experimentation before thalamocortical axons have reached the developing pallium, has allowed studies of the effects of developmental changes in cortical volume on the establishment of functional topography. Removal of the caudal one-third of the cortical sheet around the time of birth in short-tailed opossum is possible without disrupting the underlying dorsal thalamus and its relay neurons. When thalamocortical axons from these thalamic relay neurons finally reach the cortical sheet, they establish functional areas that have broadly similar rostrocaudal and mediolateral relationships to those in normal opossums (Huffman *et al.*, 1999b). In other words, all the major sensory areas are present on the reduced cortical sheet, albeit with some reduction in the size of visual cortex, which would have occupied the deleted caudal cortex. These findings indicate that the establishment of cortical functional areas depends on an interaction between thalamocortical axons rather than being precisely specified by genetic programming of progenitor cells in the telencephalic germinal zones. This interaction between ingrowing cortical afferents allows epigenetic factors such as relative activity patterns in different sensory pathways to influence the development of functional cortical topography (Larsen and Krubitzer, 2008).

Development of the hippocampal region

At the time of birth, the hippocampal primordium of all marsupials is little more than a laterally directed convexity protruding into the rostral lateral ventricle (Harman, 1997; see also atlas plate WP0/3). Like the isocortex of newborn marsupials, this hippocampal anlage is essentially two-layered: a ventricular germinal zone or neuroepithelium flanked medially by a primitive plexiform layer. In diprotodont marsupials like the brush-tailed possum, quokka and tammar wallaby, emergence of discrete hippocampal compartments like those seen in rodents (Ammon's horn ventricular zone, primary dentate neuroepithelium, hippocampal wedge and fimbrial glioepithelium; Altman and Bayer, 1990, 1995) occurs over the first two weeks of postnatal life (Harman, 1997; Hassiotis et al., 2002; see atlas plates WP12/6 and WP19/5) and dentate neuron migration begins by the end of the third postnatal week. In rodents, the fimbrial glioepithelial region is demarcated from primary dentate and Ammon's horn ventricular zone by a prominent dentate notch (Altman and Bayer, 1990), but this notch is much more subtle in the developing wallaby brain. Generation of cells in the hippocampus of the quokka occurs from just after birth (P5) until P85, with most pyramidal neurons of the cornu Ammonis fields generated before P40 and the hippocampal glial and granule cells generated after that time (Harman, 1997). Neurogenesis of cornu Ammonis, subiculum, pre- and parasubiculum in the brush-tailed possum also occurs from about birth to P46 (Sanderson and Aitkin, 1990; Sanderson and Wilson, 1997). Early-formed pyramidal neurons (P5 to P12) of CA1 to CA3 are distributed adjacent to the stratum oriens, whereas later-formed neurons are distributed in the middle of the stratum pyramidale (P21 to P32) or on the edge adjacent to the stratum radiatum (P46). Neurogenesis of the granule cells in the dentate gyrus of the brush-tailed possum also exhibits a distinct temporospatial gradient: the earliest-formed granule cells (P5, P12) are located adjacent to the molecular layer, whereas the latest-formed granule cells (P46, P68) come to lie adjacent to the hilus, i.e. laterally. Neurogenesis within the entorhinal cortex proceeds with an inside-out temporospatial gradient (P5 to P32 in the brush-tailed possum; Sanderson and Wilson, 1997).

The process of dentate-cell neurogenesis in the quokka is susceptible to the adverse effects of elevations in temperature of as little as 2°C for two hours at the peak stage of granule-cell generation (P40 to P45) (Kent and Harman, 1998), indicating the usefulness of this model system for studies of the effects of maternal hyperthermia on brain development.

Cytoarchitectural differentiation of Ammon's horn and the dentate gyrus granule cells begins at about P20 and P30, respectively, in both the tammar and quokka (Harman, 1997; see atlas plates). The presence of strong GAP-43 immunoreactivity in the developing alveus and fimbria during the second postnatal month indicates rapid growth of hippocampal afferents and efferents at this time (Hassiotis et al., 2002), but no connectional studies have been made of this system in diprotodonts.

Development of the claustrum and endopiriform nucleus

The insular claustrum is thought to derive from the lateral pallial (isocortical) neuroepithelium (Bayer and Altman, 1991; Narkiewicz and Mamos, 1990; Puelles et al., 2000), whereas neurons of the endopiriform nucleus probably arise in the ventral pallium (embryonic paleocortex) in the ventricular angle between the pallium and striatum (Bayer and Altman, 1991). Compared to other pallial derivatives, the claustrum and endopiriform nucleus of marsupials like the tammar are generated quite early, finishing neurogenesis at least one to two weeks before other pallial derivatives in a similar position (e.g. iso- and periallocortical neurons around the rhinal fissure). The claustrum and endopiriform nucleus are both strongly labelled after ^{3}H-thymidine injection at P5 in the brush-tailed possum, but the claustrum appears to have a more prolonged period of neurogenesis (peaking at P5 to P12 and ceasing around P21) than the endopiriform nucleus (completed at P9) (Sanderson and Wilson, 1997). The first dorsal claustrum neurons can be seen in position from P5 in the tammar wallaby (see atlas plate WP5/4), suggesting that claustral neurogenesis follows a similar time-course in macropods.

What remains unknown?

Naturally, most studies of the marsupial cortex have focussed on the major sensory and motor regions, because of their importance for tracing the evolution of sensory areas and the question of the sensorimotor amalgam. This attention has been at the expense of consideration of the equally behaviourally important association and periallocortical regions. On the basis of cyto- and chemoarchitecture, the iso- and periallocortex of Australian marsupials appears to approach rodents in the degree of complexity, but these few studies are based on a limited number of species and only a restricted range of neurochemical markers. More connectional studies of limbic association-cortex areas in Australian marsupials are needed.

There are several significant questions concerning the organisation of adult marsupial cortex that deserve more attention:

(1) How does the functional organisation of frontal and orbital cortex in diprotodont marsupials compare with that in rodents and primates? Has the separate evolution of complex gyrified cortical mantles in the large diprotodonts produced a functionally similar organisation of frontal association regions or have these regions evolved according to a different plan? Detailed analysis of the neurochemical and connectional characteristics of these regions in large diprotodonts is necessary to answer these questions.

(2) Does the medial occipital cortex of poly- or diprotodont marsupials have unrecognised visual function? The secondary visual area(s) in the occipital cortex of large diprotodonts is currently believed to be entirely lateral to the primary visual cortex, but it is now recognised that even small placental mammals with poor vision (like the laboratory mouse) have diverse visual areas, albeit small, medial to the primary visual cortex (van der Gucht et al., 2007). It would be surprising if the visual cortex of active, visually proficient carnivores like the dasyurids has a simpler organisational plan than rodents.

In addition, there are unresolved questions concerning isocortical development of marsupials that deserve attention. The most striking neurological difference between marsupials and placentals revolves around the difference in commissural pathways: isocortical commissural neurons situated in the parietal cortex (for example) direct their axons medially towards the corpus callosum in placental mammals, but laterally towards the internal and/or external capsule in marsupials. This important and structurally fundamental difference implies that either: (i) the guidance systems determining axonal outgrowth from commissural neurons, or (ii) the positions of chemotropic agents influencing the outgrowth of those axons, must be fundamentally different in marsupial compared to placental brains. A further question, of particular relevance to the uniqueness of Australian marsupials, concerns the mechanisms that control the pathway choices of commissural axons in diprotodonts (directing commissural axons firstly into the internal capsule and then into the fasciculus aberrans to reach the anterior commissure), as compared to polyprotodonts (directing commissural axons into the external capsule, around the striatum and into the anterior commissure). How are these fundamental differences in axonal pathway selection controlled in early pouch young marsupials?

Part III

Systems neurobiology

9

Visual system

L. D. Beazley, C. Arrese and D. M. Hunt

Summary

This chapter focusses on the malleability of visual organisation in marsupial mammals and its adaptability to ecological and evolutionary pressures. Far from being uniform or poorly developed, visual capabilities in marsupials reflect visual requirements that are essential for survival in diverse ecological niches. In addition, the retention of the sauropsid retinal prototype reflects a divergence of evolutionary trends and constraints between marsupial and placental mammals. Hence, it is not surprising that, from the configuration of visual fields to the arrangement of retinal cells, a great degree of variation is observed between the representative species of the two major taxonomic divisions of marsupials, the diprotodonts and polyprotodonts. Measurements of visual fields, cone types and retinal ganglion cell topographies reveal configurations that reflect visual priorities associated with feeding strategies, predator detection, prey catching and locomotion. Similarly, from highly reactive pupils to static apertures, the extent and speed of pupillary response are consistent with the light conditions and fluctuations to which each studied species is exposed in their respective environments. Photoreceptor dimensions and proportions resemble those found in diurnal sauropsid retinae, indicating the retention of ancestral diurnal lifestyles and photopic visual systems. The most striking difference between placental on the one hand, and marsupial and monotreme visual systems on the other, is the retention of oil droplets. The retinae of some Australian marsupials also contain three spectrally different cone classes and possess trichromatic colour vision, which was previously thought to be unique to primates within Mammalia. The visual pigment present in this third cone type is, however, uncertain. Unlike placentals, where ultraviolet-sensitive (UVS) SWS1 pigments have been retained only by a subset of rodents, UVS pigments are more widespread amongst marsupials, occurring in the honey possum (*Tarsipes rostratus*), the fat-tailed dunnart (*Sminthopsis crassicaudata*), and the grey short-tailed (*Monodelphis domestica*) and big-eared (*Didelphis aurita*) opossums. In all cases, the amino acid phenylalanine (Phe) is retained at site 86 in the SWS1 opsin. Of the species studied so far, only the quokka (*Setonix brachyurus*) and tammar wallaby (*Macropus eugenii*) possess violet-sensitive (VS) pigments resulting from the substitution of the amino acid tyrosine (Tyr) at site 86. Monotremes appear to be dichromats, but differ from all other mammals in the retention of the *SWS2* gene instead of the *SWS1* gene. Similar to placentals, marsupials possess only one melanopsin gene, *Opn4m*, which is expressed in the retina of the fat-tailed dunnart. Marsupial visual pathways show considerable differences in their cytoarchitecture and patterns of retinal terminations in the dorsal lateral geniculate nucleus, ranging from relatively simple to complex.

Introduction

In the last three decades, investigations of the mammalian visual system have provided substantial advances in the understanding of visual capabilities in relation to lifestyles and light environments. However, the majority of studies have been concerned with placental mammals, drawing inferences that were sometimes inappropriately extended to marsupial mammals. Due to the generalisation that they are strictly nocturnal, marsupials have long been considered to have limited visual capabilities, inferior to those of placentals, and this misconception still persists outside the visual sciences. The inference also overlooks the complexity of marsupial radiation, following their divergence from placental mammals 130 mya (Clemens *et al.*, 1989; Graves and Watson, 1991; Nilsson *et al.*, 2004), and their adaptation to a broad range of ecological niches and activity patterns (Johnson and Strahan, 1982; Friend and Burrows, 1983; Archer, 1984; Christensen *et al.*, 1984; Calaby, 1960). Paradoxically, early studies showed that, unlike those of placentals, marsupial retinae have retained features of their reptilian ancestors, such as oil droplets and double cones, which reflect a long history of photopic (diurnal) vision (O'Day, 1935; Walls, 1942; Ohtsuka, 1985; Jacobs, 1993; Campenhausen and Kirschfeld, 1998; Barbour *et al.*, 2002). More recent assessments of marsupial retinal and extra-retinal features have revealed their high degree of malleability in response to diverse and dynamic environments and behaviours (Rodger *et al.*, 1998; Arrese *et al.*, 2000, 2002a, 2002b, 2003, 2005, 2006b; Sumner *et al.*, 2005). Such findings, combined to the wide diversity of marsupial species and lifestyles, indicate that evolutionary pressures have shaped their visual organisation, resulting in variations and adaptations that fulfill their visual priorities.

Although the Australian, Antarctic and South American continents formed a significant southern evolutionary platform for the marsupials, phylogenetic analyses show a clear division between the Australian and American lineages (Szalay, 1982, 1994; Archer, 1984; Lee and Cockburn, 1985; Nilsson *et al.*, 2004) (Figure 9.1). Australian marsupials, isolated from severe competition and predation by large carnivores, are thought to have maintained various activity patterns encompassing diurnality (Calaby, 1960; Johnson and Strahan, 1982; Friend and Burrows, 1983; Archer, 1984; Christensen *et al.*, 1984). Their long period of evolution produced a remarkably diverse fauna, with approximately 200 species that can be divided into two phylogenetically distant groups: the polyprotodonts and diprotodonts. In contrast, South and North American marsupials, experiencing

similar evolutionary pressures to their placental counterparts, since both groups were sympatric, took refuge in nocturnality. Consequently, the highly divergent marsupial radiation (Szalay, 1982, 1994; Clemens *et al.*, 1989; de Muizon *et al.*, 1997; Cifelli and Davis, 2003) generated selective pressures that may have had differential influences on the evolution of their visual organisation.

Visual fields

The extent of an animal's visual fields scanned instantaneously without head movements delimits the portion of space in which visual control of behaviour is possible (Hughes, 1977). The relative dimensions of visual fields delimited by monocularity and/or binocularity are determined by the interaction of selective pressures that predominate in each species' lifestyles, including feeding patterns, locomotion, prey catching and predator avoidance (Walls, 1942; Polyak, 1957; Cartmill, 1972, 1974; Fite, 1973; Hughes, 1977; Lythgoe, 1979; Martin, 1999).

Predators, or animals grasping objects for manipulation at close range, have the greatest need for an accurate estimation of depth and distance (Stone, 1965; Hughes, 1977; Lythgoe, 1979; Peichl, 1992b). This ability is conferred by an extensive binocular field, where the frontal placement of the eyes maximises the field overlap and convergent inputs from the two eyes result in stereoscopic vision. By contrast, prey species tend to have eyes located to the side of the head, providing a large visual field to detect potential threats from many directions.

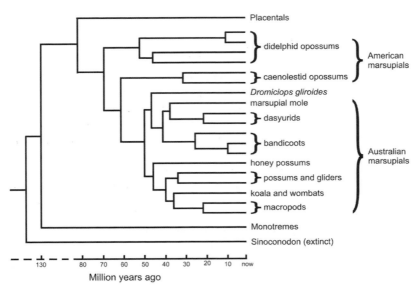

Figure 9.1 Phylogenetic tree showing the divergence of the three mammalian groups and the main marsupial phyla. Based on data from Nilsson *et al.* (2004).

Although studied in a relatively small number of marsupials, visual field measurements in marsupials have provided an insight into the manner in which behaviour is influenced by stimuli occurring at different locations in their visual space. For instance, the predator and prey status of the fat-tailed dunnart is reflected in an extensive visual field (360°), with a monocular field spanning over 270° and a binocular overlap reaching 140° (Rodger *et al.*, 1998). The frontal position of the eyes and their lateral protrusion from the head allows a panoramic view of the environment via the extensive monocular fields. As with most predators, the binocular overlap maximises accuracy when attacking and manipulating prey.

The binocular overlap of the honey possum (140°) is similar to that of the fat-tailed dunnart. However, the entire visual field is restricted (260°) by a narrow interocular space, limiting the dimensions of the monocular fields to 200° (Arrese *et al.*, 2002a). Although not a predator, the arboreal honey possum has a strong requirement for binocular vision, with specific aspects of ecology, such as strategies for feeding from flowers and locomotion, placing a premium on accurate depth perception. For instance, the minute honey possum is capable of jumping with accuracy for more than 1 m from branch to branch whilst foraging or escaping predators (Russell, 1986).

The binocular overlap of the numbat (*Myrmecobius fasciatus*: 80°; Arrese *et al.* 2000) represents a compromise between panoramic and frontal vision. Rapid running and climbing require adequate depth perception, whereas scanning of the surroundings is facilitated by the wide entire visual field (220°; Arrese *et al.*, 2000). Feeding exclusively on termites collected from subterranean galleries located by smell, the numbat is not a predator per se. However, the numbat's eyes are set relatively high on the head, so the upper visual space may be kept under surveillance, affording detection of airborne predators.

A similar binocular overlap is found in the herbivorous quokka (80°; Harman *et al.*, 1990), whereas the arboreal, carnivorous South American (big-eared) and North American (*Didelphis virginiana*) opossums that rely more strongly on an accurate estimation of depth and distance, possess an overlap of 125° (Oswaldo-Cruz *et al.*, 1979; Rapaport *et al.*, 1981).

Pupillary mobility

In terrestrial habitats, light intensity fluctuates considerably, imposing a number of constraints on the visual system of animals (Lythgoe, 1979; Martin, 1999). Furthermore, when moving between areas of different illumination, animals are exposed to rapid changes in light levels. In such conditions, visual performance is optimised when the retina adjusts rapidly to a wide range of light levels, thereby maintaining sensitivity, acuity and contrast discrimination (Lythgoe, 1979; Douglas *et al.*, 1998; Martin, 1999).

Mammals are able to regulate the amount of light reaching their retina by pupillary mobility, providing a fast and efficient response to fluctuating light levels. Generally, it is assumed that diurnal species have the most reactive pupils, whereas nocturnal species are thought to possess a wide, less mobile pupil (Walls, 1942; Tansley, 1965; Oswaldo-Cruz *et al.*, 1979; McIlwain, 1996; Warrant, 1999). By extension of this contention, marsupials were assumed to lack pupillary mobility until studies conducted in Australian species with differing lifestyles reported marked differences (Arrese *et al.*, 2000; Arrese, 2002).

A highly mobile pupil is found in the fat-tailed dunnart (Arrese, 2002), an opportunistic species that may be active during day or night, depending on seasonal and environmental conditions (Arrese *et al.*, 1999) (Figure 9.2). In its habitat of sparse grasslands or open shrublands, the dunnart also experiences rapid changes of illumination between areas of light and shade, when hunting or avoiding a predator. The rapid pupillary constriction not only reduces photoreceptor saturation in bright light, but by shielding the retina and maintaining it in a relatively dark-adapted state, also enables a faster adaptation to darkness. Consequently, when the dunnart returns to sheltered conditions, high sensitivity and contrast discrimination are rapidly restored (Frankenberg, 1979; Douglas *et al.*, 1998).

Maintenance of sensitivity and contrast discrimination is also vital to the honey possum (Arrese and Runham, 2002), a crepuscular species with activity patterns encompassing diurnal activity (Figure 9.2). Inhabiting *Banksia* heathlands and feeding exclusively on nectar and pollen, the species pursues visual tasks associated with its lifestyle in a relatively constant level of illumination. Less extensive than that of the dunnart, the pupillary mobility of the honey possum provides suitable adaptation to the minor and gradual changes in its light environment (Arrese, 2002). The response would optimise photon capture in the range of light levels experienced, whilst maintaining sensitivity. Such a mechanism is crucial for predator detection at dusk and dawn, because at low light intensity, sensitivity combined with enhanced photon capture optimises movement detection (Lythgoe, 1979; Locket, 1999).

The relative pupillary width of the nocturnal burrowing bettong (*Bettongia lesueur*) is smaller than that in the honey possum and its faster rate of constriction suggests a greater retinal sensitivity (Arrese, 2002; Figure 9.2). Confined in complex burrows during daytime and foraging only at night (Short and Turner, 1999), the species is restricted to almost constant darkness. The requirement for rapid constriction

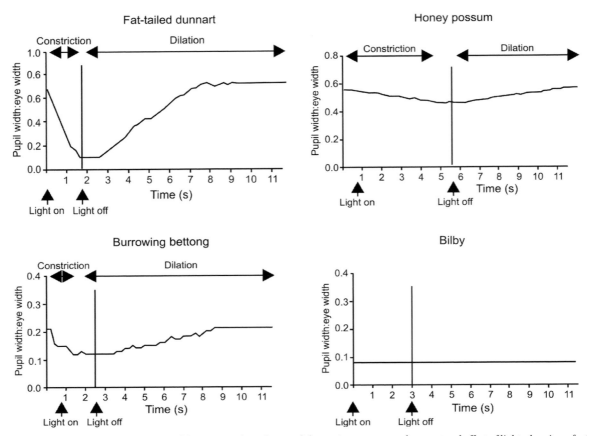

Figure 9.2 Graphic representations of the magnitude and rates of change in response to the onset and offset of light, showing a faster rate of constriction in the fat-tailed dunnart, whilst in the honey possum, constriction occurs relatively slowly, with a time of completion comparable to that for dilation. In the burrowing bettong, pupillary constriction is rapid after the onset of light, whilst dilation occurs slowly after the offset of light. No pupillary response occurs in the bilby.

may correlate with the emergence from its warren into bright moonlight.

A wide, immobile pupil is found in the diurnal numbat (Arrese *et al.*, 2000), contradicting the conception that such pupils are restricted to nocturnal animals (Figure 9.2). The absence of a pupillary response reflects the species' activity pattern in a uniform light environment. Inhabiting tall forests, with ambient light filtered by the canopy, the numbat is mostly exposed to a relatively constant level of illumination. However, peak visual performance in such a light environment might require a constant, maximum pupillary dilatation to enhance photon capture. Retinal adaptation is still possible because the numbat is unusual in possessing a cone-dominated retina (Arrese *et al.*, 2000), and may not require a responsive pupil.

The pupil of the strictly nocturnal greater bilby (*Macrotis lagotis*) is remarkably minute and remains static when exposed to varying light levels (Arrese, 2002). Active strictly during the darker hours of the night, the

bilby shelters in deep burrows by day, and hence is not exposed to any degree of light fluctuation. The species is highly susceptible to heat stress, so confinement in deep burrows, with a spiral descent that prevents light and heat infiltration, is crucial for its survival in the central deserts of Australia (Morrisson, 1962; Johnson, 1989). As reflected by the size and immobility of the pupil, the lifestyle of the bilby does not call for an adaptation to varying light levels.

Despite the lack of data for North and South American species, these studies in Australian marsupials demonstrate that Australidelphian pupils show a range of responses to changing light levels, and that the speed and extent of such responses are variable, according to the species' period(s) of activity. However, rod and cone distribution is also a determining factor. A wide, immobile pupil would be detrimental to a highly rod-dominated retina, whereas a greatly constricted pupil would be ineffectual in the case of strong cone predominance.

Anatomy and physiology of the retina

Vascular supply of the retina

Intraretinal vascular networks are seen in the retinae of those mammals that have evolved complex visual functions and require a higher level of oxygenation. Retinal thickness shows a bimodal distribution across species with avascular retinae being substantially thinner than those that are vascular (Buttery *et al.*, 1991). Avascular retinae also show heavy glycogen deposition in the Müller cells of the inner retina, short photoreceptor outer segments and lack both a tapetum and retinal taper. Avascular retinae are found in some of the primitive placental mammals, but are more common amongst the monotremes (McMenamin, 2007) and marsupials (Chase, 1982). Amongst the marsupials, the diprotodont group, including the honey possum, brush-tailed possum (*Trichosurus vulpecula*) and sugar glider (*Petaurus breviceps*) generally have avascular retinae as do some polyprotodont species, such as the Northern brown and Southern brown bandicoots (*Isoodon macrourus* and *I. obesulus*) and the bilby (Johnson, 1968; Tancred, 1981; Buttery *et al.*, 1990; Chase and Graydon, 1990; McMenamin, 2007). A vascular retina is however found in other polyprotodont species that include the Tasmanian tiger (*Thylacinus cynocephalus*), the broad-footed marsupial mouse (*Antechinus godmani*), the eastern quoll (*Dasyurus viverrinus*), the Tasmanian devil (*Sarcophilus harrisii*), the fat-tailed dunnart (*Sminthopsis crassicaudata*), the grey-bellied dunnart (*Sminthopsis griseoventer*), the numbat (*Myrmecobius fasciatus*), the North American opossum (*Didelphis virginiana*) and the Mexican mouse opossum (*Marmosa mexicana*) (O'Day, 1938a; Walls, 1939, 1942; Tancred, 1981; Buttery *et al.*, 1990; Chase and Graydon, 1990; McMenamin and Krause, 1993; Rodger *et al.*, 2001; McMenamin, 2007).

In most vertebrates, the retinal vasculature, where it exists, is arranged as an artery–capillary–vein system. What is unusual about the vasculature of marsupials is the ubiquitous presence of a distinctive end-artery system where pairs of blood vessels consisting of an artery and a corresponding vein branch in unison down to the level of the capillaries such that the arterial and venous components of each limb unite to form a terminal arteriovenous hairpin capillary or end loop (McMenamin and Krause, 1993; McMenamin, 2007; Figure 9.3). A complete system of end vessels is thereby formed that have no anastomoses. A similar system of paired vessels is seen in the vasculature of the parenchyma of the central nervous system, but not in other tissues.

Various hypotheses have been advanced to explain this unusual vascular organisation, including a role in thermoregulation and a means for optimising oxygen tension across

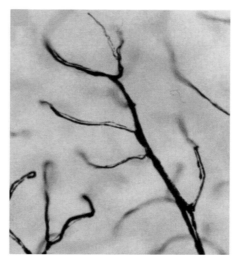

Figure 9.3 Retinal blood vessels of the North American opossum *Didelphis virginiana*. Semithin sections of carbon-filled vascular preparations. Note paired vessels with capillaries that end in blind-end loops. From McMenamin (2007), adapted by permission from BMJ Publishing Group Limited.

the retina (Walls, 1939; Craigie, 1945; Johnson, 1968; Dollery *et al.*, 1969; Chase and Graydon, 1990; Forrester *et al.*, 2002). However, the most likely reason is associated with the precocious development of marsupials (Rodger *et al.*, 2001). Vision becomes functional in marsupials much earlier than in placentals and at a stage when the vasculature is still growing across the face of the retina. The developmental study by Rodger *et al.* (2001) demonstrated that these vessels extend by interstitial growth and are open from the outset, so are able to meet the metabolic requirements of a functional retina.

Photoreceptors

Photon capture by retinal photoreceptors is the first step in the activation of the phototransduction cascade that converts light into an electrical signal for passage to the visual centres in the brain. Rod and cone photoreceptors are classified according to their morphology and function. The more light-sensitive nature of rods means they are bleached by daylight and therefore function only in dim light, whereas cones are less sensitive and provide vision in daylight with the potential for colour vision. The minimum requirement for colour vision is two spectrally distinct classes of cone photoreceptors, each containing a visual pigment with a different peak of spectral sensitivity (λ_{max}).

The number, arrangement and distribution of rods and cones across the retina are unique for each species, and reflect visual priorities associated with lifestyles and light environments. In non-primate mammals, both rods and

Table 9.1 Rod and cone dimensions (in μm), including inner-segment (IS) and outer-segment (OS) lengths, cone width at the location of oil droplets and basal width. Data from Appese *et al.* (1999, 2000, 2002a)

	Fat-tailed dunnart		Honey possum		Numbat	
	Rod	Cone	Rod	Cone	Rod	Cone
Mean total length	26.76 (± 2.20)	22 (± 1.62)	22.08 (± 1.40)	20.97 (± 2.21)	30.09 (± 0.43)	38.16 (± 1.23)
Mean IS length	9.21 (± 1.34)	16.77 (± 1.98)	9.27 (± 0.63)	17.40 (± 1.05)	18.15 (± 1.27)	31.70 (± 2.02)
Mean OS length	16.55 (± 1.89)	5.21 (± 0.85)	12.81 (± 1.11)	3.57 (± 0.48)	11.94 (± 1.03)	8.67 (± 0.16)
Width at oil droplet	–	5.02 (± 0.29)	–	4.84 (± 0.39)	–	7.4 (± 0.87)
Basal width	–	4.17 (± 0.19)	–	4.22 (± 0.06)	–	7.28 (± 0.68)

cones are long and slender, maximising light conversion into electrical activity, and thereby sensitivity in low light levels. Rods evolved to serve dim-light vision and prevail in the majority of mammals. By contrast, short and wide cones and relatively short rods, similar to those of diurnal sauropsids (reptiles), are rarely found in placental mammals, but, as discussed below, are present in marsupials, suggesting the retention of ancestral diurnal retinal features.

Cone dimensions in the fat-tailed dunnart, honey possum and numbat (Arrese *et al.*, 1999, 2000, 2002a) resemble those in diurnal sauropsid retinae, with wider basal diameters and shorter outer segments (Table 9.1). Similarly, rods in the three species are shorter and wider than those found in the majority of placentals, and are comparable to the rods of strongly diurnal species, such as the California ground squirrel (*Spermophilus beecheyi*; Long and Fisher, 1983; Kryger *et al.*, 1998) and tree shrew (*Tupaia belangeri*; Peichl *et al.*, 2000).

In many vertebrates, cone photoreceptors may consist either of single or double cells, the latter generally consisting of a large primary and a smaller accessory cell. Double cones are absent from placental retinae, but are found in monotremes (O'Day, 1938b; Young and Pettigrew, 1991), where the two components may be of equal size (twin cones), and in at least some marsupials (Ahnelt *et al.*, 1995; Arrese *et al.*, 1999). The spectral sensitivity of cone photoreceptors may also be modified by the presence of oil droplets in the distal region of the inner segment. Oil droplets contain carotenoid pigments that are spectrally matched to cone classes and act to cut off shorter wavelengths. Such coloured oil droplets are ubiquitous in reptiles and birds, and are retained as colourless droplets in marsupials (Arrese *et al.*, 2000; Arrese *et al.*, 2005) and in the platypus (*Ornithorhynchus anatinus*; O'Day, 1938b), but not in the echidna (*Tachyglossus aculeatus*; Young and Pettigrew, 1991).

The ratio of rods to cones in the retina also provides an insight into activity patterns of animals. With the major exception of the cone-rich retinae of ground and tree squirrels from the Sciuridae family of the rodents (reviewed in Hunt *et al.*, 2009a), the retinae of diurnal placental mammals remain rod-dominated, with rod-to-cone ratios of 20:1, 25:1 and 30–40:1 found in human, sheep and rabbit, respectively (Braekevelt, 1983; Vaney *et al.*, 1991; Walls, 1942). By comparison, rod-to-cone ratios in the fat-tailed dunnart (40:1) and honey possum (20:1) are considerably lower than expected for species commonly described as nocturnal, whilst the numbat shows the most extreme level of diurnal adaptation, with a cone-dominated retina with a rod-to-cone ratio of 1:13. In the nocturnal North and South American opossums, rods outnumber cones by 50–120-fold (Kolb and Wang, 1985) and 130-fold (Ahnelt *et al.*, 1995), respectively.

Retinal ganglion cell distribution in the retina

Whilst the photoreceptors absorb and transform light into a neuronal image, the retinal ganglion cells transmit the information to the brain, with their axons acting as the only link between the eye and the brain. The distribution of retinal ganglion cells, unique to each species, reflects the importance that an animal places on observing objects in specific regions of their visual field (Rodieck, 1973; Hughes, 1975, 1977, 1985; Ito and Murakami, 1984; Collin and Pettigrew, 1988a, 1988b; Collin, 1997; Shand *et al.*, 2000a, 2000b). For instance, an *area centralis*, where a dense concentric distribution of ganglion cells confers greater visual acuity for depth and distance perception, is common in predators (Walls, 1942; Stone, 1965; Tansley, 1965; Peichl, 1992a) and in arboreal species with frontally placed eyes and extensive binocular vision, such as primates. A visual streak, where the increased regional ganglion cell density extends across the retina in a position corresponding to the retinal image of an animal's horizon (Hughes, 1975; Hokoc and Oswaldo-Cruz, 1979), is found in prey species and those occurring in open habitats, whilst a concentric distribution of ganglion cells is associated with an arboreal condition (Hughes, 1975; Hebel, 1976; Hughes, 1977; Tancred, 1981; Harman *et al.*, 1999).

Retinal ganglion cell distribution in marsupials, as in placentals, shows a variety of topographic patterns. In the numbat (Arrese *et al.*, 2000), honey possum (Dunlop *et al.*, 1994), brush-tailed possum (Freeman and Tancred, 1978), tree kangaroo (*Dendrolagus doriana*; Hughes, 1975), North American (Kolb and Wang, 1985) and South American opossums (Hokoc and Oswaldo-Cruz, 1979), ganglion cells are distributed in a concentric arrangement, reflecting their arboreal habits. The density of ganglion cells forming the *area centralis* varies according to their requirements for visual acuity. In contrast, the red (*Macropus rufus*) and grey (*Macropus fuliginosus*) kangaroo (Dunlop *et al.*, 1987) and hairy-nosed wombat (*Lasiorhinus latifrons*) possess a strong visual streak correlating with their terrestrial lifestyle in an open habitat. Both specialisations are present in the fat-tailed dunnart, a predator and prey species (Arrese *et al.*, 1999), as well as in the quokka (Beazley and Dunlop, 1983), southern brown bandicoot, pademelon wallaby (*Thylogale billardierii*), Tasmanian devil and the tammar wallaby (Tancred, 1981), representing their visual priorities in relation to feeding strategies, locomotion and habitats.

Photoreceptor/retinal ganglion cell relationships

Depending on the distribution of cone types, the retina can be further divided into specialised regions responsible for visual capabilities such as colour discrimination, visual acuity and sensitivity (Ahnelt *et al.*, 1995; Hemmi and Grunert, 1999; Peichl *et al.*, 2000). These cones fall into four subtypes based on morphological criteria and into two classes based on immunocytochemistry. The nocturnal South American opossum has a rod-dominant retina with only a sparse population of cones (Ahnelt *et al.*, 1995, 1996). The antibody OS-2 that labels S-cones (Szel *et al.*, 1993) identifies cones that are present at very low densities across the entire retina. Another set of single cones is labelled by the COS-1 antibody which labels L-cones (Szel *et al.*, 1993). These cones constitute the dominant cone type in the *area centralis*. In addition, COS-1 labels a set of double cones with a principal member containing a colourless oil droplet, and a set of single cones with colourless oil droplets. These latter two cone types are concentrated in the inferior, non-tapetal half of the retina; their absence from the *area centralis* indicates a functional specialisation that may be related to vision in mesopic conditions.

The tammar wallaby was the first Australian marsupial to be studied in detail from this perspective (Hemmi and Grunert, 1999). This diprotodont species also has a preponderance of rods with a λ_{max} at 501 nm (Hemmi *et al.*, 2000). The OS-2 and COS-1 antibodies demonstrated the presence of S- and L-cones, with the distribution of these photoreceptor types and density of ganglion cells dividing the retina into three regions; a dorsal retina with low ganglion cell and low cone density, a central horizontal band with high ganglion cell and cone density, and a ventral retina with low ganglion cell numbers, but high L-cone density. The central region would be expected to give good spatial resolution, whereas the ventral region would provide high contrast sensitivity, but low acuity. Another diprotodont, the quokka, belonging to the same family as the tammar wallaby, shows

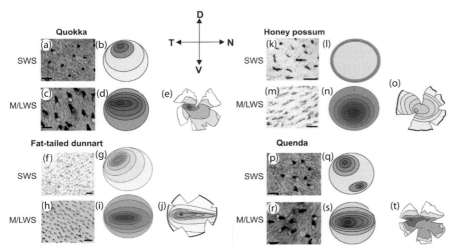

Figure 9.4 Photomicrographs of wholemounts treated with antisera labelling S- and L-cones in the quokka (a, c), fat-tailed dunnart (f, h), honey possum (k, m) and quenda (p, r). Topographies for each cone type in the quokka (b, d), dunnart (g, i), honey possum (l, n), and quenda (q, s): darker shades indicate highest densities and lighter shades, lowest densities. Retinal ganglion cell topographies in the quokka (e), dunnart (j), honey possum (o) and quenda (t), showing distributions similar to L-cones.

a similar distribution of cones with S-cones concentrated in the dorsoventral retina. With moderate to high L-cone and ganglion cell densities in the dorsal retina (Figure 9.4), this arrangement would optimise for spectral discrimination in the inferior field of view (Arrese et al., 2005).

The fat-tailed dunnart, a small polyprotodont with an arrhythmic lifestyle, also has a rod-dominated retina (Arrese et al., 1999, 2003). Cone topology is similar to the tammar wallaby with S-cones concentrated in the dorsoventral retina along with moderate numbers of L-cones to optimise for spectral discrimination in the inferiocentral visual field. The central retina has a high density of L-cones, which coincides with high ganglion cell numbers and would be expected to give enhanced visual acuity. Finally, the ventral retina shows higher convergence of a high spatial density of L-cones onto lower numbers of ganglion cells (Arrese et al., 2003), presumably to provide high contrast sensitivity.

The honey possum, a diprotodont, differs from the wallaby, quokka and dunnart in having a widespread distribution of S- and L-cones across the entire retina at spatial densities exceeding those of the other three species (Arrese et al., 2003); this would enhance spectral discrimination over the entire field. S-cones are at a higher density in the peripheral retina and may be associated with an arboreal habitat, as reported for the tarsier (Hendrickson et al., 2000) and lemur (Dkhissi-Benyahya et al., 2001). The central and mid-ventral retina have a higher proportion of cones (Arrese et al., 2002a), which coincides with a high ganglion cell density and hence lower convergence to afford good colour discrimination and acuity, important in this species for assessing the maturity of nectar-producing flowers.

The polyprotodont southern brown bandicoot or quenda, is mainly nocturnal, but may be active during the day. In this species, the density of S-cones shows separate peaks in the nasoventral and dorsotemporal retina (Arrese et al., 2005). L-cone density decreases concentrically from a horizontal streak in the mid-temporal region. Finally, the numbat, another polyprotodont, has a cone-dominated retina in keeping with its diurnal behaviour (Arrese et al., 2000).

Visual pigments

All vertebrate photopigments are based on an opsin protein that is covalently linked via a Schiff base (SB) to the chromophore (reviewed in Bowmaker, 2008; Bowmaker and Hunt, 2006). Visual pigments are found in retinal photoreceptors and are subdivided in vertebrates into a rod or Rh1 class that is generally restricted to rod photoreceptors and four different cone classes distinguished on the basis of the spectral sensitivity (λ_{max}) and amino acid sequence of

their respective opsins: longwave-sensitive (LWS) with λ_{max} 500–570 nm, middlewave-sensitive (MWS or Rh2) with λ_{max} 480–530 nm, and two shortwave-sensitive classes, SWS2 with λ_{max} 400–470 nm and SWS1 with λ_{max} 355–445 nm. In all classes, the spectral sensitivity of the pigment arises from interactions between the chromophore, 11-cis-retinal, and the amino acid residues that form the retinal-binding pocket of the opsin. In vertebrate visual pigments with λ_{max} values >385 nm, the SB is protonated, with a negatively charged residue at site 113 (usually Glu113) acting as a counter-ion to stabilise the proton of the SB (reviewed in Hunt et al., 2007).

Phylogenetic analysis based on gene sequence identity shows that the evolution of cone pigments preceded the rod pigment (Okano et al., 1992). Cone opsin sequences show an overall identity of around 40%. In contrast, the MWS cone (Rh2) and rod (Rh1) opsins show a much higher identity of around 80%, indicating a more recent separation of the Rh1 and Rh2 gene lineages and consistent with the origin of the Rh1 rod opsin gene from an ancestral duplication of the Rh2 cone opsin gene.

A nocturnal phase that is thought to mark the early evolution of the mammals around 150–200 million years ago may be the cause of the reduction in the number of cone pigment genes to only two classes. In marsupial and placental mammals, the LWS gene is paired with the SWS1 gene, with the loss of the SWS2 and Rh2 genes. The spectral sensitivity of the ancestral vertebrate SWS1 pigments peak in the ultra violet (UV) at around 360 nm (reviewed in Hunt et al., 2007), but except for some species of rodents (Yokoyama et al., 1998; Calderone and Jacobs, 1999; Jacobs et al., 2003), this is shifted in placental mammals into the violet region of the spectrum. The key residue for UV-sensitivity in all non-avian UVS pigments (the UVS pigments of birds are not ancestral and use another mechanism to shift the λ_{max} into the UV; reviewed in Hart and Hunt, 2007) is the presence of Phe at site 86. This arrangement is retained in the SWS1 pigments of muroid rodents with UVS pigments. It is substituted in all non-avian VS pigments by either Tyr (Cowing et al., 2002; Fasick et al., 2002; Carvalho et al., 2006), Val (Parry et al., 2004) or Ser (Yokoyama et al., 2005). In marsupials, the retention of Phe86 in a UVS pigment is more widespread, with examples from the diprotodonts such as the honey possum (Cowing et al., 2008) and polyprotodonts such as the fat-tailed dunnart (Strachan et al., 2004; Cowing et al., 2008), the grey short-tailed and big-eared opossums (Hunt et al., 2009b) and the quenda (Arrese et al., 2005) (Table 9.2). VS pigments possess a Phe86Tyr substitution as seen in the diprotodonts, the quokka (Arrese et al., 2005) and tammar wallaby (Deeb et al., 2003). Since Phe86 is retained by the UVS pigment of the honey possum, it would appear that the shift to VS pigments with a Phe86Tyr substitution occurred

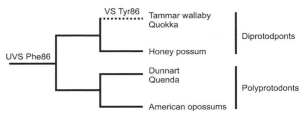

Figure 9.5 Evolution of SWS1 pigments in marsupials. The ancestral mammalian condition was UVS with Phe86. Only the Macropodiformes branch of the diprotodonts has a VS pigment (dotted line) with Tyr86.

only in the Macropodiformes lineage of kangaroos and wallabies (Figure 9.5).

The LWS pigments in marsupials shows less variability in spectral peak. In placental mammals, it is now well established that the amino-acid substitutions that are responsible for tuning the LWS pigments reside at sites 164, 261 and 269 (Merbs and Nathans, 1992; Asenjo *et al.*, 1994; Yokoyama and Radlwimmer, 2001), with polar residues present at these sites in the more longwave-shifted pigments. Site 164 is invariant with Ala in marsupials studied to date, but sites 261 and 269 vary in line with the peak sensitivities of the pigments (Table 9.2) (Deeb *et al.*, 2003; Strachan *et al.*, 2004; Arrese *et al.*, 2006a; Cowing *et al.*, 2008). Thus, the 557 and 551 nm pigments of the honey possum and quenda have Tyr261 and Thr269, the 538 nm pigment of the quokka has Tyr261 and Ala269, and the 530 nm pigments of the tammar wallaby and fat-tailed dunnart have Phe261 and Ala296 (Table 9.2). The λ_{max} for the LWS pigment of the platypus was determined by *in vitro* expression to be at 550 nm and this is consistent with the presence of Ala164, Tyr261 and Thr269 (Davies *et al.*, 2007; GenBank accession number EF050078). The λ_{max} for the LWS pigment of the echidna has not been determined spectrophotometrically, but can be inferred from the residues at the three sites, Ser180, Tyr261 and Thr269, to be around 560 nm (GenBank accession number EU636011). The use by marsupials and monotremes of the same sites for spectral tuning of LWS pigments as placental mammals (and other vertebrates) is another example of convergent evolution in the tuning of these pigments, implying that substitution at only a few amino-acid sites within the opsin protein are compatible with retaining a fully functional pigment.

An unexpected feature of the cone population in the retina of some Australian marsupials is a class of M-cones in addition to S- and L-cones. This was first reported for the fat-tailed dunnart and the honey possum (Arrese *et al.*, 2002b), and subsequently extended to the quokka and quenda (Arrese *et al.*, 2005). The presence of three cone classes offers the possibility of trichromatic colour vision and this potential was confirmed behaviourally in the fat-tailed dunnart

Table 9.2 Peak absorbance and amino-acid residues at sites 164, 261 and 269 in marsupial LWS pigments.

		Amino acid sites		
	λmax (nm)	164	261	269
Honey possum	557	Ala	Tyr	Thr
Pygmy possum	nd	Ala	Tyr	Thr
Quokka	538	Ala	Tyr	Ala
Tammar wallaby	530	Ala	Phe	Ala
Quenda	551	Ala	Tyr	Thr
Numbat	nd	Ala	Tyr	Ala
Fat-tailed dunnart	530	Ala	Phe	Ala
Human L	565	Ser	Tyr	Thr
Human M	530	Ala	Phe	Ala

nd – no data

Table 9.3 Peak spectral sensitivities of rods and M-cones.

	Peak sensitivities (nm)	
	Rod	M-cones
Fat-tailed dunnart	509	512
Honey possum	502	505
Quokka	505	502
Quenda	508	509

(Arrese *et al.*, 2006a). Interestingly, immunocytochemistry using the JH455 and JH492 antibodies which are specific for S-cones and L-cones respectively, failed to label a subset of cones in the retinae of the dunnart and honey possum (Arrese *et al.*, 2003), indicating that these cones did not express either the SWS1 or LWS pigments. Possible molecular explanations for the third cone pigment include a retained *Rh2* or *SWS2* gene or the duplication of the *LWS* gene, as has occurred in Old World primates (Nathans *et al.*, 1986). However, an extensive search of the genome of the fat-tailed dunnart in two laboratories (Strachan *et al.*, 2004; Cowing *et al.*, 2008) has failed to identify such genes. A second rod *Rh1* pigment gene was, however, identified (Cowing *et al.*, 2008); the two *Rh1* genes are identical within their coding sequences, but differ for an indel of 19 base pairs in intron 2 and four single base-pair substitutions and an indel of two base pairs in intron 3. Such major differences in intron sequences are generally associated with duplicated copies rather than alleles of a single gene. One copy is undoubtedly expressed in rod photoreceptors, but the authors have advanced the hypothesis that the second copy has acquired new regulatory sequences that target expression to a subset of cone photoreceptors. This would provide an explanation for two features of the M-cones; in all four species, the λ_{max} is very similar to the peak of the rod pigment in the same species (Table 9.3), and the photochemical properties of the M-cones, with a post-bleach build-up of photoproduct that absorbs below 430 nm, are similar to rods. It would also explain why a subset of cones did not show any labelling with the JH455 and JH492 antibodies (Arrese *et al.*, 2003). Such heterologous expression of visual pigments is not unknown. The blue sensitive cones and green rods of the tiger salamander, *Ambystoma tigrinum*, both express the same SWS2 cone pigment (Ma *et al.*, 2001), thereby providing evidence that cone pigments can function with rod transducin. The converse situation proposed for the marsupial MWS cones expressing a rod pigment may reasonably be expected to be fully functional.

The platypus and echidna, egg-laying protherian mammals belonging to the order Monotremata that diverged from the marsupial/placental around 200 mya, would appear to be dichromats. However, their colour vision system differs from that in all other mammals in that uniquely they have discarded the *SWS1* gene in favour of the *SWS2* gene (Davies *et al.*, 2007; Wakefield *et al.*, 2008), with a remnant of the *SWS1* gene retained as a non-functional pseudogene in the platypus genome (Davies *et al.*, 2007; Figure 9.6). This is then paired with the *LWS* gene to give dichromatic colour vision. The spectral peaks of the corresponding SWS2 and LWS pigments of the platypus are at 451 nm and 550 nm, respectively. The LWS peak at 550 nm is similar to that found for LWS pigments in many marsupial and placental mammals and the SWS2 peak at 451 nm is at only a marginally longer

wavelength than the violet-sensitive SWS1 pigments of some placental mammals (Hunt *et al.*, 2007). The dichromacy of the platypus is not dissimilar therefore to that of the tree squirrel *Sciurus carolinensis*, with an SWS1 pigment peaking at 440 nm (Carvalho *et al.*, 2006). The presence of the *SWS2* gene in monotremes also means that ancestral mammals prior to the protherian/therian split retained both *SWS* genes which, in combination with the *LWS* gene, would have provided the basis for trichromacy at the base of the mammalian radiation.

Melanopsin and the entrainment of circadian rhythms

A series of recent studies has shown that the detection of light for the entrainment of the circadian timing system is achieved by photon-capture by light-sensitive ganglion cells in the inner retina (Freedman *et al.*, 1999; Lucas *et al.*, 1999). These intrinsically photosensitive retinal ganglion cells (ipRGCs) project to the circadian pacemaker, the suprachiasmatic nuclei (SCh), as well as several other photoresponsive centres in the brain (Berson *et al.*, 2002; Hattar *et al.*, 2002; Sekaran *et al.*, 2003). The photosensitive pigment expressed in ipRGCs and responsible for photon-capture has been identified as melanopsin (Melyan *et al.*, 2005; Panda *et al.*, 2005; Qiu *et al.*, 2005).

Two melanopsin genes, *Opn4m* and *Opn4x*, are found in non-mammalian vertebrates, but only *Opn4m* has been described so far in placental mammals (Bellingham *et al.*, 2006), and the same situation would appear to pertain in

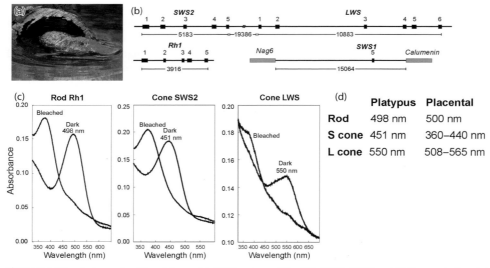

Figure 9.6 Visual pigments of the platypus (a). (b) Structure of the *Rh1*, *SWS1*, *SWS2* and *LWS* genes in the platypus genome. Intact *Rh1* and *SWS2* genes, each with five exons (black boxes), and *LWS* gene with six exons are shown. Only exon 5 of the *SWS1* pseudogene is present. Note that the *SWS2* and *LWS* genes are closely linked, with only 19.4 kb of spacer DNA. (c) *In vitro* expression and spectrophotometric analysis of the Rh1, SWS2 and LWS pigments. (d) Spectral peaks of platypus pigments compared to typical placental mammal. Adapted from Davies *et al.* (2007) with permission from Elsevier.

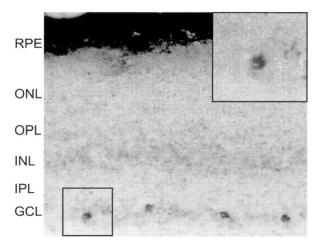

RPE

ONL

OPL

INL

IPL

GCL

Figure 9.7 *In situ* hybridisation showing retinal expression of melanopsin in adult dunnart retina. *Opn4* staining is restricted to a subset of cells in the ganglion cell layer. GCL – ganglion cell layer; IPL – inner plexiform layer; INL – inner nuclear layer; OPL – outer plexiform layer; ONL – outer nuclear layer; RPE – retinal pigmented epithelium. Adapted from Figure 3 of Pires *et al.* (2007).

marsupials (Pires *et al.*, 2007). *In silico* analysis has shown that only *Opn4m* is present in the genome of the grey opossum (Bellingham *et al.*, 2006) and is the only melanopsin to be expressed in the retina of the fat-tailed dunnart (Pires *et al.*, 2007) (Figure 9.7). For both species, sequence similarity and exon/intron structure identity confirms that this is the orthologue to the gene found in placental mammals, implying that *Opn4x* was lost from the mammalian genome prior to the marsupial/placental split. As found in placentals, *Opn4m* expression is limited to a low-density subset of ganglion cells that are irregularly distributed across the retina.

Visual pathways

The dorsal lateral geniculate nucleus of the thalamus

The visual pathways of marsupials follow an essentially similar plan to that of placental mammals, the major projections of the retina being to the dorsal lateral geniculate nucleus of the thalamus (DLG) and hence to the primary visual (striate) cortex (V1), or to the superior colliculus (SC).

Marsupials show considerable differences in the cytoarchitecture and patterns of retinal terminations in the DLG (Sanderson *et al.*, 1984; see Chapter 6 for a detailed review). In two species of rat kangaroo, *Bettongia gaimardi* and *Potorous tridactylus*, the cytoarchitecture ranges from relatively simple to complex with several well-defined cell laminae, each associated with input from a single eye. Both show a common basic pattern of three bands of retinal terminations from the contralateral eye and four from the ipsilateral eye. The wallabies and kangaroos display a more complex DLG architecture and patterning of retinal terminal bands. Bilateral retinal projections within the same DLG lamina are unusual and the number of terminal bands reaches 10 in the grey kangaroo, comprising four from the contralateral eye and six from the ipsilateral eye. The segregation of retinal inputs to the DLG in marsupials is likely to be the result of evolutionary factors that differ from those which have produced ocular segregation and complex lamination in several lines of placental mammals. The minor visual centres such as the pretectal (see Chapter 6) and accessory optic nuclei also follow a similar basic organisation in marsupial and placental mammals (Coleman and Beazley, 1988).

Organisation along the visual pathway

In order to achieve binocularity, the nerve fibres from the medial part of each retina cross at the optic chiasm to project to the opposite side of the brain, while the lateral retinal fibres continue on the same side. As a result, each half of the brain receives information about the contralateral visual field from both eyes. In placentals in general, the chiasm is homogeneous (Hoyt and Luis, 1963; Baker and Jeffery, 1989; Reese and Baker, 1990; Jeffery, 2001) with crossed and uncrossed fibres mixed before they enter the chiasm. By contrast, in marsupials fibres are gathered into fascicular groups prechiasmatically (Jeffery and Harman, 1992; Guillery and Taylor, 1993), with uncrossed fibres being restricted to the lateral region of the optic nerve and chiasm (Jeffery and Harman, 1992; Harman and Jeffery, 1995). The pattern of decussation would appear to be significantly different between placentals and marsupials, although the chiasm of a highly visual mammal, the tree shrew (*Tupaia glis*), shows some similarities to marsupials with uncrossed axons confined to lateral regions (Jeffery *et al.*, 1998). The chiasm in the tree shrew differs, however, by the presence of thick collagen bundles that serve to separate the optic fascicles.

In contrast to that of rodents, a high degree of order is maintained within the visual pathways of the quokka and dunnart (Chelvanayagam *et al.*, 1998). Axons from the nasal and temporal regions of the retina dominate the medial and lateral halves of the nerve, whereas axons from the dorsal and ventral retina exchange locations as they pass along the nerve to correct for the inversion of the dorsoventral retinal axis imposed by the lens (Dunlop *et al.*, 2000). This arrangement is maintained by decussating axons through the chiasm and throughout the tract, with nasal and ventral axons invading the SC medially, and temporal and dorsal axons invading laterally. Each retinal quadrant terminates therefore in the appropriate retinotopic sector of the SC. The organisation has implications for understanding the location of cues necessary for the establishment of retinotopic projections during development.

Reduced numbers of uncrossed retinofugal pathways and abnormally low retinal cell densities have been described in albinos of a number of placental mammals and the same developmental defect extends to the optic chiasm of marsupials, as seen in an albino wallaby of the species *Macropus rufogriseus*, where the uncrossed pathways were smaller and the cell density in the retinal ganglion cell layer in the area centralis reduced (Guillery *et al.*, 1999). Therefore, in spite of the differences in decussation between marsupials and placentals, albinism causes a similar developmental defect in both.

Higher-order visual centres

The first functional analysis of the visual cortex using evoked potential techniques was carried out in the North American opossum (Lende, 1963) and this has now been extended to several South American opossums (Christensen and Hill, 1970a, 1970b; Castro and Saraiva, 1971; Magalhaes-Rocha-Miranda *et al.*, 1973; Sousa *et al.*, 1978; Volchan *et al.*, 1988; Kahn *et al.*, 2000), the brush-tailed possum (Crewther *et al.*, 1984), the tammar wallaby (Vidyasagar *et al.*, 1992) and the northern quoll (*Dasyurus hallucatus*; Rosa *et al.*, 1999). The visual cortex contains at least two cortical areas, V1 (or striate cortex) and V2, with each organised topographically. The primary input into V1 in marsupials, as in most placental mammals, is from DLG (reviewed in Karlen and Krubitzer, 2007), with more moderate inputs from the lateral posterior nucleus (LP). In contrast, the primary input into V2 is from LP, with only sparse input from DLG.

As in placental mammals, the V1 cortical field is located at the caudal pole of the occipital lobe and contains a complete visuotopic representation of the contralateral visual field with a representation of the upper visual quadrant located caudolaterally (see Chapter 8; Figures 8.5, 8.6) and the lower visual quadrant representation located rostromedially. The horizontal meridian bisects the upper and lower visual quadrants, and the vertical meridian forms the rostrolateral boundary of V1 (reviewed in Karlen and Krubitzer, 2007). The second visual field (V2) has been mapped only in the northern quoll (Rosa *et al.*, 1999) and again as found in placentals, contains a complete representation of the visual hemifield with the vertical meridian represented at the caudomedial border of the field, adjacent to the vertical meridian border of V1. The horizontal meridian bisects the field, with the lower visual quadrant represented rostromedially and the upper visual quadrant caudolaterally.

Comparison of the anatomical composition of V1 of the marsupial species, dunnart and quokka, with macaque monkey has demonstrated that there are many shared features (Tyler *et al.*, 1998). For example, the absolute size of pyramidal neurons is remarkably similar, despite the large differences in overall area and thickness of V1, as are their specific dendritic branch patterns and the patterns of distribution of intrinsic axons. Given the evolutionary distance of over 135 million years and the very different lifestyles, the similarities would imply that these features have been evolutionarily conserved and retained from an early common ancestor.

The corpus callosum has been implicated as the major commissure mediating the interhemispheric transfer of visual information in placental mammals. As marsupials and monotremes lack a corpus callosum (Owen, 1837), such a transfer may not be possible. However, in experiments on the brush-tailed possum where a mid-sagittal transection of the optic chiasm restricted retinal input to the ipsilateral side of the brain only, it was found that high levels of interocular transfer nevertheless occurred (Robinson, 1982), the primary form of interhemispheric communication being via an enlarged anterior commissure (Martin, 1967; Ebner, 1969; Heath and Jones, 1971). Axon crossings within the anterior commissure are topographically organised (Ashwell *et al.*, 1996b) and the general pattern of formation follows closely that of the corpus callosum in placentals (Ashwell *et al.*, 1996a, 1996b; Cummings *et al.*, 1997; Shang *et al.*, 1997).

Conclusion

The representative species discussed in this chapter highlight the malleability of marsupial visual organisation in relation to ecological and evolutionary pressures. One of the striking features of marsupial vision is the presence of trichromacy in representative species from the two major marsupial lineages, which contrasts with the situation in placental mammals where trichromacy is restricted to primates. Marsupial trichromacy raises numerous questions that remain to be answered. To date, nothing is known about retinal circuitry in marsupials and knowledge of their visual pathways needs to be extended. Similarly, we are just beginning to explore the visual systems of monotremes, but have already established that colour vision, although dichromatic, has followed an alternative route to that in marsupials and placentals, with the retention of a different short-wavelength-sensitive visual pigment. In summary therefore, further studies are needed to extend our knowledge of information processing in the retina and visual cortex and the visual capabilities of these fascinating mammals.

10

Somatosensory system

L. R. Marotte, C. A. Leamey and P. M. E. Waite

Summary

The most widely studied aspect of the somatosensory system in metatherian (marsupial) mammals is that of the trigeminal pathway, in particular the pathway concerned with the whiskers, and this is reflected in the content of this chapter.

The structure and innervation of the mystacial vibrissae, or whiskers, on the snout has been described in a number of Australian species. Broadly speaking, the structure and afferent innervation is similar to that of eutherians and the American opossum. Receptors include Merkel cells and lanceolate, lamellated and free nerve endings. Intervibrissal pelage hair in several species is innervated by lanceolate and free nerve endings.

Muscle spindles in limb muscles have been well described in the brush-tailed possum (*Trichosurus vulpecula*) and resemble those of eutherian species, as do Golgi tendon organs. Three types of receptors have been identified in the joint capsules of the kowari (*Dasyuroides byrnie*): Ruffini corpuscles, lamellated corpuscles and free nerve endings. Lamellated corpuscles resembling Pacinian corpuscles are present in the interosseus region of the legs of macropods, and response properties suggest that they are responding to ground-borne vibration.

Little detailed information is available for the inputs of peripheral mechanoreceptors via the dorsal root ganglia to the dorsal column nuclei, although the medullary course and projection patterns of cervical roots in the brush-tailed possum and the potoroo (*Potorous tridactylus*) are reported to be similar to those of eutherians. In the brush-tailed possum, projections of the dorsal column nuclei have been shown to cross contralaterally in the medial lemniscus to innervate the posterior nuclear group (Po) and the ventroposterolateral nucleus (VPL) in the thalamus.

The only species for which a systematic description of the trigeminal brainstem nuclei exists is the tammar wallaby (*Macropus eugenii*). These nuclei, innervated by the central processes of the trigeminal ganglion cells comprise principalis and the spinal subnuclei (oralis, interpolaris and caudalis). Patches of succinic dehydrogenase (SDH) and cytochrome oxidase (CO) activity, thought to correlate with the mystacial vibrissae, are present in all nuclei except oralis, although they are less clear than in rodents. These patches are clearer in the brush-tailed possum.

The somatosensory thalamus of Australian diprotodont species is well differentiated, in contrast to some other marsupial species. The ventroposteromedial nucleus (VPM) and VPL, comprising the ventrobasal complex (VPL/M), are cytoarchitecturally distinct and receive tactile input from the face and head via the trigeminal system and from the trunk and limbs via the dorsal column nuclei, respectively. Together they receive a somatotopic representation of the head and body. In the brush-tailed possum and the tammar, VPM contains dorsoventrally aligned cellular aggregations corresponding to the mystacial vibrissae. The posterior nuclear group (Po) in the brush-tailed possum has been shown to receive input from the dorsal column nuclei and the spinal cord and presumably from the trigeminal system, given its cortical connections. Both VPL/M and Po in turn project to somatosensory cortex.

Up to five somatosensory cortical areas have been described in Australian marsupial species. S1 and S2 contain a complete or near complete representation of the body surface, while a parietal ventral area (PV) and two areas rostral and caudal to S1 (R and C, respectively) have been delineated in some species. In the brush-tailed possum there is a distinct barrel field in S1 in both Nissl- and SDH-stained material, with those areas mapping the face and forepaw especially clear. Such barrels of high cell density only appear to be present in the Phalangeridae. Although no barrels are present in the tammar wallaby, there are whisker-related patches of SDH and CO activity, but they are much less clear than in the possum. Somatosensory cortex has reciprocal connections with VPM, VPL and Po, and S1 has intracortical connections with both higher-order somatosensory and other cortical regions.

The major difference between eutherians and metatherians is that metatherians are born in an immature state with a large part of development taking place postnatally, and in many species it is protracted. The tammar is the species for which most is known on somatosensory development and, in particular, the trigeminal system. At birth, vibrissal follicles are already innervated, but the follicles and receptors supplying them do not resemble the adult structure until postnatal day (P) 119. The period separating these events is much longer than in the rat (3 weeks). Similarly, development of pelage hairs and their innervation takes around 200 days. Maturation of all levels of the pathway is prolonged. At birth, trigeminal afferents are distributed throughout the rostrocaudal extent of the trigeminal complex in the brainstem, but whisker-related patterns do not begin to appear until P40. In the thalamus, trigeminothalamic afferents reach VPM by P15 and electrically evoked synaptic responses are present from this time, but it is not until P52 that whisker-related patterns begin appearing and excitatory responses become dominated by non-NMDA-mediated responses. Thalamic afferents reach the cortex by P15, well ahead of descending connections to the thalamus, which first appear at P60. Whisker-related patterns characterised by the clustering of afferents in layer 4Cx, the beginning of CO patches in the newly formed layer 4Cx, and of cells and their dendrites in layers 5Cx and 6Cx that project to the thalamus, all appear around the same time, at P72–76. A fast non-NMDA-mediated thalamocortical response dominates at this time, coincident with the onset of *in vivo* responses to whisker stimulation. The long delay between the arrival of afferents and pattern formation at each level suggests that maturation of target tissue, as well as signals from the periphery may play a role in pattern formation. This is supported by findings that an increase in the non-NMDA-mediated glutamatergic response, a measure of target maturation, correlates well with pattern formation in both thalamus and cortex.

Introduction

The somatosensory system encompasses those pathways that convey information from peripheral somatosensory receptors to the cerebral cortex. Sections in this chapter describe what is known of peripheral receptors and their innervation, ascending spinal pathways and medullary relay nuclei, somatosensory thalamus and somatosensory

cortex. Finally the development of these structures and the use of the whisker pathway to study pattern development in the nervous system are described.

Peripheral receptors and their innervation

The representation of the body surface in both the spinal cord and the brain reflects the distribution and density of the peripheral receptors. These can be divided broadly into three groups on the basis of function: mechanoreceptors, which are commonly encapsulated, and nociceptors and thermoreceptors, which are free nerve endings. They are primarily present in skin, muscles or joints.

Cutaneous receptors

For Australian marsupials, most information is available for the trigeminal system and mainly for the innervation of hairy skin of the face, including the mystacial vibrissae or sinus hairs, which are aligned in rows on the snout of the animal. Scant information is available for other structures innervated by the trigeminal nerve, but is included at the end of this section. The trigeminal nerve innervates the facial skin, oral and nasal mucosa, and structures such as the vibrissae, tongue and teeth. The mystacial vibrissae, upper lip and the post-orbital skin are innervated by its maxillary branch and contain a variety of encapsulated receptors and free nerve endings.

The structure of vibrissal follicles in the brush-tailed possum (Hollis and Lyne, 1974) and the tammar (Figure 10.1a) (Marotte *et al.*, 1992), is similar to that of eutherians (Rice *et al.*, 1986) although, like the rhesus monkey (Van Horn, 1970), they also lack a ringwulst. A thickening of the mesenchymal sheath, seen in both the possum and the tammar, does not appear to be simply a homologue of the ringwulst, because, unlike the ringwulst, it is densely innervated. Afferent innervation of the follicles has been described in a number of diprotodont (brush-tailed possum (Hollis and Lyne, 1974; Loo and Halata, 1991); tammar (Loo and Halata, 1991; Marotte *et al.*, 1992)) and polyprotodont (long-nosed bandicoot *Perameles nasuta*; kowari *Dasyuroides byrnei* (Loo and Halata, 1991)) Australian marsupial species. Broadly speaking, vibrissal innervation is similar in these species, in the polyprotodont North American opossum, *Didelphus virginiana* (Loo and Halata, 1991) and in eutherians (Rice *et al.*, 1986). Receptors include Merkel endings, which give slowly adapting type I responses in glabrous and hairy skin and probably subserve the same function in the vibrissae, lanceolate (Figure 10.1b) and lamellated endings and free nerve endings (Figure 10.1b). The functions

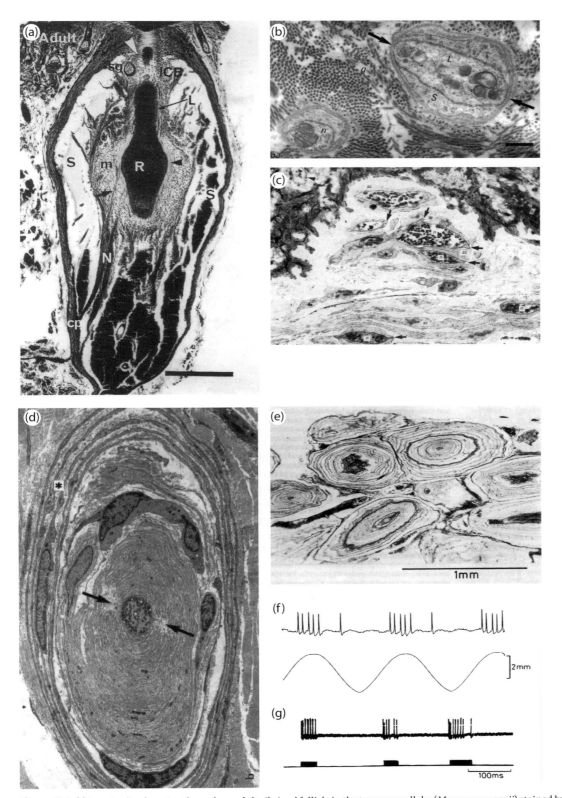

Figure 10.1 (a) Longitudinal section through an adult vibrissal follicle in the tammar wallaby (*Macropus eugenii*) stained by the Bodian method. Within the capsule (cp) the blood sinus is prominent (S). Branches of the deep vibrissal nerve (N) enter the mesenchymal thickening (m), giving rise to coarse branches (arrow) and fine branches (black arrowhead) adjacent to the root sheath (R) as well as lanceolate and lamellated (L) endings in the narrow waist region. The hair shaft in the root sheath is out of the plane of section. The inner conical body (ICB) has axons arranged circularly (white arrowhead) at the level of the sebaceous glands (sg); Scale bar = 0.5 mm. From Waite *et al.* (1994), Figure 1, with permission, copyright 1994 Wiley-Liss, Inc. (b) Electron micrograph of small lanceolate terminal (L) and free nerve ending (n) in the mesenchymal thickening of a vibrissal follicle of the tammar wallaby (*Macropus eugenii*). The terminal abuts

of the latter are not known, but lanceolate receptors are thought to be responsible for rapidly adapting responses (Iggo, 1974), and may detect acceleration of vibrissal deflection (Rice *et al.*, 1986). Free nerve endings are generally considered to mediate high threshold responses, but in the rat very few vibrissal responses are high threshold (Jacquin *et al.*, 1986) and their protected position in the wallaby follicle may indicate that they are low-threshold mechanoreceptors (Marotte *et al.*, 1992). Ruffini-like endings have been described in the rhesus monkey follicles (Halata and Munger, 1980), but were not observed in the marsupial species examined (Loo and Halata, 1991; Marotte *et al.*, 1992).

Intervibrissal pelage hairs in a number of marsupial species (Loo and Halata, 1991) are reported to be innervated by lanceolate and free nerve endings, although in the tammar they are rarely innervated (Marotte *et al.*, 1992). Loo and Halata (1991) also reported that facial guard hairs contain Merkel, lanceolate, pilo-Ruffini and free nerve endings, while in the tammar pilo-Ruffini endings are not seen (Marotte *et al.*, 1992).

Glabrous skin from the snout region in the Northern brown bandicoot (*Isoodon macrourus*) is richly innervated, having epidermal pegs with the suite of nerve endings typical of Eimer's organ: intra-epidermal nerve terminals, Merkel cells and corpuscular end-organs (Figure 10.1c), as well as free nerve endings in the dermis (Loo and Halata, 1985). The corpuscular endings have similarities to Meissner's corpuscles of the North American opossum and eutheria, such as the mole, dwarf pig and tree shrew, but differ in that they lack a capsule and the nerve terminal sometimes makes contact with the basal layer of the epidermis. Both the bandicoot and the opossum are nocturnal and the study noted that such innervation is indicative of the snout in these animals being a highly specialised organ for touch (Loo and Halata, 1985).

Intraoral receptors

Lamellated Pacinian corpuscles 'of the simple type' have been recognised in the apical region of the tongue of the tree kangaroo (*Dendrolagus ursinus*) (Kubota *et al.*, 1963), a native of New Guinea and Indonesia. Although two Australian species of kangaroo were also examined in the same study, there was no mention of the presence of corpuscles in these species. Information on the sensory innervation of the teeth is lacking for Australian marsupials, but in the South American opossum *Monodelphis domestica* three types of endings were observed: Ruffini and lamellated corpuscles, suggested to be part of the masticatory feedback system, and free nerve endings, which may act as thermo- or nociceptors (Schulze *et al.*, 1993).

Muscle and joint receptors

Muscle spindles in various limb muscles have been studied in detail for the brush-tailed possum (Jones, 1966a, 1966b) and resemble those of other mammals, including man, in both their structure and afferent and efferent innervation. Their distribution in forepaw lumbricals is not consistently related to the number of extrafusal fibres, confirming the view that it is the function of the muscle that drives the spindle content, not its size (Jones, 1966b). Golgi tendon organs, described as 'typical' were also noted near the musculotendinous junctions (Jones, 1966a).

In joint capsules of the kowari (Strasmann *et al.*, 1987), three types of receptors are seen in the fibrous layer: Ruffini corpuscles, lamellated corpuscles (Figure 10.1d) and free nerve endings. The latter two types have also been noted in the joints of the upper cervical column in *Monodelphis domestica* (Strasmann *et al.*, 1999). Lamellated corpuscles, similar to the Pacinian corpuscles of eutherian mammals, have also been described in the interosseous region

---

Caption for Figure 10.1 (Cont.)

the glassy membrane (g) while the free nerve ending lies within it. The lanceolate terminal is flanked by Schwann-cell processes (S), but is exposed to the basal lamina at each end (arrows); Bar indicates 0.5 μm. From Marotte *et al.* (1992), Figure 5a, with the permission of the Journal of Anatomy, Anatomical Society of Great Britain and Ireland and Blackwell Publishing. (c) Electron micrograph of a simple coiled corpuscle in the papillary layer under the base of an epidermal peg in the glabrous skin of the snout of the short-nosed bandicoot (*Isoodon macrourus*). Nerve terminals (n), Schwann-cell lamellae (sc) and terminal spikes (arrows) are present. An intra-epidermal nerve is indicated by an arrowhead. From Loo and Halata (1985), Figure 9, with the permission of Wiley-Blackwell. (d) Lamellated corpuscle in the joint capsule of the kowari (*Dasyuroides byrnie*), with a perineural capsule of four layers (*) and a symmetrical longitudinal cleft of the inner core (arrows); x 4000. From Strasmann *et al.* (1987), Figure 5b, copyright Springer-Verlag 1987, with the kind permission of Springer Science and Business Media. (e) Transverse section through a cluster of lamellated corpuscles in the interosseus membrane of the tammar wallaby (*Macropus eugenii*); Masson's trichrome stain. From Gregory *et al.* (1986), Figure 5B, copyright Springer-Verlag 1986, with the kind permission of Springer Science and Business Media and the authors. (f) Upper trace. Responses of one vibration-sensitive unit recorded in the interosseous nerve of the pademelon (*Thylogale billardierii*) to vibration at 12 Hz, applied where maximally effective in the interosseous region. Lower trace shows the voltage applied to the vibrator, calibrated in terms of the proportional displacement generated. From Gregory *et al.* (1986), Figure 1A, copyright Springer-Verlag 1986, with the kind permission of Springer Science and Business Media and the authors. (g) Reponses of a unit recorded in the interosseous nerve of the pademelon (*Thylogale billardierii*) to tapping lightly with the fingers on an upright mounted on the metal table supporting the animal. Lower trace shows the approximate timing of the taps. From Gregory *et al.* (1986), Figure 1B, Copyright Springer-Verlag 1986, with the kind permission of Springer Science and Business Media and the authors.

of the legs of macropod marsupials (Figure 10.1e) (Gregory *et al.*, 1986). The properties of mechanoreceptors recorded in the interosseous nerve, presumably representing the responses of these receptors, were investigated in the wallaby (*Thylogale billardierii*) (Figure 10.1f). Their functional properties are similar to eutherian Pacinian corpuscles, although less rapidly adapting, and it was speculated that they may detect ground-borne vibration (Gregory *et al.*, 1986).

Ascending spinal pathways and medullary relay nuclei

Spinothalamic tract

In eutherians, this pathway carries information on pain and temperature and also innocuous stimuli (Tracey, 2004b). Hemisection of the spinal cord in the possum revealed bilateral projections to the thalamus in the posterior nuclear group, (VPL) and the intralaminar system of nuclei (Rockel *et al.*, 1972). This is similar to that described for eutherians such as the rat (Tracey, 2004b) and more segregated than in the American opossum (Hazlett *et al.*, 1972).

Dorsal columns and nuclei

Cell bodies of peripheral mechanoreceptors in the dorsal root ganglia send their central processes to innervate dorsal column nuclei, including the gracile, cuneate and external cuneate nuclei (Tracey, 2004b). The medullary course and projection patterns of cervical dorsal root fibres have been described for the potoroo (Culberson and Albright, 1984) and in more detail for the brush-tailed possum (Culberson and Albright, 1984; Culberson, 1987), and are similar to eutherian species (Culberson and Albright, 1984). Fibres from most cervical roots separate into two as they travel rostrally into the cuneate and external cuneate nuclei. Fibres reaching the dorsal 'shell' region of the cuneate are topographically precise and mapping studies in the possum and other species (see Culberson, 1987 and below) show that this input is most extensive from cutaneous receptors in the distal extremity or forepaw. Thus distal dermatomes (C7–T1) provide a more extensive projection to the dorsal cuneate nucleus than proximal ones such as C2–C4, supplying the neck and shoulder. The ventrolateral and ventrocentral regions of the cuneate receive input from presumed low-threshold muscle afferents. The external cuneate nucleus also receives muscle afferents and in the possum this projection is large, from all cervical roots, and with a clear segmentotopic pattern, similar to that described for the

cat (Liu, 1956). Culberson (1987) noted that the extensive, well-developed cuneate system was not surprising, given the behavioural attributes of the possum, including dexterity in manipulating objects with the forepaws. However, he noted that the cells and the projections in the cuneate are not as precisely arranged as in those placental mammals with well-developed digital dexterity.

Information on the detailed somatotopic organisation of mechanosensory inputs to the cuneate/gracile nuclear complex is lacking for Australian marsupials, but has been reported for the North American opossum (*Didelphis marsupialis virginiana*; Hamilton and Johnson, 1973). Electrophysiological mapping showed that the somatotopy is similar to that in eutherians, although it was noted that the receptive fields were larger and had greater overlap compared to those seen in the racoon, for example. Consistent with the anatomical studies in the possum (above), the opossum cuneate receives a large projection from the hand, a smaller projection from the forearm and less from the upper arm and shoulder. The neck and pinnae were represented most laterally in the nucleus, whereas the gracile nucleus received inputs from the hindlimbs and feet.

In the brush-tailed possum, foci of SDH activity have been described in both the cuneate and gracile nuclei (Weller, 1983). This histochemistry is considered to reflect metabolic activity that is dependent on mitochondrial enzymic activity in dendrites and also in axon terminals (Wong-Riley, 1989), and has proved a useful marker for sensory-relay nuclei and their cortical terminations (Dawson and Killackey, 1987). Patches of high reactivity correlate with peripheral regions that have a high density of innervation (Goyal *et al.*, 1992).

Projections of the dorsal column nuclei in the brush-tailed possum have been shown to run contralaterally in the medial lemniscus to innervate Po and VPL in the thalamus (Clezy *et al.*, 1961; Rockel *et al.*, 1972), as in eutherians. These were reported to be relatively localised and discrete compared with the more widespread projections of the American opossum (Rockel *et al.*, 1972). Blumer (1963) noted in the quokka that the dorsal column projections also followed the usual mammalian plan.

Trigeminal brainstem sensory nuclei

The only systematic description of these nuclei, innervated by the central processes of trigeminal ganglion cells, is for the diprotodont Australian marsupial, the tammar (Waite *et al.*, 1994). The nuclear complex is similar cytoarchitecturally to other species (Olszewski, 1950; Torvik, 1956; Darian-Smith, 1973; Ma, 1991). The principalis nucleus (Pr5; atlas plates D23 to D25, W38 to W40), lying in the lateral pons between the trigeminal tract and the trigeminal motor nucleus, is the most rostral nucleus of the complex. It has fairly closely packed, often clustered, medium-sized cells and obliquely oriented bundles of fibres. All the spinal nuclei (Sp5 nuclear complex) feature fibre bundles running rostrocaudally. Subnucleus oralis (Sp5O; atlas plates D26, D27, W41 and W42) begins rostral to the facial nucleus, overlapping with the caudal pole of principalis rostrocaudally, and extending to the caudal pole of the facial nucleus. It is smaller with a lower density of cells, as has been described for rodents (Bates and Killackey, 1985; Ma, 1991). Subnucleus interpolaris (Sp5I; atlas plates D28 to D32 and W43 to W46) is irregularly shaped with sparse cells, in contrast to the densely packed cells described in rodents (Bates and Killackey, 1985). With both Nissl and SDH staining it appears heterogeneous, partly due to numerous fibre bundles crossing the nucleus, as well as the vagus nerve, which passes through it obliquely, as in primates. Subnucleus caudalis (Sp5C; atlas plates D32, W47 to W51) consists of a marginal layer, substantia gelatinosa and magnocellular layers.

In the tammar, all the above sensory trigeminal nuclei have higher SDH and CO activity compared to surrounding structures, as seen in other species (Belford and Killackey, 1978, 1979a, 1979b; Nomura et al., 1986; Noriega and Wall, 1991), with Pr5 showing the highest reactivity (Figure 10.2a). Particularly in rodents, patches of high reactivity reproducing the pattern of individual mystacial vibrissae on the snout, are present in Pr5, Sp5I and Sp5C (Bates and Killackey, 1985; Ma, 1991). In the tammar, patches are also seen in these nuclei, but only Pr5 (Figure 10.2a) and Sp5I display some regions where they are arranged in lines reminiscent of the rows of mystacial vibrissae. However, they are not clear enough to be identified with individual vibrissae. In the possum, SDH staining appears to more clearly reflect vibrissae-like patterns and these, as in rodents, are seen in all trigeminal subnuclei except Sp5O (Weller, 1983).

The major projection of Pr5 is contralaterally, to VPM of the thalamus, and at least in early stages of development there is no evidence of a projection to Po (Leamey et al., 1996). In the rodent, Pr5 also projects to Po (Chiaia et al., 1991), so this may be a genuine difference between the species or the projection to Po in the wallaby may form later in development. The connections of the spinal subnuclei have not been described in any Australian marsupial and there is scant information for other marsupials. In the South American opossum, Sp5C is reported to project to the inferior colliculus (Willard and Martin, 1983). In rodents, the thalamic projections of Sp5I and Sp5C include VPM, Po and ZI, like Pr5; Sp5O primarily innervates other brainstem nuclei, but also provides an input to Po and ZI. A number of other areas also receive input from the trigeminal brainstem nuclei and the reader is referred to Waite (2004) for an overview of connections in the rodent.

Somatosensory thalamus

The thalamus of the tammar is well differentiated and displays many features similar to the cat (Mayner, 1989c), while the brush-tailed possum (Rockel et al., 1972; Haight and Neylon, 1978a) appears to have features that lie between that of the tammar and the polyprotodont eastern quoll (Dasyurus viverrinus) (Haight and Neylon, 1981a, 1981b). Nevertheless, these species show more resemblance to each other than they do to the North American opossum, which has poorly differentiated thalamic nuclei (Oswaldo-Cruz and Rocha-Miranda, 1967).

Ventrobasal complex

This comprises VPM, which receives somatosensory input from the face and head via the trigeminal system, and VPL, which receives inputs from the trunk and limbs via the dorsal column nuclei and spinothalamic tracts. VPM and VPL in turn project to S1 of the somatosensory cortex (Haight and Neylon, 1978a). In the three Australian species mentioned above, VPM and VPL are cytoarchitecturally distinct (Rockel et al., 1972; Haight and Neylon, 1981a, 1981b; Mayner, 1989c; Leamey et al., 1996). In coronal section, VPL, a crescent-shaped nucleus, lies mediodorsal to the external medullary lamina and VPM is a distinct nucleus of densely packed cells, dorsal and medial to it (Figure 10.2b). Haight and Neylon (1978a) also identified a third small region in the brush-tailed possum, the posteromedial ventroposterior nucleus (VPP), which, on the basis of cortical connections, they considered part of the ventrobasal complex (Haight and Neylon, 1978b); however, this was only visible in some brains. In an early study, Dennis and Kerr (1961) noted that the hindlimb is represented lateral to the forelimb in VPL in the brush-tailed possum. In the tammar, VPM and VPL together receive a somatotopic representation of the head and body with the head having a disproportionately large representation (Faulks and Mark, 1982). The overall organisation is similar to that in eutherians (Jones, 2007) and to the North (Pubols and Pubols, 1966) and South American opossum (Didelphis marsupialis aurita) (Sousa et al., 1971).

In parts of VPM in the brush-tailed possum (Weller, 1980) and the tammar (Figure 10.2b) (Leamey et al., 1996), cells are arranged in dorsoventrally aligned bands separated by bundles of myelinated fibres, resembling the vibrissae-related barreloids in rodents (Van der Loos, 1976). VPM shows high CO activity compared to surrounding nuclei, with segmentation into bands of reactivity reflecting the arrangement seen with Nissl staining (Leamey et al., 1996). In horizontal sections these are seen as circles of reactivity, surrounded by a thin region of low activity corresponding to the myelinated

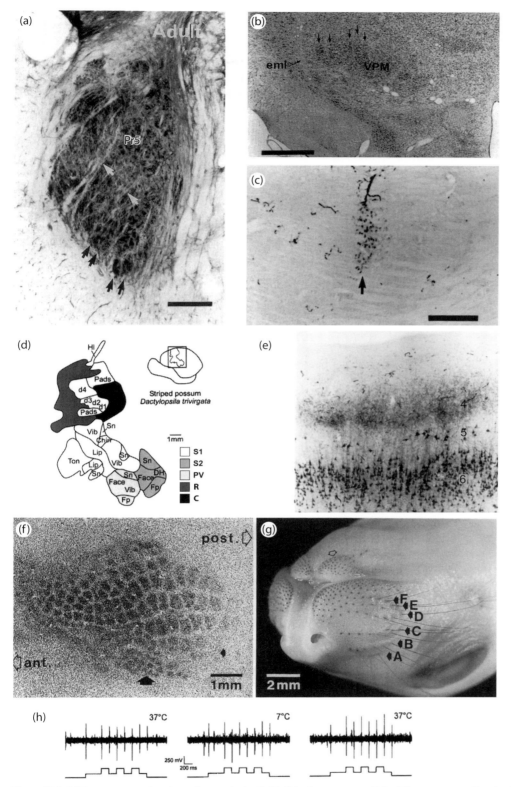

Figure 10.2 (a) Transverse section through nu. principalis (Pr5) in the tammar wallaby (*Macropus eugenii*) stained with succinic dehydrogenase. Dense patches of activity are seen aligned obliquely across the nu. (black arrows). Bundles of fibres (white arrows) also run obliquely across the nucleus; Bar = 0.5 mm. From Waite *et al*. (1994), Figure 7B, with permission, copyright 1994 Wiley-Liss, Inc. (b) Stained coronal section though the somatosensory thalamus of the tammar wallaby (*Macropus eugenii*) showing the normal histology of the ventroposteromedial nu. (VPM), which is characterised by intensely stained cells. Arrows point to the dorsoventrally aligned bands of cells in VPM; eml: external medullary lamina; bar = 2 mm. From Marotte *et al*. (1997), Figure 1b, with permission, copyright 1997 Wiley-Liss, Inc. (c) Unstained coronal section through VPM in the tammar wallaby (*Macropus eugenii*) showing the dorsoventrally aligned band of label

bundles surrounding the cell clusters (Leamey *et al.*, 1996). Anatomical and electrophysiological studies in rodents support the idea that inputs from individual facial vibrissae end on cell clusters in VPM and these in turn project to individual cortical barrels (Van der Loos, 1976; Petersen, 2003). It appears likely that a similar relationship exists in the brush-tailed possum and the tammar wallaby; electrophysiology has shown that vibrissae are represented in VPM of the tammar (Faulks and Mark, 1982) and that there is a strong correlation between vibrissae and the patches of CO or SDH activity seen in the somatosensory cortex of the tammar (Waite *et al.*, 1991) and possum (Weller, 1993). Further, a small injection of tracer in the physiologically identified area of whisker representation in somatosensory cortex of the tammar retrogradely labels a discrete band of dorsoventrally aligned cells in VPM which resembles the band seen in Nissl-stained sections (Figure 10.2c; Marotte *et al.*, 1997). Terminals within the band are also labelled, demonstrating the reciprocal connections between VPM and cortex.

VPL/M projects to the somatosensory cortex and the most detailed information on its cortical projections is known for the brush-tailed possum (Haight and Neylon, 1978b). Its field of projection is coincident with S1, as defined electrophysiologically (Haight and Weller, 1973). VPM and VPL project to S1 in a non-overlapping manner. VPL, which projects medially, apparently displays a stricter homotopy

than VPM, which projects laterally. VPP projects anteriorly and is surrounded by VPM projections caudally. VPL also has projections to S2 and it was assumed that this is also the case for VPM (Haight and Neylon, 1978b). The organisation of these projections is similar in the eastern quoll (Haight and Neylon, 1981b).

Posterior nuclear group

The nomenclature of this region in both eutherians and metatherians varies, depending on the authors, but Neylon and Haight (1983) noted that the pattern of inputs to the region is similar across species. In the brush-tailed possum (Neylon and Haight, 1983) and the eastern quoll (Haight and Neylon, 1981b) they identified a lateral and medial division of Po, as did Mayner (1989c) in the tammar. The former authors stated that the region designated as Po in their study was equivalent to the subdivision lateralis B in the study of the same species by Rockel *et al.* (1972) and corresponded fairly closely with medial Po of eutherian species.

In the brush-tailed possum, the major somatosensory projections to lateral Po found by Haight and Neylon (1981b) are from the dorsal column nuclei and the spinal cord (Rockel *et al.*, 1972). It might be expected, given its cortical connections (see below), that medial Po would receive an input from the trigeminal system. However, conclusive information is

---

Caption for Figure 10.2 (Cont.)

(arrow) resulting from a cortical deposit of horseradish peroxidase conjugated to wheat germ agglutinin (WGA-HRP); bar = 500 μm. From Marotte et al. (1997), Figure 1c, with permission, copyright 1997 Wiley-Liss, Inc. (d) Functional organisation of the somatosensory cortex in the striped possum (*Dactylopsila trivirgata*). There is a complete and inverted representation of the contralateral body surface in S1 (white); individual body areas are given to show receptive field progression. Receptive field progressions are also given for S2 and the parietal ventral area (PV). Since receptive field progressions have not been studied as extensively in the rostral (R) and caudal (C) somatosensory fields, only the outermost borders are shown. d1–4 – digits 1–4; DH – dorsal head; Fp – forepaw; Hl – hindlimb; R – rostral; Sn – snout; Ton – tongue; Vib – vibrissae. Reprinted from Karlen and Krubitzer (2007), copyright 2007 with permission from Elsevier. Based on Huffmann *et al* (1999a), copyright 1999, Wiley-Liss Inc. with the permission of Wiley. (e) Coronal section through the cortex of the tammar wallaby (*Macropus eugenii*) after an injection of WGA-HRP in VPM. Terminals in layer 4Cx are patchy. Retrogradely labelled cells are present primarily in layer 6Cx. From Mayner (1985), with the kind permission of L Mayner. (f) and g) Comparison of the arrangement of cortical barrels with the arrangement of mystacial vibrissal follicles in the brush-tailed possum (*Trichosurus vulpecula*). (f) The larger barrels are arranged in six roughly parallel, slightly curved, anteroposterior rows (ant – anterior; post – posterior). The ventral-most of these rows (small solid arrow) consists of only two or three barrels, which are more obvious in the next deeper section. Within a row the more anterior barrels are smaller than the posterior barrels and are not arranged in rows. Anterior to the short, ventral-most row of large barrels is a compact group of very small barrels (large solid arrow); thionin stain, 150 μm thick. (g) This photograph illustrates the distribution and arrangement of the possum's mystacial vibrissae, especially the six rows of large vibrissae (solid arrows A–F), and submental vibrissae adjacent to the lower lip (above, open arrow). The head, shown upside down to aid comparison with the pattern of barrels, is that of a pouch young possum estimated to be 57 days old. The position of each vibrissae follicle on the skin is marked by its pigmentation. The arrangement of follicles and the arrangement of barrels are clearly homeomorphic. From Weller (1993), copyright 1993 Wiley-Liss Inc. Reprinted with permission of John Wiley & Sons, Inc. (h) Absence of effect of S1 inactivation on an S2 neuron in the somatosensory cortex of the brush-tailed possum (*Trichosurus vulpecula*). Impulse records are shown of the response of an S2 neuron to stimulation of its receptive field on the pad of the forepaw, proximal to digit 4, with low frequency (2.7 Hz, 225 μm amplitude) rectangular pulses, superimposed for 1 s on a 1.5 s background step indentation (amplitude 200 μm). The neuron responded to each of the ON- and OFF-phases of the rectangular pulses with bursts of 2–3 action potentials during both pre- and post-cooling controls and when S1 was inactivated by cooling to 7 °C. From Coleman *et al.* (1999), Figure 5A, used with permission, The American Physiological Society.

not available for the possum and as previously mentioned in the tammar, at least at early developmental stages, there is no projection evident from Pr5 (Leamey *et al.*, 1998). In the opossum, an equivalent region to Po of the possum receives input from the trigeminal system (Walsh and Ebner, 1973), as it also does in the rat (Chiaia *et al.*, 1991). In the brush-tailed possum (Neylon and Haight, 1983), the projections of Po to the somatosensory cortex are organised in a point-to-point manner and cover most of the head and body areas, but are not as homotypically precise as the projection from VPL/M. Lateral Po projects to the body region, whereas medial Po projects to the head region. Cells in Po retrogradely labelled by injections in the head region are continuous with the label in VPM. Po also projects to posterior parietal cortex. More limited data in the eastern quoll (Haight and Neylon, 1981a) indicate that the cortical projections in this species are similar to those in the possum.

Somatosensory cortex

Up to five somatosensory areas have been described in Australian marsupial species examined so far (Figure 10.2d) (reviewed in Karlen and Krubitzer, 2007). At least two fields, S1 and S2, contain a complete or almost complete representation of the contralateral body surface. In addition, the parietal ventral area (PV) and two areas rostral (R) and caudal (C) to S1 have been delineated in some species (see below). The latter two areas may also be involved in motor and sensorimotor processing, respectively.

In the brush-tailed possum (Adey and Kerr, 1954; Elston and Manger, 1999; Haight and Weller, 1973; Huffman *et al.*, 1999a), the wallaby (*Thylogale eugenii*) (Lende, 1963), the wombat (Johnson *et al.*, 1973), and the striped possum (*Dactylopsila trivirgata*) (Figure 10.2d) and the northern quoll (*Dasyurus hallucatus*) (Huffman *et al.*, 1999a), electrophysiological mapping has shown that S1 contains an inverted map of the contralateral body surface. Most of the neurons in S1 respond to cutaneous stimulation, and Elston and Manger (1999) noted that they were non-habituating. The face and oral structures are represented laterally, progressing through the forelimbs, body and hindlimbs to the tail, which is represented medially. As in the thalamus, the head representation occupies a relatively large area, in particular the facial vibrissae. For example, in the northern quoll the representation of the snout, chin, cheek and lip vibrissae occupies 35–40% of S1. Other structures, such as digit or forepaw specialisations in some species, may also be disproportionately represented e.g. the fourth digit in the striped possum (Figure 10.2d) (Huffman *et al.*, 1999a; see Chapter 8 of this book, Figure 8.5). This organisation is

also typical of American marsupials (reviewed in Karlen and Krubitzer, 2006b) and eutherians and monotremes (reviewed in Krubitzer and Hunt, 2006).

S1 is distinguished cytoarchitectonically by a densely packed, darkly Nissl-staining layer 4Cx in the brush-tailed possum (Adey and Kerr, 1954; Elston and Manger, 1999) and tammar (Ashwell *et al.*, 2005a). In material sectioned parallel to the surface, it stains heavily for myelin, SDH and CO in the northern quoll, striped possum and fat-tailed dunnart (Huffman *et al.*, 1999a). In the brush-tailed possum there is a distinct barrel field in layer 4Cx in both Nissl- and SDH-stained material (Weller, 1972, 1993). In tangential sections, the barrels form a map of the body, with those of the face and forepaw especially clear. There are six rows of mystacial barrels (Figure 10.2f) that are homeomorphic to the six rows of large mystacial vibrissae (Figure 10.2g) and forepaw barrels that are homeomorphic to the glabrous palmer and apical digital pads. Recordings of receptive fields confirmed that the individual barrels represent specific cutaneous regions such as individual vibrissae. In coronal sections, the barrels are seen as regions of high cell density alternating with septa of low cell density where myelinated fibres are concentrated. They resemble the barrels seen in rats, in having cells throughout the barrel sides and centres, compared with those in mice, where the centres are relatively empty of cells (Woolsey and Loos, 1970; Welker and Woolsey, 1974). Weller (1972), in a large survey of Nissl-stained material in the marsupial families Didelphidae, Dasyuridae, Permamelidae, Phalangeridae, Vombatidae and Macropodidae, only found such barrels in the Phalangeridae. In the tammar, although no barrels are seen in Nissl stains, there are patches of SDH and CO activity associated with the whisker representation in the adult (Waite *et al.*, 1991), but they are more clearly seen in developing animals (Waite *et al.*, 1991; Marotte *et al.*, 1997; Mark *et al.*, 2002). These reflect patches of thalamic afferents from VPM terminating in layer 4Cx (Figure 10.2e) and corresponding clusters of cell bodies in layer 6Cx that project to the thalamus (Mayner, 1985). In the striped possum, similar barrel-like patches of CO activity have been described in the cortex, but whether barrels are apparent in Nissl-stained material was not mentioned (Huffman *et al.*, 1999a). It is not clear why neuronal aggregations are seen in phalangers, but not in the tammar (see Waite *et al.*, 1991), which are both diprotodonts, but they do not appear to be present in either Australian (Weller, 1972) or American polyprotodonts (Weller, 1972; Woolsey *et al.*, 1975; Beck *et al.*, 1996). Indeed, it is not clear why barrels or SDH/ CO patches corresponding to the whiskers are seen in some species and not in others. Woolsey *et al.* (1975), in a large survey, observed that they were not associated with whisking behaviour, they were not obscured or absent even if the behaviour of the animal was strongly dependent on another sensory modality, but they became more difficult to see as the brain size increased and we suggest this may explain their clarity in the possum compared with the tammar.

S2, which lies caudolateral to S1, contains a second non-inverted, small representation of the contralateral body surface. Electrophysiological mapping has identified this area in all species examined: brush-tailed possum (Adey and Kerr, 1954; Coleman et al., 1999; Elston and Manger, 1999; Huffman et al., 1999a), and northern quoll and striped possum (Figure 10.2d) (Huffman et al., 1999a), as well as in American opossums (Huffman et al., 1999a; Karlen and Krubitzer, 2007), although in the quoll it could not be separated from the parietal ventral area. Huffman et al. (1999a) reported only a partial representation of the body surface in the species they examined, whereas Elston and Manger (1999) believed it probably contained a complete representation in the brush-tailed possum. The face and oral structures are adjacent to these representations in S1 (Figure 10.2d). While neurons in both S1 and S2 respond to cutaneous stimulation, those in S2 often habituate and their receptive fields are larger compared to those in S1 (Elston and Manger, 1999; Huffman et al., 1999a). S2 is also architectonically distinct, with broad layers 3Cx and 5Cx and an ill-defined layer 4Cx (Adey and Kerr, 1954). In the quoll and the striped possum it stains moderately to darkly for myelin (Huffman et al., 1999a), whereas in the brush-tailed possum there is sparse myelination (Elston and Manger, 1999). Coleman et al. (1999) have shown in the brush-tailed possum that inputs can reach S2 directly from the thalamus and are not dependent on a pathway through S1 (Figure 10.2h). That is to say, S1 and S2 are organised in parallel, as they are in a range of placental mammals.

A separate, somatotopically organised field, PV, rostral to S2, has been identified in the brush-tailed possum (Elston and Manger, 1999; Huffman et al., 1999a) and the striped possum (Huffman et al., 1999a). Neurons have larger receptive fields like those of S2 and often habituate (Elston and Manger, 1999). Two additional fields rostral and caudal to S1, R and C, have also been identified in the brush-tailed possum (Elston and Manger, 1999), the quoll and the striped possum (Huffman et al., 1999a). Neurons in these areas respond to high-threshold cutaneous or deep receptors and may correspond to somatosensory areas identified in other mammals, including the Virginia opossum (Beck et al., 1996). R has been proposed to correspond to area 3a of flying foxes and primates and the rostral somatosensory area of rodents, whereas C may correspond to the posterior parietal area of squirrels and posterior parietal cortex of rodents (Karlen and Krubitzer, 2007).

As well as receiving connections from somatosensory thalamus, outlined in the previous section, somatosensory cortex has reciprocal connections with thalamic nuclei. For example, after a deposit of tracer in the whisker area of somatosensory cortex of the tammar, anterograde label coincident with retrogradely labelled cells is seen in VPM and extends into Po (Figure 10.2c) (Marotte et al., 1997). Large lesions of the cortex in the brush-tailed possum, which include somatosensory cortex, result in anterograde degeneration in VPM, VPL and Po (Rockel et al., 1972). The somatosensory cortex has also been shown to receive projections from a number of other thalamic nuclei in the brush-tailed possum and northern quoll, including the ventrolateral, ventromedial, mediodorsal, anteromedial, ventroanterior and extrageniculate nuclei, as well as parts of the centro-intralaminar complex (Ward and Watson, 1973; Haight and Neylon, 1979; Haight and Neylon, 1981a, 1981b).

Corticocortical connections have been studied in the brush-tailed possum. S1 is connected ipsilaterally with other regions of S1, with S2, PV, R, C, posterior parietal cortex, motor cortex and perirhinal cortex (Elston and Manger, 1999). This is similar to that described in the Virginia opossum, except for the connection with the motor cortex (Beck et al., 1996). Connections with other somatosensory areas are largely homotypic. Contralateral connections are similar, but less dense (Heath and Jones, 1971; Elston and Manger, 1999).

Development

The most notable difference from eutherians is that metatherians are born in an immature state and thus much of development takes place postnatally in the pouch (see Chapters 3 and 14). As for the adult, most information on the developing somatosensory system is known for the trigeminal system and the species for which most is known is the tammar.

Cutaneous receptors

Merkel cells have been described around the mouth in a number of newborn Australian species: northern quoll Dasyurus hallucatus (Figure 10.3a), brush-tailed possum, northern brown bandicoot (Gemmell et al., 1988) and stripe-faced dunnart (Gemmell and Selwood, 1994), as well as in the intervibrissal skin of the newborn tammar (Waite et al., 1994). They are also seen in the glabrous snout skin of the newborn opossum (Jones and Munger, 1985), but were not reported in the body skin of the newborn possum (Lyne et al., 1970). These studies suggest that these sensitive touch receptors assist in the journey from the urinogenital sinus to the pouch and the locating of the teat.

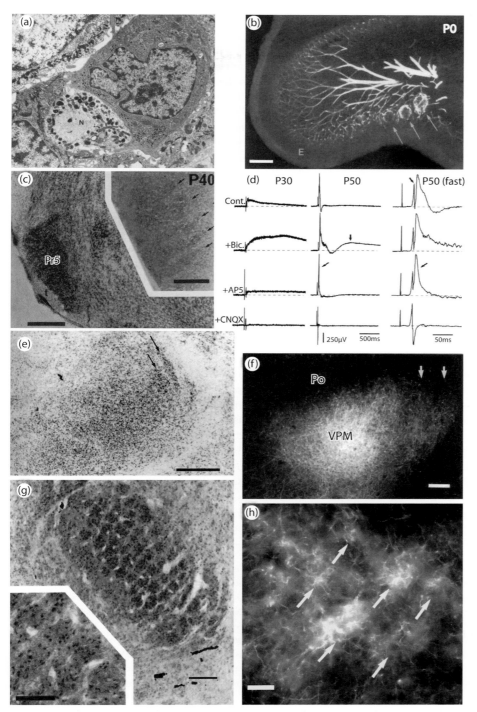

Figure 10.3 (a) A large sensory nerve (N) in close apposition to a Merkel cell (M) in the newborn quoll (*Dasyurus hallucatus*); x18 500. From Gemmell *et al*. (1988), Figure 3, copyright Springer-Verlag 1988, with the kind permission of Springer Science and Business Media. (b) to (h) tammar wallaby (*Macropus eugenii*). (b) Confocal image (four scans at 25 μm intervals) of a 100 μm section through the maxillary process on postnatal day P0, following 1,1´- dioctadecyl-3,3,3´,3´- tetramethylindocarbocyanine perchlorate (DiI) application to the trigeminal ganglion. Labelled nerve fibres can be seen branching to supply the epidermis (E) and several developing vibrissal follicles (arrows); bar = 200 μm. From Waite *et al*. (1994), Figure 3A, with permission, copyright 1994 Wiley-Liss, Inc. (c) Transverse sections through Pr5 on P40. The characteristic almond shape is apparent; bar = 0.5 mm. Inset: higher-power view of Pr5 reacted for cytochrome oxidase (CO); five oblique lines of increased reactivity (arrows) can be seen; bar = 250 μm. From Waite *et al*. (1994), Figure 12E, with permission, copyright 1994 Wiley-Liss, Inc. (d) Responses recorded in VPM following electrical stimulation of Pr5 in *in vitro* preparations of the brainstem and thalamus

Development of the vibrissal follicles and pelage hairs and their innervation has been described for the tammar (Waite *et al.*, 1994) and there is some information on follicular development in the skin of the brush-tailed possum (Lyne, 1970) and the bandicoot (Lyne, 1957). In the tammar at birth, follicles consist of a solid epidermal peg surrounded by dermal condensations. They and surrounding skin are already innervated by a dense array of trigeminal afferents (Figure 10.3b). Occasional putative Merkel cells can be seen in the epidermis. By P10, a deep vibrissal nerve is recognisable and the follicle contains a dermal papilla. By P30, a hair cone is present and by P35 hairs are seen on the skin surface. The first receptors seen in the follicle are Merkel cells around the waist region, at P48. A blood sinus can first be recognised by P63, but it is not until P119 that the inner conical body and lanceolate and lamellated receptors supplying the mesenchymal sheath and waist region are apparent and the follicle resembles the adult structure. This rate of development is much slower than that seen in rodents, where in the rat the follicle is mature in three weeks (Munger and Rice, 1986). In the tammar this takes 18 to 20 weeks, although the sequence of development is similar.

The epidermal pegs of facial pelage hairs are first seen at P30 and by P48 have increased in number with sparse axon bundles among them. By 63 days, guard hairs can be distinguished from vellus hairs and by 83 days dermal papillae are well developed and hair shafts are common. By P119, pelage hairs have increased, as have axon bundles ramifying among them and at this time innervation of guard hairs is first seen in the form of circular arrays of nerves and lanceolate endings. Innervation of vellus hairs is not obvious as it is in the adult. Development of pelage hair is not complete until around P200 and animals only start leaving the pouch for short periods around P190.

Although not compared in the same species, the development of trigeminal innervation around the mouth and mystacial skin probably precedes that of the rest of the body. Thus, in the brush-tailed possum (Lyne, 1970), the hair follicles on body skin first appear about two days after birth. Groups are formed with a single, central primary follicle, lateral primary follicles and one or two secondary follicles. Nerves are first noted in the dermis at P21 and are not seen around the primary follicles until P54. The features of follicle development are similar in the bandicoot, but occur more rapidly (Lyne, 1957). In the opossum (*Monodelphis domestica*), neurites are present in the glabrous forepaw dermis at birth, but are not seen in the epidermis until P20. From P21 to P30, innervation density increases and Merkel cells appear (Brenowitz *et al.*, 1980).

As in the tammar (Waite *et al.*, 1994), epidermal Merkel cells are also the first receptors to appear between developing vibrissae in the rat and ferret, and precede other receptors in the follicles (Munger and Rice, 1986; Mosconi and Rice, 1993). Similarly, the lag in development and innervation between vibrissal follicles and intervibrissal hairs is also seen in rats and ferrets, although there is more overlap in ferrets (Munger and Rice, 1986; Mosconi *et al.*, 1993). Overall, despite the different developmental time-courses between diverse mammalian species, they show similar sequences of maturation of follicles and

Caption for Figure 10.3 (Cont.)

at P30 and P50. Under control conditions (Cont.) at P30 (left) the post-synaptic response is prolonged, lasting in excess of 1 s. Addition of bicuculline (+Bic.) to block GABA-A receptors increases the amplitude of the response. Much of this response is NMDA mediated, as shown by the marked decrease in amplitude following addition of 2-amino-5-phosphovaleric acid (+AP5). The remainder of the post-synaptic response is blocked following addition of 6-cyano-7-nitroquinoxaline-2,3-dione (+CNQX) to block non-NMDA-mediated glutamatergic responses, leaving just the presynaptic volley. By P50 (middle, right) the control response is of much shorter duration and is dominated by the fast component. This can be distinguished from the presynaptic volley in the traces shown at a higher temporal resolution on the right. Addition of bicuculline prolonged the tail of the response, but did not affect the initial fast peak. Blockade of NMDA-mediated responses with AP5 reduced the amplitude of the slow component, but had little impact on the fast peak. Addition of CNQX completely blocked the fast peak, leaving just the presynaptic volley. This indicates that there is a change from slow, NMDA-dominated responses in the thalamus at P30 to fast, non-NMDA glutamatergic responses by P50. From Leamey *et al.* (1998), Figure 9, with permission, copyright 1998 Wiley-Liss, Inc. (e) Coronal section through VPM in a P52 wallaby stained for Nissl substance. A hint of segmentation is appearing at the lateral border of VPM (arrows); bar = 400 μm. From Leamey *et al.* (1996), Figure 5b, with permission, copyright 1996 Wiley-Liss, Inc. (f) Coronal section from the caudal half of VPM following the transport of DiI from Pr5 to the thalamus from a P50 animal. Dorsal is at the top, and medial is to the left. The fibres fill and define the borders of the VPM with no label present in the posterior nu. (Po). The fibres are no longer distributed uniformly across the nucleus as at younger ages, but rather show evidence of segregation into fingers at the lateral border (indicated by arrows); bar = 100 μm. Inset shows high power view; bar = 100 μm. From Leamey *et al.* (1998), Figure 2A, with permission, copyright 1998 Wiley-Liss, Inc. (g) Horizontal section through VPM at P75 reacted for CO. VPM is characterised by the presence of circular clusters of highly CO-positive cells; bar = 200 μm. From Leamey *et al.* (1996), Figure 6d, with permission, copyright 1996 Wiley-Liss, Inc. (h) Horizontal section through the VPM from a P75 animal following transport of DiI from Pr5. The afferents have become segregated into circular clusters (arrows). Rostral is at the top, and lateral is to the right; bar = 50 μm. From Leamey *et al.* (1998), Figure 3c, with permission, copyright 1998 Wiley-Liss, Inc.

intervibrissal skin, innervation and pattern of receptor development.

Muscle receptors

Immature muscle spindles can be recognised in forelimb and cervical musculature of the red kangaroo (*Macropus rufus*) at birth (Kubota *et al.*, 1989), whereas in the tammar they cannot be recognised by routine histology in either the fore- or hindlimbs until around P30 (Harrison, 1991; Harrison and Porter, 1992). In contrast, in the neonatal opossum (*Monodelphis domestica*) they can be seen with immunostaining for spindle fibre types in jaw, forelimb and thoracic muscle (Sciote and Rowlerson, 1998), and those authors speculated that using this method they would be recognised at birth in the tammar. The sequence of spindle development in the tammar (Harrison, 1991) is similar to that of eutherians, but occurs postnatally. Spindles are mature by around P100, prior to the time when pouch young can stand unaided, indicating that they form in the absence of feedback produced by adult forms of posture and locomotion.

Dorsal root ganglia and dorsal column nuclei

Large dorsal root ganglia are present at all levels in the newborn tammar (Harrison and Porter, 1992), and afferents, labelled from the brachial plexus or sciatic nerve, extend into the dorsal horn (Ho and Stirling, 1998). Dorsal-horn development is more advanced at cervical levels than in the lumbosacral cord (Harrison and Porter, 1992), probably relating to the greater maturity of the forelimbs needed for reaching the pouch, compared with the relatively immature hindlimbs. A rostrocaudal gradient of cord maturation has been described in the opossum (Hughes, 1973) and also occurs in eutherian species such as the rat. Distinct gracile and cuneate fasciculi are present by P17. While the histological development in the tammar is similar to that in eutherians such as the rat, it is more protracted and the tammar cord is far less mature at birth, corresponding at brachial levels to approximately E15 to E16 in the rat (Harrison and Porter, 1992).

Growth of dorsal root afferents into the cord has been studied in the South American opossum, where the first fibres, labelled by *Griffonia simplicifolia* isolectin B4, reach the lateral edge of the dorsal horn at P5 and then spread medially to reach an adult pattern of innervation by P20 (Kitchener *et al.*, 2006). CGRP-labelled afferents lag behind these slightly, reaching the lateral dorsal horn by P7 and then extending medially to give an adult pattern of labelling by P30.

Trigeminal ganglia and brainstem sensory nuclei

In the tammar, the development of the trigeminal ganglia and nuclei stretches over many months, reflecting the slow, overall pace of development (see Figure 3.6, Chapters 3 and 14). At birth, the trigeminal ganglion is well developed with peripheral processes extending to the facial skin (see above) and central afferents distributed throughout the rostrocaudal extent of the sensory trigeminal nuclear complex (Waite *et al.*, 1994). This coincides with the trigeminal root showing strong immunoreactivity for GAP-43 (Hassiotis *et al.*, 2002), a phosphoprotein which is high in growth cones and developing nerves during axon elongation (Meiri *et al.*, 1986). Reflexes involving trigeminal inputs and jaw musculature are also present at birth and are presumably involved in suckling (Waite, personal observations). All the developing nuclei except Sp5O have higher levels of SDH and CO activity, like the adult. By P40, clear rostrocaudal subdivisions within the complex have emerged (Waite *et al.*, 1994), GAP-43 expression remains high in the trigeminal root and spinal trigeminal tract (Hassiotis *et al.*, 2002) and in Pr5, five lines of CO reactivity extend obliquely across the nucleus, reminiscent of the rows of mystacial vibrissae (Figure 10.3c; Waite *et al.*, 1994). By P147, small patches of CO activity are seen ventrolaterally in Pr5, presumably corresponding to individual vibrissae, but an entire representation is not apparent (Waite *et al.*, 1994). These patches are less clear than in rats, although in both species they are more obvious in juveniles than in adult animals (Belford and Killackey, 1979b). High CO reactivity has been suggested to correlate with synaptogenesis in developing animals (Wong-Riley, 1989) and, presumably, in the rat, synaptogenesis extends over a much shorter and concentrated period of time than in the tammar. This may explain the clearer appearance of patches in the developing rat (Waite *et al.*, 1994).

Somatosensory thalamus

In the brush-tailed possum, birth of neurons in VPL/M is completed by P12, with substantial cell birth in the first postnatal week (Sanderson and Weller, 1990b). There is a gradient of neurogenesis from lateral (earlier) to medial (later), as in rodents (Angevine, 1970). Such information is not available for the tammar but, based on morphology, cell birth also appears to be primarily postnatal. At birth the diencephalon is composed of a neuroepithelial layer overlaid by a mantle zone and the boundaries of VPM cannot be clearly distinguished from VPL and surrounding nuclei on histological grounds until P47 (Leamey *et al.*, 1996), about the same time as in the brush-tailed possum (Sanderson and Weller, 1990b). The first trigeminothalamic axons from Pr5, tipped with growth cones, enter the developing VPL/M complex at P15 (Leamey *et al.*, 1998), when it is seen as a dense cellular mass separated from the dorsal lateral geniculate by the external medullary lamina (Leamey *et al.*, 1996). Afferents continue to grow into and densely fill VPM up to P40 (Leamey *et al.*, 1998). The first segmentation within the nucleus is visible in Nissl- (Figure 10.3e) and CO-reacted material laterally in VPM at P52, and fingers of reactivity similar to the whisker-related patterns seen in the adult are clear by P73–75 (Figure 10.3g; Leamey *et al.*, 1996). This segregation occurs simultaneously with the segregation of the ingrowing afferents (Figure 10.3f and h) (Leamey *et al.*, 1998). A parallel electron-microscopic and electrophysiological study (Leamey *et al.*, 1998) showed that evoked synaptic responses can be recorded in VPM from the time of afferent ingrowth; prior to P50 excitatory responses are dominated by *N*-methyl-D-aspartate (NMDA)- mediated responses (Figure 10.3d). Such responses have been shown to be important for the formation of whisker-related patterns in mice (Li *et al.*, 1994). Coincident with the onset of synaptogenesis at P40, there is a decrease in duration of the response. Synaptic transmission prior to this is presumably via transmitter release from growth cones. At P50, the response is dominated by a fast non-NMDA-mediated potential, coincident with the onset of whisker-related patterns (Figure 10.3d). The presence of functional connections prior to the onset of whisker-related patterns suggests that synaptic interactions may play a role in this pattern formation.

Projections to the diencephalon from trigeminal sensory and dorsal column nuclei also develop postnatally in the North American opossum (Martin *et al.*, 1987). In contrast, ascending connections to VPM in the rat form prenatally and the formation of whisker-related patterns takes place over a few days (Belford and Killackey, 1979a), rather than a few weeks, as seen in the tammar (Leamey *et al.*, 1996).

Somatosensory cortex

Neurogenesis in the somatosensory cortex of the brush-tailed possum is protracted and takes place in the first two postnatal months, with neurogenesis in the body region of the dorsolateral cortex lagging behind that of the head region in lateral cortex (Sanderson and Weller, 1990a). Neurogenesis finishes just before whisker-related barrels can be recognised in layer 4Cx at P65 (Sanderson and Weller, 1990a).

In the tammar, which has a similarly protracted developmental time-course to the possum, ascending input from the thalamus first reaches the cortex at P15 (Figure 10.4a, b) and afferents proceed to grow into and are widely distributed in the depth of the developing cortex without any sign of a

waiting period (Figure 10.4c, d) (Marotte *et al.*, 1997). Pre- and postsynaptic cortical responses to thalamic stimulation in a thalamocortical slice preparation can be recorded from the time of afferent arrival, but are very slow and completely mediated by NMDA receptors. Thalamocortical synapses are first seen at P28 in the cortical plate when cortical responses remain slow and are still dominated by an NMDA component (Figure 10.4e) (Leamey *et al.*, 2007). The first corticothalamic projections are seen much later, at P60, from layer 5Cx cells (Marotte *et al.*, 1997). Whisker-related patterns typified by the first hint of CO patches (Figure 10.4g) in the just-formed layer 4Cx (Figure 10.4f) (Mark *et al.*, 2002), the clustering of afferents in layer 4Cx, and the clustering of the dendrites and somata of cells in layers 5Cx and 6Cx projecting to the thalamus (Figure 10.4h) (Marotte *et al.*, 1997) all become apparent at around the same time, at P72 to P76. This is slightly earlier than the patterns are first detected with SDH staining (Waite *et al.*, 1991) and coincides with the appearance of synapses (Leamey *et al.*, 2007) and synaptic responses in layer 4Cx in response to whisker stimulation *in vivo* (Figure 10.4i; Mark *et al.*, 2002) and the domination of thalamocortical responses by a fast non-NMDA-mediated component (Figure 10.4e) (Leamey *et al.*, 2007). This component may drive pattern formation and firing of cortical neurons (Leamey *et al.*, 2007). At older ages, the responses are longer lasting with more complex sequences of activity at different depths (Mark *et al.*, 2002).

Despite the very different time-course of development of the pathway between the rat and tammar (days compared to months), afferents in the rat are also in the cortex (Catalano *et al.*, 1991, 1996) for a relatively long time prior to the onset of evoked activity *in vivo* at P2 (Verley and Axelrad, 1975). Like in the tammar, the earliest evoked activity is temporally close to the appearance of layer 4Cx (Rice *et al.*, 1985), clustering of thalamocortical afferents (Catalano *et al.*, 1996) and whisker-related patterns demonstrated histochemically (Killackey and Belford, 1979).

The whisker pathway, pattern development and plasticity

Experiments in the rodent whisker pathway have shown that the periphery is important for the development of whisker-related patterns centrally (Van der Loos and Woolsey, 1973; Weller and Johnson, 1975; Belford and Killackey, 1979b). Lesions to the whiskers or their nerve supply result in the loss of the corresponding whisker-related patch or patches and there is a critical period for this effect. Lesions made after the barrels in the cortex have formed have no effect, and the critical period for each level of the pathway correlates with the time of pattern formation at that level. Lesioning of the

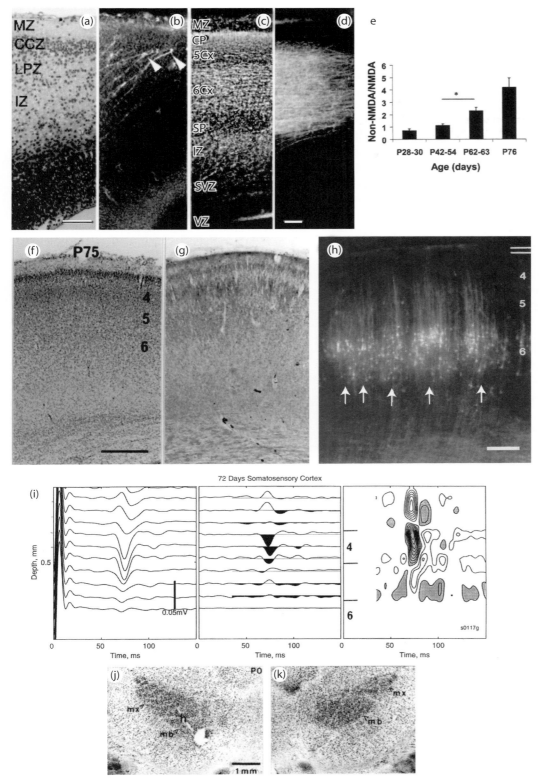

Figure 10.4 (a) to (i) tammar wallaby (*Macropus eugenii*). (a) and b) Coronal sections through the somatosensory cortex at P15. Toluidine-blue-stained plastic section showing the developing cortical layers (a) and an adjacent section showing DiI-filled thalamocortical axons (b). Axons tipped with growth cones (arrowheads) reach to the base of the CCZ of the cortical plate. MZ – marginal zone; LPZ – loose-packed zone of the cortical plate; IZ – intermediate zone; SVZ – subventricular zone; VZ – ventricular zone; bar in (a) = 100 μm also applies to b). (c) and (d) Fluorescence views of a coronal section through the somatosensory cortex at P51 exposed for bisbenzimide counterstain to show the cortical layers (c) and DiI-filled thalamocortical axons (d). Layers 5Cx and 6Cx can now be recognised at the base of the cortical

whiskers in the brush-tailed possum also prevents the formation of thalamic (Figure 10.4j and k) and cortical patterns during a postnatal critical period (Weller, 1979; Waite and Weller, 1997). The closely timed sequence of pattern formation in the brainstem, thalamus and cortex in rodents led to the suggestion that a peripherally derived signal carried by afferent fibres swept along the pathway initiating pattern formation at each level (Killackey, 1985). The periphery clearly provides an important signal for pattern formation, but the target tissue also appears to play a role. This is supported by the time-course of development of the patterns in the tammar, where the protracted development in comparison to rodents clearly separates, in time, the formation of patterns at the different levels. Thus, the onset of pattern formation is separated by weeks rather than a few days as seen in rodents (brainstem – P40, thalamus – P52, cortex – P76). There is also a long delay between the arrival of afferents (brainstem – prenatal, thalamus – P15, cortex – P15) and the onset of functional activity (Waite *et al.*, 1994; Leamey *et al.*, 1996, 1998; Marotte *et al.*, 1997) and pattern formation at each level. These findings suggest that the pattern formation is independently controlled at each level rather than by a single peripheral signal and that maturation of target cells may play a role (Waite *et al.*, 1998). This is supported by findings that an increase in the non-NMDA-mediated glutamatergic response, a measure of target maturation, correlates

well with pattern formation in both thalamus and cortex (Leamey *et al.*, 1998, 2007). For example, in the cortex, despite afferents arriving from P15, timing is controlled by the maturation of layer 4Cx cells, which are not in place until around P72 to P76.

The slow, primarily postnatal development of the whisker pathway in the tammar has been used to investigate the effects of lesions of the infraorbital nerve supplying the whiskers at various developmental times (Waite *et al.*, 2006). Lesions prior to innervation of the thalamus and cortex result in an absence of patterns at both levels, suggesting that any reinnervation of the follicles is inadequate to rescue normal pattern formation. Interestingly, lesions at times when thalamic or cortical patterns are forming or have formed indicate a single critical period for thalamus and cortex. This is despite the fact that the thalamic pattern first starts to appear around three weeks before that in the cortex and suggests a close relationship between the stabilisation of patterns in thalamus and cortex.

What remains unknown?

This chapter indicates that much about the organisation and function of somatosensory pathways in marsupials is similar to that in eutherians. However, where differences

---

Caption for Figure 10.4 (Cont.)

plate (CP); abbreviations as for (a). SP – subplate; bar in (d) = 100μm, also applies to c). (e) Contributions of *N*-methyl-D-aspartate (NMDA) and non-NMDA components of the glutamatergic postsynaptic response during development. Bar graph shows the increase in the non-NMDA to NMDA ratio during development. The ratio remains stable from P28 to P54, but increases significantly (*P <0.05) by P63. There is a further increase by P76, although this is not statistically significant. (a) to (e) from Leamey *et al.* (2007) with permission, European Journal of Neuroscience, Federation of European Neuroscience Societies and Blackwell Publishing. (f) and (g) Coronal sections through the somatosensory cortex of an animal aged P75 stained with cresyl violet (f) and cytochrome oxidase (CO) (g). (f) Layers 4Cx to 6Cx are now recognisable cytoarchitectonically beneath the compact cell zone (ccz); bar = 500 μm. (g) CO reactivity is increased in layer 4Cx and the first hints of patches are seen; bar as for (f). From Mark *et al.* (2002), Figure 1c and d. (h) Low-power fluorescent view of DiI labelling in the somatosensory cortex at P76 after a deposit of DiI in the VPM. The pair of white bars mark the boundaries of the CCZ. At this age, the first sign of clustering of retrogradely labelled cells in layers 6Cx and 5Cx and their processes is seen (arrows); bar = 200 μm. From Marotte *et al.* (1997) Figure 7d, with permission, copyright 1997 Wiley-Liss, Inc. (i) Current source density (CSD) analysis of evoked potentials recorded in the cortex of an animal aged P72. The three plots are from a single data set. Each is on the space/time domain of the recording depth in the cortex and the time after a shock to the infraorbital branch of the trigeminal nerve. The plot on the left shows averaged evoked potentials. Each trace is the average of 10 sweeps. Positive is upwards. The centre plot shows the corresponding CSD waveforms. Where currents are negative, the waveforms are filled in black. The plot on the right shows the same CSD data plotted as contours on the depth latency domain. Current sinks, indicative of excitatory synaptic activity, are shaded, giving the depth and latency. Corresponding current sources are plotted as unshaded contours. Contour levels are at 10% steps relative to the maximum sink magnitude. Boundaries of cortical laminae were determined from histology. The short latency sink indicates synaptic activity is located in layer 4Cx. From Mark *et al.* (2002), Figure 3 upper panel. (j) and (k) Results of removing all large follicles from one side of the snout of a newborn brush-tailed possum on the whisker-related pattern in the thalamus. (j) A section through the somatosensory thalamus (control side). The cell-dense part (h) represents the possum's head. Some of the cells are arranged in clusters and rows (mx). They receive sensory information from the large mystacial whisker follicles. The others (mb) receive information from the chin; Nissl stain (thionin), 100 μm thick. (k) A section through the experimental side of the somatosensory thalamus. The rows and groups of cells are missing (mx) and the region receiving information from the chin is enlarged. Nissl stain (thionin), 100 μm thick. From Waite and Weller (1997) with permission from the University of New South Wales Press.

exist, research in marsupials is frequently disadvantaged by the lack of a systematic approach in many areas. Results are often described for just one species or neural area, so that it is impossible to assess the extent to which this reflects a general difference between marsupials and eutheria or a species peculiarity. For example, the interesting observation that Pacinian corpuscles in the wallaby hindlimb are less rapidly adapting than in eutheria could reflect a significant difference in response properties related to the different locomotor style. However, as this is the only observation of hindlimb Pacinian responses in marsupials, its significance is hard to evaluate. Similarly, the extent to which whisker-related specialisations, described in the possum and tammar, are present in other species would benefit from a more systematic approach. They are not seen in Didelphidae or Dasyuridae, but their presence in Diprotodontia has only been investigated for some gliders (present), brush-tailed possum (present) and tammar (present). They may well be more widespread, especially in the smaller species such as potoroos, and the smaller possums and macropods.

There are large gaps in our understanding of function in marsupial somatosensory systems. While receptor morphology has been relatively well described, especially for facial skin, there are few studies on response properties of mechanoreceptors and none for joint or thermal receptors or nociceptors. Descriptions of responses in the dorsal column nuclei are limited to the opossum, while we know nothing about dorsal-horn physiology. The roles of cutaneous and proprioceptive inputs in co-ordinating hindlimb reflexes have not been studied, despite the synchronous activation of right and left hindlimbs needed for jumping. Similarly

for VPL/M thalamus, studies have focussed on projections and somatotopy; responses of single cells have not been described. The somatosensory cortical areas have fared rather better, both for the breadth of species investigated and the possible functions of different areas. However, the level of understanding of marsupial cortical physiology lags well behind that for eutherians.

Finally, despite the obvious ease of access to pouch young, few researchers have taken advantage of this for functional studies on development or plasticity after injury. The whisker pathway remains the only example of studies on onset of activity and the effects of early lesions. It would be of interest to know how the extended period of time in the pouch affects the development of muscle spindles, joint receptors and proprioceptive pathways. Furthermore, access to pouch young would allow studies on nerve lesions or amputations, before central connections and pathways have developed, not easily undertaken at corresponding times in utero. Similarly the potential to introduce changes within the pouch environment have not been exploited. For instance, modifications in the pouch temperature for short periods might provide interesting insights into the neural mechanisms which establish core temperature and thermoregulation. As discussed more fully in Chapter 14, access to pouch young provides unique opportunities for early functional and plasticity studies.

Acknowledgements

We would like to thank Ms Sharyn Wragg for her expert assistance with the preparation of the figures.

11

Auditory system

L. M. Aitkin and R. K. Shepherd

Summary

The first physiological study of hearing in marsupials was that of McCrady, Wever and Bray (McCrady *et al.*, 1937) who used cochlear microphonic recordings to measure the audiograms of Virginia opossums (*Didelphis virginiana*). They, and others at that time, considered marsupials to be primitive mammals, mainly because of their behaviour, method of reproduction and apparent lack of intelligence. Didelphids and other polyprotodonts may be less advanced than diprotodonts in some respects, but marsupials, as an order, have evolved independently of eutherians, so are in no evolutionary sense 'primitive' mammals.

The auditory neurobiology of the Virginia opossum and the South American short-tailed grey opossum (*Monodelphis domestica*; hereafter referred to as Monodelphis) have been studied in some detail (see Aitkin, 1998); in particular Monodelphis has received recent attention because it can be raised under laboratory conditions and has large litter sizes. With the exception of tammar wallabies (*Macropus eugenii*), Australian marsupials studied in auditory experiments are not raised in laboratories and often are wild-caught. This reduces the opportunities for study and control over the animal's age and increases the likelihood of a hearing pathology (particularly associated with the middle ear), thus increasing the difficulty of making generalisations about their auditory neurobiology.

Marsupials, especially those with large litter sizes, offer the neurobiologist opportunities to study aspects of development that occur inaccessibly *in utero* in eutherians, but accessibly in the pouch in marsupials. The use of hearing for conspecific detection and predator avoidance seems likely to be similar for eutherians and marsupials. However, the relationships between the hearing range of the adult and the sound spectra of the prehearing young are of great interest (Aitkin, 1998) and will be examined in relation to the audiograms of adult marsupials later in this review.

Morphology of the auditory systems of marsupials has long been documented – the middle ear structures (Doran and Garson, 1879; Segall, 1969; Sanchez-Villagra *et al.*, 2002), cochlea (McCrady, 1938; Fernandez and Schmidt, 1963; Gemmel and Nelson, 1992) and auditory brain structures (Stokes, 1912; Oswaldo-Cruz and Rocha-Miranda, 1968; Cowley, 1973; Morest and Winer, 1986; Aitkin, 1995; Glendenning and Masterton, 1998). Detailed examination of the auditory pathways of Australian marsupials has been restricted to two species – brush-tailed possums (*Trichosurus vulpecula*) and northern quolls or native cats (*Dasyurus hallucatus*) – although greater detail is available for the cochlear nuclei and superior olivary complexes of arboreal marsupials (Aitkin, 1996).

General structure and function of the mammalian auditory system

The external and middle ear consists of the pinna, the external auditory meatus and the middle ear (Figure 11.1a). The pinna and external auditory meatus (or ear canal) provides efficient transmission of the acoustic stimulus to the tympanic membrane including important directional cues due to the pinna (Phillips *et al.*, 1982) and protects the delicate structures of the middle ear. While the marsupial ear canal typically exhibits characteristics identical to that of eutherians, the ear canal of the feathertail glider (*Acrobates pygmaeus*) is partly occluded by a disc of bone (Segall, 1971; Aitkin and Nelson, 1992) which has the effect of reducing the sound input to the glider's tympanic membrane, with a relative enhancement near 25–30 kHz (Aitkin and Nelson, 1992).

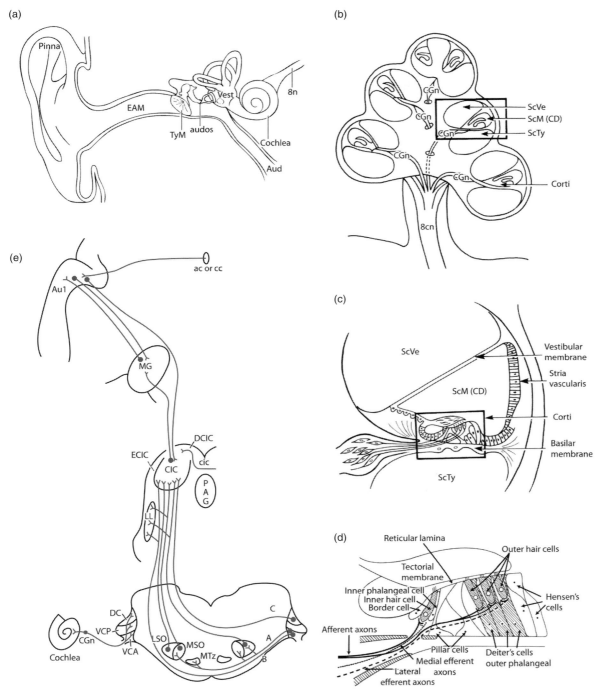

Figure 11.1 (a) Schematic diagram of the mammalian peripheral auditory system illustrating the pinna, external auditory meatus (EAM), tympanic membrane (TyM), middle ear cavity containing the ossicular chain (audos), the spiral-shaped cochlea or inner ear and the auditory nerve (8n). The marsupial peripheral auditory system exhibits only minor differences to that of eutherian mammals. Aud – auditory tube; Vest – vestibular apparatus. (b) Transverse section of the mammalian cochlea illustrating the three cochlear ducts (the scala tympani – ScTy, scala media/cochlear duct – ScM (CD) and scala vestibuli – ScVe) and the organ of Corti spiralling around the modiolus containing the cochlear ganglion (CGn) and the emerging auditory nerve (8cn). Reproduced from Ballantyne (1984), with permission of

The middle ear cavity is bound by the tympanic membrane laterally and the inner ear (or cochlea) medially (Figure 11.1a). Its prime function is to efficiently couple sound energy from the air-filled external auditory canal to the fluid-filled cochlea via the action of the middle ear ossicles (Peake and Rosowski, 1991) – the malleus, incus and stapes that connect the tympanic membrane to the oval window of the cochlea. This mechanical coupling and amplification is achieved in a manner largely related to the difference in the area of the tympanic membrane to that of the oval window of the cochlea, into which the last of the three ossicles, the stapes, inserts. A secondary pressure enhancement is due to the lever action of the ossicles.

In most respects the marsupial middle ear follows these general principles enunciated for eutherians, although the middle ears of many marsupials are not housed in the bony auditory bulla, but in soft tissue adjacent to the cochlea. The eardrum in marsupials is attached to a ring of bone called the tympanic bone, which is in turn connected by cartilage to the head (Segall, 1969).

The mammalian cochlea is a spiral-shaped labyrinth embedded within the temporal bone (Figure 11.1a, b). Sound pressure changes are converted to nerve impulses by hair cells in the coiled cochlea, which occur as rows of inner hair cells (of primary importance in sound transduction) and outer hair cells (mechanical amplifiers augmenting the sensitivity of inner hair cells near threshold) contained within the organ of Corti (Figure 11.1c, d). Unlike monotremes, both marsupial and eutherian mammals have coiled cochleae, effectively increasing the cochlear length and extending the frequency range of detection (Aitkin, 1998; Manley, 2001). The anatomy of the marsupial cochlea seems little different from those in eutheria (Fernandez and Schmidt, 1963; Aitkin, 1979, 1998; Willard and Munger,

1988); both contain a single row of inner and three rows of outer hair cells (Figure 11.1d). Again, monotremes differ in this respect, having three rows of inner and six rows of outer hair cells (Ladhams and Pickles, 1996).

In eutherians, cochlear nerve fibres arise primarily from the bases of the inner hair cells and innervate the cochlear nuclei (DC, VCA, VCP) – the first relay centre within the ascending central auditory pathway (Figure 11.1e). Cochlear nerve fibres bifurcate into anterior and posterior branches supplying the anteroventral (VCA), posteroventral (VCP) and dorsal (DC) cochlear nuclei, respectively (Irvine, 1986). Neurons from the cochlear nuclei proceed to the medial and lateral superior olives (MSO, LSO) of each side, medial nucleus of the trapezoid body (MTz) and various periolivary nuclei. Along with axons from these olivary nuclei, some cochlear nuclear afferents continue on to, mainly, the contralateral midbrain nuclei – the inferior colliculus (IC) and nuclei of the lateral lemniscus (LL; Aitkin, 1986).

Glendenning and Masterton (1998) have made a broad morphometric study of central auditory systems that include many marsupials, and Aitkin (1996) has studied the cytoarchitecture of the medullary auditory nuclei of arboreal Australian marsupials. However, very few Australian marsupials have been studied in depth, perhaps because of the difficulty of obtaining specimens to study. Even in the laboratory-raised New World marsupial, Monodelphis, there is little detailed information about the auditory pathway. In essence, all of the auditory nuclei described for eutherian mammals can be observed in the 14 Australian marsupials measured by Glendenning and Masterton (1998), but there are differences in relative sizes and disposition of the nuclei. For example Cowley (1973), in an anatomical study of the cochlear nuclei of the red

---

Caption for Figure 11.1 (Cont.)

Butterworths. (c) A magnified view of the cochlear duct in cross-section illustrating the fluid-filled scala tympani, scala media and scala vestibuli and the sensory epithelium within the organ of Corti (from Ballantyne, 1984; with permission from Butterworths). (d) A diagram of the organ of Corti containing the sensory hair cells – a single row of inner hair cells and three rows of outer hair cells. The organ of Corti is located on the basilar membrane, which separates the scala tympani from the scala media. Reproduced from Aitkin (1998), with permission from Springer. (e) Connections of the auditory pathway in a mammal, viewed in coronal sections taken at the levels of brain stem, midbrain, thalamus and cerebral cortex. Cochlear nerve fibres originating from the cochlear ganglion (GGn) branch on entering the cochlear nuclei and supply the anteroventral (VCA), posteroventral (VCP), and dorsal (DC) cochlear nuclei. Pathway A, issuing from VCA, supplies terminals to the lateral (LSO) and medial (MSO) superior olivary nuclei of the same side and the MSO and medial nu. of the trapezoid body (MTz) of the opposite side. Cells in the LSO and MSO project axons to the central nu. of the inferior colliculus (CIC); those from the MSO are largely ipsilateral (as shown), whereas those from the LSO are largely contralateral (not shown). Collateral branches supply the nuclei of the lateral lemniscus (LL). The CIC is the focus of ascending projections – the above-mentioned pathways plus pathways B (direct from VCA), C (from DC), and afferents from LL terminate here. From the inferior colliculus, including its external nucleus (ECIC) and dorsal cortex (DCIC), fibres ascend to the medial geniculate nu. of the thalamus (MG) and ultimately the primary auditory cortex (Au1). Pathways descend from the cortex to the thalamus and inferior colliculus; cross-connections exist between the auditory cortex (corpus callosum – cc – in eutherians, or anterior commissure – ac – in marsupials) and inferior colliculus (commissure of inferior colliculus, cic). PAG – periaqueductal grey. Reproduced with modification from Aitkin (1998) with permission from Springer.

kangaroo, noted that the DC was disposed medial to the restiform body with a surface on the fourth ventricle. In eutheria, the DC is on the lateral surface of the brainstem (Gates *et al.*,1996). This major relocation is also seen in the Virginia opossum (Willard and Martin, 1983), brush-tailed possum (Gates and Aitkin, 1982), northern quoll (Aitkin, 1986) and nine species of possums and gliders (Aitkin, 1996), so it may be characteristic of marsupials, although its functional significance is obscure. The huge DC of the feathertail glider (*Acrobates pygmaeus*; Aitkin and Nelson, 1989; Aitkin, 1996) has a counterpart in the eutherian mountain beaver (*Aplodontia rufa*; Merzenich *et al.*, 1973); again its functional significance is obscure.

Afferents to the superior olive from the cochlear nuclei have only been studied in the Virginia opossum (Willard and Martin, 1983, 1984). The MSO is innervated from the ventral cochlear nuclei (VCA/VCP) of both sides, the LSO from the ipsilateral VC and the MTz from the contralateral VC. The same authors have followed the ascending connections from the superior olive to the IC of Virginia opossums to find that the MSO projects to both IC, the LSO also projects bilaterally (but more strongly contralaterally), whereas the periolivary groups project ipsilaterally. In the brush-tailed possum (Aitkin and Kenyon, 1981) and northern quoll (Aitkin, 1986) the lateral nuclei of the superior olive project mainly contralaterally, whereas the medial nuclei project mainly ipsilaterally. These patterns in Australian marsupials differ from the North American opossum, but are similar to those in eutheria (Irvine, 1986).

The IC has a large central nucleus (CIC; Figure 11.1e), where most ascending afferents terminate, surrounded by pericentral and external nuclei (ECIC). In marsupials, these nuclei have been studied in Virginia opossums (Willard and Martin, 1983), brush-tailed possums (Aitkin and Kenyon, 1981) and northern quolls (Aitkin *et al.*, 1994a). In contrast to eutheria, the ECIC of the brush-tailed possum is larger than the central nucleus (Aitkin *et al.*, 1978; Aitkin and Kenyon, 1981); a similar conclusion can be drawn from examination of published anatomical material for the Virginia opossum (Oswaldo-Cruz and Rocha-Miranda, 1968; Willard and Martin, 1983). Most of the projections of the CIC terminate in the ventral division of the medial geniculate body (MGV) of the same side. Kudo and his colleagues (Kudo *et al.*, 1989) have studied the anatomy of the MG in northern quolls and report that nuclear subdivisions are very similar to those described for both eutherian species and the Virginia opossum (Morest and Winer, 1986).

The neocortex of marsupials is lissencephalic and the order lacks the corpus callosum, the distinctive band of fibres that connect the two hemispheres in eutherians; instead, the hemispheres communicate through the anterior commissure and the fasciculus aberrans (Abbie, 1939; Heath and Jones, 1971). Lende (1963) identified the auditory cortex (Au1) in the Virginia opossum as a field lying above the posterior portion of the rhinal fissure in which the largest evoked potentials to clicks clustered centrally. The auditory cortex has received more attention than most other parts of the auditory pathway of marsupials, perhaps due to its accessibility for lesion/behaviour studies made interesting because of the apparent 'primitiveness' of marsupials (Masterton *et al.*, 1969). Cortical areas responding to pure tones have been identified using microelectrode recording in brush-tailed possums (Gates and Aitkin, 1982) and northern quolls (Aitkin, 1986); these areas appear to be analogous to area Au1 in eutherians (see e.g. Merzenich and Brugge, 1973; Merzenich *et al.*, 1975).

The injection of small quantities of neuronal tracer into the central nervous system allows neuroscientists to map the complex neuronal projections radiating from a nucleus. Injections of retrograde tracers into the physiologically identified auditory cortex of brush-tailed possums has revealed input from the medial geniculate nucleus organised in a topographical manner (Aitkin and Gates, 1983). Interhemispheric connections were also demonstrated in this study. Neurons were labelled in the opposite hemisphere at corresponding sites to those where large injections of retrograde tracer were made into physiologically defined auditory cortex (Aitkin and Gates, 1983).

Kudo and his colleagues (Kudo *et al.*, 1989) made injections of retrograde tracer in mapped areas of the auditory cortex of northern quolls and found, in addition to topographically organised reciprocal ascending and descending projections to the MG, projections to three telencephalic areas, the contralateral and ipsilateral amygdala and the contralateral temporal cortex.

Hearing in marsupials

The hearing range of non-human mammals can be estimated using unconditioned motor responses ('startle responses') or by conditioning techniques in which thresholds to individual pure tones are investigated either by aversive or operant conditioning. More recently, since the demonstration that there is a close correlation between certain electrophysiological indices and behavioural thresholds, audiograms have been measured using such electrophysiological responses as otoacoustic emissions (Lonsbury-Martin *et al.*, 1987; Faulstich *et al.*, 1996; Mills and Rubel, 1996) – a measure of the motility of the outer hair cells recorded by a microphone placed in the internal auditory meatus; auditory brainstem responses (ABR) obtained by recording-evoked potentials from electrodes placed suitably near the auditory bulla and on the scalp (Aitkin *et al.*, 1996b; Cone-Wesson *et al.*, 1997; Mills and Shepherd, 2001), or thresholds for single unit responses recorded by microelectrodes from neurons in the brain (Moore and Irvine, 1979; Liberman, 1982). Unlike single unit recordings, both otoacoustic emissions and ABRs are non-invasive procedures; however, all electrophysiological techniques are performed in anaesthetised animals.

(i) Behavioural studies

Behavioural audiograms have been measured using aversive conditioning for three American marsupial species. Ravizza and his colleagues (Ravizza *et al.*, 1969) used a conditioned suppression paradigm in their study of the Virginia opossum. This species has a rather shallow insensitive audiogram with a best frequency of hearing between 16 and 32kHz at 20 dB SPL (sound pressure level). Two other opossums, Monodelphis and *Marmosa elegans*, studied using the same technique, have similarly flat, insensitive curves which are shifted towards higher frequencies (Frost and Masterton, 1994).

Reimer and Baumann (1995) used a simple operant paradigm that utilised the tendency of the Monodelphis opossum to eat small amounts of food at any one time. A tone was presented from one of two loudspeakers located at the end of two arms of a Y-maze. Food dispensers were positioned at the base of the sound sources and the animal received food if it chose the correct loudspeaker. The loudspeaker receiving the pure tone was randomised and no more than four successive trials were initiated from a speaker. The resulting audiogram, assessed using a similar range of test frequencies to those used by Frost and Masterton (1994), differed in one main respect from that study – thresholds. Those in the mid-range were some 16 dB more sensitive in the study of Reimer and Baumann, who suggested that the difference in threshold may be due to motivation (better with positive than negative reinforcement) and also, possibly, age (unspecified in the Frost and Masterton study). Subsequent ABR recordings by Reimer (1995) confirmed the shape of the behavioural audiogram and showed that the sensitivity of the mid-range diminished from young animals to achieve a value after four months similar to that of Frost and Masterton (1994). In relation to threshold sensitivity for Monodelphis, an audiogram obtained using otoacoustic emissions as a function of frequency (Faulstich *et al.*, 1996) resembles the shape of the behavioural audiogram measured by Frost and Masterton (1994), but lies close in threshold to the audiograms of Reimer (1995) and Reimer and Baumann (1995).

(ii) Electrophysiological studies

The hearing ranges of Australian marsupials have not been studied behaviourally, but electrophysiological data exist for brush-tailed possum (Aitkin *et al.*, 1978, 1984), northern quoll (Aitkin *et al.*, 1994b; Aitkin, 1998) and tammar wallaby (Cone-Wesson *et al*, 1997). More limited threshold data have been measured using ABR recordings from a few individuals in two other species – dunnart and kowari (see Aitkin, 1998). When the thresholds for activation of single units at their best or characteristic frequency (CF; frequency at which the threshold is lowest) is plotted against CF for a population of units, the resultant scattergram parallels the behavioural curve for that animal, for those species studied (Konishi, 1970; Evans, 1972; Dallos *et al.*, 1978). Extrapolating to the brush-tailed possum, single-unit thresholds obtained from the IC and auditory cortex when combined, give a scatterplot in which minimum thresholds occur between 7 and 20 kHz, rising to thresholds 40–50 dB in excess of the minima (–5 to –10 dB SPL) at 300 Hz and 40 kHz (Aitkin, 1998). A curve forming points generated by such a scatterplot would look very similar to that from a eutherian mammal, such as a cat or guinea pig.

Studies by Aitkin and his colleagues (Aitkin *et al.*, 1986b, 1994a) of northern quolls have shown that the averaged ABR audiogram drops from 40 dB SPL at 500 Hz to 10 dB at 8 kHz, rising to 60 dB at 40 kHz. The thresholds for single units of the IC at their CFs are generally more sensitive than the ABR and fall as low as 0 dB SPL at 7–12 kHz. The trends of the unit threshold distributions and averaged ABR plots are generally quite similar and again suggest auditory sensitivity similar to studied eutherians.

The diprotodont tammar wallaby has been extensively studied (Cone-Wesson *et al.*, 1997; Hill *et al.*, 1998; Liu *et al.*, 2001; Liu and Mark, 2001). The resulting ABR audiogram for the tammar wallaby defines a curve with minima of 0 dB SPL at 8–12 kHz and upper thresholds of 40 dB SPL at 1.5 kHz and 20 dB at 40 kHz (Cone-Wesson *et al.*, 1997). The plots illustrated by these authors show an audiogram similar in its central range to that of the Monodelphis opossum (Reimer, 1995), but much more sensitive than the ABR audiogram of the polyprotodont northern quoll (Aitkin *et al.*, 1996b).

Microelectrode studies of the marsupial auditory system have been carried out on the IC and auditory cortex in two species: the brush-tailed possum (Aitkin *et al.*, 1978, 1984; Gates and Aitkin, 1982) and northern quoll (Aitkin *et al.*, 1986b). Recordings have also been made in the tammar wallaby auditory brainstem (Liu *et al.*, 2001; Liu and Mark, 2001) and, in prehearing tammar wallabies, in the auditory nerve and cochlear nuclei (Gummer and Mark, 1994). The latter three studies examined developmental aspects of marsupial hearing and will be discussed later in this review. Multiunit recordings have also been made in response to free-field acoustic stimulation in the superior colliculus of tammar wallabies (Withington *et al.*, 1995).

In every mammal to date in which data are available, the CFs of units in the CIC are arranged in an orderly fashion (tonotopic organisation; Aitkin, 1986) and marsupials are no exception to this general trend (Aitkin *et al.*, 1978, 1986b). The collected CFs of the same frequency range form imaginary lines (isofrequency contours) arranged with low frequencies represented generally dorsally and high frequencies ventrally. Reimer (1993) used a functional labelling technique involving a 2-deoxyglucose (2DG) which is taken up by respiring activated neurons. If the 2DG is labelled with carbon-14 it can be used histochemically to demonstrate which neurons are active when specific tones are presented, and in Monodelphis the labelling forms bands oriented from dorsomedial to ventrolateral, similar to the orientation and trajectory of isofrequency contours observed in other species, including brush-tailed possums and northern quolls, using multiunit recording techniques (Aitkin *et al.*, 1978, 1986b).

It has been suggested that the frequency selectivity of units in the IC of brush-tailed possums is generally less sharp than that in eutherians (Aitkin *et al.*, 1978). Tuning curves are very broad in the external nucleus – some span six octaves or more – and even in the CIC no extremely sharp tuning curves are found. In association with broader tuning to tonal stimuli, units in the IC of possums are generally more strongly excited by broadband stimuli, such as white noise, than by spectrally simple stimuli such as pure tones (Aitkin *et al.*, 1984). It is interesting that the vocalisations of brush-tailed possums are spectrally very broad (Winter, 1976) and would therefore be likely to be potent activators of some units in the inferior colliculus of this species.

Another organising principle in the auditory pathway of mammals is the convergence of auditory input from each ear upon neurons in the superior olive and the vast majority in nuclei from the auditory midbrain to cortex (Aitkin, 1986; Irvine, 1986). In the brush-tailed possum, 85% of units in the IC are binaurally influenced (Aitkin *et al.*, 1978). Three principal forms of binaural interaction have been recognised. For one group, the discharge rate is influenced by the interaural time delay between binaural stimuli. For others, the response to binaural stimuli is an excitatory summation of the effects of stimulation of each ear alone. For a third group, stimuli applied to the contralateral ear are excitatory, whereas the same stimuli applied to the ipsilateral ear can reduce or inhibit the response to contralateral stimuli; these units may be sensitive to the interaural sound pressure-level difference (Irvine, 1986, 1992).

The localisation of a sound in space is made possible by the interaction of these binaural cues (interaural time and intensity differences) and the direction-dependent amplifying properties of the pinna (Coles and Hill, 1981; Semple *et al.*, 1983; Aitkin *et al.*, 1984; Coles and Guppy, 1986). The operation of neural mechanisms in the auditory brainstem using these cues results in the emergence of neural sensitivity to sound source location. Some neurons demonstrate 'spatial tuning' and respond best to sounds located at a particular locus in the horizontal and vertical planes. In the IC of possums there are neurons classified as directionally sensitive and for some of these units the same preferred location is demonstrated over a range of sound source intensities (Aitkin *et al.*, 1984).

Neural responses such as these may be responsible for some of the properties of the superior colliculus (SC) of tammar wallabies (Withington *et al.*, 1995). This nucleus is as prominent in the midbrain as the IC in most marsupial species. In addition to a largely visual input, it receives some input from the central auditory pathway. Neural responses to sound located in different spatial regions have been investigated using noise-evoked multiunit activity. There is a gradient of sound source location in the SC such that neurons in the rostral SC are 'tuned' to sounds issuing from the anterior locations, whereas sounds originating posteriorly cause greatest activity in the caudal SC. Intermediate sound source locations correlate with greatest activity in the rostrocaudal centre of the SC, suggesting a map of the spatial location of an auditory stimulus by a focus of neural activity in the SC.

Units have not been studied in the medial geniculate of marsupials, but responses in the auditory cortex have been investigated in the brush-tailed possum (Gates and Aitkin, 1982) and northern quoll (Aitkin *et al.*, 1986b). Gates and Aitkin (1982) mapped the auditory cortex of brush-tailed possums by making tangential microelectrode penetrations through the temporal cortex. They found evidence for a single cochlear representation with high CFs found dorsally and low CFs ventrally. This pattern is found irrespective of whether penetrations are made rostrally or caudally and there is no suggestion of reversals in CF or of other fields having sharply tuned neurons.

Maps with much more detail have been published for northern quolls (Aitkin *et al.*, 1986b), but only one complete auditory field, which is immediately dorsal to the rhinal fissure, has been identified. The CFs of neurons encountered in the middle layers of the auditory cortex are high (>30 kHz) at the dorsorostral limit of the field, but low (<2 kHz) ventrocaudally. Weak auditory drive is found at the edges of this field and the evidence for additional fields adjacent to this 'primary' field is inconclusive (Aitkin *et al.*, 1986b). Merzenich and Schreiner (1992) discuss the possibility that auditory fields may have evolved differently in marsupials compared with eutherians – perhaps the auditory cortical projections to the lateral amygdala of marsupials (Kudo *et al.*, 1986, 1989) may result in a secondary auditory field analogous to a secondary auditory cortical area in eutherians. The lateral amygdala is adjacent medially to the auditory cortex and it is reasonable to propose that the equivalent area, in an ancestor common to eutherians and marsupials, may have migrated to the cortical surface to form a new secondary auditory cortical field in eutherians.

Anatomical development of the auditory system in marsupials

Marsupials are born well before the visual and auditory systems have started to function, yet they can move around and feed. Presumably gravitational and tactile cues in some way facilitate the movement of the newborn from vagina to teat, but can the mother hear the young at an early age? Pouch-young quolls do not spontaneously vocalise ('isolation calls') until about P35 to P45, when they begin to move from one nipple to another (Aitkin *et al.*, 1994b, 1996b). Fourier analysis of the waveforms of calls elicited by animals at P65 or younger reveals a concentration of energy peaking at around 10 kHz. Beyond P109, the calls of the young quolls have spectra resembling those of the adult (Aitkin *et al.*, 1994b; Dempster, 1994). For quolls younger than P65, the frequencies at which peak energy occurs cluster closely around the best frequency of adult hearing (Aitkin *et al.*, 1994b, 1996b). The adult would therefore very easily detect the vocalisations of these young animals.

A similar match occurs between the frequency at which the peak energy of isolation calls made by neonatal Monodelphis occurs and the best frequency of the behavioural audiogram of the adult (Aitkin *et al.*, 1997). These data and those from the northern quoll suggest that the audiogram in any adult species should be considered not only in terms of the ecological niche of that species, but also in relation to the fostering and care of neonates (Aitkin, 1998).

The fact that so much development of the neonatal marsupial occurs outside the uterus has been a great stimulus to the physiological study of marsupial hearing (e.g. McCrady *et al.*, 1937, 1940). One problem has been to study a series of pouch young without too much diversity in genetic background. In this respect polyprotodonts, such as northern quolls and American opossums, have advantages over diprotodonts, such as tammar wallabies and brush-tailed

possums, relatively common in Australia, because the former have large litter sizes.

The ontogenetic changes in middle ear ossicles have been studied in a number of Australian and American marsupial species (Sanchez-Villagra *et al.*, 2002). There are differences among these species in terms of ossicular structure and development, and the ossification of the ossicles continues during the early days after birth and the beginning of pulmonary respiration. The basic differentiation of acousticomechanical structures in the inner ear of developing marsupials has been studied in the northern quoll by Gemmel and Nelson (1992) and in Monodelphis by Willard (1993). At birth, when the quoll first appears in the pouch, the cochlear enlargement is distinguishable from the vestibular sac. The cochlea is initially a straight tube (P3); the tip then begins to turn in a caudolateral direction to form a coil (P6); there is increasing coiling of the cochlea (P9) and by P11 there are two complete turns. At P25, when 2.5 turns are present, the organ of Corti begins to differentiate and the stapes becomes attached to the oval window. By P37, the tectorial membrane is free and by P50 the organ of Corti is fully mature (Figure 11.1b). Events are earlier in Monodelphis (weaned at P60 versus P80 for quolls). At P0, the otocyst has three-quarters of a turn, by P6, 1.5 turns and by P14 the complete spiral is formed, although hair cells remain immature (Willard, 1993).

The birth and migration of newborn neurons to form the brain (neurogenesis) has been studied for the auditory pathway in two marsupials, brush-tailed possums (Sanderson and Aitkin, 1990) and northern quolls (Aitkin *et al.*, 1991). In the polyprotodont quoll, neurons in the VC are generated prior to P3, in the superior olive at P5 to P7, and in the DC over a prolonged period. Inferior colliculus neurogenesis lags behind that in the MG, the latter taking place between P3 and P9 and the former between P7 and P22. Neurogenesis begins in the auditory cortex on P9 and is completed by about P42.

A very similar time-course occurs in the diprotodont brush-tailed possum. In the cochlear nuclei, large neurons are present by P5 and small cells appear over a more protracted period. Neurogenesis in the IC is complete by P28; that in the MG again precedes that in the midbrain (P7 to P12). It is thus possible that thalamocortical relationships are formed independently of midbrain relationships in the auditory systems of these species.

Once neurons have reached their ultimate locations, dendrites and axons grow and synaptic connections form. Although marsupial pouch young would be ideal candidates in which to study these processes, only sketchy data about the auditory system are available from the literature. Retrograde tracing studies in the Virginia opossum show that connections between the cochlear nuclei and IC are present at postnatal P15; by P22 (about 37 days after conception) these cells can be classified as being in the DC or VC (Willard and Martin, 1986). Connections appear to be established much later in northern quolls. Connections to the IC from the cochlear nuclei can be seen on P36, but results are equivocal earlier (Aitkin et al., 1986b). Targets for some of these axons, the dendritic spines, proliferate between P31 and P45 and myelin first appears in the IC of quolls at P73 (Aitkin, 1998).

Synapses are very rare in the IC of pouch-young northern quolls at P9 and increase during pouch life to reach a peak at P73, falling to adult values thereafter (Aitkin et al., 1996a). Synaptogenesis in the IC of Monodelphis increases between P26 and P30, followed by a further slower increase to the adult value at P40 (Aitkin et al., 1997).

Development of hearing in marsupials

Although startle responses (e.g. Aitkin et al., 1996b; Willard, unpublished observations) and cochlear microphonic recordings (McCrady et al., 1937, 1940) are useful in indicating whether or not young marsupials respond to sound, more accurate measures of the onset and development of auditory function are obtained using ABRs (Aitkin et al., 1996b, 1997; Reimer, 1996; Hill et al., 1998; Liu et al., 2001; Liu, 2003). In P68 quolls, responses are elicited between 1 and 16 kHz, with thresholds in excess of 55 dB (Aitkin et al., 1996b). This correlates with the time of opening of the external ear canal at P55 (Gemmel and Nelson, 1992). At P81 to P88, responses occur over the adult range at thresholds at least as low as the adult. A comparison of synaptic and neuronal packing densities suggests that, after a steady increase in the number of synapses per neuron between P9 and P63, synaptic numerical density rises steeply at about the time hearing begins (as detected by ABR recordings). Between P73 and the adult there is a pruning of both cell and synapse numbers (Aitkin et al., 1996a, 1996b).

Reimer (1996) found consistent responses to pure tones in Monodelphis at P29, when thresholds to 8 and 12 kHz were about 60 dB. At this age, middle ear structures are not mature in Monodelphis, but they are in a stage at which it is conceivable that they participate in sound transmission (Sanchez-Villagra et al., 2002). Aitkin et al. (1997) used clicks as auditory stimuli to examine the development of ABR responses in Monodelphis and found that the threshold for detection of an averaged response fell from 83 dB at P24 (the earliest age at which evoked potentials to clicks were seen in this study), to 58 dB at P40. In one study, evoked potentials were elicited at P28, in good agreement with Reimer (1996). As the animal ages, the range of effective frequencies broadens and the thresholds for these frequencies drop steadily; an adult ABR audiogram is obtained at P40 in Monodelphis (Reimer, 1996). Synaptogenesis in the IC of Monodelphis increases between P26 and P30 (Aitkin et al., 1997), as in quolls – the time when hearing first begins in this species. It is interesting that compared with normal hearing controls, studies of cats raised after a neonatal sensorineural hearing loss show that synaptogenesis in the IC is very dependent on auditory input during development (Hardie et al., 1998).

The development of auditory-evoked potentials has been studied in tammar wallabies (Hill et al., 1998; Liu et al., 2001; Liu and Mark, 2001). Spontaneous patterned neural activity can be identified in the auditory nerve and cochlear nuclei as early as P94 (Gummer and Mark, 1994). There is a sequence in development from responses in the auditory nerve at P101 to the cochlear nuclei at P110 (Liu et al., 2001). In the IC, responses first appear at P114 in its rostral pole, then caudally, and by P127 the entire IC is responsive (Liu and Mark, 2001).

What remains unknown?

There is detailed knowledge about auditory neurobiology in only a few Australian marsupials. It is clear that study should be extended laterally to species with unusual habitats or for which available evidence suggests specialised behaviours and hearing attributes. What is known suggests that the auditory systems of adult marsupials are similar in most respects to eutherians, however research into aspects of the development may offer much to all mammal species, including monotremes, marsupials and eutherians.

Acknowledgements

We thank Ms K. Smith for editorial assistance. The Bionic Ear Institute acknowledges financial support from the Victorian Government through its Operational Infrastructure Support Program.

12

Olfactory system

K. W. S. Ashwell

Summary

The olfactory pathways consist of a projection from the olfactory epithelium through the main olfactory bulb and thence to more caudal telencephalic olfactory areas, and a projection from the vomeronasal organ through the accessory olfactory bulb to the olfactory amygdala and bed nuclei of the stria terminalis. Although study of the olfactory systems of Australian marsupials has been limited, there are some findings suggestive of behaviourally important anatomical specialisations. For example, the abundant vomeronasal glands opening into the oral cavity over the tongue of the honey possum (*Tarsipes rostratus*) may represent an adaptation to facilitate sampling of odorants in that nectar feeder. Similarly, the size of the nuclei of the olfactory pathway in the marsupial mole points to a central behavioural role of the olfactory system. In diprotodont marsupials, the main elements of the olfactory system (olfactory bulb, olfactory tubercle, piriform cortex and amygdala) occupy a higher proportion of total brain weight than in common eutherian mammals (e.g. rat, rabbit and cat).

Cyto- and chemoarchitectural features of the central olfactory regions in Australian marsupials are similar to the corresponding regions in eutherian mammals. Although no recent detailed studies have been made of central connections in any Australian marsupial, it is likely that these are similar to those seen in Ameridelphians and eutherians.

Olfactory communication is just as important for Australian marsupials as other mammals and olfactory discrimination in Australian marsupials can be performed at a level of sensitivity equal to that seen in laboratory rodents. Odour-producing glands have been identified in 40 of 73 marsupial genera and sternal gland marking of territory is used extensively by many diprotodont species. Faecal and urine deposits may also be used to mark objects and distinguish sexes among both poly- and diprotodont species. Olfaction also plays an important role in assessing the palatability of foods.

Although the olfactory epithelium of many poly- and diprotodont marsupials shows some mature features at birth, the anatomical and functional maturation of the central olfactory pathways takes several weeks to months. Therefore, there is no basis for arguing that the olfactory system in newborn wallabies makes any significant contribution to pouch- or nipple-finding behaviour. In fact, full functional maturity of the olfactory system in diprotodonts is probably not achieved until close to the time that the young marsupial leaves the pouch, when olfaction becomes important for food choice.

The olfactory system has been largely ignored in Australian marsupials, despite the extraordinary anatomical specialisation seen in the marsupial mole and the striking mating-season behaviour of some small marsupial carnivores. More than any other neural system (with the possible exception of the neuroendocrine system), the behaviourally significant olfactory system is one of the most neglected areas in Australian marsupial neurobiology. Australian marsupials are potentially revealing models for exploring the nexus between olfaction, behaviour and reproductive physiology among therians.

Introduction

The sense of smell is essential for behaviourally important emotional responses, aggression, recognition of conspecifics (Shapiro *et al.*, 1996), predators and prey, neuroendocrine regulation, food selection, and reproductive and maternal functions. Furthermore, olfaction plays a key role in memory and the control of behaviour.

Olfaction is mediated through stimulation of olfactory receptor neurons by volatile chemicals. These receptor neurons are located on the upper part of the nasal septum, the cribriform plate and some of the surfaces of the endo- and ectoturbinals, which increase the surface area of the nasal cavity in many mammals. In some species (including some marsupials, see Figure 12.1a to c), olfactory neuroepithelium is also found on the ventral wall of the septum (SOO – septal organ of Masera). The olfactory neuroepithelium is a pseudostratified columnar epithelium, which is thicker than the surrounding respiratory epithelium. Olfactory receptor neurons are supported and surrounded by several types of cells (sustentacular, microvillar, globose and horizontal basal cells) and the epithelium lies on a basement membrane and an underlying lamina propria. The lamina propria contains Bowman's glands, which provide serous secretions to the mucous layer overlying the olfactory neuroepithelium. Olfactory receptor axons project to the main olfactory bulb (OB) and the output neurons of the bulb (mitral and tufted cells) project in turn to higher-order processing centres in the olfactory system and to other brain components (Figure 12.1d). The output from the OB is strongly influenced by centrifugal projections from central olfactory centres and other brain regions. The main output from the OB is to a group of structures referred to collectively as primary olfactory cortex. In rodents these are usually divided into three groups: (i) the anterior olfactory area or nucleus (often further subdivided); (ii) the medial olfactory cortex – predominantly archicortical regions (TT, IG) and the olfactory tubercle (Tu) and (iii) the lateral olfactory cortex (piriform cortex, periamygdaloid and entorhinal cortices).

In many mammals, a parallel, but anatomically and functionally separate, olfactory system is the accessory olfactory system (Figure 12.1e). In contrast to the olfactory epithelium, olfactory receptor neurons in the vomeronasal organ are exposed to non-volatile odours. Vomeronasal receptor neurons appear to respond to specific pheromone molecules at concentrations several orders of magnitude below that for the main olfactory system (see review by Shipley et al., 2004). The vomeronasal organ projects exclusively to the accessory olfactory bulb (AOB), which is usually found at the dorsocaudal rim of the OB. Since the growth of the accessory olfactory bulb is strongly influenced by gonadal steroids, sexual dimorphism of the AOB is usually significant. The AOB projects predominantly to the medial and posterior cortical nuclei of the amygdala, bed nuclei of the stria terminalis and the nucleus of the accessory olfactory tract. The secondary olfactory projections involving the amygdaloid nuclei project in turn to centres that strongly influence sexual drive, and these pathways in eutherian mammals appear to contain neurons which are strongly influenced by gonadal steroids.

Topography and histology of the marsupial nasal cavity

The nasal cavity of many mammals has folded lateral walls (maxilloturbinals, nasoturbinals, ectoturbinals, endoturbinals; Figure 12.1a to c), which provide an increased surface area for warming and moistening inspired air, slowing the nasal airflow by creating turbulence, or for supporting the olfactory sensory epithelium. Most of these folded surfaces are components of the ethmoid bone, which also carries the cribriform plate, a perforated bony plate transmitting axons from the olfactory sensory epithelium to the olfactory bulb. Macrini has found that the shape and complexity of the maxilloturbinal varies extensively between marsupial taxa (Macrini, 2007). For example, the maxilloturbinals of Vombatiformes (i.e. koala and wombats) are simple, with few scrolls and a tube-like shape caudally. As noted by Kratzing (1984a), the external appearance of the nasal region may be deceptive: the shape of the koala head would suggest a large nasal cavity, but the nasal conchae are relatively simple and extensive maxillary and frontal paranasal sinuses take up much of the air space within the head (Kratzing, 1984a). By contrast, the maxilloturbinals of *Didelphis virginiana* are quite complex (Macrini, 2007), with many branches and scrolls throughout their rostrocaudal extent. Among marsupials generally, the number of endoturbinals has been reported to range from four to six, whereas the number of ectoturbinals ranges from one to three (Macrini, 2007). Macrini concluded that the presence of five endoturbinals (not counting the nasoturbinal) and two ectoturbinals is probably the ancestral condition for marsupials.

The olfactory sensory and vomeronasal epithelium of Australian marsupials (at least those species studied to date) appears to be very similar ultrastructurally to that reported for other mammals (Kratzing, 1978, 1984a, 1984b; Schneider et al., 2008), suggesting that vomeronasal organ structure is highly conserved among therians. The vomeronasal organs of the tammar wallaby have electron-dense supporting cells and electron-lucent receptor neurons, with olfactory knob-like structures on the surface of the receptor epithelium (Schneider et al., 2008). The sensory epithelium of the tammar wallaby vomeronasal organ is thicker than in cattle, but not as thick as in the laboratory rat, which probably represents a highly specialised vomeronasal organ. Behaviour of male tammars during breeding season is suggestive of an important role for the vomeronasal organ in identifying potential mates. Male tammar wallabies retract their split upper lip when investigating the urogenital orifice of estrous females. This flehmen-like behaviour exposes the gap between the two first incisors, facilitating transfer of odorants to the oral opening of the nasopalatine duct,

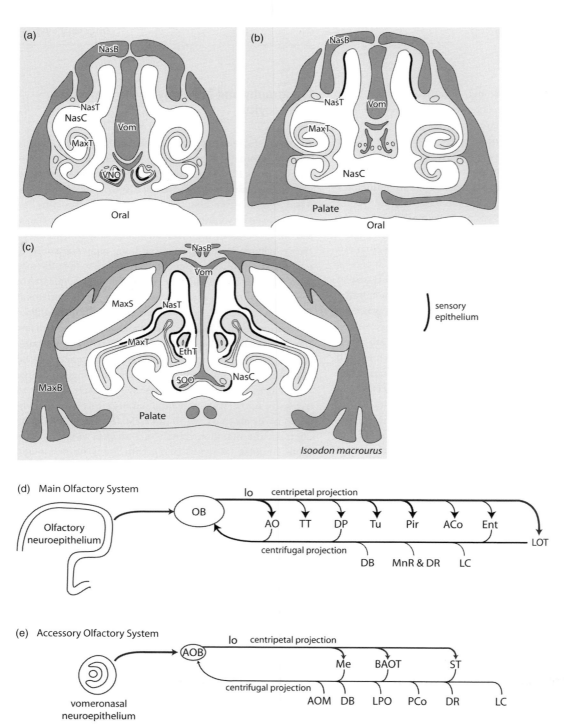

Figure 12.1 Frontal (coronal) sections through the nasal cavity of a northern brown bandicoot (*Isoodon macrourus*) have been shown at the level of the third incisor (a), the canine tooth (b) and the first molar (c). All three images have been developed from data in Kratzing (1978). Note the presence of discrete olfactory sensory epithelium in the dorsal nasal cavity, vomeronasal organ sensory epithelium (VNO) and septal olfactory organ sensory epithelium (SOO). Figures (d) and (e) show summary diagrams of the main connections (both centripetal and centrifugal) of the main and accessory olfactory systems, as known from studies of placental mammals. ACo – ant. cortical amygdaloid area; AO – anterior olfactory nu.; AOB – accessory olfactory bulb; AOM – anterior olfactory nu., medial; BAOT – bed nu. access. olfactory tract; DB – diagonal band nu.; DP – dorsal peduncular cortex; DR – dorsal raphe nu.; Ent – entorhinal cortex; EthT – ethmoid turbinate; LC – locus coeruleus; lo – lateral olfactory tract; LOT – nu. of lateral olfactory tract; LPO – lateral preoptic area; MaxB – maxillary bone (maxilla); MaxS – maxillary sinus; MaxT – maxilloturbinate; Me – medial amygdaloid nu.; MnR – median raphe nu.; NasB – nasal bone; NasC – nasal cavity; NasT – nasal turbinate; OB – olfactory bulb; Oral – oral cavity; PCo – posterior cortical amygdaloid area; Pir – piriform cortex; SOO – septal olfactory organ; ST – bed nu. stria terminalis; TT – tenia tecta; Tu – olfactory tubercle; VNO – vomeronasal organ; Vom – vomer.

which in turn leads to the vomeronasal organ (Schneider *et al.*, 2008).

A septal organ is also present in the nasal cavity of bandicoots (but not diprotodonts), and is clearly separated from the vomeronasal organ and the olfactory epithelium of the dorsal parts of the nasal cavity (Figure 12.1c). The function of the septal organ sensory epithelium (SOO) remains unknown, but its position may allow it to monitor airflow in quiet ventilation or assess the direction from which an odorant comes (Kratzing, 1978).

Kratzing also studied the types of nasal glands in four species of Australian marsupial (Kratzing, 1984b) and found some anatomical specialisations that may serve specific behaviour. For example, vomeronasal glands appear to be best developed in the nectar- and pollen-feeding honey possum (*Tarsipes rostratus*), where the associated incisive ducts open towards the oral cavity on either side of an incisive papilla. Kratzing suggested that the vomeronasal gland secretions are poured onto the dorsal surface of the spiny tongue as it moves backwards and forwards through the groove of the incisive papilla. This arrangement not only lubricates the tongue surface, but also facilitates the collection of odorants from the tongue dorsum and their transmission to the vomeronasal organ. By contrast, the duct of the vomeronasal organ opens only into the nasal cavity in the koala and agile wallaby (Kratzing, 1984b).

Another example of nasal specialisation (albeit deleterious to olfaction) is seen in the koala (Kratzing, 1984a), where the absence of glands beneath the frontal and maxillary sinus epithelium and the lack of a lateral nasal gland in adult koalas may contribute to water conservation. The sparse goblet cells appear to provide the only nasal secretion in koalas. Olfaction is probably poorer in the koala than in other Australian marsupials, since the olfactory sensory epithelium is less extensive, the septal olfactory organ absent and the vomeronasal organ relatively short (Kratzing, 1984a).

Central olfactory pathways in Australian marsupials

The centripetal (centre-seeking) and centrifugal (centre-fleeing) projections of the main and accessory olfactory pathways of a representative mammal are illustrated in Figure 12.1d, e. All the key components of the main and accessory olfactory systems can be identified in the brains of better-studied Australian marsupials (Figure 12.2, 12.3 and 12.4). Furthermore, for most Australian marsupials, the topography of the olfactory system nuclei is broadly similar to that seen in Ameridelphians. Although analysis of olfactory connections in mature Australian marsupials is limited

to a study of the brush-tailed possum using now outmoded silver staining techniques (Adey, 1953), it is likely that the central olfactory connections of Australian marsupials are similar to Ameridelphians (Haberly, 1983; Martinez-Marcos and Halpern, 2006) and rodents (Price, 1973; Haberly and Price, 1978; Schwob and Price, 1984; Martinez *et al.*, 1987). In polyprotodonts there is a topographic correspondence between the position of output neurons in the olfactory bulb and the position of centripetal axons in the lateral olfactory tract (see discussion in Johnson, 1977), but nothing is currently known about functional topographic organisation within marsupial olfactory pathways.

Perhaps the most striking olfactory system of any mammal is that possessed by the marsupial mole (Schneider, 1968)(Figure 12.2). Not only are the olfactory bulbs of this marsupial large (30.3% of the entire olfactory cortex), but the caudal parts of the forebrain are dominated by components of the olfactory pathways, in particular the anterior olfactory cortex, olfactory tubercle and piriform cortex. The surface of the olfactory tubercle occupies half of the circumference of the rostral forebrain and 21.0% of the olfactory cortex volume, suggesting that this species may have a major input of olfactory information into the ventral striatum and pallidum. Similarly, the piriform cortex in the marsupial mole occupies a greater volume of the pallium (63%) than the isocortex (29%). The proportional size of the OB in the marsupial mole is higher than that for any eutherian insectivore. Unfortunately, the secretive nature and rarity of this marsupial have precluded any detailed study of the chemoarchitecture and connectivity of this remarkable nervous system.

In other Australian marsupials the components of the olfactory pathways are not quite so prominent, but nevertheless are substantial in size. The olfactory bulbs of the brush-tailed possum and quokka occupy 3.1% and 3.4%, respectively, of the total brain weight, compared to 2.3%, 2.6% and 1.6% for the laboratory rat, rabbit and cat, respectively (Meyer, 1981). Similarly, the olfactory tubercle occupies 3.5% and 3.1% of total brain weight for the brush-tailed possum and quokka, compared to 4.0%, 2.9% and 2.1% for the laboratory rat, rabbit and cat (Meyer, 1981). Even central targets like the piriform cortex and amygdala are proportionally larger in diprotodont marsupials compared to common laboratory animals (possum and quokka – 3.8% and 3.1%, respectively, compared to 4.3%, 2.6% and 2.6% for rat, rabbit and cat, respectively). However, it should be noted that the absolute size of all these components of the olfactory system is not greater in the possum and quokka than in eutherians of comparable body size (e.g. rabbit and cat). In other words, the higher proportional size of the bulb, tubercle and piriform/amygdala in the quokka and possum is a reflection of

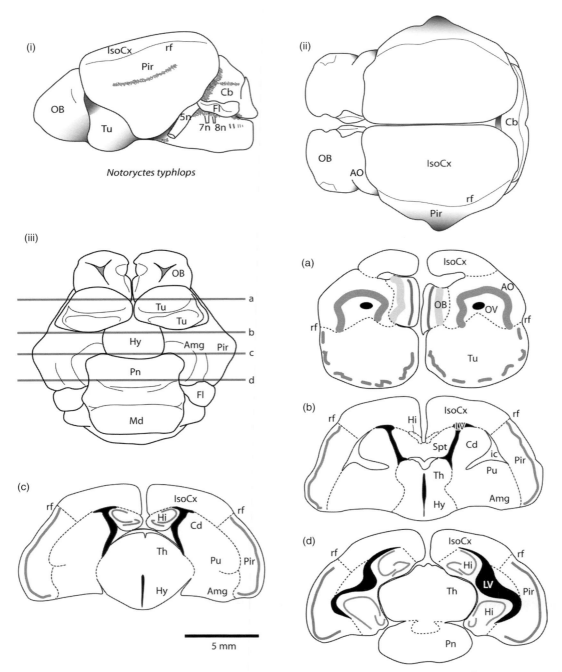

Figure 12.2 Diagrammatic representation of the extraordinary olfactory system of the marsupial mole (*Notoryctes typhlops*) as seen in a series of lateral, dorsal and ventral views (i, ii, iii, respectively), as well as a series of coronal or frontal sections (a to d) at rostrocaudal levels illustrated on the ventral view of the whole brain (iii). Note the remarkable size of the olfactory bulb, anterior olfactory nu. and piriform cortex. Diagrams (i) to (iii) have been based on illustrations in Schneider (1968). Diagrams (a) to (d) have been based on photomicrographs from the same source. 5n – trigeminal nerve; 7n – facial nerve; 8n – vestibulocochlear nerve; Amg – amygdala; AO – anterior olfactory nu.; Cb – cerebellum; Cd – caudate nu.; Fl – flocculus; Hi – hippocampus; Hy – hypothalamus; ic – internal capsule; IsoCx – isocortex; LV – lateral ventricle; Md – medulla; OB – olfactory bulb; OV – olfactory ventricle; Pir – piriform cortex; Pn – pontine nuclei; Pu – putamen; rf – rhinal fissure; Spt – septum; Th – thalamus; Tu – olfactory tubercle.

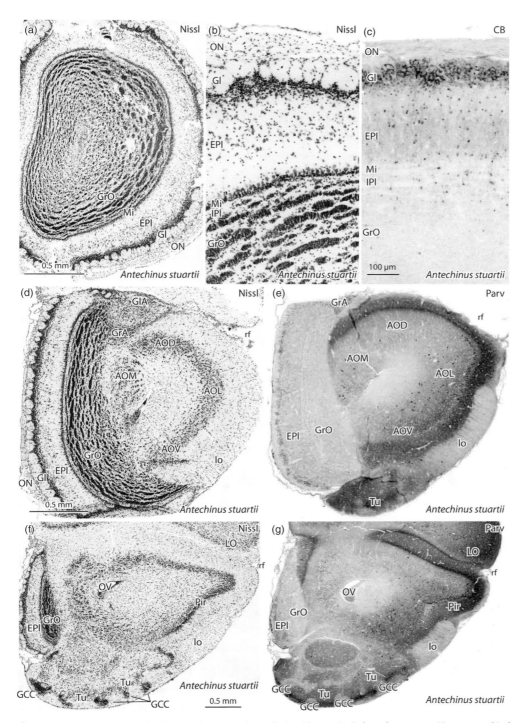

Figure 12.3 Photomicrographs of coronal sections through the olfactory bulb (a to c), anterior olfactory nu. (d, e) and rostral olfactory tubercle (f, g) in a representative polyprotodont Australian marsupial. Photomicrographs (a) and (b) show low- and high-power views of Nissl-stained sections through the olfactory bulb, while (c) is a photomicrograph of a calbindin-immunoreacted section that matches (b). Photomicrograph pairs (d), (e) and (f), (g) show matching Nissl-stained and parvalbumin-immunoreacted sections. AOD – anterior olfactory nu., dorsal; AOL – ant. olfactory nu., lateral; AOM – ant. olfactory nu., medial; AOV – ant. olfactory nu., ventral; EPl – ext. plexiform layer, olfactory bulb; GCC – granule-cell cluster; Gl – glomerular layer, olfactory bulb; GlA – glomerular layer, accessory olfactory bulb; GrA – granule cell layer, accessory olfactory bulb; GrO – granular cell layer, olfactory bulb; IPl – internal plexiform layer, olfactory bulb; LO – lateral orbital cx; lo – lateral olfactory tract; Mi – mitral cell layer, olfactory bulb; ON – olfactory nerve layer; OV – olfactory ventricle; Pir – piriform cx; rf – rhinal fissure; Tu – olfactory tubercle.

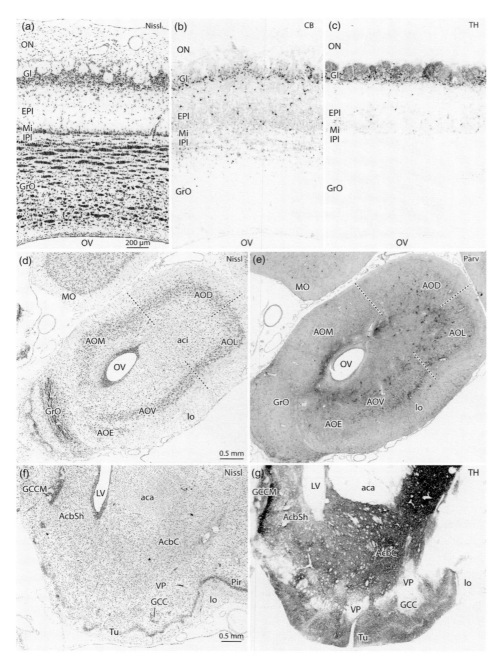

Figure 12.4 Photomicrographs of Nissl-stained and immunoreacted (calbindin – b; parvalbumin – e; tyrosine hydroxylase – c, g) adjacent coronal sections through the olfactory bulb (a to c), anterior olfactory nu. (d, e) and olfactory tubercle (f, g) of a representative diprotodont marsupial. aca – anterior commissure, anterior limb; AcbC – accumbens nu., core; AcbSh – accumbens nu., shell; aci – anterior commissure, intrabulbar limb; AOD – anterior olfactory nu., dorsal; AOE – ant. olfactory nu., external; AOL – ant. olfactory nu., lateral; AOM – ant. olfactory nu., medial; AOV – ant. olfactory nu., ventral; EPl – external plexiform layer, olfactory bulb; GCC – granule cell cluster; GCCM – granule-cell cluster, magna; Gl – glomerular layer, olfactory bulb; GrO – granular cell layer, olfactory bulb; IPl – internal plexiform layer, olfactory bulb; lo – lateral olfactory tract; LV – lateral ventricle; Mi – mitral cell layer, olfactory bulb; MO – medial orbital cortex; ON – olfactory nerve layer; OV – olfactory ventricle; Pir – piriform cortex; Tu – olfactory tubercle; VP – ventral pallidum.

the poorer development of other brain structures (cerebellum, midbrain, striatum and neocortex).

In eutherian and marsupial mammals, one of the most distinctive features of the cytoarchitectural organisation of the main olfactory bulb is the presence of a monolayer of mitral cells embedded in a thin basement layer of granule cells (Switzer and Johnson, 1977) and the mitral cell monolayer is clearly visible in both poly- and diprotodont marsupials (Figure 12.3 and Figure 12.4, respectively). The cyto- and chemoarchitectural organisation of the main olfactory bulb in Australian marsupials is very similar to that seen in eutherian mammals. Apart from the large neurons of the mitral cell layer (Mi), large neuronal bodies are also seen in the upper and middle layers of the external plexiform layer (EPl) of *Antechinus stuartii* (Figure 12.3a) and the outer third of the EPl in *Macropus eugenii* (Figure 12.4a). These large neurons are probably the external and middle populations of tufted cells, the other output neurons of the olfactory bulb. In both poly- and diprotodont marsupials, the periglomerular cells of the glomerular layer are strongly immunoreactive for calbindin and tyrosine hydroxylase (Figures 12.3c and 12.4b, c), as is seen in eutherian mammals.

The anterior olfactory nucleus is actually a laminated part of the primary olfactory cortex. It can be divided into several regions on the basis of cytoarchitecture, but the most important division is into a medial component that provides a centrifugal projection to the accessory olfactory bulb and all other divisions. In terms of both cyto- and chemoarchitecture, the AO of representative poly- and diprotodont species (Figures 12.3d, e and 12.4d, e) is similar in appearance to the AO of eutherian mammals.

The olfactory tubercle (Tu) has a superficial plexiform layer like the AO, but the architecture of the Tu is usually regarded as intermediate between a laminated cortical and massive or homogeneous striatal structure. As noted in Chapter 7, the olfactory tubercle is often considered part of the ventral striatum, despite its superficial position in adult mammals. The function of Tu in all mammals is still uncertain, but the region may allow for olfactory input to the deeper ventral striatum. Johnson noted that the Tu is exceptionally large in *Caenolestes, Perameles* and astonishingly large in *Notoryctes* (Johnson, 1977). The Tu has embedded granule-cell clusters (otherwise commonly known as islands of Calleja, although this may be a misnomer). In Australian marsupials these are present in both poly- and diprotodont species (Figure 12.3f, g; Figure 12.4f, g), but are much more prominent in the smaller polyprotodonts like *Antechinus stuartii* (Figure 12.3f, g). The function of granule-cell clusters remains unknown.

The caudal and lateral parts of the AO give way to the piriform cortex, which in turn gives way to the periamygdaloid and transitional cortices and then to the lateral entorhinal cortex. The piriform cortex is classically divided into three layers: Pir1 – the superficial plexiform or input layer, further divided into 1a, receiving afferents from the ipsilateral OB, and 1b, receiving association fibres from the AO and other parts of the primary olfactory cortex; Pir2 – the superficial compact cell layer; and Pir3 – the widest polymorphic cell layer. The piriform cortex is underlaid by the endopiriform nucleus, recognisable in all mammals (see Chapter 8).

Behavioural importance of olfaction for Australian marsupials

Olfaction plays a central role in a variety of behaviours: home-range marking, mate recognition, young recognition and food choice. Studies in a seminatural environment have shown that red kangaroos rapidly learn in an odour-cued taste-avoidance procedure (Hunt *et al.*, 1999) and the sensitivity of Australian marsupials to odorants is comparable to that seen in laboratory rodents (Croft and Eisenberg, 2006).

Pheromones are chemical messages that are released by an animal into the external environment: conspecifics perceiving the odour will exhibit a behavioural reaction specific for the odour. Pheromones may be released from glandular areas (e.g. sternal glands) or secreted in the urine and/or faeces. Odour-producing glands have been identified in most marsupial genera (Russell, 1985) and may be found on the chest (sternal glands), labia and ear, or around the urogenital sinus.

Behaviour used by conspecifics to detect the odour may include licking and sniffing the urogenital opening and/or urine, nasonasal sniffing or sniffing at some part of the head. As noted above, the vomeronasal organ is present in all marsupials. Lip-curl behaviour by marsupials has been considered as the flehmen grimace, a movement that facilitates the transfer of pheromone-laden urine or other secretions to the entry of the vomeronasal organ. This behaviour has been seen in wombats (Gaughwin, 1979) and bandicoots, but not koalas (Kratzing, 1984a). Nevertheless, some sternal gland scent marking of territory by male koalas has been reported (Kratzing, 1984a).

As with other mammals, olfaction plays a critical role for Australian marsupials in the selection of suitable foods. For example, behavioural studies with western grey kangaroos have shown that this species uses olfactory cues to avoid Myrtaceae foliage, which contains potentially toxic essential oils (Jones *et al.*, 2003). Even the (apparently) relatively poorly osmatic koala is selective in its choice of eucalypt leaves and individual trees are preferred over others. This behaviour is apparently based on olfaction, since sniffing precedes sampling in captive koalas (see Kratzing, 1984a, for discussion).

Development of olfactory pathways in Australian marsupials

A study of gross brain growth in the postnatal tammar wallaby (Renfree *et al.*, 1982) found that the olfactory lobe is the fastest-growing brain part after the cerebellum, with the olfactory lobe (presumably including the olfactory bulb, olfactory tubercle and piriform cortex, although this is not clearly defined) growing 70-fold from P30 to P180.

Studies of the development of olfactory pathways have been confined to examination of the olfactory epithelium in several species of poly- and diprotodont marsupials (Gemmell and Nelson, 1988a; Gemmell and Selwood, 1994), a study of development of electrical activity in the tammar wallaby olfactory system (Ellendorff *et al.*, 1988) and a recent analysis of development of central olfactory connections in the tammar wallaby (Ashwell *et al.*, 2008a). Tyndale-Biscoe (2005) proposed that the way in which the newborn tammar redirects its movements when it reaches the edge of the pouch and enters it might indicate a response to either the odour of the pouch or the texture of the pouch's moist hairless skin. Many of the studies of marsupial olfactory development have been motivated by the desire to determine whether the olfactory system of the newborn marsupial is sufficiently anatomically mature to contribute to the location of the pouch and/or nipple.

Gemmell and co-workers identified olfactory knobs in the nasal epithelium of the newborns of several species of diprotodont and polyprotodont marsupials (Gemmell and Nelson, 1988a; Gemmell and Rose, 1989; Gemmell and Selwood, 1994). The olfactory epithelium of the newborn marsupial is comparable in maturity to that of an E14 rat, but authors have differed on the issue of whether this is sufficiently mature to allow location of the pouch and nipple. Gemmell argued that the olfactory epithelium could be functional at birth in marsupials (Gemmell and Nelson, 1988a), but Kratzing felt that the olfactory system was not sufficiently differentiated to be of any help in finding the teat (Kratzing, 1986). In light of the functional and anatomical studies considered below, Kratzing is probably correct.

Olfactory bulb neuronal activity in developing tammar wallabies can be recorded from P39 onwards (Ellendorff *et al.*, 1988) and functional maturation of the main output neurons, the mitral cells, appears to be present by P44, but there is a long period of subsequent maturation before electrical activity of olfactory bulb neurons in this species takes on mature features (after P140). Furthermore, responses of olfactory bulb neurons to a strong-smelling stimulus (eucalyptus oil) cannot be evoked until P81. However, electrophysiological studies of this kind, which use urethane-anaesthetised animals, may not be assessing the true functional capacity of the developing olfactory system (Ashwell *et al.*, 2008a).

An anatomical study of the development of the olfactory pathways in the tammar wallaby (Ashwell *et al.*, 2008a) showed that although the olfactory system of the newborn tammar is at a comparable stage to newborn Ameridelphians (e.g. *Monodelphis domestica* and *Didelphis virginiana*; Lin *et al.*, 1988), there are several aspects of the subsequent maturation of the olfactory pathways in the wallaby which are delayed relative to Ameridelphians. For example, the emergence of glomeruli in the olfactory bulb appears to be delayed in the tammar wallaby (P25 to P54, Ashwell *et al.*, 2008a) relative to *Monodelphis domestica* (P10 to P15) (Brunjes *et al.*, 1992; Chuah *et al.*, 1997; Malun and Brunjes, 1996). Similarly, the time-course of development of the output neurons of the olfactory bulb (mitral and tufted cells) is relatively late in the wallaby compared to the opossum (P5 to after P34 in the wallaby, Ashwell *et al*, 2008a; P3 to P25 in *Monodelphis domestica*; Santacana *et al.*, 1992a, 1992b). On the other hand, both diprotodont and polyprotodont marsupials exhibit delayed maturation of olfactory pathways relative to rodents (Marchand and Bélanger, 1991; Math and Davrainville, 1980; Treloar *et al.*, 1999). Based on similarities in the level of differentiation of the olfactory epithelium and bulb (Ashwell *et al.*, 2008a), the newborn tammar wallaby appears to be at a comparable stage of development to that seen in the rat at an early fetal age (E14) (Ashwell and Paxinos, 2008; De Carlos *et al.*, 1996). Furthermore, with respect to the structural development of the central olfactory pathways, both *Macropus eugenii* and *Monodelphis domestica* are at least one week (and often two or three weeks) behind the laboratory rat (Ashwell *et al.*, 2008a).

The main and accessory olfactory pathways are not the only connections between the olfactory epithelium and telencephalon during early development. As in eutherian mammals, there is an extensive terminal nerve projection from the olfactory epithelium to the rostral forebrain of the newborn wallaby (Ashwell *et al.*, 2008a). This pathway begins to degenerate by the end of the first postnatal week and disappears by the end of the second postnatal week. The transient terminal nerve projection is probably providing an avenue for entry of luteinising hormone releasing hormone (LHRH)+ and FMRFamide+ axons and neurons into the developing forebrain as has been reported for other therian mammals at a similar developmental stage (Cummings and Brunjes, 1995; Malz and

Schwanzel-Fukuda *et al.*, 1988; Zheng *et al.*, 1990; Tarozzo *et al.*, 1994; Kuhn, 2002).

As regards the question of whether the olfactory system of the newborn Australian marsupial contributes to location of either the pouch entrance or the nipple, several lines of evidence exclude this possibility at least for the tammar wallaby (Ashwell *et al.*, 2008a). Firstly, although olfactory receptor neurons express some immunoreactivity to GAP-43 at the time of birth, mature receptor neuron morphology does not appear until P5 to P12. Secondly, although some degree of outgrowth of olfactory receptor neuron axons into the forebrain is present at birth, the degree of differentiation of the olfactory bulb is poor at this age. In particular, the olfactory bulb of the tammar wallaby does not begin to exhibit lamination until the end of the second postnatal week and recognisable sensory glomeruli do not form until towards the end of the fourth postnatal week. Thirdly, central projections of the olfactory bulb (along the lateral olfactory tract) are not established until at least the second week of postnatal life and bulb output neurons cannot be retrogradely labelled from the lateral olfactory tract until after P15. Fourthly, the olfactory tubercle and piriform cortex do not differentiate as distinct neuronal targets for the olfactory bulb efferents until the third postnatal week. Finally, there are no apparent connections (either direct or indirect) between rostral forebrain olfactory centres and the brainstem at birth. Therefore, there is no evidence to support the hypothesis that olfaction contributes to behaviour in this species at birth.

The components of the primary olfactory cortex also engage in commissural connections through the paleocortical component of the anterior commissure. In the tammar wallaby, commissural connections from the pyramidal neurons of the piriform cortex are the first to cross the midline, from about P14 (Ashwell *et al.*, 1996a, 1996b; Shang *et al.*, 1997). The neurons contributing to the first commissural connections are located in the rostral piriform cortex close to the rhinal fissure (Shang *et al.*, 1997). This means that piriform pyramidal neurons contributing to the commissural pathways in this wallaby

begin to grow their axons across the midline before they receive the first projections from rostral parts of the olfactory system and well before any significant functional input from the olfactory bulb (Ellendorff *et al.*, 1988). In fact, outgrowth of commissural axons must occur within only a few days of piriform neurons leaving the mitotic cycle and coincidentally with the establishment of piriform cortex lamination.

What remains unknown?

Probably the most interesting aspect of the olfactory system concerns its functional and anatomical specialisations, hinted at by findings to date in some Australian marsupials. In particular, the extraordinary size of the olfactory regions in the marsupial mole is indicative of a central behavioural role of the olfactory system for that marsupial, and yet nothing is known about functional olfactory specialisations in the mole or how this strange creature uses olfaction while seeking food. The extensive olfactory tubercle in the marsupial mole suggests that this mammal could provide new insights into the role of the tubercle in influencing the ventral striatum and brain reward pathways.

In eutherian mammals, components of the accessory olfactory system and its targets (accessory olfactory bulb, anterior olfactory nucleus, medial amygdaloid nucleus, medial preoptic area) are strongly influenced by circulating gonadal steroids and exhibit significant sexual dimorphism. Yet, little is known about how these brain regions are influenced by hormones in Australian marsupials. In particular, those species of Australian polyprotodont marsupials that exhibit male semelparity (e.g. *Antechinus stuartii*) are likely to show significant correlation between hormonal influences and brain structure, but this has not been studied in sufficient detail to explain the striking behaviour of males in these species. These sorts of study have enormous potential for illuminating the evolution of behavioural control in therians.

Motor system and spinal cord

K. W. S. Ashwell

Summary

The grey matter of the spinal cords of all mammals can be divided on the basis of similar criteria into 10 cytoarchitectonic regions known as Rexed's laminae. Rexed's laminae 1 to 6 correspond with the dorsal horn, lamina 7 largely corresponds with the intermediate grey matter and laminae 8 to 9 are found within the ventral horn. Lamina 10 surrounds the central canal.

Two descending motor pathways to the spinal cord have been studied in Australian marsupials. The corticospinal tract allows relatively direct activation of spinal-cord motorneurons by cortical activity. In Australian marsupials, most fibres run in the most ventral part of the dorsal white matter, although a few lateral corticospinal fibres are present in the most medial part of the lateral white matter in the brush-tailed possum and potoroo. In the relatively dextrous brush-tailed possum, the corticospinal tract has a robust projection to cervical and upper thoracic segmental levels, but the pathway is not so well developed in macropods. The rubrospinal pathway in the brush-tailed possum is similar to that in the North American opossum, and the pathway extends to many more segmental levels than the corticospinal tract. The relationship between corticospinal projection and forelimb manipulative behaviour warrants further investigation in Australian marsupials.

Immature marsupials exhibit characteristic rhythmic movements of the forelimbs that are clearly critical for climbing from the urogenital sinus to the pouch, but may also contribute to initiation of delivery and/or stimulation of lactation and milk ejection. Rhythmic forelimb movements involve reaching and digital extension alternating with retraction and grasping. These movements are stimulated by a drop in ambient temperature and are dependent on rhythm generators in the ipsilateral spinal cord and a pattern-co-ordinating pathway, which involve both glycinergic and GABAergic interneurons.

As might be expected, given the need to climb to the pouch on the day of birth, the brachial levels of the spinal cord (caudal cervical and upper thoracic segments) of newborn marsupials are more advanced than lumbar or sacral levels. Nevertheless, even the brachial levels of the spinal cord of newborn marsupials appear surprisingly histologically immature, given their critical role in reaching the safety of the pouch. In the tammar wallaby, motorneuron cell death occurs in lumbar spinal cord in two phases (phase I from birth to P40 and phase II from about P90 to P150), and both phases together account for approximately 70% loss of the motorneurons generated before birth.

Although not directly studied in Australian marsupials, it is likely that dorsal-column pathways develop from the time of birth or perhaps slightly before, whereas the corticospinal tract develops from about P20 and may retain some plasticity into adult life, at least as suggested by the sustained levels of GAP-43 immunoreactivity in this pathway.

Introduction

Motor pathways in mammals involve areas in the cerebral cortex (primary motor and supplementary motor cortex), basal ganglia (caudate and putamen, globus pallidus, substantia nigra), so-called 'motor' thalamic nuclei (VA, VL), the cerebellum and precerebellar nuclei, brainstem reticular formation and the spinal cord. Several of these elements are considered elsewhere, and they will only be discussed here in the context of functional studies of their role in motor control.

Very little is currently known about the development of the spinal cord in Australian marsupials and what information is available is largely concerned with development of motor and spinal reflex systems. Therefore this chapter will focus on the development of the spinal-cord motor nuclei and connections, with brief consideration of other systems in the spinal cord.

Spinal-cord gross anatomy and relative area of the white matter

The spinal cord begins at the foramen magnum of the skull base and may extend for a variable distance down the vertebral column within the vertebral canal. In general, the length of the spinal cord relative to the vertebral column appears to be linked to flexibility of the vertebral column. In mammals, where flexion of the vertebral column is behaviourally important and highly developed, e.g. the short-beaked echidna (*Tachyglossus aculeatus*), the spinal cord is relatively short, extending no further than the mid-thoracic levels (Ashwell and Zhang, 1997), whereas in mammals where vertebral flexion is minimal (e.g. the platypus, ungulates, carnivores), the spinal cord extends to caudal lumbar or sacral vertebral levels (Ariëns-Kappers *et al.*, 1960). In macropods, the spinal cord extends to caudal lumbar levels, whereas in the more flexible possums the spinal cord ends at the upper lumbar level. Shortening of the spinal cord (with concomitant extension of dorsal and ventral roots) may be an adaptation to promote flexibility of the vertebral column, since the more robust dorsal and ventral roots are likely to be more resilient during stretching than the delicate internal structures of the spinal cord.

The spinal cord presents two enlargements: one at caudal cervical to upper thoracic levels (brachial or cervical enlargement), the other at caudal lumbar to upper sacral levels (lumbosacral enlargement), the segmental levels supplying the fore- and hindlimbs, respectively. The relative size of these spinal-cord enlargements reflects the behavioural importance of fore- and hindlimbs and the expansion is primarily due to provision of motorneurons and sensory cells for controlling and processing information from the limbs, respectively. The increased number of ascending pathways from cervical levels also contributes to enlargement of the cervical spinal cord in mammals. In most Australian marsupials (as in most eutherians), the brachial (cervical) enlargement is the greater in size and this is true of dasyuromorphia and peramelemorphia, but macropods have a larger lumbosacral enlargement for robust supply of the hindlimbs and tail.

The relative sizes of brachial and lumbosacral enlargements often leave their mark on vertebral anatomy (Figure 13.1) and this in turn reflects locomotor habits. The vertebral-canal area is large where the spinal cord is widest

and narrow at thoracic segmental levels. As a measure of this, the spinal quotient (SQ) is the ratio of the vertebral-canal area at fore- or hindlimb/tail associated vertebral levels (respectively, cervical vertebra 5 to thoracic vertebra 1, and lumbar vertebra 1 to sacral vertebra 1) to the average canal area in mid-thoracic levels (thoracic vertebrae 2 to 12), where the spinal cord serves only visceral function and axial musculature. Quadrupedal cursorial marsupials like the thylacine and Tasmanian devil (Figure 13.1a, b) have values for forelimb SQ that are a little below 2.0 and around twice the hindlimb SQ. On the other hand, macropods (Figure 13.1c, d) have forelimb SQs that are similar to or below hindlimb SQs, reflecting the smaller muscle mass within the forelimb of the essentially bipedal kangaroos and wallabies, and the expansion of the lumbosacral segments to serve the large hindlimbs and tail. Conversely, arboreal or fossorial marsupials (Figure 13.1e to g) have values for forelimb SQs well over 2.0. A large spinal cord at cervical and upper thoracic segments is required for the control of muscles for climbing or digging, as well as processing proprioceptive information from the forelimbs.

The relative size of white matter (i.e. ascending and descending fibres) to grey matter (neuronal cell bodies, dendrites and unmyelinated axons) shows a clear relationship with body size in all mammals and this is no different in marsupials. Larger mammals have a much higher cross-sectional area of the spinal cord devoted to white matter than small mammals. This appears to arise because grey matter area increases as the square of body weight, whereas white matter area increases as the cube of body weight. In the case of large macropods, dorsal column area (mainly devoted to conveying discriminative touch and proprioceptive information to the brainstem, but also including the descending dorsal corticospinal tract; see below) is comparable to carnivores and ungulates of similar body weight (Ariëns-Kappers *et al.*, 1960).

Spinal-cord grey matter topography in Australian marsupials

The cytoarchitectonic scheme originally developed by Rexed for the cat (Rexed 1952, 1954) is widely accepted in the neuroscience literature and has been applied to a wide range of eutherians (Grant and Koerber, 2004), marsupials (Martin and Dom, 1970; Martin *et al.*, 1970) and monotremes (Ashwell and Zhang, 1997). The scheme is based on the appearance and arrangement of neuronal and glial cell bodies, as seen in Nissl stains, and divides the grey matter of the spinal cord into 10 cytoarchitectonic zones (Figure 13.2). The dorsal horn contains Rexed's laminae 1 to 5 or 6,

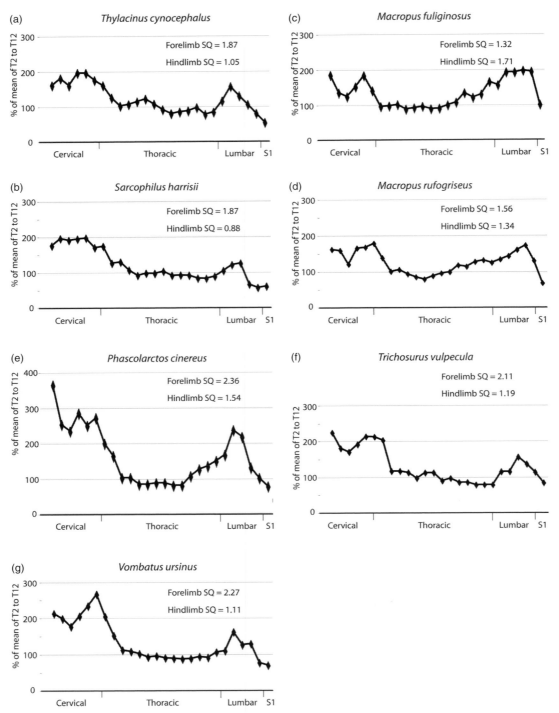

Figure 13.1 Area of the vertebral canal at different vertebral levels compared to the mid-thoracic levels in a variety of Australian marsupials ((a) – thylacine; (b) – Tasmanian devil; (c) – western grey kangaroo; (d) – red-necked wallaby; (e) – koala; (f) – brush-tailed possum; (g) – wombat). Forelimb and hindlimb spinal quotients, i.e. the ratios of mean vertebral canal area for forelimb (fifth cervical to first thoracic vertebrae, inclusive) and hindlimb (first lumbar to first sacral vertebrae, inclusive) levels to mid-thoracic levels (thoracic vertebrae 2 to 12), are given for each species. Note that spinal-cord segmental levels supplying the hindlimb and tail are L3 to coccygeal, but that these levels and associated nerve roots lie alongside the lumbar vertebrae because the spinal cord is shorter than the vertebral column for these mammals.

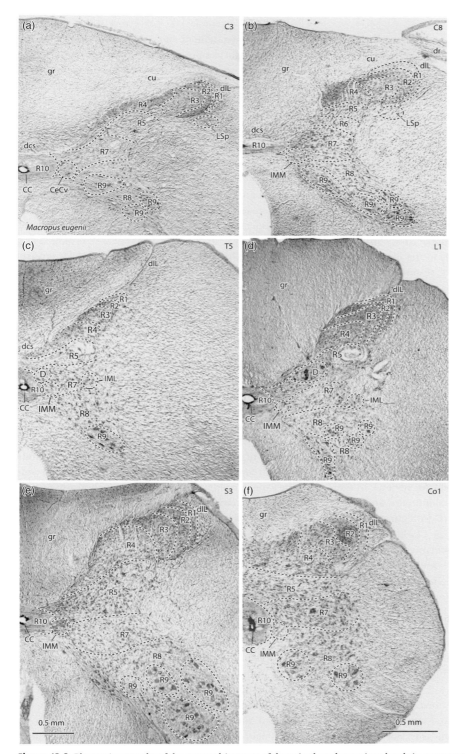

Figure 13.2 Photomicrographs of the cytoarchitecture of the spinal cord at various levels in a representative diprotodont Australian marsupial (the tammar wallaby – *Macropus eugenii*). The midline is to the left in all photomicrographs. Dashed lines and labels R1 to R10 indicate the laminae of Rexed. The scale bar in (e) applies to (a) to (e), but the scale in (f) applies only to that photomicrograph. CC – central canal; CeCv – central cervical nu.; cu – cuneate fasciculus; D – dorsal nu. of the spinal cord; dcs – dorsal corticospinal tract; dlL – dorsolateral tract of Lissauer; dr – dorsal root; gr – gracile fasciculus; IML – intermediolateral cell column; IMM – intermediomedial cell column; LSp – lateral spinal nu.

the intermediate zone largely corresponds to Rexed's laminae 7, while the ventral horn includes Rexed's laminae 8 and 9 (and some of lamina 7). The area around the central canal is Rexed's lamina 10.

Lamina 1 is also known as the marginal zone and forms a thin rim along the dorsal and dorsolateral edges of the dorsal horn (Figure 13.2). As in the rat, most of the constituent cells are small. A region of longitudinally running fibres, known as the dorsolateral fasciculus (dlL) or Lissauer's zone, surrounds the external and lateral margin of lamina 1. Lamina 2 is also known as the substantia gelatinosa and immediately underlies lamina 1. In all mammals it is characterised in Nissl-stained preparations by the presence of many small round cell bodies with sparse Nissl substance (rough endoplasmic reticulum). Cells in the outer part of lamina 2 are primarily nociceptive (i.e. responsive to painful stimuli), whereas those in the less compact inner zone of lamina 2 respond to brush stimuli of the skin surface. Lamina 3 has a cytoarchitectonic appearance similar to 2, but, as in the rat, has a wider range of cell sizes and is less compact. Lamina 4 forms the base of the head of the dorsal horn and becomes continuous with the contralateral lamina 4 through the dorsal commissure at caudal spinal cord levels (Figure 13.2f). The cells in lamina 4 are multipolar and larger than those in lamina 3. Lamina 5 forms the neck of the dorsal horn and contains neurons that contribute axons to many ascending pathways (to the lateral cervical nucleus of the spinal cord, dorsal column nuclei, brainstem reticular formation, midbrain, cerebellum and thalamus; see Grant and Koerber, 2004, for review). In the thoracic and upper lumbar segments, laminae 4 and 5 are interrupted by the nucleus dorsalis (D in Figure 13.2c, d; also known as the nucleus thoracis or Clarke's column). The nucleus dorsalis receives proprioceptive and cutaneous information from the lower limb and projects to the cerebellum via the dorsal spinocerebellar tract. It is clearly present in both poly- and diprotodonts (Johnson, 1977; Terman et al., 1998). In eutherians, the neurons of lamina 4 have widespread dendritic fields, but this has not been specifically studied in Australian marsupials. Lamina 6 forms the base of the dorsal horn and in most mammals is present only in cervical (Figure 13.2b), lower lumbar and upper sacral levels.

Lamina 7 corresponds to the intermediate zone of the spinal cord at all segmental levels and also extends into the ventral horn in some levels (e.g. mid-cervical to caudal lumbar levels). As in rodents, it has a lighter and more homogeneous appearance in Nissl-stained sections of marsupial spinal cord compared to adjacent laminae. In thoracic and upper lumbar levels (Figure 13.2c, d) it includes an intermediolateral cell column (IML) that contains the cell bodies of preganglionic sympathetic neurons. An intermediomedial cell column (IMM) for processing sensory information from visceral structures is visible from cervical to sacral levels. Around the lumbosacral junction, lamina 7 includes a sacral parasympathetic nucleus containing the cell bodies of preganglionic parasympathetic neurons. It also includes a central cervical nucleus at cervical levels (Figure 13.2a), which projects to the cerebellum and vestibular nuclei in rodents (Grant and Koerber, 2004).

Lamina 8 is located in the ventral or ventromedial part of the ventral horn of the spinal cord. It is cytoarchitecturally more heterogeneous than lamina 7 and contains commissurally projecting neurons. In eutherians, neurons located in lamina 8 project to the brainstem reticular formation and thalamus (see Grant and Koerber, 2004). Lamina 9 consists of large neurons around the ventral and lateral edges of the ventral horn. Most of these large neurons are α motorneurons projecting axons through the ventral roots, but smaller γ motorneurons supplying the intrafusal fibres of the muscle spindles are also present. Lamina 9 neurons are more numerous at lower cervical, upper thoracic and lumbosacral levels (Figure 13.2b, d and e), where motorneurons form a lateral motor column for supply of the limb musculature. A medial motor column is present on each side throughout the entire length of the spinal cord and contains motorneurons for supply of axial musculature (e.g. cervical, intercostal, abdominal and pelvic musculature).

Lamina (or, speaking more precisely, area) 10 surrounds the central canal. In eutherians, neurons in this region contribute axons to pathways projecting to the brainstem reticular formation, amygdala, hypothalamus and thalamus and may be involved in visceral pain sensation. The central canal contains the delicate Reissner's fibre, which is secreted by the subcommissural organ and extends the entire length of the central canal of the spinal cord in all therians (Tulsi, 1982). The role of Reissner's fibre remains obscure, but may involve clearing debris from the central canal or sensing cerebrospinal fluid pressure (see discussion in Tulsi, 1982).

Some neuronal cell bodies are located outside the grey matter of the spinal cord in all mammals (Ashwell and Zhang, 1997; Grant and Koerber, 2004). These include neurons in the lateral spinal nucleus (LSp), located along the lateral edge of the dorsal horn at all segmental levels; and the lateral cervical nucleus, located lateral to the lateral spinal nucleus at upper cervical segmental levels. In rodents, neurons of the lateral spinal nucleus are responsive to stimulation of subcutaneous and deep structures, whereas neurons of the lateral cervical nucleus respond to hair movement and noxious stimuli (see Grant and Koerber, 2004 for review of these structures in the laboratory rat).

Descending projections to the spinal cord in Australian marsupials

The course and sites of termination of corticospinal and rubrospinal tracts have been examined in Australian marsupials and may be compared to those in eutherians.

Corticospinal tract

The course and pattern of termination of corticospinal tract axons in mammals has been considered to fall into four categories (Ariëns-Kappers et al., 1960; Towe, 1973). Group 1 mammals (including all marsupials studied to date, e.g. didelphids, macropods and phalangerids) have corticospinal tract axons reaching only the cervical and thoracic levels and terminating mainly in the dorsal part of the contralateral grey matter (Rexed's lamina 4 and medial laminae 5 and 6). Group 2 mammals (including rodents) have corticospinal tract axons running the entire length of the spinal cord and terminating mainly in the contralateral dorsal horn and laminae 5, 6 and 7. Groups 3 and 4 (including primates) have corticospinal axons terminating extensively on lamina 9 motorneurons, with differing degrees of control. The corticospinal tract is smaller in Australian polyprotodonts like dunnarts and antechinus (Figure 13.3a, b) than in rodents of similar brain and/or body weight. On the other hand, the relative size of the corticospinal tract in phalangerids like the brush-tailed possum is comparable to dextrous carnivores or small primates of similar brain weight.

In all mammals studied, the neuronal cell bodies giving rise to the corticospinal tract axons are located in layer 5Cx of the relevant cortical regions. In eutherians, most corticospinal neurons are located in the primary motor cortex and in forelimb and hindlimb parts of the primary somatosensory cortex (Tracey, 2004a). Some additional corticospinal neurons are located in supplementary motor, secondary somatosensory and visual association areas. The corticospinal tract plays a role either in direct control of movement by its terminations on motorneurons themselves, or indirectly by its termination on interneurons in the intermediate grey matter (lamina 7) and ventral horn (lamina 8). In the brush-tailed possum (Trichosurus vulpecula), most corticospinal influence is mediated by interneurons (Martin et al., 1970; Rees and Hore, 1970; Hore and Porter, 1971, 1972; Aoki and McIntyre, 1973), but there is, nevertheless, close physiological coupling between corticospinal axons and forelimb muscle motorneurons (Hore and Porter, 1971).

The corticospinal tract of the arboreal brush-tailed possum (Trichosurus vulpecula) arises mainly from the immediately postorbital cortex (i.e. immediately caudal to the α/α' sulcus) and the parietal cortex (i.e. middle of the rostrocaudal extent of the cortex)(Martin et al., 1970). The first of these sites corresponds approximately with the cortical position of corticospinal tract neurons as seen in Monodelphis domestica and Didelphis virginiana (Nudo and Masterton, 1990). There does not appear to be any significant corticospinal projection from the occipital or caudal temporal cortex in the brush-tailed possum. Corticospinal fibres descend through the ipsilateral pyramid (py) and most fibres decussate in the caudal medullary pyramidal decussation. By the cervical spinal cord level, most corticospinal fibres are located in a large dorsal corticospinal tract (dcs), immediately ventral to the dorsal columns, with some fibres in a much smaller lateral corticospinal bundle in the most medial part of the lateral funiculus. At cervical levels, most corticospinal fibres of the brush-tailed possum terminate in the medial parts of Rexed's laminae 3 to 6, although a smaller portion extends into the lateral parts of the dorsal horn (Figure 13.3c). Of particular functional significance, some corticospinal fibres terminate in lamina 7 and 8 in close proximity to motorneurons of lamina 9 and this seems to be better developed than in Didelphis virginiana (Martin et al., 1970). Termination of corticospinal fibres on ventral horn motorneurons is an important feature of those eutherians with the ability to perform rapid, co-ordinated movements of distal musculature of the forelimb (e.g. primates like the cebus monkey), but is not prominent in rodents (compare Figure 13.3d to g). The larger size of the pyramidal tract and greater extension of the corticospinal terminations into the ventral horn of the brush-tailed possum compared to Didelphis virginiana is consistent with the superior forelimb dexterity shown by the former (but see below). Although the corticospinal tract may be traced further caudally in the brush-tailed possum, it rapidly declines in size and termination at thoracic levels (where lamina 6 is absent) is confined to the medial parts of laminae 4, 5 and 7. An important function of the corticospinal tract in the brush-tailed possum is to bring about appropriately timed release of the forelimb grasp during arboreal locomotion (Aoki and McIntyre, 1973).

Despite the restriction of the corticospinal tract terminals to cervicothoracic levels in the brush-tailed possum, Hore and Porter (1971) and Aoki and McIntyre (1976) showed that pathways exist for cortical activity to indirectly affect lumbosacral segment motorneurons. This is manifested by inhibition of hindlimb flexor digitorum longus muscle motorneurons after corticospinal (pyramidal) tract activation, in a mechanism that allows release of the hindlimb grip when the forelimb grasp is activated. This function would come into play when the possum moves from a habitual posture to continue climbing or descent (Hore and Porter, 1971). These pathways may depend on long propriospinal pathways descending from cervical to lumbosacral levels.

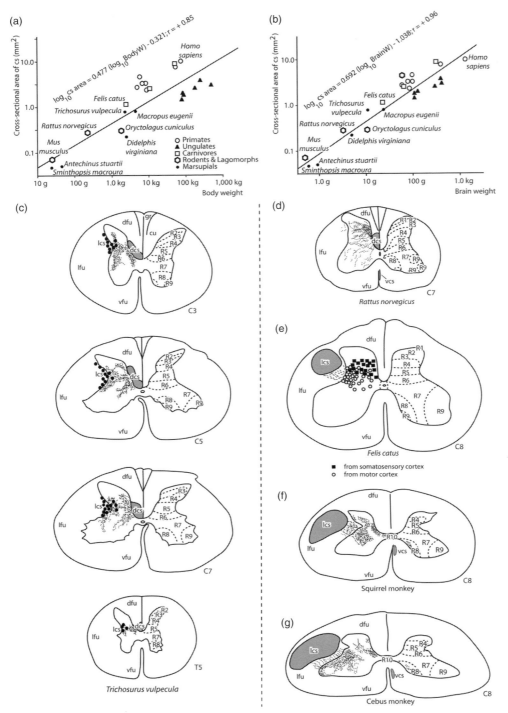

Figure 13.3 Plots of cross-sectional area of the corticospinal tract (as measured at the level of the inferior olivary complex) against body weight (a) and brain weight (b) for a variety of therian mammals. Note that the cs cross-sectional area is lower in marsupials compared to many eutherians when considered relative to body weight (but with the noticeable exception of the relatively dextrous brush-tailed possum, *Trichosurus vulpecula*). On the other hand, cs cross-sectional area is comparable across all therians when related to brain weight. Data for cross-sectional area of cs are drawn from Towe (1973) and measurements by the author. All measurements are taken from fixed, dehydrated histological sections. Diagrams of cross-sections through the spinal cord of the brush-tailed possum are shown in (c). Representative

The corticospinal tract of the ground-dwelling small macropod, the potoroo (*Potorous apicalis*), also arises mainly from cortex immediately caudal to the α sulcus, as well as adjacent parietal cortex (Martin *et al.*, 1972). As in the brush-tailed possum, corticospinal fibres decussate at the caudal medullary pyramidal decussation and give rise to a large dorsal corticospinal tract extending as far caudally as the twelfth thoracic segment and a smaller lateral corticospinal tract extending as far as the eighth thoracic segment. Corticospinal fibres in the potoroo ramify in (mainly) medial laminae 3 to 6, with some fibres distributed to the lateral parts of the same laminae. There is some evidence that fibres from motor cortical regions are distributed to deeper laminae (5, 6), whereas fibres from somatosensory cortex are distributed to the more superficial laminae (3, 4). Very few fibres enter lamina 7, in contrast to the brush-tailed possum. The corticospinal tract of other diprotodonts, e.g. the quokka *Setonix brachyurus* and large kangaroos, are broadly similar to the potoroo in terms of site of origin, course, rostrocaudal extent and laminar regions of termination (Watson, 1971; Watson and Freeman, 1977).

Rubrospinal tract

The rubrospinal tract arises from neurons in both the parvicellular (small-celled) and magnocellular (large-celled) parts of the red nucleus (for review see Tracey, 2004a; see also Chapter 4). In eutherians there is a myotopic organisation of the red nucleus such that the ventrolateral part of the nucleus projects to the lumbar spinal cord, whereas the dorsomedial parts project to the cervical spinal cord (Pompeiano and Brodal, 1957). Most rubrospinal axons in rodents terminate in the contralateral laminae 5 to 7, where they contact both excitatory and inhibitory interneurons to indirectly influence motorneuron activity. In rats, lesions of the rubrospinal tract impair skilled reaching movements (for discussion see Tracey, 2004a).

In contrast to the findings described above for the corticospinal tract, the course and termination of the rubrospinal tract in the brush-tailed possum (Figure 13.4a) is very similar to that in *Didelphis virginiana* (Figure 13.4b) (Warner and Watson, 1972). Rubrospinal terminals appear to be distributed predominantly to the lateral parts of laminae 3 to 7, in contrast to the distribution of corticospinal terminals mainly to medial laminae 5 and 6. The rubrospinal tract also extends much further caudally (as far as coccygeal segments) than the corticospinal tract (see above) and, unlike the corticospinal tract, rubrospinal fibres terminate heavily in the lumbosacral enlargement. The pattern of rubrospinal tract termination is essentially similar in both the brush-tailed possum and North American opossum to that in the domestic cat (Figure 13.4c; Nyberg-Hansen and Brodal, 1964; Hongo *et al.*, 1969), indicating that the pattern of terminations for this tract may be highly conserved across therians.

Skilled movements in marsupial and eutherian mammals

Studies of manual dexterity in brush-tailed possums indicate that these phalangerids have the ability to quickly solve problems requiring lateralised forelimb use, including manipulation with the forepaw. In most animals (over 95%) there appears to be a preferred forelimb, with the side of preference roughly equally divided between left and right (Megirian *et al.*, 1977). Once a preference for a particular forepaw is developed during training, cortical lesions and anaesthesia of the forelimb appear to be relatively ineffective in causing transfer of preference to the other forepaw. The lack of effect of cortical lesions suggests that descending pathways from the cortex exert relatively little effect on the maintenance of forepaw preference.

One might expect that the more segmentally and laminar extensive the penetration of corticospinal fibres, the more

---

Caption for Figure 13.3 (Cont.)
eutherians showing the position and distribution of corticospinal fibres at cervicothoracic junction levels are illustrated in (d) to (g). In the brush-tailed possum, corticospinal fibres are mainly located in the dorsal corticospinal tract (dcs – solid filled region) within the dorsal funiculus (dfu) with sporadic large fibres in the lateral funiculus (lfu) (black dots indicate the labelled lateral corticospinal tract – lcs). In the laboratory rat, the corticospinal fibres are predominantly in the dcs with a few fibres in the ventral corticospinal tract (vcs) within the ventral funiculus (vfu). In the domestic cat, the bulk of corticospinal tract fibres descend in the lateral funiculus. Fibres from the somatosensory cortex are distributed to the dorsal horn, whereas fibres from neurons in the motor cortex are distributed predominantly to the intermediate grey. In primates, most corticospinal tract fibres run in the lateral funiculus with a small ventral corticospinal tract. Monkeys with good digital control have corticospinal tract fibres distributed to the lateral motor column. Diagrams for the brush-tailed possum have been based on findings in Martin *et al.* (1970). The diagram for the laboratory rat is based on Tracey (2004b). The diagrams for the cat and monkeys have been based on data in Nyberg-Hansen and Brodal (1963) and Bortoff and Strick (1993), respectively. cu – cuneate fasciculus; dcs – dorsal corticospinal tract; dfu – dorsal funiculus; gr – gracile fasciculus; lcs – lateral corticospinal tract; lfu – lateral funiculus; R1 to R10 – Rexed's laminae 1 to 10; vcs – ventral corticospinal tract; vfu – ventral funiculus.

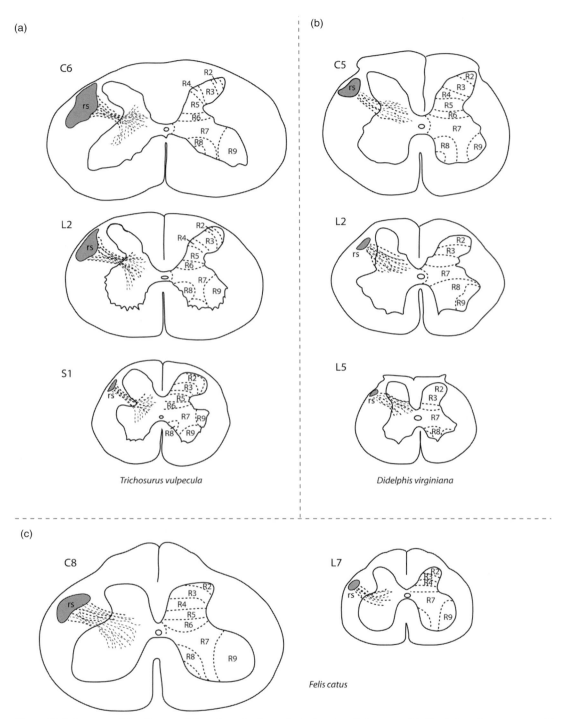

Figure 13.4 Course and distribution of the rubrospinal tract in the brush-tailed possum (a), North American opossum (b) and domestic cat (c). Note the similar distribution of terminals in all three species. Abbreviations R1 to R10 indicate Rexed's laminae. Diagrams for the brush-tailed possum have been based on data in Warner and Watson (1972). The diagrams for *Didelphis virginiana* have been based on data in Martin and Dom (1970), whereas the diagrams for the cat have been based on data in Nyberg-Hansen and Brodal (1964). R2 to R9 – Rexed's lamina 2 to 9; rs – rubrospinal tract.

direct the control exerted by the cerebral cortex over limb musculature, but the reality may be more complex. Heffner and Masterton (1975) reported that variation in digital dexterity among mammals corresponds most closely with variation in the placement of corticospinal terminals within the spinal-cord grey matter and less closely with variation in the size of the tract itself, but later studies (Iwaniuk et al., 1999) have questioned whether the inclusion of large numbers of primates in the Heffner and Masterton data set may have biased the sample and confounded the analysis.

Iwaniuk and colleagues have developed a behavioural index of forelimb dexterity and assessed the relationship between dexterity and measures of brain size in 18 species of American and Australian marsupials (Iwaniuk et al., 1999). Proximal forelimb dexterity (i.e. excluding the forepaws) is high in Caluromys derbianus of the American marsupials and in all the examined species of Petauridae, Pseudocheiridae and Phalangeridae (including Trichosurus vulpecula). By contrast, proximal dexterity is poor in most of the macropods. Distal forelimb dexterity (i.e. only the forepaws) is also best in Caluromys derbianus, the dasyurid Dasyurus hallucatus and the above-mentioned possums. The degree of distal dexterity varies widely across macropods, being high for Lumholtz's tree kangaroo (Dendrolagus lumholtzi) and poor for the red kangaroo (Macropus rufus). Iwaniuk and colleagues found that distal dexterity was positively correlated with size of the cerebral cortex, as would be expected, given the importance of a large isocortex for an extensive corticospinal tract or other indirect descending motor pathways. The size of the cerebellum may also be correlated with proximal limb dexterity, probably as a reflection of spino- and pontocerebellum development (see Chapter 5).

In mammals, direct termination of corticospinal axons on lamina 9 motorneurons (as seen in primates) is often considered to be directly related to the ability to perform fine independent finger movements to manipulate objects, but note that Iwaniuk and Whishaw (2000) have highlighted the potential for synergistic action of multiple descending pathways (e.g. tectospinal and reticulospinal tracts, as well as the more usually considered corticospinal and rubrospinal tracts) to contribute to skilled forelimb movements in non-mammalian tetrapods. The limited segmental projection and laminar termination of corticospinal axons in marsupials (group I discussed above) might be expected to limit the amount of fine forepaw manipulation that they can perform. On the other hand, behavioural studies of forepaw manipulative capacity in Ameridelphians (e.g. Monodelphis domestica) suggest that these skills are better than might be expected (Ivanco et al., 1996). Although the food-handling behaviour of Monodelphis is less complex than that seen in rodents, the marsupials are capable of ballistic single

forelimb movements to grasp prey and can use a single limb to take food from a shelf (Ivanco et al., 1996). The restriction of corticospinal terminals to mainly the dorsal horn in marsupials translates to an inability to perform rotatory movements of the forelimb during prey capture and manipulation, and limitations in digital manipulative ability. Forepaw movements of Monodelphis are also less plastic and individualistic than rodents (Ivanco et al., 1996).

Correlation of the degree of skilled forepaw movements with the anatomy and physiology of descending motor pathways has not been extensively studied in Australian marsupials. Behavioural studies indicate that some dasyurid marsupials catch and consume crickets in a similar fashion to Ameridelphians, i.e. a ballistic single paw grasp followed by a killing bite, whereas others are more selective, removing wings and legs before consuming the abdomen. Some diprotodonts exhibit more complex manipulative behaviour, e.g. some tree kangaroos (Dendrolagus matschiei) use a single primate-like prehensile movement when reaching for, and transferring leaves to, the mouth (Ivanco et al., 1996). Clearly this is a topic that needs further study to establish the relationship between the anatomy and physiology of descending motor pathways and behaviour in Australian marsupials.

Propriospinal tracts

Propriospinal tracts interconnect different segmental levels of the spinal cord; they play important roles in the coordination of fore- and hindlimb movements and perhaps nociception. Aoki and McIntyre (1976) showed that there are descending long propriospinal reflex linkages between the fore- and hindlimb in the brush-tailed possum, which are similar to those in the cat. As an arboreal mammal, the brush-tailed possum needs a mechanism to inhibit and release hindlimb grip when the forepaw is stimulated during climbing. This reflex is spared after pyramidal-tract transection in the medulla, indicating that propriospinal pathways within the cord are most likely responsible for its mediation. Nevertheless, the reflex is reduced in magnitude and duration after medullary transection, suggesting that hindlimb release may be at least modulated by connections through the brainstem (i.e. spinobulbospinal activity).

Muscle spindles and their innervation

Structure and function of muscle spindles in Australian marsupials have received relatively little attention. Morphologically, muscle spindles are broadly similar across all mammals (Voss, 1963; Jones, 1966a, 1966b). In the brush-tailed possum, the small forepaw lumbricals have one muscle spindle for every 100 to 150 extrafusal fibres, comparable to a ratio of 1 to 200 for human first lumbricals (Jones, 1966a). Intrafusal muscle fibres of pedal lumbricals in the brush-tailed possum are of both nuclear bag and nuclear chain types as described for eutherians (Jones, 1966b). The morphology of both muscle spindles, sensory nerve endings and their associated group Ia and II afferents are similar to those reported for eutherians (Jones, 1966b).

Motor behaviour of pouch-young marsupials

Reaching and grasping movements by newborn marsupials are seen in all poly- and diprotodont species studied so far, and occur well before any myelination of descending pathways is evident (Langworthy, 1928). Wide side-to-side movements of the head also occur and may be accompanied by gross contractions of the trunk and abdominal musculature.

Movement of the perinatal and pouch-young marsupial may serve several critically important functions. Prior to birth, forelimb movements may assist in the rupture of the extra-embryonic membranes to initiate delivery. Secondly, in the first few hours after birth, rhythmic stepping and grasping movements of the forelimbs allow the young to climb from the urogenital sinus to the pouch and nipple. Thirdly, movement of the pouch young shortly after attachment to the teat produces traction, which elongates the teat (Hartman, 1920). Finally, throughout the entire postnatal period, grasping and forelimb movements against the nipple may stimulate milk production and ejection and delay reactivation of other embryos in diapause (Renfree, 1979; Ho, 1997). Newborn brush-tailed possums and tammar wallabies have short, wide forelimb bones with high muscle mass per unit limb length to compensate for the relatively poor leverage of neonatal forelimb bones (Lentle *et al.*, 2006).

Of the Australian marsupials, newborn motor behaviour has been best studied in the tammar wallaby (*Macropus eugenii*). When removed from the pouch, young wallabies display clock-like alternating forelimb movements: i.e. when one limb is extended to the most rostral position, the contralateral limb is retracted. These stepping movements occur in conjunction with the grasping of fur by the digits of the pronated forelimbs. During each step cycle, the digits are extended when the limb is protracting and straightening and then closed during flexion and adduction. In newborn animals, the stepping frequency is about 0.1 to 0.2 Hz, but the stepping frequency and pattern regularity increase between P1 and P42 to plateau at about 0.4 Hz (Ho, 1997). The initiation of stepping movements is strongly linked to ambient temperature, in that reduction of temperature from 37°C to room temperature immediately elicits forelimb movements, whereas return to 37°C suppresses both the amplitude and frequency of the forelimb movements (Ho, 1997). Forelimb activity persists until P140, by which stage the forelimbs are actively used to hold onto surrounding objects to maintain balance whenever the young wallaby is removed from the pouch.

The hindlimbs of pouch-young marsupials exhibit large developmental differences from the forelimbs. At birth in the tammar wallaby, the hindlimbs are little more than limb buds. Furthermore, the hindlimbs show no active movement for the first four to five weeks of pouch life, although they are capable of withdrawal reflexes in response to noxious stimuli (Ho, 1997).

Development of the spinal cord and motor connections

Climbing from the urogenital sinus to the pouch naturally requires some functional maturity of those neural systems controlling forelimb and perhaps upper axial musculature, but appears to not require mature hindlimb circuitry and musculature. In all marsupials studied, levels of the spinal cord concerned with control of the forelimbs are more advanced at birth both structurally and functionally (Lavallée and Pflieger, 2009) than those controlling the hindlimb, although it is surprising how immature the brachial levels of the spinal cord are in newborn marsupials, given the demands of the climb to the pouch.

Of all Australian marsupials, spinal cord and motor development has been most studied in the tammar (*Macropus eugenii*). Even at birth in the tammar, the spinal cord as a whole is still very immature, with a dorsoventrally deep central canal flanked by proliferative neuroepithelium (Harrison and Porter, 1992); histologically, the newborn tammar spinal cord is similar in appearance to that of an E(embryonic day)15 or E16 rodent. Nevertheless, production of neurons is sufficiently advanced in the brachial levels (i.e. caudal cervical and upper thoracic segmental levels) for formation of the dorsal horn and lateral motor columns to be underway, whereas the lateral motor columns in the lumbar and sacral levels are still poorly developed at this stage. This stage of development in the lumbar spinal cord corresponds with invasion of the muscle masses of the thigh and shank by motor nerves (Comans *et al.*, 1987). Structural maturation of the cervical spinal cord proceeds quite rapidly, such that by P17 the central canal has greatly reduced in size and dorsal horns and lateral motor columns are fully developed. Distinct dorsal columns (fasciculus gracilis and cuneatus) can also be distinguished by the middle of the third postnatal week. By P34, the spinal cord at all levels has an essentially mature appearance with motorneurons in the lateral motor column taking on a more mature polygonal profile.

As in other vertebrates, motorneurons in the lateral motor columns of the tammar spinal cord undergo developmental cell death. In the case of the lumbar spinal cord this appears to occur in two phases: the first from birth to P40, resulting in the loss of 59% of the motorneurons originally generated; the second from P90 to P150 and accounting for a further 24% of motorneurons (Comans *et al.*, 1987). It is possible that the two phases of motorneuron death correspond to losses of α and γ motorneurons, respectively (Comans *et al.*, 1987), but this hypothesis has never been tested. As in other tetrapods, removal of muscle mass by amputation before the period of naturally occurring neuronal death increases the number of motorneurons lost during development (Comans *et al.*, 1988), indicating the developmental dependence of motorneurons on their peripheral target for support and survival. This may be to such an extent that the relevant column of motorneurons is absent from the mature spinal cord following complete amputation.

As might be expected, maturation of muscle afferents to the newborn tammar spinal cord is much more advanced in the brachial segmental levels than lumbar levels. At birth, afferent fibres associated with forelimb muscles have already invaded the spinal cord and entered the ventral horn, close to the motorneuron cell bodies, whereas afferents from hindlimb muscles have reached only the intermediate zone of the spinal cord (Ho and Stirling, 1998). By the fourth postnatal week, muscle afferent innervation in both brachial and lumbar segmental levels is morphologically similar. Functional maturation of stretch reflexes corresponds temporally with the structural maturation of the pathways: afferent discharges from stretching the biceps brachii of the forelimb can be recorded at birth, but stretching the gastrocnemius of the hindlimb does not elicit afferent discharges until P4. As noted above for the structural maturation, spinal reflexes become much more similar in brachial and lumbar levels by the third postnatal week. As for eutherians, the end of the rapid phase (i.e. phase I) motorneuron death corresponds with the establishment of proprioceptive afferent projections from the limbs (Ho and Stirling, 1998).

Maturation of muscle cells and spindles in the tammar follows a similarly protracted course to the dorsal and ventral horns of the spinal cord. At birth, forelimb and axial muscles contain primary myotubes at a relatively advanced, uniform level of maturity, but primary myotubes have only just begun to emerge in the hindlimb musculature (Harrison and Porter, 1992). Myofibres appear in the forelimb muscles at P7, but muscle spindles do not develop

in either forelimb or hindlimb musculature until after the third postnatal week (Harrison, 1991). Although the general morphology and overall sequence of muscle-spindle development in the hindlimb of the tammar is similar to that seen in eutherians, the timing of muscle-spindle development in marsupials is remarkably delayed relative to that seen in eutherians. In all eutherians studied (e.g. mouse, rat, cat, pig, sheep and human) hindlimb muscle-spindle development commences during mid- or late gestation and the adult complement of intrafusal muscle fibres is present by birth, whereas in the tammar wallaby the major components of hindlimb muscle spindles are not assembled until P50 and do not reach mature morphology until P100 (see Harrison, 1991, for discussion).

Findings in the tammar are broadly similar to those in Ameridelphian marsupials, e.g. *Monodelphis domestica* and *Didelphis virginiana*, although the motor columns of the caudal spinal cord of *Didelphis* may be less mature at birth than in the tammar or *Monodelphis*. Surprisingly, some authors have reported well-developed lateral motor columns and dorsal columns and the presence of immature muscle spindles in the forelimb and cervical musculature (longus colli, longus capitis, anterior and lateral rectus capitis) of the red kangaroo (*Macropus rufus*) at birth (Kubota *et al.*, 1989), even though there is no apparent difference in the functional maturation of the motor systems of newborn red kangaroos relative to tammars.

Control of rhythmic motor activity and interlimb co-ordination in pouch-young wallabies

Motorneurons of marsupial spinal cord exhibit spontaneous activity very soon after birth (Lavallée and Pflieger, 2009). *In vitro* experiments with the isolated brainstem and spinal cord (i.e. caudal to the pontine flexure) of tammar wallabies (Ho, 1997, 1998) show that rhythmic forelimb activity can be elicited even in the absence of the midbrain and forebrain. The neural network controlling rhythmic forelimb movements can be divided into two components: a rhythm generator within each side of the spinal cord and a pattern-co-ordinating pathway, which involves both glycinergic and GABAergic interneurons (Ho, 1997).

Development of ascending spinal-cord pathways

Very little is known about the maturation of either ascending or descending spinal-cord pathways in Australian marsupials. The robust development of muscle afferents and the strong labelling of the same within the cervical segment spinal cord of the tammar wallaby (Ho, 1998) suggest that dorsal column pathways have at least begun development by the time of birth. A clue to the time-course of development of some pathways may be obtained by examining changes in the level of the calmodulin-binding phosphoprotein GAP-43 that is enriched in growing axons and is present in high concentrations during the period of axonal elongation. Certainly, the high levels of GAP-43 immunoreactivity within the dorsal columns of the tammar at birth (Hassiotis et al., 2002) suggest that at least some fibres of the fasciculus gracilis and cuneatus are actively growing at birth and continue to do so until the end of the third postnatal month.

Studies in Ameridelphians (Martin et al., 1983; Qin et al., 1993) indicate that axons from caudal levels of the spinal cord (i.e. caudal thoracic and below) reach the hindbrain by P7. These first afferents from the spinal cord are distributed to the lateral reticular nucleus and inferior olivary complex (see Chapter 5 on precerebellar nuclei) and developing cerebellum. Ascending projections from Clarke's column (nucleus dorsalis) reach the cerebellum of the Virginia opossum by about P20, but dorsal column afferents do not reach the gracile nucleus until about P30. Development of projections from caudal spinal cord to the brainstem of Monodelphis domestica is similar to the Virginia opossum, in that the first ascending projections appear to be to the lateral reticular nucleus and cerebellum at P3, but ascending dorsal column pathways to the nucleus gracilis appear as early as P7 (Qin et al., 1993). The early spinal projections to precerebellar nuclei coincide quite closely with the end of migration of those neurons from the rhombic lip region, but it remains to be seen whether any behaviourally significant reflexes are mediated through these nuclei in young marsupials.

Development of descending spinal-cord pathways

Even at birth, there is a robust projection to the cervical spinal cord from the medial reticular formation of the brainstem (i.e. gigantocellular and ventral gigantocellular reticular nuclei) in the tammar wallaby (McCluskey et al., 2008). The first axons to grow into the spinal cord of the Virginia opossum are also from the gigantocellular reticular nucleus and ventral gigantocellular reticular nucleus (from P5; Martin et al., 1988), but whether these projections are present at birth in opossums is still uncertain. Certainly, in the spinal cord of newborn Didelphis virginiana, there is a substantial population of tyrosine-hydroxylase-immunoreactive axons, which reach as far as the lumbosacral spinal cord (Pindzola et al., 1990). The pattern of projections from the brainstem reticular formation to the spinal cord is essentially similar

in the early pouch-young opossum and tammar wallaby to that seen in adults, suggesting that developing medullary reticulospinal pathways are more stereotyped than developing forebrain pathways (Martin et al., 1988) although they retain plasticity until the end of the second week of postnatal life (Wang et al., 1994).

Development of descending projections to more caudal spinal cord (i.e. thoracic segmental levels) is at about the same time as for cervical levels. Studies in Monodelphis domestica and Didelphis virginiana indicate that the first axons from brainstem nuclei reach the thoracic spinal cord at around the time of birth (Wang et al., 1992; Martin et al., 1993). The earliest terminals in lumbar spinal cord are consistently from medial magnocellular reticular formation, raphe nuclei and locus coeruleus (Wang et al., 1992; Martin et al., 1993), with some animals showing evidence of projections from the lateral vestibular nucleus. Sources of axons to the thoracic spinal cord at P5 in Didelphis virginiana include the hypothalamus, red nucleus, pontine and medullary reticular formation, locus coeruleus, raphe nuclei of the caudal pons and medulla, spinal trigeminal nucleus, inferior, medial and lateral vestibular nuclei, accessory oculomotor nuclei and the interstitial nucleus of Cajal (Cabana and Martin, 1984). Dorsal column nuclei project to the thoracic spinal cord from P14 and axons from the nucleus ambiguus, deep cerebellar nuclei and superior colliculus (intermediate and deep layers) reach the thoracic spinal cord by P17. In many respects the pattern of early projections to the spinal cord in marsupials has similarities to eutherians (albeit at embryonic stages for the latter). In both marsupials and eutherians, the first projections to the cervical spinal cord are from the medial reticular formation (i.e. large-celled neuronal groups – gigantocellular and ventral gigantocellular nuclei). These are established around the time of birth in marsupials (McCluskey et al., 2008) and at E13 or E14 in rodents (Auclair et al., 1993). Projections from the lateral vestibular nucleus are also established early, i.e. in the birth to P5 stage in marsupials (Wang et al., 1992; Martin et al., 1993; McCluskey et al., 2008) and by E14 in the laboratory rat (Auclair et al., 1993).

Although the development of descending cortical pathways to the spinal cord has not been specifically studied in Australian marsupials, the time-course of the rise and fall of GAP-43 immunoreactivity in the tammar wallaby pyramidal tract (Hassiotis et al., 2002) suggests that the corticospinal tracts develop from about P20 onwards. The pyramidal tract retains some GAP-43 immunoreactivity into adult life, suggesting that some plasticity may be retained in this pathway into adult life, but this has never been analysed functionally.

What remains unknown?

The range of corticospinal-tract anatomy and physiology in Australian marsupials, and how this influences or limits their behaviour, is largely unexplored. Studies to date have been focussed on smaller polyprotodont and selected diprotodont species, but the range of observed manipulative behaviour in Australian marsupials may reflect more extensive corticospinal or other descending motor projections than hitherto appreciated.

The timing of the development of the ascending and descending pathways in Australian marsupials has not been directly studied in detail and there are currently very little data available on how these pathways contribute to the behaviour of early pouch-young Australian marsupials.

Part IV

Australian marsupials as models

14

Australian marsupials as models of brain development

L. R. Marotte, P. M. E. Waite and C. A. Leamey

Summary

Australian marsupials are eminently suitable as models for studying the development of the mammalian nervous system. Despite the fact that metatherian (marsupial) mammals diverged from eutherian mammals around 135 million years ago, their nervous systems are remarkably similar. Here, we use the visual system as an example to highlight the high degree of similarity. In contrast, their mode of reproduction is dramatically different: marsupials are born at a very early stage and complete a protracted development in the pouch, allowing unrivalled access for experimental manipulation that would be either extremely difficult or impossible in eutherian mammals. Some insights into the development of the nervous system gained from such research are described. Finally, possible directions for the future are discussed, in particular the so-far unexploited, but unrivalled, suitability of the marsupial model for physiological analysis of the developing nervous system in the unanaesthetised animal and the feasibility of manipulating levels of molecules in subregions of the developing nervous system to investigate their function.

Introduction

Although the Metatheria diverged from the Eutheria around 135 million years ago, the structure of their nervous systems is remarkably similar. The differences between the groups are no greater than differences seen within the groups. The only major exception to this is the fact that marsupials lack a corpus callosum and cortical connections are made through an expanded anterior commissure. In the more advanced Australian diprotodonts, which include the kangaroos, wallabies and possums, a fibre

bundle, the fasciculus aberrans, passes from the internal capsule to cross in the dorsal part of the commissure and connects more dorsal neocortical areas (Heath and Jones, 1971; Ashwell *et al.*, 1996b). Although the route taken by isocortical commissural axons differs, other features, such as their topography in the commissure and developmental overproduction (Ashwell *et al.*, 1996b) is similar to that described for eutherian species. For example, the pattern of inter-hemispheric visual connections in the adult (Heath and Jones, 1971; Crewther *et al.*, 1984; Sheng *et al.*, 1990), as well as their developmental pattern, which shows an early exuberance followed by restriction (Sheng *et al.*, 1990), is remarkably similar between the groups.

The nervous system of Australian marsupials is typically mammalian

To illustrate the similarities between eutherian and metatherian species, some of the features of the visual system will be discussed briefly. The visual system has been intensively studied in the diprotodont tammar wallaby (*Macropus eugenii*) (Figure 14.1b). This species has more features in common with eutherians such as the cat and monkey than more commonly used laboratory animals such as rodents. The tammar is a highly visual animal which is active during the day, as well as at dawn and dusk (Wye-Dvorak *et al.*, 1987; Hemmi and Mark, 1998; Hemmi, 1999), unlike rodents, which are nocturnal. It has colour vision (Hemmi, 1999) and a sizeable binocular field (Wye-Dvorak *et al.*, 1987; Wimborne *et al.*, 1999). Its retina has a well-developed visual streak containing an area centralis for high acuity vision (Wong *et al.*, 1986), and the dorsal lateral geniculate nucleus (DLG) is laminated with interleaving eye-specific terminal bands (Wye-Dvorak, 1984) like that of the cat (Hickey and Guillery, 1974) and

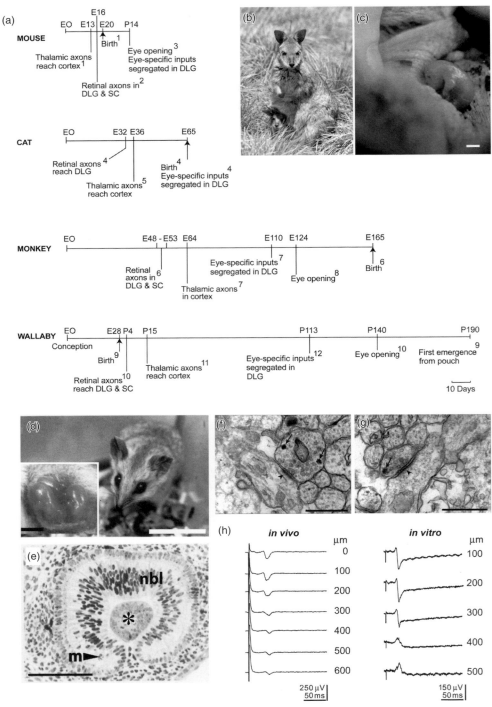

Figure 14.1 (a) Time lines comparing selected milestones in visual development in three eutherian species used in developmental studies with that of the marsupial wallaby (*Macropus eugenii*). Note the protracted and postnatal development in the wallaby compared with the largely prenatal development in the eutherian species. Birth is indicated by an arrow for each species. [1] (Auladell *et al.*, 2000); [2] (Godement *et al.*, 1984); [3] (Jaubert-Miazza *et al.*, 2005); [4] (Shatz, 1983); [5] (Ghosh and Shatz, 1992); [6] (Meissirel *et al.*, 1997); [7] (Rakic, 1977); [8] (Hendrickson and Provis, 2006); [9] (Tyndale-Biscoe and Janssens, 1988); [10] (Ding and Marotte, 1996); [11] (Sheng *et al.*, 1990); [12] (Marotte, 1990). E: embryonic day. P: postnatal day. DLG: dorsal lateral geniculate nu. SC: superior colliculus. (b) An adult tammar wallaby (*Macropus eugenii*) with a large

monkey (Wiesel *et al.*, 1974). The adult tammar weight of 5 to 7 kg, and thus the size of its brain is highly amenable to physiological studies. Response properties of cells and functional segregation in the DLG are similar to those described in cats (Henry and Mark, 1992), while response properties of cortical cells have similar characteristics to those in cats and primates (Vidyasagar *et al.*, 1992; Ibbotson and Mark, 2003). Further, aspects of visual development also appear to have more in common with carnivores than with rodents. The development of topographic projections from the retina to the superior colliculus in the wallaby, where retinal axons grow into the SC in coarse retinotopic order with little branching and then form terminal arbors in their retinotopically correct positions (Ding and Marotte, 1996, 1997), is similar to that described in carnivores (Chalupa *et al.*, 1996; Chalupa and Snider, 1998) and in contrast to the rat, where axons are reported to branch and arborise more widely in the initial stages of development of the projection (Simon and O'Leary, 1992).

Mode of reproduction: most development occurs postnatally in the pouch

While the marsupial nervous system is that of a typical mammal, its mode of reproduction is dramatically different to eutherians. It is the combination of these two features that make it an extremely useful model for the study of early development of the brain and its connections. The young are born very early in development, at a stage when most parts of the nervous system are extremely immature, and complete their maturation in the pouch attached to the teat for a protracted period. They are hairless, the hindlimbs are mere paddles and the eyes and ears are closed. The previously mentioned tammar wallaby, for example, the species most used for developmental studies, weighs just 350–400 mg or 0.01% of its mother's weight when it is born after 28 days of gestation (Figure 14.1c) and it does not first leave the pouch until approximately postnatal day (P) 190 (Tyndale-Biscoe and Janssens, 1988).

The visual system demonstrates these features of protracted, postnatal development. The time from conception to eye opening (around 168 days) is similar to that of the human (182 days) (Mark and Marotte, 1992). Figure. 14.1a shows a comparison of several visual developmental milestones in the wallaby with three eutherian species, the laboratory mouse, cat and macaque. Almost the entire development of the visual system occurs postnatally in the wallaby. At birth, retinal ganglion cell axons have only just left the eye and first reach the superior colliculus (SC) and DLG a few days later, with thalamocortical axons reaching the visual cortex about 10 days after that. In contrast, in the eutherian species these events occur prenatally. Similarly, although segregation of the eye-specific regions within the DLG occurs in all species before or by the time of eye opening, this occurs postnatally in the wallaby, nearly four months after birth, but prenatally in the three eutherian species. Other aspects of visual development, such as the birth of neurons which will form the cortical layers (Marotte and Sheng, 2000) and initiation of corticogeniculate and corticocollicular connections (Sheng *et al.*, 1990), show a similar contrast in timing relative to birth when compared to eutherians e.g. cat (Anker 1977; Luskin and Shatz 1985). Another species of wallaby, the quokka (*Setonix brachyurus*), also used for studies of visual development, shows a similar, protracted postnatal pattern of development (Beazley and Dunlop,

---

Caption for Figure 14.1 (Cont.)

furred young in the pouch. Its height in this posture is approximately 0.7 m. (c) A tammar wallaby pouch young on the day of birth, attached to the teat in the mother's pouch. The forelimbs are well developed, but the hindlimbs are still paddle-like. The developing eye is covered by skin. The human thumb in the picture gives an idea of its size; bar = 5mm. (d) Adult fat-tailed dunnart (*Sminthopsis crassicaudata*); bar = 2 cm. Inset shows a ventral view of the pouch with young aged seven days; bar = 3 mm. From Dunlop *et al.*, (1997). Copyright 1997, Alan. R. Liss, Inc. Reprinted with permission of John Wiley and Sons, Inc. (e) Frontal semithin (0.5-µm thick) plastic section stained with toluidine blue through the eye of the dunnart on the day of birth. Ventral is down. The lens (*) is surrounded by the retina, which consists of neuroblast cells (nbl). There is no ganglion cell layer apparent and no axons have entered the optic stalk ventrally; m indicates a mitotic figure; bar = 250 µm. From Dunlop *et al.*, (1997). Copyright 1997, Alan. R. Liss, Inc. Reprinted with permission of John Wiley & Sons, Inc. (f) and g) Retinocollicular synapses in the wallaby, identified by labelling from the eye with horseradish peroxidase conjugated to wheat germ agglutinin, on P14. Arrows indicate electron-dense reaction product and arrowheads indicate synaptic contacts; bars = 0.5 µm. (h) *In vivo* vs. *in vitro* response differences in the superior colliculus of the wallaby. Responses recorded at different depths in response to stimulation of the optic nerve in an animal on P32. Each trace is an average of five sweeps at a frequency of stimulation of 0.03 Hz. Responses recorded *in vivo* show a non-reversing waveform with depth indicative of presynaptic activity only, representing the response of optic axons. *In vitro* there was a clear reversal of the potential with depth, indicative of a postsynaptic response. (f) to (h) from Flett *et al.* (2006). Copyright: the authors (2006). Journal Compilation. Copyright: Federation of European Neuroscience Societies and Blackwell Publishing Ltd.

1983; Dunlop and Beazley, 1985; Harman and Beazley, 1986, 1987) as does the possum (*Trichosurus vulpecula*) (Sanderson *et al.*, 1982; Sanderson and Aitkin, 1990). In the polyprotodont marsupial mouse, the fat-tailed dunnart (*Sminthopsis crassicaudata*) (Figure 14.1d), the visual system is even more immature at birth. The ganglion cell layer is yet to develop and no optic axons have left the eye (Figure 14.1e). Axons do not reach the primary visual centres until P15 (Dunlop *et al.*, 1997). In addition, this species is multiparous (Figure 14.1d, inset), unlike the tammar and the quokka.

The somatosensory (Waite *et al.*, 1998) and auditory (Hill *et al.*, 1998) systems show similar protracted, largely postnatal development. The exceptions are pathways that aid the pouch young in their climb to the pouch and attachment to the teat.

This mode of development means that the marsupial young are accessible for experimental manipulation in the pouch at very early stages of development, which in eutherians would occur prenatally. Pouch young are robust and can be removed from the mother's teat in the pouch from the day of birth onwards, anaesthetised, operated on and returned to the teat for varying survival times with very low mortality. This allows a variety of procedures that would be impossible or extremely difficult to carry out in placentals, such as the use of sensitive *in vivo* tracers to map neuronal connections (Sheng *et al.*, 1990, 1991; Pearce and Marotte, 2003), microsurgical procedures (Dunlop *et al.*, 2007), including eye rotations (Hoffmann *et al.*, 1995; Marotte and Mark, 1988), and *in vivo* electrophysiological recording (Pearce *et al.*, 2000; Mark *et al.*, 2002; Flett *et al.*, 2006), including sequential recordings in the same animal (Hill *et al.*, 1998). The latter is not feasible in eutherians, where *in vitro* slice preparations must be used for making recordings. This means that all inputs and outputs have been removed, a situation quite removed from the *in vivo* situation and one which is likely to affect the responses obtained. The commonly used species for developmental studies, the tammar and the quokka, also breed well in captivity (Mark and Marotte, 1992). The endocrinology of the tammar, in particular, is well understood and the ability to produce timed births throughout a large part of the year is routine (Tyndale-Biscoe *et al.*, 1974; Tyndale-Biscoe and Hinds, 1984). The protracted development means that events that overlap in rapidly developing species such as rodents may be separated in time so that an underlying sequence is revealed (Mark and Marotte, 1992), which can help to elucidate developmental mechanisms (Waite *et al.*, 1998; Waite *et al.*, 2006). Experimental intervention may then allow testing of possible mechanisms (Waite *et al.*, 2006; Dunlop *et al.*, 2007).

Some insights on development of the nervous system gained from marsupials

Retinal development

The protracted, postnatal development of the retina in the quokka not only enables detailed analysis of the birth dates of cell types over the entire time-course of development, but also allows the unambiguous identification of ganglion cells in the ganglion cell layer by retrograde labelling from visual centres and optic tracts (Harman and Beazley, 1989). Distinguishing ganglion cells from, for example, displaced amacrine cells, has been a difficulty in previous studies in eutherian species (Wong and Hughes, 1987). The study in the quokka remains one of the few that describe the complete sequence of retinal cell genesis (see Rapaport, 2006) and was the first to reveal that cell genesis occurred in two phases. Ganglion cells, amacrine cells, including displaced amacrines, horizontal cells and cones are produced in the first phase, while amacrine cells, bipolar and horizontal cells and cones are produced in the second phase. In contrast to the first phase, where there is a central to peripheral wave of cell genesis, in the second phase the sequence is from mid-temporal to peripheral retina. This led the authors to suggest that the later asymmetrical cell addition could contribute to the establishment of ganglion-cell density gradients and further, that the two phases go some way in explaining the density gradients of some of the cell types in the retina (Harman and Beazley, 1987; Harman *et al.*, 1992; Rapaport *et al.*, 2004). Cell genesis in two phases has since been confirmed in the rat (Rapaport *et al.*, 2004) and monkey (La Vail *et al.*, 1991; Rapaport *et al.*, 1996).

Retinocollicular development

Spontaneous retinal activity is suggested to play a role in the development of the topographic map made by retinal axons in the superior colliculus. However, the basic information regarding when the pathway is first capable of transmitting this activity to the colliculus was lacking because the prenatal development of the initial stages in eutherian mammals necessitated the use of *in vitro* preparations. Further, the reported onset of retinocollicular synaptic transmission *in vitro* (Reece and Lim, 1998) was earlier than that seen *in vivo* (Lim and Ho, 1997) in the rat. The entirely postnatal development of the wallaby retinocollicular projection allows a direct comparison of *in vivo* vs. *in vitro* onset in the same animals (Flett *et al.*, 2006). Retinal ganglion cell synapses are present in the colliculus (Figure 14.1f and g) at the time that the first postsynaptic responses can be recorded *in vitro*. Recordings confirm that there is indeed a discrepancy in the time of onset of synaptic transmission, with the onset *in vivo* occurring some three weeks later than that detected *in vitro*, at around the time axons begin arborising. Figure 14.1h shows an example of the *in vivo* vs. *in vitro* discrepancy at P32. This appears to be unrelated to anaesthetic and could indicate a suppressive mechanism to prevent excessive activity from the retina reaching the colliculus when inhibitory mechanisms are poorly developed there (Flett *et al.*, 2006). Alternatively, if the large increase in synapses seen in rat hippocampal slices *in vitro* compared to perfusion-fixed hippocampus (Kirov *et al.*, 1999) is also occurring in the wallaby *in vitro* preparations, this may increase postsynaptic responses to detectable levels. These findings highlight possible problems when drawing conclusions from *in vitro* preparations and the need, ideally, to determine the onset in unanaesthetised animals (see future directions).

Cortical development

The accessibility of developing pouch young for sensitive neuronal tracing methods, including *in vivo* methods, as well as *in vivo* electrophysiological recording has been exploited to investigate thalamocortical development. The anatomical development of connections from the thalamus in visual and somatosensory cortex at both the light- and electron-microscopic level and their functional development revealed a quite different mode of axonal ingrowth to that previously described for carnivores and primates. In the latter species, thalamic axons are said to undergo a waiting period in the subplate before growing into the developing cortex, and connections made there are postulated to play an important role in directing thalamocortical connectivity. However, the prenatal development of connections precluded *in vivo* recording. Although in isolated *in vitro* slices it has been shown that connections are made onto subplate cells early in development, the source of these connections remains unknown (Friauf and Shatz, 1991). Thalamic axons have been shown to make connections on subplate cells, but this was demonstrated at a relatively late stage when these axons are already synapsing with cells in layer 4Cx, their main input layer in the adult (Herrmann *et al.*, 1994). What is happening at earlier stages is not known. In contrast, in the wallaby, thalamic axons grow into the developing cortex without any sign of a waiting period in the subplate and the first synapses they make are with cells residing in developing layer 6Cx (Pearce and Marotte, 2003), as identified by birthdating studies (Marotte and Sheng, 2000), and one of their normal targets in the adult cortex (Sheng *et al.*, 1991). Their first detectable functional connections, identified by *in vivo* current-source density analysis after optic-nerve

stimulation, are also made with cells in developing layer 6Cx and axons continued to grow into the developing layers as they differentiated at the base of the cortical plate (Pearce *et al.*, 2000). It does not appear that this mode of development is peculiar to marsupials because thalamocortical axons in rodents also show no waiting period in either visual (Kageyama and Robertson, 1993) or somatosensory cortex (Catalano *et al.*, 1991). These apparent differences in thalamocortical development remain to be resolved.

In both visual and somatosensory cortex there is a lag between the presence of thalamocortical synapses and the onset of synaptic responses to peripheral stimulation *in vivo*. *In vitro* pharmacological investigations in a thalamocortical slice preparation (Leamey *et al.*, 2007) allowed a dissection of chemical transmission in developing thalamocortical connections and revealed possible contributions not only to the above discrepancy, but to the lag between the onset of synaptic responses and the first identification of synapses. These studies highlight the advantages of being able to use both *in vivo* and *in vitro* approaches to analyse developing pathways.

Auditory development

The ability to record *in vivo* at relatively early stages of auditory development, prior to the onset of sound-evoked responses, has been exploited to investigate spontaneous neural activity in the auditory system of the tammar wallaby (Gummer and Mark, 1994). It was known at the time that such activity is present in synchronised, rhythmic bursts in isolated developing retina (Meister *et al.*, 1991; Wong *et al.*, 1993) and in the developing somatosensory system (Fitzgerald, 1987), and spontaneous activity was believed to be important for the refinement of visual connections (Shatz, 1990). However the description of rhythmic bursting in the eighth nerve and cochlear nucleus prior to hearing in the tammar extended this to the developing mammalian auditory system. Spontaneous activity has been implicated in a number of developmental auditory processes (reviewed in Friauf and Lohmann, 1999), but probably the most convincing evidence for a role prior to hearing is in the development of tonotopic organisation in the auditory brainstem of mice (Leao *et al.*, 2006).

Auditory brainstem responses (ABR) of the adult wallaby (Cone-Wesson *et al.*, 1997) are similar to those of other mammals, so that their protracted postnatal development provides 'a slow-motion picture' of auditory aspects of mammalian development (Liu and Mark, 2001). A detailed study of the development of auditory brainstem responses in this species, including repeated examination of responses over time in the same animals (Hill *et al.*, 1998) has been carried out, as well as studies to relate the components of the ABR during development to inferior collicular (Liu and Mark, 2001) and focal responses of auditory nerve and cochlear nucleus responses (Liu *et al.*, 2001).

Somatosensory development

Similar to the 'barrels' reported in the somatosensory pathway of some eutherian species (Van der Loos and Woolsey, 1973), the somatosensory system of the tammar is also characterised by the presence of whisker-related patterns (Waite *et al.*, 1991, 1994; Leamey *et al.*, 1996; Marotte *et al.*, 1997). The protracted postnatal developmental pattern of the tammar has been utilised to investigate the anatomical and functional changes which underlie the formation of these patterns; this is difficult in eutherian species due to their rapid, predominantly intrauterine, maturation. The clear temporal separation of pattern formation in brainstem, thalamus and cortex indicates that pattern formation is triggered independently at each level of the pathway (Waite *et al.*, 1998). Further, the maturation of the target tissue, as indicated by an increase in the non-NMDA-mediated glutamatergic response, correlated well with pattern formation in both thalamus and cortex (Leamey *et al.*, 1998, 2007), suggesting that this, together with peripherally related signals, may be a driving force for the clustering of functionally related cells and afferents. The somatosensory system and its development are discussed in more detail in Chapter 10.

Future directions

The protracted developmental timecourse is likely to provide further opportunities for elucidating mechanisms of axonal guidance and pathway development. For example, Neveu and co-workers (Neveu *et al.*, 2006; Neveu and Jeffery, 2007) have suggested that the optic chiasm of marsupials may be a more appropriate model than the rodent chiasm to study mechanisms controlling the development of the chiasm that are applicable to humans. In humans (Neveu *et al.*, 2006), uncrossed axons are confined laterally as they approach the chiasm, similar to the pattern in marsupials (Jeffery and Harman, 1992; Harman and Jeffery, 1995) and unlike that seen in rodents (Baker and Jeffery, 1989; Baker, 1990), where midline interactions occur. Further, early eye removal in marsupials does not affect the crossing pattern of projections from the remaining eye (Taylor and Guillery, 1995), a similar outcome to that seen in humans when one eye fails to develop (Neveu *et al.*, 2006).

The ability to carry out *in vivo* physiological investigations on the intact animal at the earliest developmental stages is a great advantage of this model system, especially in combination with *in vitro* pharmacological investigations to further dissect mechanisms of interest. What is yet to be exploited is the unrivalled suitability of the marsupial model for physiological analysis of the developing nervous system in the unanaesthetised animal. An investigation of spontaneous activity in the developing visual system is one area that would be fruitful. Such activity, arising in the retina, is claimed to play a pivotal role in shaping connections in developing visual centres (Flett *et al.*, 2006; Huberman *et al.*, 2008). Recent studies, however, raise questions regarding the relative roles of activity-dependent vs. axon guidance cues (Huberman *et al.*, 2003, 2005; Sun *et al.*, 2008) which could be addressed by recording with implanted electrodes in awake pouch young.

Another area suitable for exploitation is the ability to manipulate the levels of molecules of interest at chosen time points to test their developmental role, using *in vivo* electroporation of DNA constructs for up regulation, or siRNA for down regulation. Gene knockouts in the mouse have been a useful tool for such studies, but they remove expression of the gene in all tissues over the entire developmental time period, which can confound interpretation of the results; for example, it is difficult to dissect the effects in one tissue vs. another (Grubb *et al.*, 2003). They can also be lethal. Conditional knockouts, which allow alteration of gene expression levels in specific tissues, avoid some of these difficulties, but are expensive and time-consuming to produce and depend on a tissue-specific marker amenable to genetic manipulation; their use so far is not widespread. Intrauterine surgery and the size of embryos of common laboratory mammals mean that electroporation is simply not feasible in most cases. By contrast in the metatherian model, levels of the desired molecule can be manipulated at the appropriate time and in the tissue of choice or within subregions of this tissue. This is particularly useful if gradients of a molecule are under investigation, for example those involved in topographic mapping in the visual system.

Acknowledgements

This chapter is dedicated to Richard F. Mark (1934–2003) who championed the use of Australian marsupials as models for the early development of the nervous system and who, together with his colleagues and students, did much of the work on which this chapter is based. We would like to thank Dr L. Aitkin and Professor S. Dunlop for their comments on the chapter and Ms S. Wragg for preparation of the figure.

15

Marsupials as models for development, ageing and disease: a neurobiological and comparative context

B. M. McAllan and S. J. Richardson

Introduction

Marsupials are an under-recognised group of potential animal models for many biological processes, such as development, ageing, regeneration and disease. Marsupials are born at a very undeveloped stage compared to eutherian ('placental') mammals and are therefore excellent models for analysis of developmental systems. Furthermore, the South American grey short-tailed opossum (*Monodelphis domestica*) genome has been sequenced and is being annotated, and the tammar wallaby (*Macropus eugenii*) genome is currently being sequenced. Annotation of these genomes will result in more powerful exploitation of molecular analyses of these species as models for normal biological processes and for disease. The increasing number of marsupials whose life histories are being elucidated has revealed an increasing number of species that are natural models for specific diseases in humans. The advantage of this is that it allows the disease process to be studied in a natural model compared to transgenic mice models, which often do not present an accurate phenotype of the human disease.

The 'other' mammals: marsupials and monotremes

The Prototheria (monotremes) and Metatheria (marsupials) differ from Eutheria ('true' mammals, often called placental mammals) by their means of reproduction. The mammalian features of hair, two dentary bones making up the lower jaw, three middle ear bones and suckling their young using mammary tissue modified from apical glands of the skin are shared between all three groups (Blackburn, 1993; Augee *et al.*, 2008). It appears that the common ancestor of the Theria was still oviparous, similar to the extant monotremes, and that viviparity occurred independently in the phylogeny of Metatheria and Eutheria (Zeller, 1999).

Monotremes

The subclass Prototheria is represented today by only one group of mammals, the order Monotremata (monotremes). Monotremes diverged from other mammalian lineages approximately 200 million years ago, and there are only five extant species of monotremes, the best known of which are the platypus (*Ornithorhynchus anatinus*), and the short-beaked echidna (*Tachyglossus aculeatus*; Augee *et al.*, 2008). Monotremes can be identified anatomically from other mammalian subclasses by their retention of interclavicle, coracoid and precoracoid bones in the shoulder girdle, and ectopterygoid bones at the base of the skull (Augee *et al.*, 2008). Monotremes have a true cloaca, in contrast to marsupials, which have a simple external opening for excretory and reproductive functions (known as a urogenital sinus; Temple-Smith and Grant, 2001). The monotremes differ in their mode of reproduction from other mammals by laying eggs and also producing milk for their altricial young (Temple-Smith and Grant, 2001). Monotremes feed their young through lactational pores, which duct directly to the surface of the brood area (Blackburn, 1993). Lactation occurs for 3–4 months for platypus and 5–6 months for the short-beaked echidnas (Beard and Grigg, 2000; Temple-Smith and Grant, 2001).

Marsupials

Marsupial mammals diverged from eutherian (placental) mammals about 125–130 mya, although some debate exists about the exact timing, with dates of between 125–180 mya suggested (Murphy *et al.*, 2004; Renfree, 2006). There are about 270 species of extant marsupial mammals, which are found in Australia, South America and Papua New Guinea.

A main difference in the biology between the marsupials and other mammalian groups is that marsupial neonates are born in an extremely altricial state (similar to monotremes) after a relatively short gestation period for body size, but development occurs postnatally in the pouch or pouch area, as seen in some didelphids ('lactation over gestation' option; Renfree 1983, 2006). The marsupials do have a functional placenta, with most having a chorio-vitelline placenta, whereas the bandicoots have an invasive chorio-allantoic placenta (Freyer *et al.*, 2003, 2007). Thus it is incorrect to use the term 'placentals' to separate the eutherian mammals from marsupial mammals. The gestation period can be as short as 10.5 days (*Sminthopsis macroura*, the stripe-faced dunnart, Selwood and Woolley, 1991), with some didelphids and dasyurids having gestational lengths of 12 to 16 days (see McAllan, 2003), but most species demonstrate a gestational length of about 27 to 35 days (e.g. *Macropus eugenii*, the tammar wallaby, Tyndale-Biscoe and Renfree, 1987). Marsupial neonates weigh as little as 5 mg (*Tarsipes rostratus*, the honey possum, Renfree 1979), and are undeveloped compared to eutherian mammals, with most organ development occurring postbirth, during the long, frequently complex, lactational period in the pouch.

The altricial development at birth includes the partially developed brain, with somatosensory and brainstem regulatory regions well developed, but with other cerebral and limbic regions and the endocrine system remaining very undeveloped (see Chapters 3, 4, 8, 9, 10, 11 and 12 this volume; Reynolds *et al.*, 1985; Gemmell and Nelson 1988a, 1988b; Nelson, 1992; Nelson *et al.*, 2003). The marsupial neonate has relatively well-developed digestive, respiratory and circulatory systems, although lower limbs and musculoskeletal structures of all but those needed to climb to the pouch after birth are poorly developed (Shaw and Renfree, 2006). Marsupials are born retaining a functional mesonephric kidney, and the scrotum and primordial mammary glands are present at birth, even though the testes and ovaries do not develop until well after birth (Shaw *et al.*, 1990; Shaw and Renfree, 2006).

The extremely undeveloped neonate has provided the opportunity for marsupials to be used as models for developmental mechanisms, including the role of hypoxia and skin respiration in marsupial neonates paralleling the developing respiratory system and prematurity in eutherians (e.g. Frappell and McFarlane, 2006), and development of thermogenesis and metabolic activities (e.g. Geiser *et al.*, 2006). Part of their appeal as a model system has been the access to pouch young without having to sacrifice the mother, and the ability to follow developmental mechanisms in the same individuals (or litter mates) throughout life. What has not always been appreciated is their potential as model systems for life history and ageing processes.

Features of marsupial life histories

Marsupials range in size from small carnivores (*Planigale gilesi* Giles' planigale, 5 g) to large herbivores (*Macropus giganteus*, grey kangaroo, up to 85 kg; van Dyck and Strahan, 2008). Their social organisation includes strictly solitary and territorial (outside the breeding season, e.g. *Phascogale calura*, the red-tailed phascogale; and most small macropods) through to large more complex social structures (e.g. *Macropus rufus*, red kangaroo; *Petrogale persephone*, the Proserpine rock-wallaby). Known life expectancy is from 11.5 months (*Antechinus* sp. males; but not females, whose life expectancy ranges from 1.5 to 3 years) to about 30 years (many species of the larger kangaroos; van Dyck and Strahan, 2008). The range of life-history circumstances found in marsupials can often be overlooked when searching for accurate predictors for behavioural and physiological models for both human and non-human biological adaptations.

A life-history feature of all marsupials is the short gestation period followed by a long lactational phase, which contrasts to eutherian mammals, where, relative to body size, the longer gestational phase is followed by a shorter lactational phase. This difference has frequently been interpreted as a negative, even primitive adaptation persisting in marsupials (Harvey *et al.*, 1991; McNab, 2006) and few comparative studies have addressed the differences in a systematic manner (Lovegrove and Haines 2004; Lovegrove, 2009). Most studies focus on the metabolic contrasts between the groups, frequently relying on data that are inconsistent or incomplete (Lovegrove and Haines, 2004; White and Seymour, 2004; White *et al.*, 2009). It has been cautioned that, because of the sporadic data collected for all mammalian groups, data should be considered in body-size clades before looking for evolutionary trends and mechanisms (for discussion see Lovegrove and Haines, 2004).

What is confirmed is that the normothermic marsupial body temperatures are generally slightly, but not significantly, lower than in eutherian mammals (White and Seymour, 2004; Withers *et al.*, 2006) and their mass-specific-based metabolic rate is lower and thus considered to be 'poorer' (Harvey *et al.*, 1991). The contention that they are 'less able' warm-blooded organisms has been seen as a facile argument by some physiologists, because birds thermoregulate at higher temperatures than eutherian mammals, and yet are not seen as superior to mammals (van Dyck and Strahan 2008). However, when data for metabolism are matched for life-history traits (herbivory/ carnivory; torpor use/no torpor use; litter size; habitat locality) there is little difference between the mammalian groups (Lovegrove, 2003; Geiser, 2004; Withers *et al.*, 2006; Clarke and Rothery, 2008; Frappell, 2008). For example, when climatic indicators are used to determine differences in basal metabolism, the main determinates are body mass, ambient temperature, rainfall patterns and latitude, not phylogeny (Lovegrove, 2003).

Current models for ageing and disease

Ageing can be described as the progressive deterioration of physiological function that occurs as adults grow older, and this eventually leads to decreased health, a decline in general welfare, and an increasing incidence of degenerative disease, functional loss and death (Austad, 2005). In recent years there has been prolific use of transgenic animal models for human disease, with many aficionados adhering to the notion that knocking out (or in) of specific genes accurately reflects a disease state. This is commonly observed in models for Alzheimer's disease (AD). The reproduction of AD pathology in transgenic mice is based on the transfection of mutant amyloid precursor protein (APP), presenilins (PS) and tau (for formation of neurofibrillary tangles) genes, and even when combined, these transgenic mice produce only some of the pathological changes seen in AD (Götz *et al.*, 2007; Woodruff-Pak, 2008; Kokjohn and Roher, 2009). Moreover, no specific mutations in the β-amyloid peptide (Aβ) explain the massive pathological amyloid deposition in the most common sporadic form of AD. Accumulation of both Aβ and tau (as neurofibrillary tangles) are evident in other neurodegenerative disorders, and these features are widely seen in aged, non-demented individuals (Kokjohn and Roher, 2009). In spite of significant advances made with the initial transgenic mouse models of AD, these models of over-expression of human APP appear to be incomplete because they fail to express the full array of human AD neuropathology (see Woodruff-Pak, 2008). Contention concerns the precise phenotype of transgenic animals and their relevance to human disease states, because the altered gene can transfer more than the specific defect, and for some mouse phenotypes this can mean lowered fertility and altered neurohormonal pathway responses, resulting in a cascade of unexpected, even unwanted, side effects (Spires and Hyman, 2005; Melrose *et al.*, 2006; Götz *et al.* 2007; Heng *et al.*, 2008; Liu *et al.*, 2008). The issues of concern mostly relate to accuracy of the models for the disease under question, but sometimes the ethics of transgenic animal generation for disease elucidation have been questioned (Lavery, 2000; Olsen *et al.*, 2007). Some authors are disquieted by the notion that pain, distress and poor quality of life are genetically modified with the 'end justifying the means' for human wellbeing (Olsen *et al.*, 2007). Marsupials are under-recognised animals as models for disease states

(e.g. neurodegenerative diseases) and as models for senescence, and this may be related to the paucity of knowledge about them outside of Australia. Most models are based on invertebrates or rodents, with few natural models for disease studied (Woodruff-Pak, 2008; Kokjohn and Roher, 2009).

Ageing in mammals: marsupials compared to eutherians

The variable ageing rates in metazoans has been known for some time (see Austad, 1997, for discussion), and there is a >10 000-fold variation in longevity in metazoan organisms with confirmed life expectancies ranging from a few days (rotifers, in Austad, 2005) to more than 200 years (deep sea fish, Cailliet *et al.*, 2001; bowhead whales, in Garde *et al.*, 2007). Considerable data has accumulated about relative body size and longevity (Calder, 1984), which has resulted in two of the tenets of the theories of ageing. These are: (i) that smaller mammalian species have shorter lives than larger mammals and (ii) that ageing is inevitable, with all species experiencing programmed death (Calder, 1984; Weinert and Timiras, 2003). However, much of the ageing data rely on age at death, and the assumption that longevity and senescence are related (Austad and Fischer, 1991). Mammals with dissimilar life histories may be differentially represented. Thus, mammals that die young, but in which mortality is caused by activities unrelated to ageing, may not be part of the 'live fast – die young' organismal grouping, but simply represent mammals that have continued exposure to adverse environmental conditions such as heavy predation (e.g. bank vole *Clethrionomys glareolus*, Selman *et al.*, 2008; Speakman, 2005a). Other mammals may live for long periods of time, but senescence is swift and cataclysmic when it occurs (e.g. naked mole-rat, *Heterocephalus glaber*, Austad and Fischer, 1991; Speakman 2005a).

The hypotheses concerning the evolutionary benefits of a long life have been widely discussed (Austad and Fischer, 1991; Speakman, 2005a, 2005b; Magalhães *et al.*, 2007; Selman *et al.*, 2008), including the negative evolutionary effects of reduced mortality, the restriction of breeding to older, experienced mammals, and thus limitation of genetic diversity (Goldsmith, 2004). Conversely, an advantage of a long life could include the selection for immunocompetent survivors of disease, and imparting of learned knowledge of predator avoidance, food collection and parenting behaviours to offspring or other closely related members of the species (Weinert and Timiras, 2003; Goldsmith, 2004). These factors are important in epidemiological and social studies of human history (Caruso *et al.*, 2000; Mountz *et al.*, 2002).

Attempts to clarify the effects of different life-history strategies on ageing, and to compartmentalise the strategies to optimise models for ageing, have not been able to determine a clear pattern, except that volant (flying) mammals live longer than non-volant mammals (Read and Harvey, 1989; Austad and Fischer, 1991; Phelan and Austad, 1994; Speakman, 2005a, 2005b, 2008). Calculations of ageing and relative ageing among species are often fraught with the complications of data collected from captive vs. wild species, and the nature of differences between very closely related species (Speakman, 2005a, 2005b, 2008). Our own unpublished data has shown that in captivity the fat-tailed dunnart (*Sminthopsis crassicaudata*), a 12 to 15 g carnivorous marsupial, can regularly live as long as four years, and both sexes can reproduce in their third year of life (Table 15.1). Data on life expectancy from wild dunnarts is not available. Published data from other small species of both marsupials and eutherian mammals suggest that this level of longevity is unusually long for a small non-volant mammal (Gorbunova et al., 2008).

Data on ageing in marsupials have been collected almost exclusively from studies on wild animals, and the data are then compared with data from captive rodents (Austad and Fischer, 1991; Austad, 1997). In contrast to captive rodents, data are drawn from the wild for some of the longest-lived small eutherian mammals, the bats, with longevity records of 41 of the 50 species in one study drawn from animals living in the wild (Austad and Fischer, 1991). No correlation has been found with body size, reproduction rate and longevity in bats, and while use of hibernation may extend the life span of bats, this is only by an average of six years (Read and Harvey, 1989; Wilkinson and South, 2002). Data for marsupials are as intriguing. No correlation between brain size, body mass or metabolism and longevity have been found for marsupials, and their longevity has been unfavourably compared with bats. However, longevity data from the mountain pygmy possum (*Burramys parvus*, 50 g), suggests they can compete with some bats for longevity. Data collected from the wild demonstrate that they can live longer than 10 years, including reproducing in their 10th year (Broome, 2001). The comparison of the marsupial life span with other mammals with similar life-history strategies is more complicated (Austad and Fischer, 1991; Austad 2005; Speakman, 2005a, 2005b). Calculation of the longevity quotient (LQ = actual life span/predicted life span from non-flying eutherian mammalian regression) has determined that marsupials are less long-lived than other mammals (Austad and Fischer, 1991). However, more recent and extensive data analysis has concluded that there are few differences related directly to marsupial phylogeny, and that many of the published longevity analyses are inherently flawed (Speakman, 2005a; and see Speakman, 2005b, for discussion). It would appear that, rather than make broad phylogenetic assumptions, ageing and longevity data need to be assessed on a case-by-case basis.

Longevity data from one species of carnivorous marsupial, *Antechinus stuartii*, the brown marsupial 'shrew', are perhaps the most intriguing. The males of this, and all members of the genus, die from a catastrophic life-history event, male 'die-off', which is driven by rising circulating concentrations of testosterone and cortisol – even though these concentrations are physiologically not high for eutherian mammals (for review see Naylor et al., 2008). Males (35 to 40 g) live for 11.5 months, fulfilling the 'living fast – dying young' paradigm as proposed by Promislow and Harvey (1990), but females (20 to 25 g) do not, living for 18 months to three years (76 to 252 weeks), and regularly reproducing in their second and third years of life in both the wild and in captivity (Table 15.1, McAllan et al., 1991; Dickman 1986; Fisher et al., 2006).

Two important facts, often overlooked by longevity analysers, must be considered before dismissing these mammals as 'longevity losers'. Firstly, males are otherwise robust until the last few weeks of life and, in the wild, the entire male cohort dies within a two-week period, irrespective of mating success, social stress or environmental conditions (see Naylor et al., 2008). In captivity, male *Antechinus* can survive male 'die-off' if given enough food daily to surmount the negative nitrogen balance experienced during the mating and 'die-off' period, and can live for at least another year, as might be predicted for their size (McAllan, 2009). While (in contrast to second-year females) the males can never reproduce again due to the complete and irreversible collapse of the seminiferous germinal epithelium, they do recover physiologically, although some behavioural disturbances can remain in some individuals (McAllan, 2009). Thus, *Antechinus* offer a unique opportunity as an 'unmodified' animal model to look at the effects of gene-signalling of senescence events of ageing. Humans now live extended life spans compared to a few generations ago, often living much longer than their reproductive 'life span', similar to captive male *Antechinus*.

Secondly, although female *Antechinus* have also been unfavourably compared to the longest-lived laboratory mouse whose longevity record is four years (30 to 40 g, mostly derived from the North American *Mus musculus musculus*, (see Austad, 2005), the published and analysed longevity data from *Antechinus* are from wild animals. Wild-type laboratory mice are often thought to live without the stress of predation, parasites and pathogens, and in optimum food, water and thermal conditions (Miller et al., 2002; Harper, 2008). In wild (feral) populations of house mice from Australia (14 to 18 g, mostly derived from the

Table 15.1 Timelines for life history and neurological events for four species of mammals, comparing marsupial and eutherian events. Data are from two laboratory established mammals, *Sminthopsis crassicaudata* and *Mus musculus musculus*, and two Australian wild-caught mammals, *Antechinus stuartii* and *Mus musculus domesticus*. Data are gestational length, days to weaning from birth, age at sexual maturity from birth, and the longevity record for that species. Data from both sexes are shown as there are some sex differences in most species. Neurological developmental data are in days post-conception (PC: gestation and, for some species, plus days after birth). Amyloid depositional data includes information as to where plaques are known to be developing and include data for when dementia (D) is known, and when no dementia is present (ND) and the Braak and Braak (1991) staging (from 0–VI). Mouse amyloid presentation data are from non-transgenic mice. If information is not known a '-' is used NT – no tangles. Data are from Marlow (1961), Braak and Braak (1991), Ashwell *et al.* (1996a), Dunlop *et al.* (1997), Darlington *et al.* (1999), Westman *et al.* (2002), Ernest (2003).

Species	Gestation length (days)	Age at weaning (days)	Age at sexual maturity (days)	Reproductive life span (range in days)	Longevity record (days)	Axons appear in the optic stalk (days)	Optic axons invade visual centres (days)	Superior colliculus segregation (days)	Eyes open (days)	Amyloid deposition (days)
Sminthopsis crassicaudata Male body mass 15–18 g	12	70	150	150 – >1095	≈1460	PC 16	PC 28.5	PC 63.5	PC 58.5	–
Sminthopsis crassicaudata Female body mass 13–16 g	12	70	90	120 – >1095	≈1460	PC 16	PC 28.5	PC 63.5	PC 58.5	–
Mus musculus Male body mass 30–35 g	20	45	72	72 – 500	≈1460	PC 12.3	PC 15.5	PC 24	PC 30	>365 (NT)
Mus musculus Female body mass 30–35 g	20	45	72	72 – 350	≈1460	PC 12.3	PC 15.5	PC 24	PC 30	>365 (NT)
Antechinus stuartii Male body mass 30–40 g	27	100	275	275 – 350	350	–	–	–	PC 90	≈275
Antechinus stuartii Female body mass 20–25 g	27	100	275	275 – ≈1095	≈1095	–	–	–	PC 90	> ≈275
Mus musculus domesticus Male body mass 15–18 g	19	45	72	≈72 – 225	≈500	–	–	–	–	–
Mus musculus domesticus Female body mass 13–16 g	19	45	72	≈72 – 225	≈500	–	–	–	–	–
Homo sapiens Female body mass 50–80 kg	280	≈350	≈4380–6205	4380 – 17 520	≈44 530	PC 51	PC 60	PC 175	PC 182	17 155 (D VI) >14 600 (ND II)
Homo sapiens Male body mass 60–90 kg	280	≈350	≈4745–6570	≈4745 – 29 200	≈43 800	PC 51	PC 60	PC 175	PC 182	20 440 (D VI) >14 600 (ND II)

European house mouse *Mus musculus domesticus*) the average life expectancy is 32 weeks, with occasional animals living as long as 75 weeks (Sutherland *et al.*, 2004), that is, the longest-lived feral mouse lives as long as the 'least successful' *Antechinus* female, who survive long enough only to wean their young. Moreover, when Australian feral house mice are held in captivity they can survive for more than 50 weeks of age, however their growth and survivorship can be significantly affected by ambient temperature (McAllan *et al.*, 2008b). House mouse longevity data are also confounded by strain, and hybrid analysis suggests that there is considerable variability that may negatively affect their usefulness as models for ageing humans (Table 15.1, Phelan and Austad 1994; Miller *et al.*, 1999; Miller *et al.*, 2002; Flurkey *et al.*, 2007; Harper, 2008). This information suggests that the opinions on longevity data for marsupials and widely used laboratory mammals, as well as the mechanisms of ageing in all mammals, do need to be revisited in a systematic manner.

Parallels between models for disease and ageing and humans

The limitations of the house mice in the laboratory may not make them a suitable model for ageing processes. The reproductive potential of mice makes them an attractive laboratory animal, but their rapid turnover postreproduction does not mirror the life-history strategy adopted by humans, where postreproductive life span can be as long as, or longer than, the reproductive life span. Moreover, house mice continue to grow throughout life (Singleton *et al.*, 2001) and this is not true for humans, where musculoskeletal growth slows around puberty, ceasing completely three to five years after puberty (Eshet *et al.*, 2004; Nilsson and Baron, 2004). Similarly to the situation for humans, in dasyurids such as *A. stuartii* and *S. macroura*, growth stabilises at the onset of reproductive maturity (Woolley, 1990a, 1990b; McAllan *et al.*, 1991), offering parallels to this aspect of the life history of humans. Besides their reproductive success, the use of mice as models for ageing has relied on the genetic modification of mice, especially transgenic models for illness. For example, the genes for hormonal changes such as are found in transgenic mice that either over-express or under-express growth hormone have been used to investigate the relationship between growth hormone and insulin-like growth factor 1, and longevity (Carter *et al.*, 2002). However, hormones that have a direct bearing on organisational development usually result in mutant and transgenic mammals that develop outside the hormonal pattern of normal mammals, which can exhibit significant differences in the

development of many organs and tissues, and can include neurological and behavioural defects, impaired growth and delayed puberty (Carter *et al.*, 2002).

Thus, in comparison to other mammalian models, the significant parallels to the growth and longevity patterns in humans and dasyurids makes the members of the genera *Sminthopsis* and *Antechinus* attractive mammalian models for ageing (see Table 15.1). One possible disadvantage of using dasyurid marsupials as models, however, may be the perceived slower reproductive life history and thus their slow replacement in the laboratory, although this has been argued against by some authors (Harman *et al.*, 2003a). Researchers must trade off the significant advantages of appropriate models for the disease or ageing profile under questions, and the convenience of a fecund unsuitable laboratory model.

The ageing marsupial brain

Current knowledge about the marsupial brain as it ages is limited, as most studies have focussed on neonatal development rather than longitudinal decline (e.g. Leamey *et al.*, 1998; Sheng *et al.*, 1990; see Chapters 3, 10, 11, 12 and 14 of this book). Like much of the work on marsupial biology, data on the ageing marsupial brain are patchy and are limited to a few species (*Sminthopsis crassicaudata*, Harman and Moore, 1999; *Monodelphis domestica*, Grabiec *et al.*, 2009). While this might be seen as a negative factor in the quest for marsupial models for disease, it must be remembered that very few of the >4000 species of eutherian mammals, or indeed of any non-mammalian vertebrates or invertebrates, are used as models for human disease. Besides transgenic mice, experimental models for ageing and the brain have included the worm *Caenorhabditis elegans* (Gill, 2006; Kleeman and Murphy, 2009), the fruit fly *Drosophila melanogaster* (Zoghbi and Botas, 2002; Bonini and Fortini, 2003), the clawed toad *Xenopus* sp. (Lee *et al.*, 2005, Endo *et al.*, 2007), the chick embryo (Carrodeguas *et al.*, 2005; Lee *et al.*, 2005), the kokanee salmon, *Oncorhynchus nerka* (Maldonado *et al.*, 2002), the tree shrew, *Tupiaia glis* (Pawlik *et al.*, 1999; Primmer, 2002) and the macaque *Macaca mulatta* (Tayebati, 2005). Advantages to some of these models for ageing are the short generational times (e.g. *C. elegans*, *D. melanogaster*), ease of care in the laboratory (e.g. *C. elegans*, *D. melanogaster*, *Xenopus* sp., chick embryo), appropriateness of the models to the mechanism under study (e.g. kokanee salmon, tree shrews), and phylogenetic closeness to humans (tree shrews, macaques). Disadvantages are the long generational times (tree shrews, macaques), phylogenetic distance from humans (e.g. *C. elegans*, *D. melanogaster*, *Xenopus* sp., kokanee salmon, chick embryo) and the ethical issues surrounding primate studies.

Marsupial brain development: comparison with eutherian brains

The brain of marsupials develops slowly in comparison to similarly sized eutherians (Darlington *et al.*, 1999), but continues to develop long after the brain has stopped developing in eutherians (Finlay *et al.*, 1998). The poorer nutrient transfer via lactation instead of longer exposure to placental transfer of nutrients has been considered a major reason why this has occurred, although significant variability in neurogenic development exists amongst eutherian mammals, which can confound this simple argument (Finlay *et al.* 1998; Darlington *et al.*, 1999).

Studies of the sensory system have shown that the visual system takes on the adult configuration by weaning (Beazley and Dunlop 1983; Sheng *et al.*, 1990), and the visual cortex in marsupials contains at least two topographically organised cortical areas, V1 and V2 (Karlen and Krubitzer, 2007; see Chapters 8 and 9 of this book). Electrophysiological recording data and architectonic analysis have determined that at least two unimodal somatosensory fields are present in the neocortex of adult marsupials, S1 and S2 (primary and secondary somatosensory areas, respectively) along with R, C and PV fields in some cases (see Chapters 8 and 10 of this book). The information on the organisation and connections of auditory cortex in marsupials are scant, however existing data indicate that there are features of organisation that are similar to that described in eutherian mammals, and, again similar to the situation for eutherians, the cortical magnification of specific frequencies appears to be related to specific behaviours and life-history activities such as prey detection and predator avoidance (Kudo *et al.*, 1989; Aitkin, 1995; see Chapter 11 of this book).

Motor-cortex mapping has demonstrated that there are two distinct types of organisation, which may indicate that the motor cortex has evolved differently in separate phylogenetic lineages. Neuroanatomical and connectional data support the observation that the two types of motor-cortex organisation that exist within marsupials include one where the somatosensory and motor-cortical information overlaps in the sensorimotor amalgam (e.g. the quokka, *Setonix brachyurus*, red kangaroo *Macropus rufus* and opossum *Didelphis virginiana*) and another where the sensory and motor cortices are separated (e.g. wombat *Vombatus ursinus*, Tasmanian pademelon *Thylogale billardierii* and brush-tailed possum *Trichosurus vulpecula*; Haight and Neylon, 1979; Karlen and Krubitzer, 2007; Neylon and Haight, 1983).

A major point is that the cortical structure and function of the neocortex is essentially similar between eutherians and marsupials, and that any differences are related to life-history strategies, not necessarily phylogeny (Kornack, 2000; Karlen and Krubitzer, 2007). The neocortex is essentially a mammalian structure, and differs markedly across all mammalian lineages, with relative and absolute expansions in size occurring independently in several different lineages, marsupial and eutherian alike, with the largest expansion occurring in the primates (60–80% of total brain mass; Kornack, 2000).

Adult neurogenesis in marsupial brains

Coupled with the neocortical development in mammals is the capability of neurogenesis in the adult neocortex of many mammals. The ability and rates of adult neurogenesis vary considerably between species, with some species showing no neurogenic activity in the neocortex (cat, rodents, in Doetsch and Scharff, 2001), and others showing considerable neurogenesis (e.g. some primates, although other primates show none; Gould *et al.*, 1999a; Doetsch and Scharff, 2001; Doetsch, 2003a). All mammals studied so far, including marsupials, show new neurons which present themselves in the adult olfactory bulb and hippocampus (Doetsch and Scharff, 2001; Harman *et al.*, 2003a; Grabiec *et al.*, 2009). Studies on eutherians have shown that there are neural stem cells and progenitor cells which migrate from the ventricular areas to the sites of neurogenesis (Doestch and Scharff, 2001). The adult neurogenesis occurs in the inner rim of the dentate gyrus (subgranular layer SGL) and in the subventricular zone (SVZ), with migration of new cells occurring from these areas.

The SVZ is a layer of dividing cells that are found along the lateral walls of the lateral ventricle, and the new neurons are generated along this region and then form a chain of migrating neuroblasts which band together to form the rostral migrating stream, leading to the olfactory bulb (Doestch and Scharff, 2001; Doetsch, 2003b). Once in the olfactory bulb, the cells then differentiate into the granule and periglomerular cells, both of which are inhibitory neurons (Doestch and Scharff, 2001; Doetsch, 2003b). The SGL is the layer of cells which lies between the granular layer of the dentate gyrus and the hilus and in rodent models the new neurons generated migrate only a short distance to form the granular layer of the dentate gyrus (Doestch and Scharff, 2001; Doetsch, 2003b). Neurogenesis in the SGL occurs in foci closely associated with blood vessels. The foci contain SGL astrocytes, which have basal processes found under the blades of the dentate gyrus and also have an apical process, which penetrates into the granule-cell layer. There are also dividing immature neuronal cells, which have begun to express markers of neuronal differentiation, newly generated neurons and endothelial cells (Doetsch, 2003a). The observation that the generation of new functional neurons occurs throughout life in many primate species as well as other non-primate mammals suggests that this is a universal mammalian phenomenon (Gould *et al.*, 1999a). The neuronal stem cells differentiate into viable new neurons and are believed to be important in memory formation, learning and retention of the discriminatory sense of smell (Doetsch, 2003b; Doetsch and Hen, 2005). Adult neurogenesis in mammals is influenced by a range of hormones (see Richardson *et al.*, 2007).

In marsupials there is evidence of adult neurogenesis in the grey short-tailed opossum (*Monodelphis domestica*) and the fat-tailed dunnart (*Sminthopsis crassicaudata*), the two species that have been comprehensively studied thus far, confirming that the neurogenesis in the SVZ and SGL is a common mammalian trait (Gould *et al.*, 1999a). In the fat-tailed dunnart, tritiated thymidine labelling techniques have been used to identify dividing cells in the ventricular wall of the lateral ventricles, the olfactory bulbs and dentate gyrus (Harman *et al.*, 2003a). Similar to rodents, no new dividing cells appear in neocortical areas. There was, however, immunohistochemical staining revealing cells of glial divisionary origin (glial fibrillary acidic protein, GFAP, and polysialic acid-linked neural-cell adhesion molecule, PSA-NCAM) in thalamic areas, hypothalamic areas, rostral midbrain and piriform cortex, indicating that some activity occurs during adulthood, but this was not detected by using tritiated thymidine labelling (Harman *et al.*, 2003a). Similar results have been found for *Monodelphis domestica*, where labelled neurons have been found in similar regions, with a similar rate of decline over the course of a month after injection (Grabiec *et al.*, 2009). In both species neurogenesis still occurred in old animals, but cell counts are lower than for younger animals (Harman *et al.*, 2003a; Grabiec *et al.*, 2009).

The role of neurogenesis in adult learning and memory can be tested using behavioural strategies (Bonney and Wynne, 2003, 2004). Marsupials are capable of configural learning, where an animal can discriminate compound stimuli not only on the basis of the reinforcement outcomes associated with each of the elements of which they are made, but also by relying on the context or configuration in which a stimulus appears (Bonney and Wynne, 2003). They are also capable of discriminate learning, where an animal must discriminate between two objects (Bonney and Wynne, 2004). Because of their ability to perform these standard learning tasks, they make ideal models for following ageing processes in memory and learning.

One aspect of the role of the SVZ in adult neurogenesis is its association with the choroid plexus of the lateral ventricles (Doetsch, 2003a). The various choroid plexuses in the opossum appear in the same order as those in eutherian mammals, namely first in the fourth ventricle, then the lateral ventricles and finally in the third ventricle (Ek *et al.*, 2003). The opossum choroid plexuses also go through the same developmental ultrastructural stages as eutherian choroid plexuses, with the exception that they lack the glycogen-rich stage. The choroid plexuses in the opossum stop growing at P65 (Ek *et al.*, 2003), and, similar to eutherian mammals, act as a barrier to small molecules in the adult (Ek *et al.*, 2006). Because the opossum choroid plexuses are developmentally, structurally and functionally very similar to those in eutherian species, experiments can be carried out on early stages of brain development with minimal intervention compared to analysing equivalent stages of brain development in eutherian mammals.

These results for marsupials are similar to those seen for many eutherian mammals (Gould *et al.*, 1999a, 1999b), and parallel the conservative retention of some structures, such as the similarities between macaques and the dunnart, the quokka (*Setonix brachyurus*) and the northern quoll (*Dasyurus hallucatus*) in the visual cortical structures (Tyler *et al.*, 1998; Rosa *et al.*, 1999). Coupled with the comprehensive data on the development of the nervous system in marsupials (see Chapters 3 and 14 of this book), this means that marsupials can be used to follow the complete life history of neurogenic patterns. The easy access of marsupials from 'embryonic' stages of postnatal development in the pouch complemented by access to data generated from adults, especially from the small dunnart and opossum species, makes them an excellent laboratory tool for this field of study.

Marsupial models for regeneration are age dependent

Marsupials have long been used as models for neurogenic regeneration after trauma. Two species of opossum have been used as a model for spinal lesions for some time (*Didelphis virginiana*, Martin *et al.*, 2000; *Monodelphis domestica*, Lane *et al.*, 2007), with most studies focussing on the severance of the spinal cord at early postnatal stages. This is because regeneration and axonal growth across the lesion occurs in neonate marsupials, but is lost by adulthood (Lane *et al.*, 2007). Such studies follow regeneration ability during development and ageing of these species.

The use of neural stem cells in the developing and ageing visual system has also been investigated (Sakaguchi *et al.*,

2005). Retinal ganglion cells do not disappear during life in the quokka, although there does seem to be a change in the lipofuscin content of some cells (Harman and Moore, 1999). Lipofuscin is a chemically and morphologically polymorphous waste material, originating from a variety of intracellular structures, which accumulates in the lysosomes of most ageing eukaryotic cells. Lipofuscin is considered to be a marker of ageing because the amount of lipofuscin increases with age (Xu *et al.*, 2008). These results are similar to those found for the rat (Harman *et al.*, 2003b), and both studies challenge the tenet that neural degeneration always occurs throughout the mammalian nervous system (Harman *et al.*, 2003a, 2003b). Progenitor cells have been transplanted by intraocular injection into the eyes of recipients of different developing and adult ages, with the best morphologic differentiation and integration only observed after transplantation into the youngest (P5 to P10) host retinas three to four weeks post transplant. While grafted cells are capable of surviving in the older host retinas, little integration is observed (Sakaguchi *et al.*, 2005). The reasons for this lack of integration are yet to be explored. The dunnarts have a visual system most similar to the primates, with three cone pigments, whereas other non-primate mammals have only two (Cowing *et al.*, 2008). Thus, dunnarts will make excellent models for studying the processes of retinal graft integration; indeed they are more suitable than any other non-primate model because of similarities in their visual system, and also could supersede non-human primate models because of their relative ease of care in the laboratory compared to primates.

Marsupial models for the effects of stress hormones and sex hormones on the nervous system

Besides lesion and retinal studies, the marsupials make an ideal group on which to study the effects of stress and the nervous system. Two genera of the family Dasyuridae are possible models, *Antechinus* and *Sminthopsis*, allowing study of the sex hormones in relation to physiological changes associated with ageing. The synchronised cycle of *Antechinus* sp. is typified by the precise onset of the secretion of sex hormones, and in males this is followed by a physiologically significant rise in the stress hormone, cortisol (Barnett, 1973; Bradley *et al.*, 1980; McDonald *et al.*, 1981). Whereas plasma cortisol concentrations have been shown to rise transiently in females, their physiology remains relatively resistant to the effects of increased cortisol (Barnett, 1973; Barker *et al.*, 1978; McDonald *et al.*, 1981). Many components of the ageing process are expressed differently between male and female humans, and the role of sex hormones has been explored

frequently, but the precise roles of testosterone or oestrogen in ageing are still undecided (Ljubojevic *et al.*, 2004; Park *et al.*, 2004; Verzola *et al.*, 2004). Moreover, a stress component to the ageing process is hypothesised to interact with the sex-steroid differences between males and females in mammals (Kudielka and Kirschbaum, 2005), and this too can be readily explored in *Antechinus* sp. Thus, dasyurid marsupial biology may offer some insights into inter-relations between ageing and hormonal changes.

The complex interactions between brain development and ageing of the brain have been linked to the alterations in circulating sex steroid and cortisol concentrations. Elevated circulating cortisol and altered sex-steroid profiles are known to occur in neurodegenerative diseases such as Alzheimer's disease (AD; de Leon *et al.*, 1997; McEwen, 1999). These diseases are hypothesised to have an upregulated hypothalamic–pituitary–adrenal axis, a dysregulated hypothalamic–pituitary–gonadal axis, and dysregulated metabolic activities (Murialdo *et al.*, 2001; Tanzi *et al.*, 2004; Watson and Craft, 2004). A problem impeding investigation into the causes of AD has been the lack of suitable animal models (Stoothoff and Johnson, 2005). AD is characterised by neuronal death, oxidative damage, hyperphosphorylation and intra-cellular deposition of the microtubule-associated protein tau, and extracellular deposition of β-amyloid (Aβ), although mechanisms are unclear (Schroeder and Koo, 2005). Some animals naturally produce Aβ plaques and neurofibrillary tangles, including the lamprey *Petromyzon marinus*, the worm *Caenorhabditis elegans*, the fruit fly *Drosophila melanogaster*, the clawed toad *Xenopus* sp., chick embryos, kokanee salmon and tree shrews (Pawlik *et al.*, 1999; Apelt *et al.*, 2004; Götz *et al.*, 2004, 2007; Schwab *et al.*, 2004; Carrodeguas *et al.*, 2005; Lee *et al.*, 2005). Many transgenic mice have been created that produce Aβ, but the limitations of transgenic mice are that not all pathologies and behavioural defects are seen (Götz *et al.*, 2004, 2007; Schwab *et al.*, 2004; Lee *et al.*, 2005).

As a model for AD, mice have been genetically programmed to express genes that cause aspects of AD pathology. Because these genes are not naturally expressed in mice, transgenic models limit the ability to investigate the natural stimuli of AD pathology (Schwab *et al.*, 2004). In addition, mutant genes may function differently in transgenic mice than in the mammals that produce them naturally (Schwab *et al.*, 2004). The significant modification of normal physiology and behaviour by transgenic actions in mice has led some authors to suggest that transgenic mice are of limited use in studies on ageing and their related disorders (Carter *et al.*, 2002; Schwab *et al.*, 2004). Others caution the reliance on their use unless specific care is taken to

Figure 15.1 (a) Aligned nucleotide sequences of human and antechinus Aβ and flanking regions of APP (Sequences aligned with *ClustalW*). (b) Aligned protein sequences of mammalian Aβ and flanking regions of APP. (Sequences aligned with *ClustalW*). Residues are numbered from the β-secretase site. Residues conserved across species are marked (*); substitutions with very similar residues are marked (:); and similar residues with (.).

ensure that the phenotype is accurately determined for the focus of the study (Melrose *et al.*, 2006; Heng *et al.*, 2008). Aβ formation is not a part of the natural ageing process in rodents, although it is in non-human primates and many other non-primate mammals (Fiala, 2007). Indeed, it has been found that learning and memory deficits found in aged wild-type and mutant mice are unrelated to the formation of Aβ plaques (Gruart *et al.*, 2008). Moreover, the presence of Aβ may be part of the ageing process in many mammals, not specifically related to the onset of AD and inter-relation between ageing and Aβ is yet to be fully determined (Pawlik *et al.*, 1999; Stoothoff and Johnson, 2005).

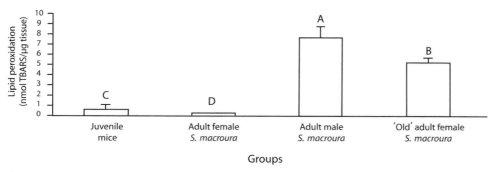

Figure 15.2 Lipid peroxidation in brains from *Antechinus* compared with mice and *Sminthopsis*. Different letters indicate significantly different from other groups ($P < 0.05$).

Marsupial mammals may give us insights into AD pathologies and the biology of ageing. We have demonstrated the presence of Aβ (McAllan *et al.*, 2006), and cloned and sequenced the Aβ portion of the APP gene in the marsupial shrew, *Antechinus stuartii* (Naylor, 2005, Figure 15.1). The derived amino-acid sequence has four residues, which differ from those in eutherian species. These residues may affect the processing of APP, and thereby shed light on the mechanism of Aβ production. This situation of naturally occurring amyloid in the brains of *Antechinus* is in strong contrast to mouse models of amyloidosis, where amyloidosis has to be artificially induced, usually by transgenesis.

Another aspect of AD pathology is the reactive-oxygen-species-mediated damage, which is usually manifested as lipid peroxidation, and which can be measured by reacting brain homogenates with thiobarbituric acid (Pratico *et al.*, 2000). We have used lipid peroxidation assays to measure the amount of thiobarbituric-acid-reactive substances (TBARS) in juvenile mice, adult female *S. macroura*, juvenile male *A. stuartii* and adult *A. stuartii* of both sexes. Our preliminary data showed significantly more lipid peroxidation in brains of male *Antechinus* than those of mice and *Sminthopsis* (Figure 15.2). Lipid peroxidation values from *Antechinus* are 10-fold higher than those from juvenile mouse and adult *Sminthopsis* controls (Naylor, 2005; Naylor *et al.*, 2008). Lipid peroxidation in brain homogenates from adult males is also significantly higher than that from adult female *Antechinus*. This data further strengthens the argument for using *Antechinus* as a model for AD.

It has also been found that stress can affect neurogenesis in the adult fat-tailed dunnart (Harman *et al.*, 2003a), where a relatively transient stressor was seen to significantly decrease neurogenesis. A study using the serotonin receptor 1a (5-HT1A) blocker buspirone, used in anti-anxiety treatment, significantly increased neurogenesis in the *M. domestica* in both young and aged animals (Grabiec *et al.*, 2009).

Thus, the potential of marsupials as animal models for the effects of stress on the nervous system seems significant and exciting.

Marsupial models for neurodegeneration

Marsupials can be used as natural models for neurodegenerative human diseases. The power of unmodified models is that natural site-directed mutagenesis or transgenesis has occurred. Four examples where marsupials could be used as models for neurodegenerative human diseases are: (i) Transthyretin familial amyloidotic polyneuropathy; (ii) Transthyretin senile systemic amyloidosis; (iii) naturally occurring cerebral amyloid and (iv) tau phosphorylation. The third example, naturally occurring cerebral amyloid, has been discussed previously, as it has a clear relationship (causal or otherwise) on the ageing of the brain and development of AD. Thus, exploitation of marsupial species may provide powerful insights into several neurodegenerative diseases in humans.

Transthyretin familial amyloidotic polyneuropathy

Transthyretin (TTR) is a homotetrameric protein in the blood and/or cerebrospinal fluid of vertebrates at some stage during the life cycle (see Richardson, 2007; Richardson *et al.*, 2007). TTR transports thyroid hormones via the blood and cerebrospinal fluid to target cells in the body and brain. In humans, TTR can change from being a soluble protein to forming insoluble aggregates (amyloid) that deposit on cell membranes, thereby disrupting normal cell function. There are two forms of TTR amyloid: familial amyloidotic polyneuropathy (FAP) and senile systemic amyloidosis (SSA). FAP is caused by a point mutation in TTR. TTR FAP is an autosomal hereditary polyneuropathy that initially manifests in nerves, but later is apparent in many organs, and more than 100 point mutations that result in its formation have been identified (Benson, 2009).

TTR (or its gene) has been sequenced from more than 20 species. The evolutionary changes in TTR are clustered around the N-terminal regions, whereas the point mutations in human TTR that result in disease (amyloidosis) are evenly spread throughout the molecule (see Richardson, 2007). In general, the amino-acid residues that occur as point mutations in human TTR and result in FAP are not found in that position in other species. One notable exception is at amino-acid position 68. In wild-type human TTR, residue 68 is Ile, but if it is mutated to Leu, TTR amyloidosis results (Almeida *et al.*, 1991). However, Leu is the normal amino acid in this position in several marsupial TTRs (see Richardson, 2007). These marsupials are apparently without symptoms of amyloidosis. Elucidation of the three-dimensional structure of a marsupial TTR with Leu68 would reveal how Leu is accommodated at position 68 in a marsupial TTR, but not in a human TTR. This could give valuable insights into a mechanism of TTR amyloid formation in FAP.

Transthyretin senile systemic amyloidosis

Senile systemic amyloidosis (SSA) involves wild-type TTR, which deposits in the heart. At least 20% of people over the age of 70 have SSA. The reason for wild-type TTR to form SSA is unknown. It is generally believed that TTR dissociates from tetramers into dimers, and from dimers into monomers before forming amyloid fibrils (Zhang and Kelly, 2003). Most studies have used abnormally low pH (e.g. pH 4), which is not physiologically realistic, to induce human TTR amyloid formation (e.g. Zhang and Kelly, 2003). However, we have demonstrated that under physiologically acidic (pH 6.5) conditions that cause the human TTR dimers to dissociate into monomers, tammar wallaby TTR dimers remain intact, thereby halting the progression that leads to amyloid formation (Altland and Richardson, 2009). Our preliminary data suggest that specific histidine residues are responsible for the stability of wallaby TTR compared to human TTR. Structural data on key residues in wallaby TTR could give insight into the mechanism of wild-type human TTR SSA amyloid formation.

Tau phosphorylation

Marsupial models can also be considered for other aspects of the development of AD besides Aβ involvement, i.e. that of tau phosphorylation. In the normal mature neuron, the promotion of the assembly and stability of microtubules is performed by microtubule-associated proteins (MAP), including the protein tau. The biological activity of tau in promoting assembly and stability of microtubules is regulated by its degree of phosphorylation. Hyper-phosphorylation of tau depresses microtubule assembly and the ability of tau to bind to microtubules (Wong *et al.*, 2002; Iqbal *et al.*, 2005; Lee *et al.*, 2005). In AD, tau is excessively phosphorylated, leading to aggregations in cell bodies and axons (neurofibrillary tangles), which contribute to neuronal death (Wong *et al.*, 2002; Lee *et al.*, 2005). Mammals using torpor can regulate the phosphorylation of tau (Härtig *et al.*, 2007). The reversibility of the formation and degradation of paired helical filament (PHF)-like tau observed during arousal from torpor occurs in hippocampal and cortical regions, which are affected in AD (Härtig *et al.*, 2007). In such cases, hyperphosphorylation of tau is not associated with neurofibrillary tangle formation. Many marsupial species use torpor as part of their life-history cycle of energy budgeting, and species include *Sminthopsis crassicaudata* and *S. macroura* (Geiser, 2004; Geiser *et al.*, 2006). If *Sminthopsis* spp. tau is phosphorylated whilst in torpor, and dephosphorylated during arousal, this naturally occurring reversible PHF-like phosphorylation of tau will be a model for studying the mechanism of tau phosphorylation in AD.

Where to from here?

We have demonstrated that marsupials offer the potential to act as significant and instructive models for specific neurological aspects of human ageing and disease. We have also listed the reasons why specific, appropriate animal models for ageing and disease, both marsupial and eutherian, should be used which could markedly enhance the potential outcomes. Besides further development of the studies outlined in the review, especially those of stress and ageing (see Pardon and Rattray, 2008), there are several potential areas of study in which work on marsupials could contribute.

In studies on ageing, the calcium-dependent proteases (calpains) have been implicated in degenerative processes in both muscles and neurons, and disturbances to calcium homeostasis are believed to be the result (Baudry *et al.*, 1986; Nixon 2003; Hajieva *et al.*, 2009). Over-activation of the proteolytic enzymes results in unwanted cleavage of key structural elements, which can also lead to generation of cytotoxic proteolytic fragments, and also promote inappropriate activation or inactivation of regulatory enzymes and signalling molecules (Nixon, 2003; Hinman and Abraham, 2007; Raynaud and Marcilhac, 2007; Hajieva *et al.*, 2009). Calpain activation has been implicated in such diseases as cataracts, neuronal apoptosis, arthritis, neurodegenerative disorders, including AD, and in the development of type II diabetes (Nixon, 2003; Raynaud and Marcilhac, 2007). Ageing bats show reduced calpain activity compared to mice and this has been hypothesised to be a causal factor in the longevity of bats compared to rodents (Baudry *et al.*, 1986). Thus the developmental advantages of marsupials such as *Sminthopsis* sp. and *Antechinus* sp. (see Table 15.1) will allow full exploration of the role of calpains throughout their life history. The roles of stress and changes in calpain activity can also be studied.

Other neurodegenerative changes that can occur include alterations in the white matter, programmed cell death, and also telomerase activity (Lossi *et al.*, 2005; Hinman and Abraham, 2007; Ricklefs, 2008; Gorbanova and Selanov, 2009). Studies on these topics are mostly limited to *in vitro* or rodent models, and the rodent models do not necessarily demonstrate similar changes to humans (Lossi *et al.*, 2005; Hinman and Abraham, 2007; Ricklefs, 2008; Gorbanova and Selanov, 2009). The potential of marsupials to fill some of these roles as models is considerable, and with the publication of the opossum genome this potential is increased, as has been already discovered (Santangelo *et al.*, 2007; Samollow, 2008). Clearly the use of marsupials in research on ageing and disease is an exciting and under-used frontier.

Part V

Atlases

High-resolution files of all brain atlases are available at www.cambridge.org/9780521519458.

16

Atlas of the brain of the stripe-faced dunnart
(*Sminthopsis macroura*)

K. W. S. Ashwell, B. M. McAllan and J. K. Mai

Introduction

The stripe-faced dunnart is a small dasyurid marsupial carnivore widespread throughout the arid and semi-arid regions of inland central and northern Australia. It derives its common name from the prominent line of dark hair running from between the eyes to the occipital region. The specific name refers to the (long) tail, which acts as a fat reservoir when food supplies are good, and is longer (by 1.25 times) than the head and body length combined. These dunnarts are naturally found in low shrubland and tussock grasslands and are probably strictly nocturnal. Both sexes reach adult body weights of 15 to 25 g. Stripe-faced dunnarts have the shortest gestation known for any marsupial (11 days). They can be easily maintained in colonies, with litter sizes of up to eight offspring (the maximum number of teats in the pouch).

Methods

Ethics, anaesthesia and perfusion

All experimental procedures were approved by the Animal Ethics Experimentation Committee of the University of Sydney, conform to NIH principles of laboratory animal care and were carried out according to the ethical guidelines of the National Health and Medical Research Council (Australia).

A mature female dunnart (body weight of 27.2 g) was deeply anaesthetised by intraperitoneal injection of sodium phenobarbital (35 mg/kg) and the thoracic cavity opened by paramedian incisions through the anterior rib cage. Transcardial perfusion was commenced with 0.9% saline for 10 minutes, followed by 4% paraformaldehyde in 0.1M phosphate buffer (pH 7.4) for 20 minutes. The carcass was allowed to rest for 1 hour before the brain was removed from the skull and weighed (0.392 g). The brain was embedded in paraffin and sectioned in the coronal plane at a thickness of 14 μm. Sections at intervals of 140 μm were stained with cresyl violet and coverslipped with DePeX. Note that some transverse stretching of D11 has slightly distorted that section.

Photomicrography and delineation

Sections at intervals of 280 μm were photographed at a magnification of 100× with the aid of a *.slide* photomicrographic system (*Olympus* tiling microscope) in the Department of Anatomy at the Heinrich Heine University in Düsseldorf, Germany. Images were photomerged and saved as *Adobe Photoshop CS2* files with sufficient resolution to retain fine cytoarchitectural detail to a scale of less than 2 μm.

Photoshop files were placed in *Adobe Illustrator CS2* and tracings made of tracts and nuclei. Professors George Paxinos and Charles Watson have kindly given permission to use the nomenclature and abbreviation systems which they have applied in their numerous atlases of placental mammal brains and these have been applied to all the atlases in this book with appropriate modifications specific to marsupial brains.

Each plate of the coronal sections through this non-stereotaxic atlas of the dunnart brain shows a scale in mm, but note that these measurements are of dehydrated brain. Fresh or frozen section material would be approximately 150% of the size depicted. The line diagram of each plate also depicts a small finder diagram in the top right-hand corner, which shows the position of the section and the distance in mm from the rostral tip of the olfactory bulb. These rostrocaudal distances also refer to a dehydrated brain and need to be corrected accordingly if one intends drawing comparisons with frozen material.

A common index of abbreviations used in the three atlases (Chapters 16 to 18) and many of the text chapter figures has been included after Chapter 18. Atlas plate files are available at www.cambridge.org/9780521519458.

D1 & D2

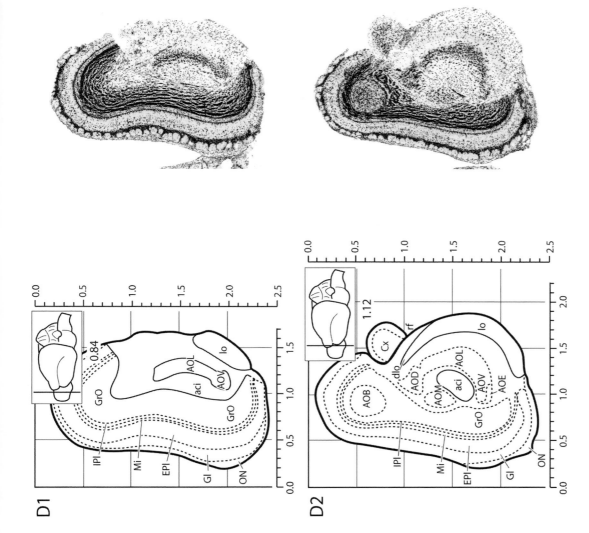

aci   ant. commissure, intrabulbar part
AOB   accessory olfactory bulb
AOD   anterior olfactory nu., dorsal
AOE   anterior olfactory nu., external
AOL   anterior olfactory nu., lateral
AOM anterior olfactory nu., medial
AOV   anterior olfactory nu., ventral
Cx   cerebral cortex
dlo   dorsal lateral olfactory tract
EPl   ext. plexiform layer olfactory bulb
Gl   glomerular layer olfactory bulb
GrO   granular cell layer olfactory bulb
IPl   internal plexiform layer olfactory bulb
lo   lateral olfactory tract
Mi   mitral cell layer olfactory bulb
ON   olfactory nerve layer
rf   rhinal fissure

D3 & D4

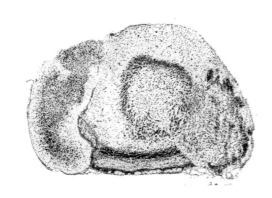

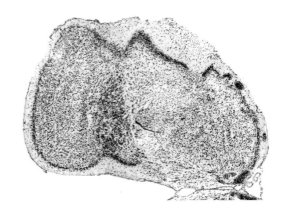

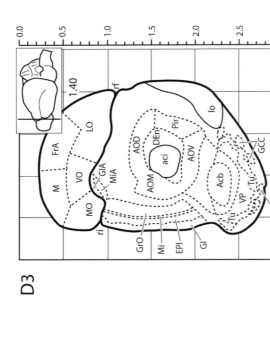

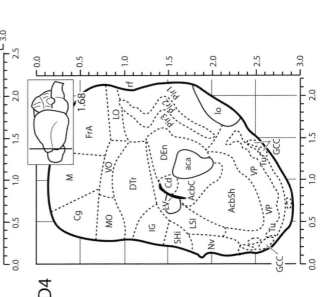

D3

D4

aca anterior commissure, anterior limb
Acb accumbens nucleus
AcbC accumbens nucleus, core
AcbSh accumbens nucleus, shell
aci anterior commissure, intrabulbar part
AOD anterior olfactory nucleus, dorsal
AOM anterior olfactory nucleus, medial
AOV anterior olfactory nucleus, ventral
Cd caudate nucleus
Cg cingulate cortex
DEn dorsal endopiriform nucleus
DTr dorsal transition zone (of cortex)
EPl external plexiform layer olfactory bulb
FrA frontal association cortex
GCC granule cell clusters
Gl glomerular layer olfactory bulb
GlA glomerular layer accessory olfactory bulb
GrO granular cell layer olfactory bulb
IG indusium griseum
LO lateral orbital cortex
lo lateral olfactory tract
LSI lateral septal nucleus, intermediate
LV lateral ventricle
M motor cortex (unspecified)
Mi mitral cell layer olfactory bulb
MiA mitral cell layer accessory olfactory bulb
MO medial orbital cortex
Nv navicular postolfactory nucleus
Pir piriform cortex
Pir1 piriform cortex, layer 1
Pir2 piriform cortex, layer 2
Pir3 piriform cortex, layer 3
rf rhinal fissure
ri rhinal incisure
SHi septohippocampal nucleus
Tu olfactory tubercle
VO ventral orbital cortex
VP ventral pallidum

D5

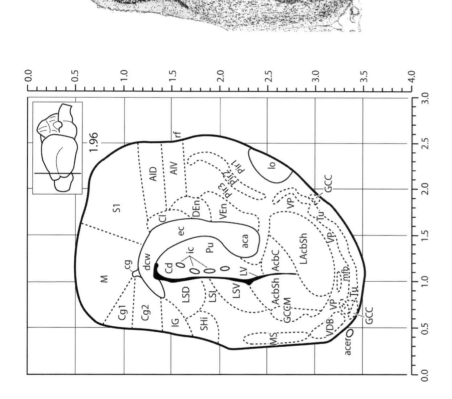

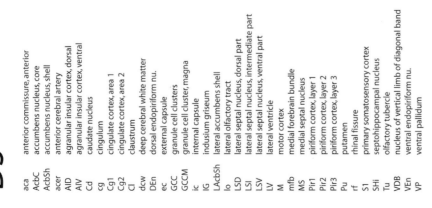

aca	anterior commissure, anterior
AcbC	accumbens nucleus, core
AcbSh	accumbens nucleus, shell
acer	anterior cerebral artery
AID	agranular insular cortex, dorsal
AIV	agranular insular cortex, ventral
Cd	caudate nucleus
cg	cingulum
Cg1	cingulate cortex, area 1
Cg2	cingulate cortex, area 2
Cl	claustrum
dcw	deep cerebral white matter
DEn	dorsal endopiriform nu.
ec	external capsule
GCC	granule cell clusters
GCCM	granule cell cluster, magna
ic	internal capsule
IG	indusium griseum
LAcbSh	lateral accumbens shell
lo	lateral olfactory tract
LSD	lateral septal nucleus, dorsal part
LSI	lateral septal nucleus, intermediate part
LSV	lateral septal nucleus, ventral part
LV	lateral ventricle
M	motor cortex
mfb	medial forebrain bundle
MS	medial septal nucleus
Pir1	piriform cortex, layer 1
Pir2	piriform cortex, layer 2
Pir3	piriform cortex, layer 3
Pu	putamen
rf	rhinal fissure
S1	primary somatosensory cortex
SHi	septohippocampal nucleus
Tu	olfactory tubercle
VDB	nucleus of vertical limb of diagonal band
VEn	ventral endopiriform nu.
VP	ventral pallidum

D6

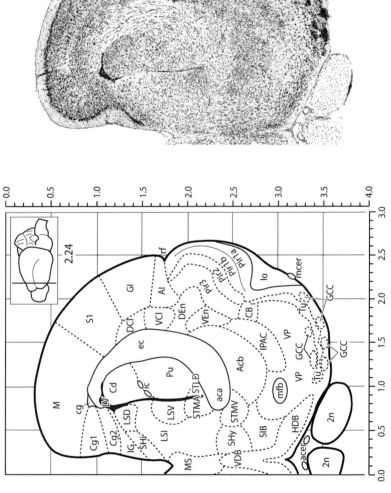

2.24

2n	optic nerve	DCl	dorsal part of claustrum
aca	anterior commissure, ant. part	DEn	dorsal endopiriform nu.
Acb	accumbens nucleus	ec	external capsule
acer	anterior cerebral artery	GCC	granule cell clusters
AI	agranular insular cortex	GI	granular insular cortex
CB	cell bridges of ventral striatum	HDB	nucleus horiz. limb diagonal band
Cd	caudate nucleus	ic	internal capsule
cg	cingulum	IG	indusium griseum
Cg1	cingulate cortex, area 1	IPAC	interstitial nu. post. limb ant. comm.
Cg2	cingulate cortex, area 2	lo	lateral olfactory tract

LSD	lateral septal nucleus, dorsal part	STMV	bed nu. stria terminalis, med. div., vent.
LSI	lateral septal nucleus, intermed.	Tu	olfactory tubercle
LSV	lateral septal nucleus, vent. part	VCl	ventral part of claustrum
LV	lateral ventricle	VDB	nucleus vertical limb diagonal band
M	motor cortex	VEn	ventral endopiriform nu.
mcer	middle cerebral artery	VP	ventral pallidum
mfb	medial forebrain bundle		
MS	medial septal nucleus		
Pir1a	piriform cortex, layer 1a		
Pir1b	piriform cortex, layer 1b		

Pir2	piriform cortex, layer 2	
Pir3	piriform cortex, layer 3	
Pu	putamen	
rf	rhinal fissure	
S1	primary somatosensory cortex	
SHi	septohippocampal nucleus	
SHy	septohypothalamic nucleus	
SIB	substantia innominata, basal	
STLD	bed nu. stria terminalis, lat. div., dorsal	
STMA	bed nu. stria terminalis, med. div., ant.	

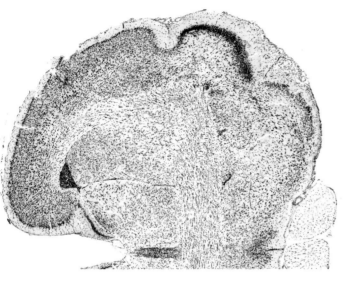

D7

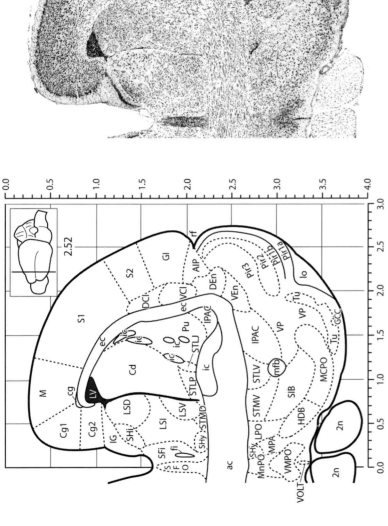

2n	optic nerve	
ac	anterior commissure	
AIP	agranular insular cx, posterior	
Cd	caudate nucleus	
cg	cingulum	
Cg1	cingulate cortex, area 1	
Cg2	cingulate cortex, area 2	
DCl	dorsal part of claustrum	
DEn	dorsal endopiriform nu.	
ec	external capsule	
fi	fimbria of the hippocampus	

GCC	granule cell cluster	
GI	granular insular cortex	
HDB	nu. horiz. limb diagonal band	
ic	internal capsule	
IG	indusium griseum	
IPAC	interstitial nu. post. limb ac	
Io	lateral olfactory tract	
LPO	lateral preoptic area	
LSD	lateral septal nu, dorsal part	
LSI	lateral septal nu., intermediate	
LSV	lateral septal nu, ventral part	

LV	lateral ventricle	
M	motor cortex	
MCPO	magnocellular preoptic nu.	
mfb	medial forebrain bundle	
MnPO	median preoptic nu.	
MPA	medial preoptic area	
Pir1a	piriform cortex, layer 1a	
Pir1b	piriform cortex, layer 1b	
Pir2	piriform cortex, layer 2	
Pir3	piriform cortex, layer 3	
Pu	putamen	

rf	rhinal fissure	
S1	primary somatosensory cx	
S2	secondary somatosensory cx	
SFi	septofimbrial nucleus	
SFO	subfornical organ	
SHi	septohippocampal nucleus	
SHy	septohypothalamic nucleus	
SIB	substantia innominata, basal	
STLJ	bed nu. st, lat. div., juxtacap.	
STLP	bed nu. st, lat. div., posterior	
STLV	bed nu. st, lat. div., ventral	

STMD	bed nu. st, med. div., dorsal	
STMV	bed nu. st, med. div., ventral	
Tu	olfactory tubercle	
VCl	ventral part of claustrum	
VEn	ventral endopiriform nu.	
VMPO	ventromedial preoptic nu.	
VOLT	vascular organ lamina terminalis	
VP	ventral pallidum	

D8

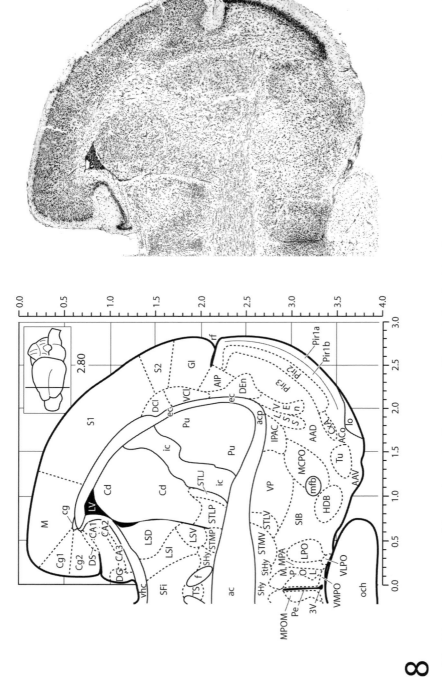

3V	3rd ventricle	CxA	cortex-amyg. transition	LSI	lateral septal nu., intermed.
AAD	anterior amyg. area, dorsal	DCl	dorsal part of claustrum	LSS	lateral stripe striatum
AAV	anterior amyg. area, ventral	DEn	dorsal endopiriform nu.	LSV	lateral septal nu., vent. part
ac	anterior commissure	DG	dentate gyrus	LV	lateral ventricle
ACo	ant. cortical amygdaloid area	DS	dorsal subiculum	M	motor cortex
acp	anterior comm., posterior limb	ec	external capsule	MCPO	magnocellular preoptic nu.
AIP	agranular insular area	f	fornix	mfb	medial forebrain bundle
CA1	field CA1 of hippocampus	GI	granular insular cortex	MPA	medial preoptic area
CA2	field CA2 of hippocampus	HDB	nu. horiz. limb diag. band	MPOL	medial preoptic nu., lateral
CA3	field CA3 of hippocampus	ic	internal capsule	MPOM	medial preoptic nu., medial
Cd	caudate nucleus	IPAC	interstitial nu. post. limb ac	och	optic chiasm
cg	cingulum	lo	lateral olfactory tract	Pe	perivent. hypothal. nu.
Cg1	cingulate cortex, area 1	LPO	lateral preoptic area	Pir1a	piriform cortex, layer 1a
Cg2	cingulate cortex, area 2	LSD	lateral septal nu., dorsal part	Pir1b	piriform cortex, layer 1b

Pir2	piriform cortex, layer 2	STMV	bed nu. st. med. div., ventral
Pir3	piriform cortex, layer 3	TS	triangular septal nu.
Pu	putamen	Tu	olfactory tubercle
rf	rhinal fissure	VCl	ventral part of claustrum
S1	primary somatosensory cx	VEn	ventral endopiriform nu.
S2	secondary somatosensory cx	vhc	ventral hippocamp. comm.
SFi	septofimbrial nu.	VLPO	ventrolat. preoptic nu.
SHy	septohypothal. nu.	VMPO	ventromed. preoptic nu.
SIB	substantia innominata, basal	VP	ventral pallidum
StHy	striohypothalamic nu.		
STLJ	bed nu. st, lat. div., juxtacap.		
STLP	bed nu. st. lat. div., posterior		
STLV	bed nu. st. lat. div., ventral		
STMP	bed nu. st. med. div., post.		

D9

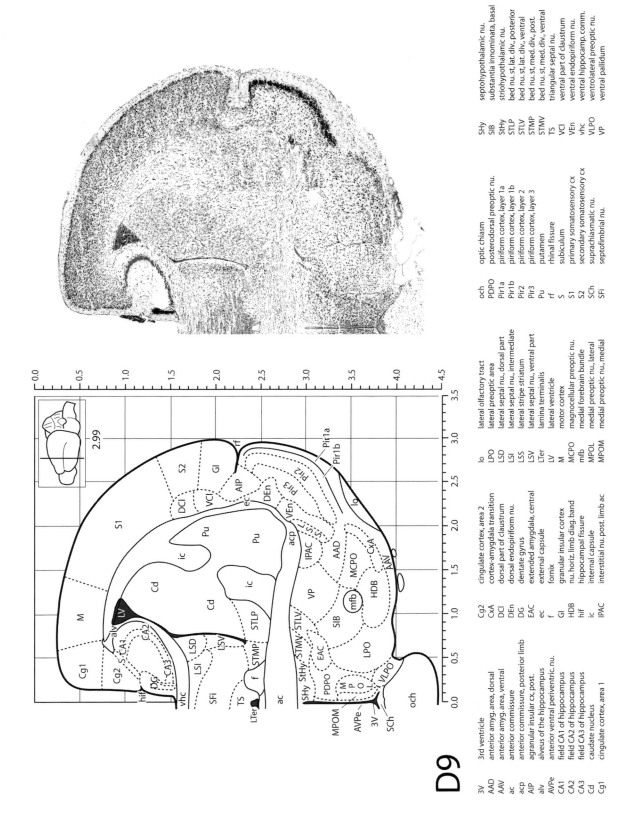

2.99

3V	3rd ventricle
AAD	anterior amyg. area, dorsal
AAV	anterior amyg. area, ventral
ac	anterior commissure
acp	anterior commissure, posterior limb
AIP	agranular insular cx, post.
alv	alveus of the hippocampus
AVPe	anterior ventral periventric. nu.
CA1	field CA1 of hippocampus
CA2	field CA2 of hippocampus
CA3	field CA3 of hippocampus
Cd	caudate nucleus
Cg1	cingulate cortex, area 1

Cg2	cingulate cortex, area 2
CxA	cortex-amygdala transition
DCl	dorsal part of claustrum
DEn	dorsal endopiriform nu.
DG	dentate gyrus
EAC	extended amygdala, central
ec	external capsule
f	fornix
GI	granular insular cortex
HDB	nu. horiz. limb diag. band
hif	hippocampal fissure
ic	internal capsule
IPAC	interstitial nu. post. limb ac

lo	lateral olfactory tract
LPO	lateral preoptic area
LSD	lateral septal nu., dorsal part
LSI	lateral septal nu., intermediate
LSS	lateral stripe striatum
LSV	lateral septal nu., ventral part
LTer	lamina terminalis
LV	lateral ventricle
M	motor cortex
MCPO	magnocellular preoptic nu.
mfb	medial forebrain bundle
MPOL	medial preoptic nu., lateral
MPOM	medial preoptic nu., medial

och	optic chiasm
PDPO	posterodorsal preoptic nu.
Pir1a	piriform cortex, layer 1a
Pir1b	piriform cortex, layer 1b
Pir2	piriform cortex, layer 2
Pir3	piriform cortex, layer 3
Pu	putamen
rf	rhinal fissure
S	subiculum
S1	primary somatosensory cx
S2	secondary somatosensory cx
SCh	suprachiasmatic nu.
SFi	septofimbrial nu.

SHy	septohypothalamic nu.
SIB	substantia innominata, basal
StHy	striohypothalamic nu.
STLP	bed nu. st, lat. div., posterior
STLV	bed nu. st, lat. div., ventral
STMP	bed nu. st, med. div., post.
STMV	bed nu. st, med. div., ventral
TS	triangular septal nu.
VCl	ventral part of claustrum
VEn	ventral endopiriform nu.
vhc	ventral hippocamp. comm.
VLPO	ventrolateral preoptic nu.
VP	ventral pallidum

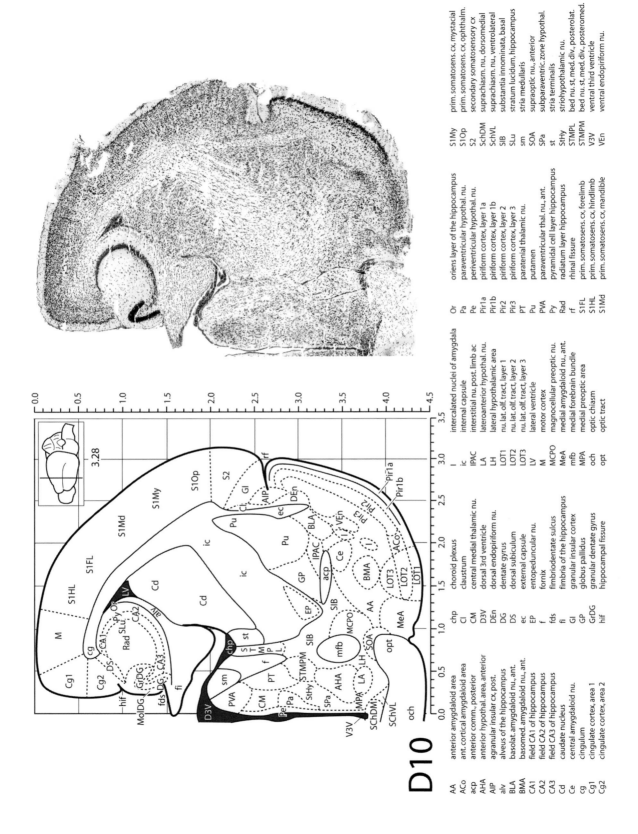

D10

AA	anterior amygdaloid area	I	intercalated nuclei of amygdala	Or	oriens layer of the hippocampus	S1My	prim. somatosens. cx, mystacial
ACo	ant. cortical amygdaloid area	ic	internal capsule	Pa	paraventricular hypothal. nu.	S1Op	prim. somatosens. cx, ophthalm.
acp	anterior comm., posterior	IPAC	interstitial nu. post. limb ac	Pe	periventricular hypothal. nu.	S2	secondary somatosensory cx
AHA	anteroanterior hypothal. area, anterior	LA	lateroanterior hypothal. nu.	Pir1a	piriform cortex, layer 1a	SchDM	suprachiasm. nu., dorsomedial
AIP	agranular insular cx, post.	LH	lateral hypothalamic area	Pir1b	piriform cortex, layer 1b	SchVL	suprachiasm. nu., ventrolateral
alv	alveus of the hippocampus	LOT1	nu. lat. olf. tract, layer 1	Pir2	piriform cortex, layer 2	SIB	substantia innominata, basal
BLA	basolat. amygdaloid nu., ant.	LOT2	nu. lat. olf. tract, layer 2	Pir3	piriform cortex, layer 3	SLu	stratum lucidum, hippocampus
BMA	basomed. amygdaloid nu., ant.	LOT3	nu. lat. olf. tract, layer 3	PT	paratenial thalamic nu.	sm	stria medullaris
CA1	field CA1 of hippocampus	LV	lateral ventricle	Pu	putamen	SOA	supraoptic nu., anterior
CA2	field CA2 of hippocampus	M	motor cortex	PVA	paraventricular thal. nu., ant.	SPa	subparaventric. zone hypothal.
CA3	field CA3 of hippocampus	MCPO	magnocellular preoptic nu.	Py	pyramidal cell layer hippocampus	st	stria terminalis
Cd	caudate nucleus	MeA	medial amygdaloid nu., ant.	Rad	radiatum layer hippocampus	StHy	striohypothalamic nu.
Ce	central amygdaloid nu.	mfb	medial forebrain bundle	rf	rhinal fissure	STMPL	bed nu. st, med. div., posterolat.
cg	cingulum	MPA	medial preoptic area	S1FL	prim. somatosens. cx, forelimb	STMPM	bed nu. st, med. div., posteromed.
Cg1	cingulate cortex, area 1	och	optic chiasm	S1HL	prim. somatosens. cx, hindlimb	V3V	ventral third ventricle
Cg2	cingulate cortex, area 2	opt	optic tract	S1Md	prim. somatosens. cx, mandible	VEn	ventral endopiriform nu.
chp	choroid plexus						
Cl	claustrum						
CM	central medial thalamic nu.						
D3V	dorsal 3rd ventricle						
DEn	dorsal endopiriform nu.						
DG	dentate gyrus						
DS	dorsal subiculum						
ec	external capsule						
EP	entopeduncular nu.						
f	fornix						
fds	fimbriodentate sulcus						
fi	fimbria of the hippocampus						
GI	granular insular cortex						
GP	globus pallidus						
GrDG	granular dentate gyrus						
hif	hippocampal fissure						

D11

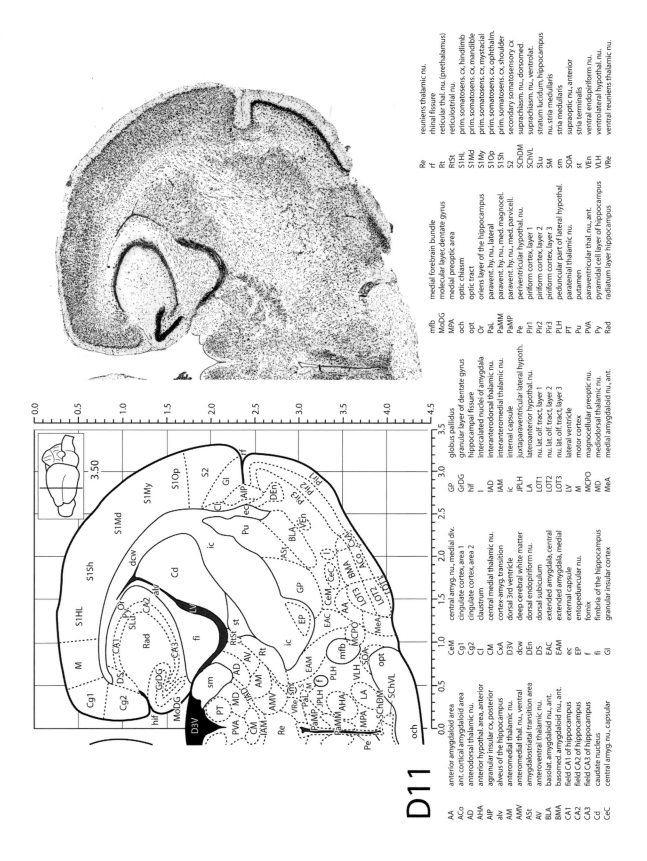

3.50

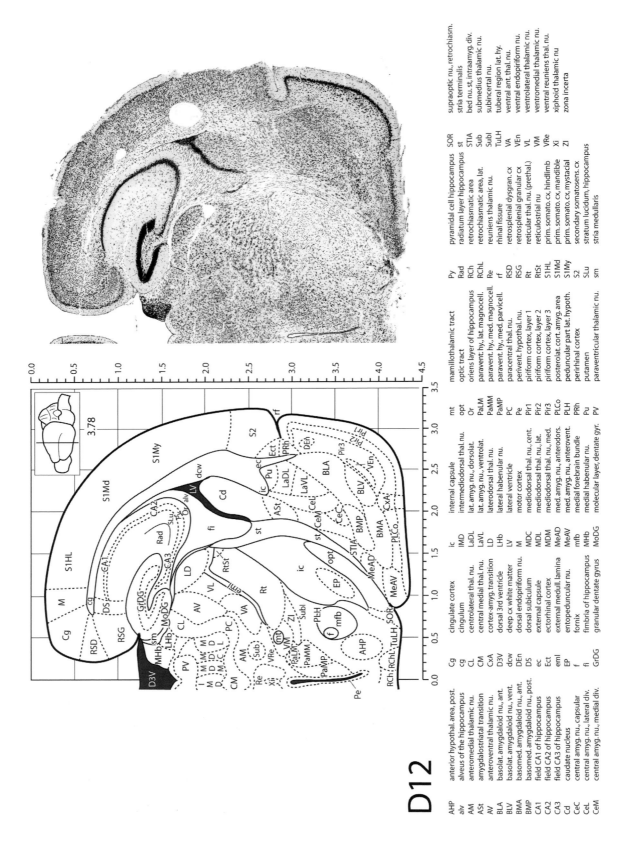

D12

AHP anterior hypothal. area, post.
alv alveus of the hippocampus
AM anteromedial thalamic nu.
ASt amygdalostriatal transition
AV anteroventral thalamic nu.
BLA basolat. amygdaloid nu., ant.
BLV basolat. amygdaloid nu., vent.
BMA basomed. amygdaloid nu., ant.
BMP basomed. amygdaloid nu., post.
CA1 field CA1 of hippocampus
CA2 field CA2 of hippocampus
CA3 field CA3 of hippocampus
Cd caudate nucleus
CeC central amyg. nu., capsular
CeL central amyg. nu., lateral div.
CeM central amyg. nu., medial div.

Cg cingulate cortex
cg cingulum
CL centrolateral thal. nu.
CM central medial thal. nu.
CxA cortex-amyg. transition
D3V dorsal 3rd ventricle
dcw deep cx white matter
DEn dorsal endopiriform nu.
DS dorsal subiculum
ec external capsule
Ect ectorhinal cortex
eml external medull. lamina
EP entopeduncular nu.
f fornix
fi fimbria of hippocampus
GrDG granular dentate gyrus

ic internal capsule
IMD intermediodorsal thal. nu.
LaDL lat. amyg. nu., dorsolat.
LaVL lat. amyg. nu., ventrolat.
LD laterodorsal thal. nu.
LHb lateral habenular nu.
LV lateral ventricle
M motor cortex
MDC mediodorsal thal. nu., cent.
MDL mediodorsal thal. nu., lat.
MDM mediodorsal thal. nu., med.
MeAD med. amyg. nu., anterodors.
MeAV med. amyg. nu., anterovent.
mfb medial forebrain bundle
MHb medial habenular nu.
MoDG molecular layer, dentate gyr.

mt mamillothalamic tract
opt optic tract
Or oriens layer of hippocampus
PaLM paravent. hy., lat. magnocell.
PaMM paravent. hy., med. magnocell.
PaMP paravent. hy., med. parvicell.
PC paracentral thal. nu.
Pe perivent. hypothal. nu.
Pir1 piriform cortex, layer 1
Pir2 piriform cortex, layer 2
Pir3 piriform cortex, layer 3
PLCo posterolat. cort. amyg. area
PLH peduncular part lat. hypoth.
PRh perirhinal cortex
Pu putamen
PV paraventricular thalamic nu.

Py pyramidal cell hippocampus
Rad radiatum layer hippocampus
RCh retrochiasmatic area
RChL retrochiasmatic area, lat.
Re reuniens thalamic nu.
rf rhinal fissure
RSD retrosplenial dysgran. cx
RSG retrosplenial granular cx
Rt reticular thal. nu. (prethal.)
RtSt reticulostrial nu
S1HL prim. somato. cx, hindlimb
S1Md prim. somato. cx, mandible
S1My prim. somato. cx, mystacial
S2 secondary somatosens. cx
SLu stratum lucidum, hippocampus
sm stria medullaris

SOR supraoptic nu., retrochiasm.
st stria terminalis
STIA bed nu. st, intraamyg. div.
Sub submedius thalamic nu.
Subl subincertal nu.
TuLH tuberal region lat. hy.
VA ventral ant. thal. nu.
VEn ventral endopiriform nu.
VL ventrolateral thalamic nu.
VM ventromedial thalamic nu.
VRe ventral reuniens thal. nu.
Xi xiphoid thalamic nu
ZI zona incerta

D13

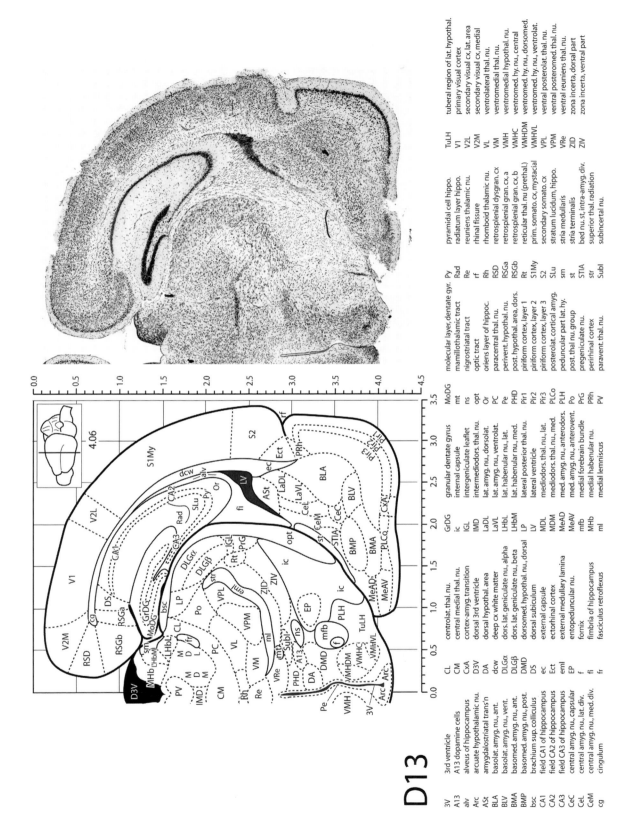

4.06

D14

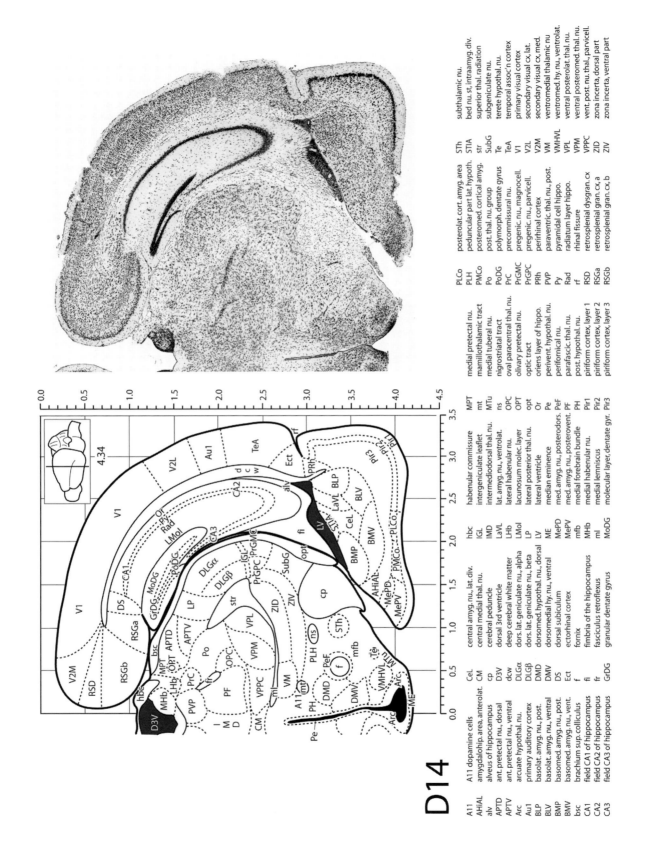

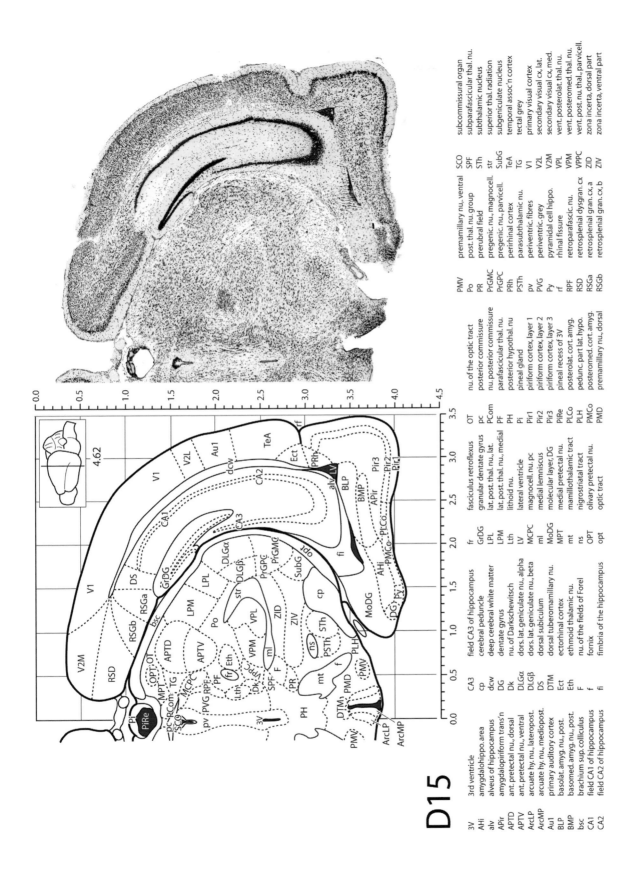

D15

3V — 3rd ventricle
AHi — amygdalohippo. area
alv — alveus of hippocampus
APir — amygdalopiriform trans'n
APTD — ant. pretectal nu., dorsal
APTV — ant. pretectal nu., ventral
ArcLP — arcuate hy. nu., lateropost.
ArcMP — arcuate hy. nu., mediopost.
Au1 — primary auditory cortex
BLP — basolat. amyg. nu., post.
BMP — basomed. amyg. nu., post.
bsc — brachium sup. colliculus
CA1 — field CA1 of hippocampus
CA2 — field CA2 of hippocampus

CA3 — field CA3 of hippocampus
cp — cerebral peduncle
dcw — deep cerebral white matter
DG — dentate gyrus
Dk — nu. of Darkschewitsch
DLGα — dors. lat. geniculate nu., alpha
DLGβ — dors. lat. geniculate nu., beta
DS — dorsal subiculum
DTM — dorsal tuberomamillary nu.
Ect — ectorhinal cortex
Eth — ethmoid thalamic nu.
F — nu. of the fields of Forel
f — fornix
fi — fimbria of the hippocampus

fr — fasciculus retroflexus
GrDG — granular dentate gyrus
LPL — lat. post. thal. nu., lat.
LPM — lat. post. thal. nu., medial
Lth — lithoid nu.
LV — lateral ventricle
MCPC — magnocell. nu. pc
ml — medial lemniscus
MoDG — molecular layer, DG
MPT — medial pretectal nu.
mt — mamillothalamic tract
ns — nigrostriatal tract
OPT — olivary pretectal nu.
opt — optic tract

OT — nu. of the optic tract
pc — posterior commissure
PCom — nu. posterior commissure
PF — parafascicular thal. nu.
PH — posterior hypothal. nu
Pi — pineal gland
Pir1 — piriform cortex, layer 1
Pir2 — piriform cortex, layer 2
Pir3 — piriform cortex, layer 3
PiRe — pineal recess of 3V
PLCo — posterolat. cort. amyg.
PLH — peduncl. part lat. hypo.
PMCo — posteromed. cort. amyg.
PMD — premamillary nu., dorsal

PMV — premamillary nu., ventral
Po — post. thal. nu. group
PR — prerubral field
PrGMC — pregenic. nu., magnocell.
PrGPC — pregenic. nu., parvicell.
PRh — perirhinal cortex
PSTh — parasubthalamic nu.
pv — periventric. fibres
PVG — periventric. grey
Py — pyramidal cell hippo.
rf — rhinal fissure
RPF — retroparafascic. nu.
RSD — retrosplenial dysgran. cx
RSGa — retrosplenial gran. cx, a
RSGb — retrosplenial gran. cx, b

SCO — subcommissural organ
SPF — subparafascicular thal. nu.
STh — subthalamic nucleus
str — superior thal. radiation
SubG — subgeniculate nucleus
TeA — temporal assoc'n cortex
TG — tectal grey
V1 — primary visual cortex
V2L — secondary visual cx. lat.
V2M — secondary visual cx, med.
VPL — vent. posterolat. thal. nu.
VPM — vent. posteromed. thal. nu.
VPPC — vent. post. nu. thal., parvicell.
ZID — zona incerta, dorsal part
ZIV — zona incerta, ventral part

D16

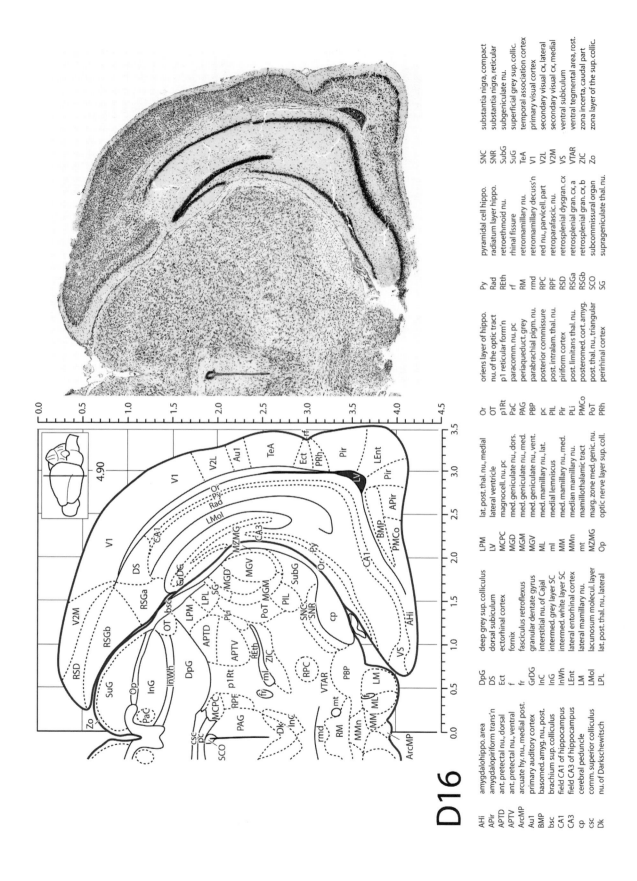

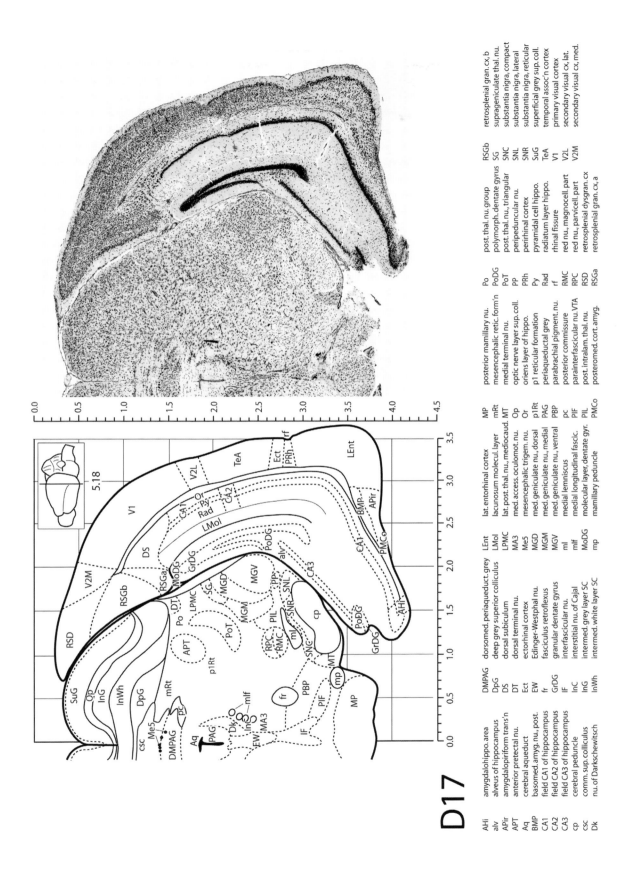

0.0
0.5
1.0
1.5
2.0
2.5
3.0
3.5
4.0
4.5

5.18

D17

AHi	amygdalohippo. area
alv	alveus of hippocampus
APir	amygdalopiriform trans'n
APT	anterior pretectal nu.
Aq	cerebral aqueduct
BMP	basomed.amyg.nu, post.
CA1	field CA1 of hippocampus
CA2	field CA2 of hippocampus
CA3	field CA3 of hippocampus
cp	cerebral peduncle
csc	comm.sup.colliculus
Dk	nu. of Darkschewitsch

DMPAG	dorsomed.periaqueduct.grey
DpG	deep grey superior colliculus
DS	dorsal subiculum
DT	dorsal terminal nu.
Ect	ectorhinal cortex
EW	Edinger-Westphal nu.
fr	fasciculus retroflexus
GrDG	granular dentate gyrus
IF	interfascicular nu.
InC	interstitial nu. of Cajal
InG	intermed. grey layer SC
InWh	intermed. white layer SC

LEnt	lat. entorhinal cortex
LMol	lacunosum molecul. layer
LPMC	lat.post.thal.nu, mediocaud.
MA3	med.access. oculomot. nu.
Me5	mesencephalic trigem. nu.
MGD	med. geniculate nu, dorsal
MGM	med. geniculate nu, medial
MGV	med. geniculate nu, ventral
ml	medial lemniscus
mlf	medial longitudinal fascic.
MoDG	molecular layer, dentate gyr.
mp	mamillary peduncle

MP	
mRt	mesencephalic retic. form'n
MT	medial terminal nu.
Op	optic nerve layer sup. coll.
Or	oriens layer of hippo.
p1Rt	p1 reticular formation
PAG	periaqueductal grey
PBP	parabrachial pigment. nu.
pc	posterior commissure
PIF	parainterfascicular nu.VTA
PIL	post.intralam.thal.nu.
PMCo	posteromed. cort.amyg.

Po	posterior mamillary nu.
PoDG	polymorph. dentate gyrus
PoT	post. thal. nu., triangular
PP	peripeduncular nu.
PRh	perirhinal cortex
Py	pyramidal cell hippo.
Rad	radiatum layer hippo.
rf	rhinal fissure
RMC	red nu, magnocell. part
RPC	red nu, parvicell. part
RSD	retrosplenial dysgran. cx
RSGa	retrosplenial gran. cx, a

RSGb	retrosplenial gran.cx, b
SG	suprageniculate thal. nu.
SNC	substantia nigra, compact
SNL	substantia nigra, lateral
SNR	substantia nigra, reticular
SuG	superficial grey sup. coll.
TeA	temporal assoc'n cortex
V1	primary visual cortex
V2L	secondary visual cx, lat.
V2M	secondary visual cx, med.

D18

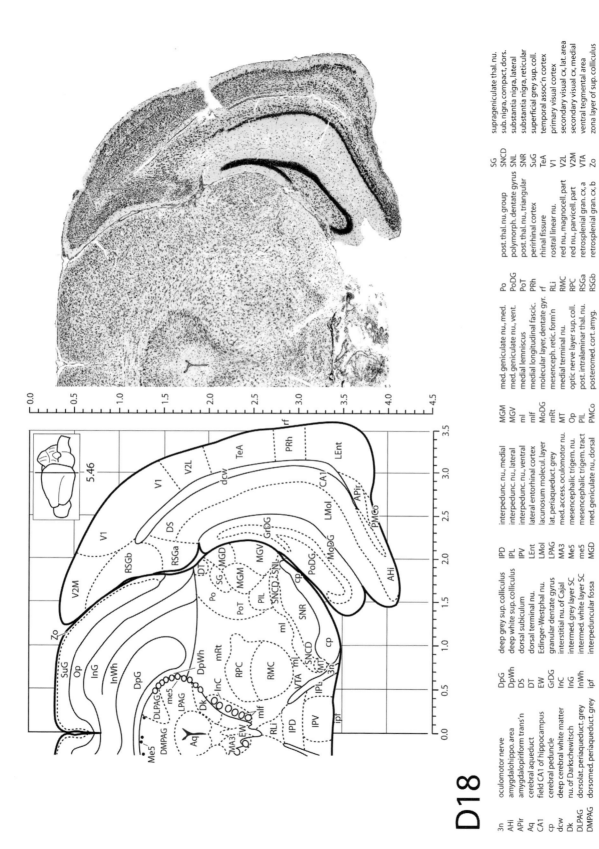

3n	oculomotor nerve	DpG	deep grey sup. colliculus
AHi	amygdalohippo. area	DpWh	deep white sup. colliculus
APir	amygdalopiriform trans'n	DS	dorsal subiculum
Aq	cerebral aqueduct	DT	dorsal terminal nu.
CA1	field CA1 of hippocampus	EW	Edinger-Westphal nu.
cp	cerebral peduncle	GrDG	granular dentate gyrus
dcw	deep cerebral white matter	InC	interstitial nu. of Cajal
Dk	nu. of Darkschewitsch	InG	intermed. grey layer SC
DLPAG	dorsolat. periaqueduct. grey	InWh	intermed white layer SC
DMPAG	dorsomed. periaqueduct. grey	ipf	interpeduncular fossa

IPD	interpedunc. nu., dorsal	MGM	med. geniculate nu., med.
IPL	interpedunc. nu., lateral	MGV	med. geniculate nu., vent.
IPV	interpedunc. nu., ventral	ml	medial lemniscus
LEnt	lateral entorhinal cortex	mlf	medial longitudinal fascic.
LMol	lacunosum molecul. layer	MoDG	molecular layer, dentate gyr.
LPAG	lat. periaqueduct. grey	mRt	mesenceph. retic. form'n
MA3	med. access. oculomotor nu.	MT	medial terminal nu.
Me5	mesencephalic trigem. nu.	Op	optic nerve layer sup. coll.
me5	mesencephalic trigem. tract	PIL	post. intralaminar thal. nu.
MGD	med. geniculate nu., dorsal	PMCo	posteromed. cort. amyg.

Po	post. thal. nu. group	SG	suprageniculate thal. nu.
PoDG	polymorph. dentate gyrus	SNCD	sub. nigra, compact, dors.
PoT	post. thal. nu., triangular	SNL	substantia nigra, lateral
PRh	perirhinal cortex	SNR	substantia nigra, reticular
rf	rhinal fissure	SuG	superficial grey sup. coll.
RLi	rostral linear nu.	TeA	temporal assoc'n cortex
RMC	red nu., magnocell. part	V1	primary visual cortex
RPC	red nu., parvicell. part	V2L	secondary visual cx, lat. area
RSGa	retrosplenial gran.cx, a	V2M	secondary visual cx, medial
RSGb	retrosplenial gran. cx, b	VTA	ventral tegmental area
		Zo	zona layer of sup. colliculus

D19

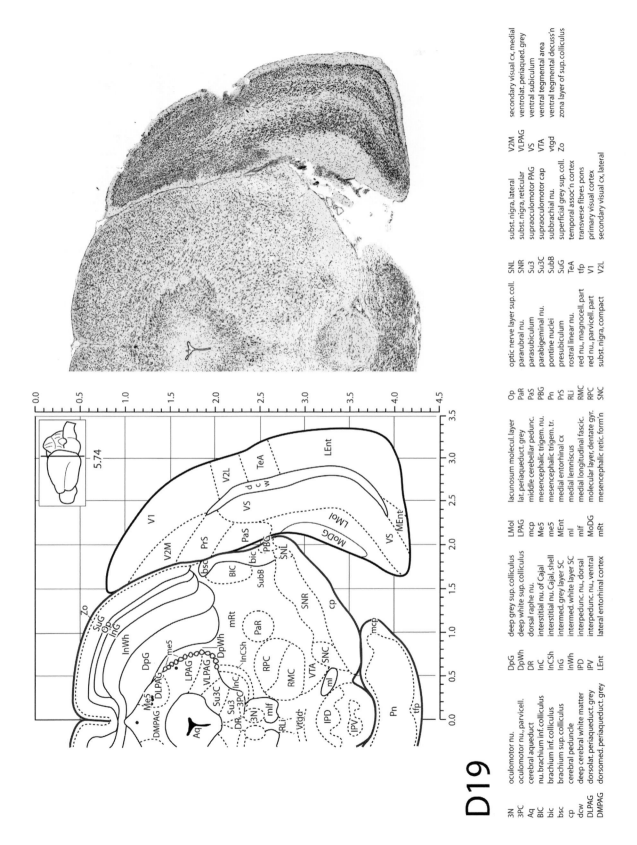

3N	3PC	oculomotor nu.
3PC		oculomotor nu., parvicell.
Aq		cerebral aqueduct
BIC		nu. brachium inf. colliculus
bic		brachium inf. colliculus
bsc		brachium sup. colliculus
cp		cerebral peduncle
dcw		deep cerebral white matter
DLPAG		dorsolat. periaqueduct. grey
DMPAG		dorsomed. periaqueduct. grey

DpG	deep grey sup. colliculus
DpWh	deep white sup. colliculus
DR	dorsal raphe nu.
InC	interstitial nu. of Cajal
InCSh	interstitial nu. Cajal, shell
InG	intermed. grey layer SC
InWh	intermed. white layer SC
IPD	interpedunc. nu., dorsal
IPV	interpedunc. nu., ventral
LEnt	lateral entorhinal cortex

LMol	lacunosum molecul. layer
LPAG	lat. periaqueduct. grey
mcp	middle cerebellar pedunc.
Me5	mesencephalic trigem. nu.
me5	mesencephalic trigem. tr.
MEnt	medial entorhinal cx
ml	medial lemniscus
mlf	medial longitudinal fascic.
MoDG	molecular layer, dentate gyr.
mRt	mesencephalic retic. form'n

Op	optic nerve layer sup. coll.
PaR	pararubral nu.
PaS	parasubiculum
PBG	parabigeminal nu.
Pn	pontine nuclei
PrS	presubiculum
RLi	rostral linear nu.
RMC	red nu., magnocell. part
RPC	red nu., parvicell. part
SNC	subst. nigra, compact

SNL	subst. nigra, lateral
SNR	subst. nigra, reticular
Su3	supraoculomotor PAG
Su3C	supraoculomotor cap
SubB	subbrachial nu.
SuG	superficial grey sup. coll.
TeA	temporal assoc'n cortex
tfp	transverse fibres pons
V1	primary visual cortex
V2L	secondary visual cx, lateral

V2M	secondary visual cx, medial
VLPAG	ventrolat. periaqued. grey
VS	ventral subiculum
VTA	ventral tegmental area
vtgd	ventral tegmental decuss'n
Zo	zona layer of sup. colliculus

D20

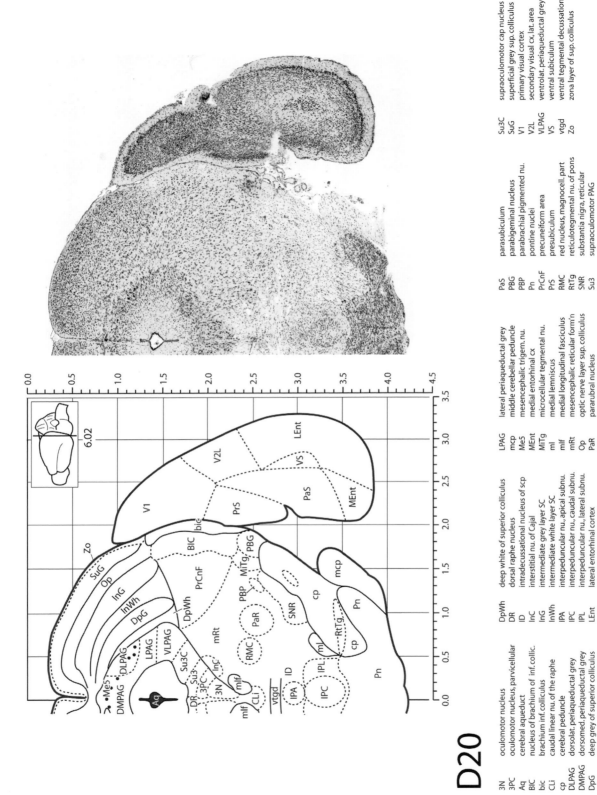

3N	oculomotor nucleus	DpWh	deep white of superior colliculus
3PC	oculomotor nucleus, parvicellular	DR	dorsal raphe nucleus
Aq	cerebral aqueduct	ID	intradecussational nucleus of scp
BIC	nucleus of brachium of inf. collic.	InC	interstitial nu. of Cajal
bic	brachium inf. colliculus	InG	intermediate grey layer SC
CLi	caudal linear nu. of the raphe	InWh	intermediate white layer SC
cp	cerebral peduncle	IPA	interpeduncular nu., apical subnu.
DLPAG	dorsolat. periaqueductal grey	IPC	interpeduncular nu., caudal subnu.
DMPAG	dorsomed. periaqueductal grey	IPL	interpeduncular nu., lateral subnu.
DpG	deep grey of superior colliculus	LEnt	lateral entorhinal cortex

LPAG	lateral periaqueductal grey	PaS	parasubiculum
mcp	middle cerebellar peduncle	PBG	parabigeminal nucleus
Me5	mesencephalic trigem. nu.	PBP	parabrachial pigmented nu.
MEnt	medial entorhinal cx	Pn	pontine nuclei
MiTg	microcellular tegmental nu.	PrCnF	precuneiform area
ml	medial lemniscus	PrS	presubiculum
mlf	medial longitudinal fasciculus	RMC	red nucleus, magnocell. part
mRt	mesencephalic reticular form'n	RtTg	reticulotegmental nu. of pons
Op	optic nerve layer sup. colliculus	SNR	substantia nigra, reticular
PaR	pararubral nucleus	Su3	supraoculomotor PAG

Su3C	supraoculomotor cap nucleus
SuG	superficial grey sup. colliculus
V1	primary visual cortex
V2L	secondary visual cx, lat. area
VLPAG	ventrolat. periaqueductal grey
VS	ventral subiculum
vtgd	ventral tegmental decussation
Zo	zona layer of sup. colliculus

D21

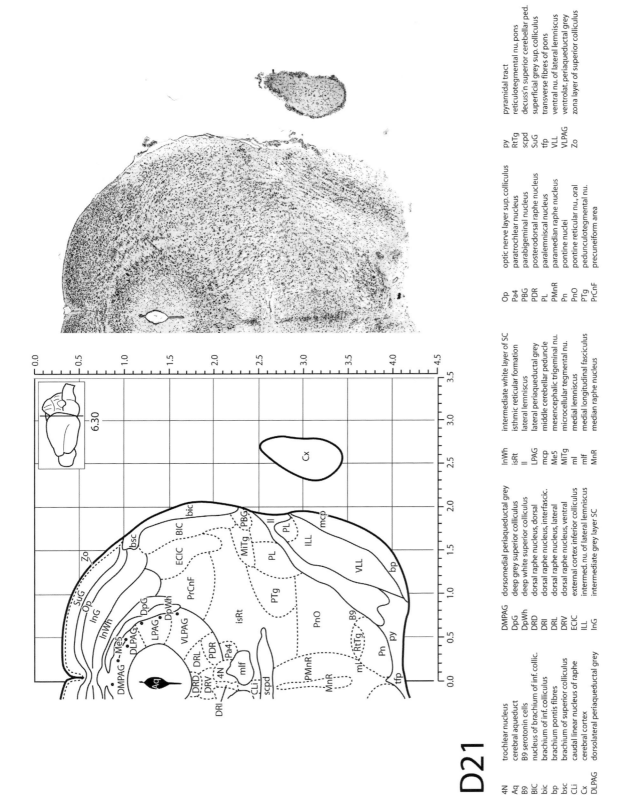

4N	trochlear nucleus	DMPAG	dorsomedial periaqueductal grey	InWh	intermediate white layer of SC
Aq	cerebral aqueduct	DpG	deep grey superior colliculus	isRt	isthmic reticular formation
B9	B9 serotonin cells	DpWh	deep white superior colliculus	ll	lateral lemniscus
BIC	nucleus of brachium of inf. collic.	DRD	dorsal raphe nucleus, dorsal	LPAG	lateral periaqueductal grey
bic	brachium of inf. colliculus	DRI	dorsal raphe nucleus, interfascic.	mcp	middle cerebellar peduncle
bp	brachium pontis fibres	DRL	dorsal raphe nucleus, lateral	Me5	mesencephalic trigeminal nu.
bsc	brachium of superior colliculus	DRV	dorsal raphe nucleus, ventral	MiTg	microcellular tegmental nu.
CLi	caudal linear nucleus of raphe	ECIC	external cortex inferior colliculus	ml	medial lemniscus
Cx	cerebral cortex	ILL	intermed. nu. of lateral lemniscus SC	mlf	medial longitudinal fasciculus
DLPAG	dorsolateral periaqueductal grey	InG	intermediate grey layer SC	MnR	median raphe nucleus

Op	optic nerve layer sup. colliculus	py	pyramidal tract
Pa4	paratrochlear nucleus	RtTg	reticulotegmental nu. pons
PBG	parabigeminal nucleus	scpd	decuss'n superior cerebellar ped.
PDR	posterodorsal raphe nucleus	SuG	superficial grey sup. colliculus
PL	paralemniscal nucleus	tfp	transverse fibres of pons
PMnR	paramedian raphe nucleus	VLL	ventral nu. of lateral lemniscus
Pn	pontine nuclei	VLPAG	ventrolat. periaqueductal grey
PnO	pontine reticular nu., oral	Zo	zona layer of superior colliculus
PTg	pedunculotegmental nu.		
PrCnF	precuneiform area		

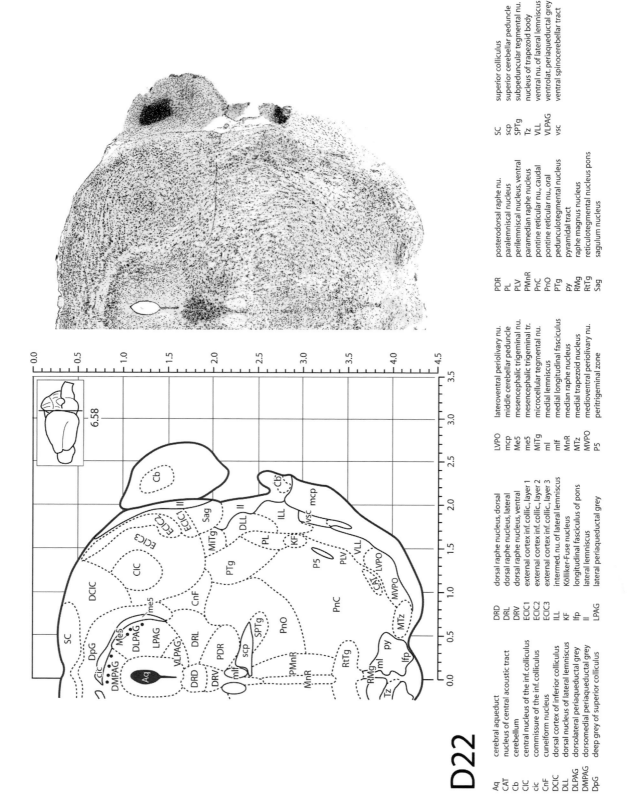

D22

6.58

Aq	cerebral aqueduct	
CAT	nucleus of central acoustic tract	
Cb	cerebellum	
CIC	central nucleus of the inf. colliculus	
cic	commissure of the inf. colliculus	
CnF	cuneiform nucleus	
DCIC	dorsal cortex of inferior colliculus	
DLL	dorsal nucleus of lateral lemniscus	
DLPAG	dorsolateral periaqueductal grey	
DMPAG	dorsomedial periaqueductal grey	
DpG	deep grey of superior colliculus	

DRD	dorsal raphe nucleus, dorsal
DRL	dorsal raphe nucleus, lateral
DRV	dorsal raphe nucleus, ventral
ECIC1	external cortex inf. collic., layer 1
ECIC2	external cortex inf. collic., layer 2
ECIC3	external cortex inf. collic., layer 3
ILL	intermed. nu. of lateral lemniscus
KF	Kölliker-Fuse nucleus
lfp	longitudinal fasciculus of pons
ll	lateral lemniscus
LPAG	lateral periaqueductal grey

LVPO	lateroventral periolivary nu.
mcp	middle cerebellar peduncle
Me5	mesencephalic trigeminal nu.
me5	mesencephalic trigeminal tr.
MiTg	microcellular tegmental nu.
ml	medial lemniscus
mlf	medial longitudinal fasciculus
MnR	median raphe nucleus
MTz	medial trapezoid nucleus
MVPO	medioventral periolivary nu.
P5	peritrigeminal zone

PDR	posterodorsal raphe nu.
PL	paralemniscal nucleus
PLV	perilemniscal nucleus, ventral
PMnR	paramedian raphe nucleus
PnC	pontine reticular nu., caudal
PnO	pontine reticular nu., oral
PTg	pedunculotegmental nucleus
py	pyramidal tract
RMg	raphe magnus nucleus
RtTg	reticulotegmental nucleus pons
Sag	sagulum nucleus

SC	superior colliculus
scp	superior cerebellar peduncle
SPTg	subpeduncular tegmental nu.
Tz	nucleus of trapezoid body
VLL	ventral nu. of lateral lemniscus
VLPAG	ventrolat. periaqueductal grey
vsc	ventral spinocerebellar tract

D23

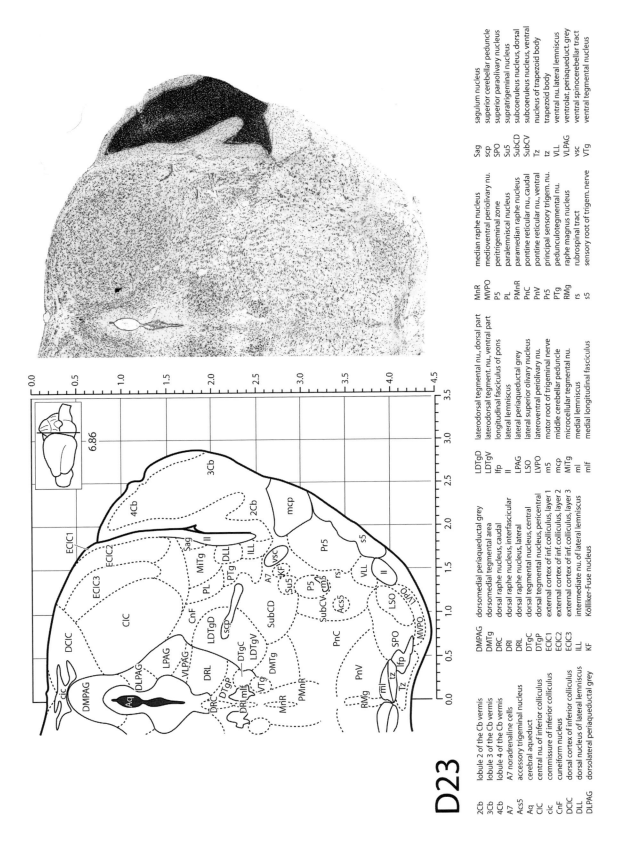

6.86

2Cb	lobule 2 of the Cb vermis
3Cb	lobule 3 of the Cb vermis
4Cb	lobule 4 of the Cb vermis
A7	A7 noradrenaline cells
Acs5	accessory trigeminal nucleus
Aq	cerebral aqueduct
CIC	central nu. of inferior colliculus
cic	commissure of inferior colliculus
CnF	cuneiform nucleus
DCIC	dorsal cortex of inferior colliculus
DLL	dorsal nucleus of lateral lemniscus
DLPAG	dorsolateral periaqueductal grey

DMPAG	dorsomedial periaqueductal grey
DMTg	dorsomedial tegmental area
DRC	dorsal raphe nucleus, caudal
DRI	dorsal raphe nucleus, interfascicular
DRL	dorsal raphe nucleus, lateral
DTgC	dorsal tegmental nucleus, central
DTgP	dorsal tegmental nucleus, pericentral
ECIC1	external cortex of inf. colliculus, layer 1
ECIC2	external cortex of inf. colliculus, layer 2
ECIC3	external cortex of inf. colliculus, layer 3
ILL	intermediate nu. of lateral lemniscus
KF	Kölliker-Fuse nucleus

LDTgD	laterodorsal tegmental nu., dorsal part
LDTgV	laterodorsal tegment. nu., ventral part
lfp	longitudinal fasciculus of pons
ll	lateral lemniscus
LPAG	lateral periaqueductal grey
LSO	lateral superior olivary nucleus
LVPO	lateroventral periolivary nu.
m5	motor root of trigeminal nerve
mcp	middle cerebellar peduncle
MiTg	microcellular tegmental nu.
ml	medial lemniscus
mlf	medial longitudinal fasciculus

MnR	median raphe nucleus
MVPO	medioventral periolivary nu.
P5	peritrigeminal nucleus
PL	paralemniscal zone
PMnR	paramedian raphe nucleus
PnC	pontine reticular nu., caudal
PnV	pontine reticular nu., ventral
Pr5	principal sensory trigem. nu.
PTg	pedunculotegmental nu.
RMg	raphe magnus nucleus
rs	rubrospinal tract
s5	sensory root of trigem. nerve

Sag	sagulum nucleus
scp	superior cerebellar peduncle
SPO	superior paraolivary nucleus
Su5	supratrigeminal nucleus
SubCD	subcoeruleus nucleus, dorsal
SubCV	subcoeruleus nucleus, ventral
Tz	nucleus of trapezoid body
tz	trapezoid body
VLL	ventral nu. lateral lemniscus
VLPAG	ventrolat. periaqueduct. grey
vsc	ventral spinocerebellar tract
VTg	ventral tegmental nucleus

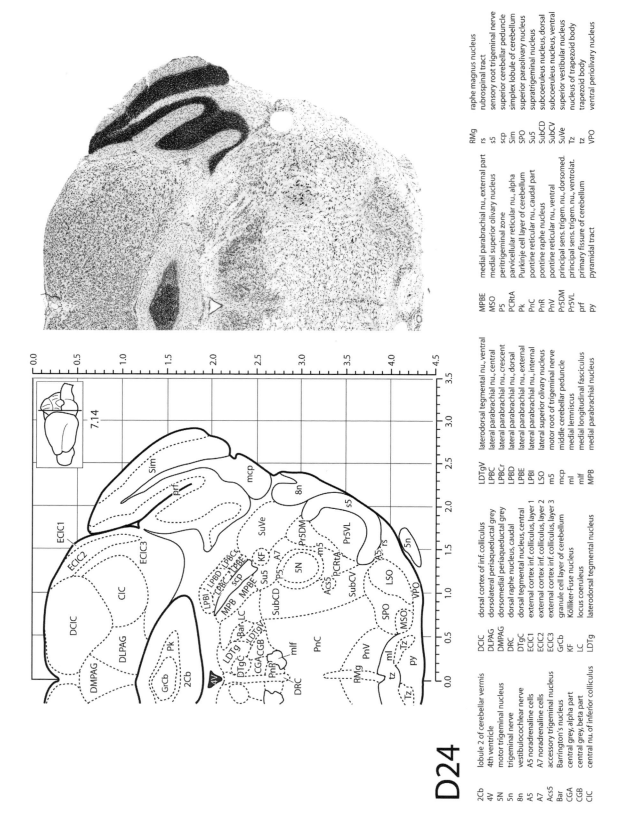

D24

2Cb	lobule 2 of cerebellar vermis	DCIC	dorsal cortex of inf. colliculus	LDTgV	laterodorsal tegmental nu., ventral	MPBE	medial parabrachial nu., external part	RMg	raphe magnus nucleus
4V	4th ventricle	DLPAG	dorsolateral periaqueductal grey	LPBC	lateral parabrachial nu., central	MSO	medial superior olivary nucleus	rs	rubrospinal tract
5N	motor trigeminal nucleus	DMPAG	dorsomedial periaqueductal grey	LPBCr	lateral parabrachial nu., crescent	P5	peritrigeminal zone	s5	sensory root trigeminal nerve
5n	trigeminal nerve	DRC	dorsal raphe nucleus, caudal	LPBD	lateral parabrachial nu., dorsal	PCRtA	parvicellular reticular nu., alpha	scp	superior cerebellar peduncle
8n	vestibulocochlear nerve	DTgC	dorsal tegmental nucleus, central	LPBE	lateral parabrachial nu., external	Pk	Purkinje cell layer of cerebellum	Sim	simplex lobule of cerebellum
A5	A5 noradrenaline cells	ECIC1	external cortex inf. colliculus, layer 1	LPBI	lateral parabrachial nu., internal	PnC	pontine reticular nu., caudal part	SPO	superior paraolivary nucleus
A7	A7 noradrenaline cells	ECIC2	external cortex inf. colliculus, layer 2	LSO	lateral superior olivary nucleus	PnR	pontine raphe nucleus	Su5	supratrigeminal nucleus
Acs5	accessory trigeminal nucleus	ECIC3	external cortex inf. colliculus, layer 3	m5	motor root of trigeminal nerve	PnV	pontine reticular nu., ventral	SubCD	subcoeruleus nucleus, dorsal
Bar	Barrington's nucleus	GrCb	granule cell layer of cerebellum	mcp	middle cerebellar peduncle	Pr5DM	principal sens. trigem. nu., dorsomed.	SubCV	subcoeruleus nucleus, ventral
CGA	central grey, alpha part	KF	Kölliker-Fuse nucleus	ml	medial lemniscus	Pr5VL	principal sens. trigem. nu., ventrolat.	SuVe	superior vestibular nucleus
CGB	central grey, beta part	LC	locus coeruleus	mlf	medial longitudinal fasciculus	prf	primary fissure of cerebellum	Tz	nucleus of trapezoid body
CIC	central nu. of inferior colliculus	LDTg	laterodorsal tegmental nucleus	MPB	medial parabrachial nucleus	py	pyramidal tract	tz	trapezoid body
								VPO	ventral periolivary nucleus

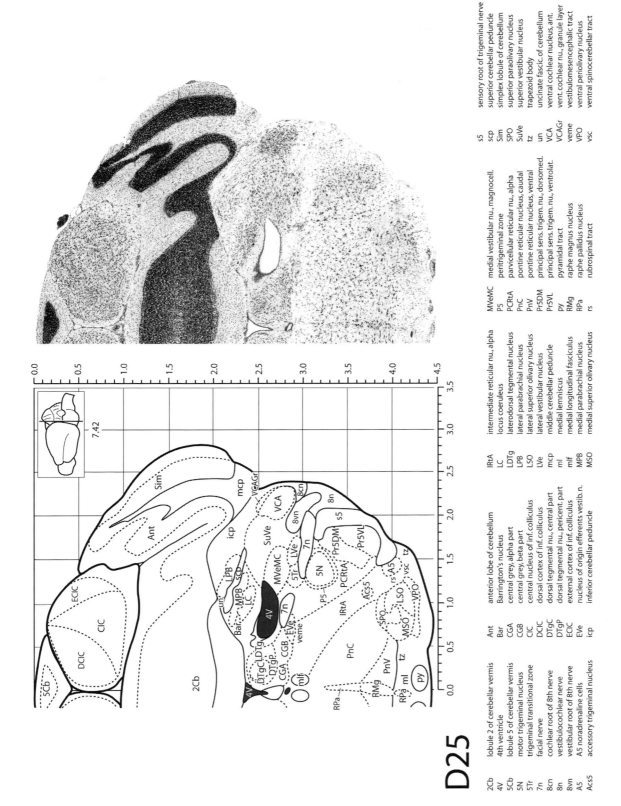

D25

2Cb	lobule 2 of cerebellar vermis	Ant	anterior lobe of cerebellum	IRtA
4V	4th ventricle	Bar	Barrington's nucleus	LC
5Cb	lobule 5 of cerebellar vermis	CGA	central grey, alpha part	LDTg
5N	motor trigeminal nucleus	CGB	central grey, beta part	LPB
5Tr	trigeminal transitional zone	CIC	central nucleus of inf. colliculus	LSO
7n	facial nerve	DCIC	dorsal cortex of inf. colliculus	LVe
8cn	cochlear root of 8th nerve	DTgC	dorsal tegmental nu., central part	mcp
8n	vestibulocochlear nerve	DTgP	dorsal tegmental nu., pericent. part	ml
8vn	vestibular root of 8th nerve	ECIC	external cortex of inf. colliculus	mlf
A5	A5 noradrenaline cells	EVe	nucleus of origin efferents vestib. n.	MPB
Acs5	accessory trigeminal nucleus	icp	inferior cerebellar peduncle	MSO

IRtA	intermediate reticular nu., alpha	MVeMC	medial vestibular nu., magnocell.	s5
LC	locus coeruleus	P5	peritrigeminal zone	scp
LDTg	laterodorsal tegmental nucleus	PCRtA	parvicellular reticular nu., alpha	Sim
LPB	lateral parabrachial nucleus	PnC	pontine reticular nucleus, caudal	SPO
LSO	lateral superior olivary nucleus	PnV	pontine reticular nucleus, ventral	SuVe
LVe	lateral vestibular nucleus	Pr5DM	principal sens. trigem. nu., dorsomed.	tz
mcp	middle cerebellar peduncle	Pr5VL	principal sens. trigem. nu., ventrolat.	un
ml	medial lemniscus	py	pyramidal tract	VCA
mlf	medial longitudinal fasciculus	RMg	raphe magnus nucleus	VCAGr
MPB	medial parabrachial nucleus	RPa	raphe pallidus nucleus	veme
MSO	medial superior olivary nucleus	rs	rubrospinal tract	VPO
				vsc

s5	sensory root of trigeminal nerve
scp	superior cerebellar peduncle
Sim	simplex lobule of cerebellum
SPO	superior paraolivary nucleus
SuVe	superior vestibular nucleus
tz	trapezoid body
un	uncinate fascic. of cerebellum
VCA	ventral cochlear nucleus, ant.
VCAGr	vent. cochlear nu., granule layer
veme	vestibulomesencephalic tract
VPO	ventral periolivary nucleus
vsc	ventral spinocerebellar tract

D26

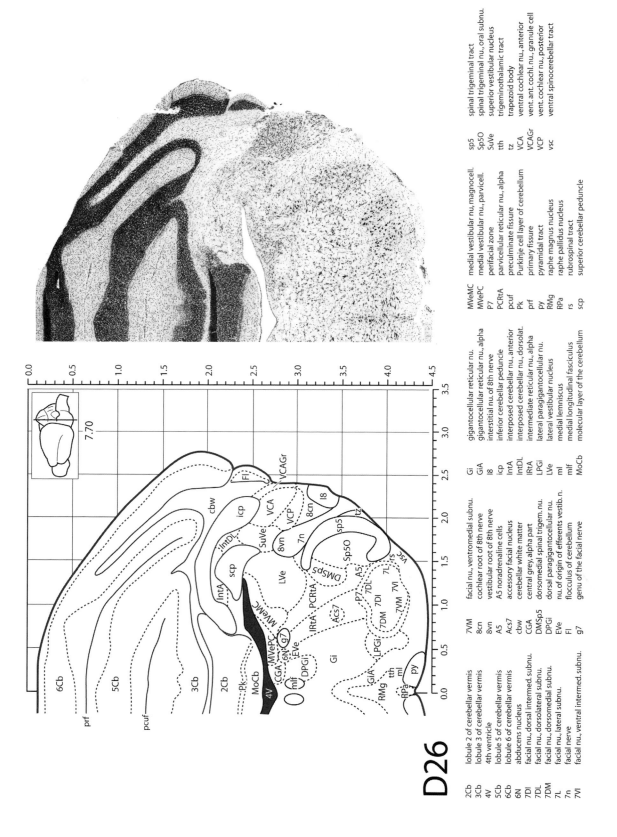

7.70

2Cb	lobule 2 of cerebellar vermis	7VM	facial nu., ventromedial subnu.
3Cb	lobule 3 of cerebellar vermis	8cn	cochlear root of 8th nerve
4V	4th ventricle	8vn	vestibular root of 8th nerve
5Cb	lobule 5 of cerebellar vermis	A5	A5 noradrenaline cells
6Cb	lobule 6 of cerebellar vermis	Acs7	accessory facial nucleus
6N	abducens nucleus	cbw	cerebellar white matter
7DI	facial nu., dorsal intermed. subnu.	CGA	central grey, alpha part
7DL	facial nu., dorsolateral subnu.	DMSp5	dorsomedial spinal trigem. nu.
7DM	facial nu., dorsomedial subnu.	DPGi	dorsal paragigantocellular nu.
7L	facial nu., lateral subnu.	EVe	nu. of origin of efferents vestib. n.
7n	facial nerve	Fl	flocculus of cerebellum
7VI	facial nu., ventral intermed. subnu.	g7	genu of the facial nerve

Gi	gigantocellular reticular nu.	MVeMC	medial vestibular nu., magnocell.	sp5	spinal trigeminal tract
GiA	gigantocellular reticular nu., alpha	MVePC	medial vestibular nu., parvicell.	SpSO	spinal trigeminal nu., oral subnu.
I8	interstitial nu. of 8th nerve	P7	perifacial zone	SuVe	superior vestibular nucleus
icp	inferior cerebellar peduncle	PCRtA	parvicellular reticular nu., alpha	tth	trigeminothalamic tract
IntA	interposed cerebellar nu., anterior	pcuf	preculminate fissure	tz	trapezoid body
IntDL	interposed cerebellar nu., dorsolat.	Pk	Purkinje cell layer of cerebellum	VCA	ventral cochlear nu., anterior
IRtA	intermediate reticular nu., alpha	prf	primary fissure	VCAGr	vent. ant. cochl. nu., granule cell
LPGi	lateral paragigantocellular nu.	py	pyramidal tract	VCP	vent. cochlear nu., posterior
LVe	lateral vestibular nucleus	RMg	raphe magnus nucleus	vsc	ventral spinocerebellar tract
ml	medial lemniscus	RPa	raphe pallidus nucleus		
mlf	medial longitudinal fasciculus	rs	rubrospinal tract		
MoCb	molecular layer of the cerebellum	scp	superior cerebellar peduncle		

D27

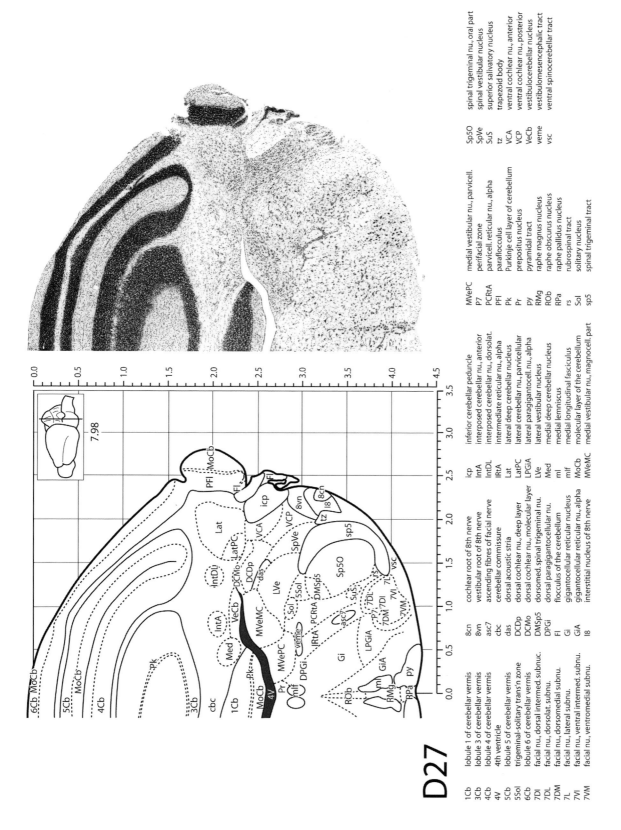

7.98

1Cb	lobule 1 of cerebellar vermis
3Cb	lobule 3 of cerebellar vermis
4Cb	lobule 4 of cerebellar vermis
4V	4th ventricle
5Cb	lobule 5 of cerebellar vermis
5Sol	trigeminal-solitary trans'n zone
6Cb	lobule 6 of cerebellar vermis
7DI	facial nu., dorsal intermed. subnuc.
7DL	facial nu., dorsolat. subnu.
7L	facial nu., lateral subnu.
7VI	facial nu., ventral intermed. subnu.
7VM	facial nu., ventromedial subnu.
8cn	cochlear root of 8th nerve
8vn	vestibular root of 8th nerve
asc7	ascending fibres of facial nerve
cbc	cerebellar commissure
das	dorsal acoustic stria
DCDp	dorsal cochlear nu., deep layer
DCMo	dorsal cochlear nu., molecular layer
DMSp5	dorsomed. spinal trigeminal nu.
DPGi	dorsal paragigantocellular nu.
Fl	flocculus of the cerebellum
Gi	gigantocellular reticular nucleus
GiA	gigantocellular reticular nu., alpha
I8	interstitial nucleus of 8th nerve
icp	inferior cerebellar peduncle
IntA	interposed cerebellar nu., anterior
IntDL	interposed cerebellar nu., dorsolat.
IRtA	intermediate reticular nu., alpha
Lat	lateral deep cerebellar nucleus
LatPC	lateral cerebellar nu., parvicellular
LPGiA	lateral paragigantocell. nu., alpha
LVe	lateral vestibular nucleus
Med	medial deep cerebellar nucleus
ml	medial lemniscus
mlf	medial longitudinal fasciculus
MoCb	molecular layer of the cerebellum
MVeMC	medial vestibular nu., magnocell. part
MVePC	medial vestibular nu., parvicell.
P7	perifacial zone
PCRtA	parvicell. reticular nu., alpha
PFl	paraflocculus
Pk	Purkinje cell layer of cerebellum
Pr	prepositus nucleus
py	pyramidal tract
RMg	raphe magnus nucleus
ROb	raphe obscurus nucleus
RPa	raphe pallidus nucleus
rs	rubrospinal tract
Sol	solitary nucleus
sp5	spinal trigeminal tract
SpSO	spinal trigeminal nu., oral part
SpVe	spinal vestibular nucleus
SuS	superior salivatory nucleus
tz	trapezoid body
VCA	ventral cochlear nu., anterior
VCP	ventral cochlear nu., posterior
VeCb	vestibulocerebellar nucleus
veme	vestibulomesencephalic tract
vsc	ventral spinocerebellar tract

D28

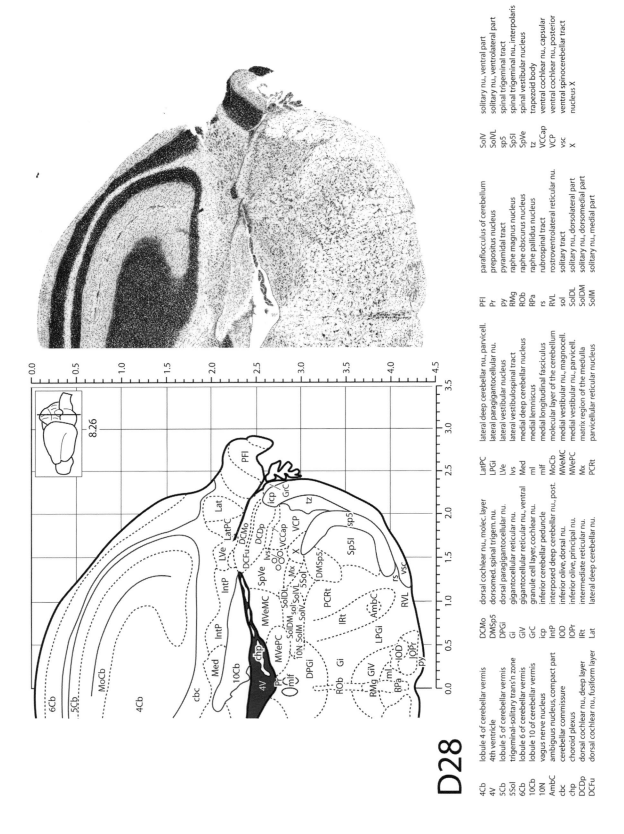

8.26

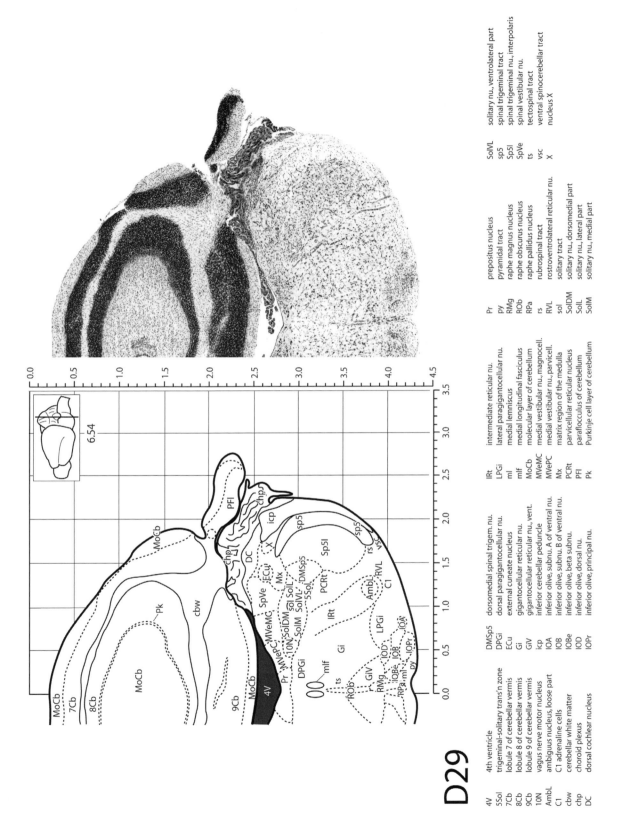

D29

6.54

4V	4th ventricle	DMSp5	dorsomedial spinal trigem. nu.	IRt	intermediate reticular nu.	Pr	prepositus nucleus	SolVL	solitary nu., ventrolateral part
5Sol	trigeminal-solitary trans'n zone	DPGi	dorsal paragigantocellular nu.	LPGi	lateral paragigantocellular nu.	py	pyramidal tract	sp5	spinal trigeminal tract
7Cb	lobule 7 of cerebellar vermis	ECu	external cuneate nucleus	ml	medial lemniscus	RMg	raphe magnus nucleus	Sp5I	spinal trigeminal nu., interpolaris
8Cb	lobule 8 of cerebellar vermis	Gi	gigantocellular reticular nu.	mlf	medial longitudinal fasciculus	ROb	raphe obscurus nucleus	SpVe	spinal vestibular nu.
9Cb	lobule 9 of cerebellar vermis	GiV	gigantocellular reticular nu., vent.	MoCb	molecular layer of cerebellum	RPa	raphe pallidus nucleus	ts	tectospinal tract
10N	vagus nerve motor nucleus	icp	inferior cerebellar peduncle	MVeMC	medial vestibular nu., magnocell.	rs	rubrospinal tract	vsc	ventral spinocerebellar tract
AmbL	ambiguus nucleus, loose part	IOA	inferior olive, subnu. A of ventral nu.	MVePC	medial vestibular nu., parvicell.	RVL	rostroventrolateral reticular nu.	X	nucleus X
C1	C1 adrenaline cells	IOB	inferior olive, subnu. B of ventral nu.	Mx	matrix region of the medulla	sol	solitary tract		
cbw	cerebellar white matter	IOBe	inferior olive, beta subnu.	PCRt	parvicellular reticular nucleus	SolDM	solitary nu., dorsomedial part		
chp	choroid plexus	IOD	inferior olive, dorsal nu.	PFl	paraflocculus of cerebellum	SolL	solitary nu., lateral part		
DC	dorsal cochlear nucleus	IOPr	inferior olive, principal nu.	Pk	Purkinje cell layer of cerebellum	SolM	solitary nu., medial part		

D30

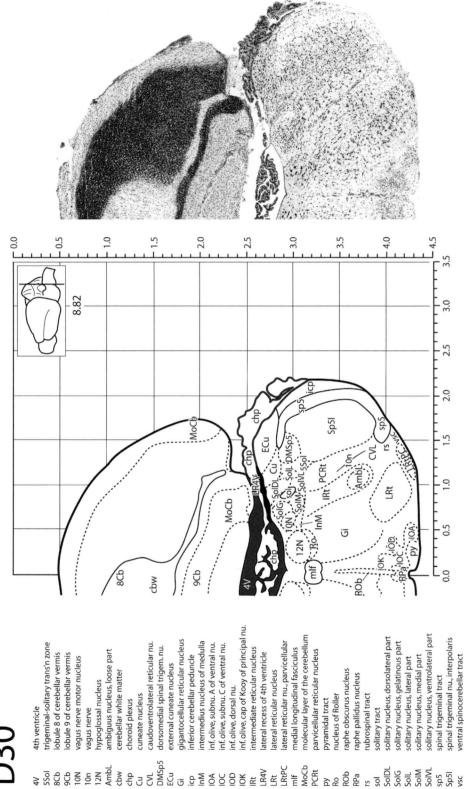

8.82

4V	4th ventricle
5Sol	trigeminal-solitary trans'n zone
8Cb	lobule 8 of cerebellar vermis
9Cb	lobule 9 of cerebellar vermis
10N	vagus nerve motor nucleus
10n	vagus nerve
12N	hypoglossal nucleus
AmbL	ambiguus nucleus, loose part
cbw	cerebellar white matter
chp	choroid plexus
Cu	cuneate nucleus
CVL	caudoventrolateral reticular nu.
DMSp5	dorsomedial spinal trigem. nu.
ECu	external cuneate nucleus
Gi	gigantocellular reticular nucleus
icp	inferior cerebellar peduncle
InM	intermedius nucleus of medulla
IOA	inf. olive, subnu. A of ventral nu.
IOC	inf. olive, subnu. C of ventral nu.
IOD	inf. olive, dorsal nu.
IOK	inf. olive, cap of Kooy of principal nu.
IRt	intermediate reticular nucleus
LR4V	lateral recess of 4th ventricle
LRt	lateral reticular nucleus
LRtPC	lateral reticular nu., parvicellular
mlf	medial longitudinal fasciculus
MoCb	molecular layer of the cerebellum
PCRt	parvicellular reticular nucleus
py	pyramidal tract
Ro	nucleus of Roller
ROb	raphe obscurus nucleus
RPa	raphe pallidus nucleus
rs	rubrospinal tract
sol	solitary tract
SolDL	solitary nucleus, dorsolateral part
SolG	solitary nucleus, gelatinous part
SolL	solitary nucleus, lateral part
SolM	solitary nucleus, medial part
SolVL	solitary nucleus, ventrolateral part
sp5	spinal trigeminal tract
Sp5I	spinal trigeminal nu., interpolaris
vsc	ventral spinocerebellar tract

D31 & D32

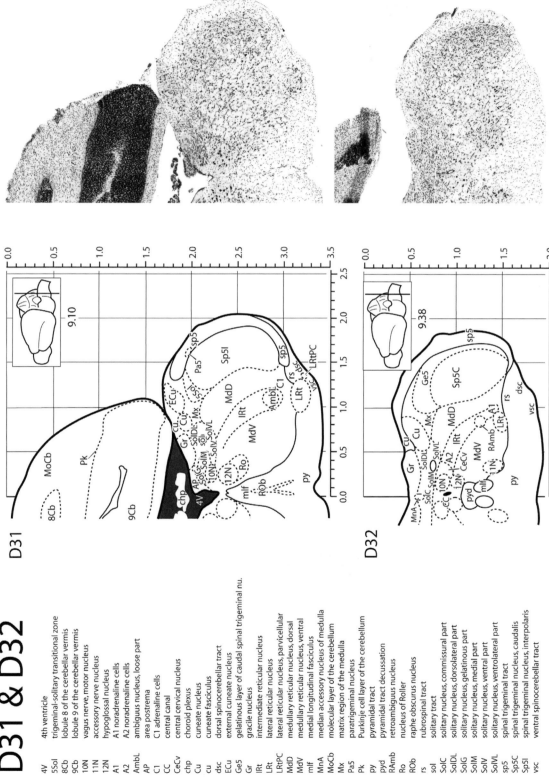

4V	4th ventricle
5Sol	trigeminal-solitary transitional zone
8Cb	lobule 8 of the cerebellar vermis
9Cb	lobule 9 of the cerebellar vermis
10N	vagus nerve, motor nucleus
11N	accessory nerve nucleus
12N	hypoglossal nucleus
A1	A1 noradrenaline cells
A2	A2 noradrenaline cells
AmbL	ambiguus nucleus, loose part
AP	area postrema
C1	C1 adrenaline cells
CC	central canal
CeCv	central cervical nucleus
chp	choroid plexus
Cu	cuneate nucleus
cu	cuneate fasciculus
dsc	dorsal spinocerebellar tract
ECu	external cuneate nucleus
Ge5	gelatinous layer of caudal spinal trigeminal nu.
Gr	gracile nucleus
IRt	intermediate reticular nucleus
LRt	lateral reticular nucleus
LRtPC	lateral reticular nucleus, parvicellular
MdD	medullary reticular nucleus, dorsal
MdV	medullary reticular nucleus, ventral
mlf	medial longitudinal fasciculus
MnA	median accessory nucleus of medulla
MoCb	molecular layer of the cerebellum
Mx	matrix region of the medulla
Pa5	paratrigeminal nucleus
Pk	Purkinje cell layer of the cerebellum
py	pyramidal tract
pyd	pyramidal tract decussation
RAmb	retroambiguus nucleus
Ro	nucleus of Roller
ROb	raphe obscurus nucleus
rs	rubrospinal tract
sol	solitary tract
SolC	solitary nucleus, commissural part
SolDL	solitary nucleus, dorsolateral part
SolG	solitary nucleus, gelatinous part
SolM	solitary nucleus, medial part
SolV	solitary nucleus, ventral part
SolVL	solitary nucleus, ventrolateral part
sp5	spinal trigeminal tract
Sp5C	spinal trigeminal nucleus, caudalis
Sp5I	spinal trigeminal nucleus, interpolaris
vsc	ventral spinocerebellar tract

17

Stereotaxic atlas of the brain of the tammar wallaby (*Macropus eugenii*)

K. W. S. Ashwell and L. R. Marotte

Introduction

The tammar wallaby (or just the tammar) is one of the smallest species of *Macropus*, averaging 7.5 kg for adult males (range of 5 to 9 kg) and 5.5 kg for adult females (range of 4 to 6 kg). The natural population of *M. eugenii* was originally found in Western and South Australia and on 10 or more offshore islands, where the tammar inhabits coastal scrub and dry sclerophyll forest. The species can be easily maintained in colonies, making it a convenient experimental animal for neuroscience studies.

Methods

Ethics, anaesthesia and stereotaxic procedures

All experimental procedures were approved by the Animal Ethics and Experimentation Committee of the Australian National University, conform to NIH principles of laboratory animal care and were carried out according to the ethical guidelines of the National Health and Medical Research Council (Australia).

A 2.5-year-old male wallaby weighing 5.5 kg was sedated by intramuscular injection of 0.4 ml of xylazine (8 mg/kg) and the tail vein cannulated. Anaesthesia was induced with an initial dose of 2 ml 5% sodium thiopentone intravenously and maintained with 0.5 ml aliquots as required. The head was placed in a stereotaxic frame such that the plane including both external acoustic meatuses and the medial angles of each eye was depressed 10° from the horizontal (Figure 17.1a). This allows the forward view of the wallaby to be unobstructed by the nose-bar in visual system experiments. The medial angle of the eye approximately corresponds to the lachrymal foramen on the wallaby skull. Bregma (Figure 17.1b) was observed on the skull and the stereotaxic

co-ordinates noted. Burr holes were drilled through the skull at 3 mm on either side of the midline at bregma and 3 mm on either side of the midline at points 8 mm and 16 mm caudal to bregma. An 18-gauge needle was lowered into each burr hole at the above co-ordinates to a depth of 20 mm for the paired insertions at bregma and to a depth of 18 mm for the remaining pairs of insertion sites.

Perfusion and blocking

The needle was removed after insertion and the animal deeply anaesthetised with 5% thiopentone before the thoracic cage was opened. Sodium nitrite (1.0%) and heparin sulfate (0.6 mg) were injected into the left ventricle before the right atrium and left ventricle were incised. A cannula was inserted into the ascending aorta and secured there by a ligature before perfusion was commenced with room temperature (24 °C) 0.9% saline (200 ml) followed by 1600 ml of room temperature 4% paraformaldehyde in 0.1M phosphate buffer (PB pH 7.4) over 30 minutes. One hour was allowed after cessation of perfusion before the animal was decapitated at the cervical vertebra 2 (axis) level and the dorsal atlas and dorsal occipital and parietal bones removed to expose the brain. Care was taken to keep the part of the skull with bregma intact and undisplaced. The head was immersed overnight in the fixative at room temperature.

The following morning, the head was re-inserted into the stereotaxic frame, making sure that all adjustments were identical to the previous day and that the plane including the medial angles of the eyes and external acoustic meatuses was still tilted 10° below the horizontal. The position of bregma was rechecked and found to be within ±0.2 mm of the rostrocaudal position noted the previous day. A scalpel blade oriented in the coronal plane was mounted on the stereotaxic arm and inserted into the brain at a point 10 mm caudal to bregma, thereby partially dividing the

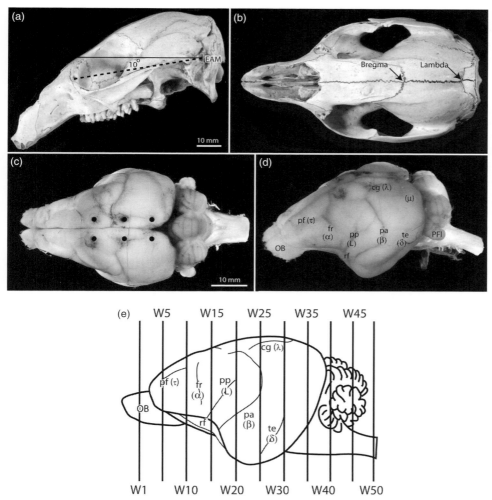

Figure 17.1 Photographs (a) and (b) show lateral and dorsal views of a representative *Macropus eugenii* skull, indicating the orientation of the skull in the stereotaxic frame and the positions of the external acoustic meatus (EAM, a), bregma and lambda (b). The dashed line indicates the plane containing the EAM and the lachrymal foramen at the medial angle of the orbit, whereas the continuous horizontal line depicts the horizontal plane. Photograph (c) shows the dorsal surface of the brain depicted in subsequent plates, with the position of vertical needles marked (●). Photograph (d) shows the lateral view of the atlas brain with major sulci denoted. The final figure (e) illustrates the position of coronal sections on a lateral view of the wallaby brain.

brain into rostral and caudal halves. The brain was removed from the skull, taking care to retain the paraflocculus and olfactory bulbs. The brain was photographed (Figure 17.1c, d), weighed and measured. The rostrocaudal length of the cerebral hemisphere was 39.7 mm and the dorsoventral height of the hemisphere was 27.5 mm. The weight of the brain at this stage was 19.4 g.

Histology and photomicrography

The brain was separated into rostral and caudal parts along the plane cut previously during blocking and fixed by immersion in the above fixative for a further five weeks before being submerged in 30% sucrose in 0.1M PB for 3 days at 4 °C. When the brain parts had sunk they were embedded in albumin gelatine hardened with glutaraldehyde before being mounted on a cryostat chuck with *Tissuetek* mounting medium and frozen. Coronal sections 50 μm thick were cut parallel to the blocking incision and series of every 22nd section retained for mounting from 0.1M PBS onto gelatinised glass slides. These sections were then stained with cresyl violet and coverslipped with *DePeX*.

A series of sections at intervals of 1.1 mm were photographed with the aid of an *Olympus Provis* microscope equipped with a *SPOT* photomicrography system in the laboratory of Professor George Paxinos. In most cases the right side of the brain has been illustrated, except for plate W29, for which the right half of the section was incomplete because of the blocking transection. For this level, the left half of the section has been shown, mirror-imaged to match the orientation of the other plates. The features visible in the left half of this section are essentially identical to those in the damaged right half. Sections were photographed at a magnification of 20× and merged with *Adobe Photoshop CS2*. The large size of sections in the middle of the series (W20 to W30, Figure 17.1e) made photomerging extremely difficult, even with a high-powered computer and most of these sections have been represented as two or three montaged components. Images were saved as files with sufficient resolution to retain individual neurons as distinct points (approximately 5 μm resolution) even over sections of 27 mm width.

Illustrations, delineations and nomenclature

Photoshop images were placed in *Adobe Illustrator CS2* and the outlines of significant boundaries traced directly on the images. Nomenclature for tracts and nuclei is identical to that used for the dunnart atlas, except for sulci and features specific for macropods. The nomenclature for cortical sulci includes both the Greek letter system of Ziehen (1897) and Johnson (1977) and a two-letter system based on the nomenclature of Mayner (1989a).

All atlas figures have scales in mm indicating distance from the midline and vertical depth. Please note that the vertical depth is not zeroed at bregma. Researchers using the stereotaxic co-ordinates are advised to measure depth of penetration from the pial surface.

Acknowledgements

We would like to thank Ms S. Wragg for photography of the skull and brain. Atlas plate files are available at www.cambridge.org/9780521519458.

W1, W2 & W3

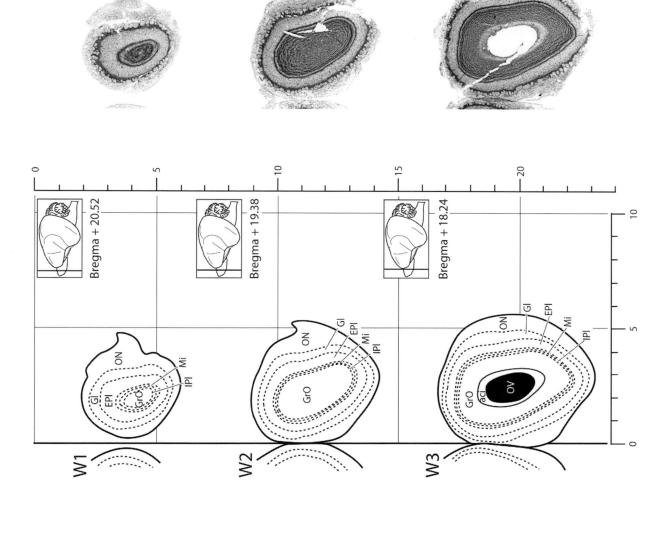

aci anterior commissure, intrabulbar limb
EPl external plexiform layer of olfactory bulb
Gl   glomerular layer of olfactory bulb
GrO granular cell layer of olfactory bulb
IPl internal plexiform layer of olfactory bulb
Mi   mitral cell layer of olfactory bulb
ON   olfactory nerve layer of olfactory bulb
OV   olfactory ventricle

Bregma + 20.52

Bregma + 19.38

Bregma + 18.24

W4 & W5

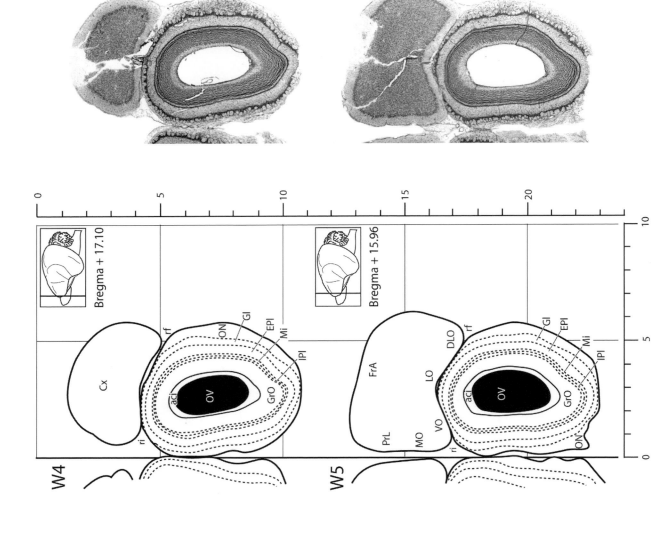

aci anterior commissure, intrabulbar limb
Cx cerebral cortex (unspecified)
DLO dorsolateral orbital cortex
EPl external plexiform layer of olfactory bulb
FrA frontal association cortex
Gl glomerular layer of olfactory bulb
GrO granular cell layer of olfactory bulb
IPl internal plexiform layer of olfactory bulb
LO lateral orbital cortex
Mi mitral cell layer of olfactory bulb
MO medial orbital cortex
ON olfactory nerve layer of olfactory bulb
OV olfactory ventricle
PrL prelimbic cortex
rf rhinal fissure
ri rhinal incisure
VO ventral orbital cortex

Bregma + 17.10

Bregma + 15.96

W6 & W7

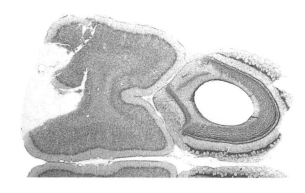

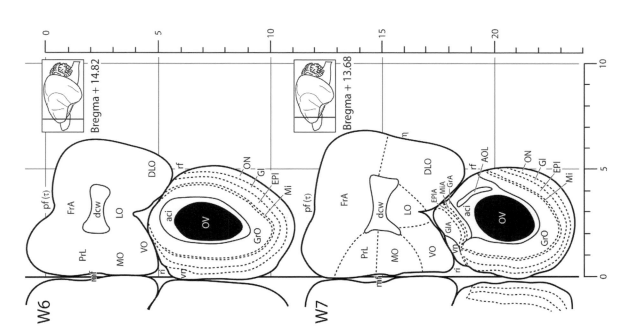

aci anterior commissure, intrabulbar limb
AOL anterior olfactory nucleus, lateral part
dcw deep cerebral white matter
DLO dorsolateral orbital cortex
EPI external plexiform layer of olfactory bulb
EPlA external plexiform layer of accessory olfactory bulb
η sulcus η of cerebral cortex
FrA frontal association cortex
Gl glomerular layer of olfactory bulb
GlA glomerular layer, accessory olfactory bulb
GrA granule cell layer, accessory olfactory bulb
GrO granular cell layer of olfactory bulb
LO lateral orbital cortex
mf medial frontal sulcus
Mi mitral cell layer of olfactory bulb
MiA mitral cell layer, accessory olfactory bulb
MO medial orbital cortex
ON olfactory nerve layer of olfactory bulb
OV olfactory ventricle
pf (τ) prefrontal (τ) sulcus
PrL prelimbic cortex
rf rhinal fissure
ri rhinal incisure
vn vomeronasal nerve layer, accessory olfactory bulb
VO ventral orbital cortex

W8

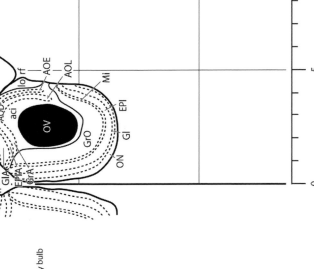

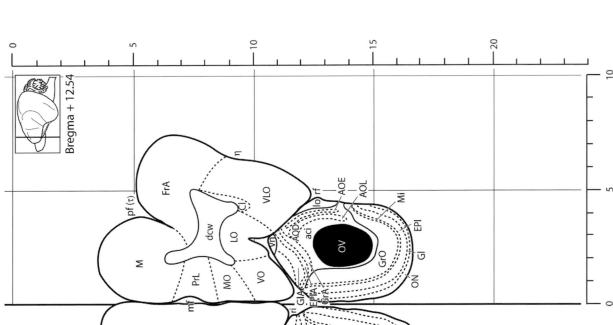

Bregma + 12.54

aci	anterior commissure, intrabulbar limb
AOD	anterior olfactory nucleus, dorsal part
AOE	anterior olfactory nucleus, external part
AOL	anterior olfactory nucleus, lateral part
Cl	claustrum
dcw	deep cerebral white matter
η	sulcus η of cerebral cortex
EPl	external plexiform layer of olfactory bulb
EPlA	external plexiform layer, accessory olfactory bulb
FrA	frontal association cortex
Gl	glomerular layer of olfactory bulb
GlA	glomerular layer, accessory olfactory bulb
GrA	granule cell layer, accessory olfactory bulb
GrO	granular cell layer of olfactory bulb
LO	lateral orbital cortex
lo	lateral olfactory tract
M	motor cortex
mf	medial frontal sulcus
Mi	mitral cell layer of olfactory bulb
MO	medial orbital cortex
ON	olfactory nerve layer of olfactory bulb
OV	olfactory ventricle
pf (τ)	prefrontal (τ) sulcus
PrL	prelimbic cortex
rf	rhinal fissure
ri	rhinal incisure
vn	vomeronasal nerve layer of accessory olfactory bulb
VLO	ventrolateral orbital cortex
VO	ventral orbital cortex

W9

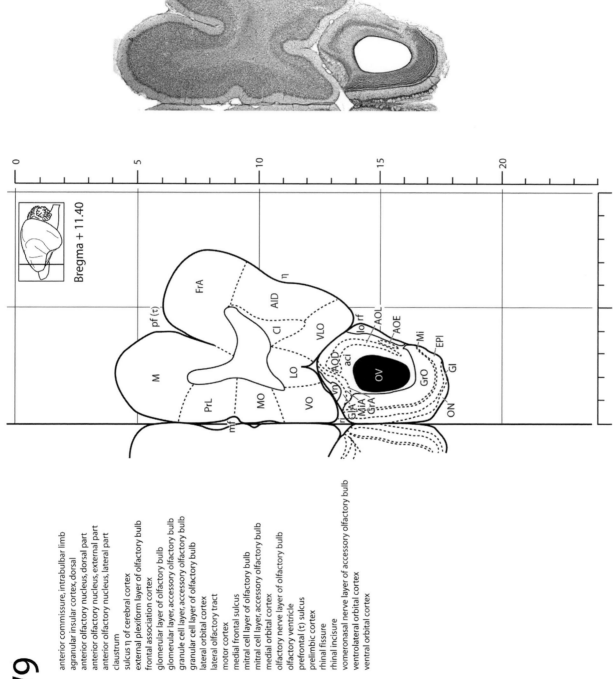

Bregma + 11.40

aci	anterior commissure, intrabulbar limb
AID	agranular insular cortex, dorsal
AOD	anterior olfactory nucleus, dorsal part
AOE	anterior olfactory nucleus, external part
AOL	anterior olfactory nucleus, lateral part
Cl	claustrum
η	sulcus η of cerebral cortex
EPl	external plexiform layer of olfactory bulb
FrA	frontal association cortex
Gl	glomerular layer of olfactory bulb
GlA	glomerular layer, accessory olfactory bulb
GrA	granule cell layer, accessory olfactory bulb
GrO	granule cell layer of olfactory bulb
LO	lateral orbital cortex
lo	lateral olfactory tract
M	motor cortex
mf	medial frontal sulcus
Mi	mitral cell layer of olfactory bulb
MiA	mitral cell layer, accessory olfactory bulb
MO	medial orbital cortex
ON	olfactory nerve layer of olfactory bulb
OV	olfactory ventricle
pf (τ)	prefrontal (τ) sulcus
PrL	prelimbic cortex
rf	rhinal fissure
ri	rhinal incisure
vn	vomeronasal nerve layer of accessory olfactory bulb
VLO	ventrolateral orbital cortex
VO	ventral orbital cortex

W10

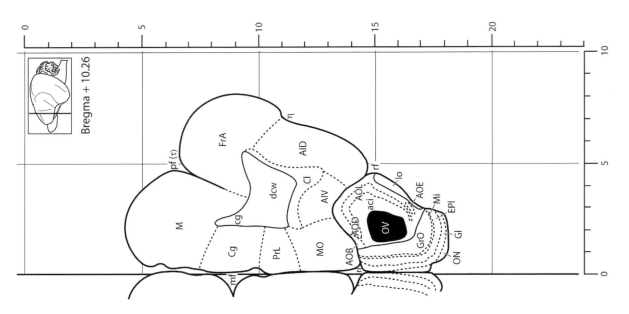

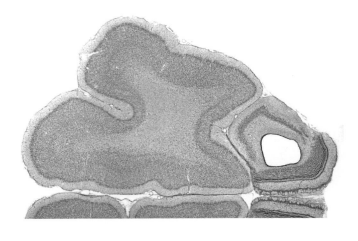

Bregma + 10.26

W11

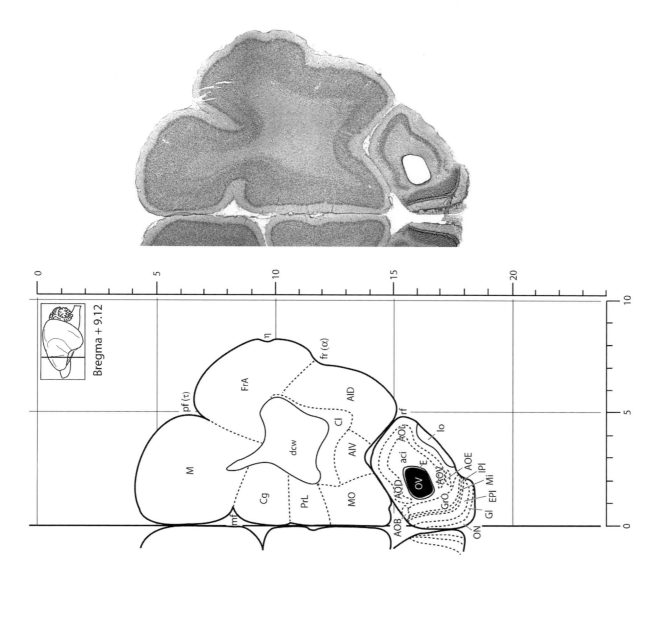

Bregma + 9.12

aci	anterior commissure, intrabulbar limb
AID	agranular insular cortex, dorsal
AIV	agranular insular cortex, ventral
AOB	accessory olfactory bulb
AOD	anterior olfactory nucleus, dorsal part
AOE	anterior olfactory nucleus, external part
AOL	anterior olfactory nucleus, lateral part
AOV	anterior olfactory nucleus, ventral part
Cg	cingulate cortex
Cl	claustrum
dcw	deep cerebral white matter
η	sulcus η of cerebral cortex
E	ependymal and subependymal layers
EPl	external plexiform layer of olfactory bulb
fr (α)	frontal (α) sulcus
FrA	frontal association cortex
Gl	glomerular layer of olfactory bulb
GrO	granular cell layer of olfactory bulb
IPl	internal plexiform layer of olfactory bulb
lo	lateral olfactory tract
M	motor cortex
mf	medial frontal sulcus
Mi	mitral cell layer of olfactory bulb
MO	medial orbital cortex
ON	olfactory nerve layer of olfactory bulb
OV	olfactory ventricle
pf (τ)	prefrontal (τ) sulcus
PrL	prelimbic cortex
rf	rhinal fissure

W12

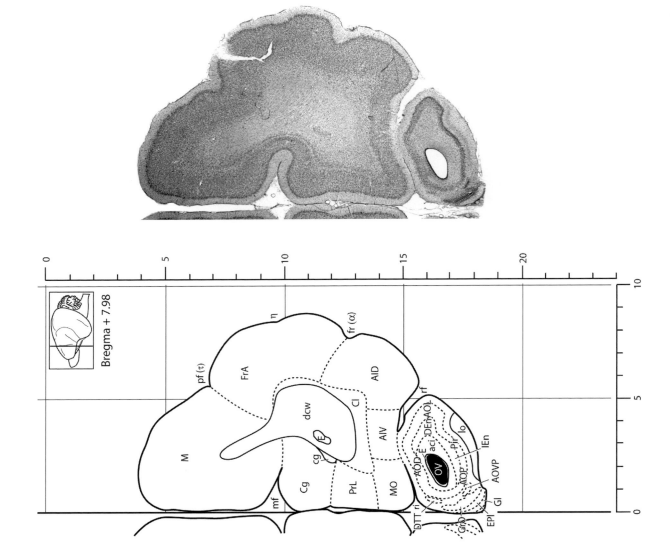

Bregma + 7.98

aci — anterior commissure, intrabulbar limb
AID — agranular insular cortex, dorsal
AIV — agranular insular cortex, ventral
AOD — anterior olfactory nucleus, dorsal part
AOL — anterior olfactory nucleus, lateral part
AOP — anterior olfactory nucleus, posterior part
AOVP — anterior olfactory area, ventroposterior part
Cg — cingulate cortex
cg — cingulum
Cl — claustrum
dcw — deep cerebral white matter
DEn — dorsal endopiriform nucleus
DTT — dorsal tenia tecta
η — sulcus η of cerebral cortex
E — ependymal and subependymal layers
EPl — external plexiform layer of olfactory bulb
fr (α) — frontal (α) sulcus
FrA — frontal association cortex
Gl — glomerular layer of olfactory bulb
GrO — granular cell layer of olfactory bulb
IEn — intermediate endopiriform nucleus
lo — lateral olfactory tract
M — motor cortex
mf — medial frontal sulcus
MO — medial orbital cortex
OV — olfactory ventricle
pf (τ) — prefrontal (τ) sulcus
Pir — piriform cortex
PrL — prelimbic cortex
rf — rhinal fissure
ri — rhinal incisure

W13

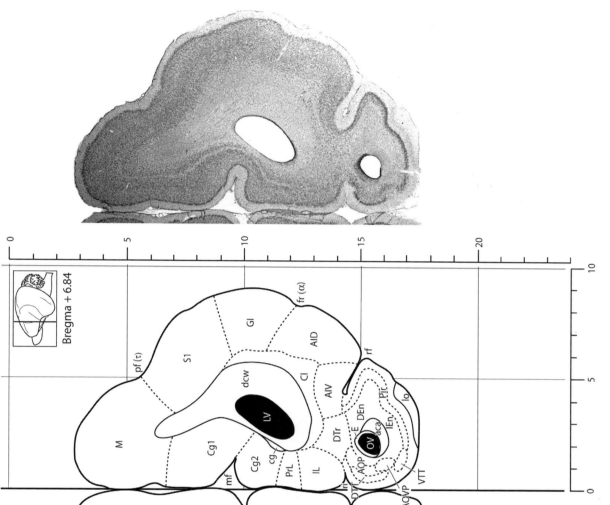

Bregma + 6.84

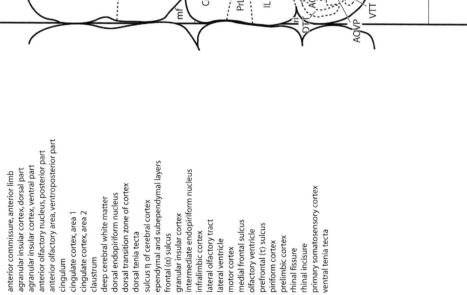

aca   anterior commissure, anterior limb
AID   agranular insular cortex, dorsal part
AIV   agranular insular cortex, ventral part
AOP   anterior olfactory nucleus, posterior part
AOVP   anterior olfactory area, ventroposterior part
cg   cingulum
Cg1   cingulate cortex, area 1
Cg2   cingulate cortex, area 2
Cl   claustrum
dcw   deep cerebral white matter
DEn   dorsal endopiriform nucleus
DTr   dorsal transition zone of cortex
DTT   dorsal tenia tecta
η   sulcus η of cerebral cortex
E   ependymal and subependymal layers
fr (α)   frontal (α) sulcus
GI   granular insular cortex
IEn   intermediate endopiriform nucleus
IL   infralimbic cortex
lo   lateral olfactory tract
LV   lateral ventricle
M   motor cortex
mf   medial frontal sulcus
OV   olfactory ventricle
pf (τ)   prefrontal (τ) sulcus
Pir   piriform cortex
PrL   prelimbic cortex
rf   rhinal fissure
ri   rhinal incisure
S1   primary somatosensory cortex
VTT   ventral tenia tecta

W14

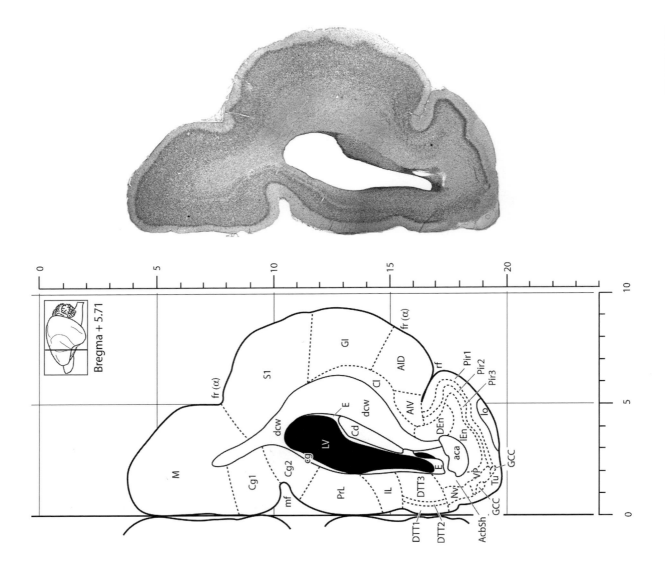

Bregma + 5.71

aca   anterior commissure, anterior limb
AcbSh   accumbens nucleus, shell
AID   agranular insular cortex, dorsal part
AIV   agranular insular cortex, ventral part
Cd   caudate nucleus
cg   cingulum
Cg1   cingulate cortex, area 1
Cg2   cingulate cortex, area 2
Cl   claustrum
dcw   deep cerebral white matter
DEn   dorsal endopiriform nucleus
DTT1   dorsal tenia tecta, layer 1
DTT2   dorsal tenia tecta, layer 2
DTT3   dorsal tenia tecta, layer 3
E   ependymal and subependymal layers
fr (α)   frontal (α) sulcus
GCC   granule cell cluster
GI   granular insular cortex
IEn   intermediate endopiriform nucleus
IL   infralimbic cortex
lo   lateral olfactory tract
LV   lateral ventricle
M   motor cortex
mf   medial frontal sulcus
Nv   navicular postolfactory nucleus
Pir1   piriform cortex, layer 1
Pir2   piriform cortex, layer 2
Pir3   piriform cortex, layer 3
PrL   prelimbic cortex
rf   rhinal fissure
S1   primary somatosensory cortex
Tu   olfactory tubercle
VP   ventral pallidum

W15

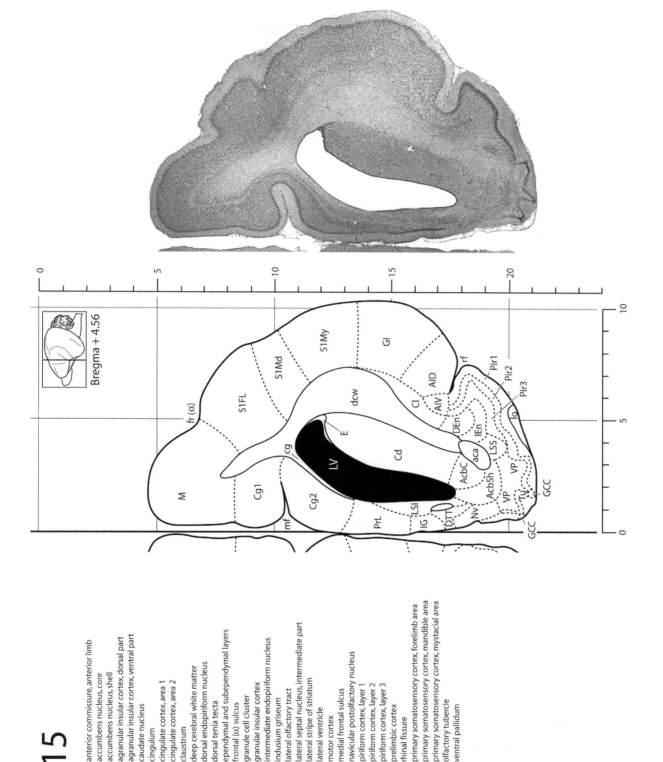

Bregma + 4.56

aca anterior commissure, anterior limb
AcbC accumbens nucleus, core
AcbSh accumbens nucleus, shell
AID agranular insular cortex, dorsal part
AIV agranular insular cortex, ventral part
Cd caudate nucleus
cg cingulum
Cg1 cingulate cortex, area 1
Cg2 cingulate cortex, area 2
Cl claustrum
dcw deep cerebral white matter
DEn dorsal endopiriform nucleus
DTT dorsal tenia tecta
E ependymal and subependymal layers
fr (α) frontal (α) sulcus
GCC granule cell cluster
GI granular insular cortex
IEn intermediate endopiriform nucleus
IG indusium griseum
lo lateral olfactory tract
LSI lateral septal nucleus, intermediate part
LSS lateral stripe of striatum
LV lateral ventricle
M motor cortex
mf medial frontal sulcus
Nv navicular postolfactory nucleus
Pir1 piriform cortex, layer 1
Pir2 piriform cortex, layer 2
Pir3 piriform cortex, layer 3
PrL prelimbic cortex
rf rhinal fissure
S1FL primary somatosensory cortex, forelimb area
S1Md primary somatosensory cortex, mandible area
S1My primary somatosensory cortex, mystacial area
Tu olfactory tubercle
VP ventral pallidum

W16

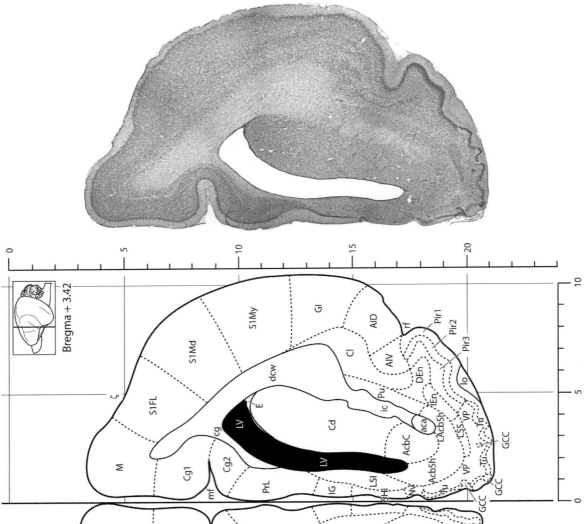

Bregma + 3.42

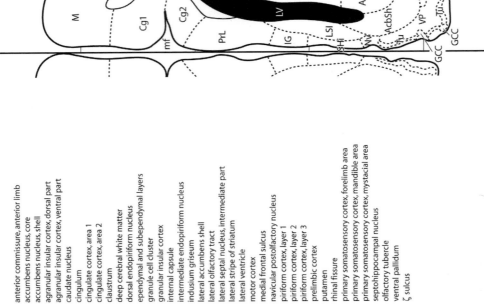

aca anterior commissure, anterior limb
AcbC accumbens nucleus, core
AcbSh accumbens nucleus, shell
AID agranular insular cortex, dorsal part
AIV agranular insular cortex, ventral part
Cd caudate nucleus
cg cingulum
Cg1 cingulate cortex, area 1
Cg2 cingulate cortex, area 2
Cl claustrum
dcw deep cerebral white matter
DEn dorsal endopiriform nucleus
E ependymal and subependymal layers
GCC granule cell cluster
GI granular insular cortex
ic internal capsule
IEn intermediate endopiriform nucleus
IG indusium griseum
LAcbSh lateral accumbens shell
lo lateral olfactory tract
LSI lateral septal nucleus, intermediate part
LSS lateral stripe of striatum
LV lateral ventricle
M motor cortex
mf medial frontal sulcus
Nv navicular postolfactory nucleus
Pir1 piriform cortex, layer 1
Pir2 piriform cortex, layer 2
Pir3 piriform cortex, layer 3
PrL prelimbic cortex
Pu putamen
rf rhinal fissure
S1FL primary somatosensory cortex, forelimb area
S1Md primary somatosensory cortex, mandible area
S1My primary somatosensory cortex, mystacial area
SHi septohippocampal nucleus
Tu olfactory tubercle
VP ventral pallidum
ζ ζ sulcus

W17

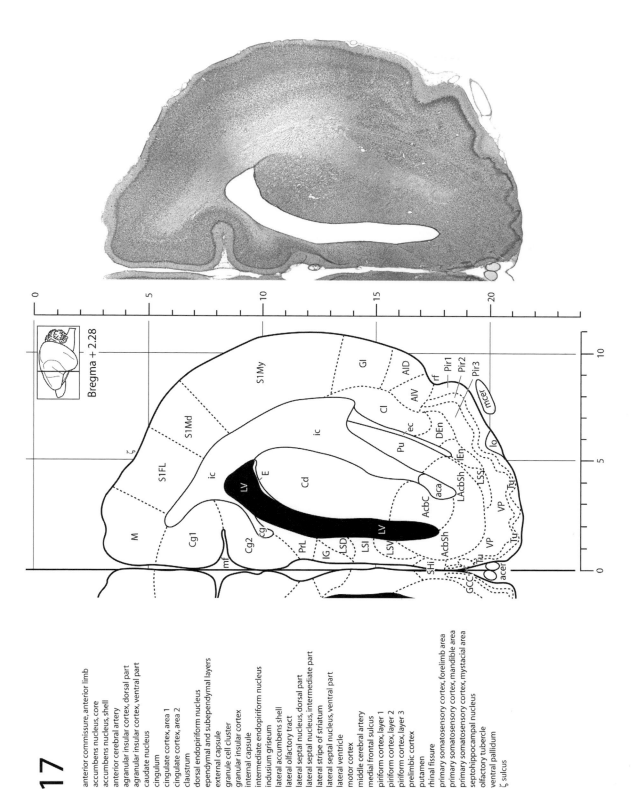

Bregma + 2.28

aca	anterior commissure, anterior limb
AcbC	accumbens nucleus, core
AcbSh	accumbens nucleus, shell
acer	anterior cerebral artery
AID	agranular insular cortex, dorsal part
AIV	agranular insular cortex, ventral part
Cd	caudate nucleus
cg	cingulum
Cg1	cingulate cortex, area 1
Cg2	cingulate cortex, area 2
Cl	claustrum
DEn	dorsal endopiriform nucleus
E	ependymal and subependymal layers
ec	external capsule
GCC	granule cell cluster
GI	granular insular cortex
ic	internal capsule
IEn	intermediate endopiriform nucleus
IG	indusium griseum
LAcbSh	lateral accumbens shell
lo	lateral olfactory tract
LSD	lateral septal nucleus, dorsal part
LSI	lateral septal nucleus, intermediate part
LSS	lateral stripe of striatum
LSV	lateral septal nucleus, ventral part
LV	lateral ventricle
M	motor cortex
mcer	middle cerebral artery
mf	medial frontal sulcus
Pir1	piriform cortex, layer 1
Pir2	piriform cortex, layer 2
Pir3	piriform cortex, layer 3
PrL	prelimbic cortex
Pu	putamen
rf	rhinal fissure
S1FL	primary somatosensory cortex, forelimb area
S1Md	primary somatosensory cortex, mandible area
S1My	primary somatosensory cortex, mystacial area
SHi	septohippocampal nucleus
Tu	olfactory tubercle
VP	ventral pallidum
ζ	ζ sulcus

W18

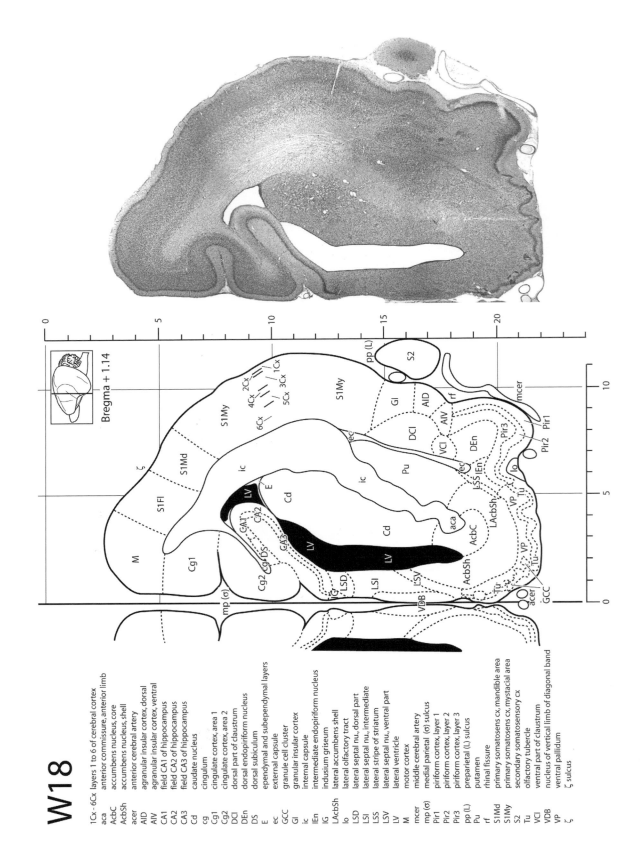

Bregma + 1.14

1Cx - 6Cx   layers 1 to 6 of cerebral cortex
aca   anterior commissure, anterior limb
AcbC   accumbens nucleus, core
AcbSh   accumbens nucleus, shell
acer   anterior cerebral artery
AID   agranular insular cortex, dorsal
AIV   agranular insular cortex, ventral
CA1   field CA1 of hippocampus
CA2   field CA2 of hippocampus
CA3   field CA3 of hippocampus
Cd   caudate nucleus
cg   cingulum
Cg1   cingulate cortex, area 1
Cg2   cingulate cortex, area 2
DCl   dorsal part of claustrum
DEn   dorsal endopiriform nucleus
DS   dorsal subiculum
E   ependymal and subependymal layers
ec   external capsule
GCC   granule cell cluster
GI   granular insular cortex
ic   internal capsule
IEn   intermediate endopiriform nucleus
IG   indusium griseum
LAcbSh   lateral accumbens shell
lo   lateral olfactory tract
LSD   lateral septal nu., dorsal part
LSI   lateral septal nu., intermediate
LSS   lateral stripe of striatum
LSV   lateral septal nu., ventral part
LV   lateral ventricle
M   motor cortex
mcer   middle cerebral artery
mp (σ)   medial parietal (σ) sulcus
Pir1   piriform cortex, layer 1
Pir2   piriform cortex, layer 2
Pir3   piriform cortex, layer 3
pp (L)   preparietal (L) sulcus
Pu   putamen
rf   rhinal fissure
S1Md   primary somatosens cx, mandible area
S1My   primary somatosens cx, mystacial area
S2   secondary somatosensory cx
Tu   olfactory tubercle
VCl   ventral part of claustrum
VDB   nucleus of vertical limb of diagonal band
VP   ventral pallidum
ζ   ζ sulcus

W19

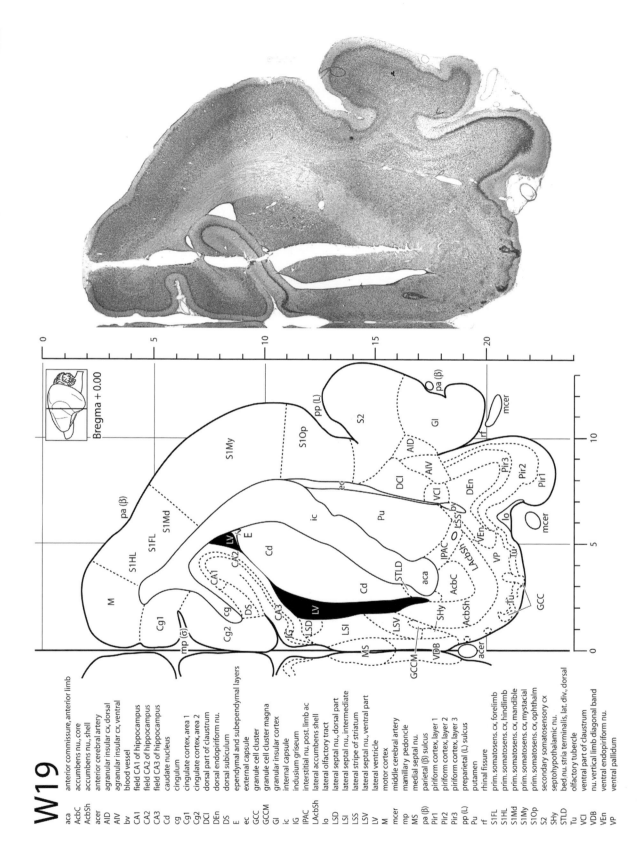

Bregma + 0.00

aca anterior commissure, anterior limb
AcbC accumbens nu., core
AcbSh accumbens nu., shell
acer anterior cerebral artery
AID agranular insular cx, dorsal
AIV agranular insular cx, ventral
bv blood vessel
CA1 field CA1 of hippocampus
CA2 field CA2 of hippocampus
CA3 field CA3 of hippocampus
Cd caudate nucleus
cg cingulum
Cg1 cingulate cortex, area 1
Cg2 cingulate cortex, area 2
DCl dorsal part of claustrum
DEn dorsal endopiriform nu.
DS dorsal subiculum
E ependymal and subependymal layers
ec external capsule
GCC granule cell cluster
GCCM granule cell cluster magna
GI granular insular cortex
ic internal capsule
IG indusium griseum
IPAC interstitial nu. post. limb ac
LAcbSh lateral accumbens shell
lo lateral olfactory tract
LSD lateral septal nu., dorsal part
LSI lateral septal nu., intermediate
LSS lateral stripe of striatum
LSV lateral septal nu., ventral part
LV lateral ventricle
M motor cortex
mcer middle cerebral artery
mp mamillary pedoncle
MS medial septal nu.
pa (β) parietal (β) sulcus
Pir1 piriform cortex, layer 1
Pir2 piriform cortex, layer 2
Pir3 piriform cortex, layer 3
pp (L) preparietal (L) sulcus
Pu putamen
rf rhinal fissure
S1FL prim. somatosens. cx, forelimb
S1HL prim. somatosens. cx, hindlimb
S1Md prim. somatosens. cx, mandible
S1My prim. somatosens. cx, mystacial
S1Op prim. somatosens. cx, ophthalm
S2 secondary somatosensory cx
SHy septohypothalamic nu.
STLD bed nu. stria terminalis, lat. div., dorsal
Tu olfactory tubercle
VCl ventral part of claustrum
VDB nu. vertical limb diagonal band
VEn ventral endopiriform nu.
VP ventral pallidum

W20

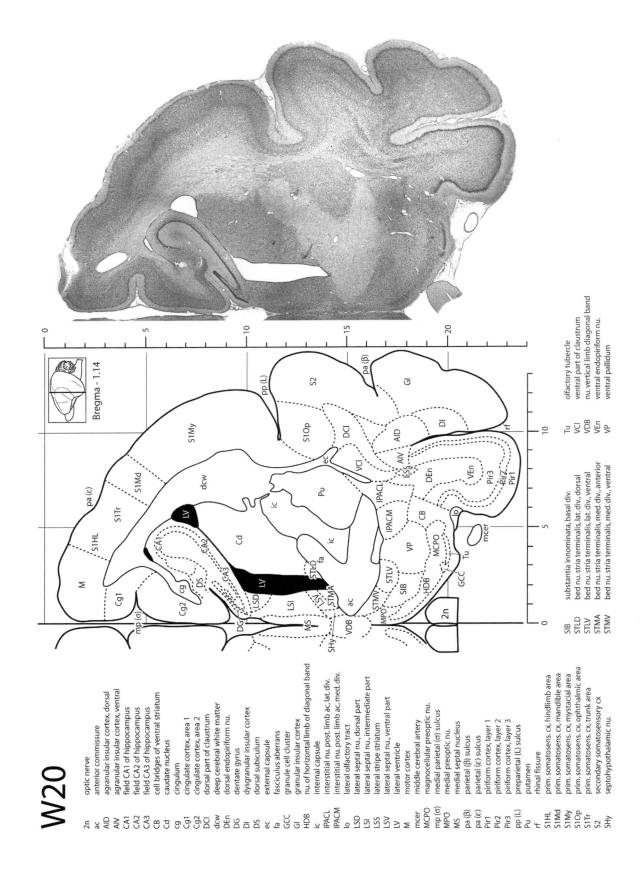

Bregma - 1.14

2n	optic nerve	
ac	anterior commissure	
AID	agranular insular cortex, dorsal	
AIV	agranular insular cortex, ventral	
CA1	field CA1 of hippocampus	
CA2	field CA2 of hippocampus	
CA3	field CA3 of hippocampus	
CB	cell bridges of ventral striatum	
Cd	caudate nucleus	
cg	cingulum	
Cg1	cingulate cortex, area 1	
Cg2	cingulate cortex, area 2	
DCl	dorsal part of claustrum	
dcw	deep cerebral white matter	
DEn	dorsal endopiriform nu.	
DG	dentate gyrus	
DI	dysgranular insular cortex	
DS	dorsal subiculum	
ec	external capsule	
fa	fasciculus aberrans	
GCC	granule cell cluster	
GI	granular insular cortex	
HDB	nu. of horizontal limb of diagonal band	
ic	internal capsule	
IPACL	interstitial nu. post. limb ac, lat. div.	
IPACM	interstitial nu. post. limb ac, med. div.	
lo	lateral olfactory tract	
LSD	lateral septal nu., dorsal part	
LSI	lateral septal nu., intermediate part	
LSS	lateral stripe striatum	
LSV	lateral septal nu., ventral part	
LV	lateral ventricle	
M	motor cortex	
mcer	middle cerebral artery	
MCPO	magnocellular preoptic nu.	
mp (σ)	medial parietal (σ) sulcus	
MPO	medial preoptic nu.	
MS	medial septal nucleus	
pa (β)	parietal (β) sulcus	
pa (ε)	parietal (ε) sulcus	
Pir1	piriform cortex, layer 1	
Pir2	piriform cortex, layer 2	
Pir3	piriform cortex, layer 3	
pp (L)	preparietal (L) sulcus	
Pu	putamen	
rf	rhinal fissure	
S1HL	prim. somatosens. cx, hindlimb area	
S1Md	prim. somatosens. cx, mandible area	
S1My	prim. somatosens. cx, mystacial area	
S1Op	prim. somatosens. cx, ophthalmic area	
S1Tr	prim. somatosens. cx, trunk area	
S2	secondary somatosensory cx	
SHy	septohypothalamic nu.	

SIB	substantia innominata, basal div.	
STLD	bed nu. stria terminalis, lat. div, dorsal	
STLV	bed nu. stria terminalis, lat. div, ventral	
STMA	bed nu. stria terminalis, med. div., anterior	
STMV	bed nu. stria terminalis, med. div, ventral	

Tu	olfactory tubercle
VCl	ventral part of claustrum
VDB	nu. vertical limb diagonal band
VEn	ventral endopiriform nu.
VP	ventral pallidum

W21

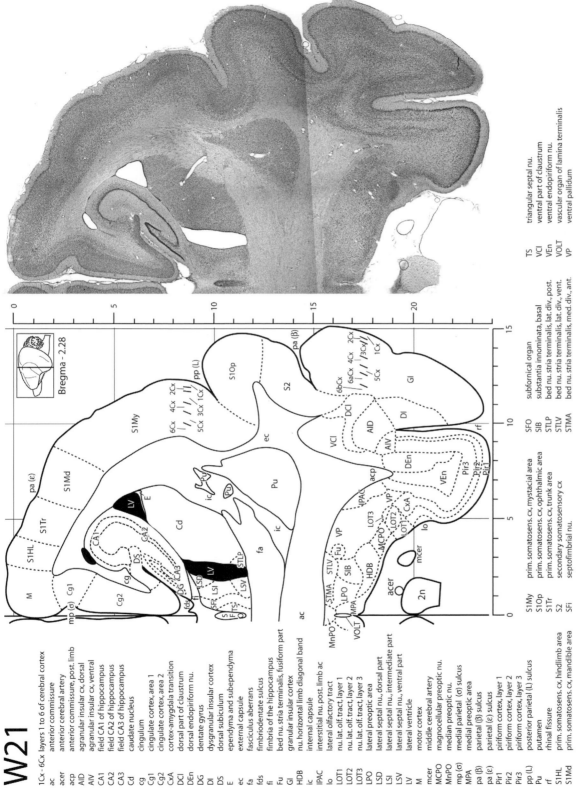

Bregma - 2.28

1Cx - 6Cx layers 1 to 6 of cerebral cortex
ac anterior commissure
acer anterior cerebral artery
acp anterior commissure, post. limb
AID agranular insular cx, dorsal
AIV agranular insular cx, ventral
CA1 field CA1 of hippocampus
CA2 field CA2 of hippocampus
CA3 field CA3 of hippocampus
Cd caudate nucleus
cg cingulum
Cg1 cingulate cortex, area 1
Cg2 cingulate cortex, area 2
CxA cortex-amygdala transition
DCl dorsal part of claustrum
DEn dorsal endopiriform nu.
DG dentate gyrus
DI dysgranular insular cortex
DS dorsal subiculum
E ependyma and subependyma
ec external capsule
fa fasciculus aberrans
fds fimbriodentate sulcus
fi fimbria of the hippocampus
Fu fimbria of the hippocampus, fusiform part
GI granular insular cortex
HDB nu. horizontal limb diagonal band
ic internal capsule
IPAC interstitial nu. post. limb ac
lo lateral olfactory tract
LOT1 nu. lat. olf. tract, layer 1
LOT2 nu. lat. olf. tract, layer 2
LOT3 nu. lat. olf. tract, layer 3
LPO lateral preoptic area
LSD lateral septal nu., dorsal part
LSI lateral septal nu., intermediate part
LSV lateral septal nu., ventral part
LV lateral ventricle
M motor cortex
mcer middle cerebral artery
MCPO magnocellular preoptic nu.
MnPO median preoptic nu.
mp (σ) medial parietal (σ) sulcus
MPA medial preoptic area
pa (β) parietal (β) sulcus
pa (ε) parietal (ε) sulcus
Pir1 piriform cortex, layer 1
Pir2 piriform cortex, layer 2
Pir3 piriform cortex, layer 3
pp (L) posterior parietal (L) sulcus
Pu putamen
rf rhinal fissure
S1HL prim. somatosens.cx, hindlimb area
S1Md prim. somatosens.cx, mandible area

S1My prim. somatosens. cx, mystacial area
S1Op prim. somatosens. cx, ophthalmic area
S1Tr prim. somatosens.cx, trunk area
S2 secondary somatosensory cx
SFi septofimbrial nu.

SFO subfornical organ
SIB substantia innominata, basal
STLP bed nu. stria terminalis, lat. div., post.
STLV bed nu. stria terminalis, lat. div., vent.
STMA bed nu. stria terminalis, med. div., ant.

TS triangular septal nu.
VCl ventral part of claustrum
VEn ventral endopiriform nu.
VOLT vascular organ of lamina terminalis
VP ventral pallidum

W22

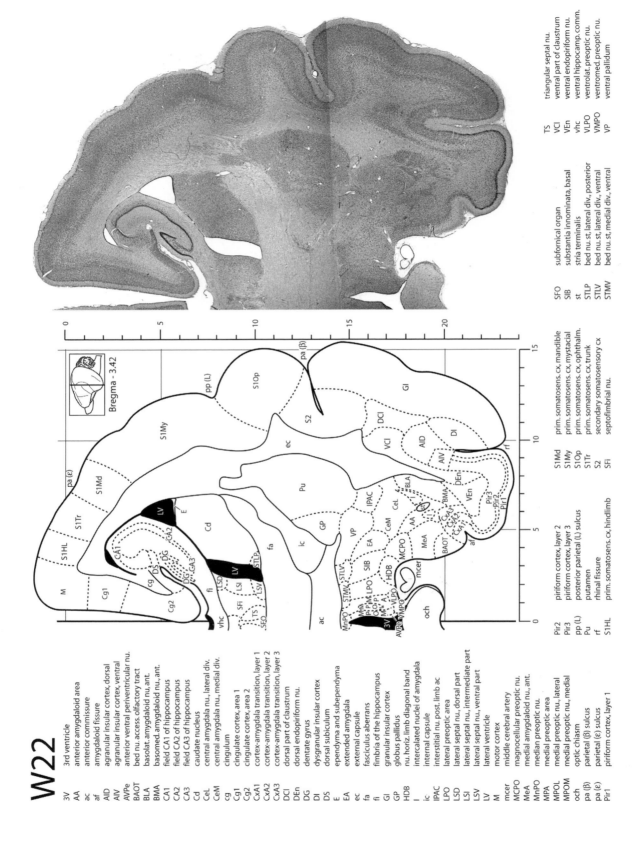

Bregma -3.42

3V	3rd ventricle		Pir2	piriform cortex, layer 2
AA	anterior amygdaloid area		Pir3	piriform cortex, layer 3
ac	anterior commissure		pp (L)	posterior parietal (L) sulcus
af	amygdaloid fissure		Pu	putamen
AID	agranular insular cortex, dorsal		rf	rhinal fissure
AIV	agranular insular cortex, ventral		S1HL	prim. somatosens. cx, hindlimb
AVPe	anterior ventral periventricular nu.		S1Md	prim. somatosens. cx, mandible
BAOT	bed nu. access. olfactory tract		S1My	prim. somatosens. cx, mystacial
BLA	basolat. amygdaloid nu., ant.		S1Op	prim. somatosens. cx, ophthalm.
BMA	basomed. amygdaloid nu., ant.		S1Tr	prim. somatosens. cx, trunk
CA1	field CA1 of hippocampus		S2	secondary somatosensory cx
CA2	field CA2 of hippocampus		SFi	septofimbrial nu.
CA3	field CA3 of hippocampus			
Cd	caudate nucleus			
CeL	central amygdala nu., lateral div.			
CeM	central amygdala nu., medial div.			
cg	cingulum			
Cg1	cingulate cortex, area 1			
Cg2	cingulate cortex, area 2			
CxA1	cortex-amygdala transition, layer 1			
CxA2	cortex-amygdala transition, layer 2			
CxA3	cortex-amygdala transition, layer 3			
DCl	dorsal part of claustrum			
DEn	dorsal endopiriform nu.			
DG	dentate gyrus			
DI	dysgranular insular cortex			
DS	dorsal subiculum			
E	ependyma and subependyma			
EA	extended amygdala			
ec	external capsule			
fa	fasciculus aberrans			
fi	fimbria of the hippocampus			
GI	granular insular cortex			
GP	globus pallidus			
HDB	nu. horiz. limb diagonal band			
I	intercalated nuclei of amygdala			
ic	internal capsule			
IPAC	interstitial nu. post. limb ac			
LPO	lateral preoptic area			
LSD	lateral septal nu., dorsal part			
LSI	lateral septal nu., intermediate part			
LSV	lateral septal nu., ventral part			
LV	lateral ventricle			
M	motor cortex			
mcer	middle cerebral artery			
MCPO	magnocellular preoptic nu.			
MeA	medial amygdaloid nu., ant.			
MnPO	median preoptic nu.			
MPA	medial preoptic area			
MPOL	medial preoptic nu., lateral			
MPOM	medial preoptic nu., medial			
och	optic chiasm			
pa (β)	parietal (β) sulcus			
pa (ε)	parietal (ε) sulcus			
Pir1	piriform cortex, layer 1			

SFO	subfornical organ		TS	triangular septal nu.
SIB	substantia innominata, basal		VCl	ventral part of claustrum
st	stria terminalis		VEn	ventral endopiriform nu.
STLP	bed nu. st, lateral div., posterior		vhc	ventral hippocamp. comm.
STLV	bed nu. st, lateral div., ventral		VLPO	ventrolat. preoptic nu.
STMV	bed nu. st, medial div., ventral		VMPO	ventromed. preoptic nu.
			VP	ventral pallidum

W23

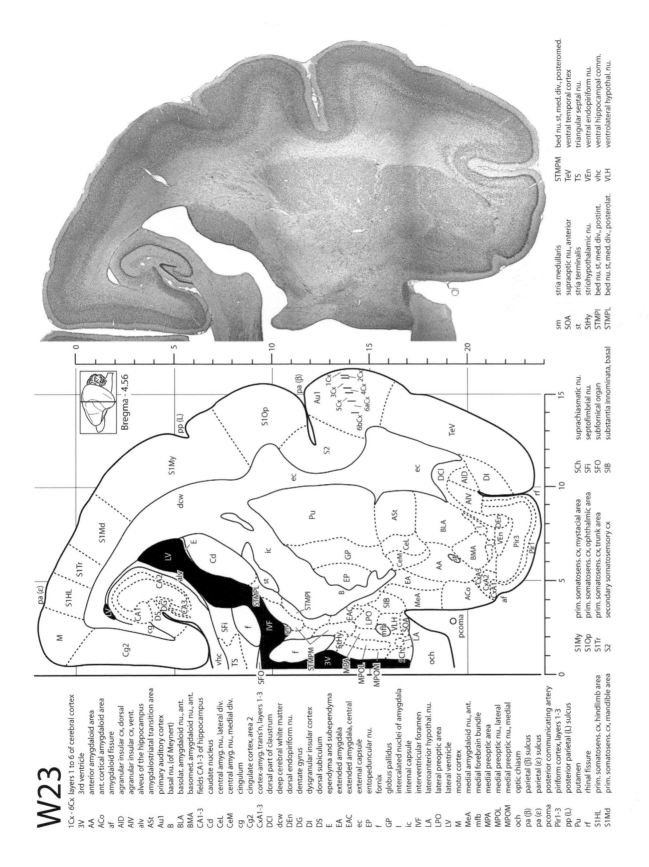

Bregma - 4.56

1Cx - 6Cx layers 1 to 6 of cerebral cortex
3V   3rd ventricle
AA   anterior amygdaloid area
ACo   ant. cortical amygdaloid area
af   amygdaloid fissure
AID   agranular insular cx, dorsal
AIV   agranular insular cx, vent.
alv   alveus of the hippocampus
ASt   amygdalostriatal transition area
Au1   primary auditory cortex
B   basal nu. (of Meynert)
BLA   basolat. amygdaloid nu., ant.
BMA   basomed. amygdaloid nu., ant.
CA1-3   fields CA1-3 of hippocampus
Cd   caudate nucleus
CeL   central amyg. nu., lateral div.
CeM   central amyg. nu., medial div.
cg   cingulum
Cg2   cingulate cortex, area 2
CxA1-3   cortex-amyg. trans'n, layers 1-3
DCl   dorsal part of claustrum
dcw   deep cerebral white matter
DEn   dorsal endopiriform nu.
DG   dentate gyrus
DI   dysgranular insular cortex
DS   dorsal subiculum
E   ependyma and subependyma
EA   extended amygdala
EAC   extended amygdala, central
ec   external capsule
EP   entopeduncular nu.
f   fornix
GP   globus pallidus
I   intercalated nuclei of amygdala
ic   internal capsule
IVF   interventricular foramen
LA   lateroanterior hypothal. nu.
LPO   lateral preoptic area
LV   lateral ventricle
M   motor cortex
MeA   medial amygdaloid nu., ant.
mfb   medial forebrain bundle
MPA   medial preoptic area
MPOL   medial preoptic nu., lateral
MPOM   medial preoptic nu., medial
och   optic chiasm
pa (β)   parietal (β) sulcus
pa (ε)   parietal (ε) sulcus
pcoma   posterior communicating artery
Pir1-3   piriform cortex, layers 1-3
pp (L)   posterior parietal (L) sulcus
Pu   putamen
rf   rhinal fissure
S1HL   prim. somatosens. cx, hindlimb area
S1Md   prim. somatosens. cx, mandible area

S1My   prim. somatosens. cx, mystacial area
S1Op   prim. somatosens. cx, ophthalmic area
S1Tr   prim. somatosens. cx, trunk area
S2   secondary somatosensory cx

SCh   suprachiasmatic nu.
SFi   septofimbrial nu.
SFO   subfornical organ
SIB   substantia innominata, basal

sm   stria medullaris
SOA   supraoptic nu., anterior
st   stria terminalis
StHy   striohypothalmic nu.
STMPl   bed nu. st, med. div., postint.
STMPL   bed nu. st, med. div., posterolat.

STMPM   bed nu. st, med. div., posteromed.
TeV   ventral temporal cortex
TS   triangular septal nu.
VEn   ventral endopiriform nu.
vhc   ventral hippocampal comm.
VLH   ventrolateral hypothal. nu.

W24

Bregma - 5.70

Abbreviations:

3V — 3rd ventricle
ACo — ant. cortical amygdaloid area
AD — anterodorsal thalamic nu.
AHD — ant. hypothal. area, dorsal
AHC — ant. hypothal. area, central
AIP — agranular insular cx, post.
AIV — agranular insular cx, vent.
alv — alveus of the hippocampus
AM — anteromedial thalamic nu.
ANS — access. neurosecretory nu.
ASt — amygdalostriatal trans'n area
Au1 — primary auditory cortex
AVDM — ant. vent. thal. nu., dorsomed.
AVVL — ant. vent. thal. nu., ventrolat.
B — basal nu. (of Meynert)
BLA — basolat. amyg. nu., ant.
BMA — basomed. amyg. nu., ant.
CA1-3 — fields CA1-3 of hippocamp.
Cd — caudate nucleus
CeL — central amyg. nu., lat. div.
CeM — central amyg. nu., med. div.
cg — cingulum
Cg2 — cingulate cortex, area 2
CxA — cortex-amyg. transition
D3V — dorsal 3rd ventricle
dcw — deep cerebral white matter
DEn — dorsal endopiriform nu.
DG — dentate gyrus
DS — dorsal subiculum
EA — extended amygdala
EAM — extended amygdala, med.
ec — external capsule
EP — entopeduncular nu.
ESO — episupraoptic nu.
f — fornix
fi — fimbria of hippocampus
GP — globus pallidus
I — intercalated nuclei amyg.
IAD — interanterodorsal thal. nu.
ic — internal capsule
La — lateral amygdaloid nu.
LV — lateral ventricle
MeA — medial amygdala
mfb — medial forebrain bundle
opt — optic tract
pa (β) — parietal (β) sulcus
pa (ε) — parietal (ε) sulcus
PaL — paravent. hypoth. nu., lateral
PaM — paravent. hy. nu., magnocell.
PC — paracentral thalamic nu.
Pe — periventricular hypothal. nu.
Pir1-3 — piriform cortex, layers 1-3
PLH — peduncular part of lateral hypoth.
PT — paratenial thalamic nu.

Pu — putamen
PVA — paraventricular thal. nu., ant.
RCh — retrochiasmatic area
RChL — retrochiasmatic area, lateral
rf — rhinal fissure
Rt — reticular thal. nu. (prethalamus)
RtSt — reticulostrial nu.
S1HL — prim. somatosens. cx, hindlimb

S1Md — prim. somato. cx, mandible
S1My — prim. somatosens. cx, mystacial
S1Op — prim. somatosens. cx, ophthalm.
S1Tr — prim. somatosens. cx, trunk
S2 — secondary somatosensory cx
sm — stria medullaris
SOA — supraoptic nu., anterior

sod — supraoptic decussation
st — stria terminalis
StHy — striohypothalamic nu.
STIA — bed nu. st, intraamyg. div.
TeV — ventral temporal cortex
VEn — ventral endopiriform nu.
vhc — ventral hippocamp. comm.

W25

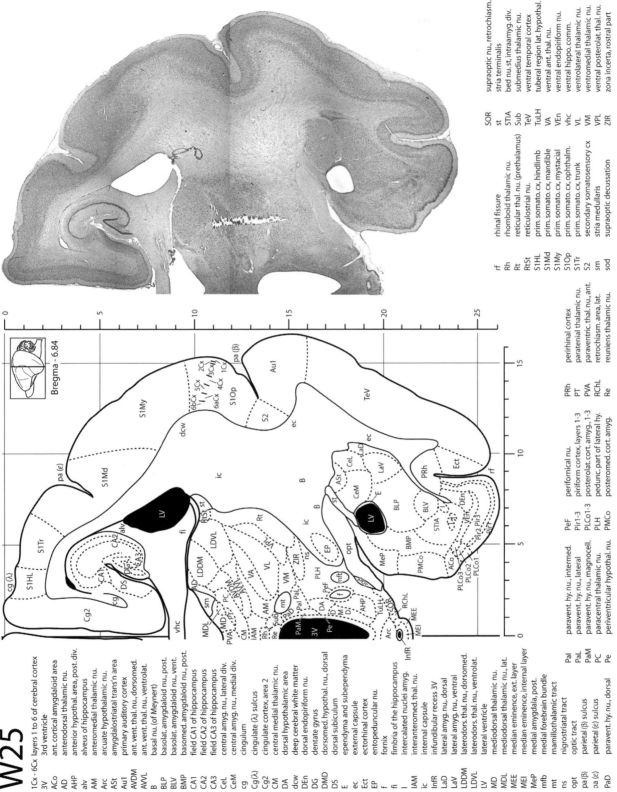

Bregma - 6.84

W26

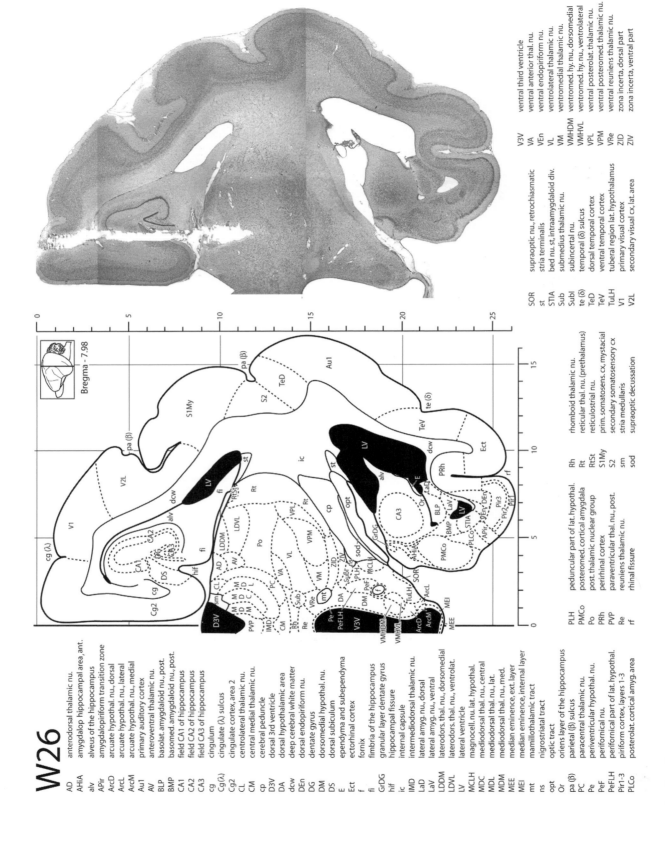

Bregma - 7.98

W27

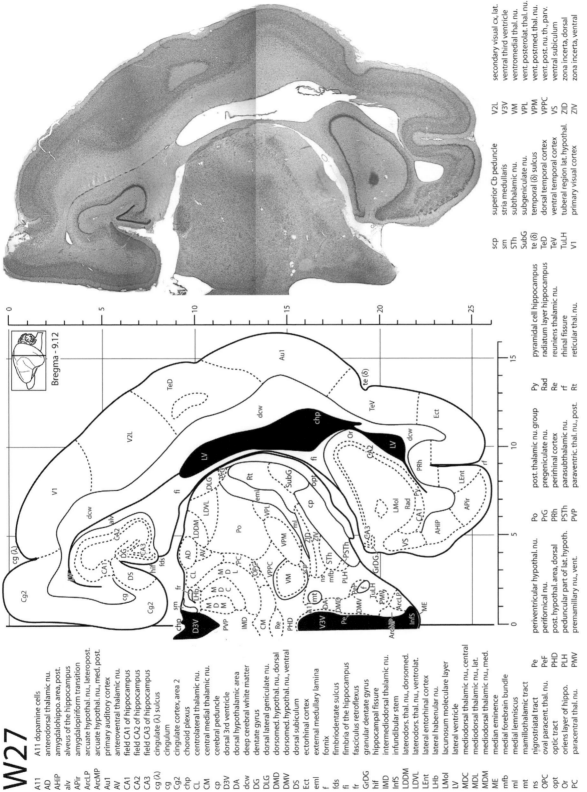

Bregma - 9.12

A11	A11 dopamine cells	
AD	anterodorsal thalamic nu.	
AHiP	amygdalohippo. area, post.	
alv	alveus of the hippocampus	
APir	amygdalopiriform transition	
ArcLP	arcuate hypothal. nu., lateropost.	
ArcMP	arcuate hypothal. nu., med. post.	
Au1	primary auditory cortex	
AV	anteroventral thalamic nu.	
CA1	field CA1 of hippocampus	
CA2	field CA2 of hippocampus	
CA3	field CA3 of hippocampus	
cg (λ)	cingulate (λ) sulcus	
cg	cingulum	
Cg2	cingulate cortex, area 2	
chp	choroid plexus	
CL	centrolateral thalamic nu.	
CM	central medial thalamic nu.	
cp	cerebral peduncle	
D3V	dorsal 3rd ventricle	
DA	dorsal hypothalamic area	
dcw	deep cerebral white matter	
DG	dentate gyrus	
DLG	dorsal lateral geniculate nu.	
DMD	dorsomed. hypothal. nu., dorsal	
DMV	dorsomed. hypothal. nu., ventral	
DS	dorsal subiculum	
Ect	ectorhinal cortex	
eml	external medullary lamina	
f	fornix	
fds	fimbriodentate sulcus	
fi	fimbria of the hippocampus	
fr	fasciculus retroflexus	
GrDG	granular dentate gyrus	
hif	hippocampal fissure	
IMD	intermediodorsal thalamic nu.	
InfS	infundibular stem	
LDDM	laterodors. thal. nu., dorsomed.	
LDVL	laterodors. thal. nu., ventrolat.	
LEnt	lateral entorhinal cortex	
LHb	lateral habenular nu.	
LMol	lacunosum moleculare layer	
LV	lateral ventricle	
MDC	mediodorsal thalamic nu., central	
MDL	mediodorsal thalamic nu., lat.	
MDM	mediodorsal thalamic nu., med.	
ME	median eminence	
mfb	medial forebrain bundle	
ml	medial lemniscus	
mt	mamillothalamic tract	
ns	nigrostriatal tract	Pe	periventricular hypothal. nu.	Po	post. thalamic nu. group	Py	pyramidal cell hippocampus
OPC	oval paracent. thal. nu.	PeF	perifornical nu.	PrG	pregeniculate nu.	Rad	radiatum layer hippocampus
opt	optic tract	PHD	post. hypothal. area, dorsal	PRh	perirhinal cortex	Re	reuniens thalamic nu.
Or	oriens layer of hippo.	PLH	peduncular part of lat. hypoth.	PSTh	parasubthalamic nu.	rf	rhinal fissure
PC	paracentral thal. nu.	PMV	premamillary nu., vent.	PVP	paraventric. thal. nu., post.	Rt	reticular thal. nu.

scp	superior Cb peduncle	V2L	secondary visual cx. lat.
sm	stria medullaris	V3V	ventral third ventricle
STh	subthalamic nu.	VM	ventromedial thal. nu.
SubG	subgeniculate nu.	VPL	vent. posterolat. thal. nu.
te (δ)	temporal (δ) sulcus	VPM	vent. postmed. thal. nu.
TeD	dorsal temporal cortex	VPPC	vent. post. nu. th., parv.
TeV	ventral temporal cortex	VS	ventral subiculum
TuLH	tuberal region lat. hypothal.	ZID	zona incerta, dorsal
V1	primary visual cortex	ZIV	zona incerta, ventral

W28

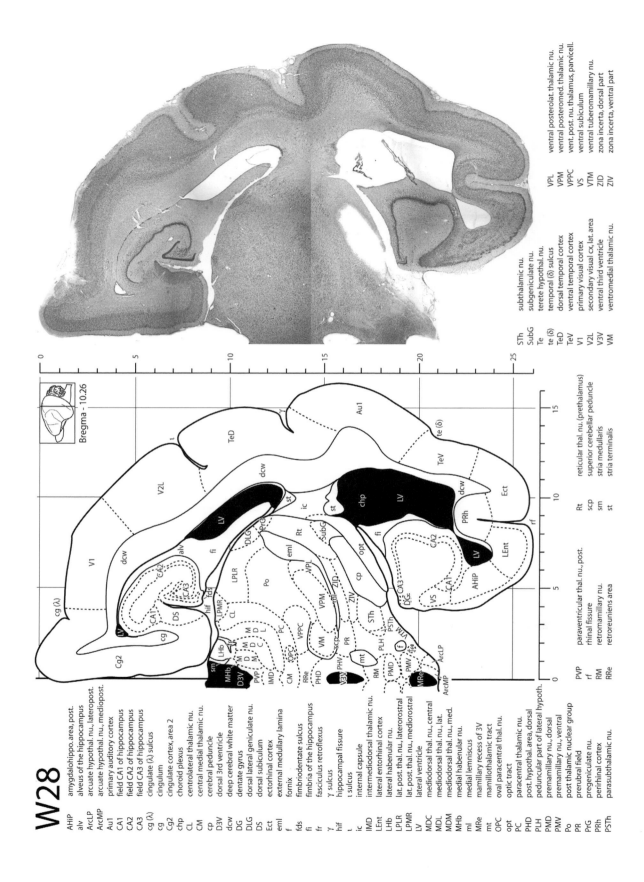

Bregma - 10.26

AHiP	amygdalohippo. area, post.
alv	alveus of the hippocampus
ArcLP	arcuate hypothal. nu., lateropost.
ArcMP	arcuate hypothal. nu., mediopost.
Au1	primary auditory cortex
CA1	field CA1 of hippocampus
CA2	field CA2 of hippocampus
CA3	field CA3 of hippocampus
cg	cingulum
cg (λ)	cingulate (λ) sulcus
Cg2	cingulate cortex, area 2
chp	choroid plexus
CL	centrolateral thalamic nu.
CM	central medial thalamic nu.
cp	cerebral peduncle
D3V	dorsal 3rd ventricle
dcw	deep cerebral white matter
DG	dentate gyrus
DLG	dorsal lateral geniculate nu.
DS	dorsal subiculum
Ect	ectorhinal cortex
eml	external medullary lamina
f	fornix
fds	fimbriodentate sulcus
fi	fimbria of the hippocampus
fr	fasciculus retroflexus
γ	γ sulcus
hif	hippocampal fissure
ι	ι sulcus
ic	internal capsule
IMD	intermediodorsal thalamic nu.
LEnt	lateral entorhinal cortex
LHb	lateral habenular nu.
LPLR	lat. post. thal. nu., laterorostral
LPMR	lat. post. thal. nu., mediorostral
LV	lateral ventricle
MDC	mediodorsal thal. nu., central
MDL	mediodorsal thal. nu., lat.
MDM	mediodorsal thal. nu., med.
MHb	medial habenular nu.
ml	medial lemniscus
MRe	mamillary recess of 3V
mt	mamillothalamic tract
OPC	oval paracentral thal. nu.
opt	optic tract
PC	paracentral thalamic nu.
PHD	post. hypothal. area, dorsal
PLH	peduncular part of lateral hypoth.
PMD	premamillary nu., dorsal
PMV	premamillary nu., ventral
Po	post thalamic nuclear group
PR	prerubral field
PrG	pregeniculate nu.
PRh	perirhinal cortex
PSTh	parasubthalamic nu.

PVP	paraventricular thal. nu., post.
rf	rhinal fissure
RM	retromamillary nu.
RRe	retroreuniens area

Rt	reticular thal. nu. (prethalamus)
scp	superior cerebellar peduncle
sm	stria medullaris
st	stria terminalis

STh	subthalamic nu.
SubG	subgeniculate nu.
Te	terete hypothal. nu.
te (δ)	temporal (δ) sulcus
TeD	dorsal temporal cortex
TeV	ventral temporal cortex
V1	primary visual cortex
V2L	secondary visual cx, lat. area
V3V	ventral third ventricle
VM	ventromedial thalamic nu.

VPL	ventral posterolat. thalamic nu.
VPM	ventral posteromed. thalamic nu.
VPPC	vent. post. nu. thalamus, parvicell.
VS	ventral subiculum
VTM	ventral tuberomamillary part
ZID	zona incerta, dorsal part
ZIV	zona incerta, ventral part

W29

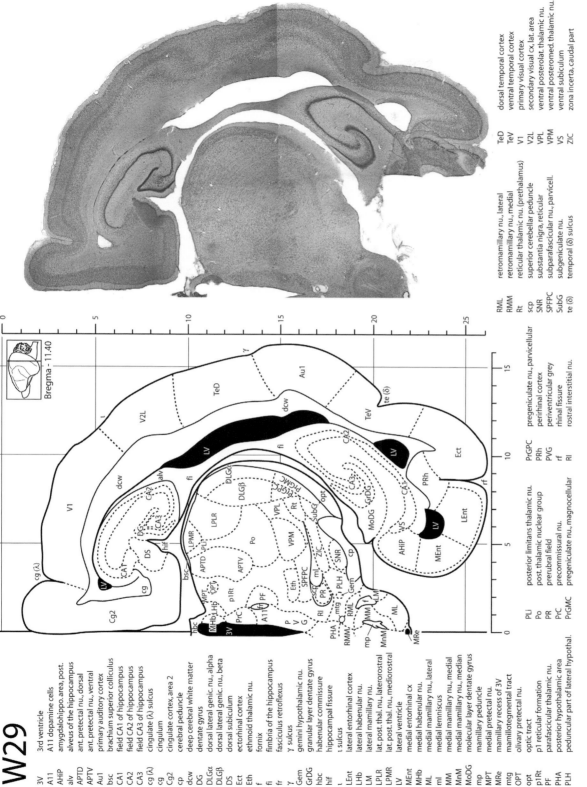

Bregma - 11.40

3V	3rd ventricle
A11	A11 dopamine cells
AHiP	amygdalohippo. area, post.
alv	alveus of the hippocampus
APTD	ant. pretectal nu., dorsal
APTV	ant. pretectal nu., ventral
Au1	primary auditory cortex
bsc	brachium superior colliculus
CA1	field CA1 of hippocampus
CA2	field CA2 of hippocampus
CA3	field CA3 of hippocampus
cg (λ)	cingulate (λ) sulcus
cg	cingulum
Cg2	cingulate cortex, area 2
cp	cerebral peduncle
dcw	deep cerebral white matter
DG	dentate gyrus
DLGα	dorsal lateral genic. nu., alpha
DLGβ	dorsal lateral genic. nu., beta
DS	dorsal subiculum
Ect	ectorhinal cortex
Eth	ethmoid thalamic nu.
f	fornix
fi	fimbria of the hippocampus
fr	fasciculus retroflexus
γ	γ sulcus
Gem	gemini hypothalamic nu.
GrDG	granular layer dentate gyrus
hbc	habenular commissure
hif	hippocampal fissure
ι	ι sulcus
LEnt	lateral entorhinal cortex
LHb	lateral habenular nu.
LM	lateral mamillary nu.
LPLR	lat. post. thal. nu., laterorostral
LPMR	lat. post. thal. nu., mediorostral
LV	lateral ventricle
MEnt	medial entorhinal cx
MHb	medial habenular nu.
ML	medial mamillary nu., lateral
ml	medial lemniscus
MM	medial mamillary nu., medial
MnM	medial mamillary nu., median
MoDG	molecular layer dentate gyrus
mp	mamillary peduncle
MPT	medial pretectal nu.
MRe	mamillary recess of 3V
mtg	mamillotegmental tract
OPT	olivary pretectal nu.
opt	optic tract
p1Rt	p1 reticular formation
PF	parafascicular thalamic nu.
PHA	posterior hypothalamic area
PLH	peduncular part of lateral hypothal.

PLi	posterior limitans thalamic nu.
Po	post. thalamic nuclear group
PR	prerubral field
PrC	precommissural nu.
PrGMC	pregeniculate nu., magnocellular
PrGPC	pregeniculate nu., parvicellular
PRh	perirhinal cortex
PVG	periventricular grey
rf	rhinal fissure
RI	rostral interstitial nu.

RML	retromamillary nu., lateral
RMM	retromamillary nu., medial
Rt	reticular thalamic nu. (prethalamus)
scp	superior cerebellar peduncle
SNR	substantia nigra, reticular
SPFPC	subparafascicular nu., parvicell.
SubG	subgeniculate nu.
te (δ)	temporal (δ) sulcus

TeD	dorsal temporal cortex
TeV	ventral temporal cortex
V1	primary visual cortex
V2L	secondary visual cx, lat. area
VPL	ventral posterolat. thalamic nu.
VPM	ventral posteromed. thalamic nu.
VS	ventral subiculum
ZIC	zona incerta, caudal part

W30

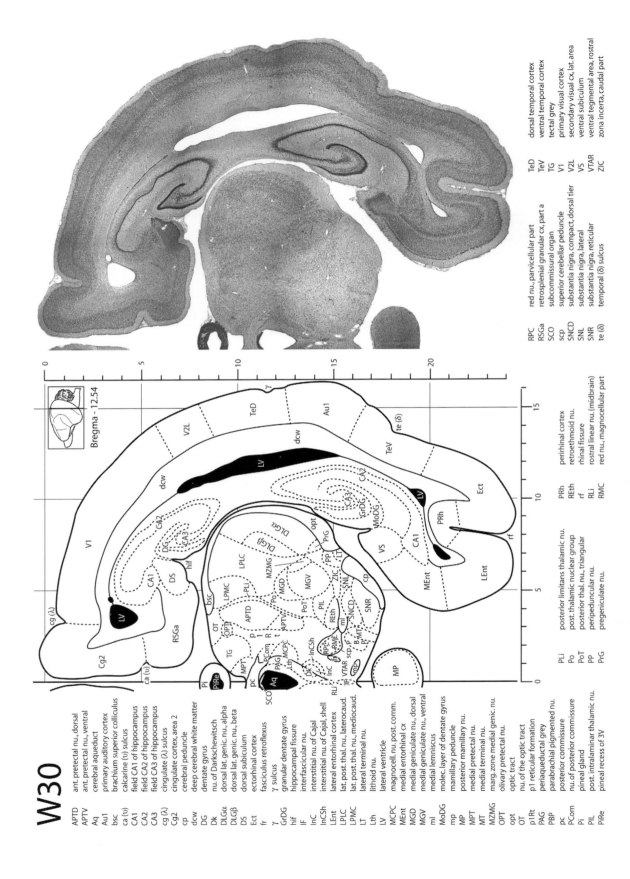

Bregma - 12.54

APTD ant. pretectal nu., dorsal
APTV ant. pretectal nu., ventral
Aq cerebral aqueduct
Au1 primary auditory cortex
bsc brachium superior colliculus
ca (b) calcarine (b) sulcus
CA1 field CA1 of hippocampus
CA2 field CA2 of hippocampus
CA3 field CA3 of hippocampus
cg (λ) cingulate (λ) sulcus
Cg2 cingulate cortex, area 2
cp cerebral peduncle
dcw deep cerebral white matter
DG dentate gyrus
Dk nu. of Darkschewitsch
DLGα dorsal lat. genic. nu., alpha
DLGβ dorsal lat. genic. nu., beta
DS dorsal subiculum
Ect ectorhinal cortex
fr fasciculus retroflexus
γ γ sulcus
GrDG granular dentate gyrus
hif hippocampal fissure
IF interfascicular nu.
InC interstitial nu. of Cajal
InCSh interstitial nu. of Cajal, shell
LEnt lateral entorhinal cortex
LPLC lat. post. thal. nu., laterocaud.
LPMC lat. post. thal. nu., mediocaud.
LT lateral terminal nu.
Lth lithoid nu.
LV lateral ventricle
MCPC magnocell. nu. post. comm.
MEnt medial entorhinal cx
MGD medial geniculate nu., dorsal
MGV medial geniculate nu., ventral
ml medial lemniscus
MoDG molec. layer of dentate gyrus
mp mamillary peduncle
MP posterior mamillary nu.
MPT medial pretectal nu.
MT medial terminal nu.
MZMG marg. zone medial genic. nu.
OPT olivary pretectal nu.
opt optic tract
OT nu. of the optic tract
p1Rt p1 reticular formation
PAG periaqueductal grey
PBP parabrachial pigmented nu.
pc posterior commissure
PCom nu. of posterior commissure
Pi pineal gland
PIL post. intralaminar thalamic nu.
PiRe pineal recess of 3V

PLi posterior limitans thalamic nu.
Po post. thalamic nuclear group
PoT posterior thal. nu., triangular
PP peripeduncular nu.
PrG pregeniculate nu.

PRh perirhinal cortex
REth retroethmoid nu.
rf rhinal fissure
RLi rostral linear nu. (midbrain)
RMC red nu, magnocellular part

RPC red nu, parvicellular part
RSGa retrosplenial granular cx, part a
SCO subcommissural organ
scp superior cerebellar peduncle
SNCD substantia nigra, compact, dorsal tier
SNL substantia nigra, lateral
SNR substantia nigra, reticular
te (δ) temporal (δ) sulcus

TeD dorsal temporal cortex
TeV ventral temporal cortex
TG tectal grey
V1 primary visual cortex
V2L secondary visual cx, lat. area
VS ventral subiculum
VTAR ventral tegmental area, rostral
ZIC zona incerta, caudal part

W31

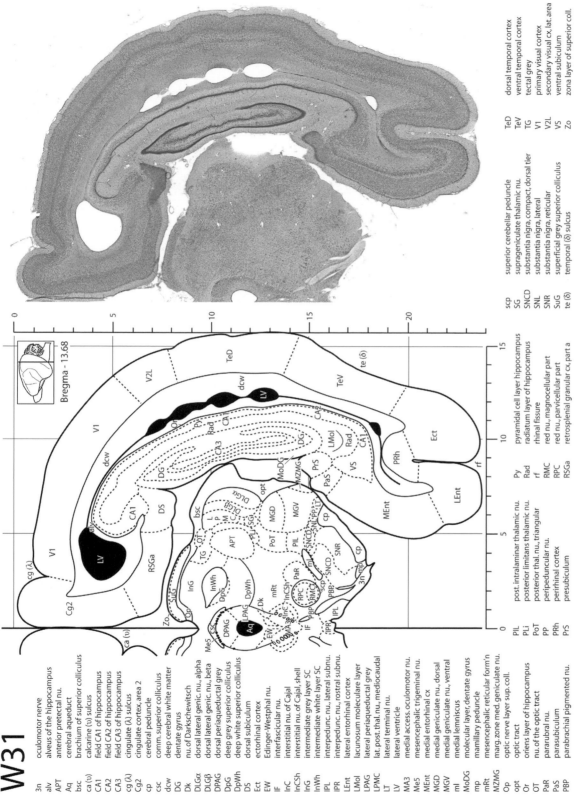

Bregma - 13.68

3n	oculomotor nerve
alv	alveus of the hippocampus
APT	anterior pretectal nu.
Aq	cerebral aqueduct
bsc	brachium of superior colliculus
ca (υ)	calcarine (υ) sulcus
CA1	field CA1 of hippocampus
CA2	field CA2 of hippocampus
CA3	field CA3 of hippocampus
cg (λ)	cingulate (λ) sulcus
Cg2	cingulate cortex, area 2
cp	cerebral peduncle
csc	comm. superior colliculus
dcw	deep cerebral white matter
DG	dentate gyrus
Dk	nu. of Darkschewitsch
DLGα	dorsal lateral genic. nu., alpha
DLGβ	dorsal lateral genic. nu., beta
DPAG	dorsal periaqueductal grey
DpG	deep grey superior colliculus
DpWh	deep white superior colliculus
DS	dorsal subiculum
Ect	ectorhinal cortex
EW	Edinger-Westphal nu.
IF	interfascicular nu.
InC	interstitial nu. of Cajal
InCSh	interstitial nu. of Cajal, shell
InG	intermediate grey layer SC
InWh	intermediate white layer SC
IPL	interpedunc. nu., lateral subnu.
IPR	interpedunc. nu., rostral subnu.
LEnt	lateral entorhinal cortex
LMol	lacunosum moleculare layer
LPAG	lateral periaqueductal grey
LPMC	lat. post.thal. nu., mediocaudal
LT	lateral terminal nu.
LV	lateral ventricle
MA3	medial access. oculomotor nu.
Me5	mesencephalic trigeminal nu.
MEnt	medial entorhinal cx
MGD	medial geniculate nu., dorsal
MGV	medial geniculate nu., ventral
ml	medial lemniscus
MoDG	molecular layer, dentate gyrus
mp	mamillary peduncle
mRt	mesencephalic reticular form'n
MZMG	marg. zone med. geniculate nu.
Op	optic nerve layer sup. coll.
opt	optic tract
Or	oriens layer of hippocampus
OT	nu. of the optic tract
PaR	pararubral nu.
PaS	parasubiculum
PBP	parabrachial pigmented nu.

PIL	post. intralaminar thalamic nu.
PLi	posterior limitans thalamic nu.
PoT	posterior thal. nu., triangular
PP	peripeduncular nu.
PRh	perirhinal cortex
PrS	presubiculum

Py	pyramidal cell layer hippocampus
Rad	radiatum layer of hippocampus
rf	rhinal fissure
RMC	red nu., magnocellular part
RPC	red nu., parvicellular part
RSGa	retrosplenial granular cx, part a

scp	superior cerebellar peduncle
SG	suprageniculate thalamic nu.
SNCD	substantia nigra, compact, dorsal tier
SNL	substantia nigra, lateral
SNR	substantia nigra, reticular
SuG	superficial grey superior colliculus
te (δ)	temporal (δ) sulcus

TeD	dorsal temporal cortex
TeV	ventral temporal cortex
TG	tectal grey
V1	primary visual cortex
V2L	secondary visual cx, lat. area
VS	ventral subiculum
Zo	zona layer of superior coll.

W32

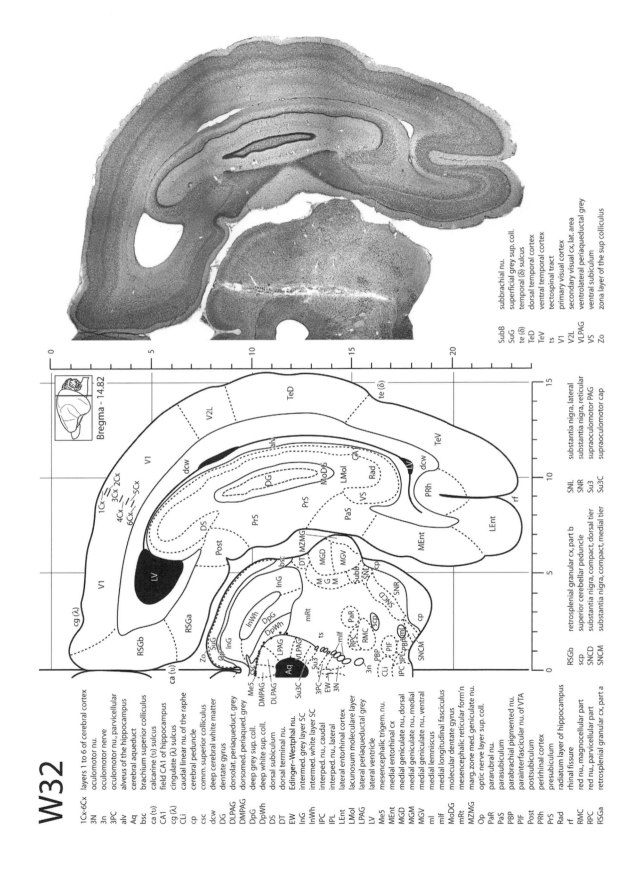

Bregma - 14.82

W33

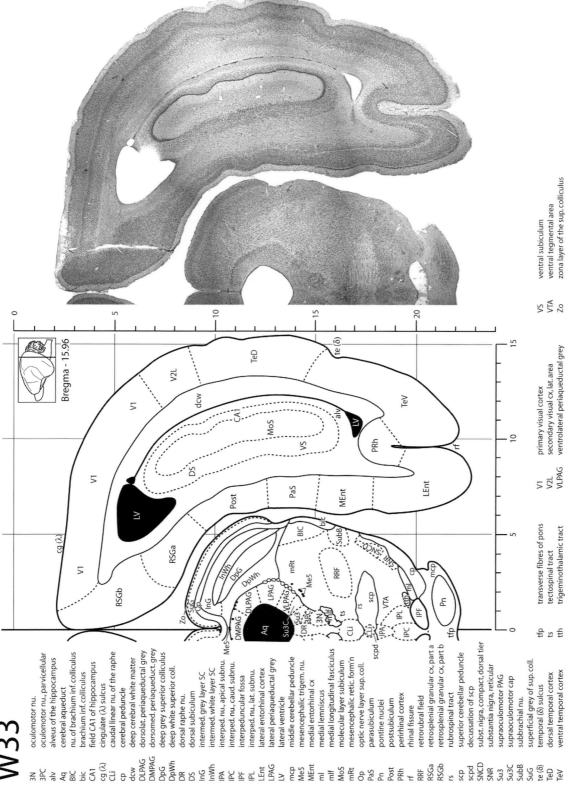

Bregma - 15.96

3N	oculomotor nu.
3PC	oculomotor nu, parvicellular
alv	alveus of the hippocampus
Aq	cerebral aqueduct
BIC	nu. of brachium inf. colliculus
bic	brachium inf. colliculus
CA1	field CA1 of hippocampus
cg (λ)	cingulate (λ) sulcus
CLi	caudal linear nu. of the raphe
cp	cerebral peduncle
dcw	deep cerebral white matter
DLPAG	dorsolat. periaqueductal grey
DMPAG	dorsomed. periaqueduct. grey
DpG	deep grey superior colliculus
DpWh	deep white superior coll.
DR	dorsal raphe nu.
DS	dorsal subiculum
InG	intermed. grey layer SC
InWh	intermed. white layer SC
IPA	interped.nu, apical subnu.
IPC	interped.nu., caud. subnu.
IPF	interpeduncular fossa
IPL	interped.nu., lat. subnu.
LEnt	lateral entorhinal cortex
LPAG	lateral periaqueductal grey
LV	lateral ventricle
mcp	middle cerebellar peduncle
Me5	mesencephalic trigem. nu.
MEnt	medial entorhinal cx
ml	medial lemniscus
mlf	medial longitudinal fasciculus
MoS	molecular layer subiculum
mRt	mesencephalic retic. form'n
Op	optic nerve layer sup. coll.
PaS	parasubiculum
Pn	pontine nuclei
Post	postsubiculum
PRh	perirhinal cortex
rf	rhinal fissure
RRF	retrorubral field
RSGa	retrosplenial granular cx, part a
RSGb	retrosplenial granular cx, part b
rs	rubrospinal tract
scp	superior cerebellar peduncle
scpd	decussation of scp
SNCD	subst. nigra, compact, dorsal tier
SNR	substantia nigra, reticular
Su3	supraoculomotor PAG
Su3C	supraoculomotor cap
SubB	subbrachial nu.
SuG	superficial grey of sup.coll.
te (δ)	temporal (δ) sulcus
TeD	dorsal temporal cortex
TeV	ventral temporal cortex

tfp	transverse fibres of pons
ts	tectospinal tract
tth	trigeminothalamic tract

V1	primary visual cortex
V2L	secondary visual cx, lat. area
VLPAG	ventrolateral periaqueductal grey

VS	ventral subiculum
VTA	ventral tegmental area
Zo	zona layer of the sup. colliculus

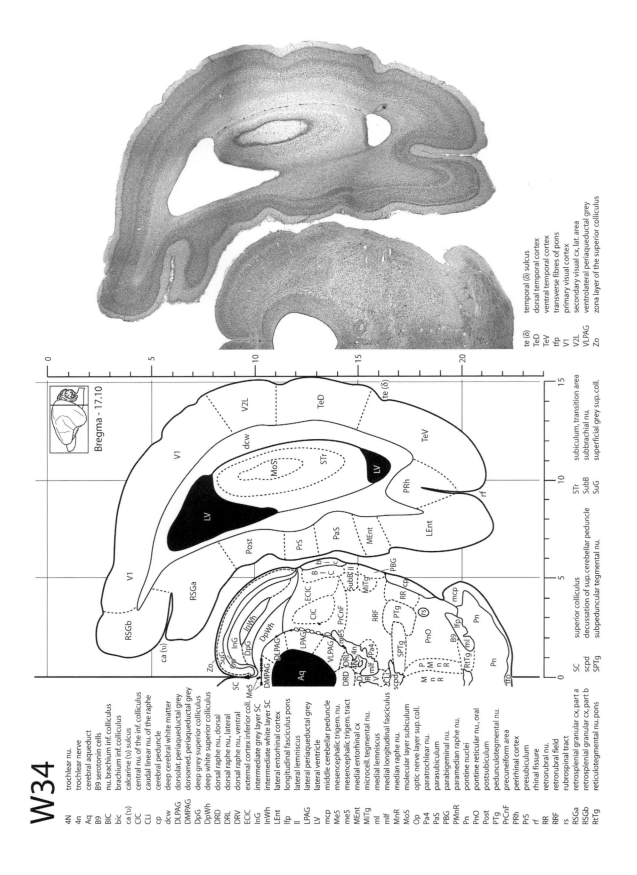

W34

4N	trochlear nu.
4n	trochlear nerve
Aq	cerebral aqueduct
B9	B9 serotonin cells
BIC	nu. brachium inf.colliculus
bic	brachium inf.colliculus
ca (v)	calcarine (v) sulcus
CIC	central nu. of the inf. colliculus
CLi	caudal linear nu. of the raphe
cp	cerebral peduncle
dcw	deep cerebral white matter
DLPAG	dorsolat. periaqueductal grey
DMPAG	dorsomed. periaqueductal grey
DpG	deep grey superior colliculus
DpWh	deep white superior colliculus
DRD	dorsal raphe nu, dorsal
DRL	dorsal raphe nu, lateral
DRV	dorsal raphe nu, ventral
ECIC	external cortex inferior coll.
InG	intermediate grey layer SC
InWh	intermediate white layer SC
LEnt	lateral entorhinal cortex
lfp	longitudinal fasciculus pons
ll	lateral lemniscus
LPAG	lateral periaqueductal grey
LV	lateral ventricle
mcp	middle cerebellar peduncle
Me5	mesencephalic trigem. nu.
me5	mesencephalic trigem. tract
MEnt	medial entorhinal cx
MiTg	microcell. tegmental nu.
ml	medial lemniscus
mlf	medial longitudinal fasciculus
MnR	median raphe nu.
MoS	molecular layer subiculum
Op	optic nerve layer sup.coll.
Pa4	paratrochlear nu.
PaS	parasubiculum
PBG	parabigeminal nu.
PMnR	paramedian raphe nu.
Pn	pontine nuclei
PnO	pontine reticular nu, oral
Post	postsubiculum
PTg	pedunculotegmental nu.
PrCnF	precuneiform area
PRh	perirhinal cortex
PrS	presubiculum
rf	rhinal fissure
RR	retrorubral nu.
RRF	retrorubral field
rs	rubrospinal tract
RSGa	retrosplenial granular cx, part a
RSGb	retrosplenial granular cx, part b
RtTg	reticulotegmental nu. pons

SC	superior colliculus
scpd	decussation of sup. cerebellar peduncle
SPTg	subpeduncular tegmental nu.
STr	subiculum, transition area
SubB	subbrachial nu.
SuG	superficial grey sup.coll.

te (δ)	temporal (δ) sulcus
TeD	dorsal temporal cortex
TeV	ventral temporal cortex
tfp	transverse fibres of pons
V1	primary visual cortex
V2L	secondary visual cx, lat. area
VLPAG	ventrolateral periaqueductal grey
Zo	zona layer of the superior colliculus

Bregma - 17.10

W33

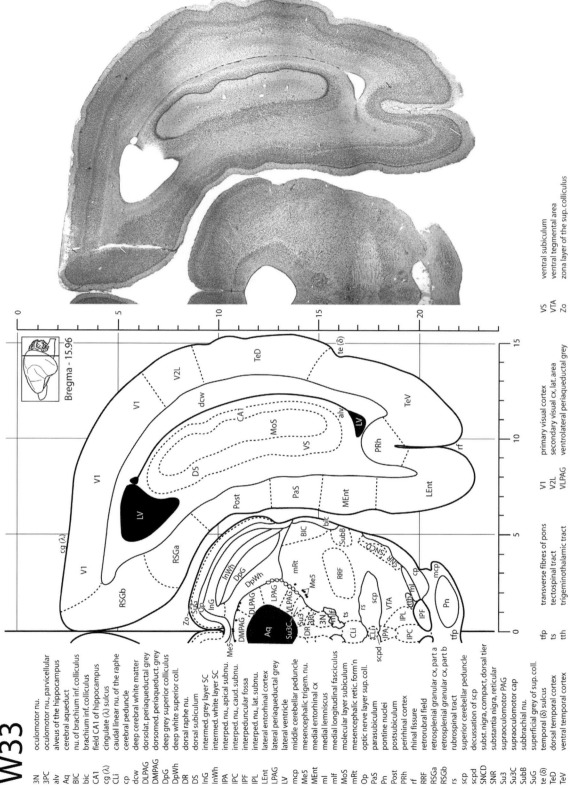

Bregma - 15.96

W34

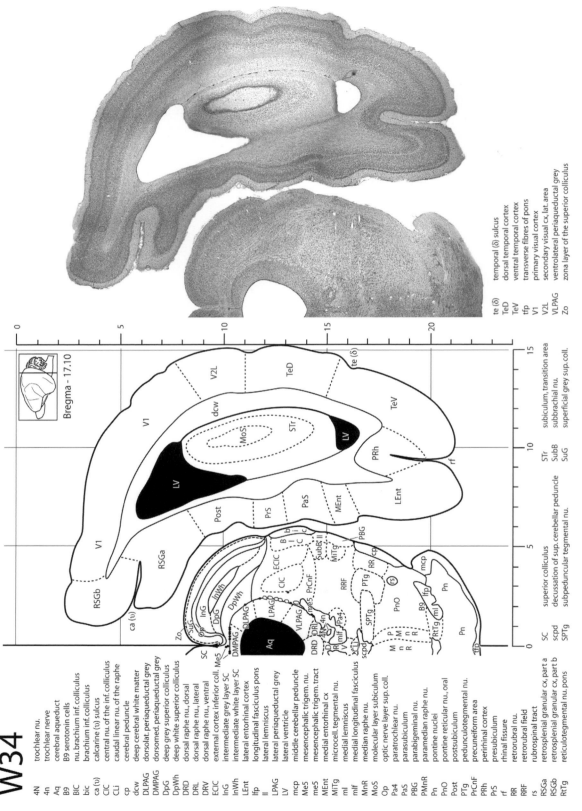

Bregma - 17.10

4N	trochlear nu.
4n	trochlear nerve
Aq	cerebral aqueduct
B9	B9 serotonin cells
BIC	nu. brachium inf. colliculus
bic	brachium inf. colliculus
ca (υ)	calcarine (υ) sulcus
CIC	central nu. of the inf. colliculus
CLi	caudal linear nu. of the raphe
cp	cerebral peduncle
dcw	deep cerebral white matter
DLPAG	dorsolat. periaqueductal grey
DMPAG	dorsomed. periaqueductal grey
DpG	deep grey superior colliculus
DpWh	deep white superior colliculus
DRD	dorsal raphe nu., dorsal
DRL	dorsal raphe nu., lateral
DRV	dorsal raphe nu., ventral
ECIC	external cortex inferior coll.
InG	intermediate grey layer SC
InWh	intermediate white layer SC
LEnt	lateral entorhinal cortex
lfp	longitudinal fasciculus pons
ll	lateral lemniscus
LPAG	lateral periaqueductal grey
LV	lateral ventricle
mcp	middle cerebellar peduncle
Me5	mesencephalic trigem. nu.
me5	mesencephalic trigem. tract
MEnt	medial entorhinal cx
MiTg	microcell. tegmental nu.
ml	medial lemniscus
mlf	medial longitudinal fasciculus
MnR	median raphe nu.
MoS	molecular layer subiculum
Op	optic nerve layer sup.coll.
Pa4	paratrochlear nu.
PaS	parasubiculum
PBG	parabigeminal nu.
PMnR	paramedian raphe nu.
Pn	pontine nuclei
PnO	pontine reticular nu., oral
Post	postsubiculum
PTg	pedunculotegmental nu.
PrCnF	precuneiform area
PRh	perirhinal cortex
PrS	presubiculum
rf	rhinal fissure
RR	retrorubral nu.
RRF	retrorubral field
rs	rubrospinal tract
RSGa	retrosplenial granular cx, part a
RSGb	retrosplenial granular cx, part b
RtTg	reticulotegmental nu. pons

SC	superior colliculus
scpd	decussation of sup. cerebellar peduncle
SPTg	subpeduncular tegmental nu.
STr	subiculum, transition area
SubB	subbrachial nu.
SuG	superficial grey sup. coll.

te (δ)	temporal (δ) sulcus
TeD	dorsal temporal cortex
TeV	ventral temporal cortex
tfp	transverse fibres of pons
V1	primary visual cortex
V2L	secondary visual cx, lat. area
VLPAG	ventrolateral periaqueductal grey
Zo	zona layer of the superior colliculus

W35

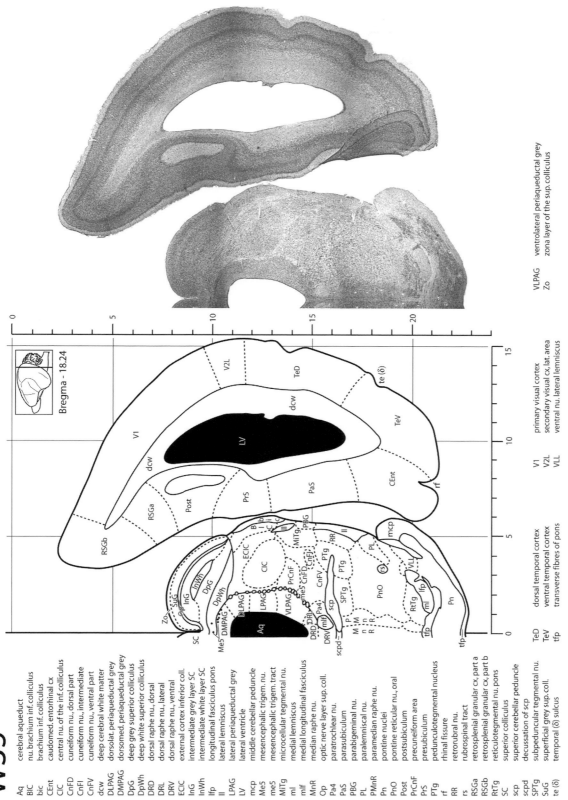

Bregma - 18.24

Abbreviation	Full name
Aq	cerebral aqueduct
BIC	nu.brachium inf. colliculus
bic	brachium inf. colliculus
CEnt	caudomed. entorhinal cx
CIC	central nu. of the inf. colliculus
CnFD	cuneiform nu., dorsal part
CnFI	cuneiform nu., intermediate
CnFV	cuneiform nu., ventral part
dcw	deep cerebral white matter
DLPAG	dorsolat. periaqueductal grey
DMPAG	dorsomed. periaqueductal grey
DpG	deep grey superior colliculus
DpWh	deep white superior colliculus
DRD	dorsal raphe nu., dorsal
DRL	dorsal raphe nu., lateral
DRV	dorsal raphe nu., ventral
ECIC	external cortex inferior coll.
InG	intermediate grey layer SC
InWh	intermediate white layer SC
lfp	longitudinal fasciculus pons
ll	lateral lemniscus
LPAG	lateral periaqueductal grey
LV	lateral ventricle
mcp	middle cerebellar peduncle
Me5	mesencephalic trigem. nu.
me5	mesencephalic trigem. tract
MiTg	microcellular tegmental nu.
ml	medial lemniscus
mlf	medial longitudinal fasciculus
MnR	median raphe nu.
Op	optic nerve layer sup. coll.
Pa4	paratrochlear nu.
PaS	parasubiculum
PBG	parabigeminal nu.
PL	paralemniscal nu.
PMnR	paramedian raphe nu.
Pn	pontine nuclei
PnO	pontine reticular nu., oral
Post	postsubiculum
PrCnF	precuneiform area
PrS	presubiculum
PTg	pedunculotegmental nucleus
rf	rhinal fissure
RR	retrorubral nu.
rs	rubrospinal tract
RSGa	retrosplenial granular cx, part a
RSGb	retrosplenial granular cx, part b
RtTg	reticulotegmental nu. pons
SC	superior colliculus
scp	superior cerebellar peduncle
scpd	decussation of scp
SPTg	subpeduncular tegmental nu.
SuG	superficial grey sup. coll.
te (δ)	temporal (δ) sulcus

Abbreviation	Full name
TeD	dorsal temporal cortex
TeV	ventral temporal cortex
tfp	transverse fibres of pons

Abbreviation	Full name
V1	primary visual cortex
V2L	secondary visual cx, lat. area
VLL	ventral nu. lateral lemniscus

Abbreviation	Full name
VLPAG	ventrolateral periaqueductal grey
Zo	zona layer of the sup. colliculus

W36

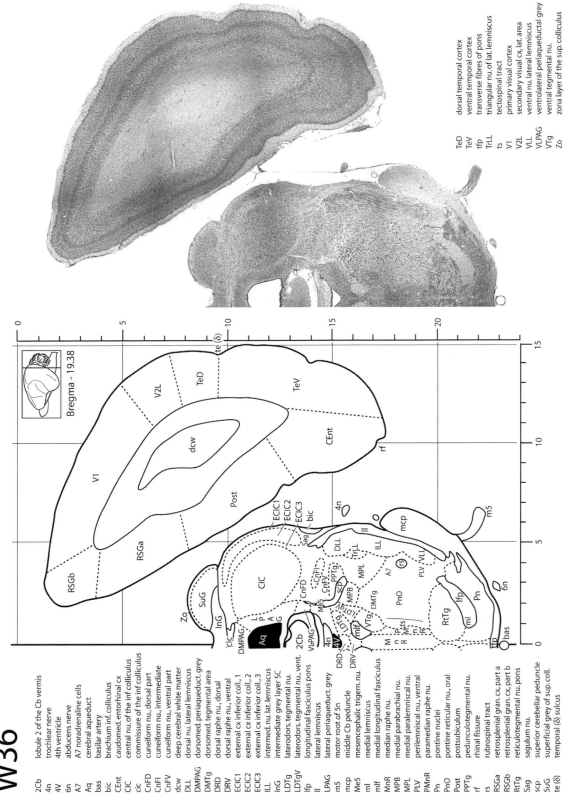

Bregma - 19.38

te (δ)

2Cb	lobule 2 of the Cb vermis
4n	trochlear nerve
4V	4th ventricle
6n	abducens nerve
A7	A7 noradrenaline cells
Aq	cerebral aqueduct
bas	basilar artery
bic	brachium inf. colliculus
CEnt	caudomed. entorhinal cx
CIC	central nu. of the inf colliculus
cic	commissure of the inf colliculus
CnFD	cuneiform nu., dorsal part
CnFI	cuneiform nu., intermediate
CnFV	cuneiform nu., ventral part
dcw	deep cerebral white matter
DLL	dorsal nu. lateral lemniscus
DMPAG	dorsomed. periaqueduct. grey
DMTg	dorsomed. tegmental area
DRD	dorsal raphe nu., dorsal
DRV	dorsal raphe nu., ventral
ECIC1	external cx inferior coll., 1
ECIC2	external cx inferior coll., 2
ECIC3	external cx inferior coll., 3
ILL	intermed. nu. lat. lemniscus
InG	intermediate grey layer SC
LDTg	laterodors. tegmental nu.
LDTgV	laterodors. tegmental nu., vent.
lfp	longitudinal fasciculus pons
ll	lateral lemniscus
LPAG	lateral periaqueduct. grey
m5	motor root of 5n
mcp	middle Cb peduncle
Me5	mesencephalic trigem. nu.
ml	medial lemniscus
mlf	medial longitudinal fasciculus
MnR	median raphe nu.
MPB	medial parabrachial nu.
MPL	medial paralemniscal nu.
PLV	perilemniscal nu., ventral
PMnR	paramedian raphe nu.
Pn	pontine nuclei
PnO	pontine reticular nu., oral
Post	postsubiculum
PPTg	pedunculotegmental nu.
rf	rhinal fissure
rs	rubrospinal tract
RSGa	retrosplenial gran. cx, part a
RSGb	retrosplenial gran. cx, part b
RtTg	reticulotegmental nu. pons
Sag	sagulum nu.
scp	superior cerebellar peduncle
SuG	superficial grey of sup. coll.
te (δ)	temporal (δ) sulcus
TeD	dorsal temporal cortex
TeV	ventral temporal cortex
tfp	transverse fibres of pons
TrLL	triangular nu. of lat. lemniscus
ts	tectospinal tract
V1	primary visual cortex
V2L	secondary visual cx, lat. area
VLL	ventral nu. lateral lemniscus
VLPAG	ventrolateral periaqueductal grey
VTg	ventral tegmental nu.
Zo	zona layer of the sup. colliculus

W37

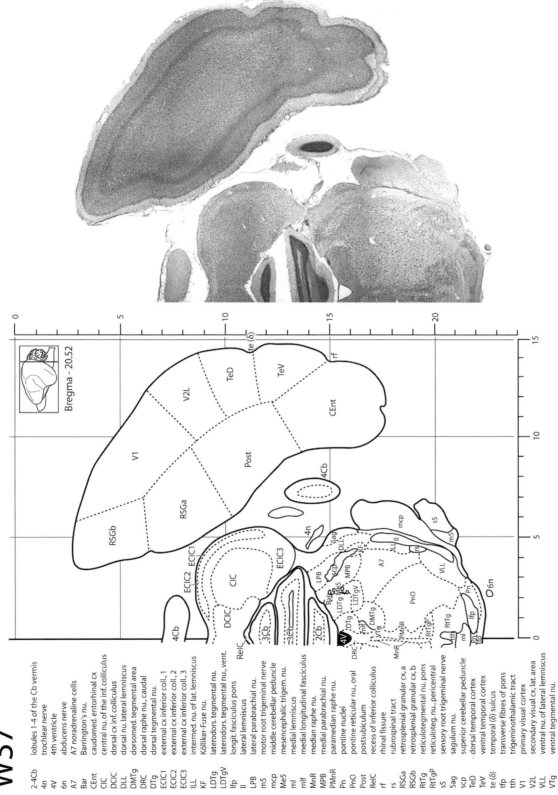

Bregma - 20.52

2-4Cb	lobules 1-4 of the Cb vermis
4n	trochlear nerve
4V	4th ventricle
6n	abducens nerve
A7	A7 noradrenaline cells
Bar	Barrington's nu.
CEnt	caudomed. entorhinal cx
CIC	central nu. of the inf. colliculus
DCIC	dorsal cx inf. colliculus
DLL	dorsal nu. lateral lemniscus
DMTg	dorsomed. tegmental area
DRC	dorsal raphe nu., caudal
DTg	dorsal tegmental nu.
ECIC1	external cx inferior coll., 1
ECIC2	external cx inferior coll., 2
ECIC3	external cx inferior coll., 3
ILL	intermed. nu. of lat. lemniscus
KF	Kölliker-Fuse nu.
LDTg	laterodors. tegmental nu.
LDTgV	laterodors. tegmental nu., vent.
lfp	longit. fasciculus pons
ll	lateral lemniscus
LPB	lateral parabrachial nu.
m5	motor root trigeminal nerve
mcp	middle cerebellar peduncle
Me5	mesencephalic trigem. nu.
ml	medial lemniscus
mlf	medial longitudinal fasciculus
MnR	median raphe nu.
MPB	medial parabrachial nu.
PMnR	paramedian raphe nu.
Pn	pontine nuclei
PnO	pontine reticular nu., oral
Post	postsubiculum
RelC	recess of inferior colliculus
rf	rhinal fissure
rs	rubrospinal tract
RSGa	retrosplenial granular cx, a
RSGb	retrosplenial granular cx, b
RtTg	reticulotegmental nu. pons
RtTgP	reticuloteg. nu., pericentral
s5	sensory root trigeminal nerve
Sag	sagulum nu.
scp	superior cerebellar peduncle
TeD	dorsal temporal cortex
TeV	ventral temporal cortex
te (δ)	temporal (δ) sulcus
tfp	transverse fibres of pons
tth	trigeminothalamic tract
V1	primary visual cortex
V2L	secondary visual cx, lat. area
VLL	ventral nu. of lateral lemniscus
VTg	ventral tegmental nu.

W38

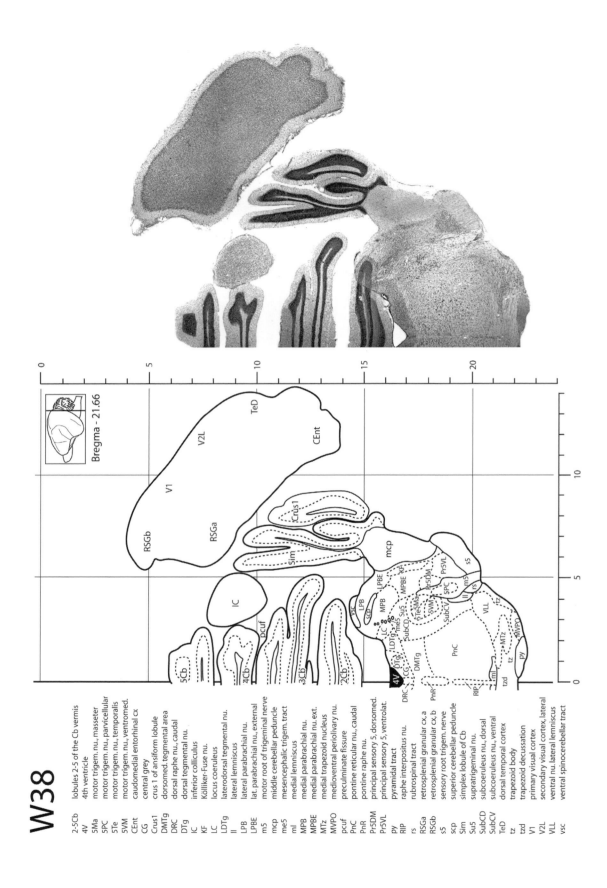

Bregma - 21.66

2-5Cb	lobules 2-5 of the Cb vermis
4V	4th ventricle
5Ma	motor trigem. nu., masseter
5PC	motor trigem. nu., parvicellular
5Te	motor trigem. nu., temporalis
5VM	motor trigem. nu., ventromed.
CEnt	caudomedial entorhinal cx
CG	central grey
Crus1	crus 1 of ansiform lobule
DMTg	dorsomed. tegmental area
DRC	dorsal raphe nu., caudal
DTg	dorsal tegmental nu.
IC	inferior colliculus
KF	Kölliker-Fuse nu.
LC	locus coeruleus
LDTg	laterodorsal tegmental nu.
ll	lateral lemniscus
LPB	lateral parabrachial nu.
LPBE	lat. parabrachial nu., external
m5	motor root of trigeminal nerve
mcp	middle cerebellar peduncle
me5	mesencephalic trigem. tract
ml	medial lemniscus
MPB	medial parabrachial nu.
MPBE	medial parabrachial nu. ext.
MTz	medial trapezoid nucleus
MVPO	medioventral periolivary nu.
pcuf	preculminate fissure
PnC	pontine reticular nu., caudal
PnR	pontine raphe nu.
Pr5DM	principal sensory 5, dorsomed.
Pr5VL	principal sensory 5, ventrolat.
py	pyramidal tract
RIP	raphe interpositus nu.
rs	rubrospinal tract
RSGa	retrosplenial granular cx, a
RSGb	retrosplenial granular cx, b
s5	sensory root trigem. nerve
scp	superior cerebellar peduncle
Sim	simplex lobule of Cb
Su5	supratrigeminal nu.
SubCD	subcoeruleus nu., dorsal
SubCV	subcoeruleus nu., ventral
TeD	dorsal temporal cortex
tz	trapezoid body
tzd	trapezoid decussation
V1	primary visual cortex
V2L	secondary visual cortex, lateral
VLL	ventral nu. lateral lemniscus
vsc	ventral spinocerebellar tract

W39

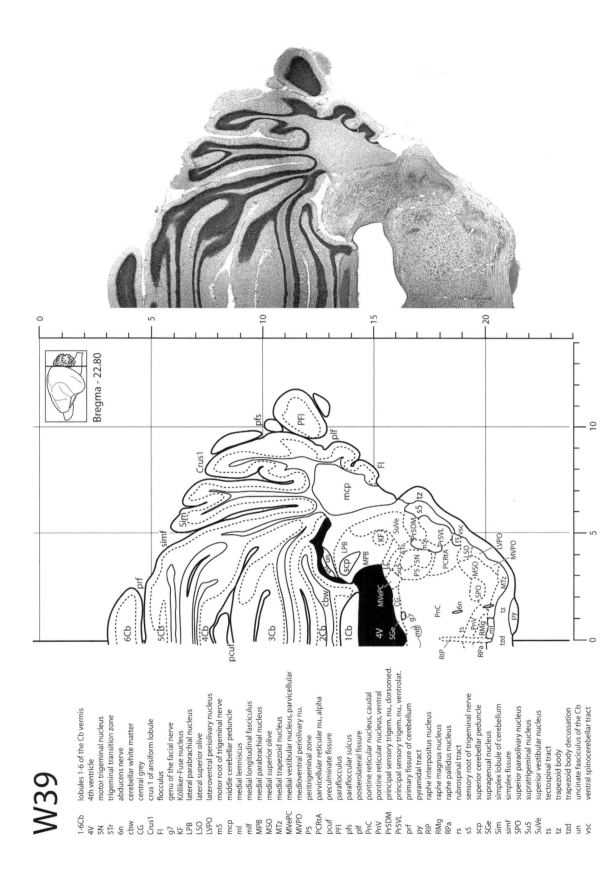

Bregma - 22.80

1-6Cb	lobules 1-6 of the Cb vermis
4V	4th ventricle
5N	motor trigeminal nucleus
5Tr	trigeminal transition zone
6n	abducens nerve
cbw	cerebellar white matter
CG	central grey
Crus1	crus 1 of ansiform lobule
Fl	flocculus
g7	genu of the facial nerve
KF	Kölliker-Fuse nucleus
LPB	lateral parabrachial nucleus
LSO	lateral superior olive
LVPO	lateroventral periolivary nucleus
m5	motor root of trigeminal nerve
mcp	middle cerebellar peduncle
ml	medial lemniscus
mlf	medial longitudinal fasciculus
MPB	medial parabrachial nucleus
MSO	medial superior olive
MTz	medial trapezoid nucleus
MVePC	medial vestibular nucleus, parvicellular
MVPO	medioventral periolivary nu.
P5	peritrigeminal zone
PCRtA	parvicellular reticular nu., alpha
pcuf	preculminate fissure
PFl	paraflocculus
pfs	parafloccular sulcus
plf	posterolateral fissure
PnC	pontine reticular nucleus, caudal
PnV	pontine reticular nucleus, ventral
Pr5DM	principal sensory trigem. nu., dorsomed.
Pr5VL	principal sensory trigem. nu., ventrolat.
prf	primary fissure of cerebellum
py	pyramidal tract
RIP	raphe interpositus nucleus
RMg	raphe magnus nucleus
RPa	raphe pallidus nucleus
rs	rubrospinal tract
s5	sensory root of trigeminal nerve
scp	superior cerebellar peduncle
SGe	supragenual nucleus
Sim	simplex lobule of cerebellum
simf	simplex fissure
SPO	superior paraolivary nucleus
Su5	supratrigeminal nucleus
SuVe	superior vestibular nucleus
ts	tectospinal tract
tz	trapezoid body
tzd	trapezoid body decussation
un	uncinate fasciculus of the Cb
vsc	ventral spinocerebellar tract

W40

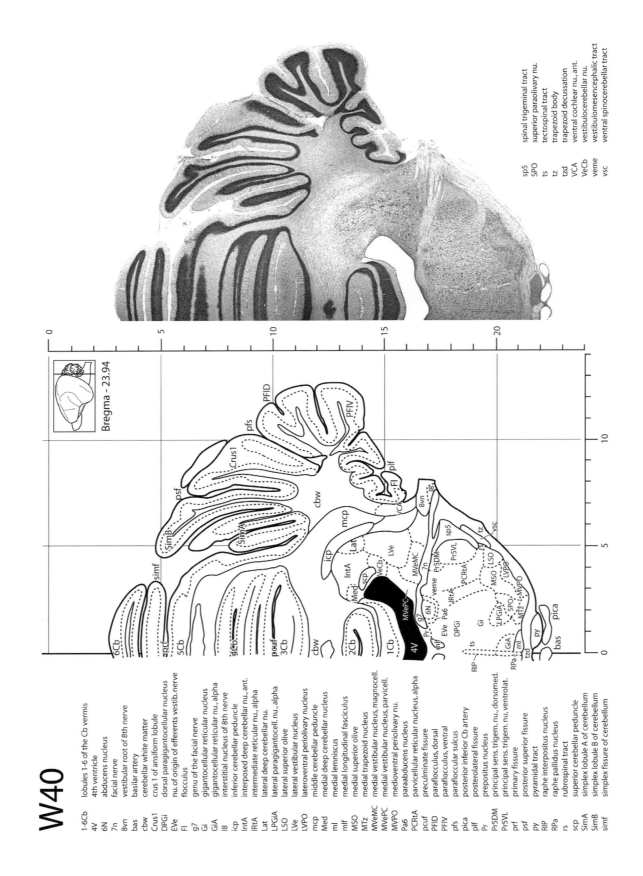

Bregma - 23.94

1-6Cb	lobules 1-6 of the Cb vermis
4V	4th ventricle
6N	abducens nucleus
7n	facial nerve
8vn	vestibular root of 8th nerve
bas	basilar artery
cbw	cerebellar white matter
Crus1	crus 1 of ansiform lobule
DPGi	dorsal paragigantocellular nucleus
EVe	nu. of origin of efferents vestib. nerve
Fl	flocculus
g7	genu of the facial nerve
Gi	gigantocellular reticular nucleus
GiA	gigantocellular reticular nu., alpha
I8	interstitial nucleus of 8th nerve
icp	inferior cerebellar peduncle
IntA	interposed deep cerebellar nu., ant.
IRtA	intermediate reticular nu., alpha
Lat	lateral deep cerebellar nu.
LPGiA	lateral paragigantocell. nu., alpha
LSO	lateral superior olive
LVe	lateral vestibular nucleus
LVPO	lateroventral periolivary nucleus
mcp	middle cerebellar peduncle
Med	medial deep cerebellar nucleus
ml	medial lemniscus
mlf	medial longitudinal fasciculus
MSO	medial superior olive
MTz	medial trapezoid nucleus
MVeMC	medial vestibular nucleus, magnocell.
MVePC	medial vestibular nucleus, parvicell.
MVPO	medioventral periolivary nu.
Pa6	paraabducens nucleus
PCRtA	parvicellular reticular nucleus, alpha
pcuf	preculminate fissure
PFID	paraflocculus, dorsal
PFIV	paraflocculus, ventral
pfs	parafloccular sulcus
pica	posterior inferior Cb artery
plf	posterolateral fissure
Pr	prepositus nucleus
Pr5DM	principal sens. trigem. nu., dorsomed.
Pr5VL	principal sens. trigem. nu, ventrolat.
prf	primary fissure
psf	posterior superior fissure
py	pyramidal tract
RIP	raphe interpositus nucleus
RPa	raphe pallidus nucleus
rs	rubrospinal tract
scp	superior cerebellar peduncle
SimA	simplex lobule A of cerebellum
SimB	simplex lobule B of cerebellum
simf	simplex fissure of cerebellum
sp5	spinal trigeminal tract
SPO	superior paraolivary nu.
ts	tectospinal tract
tz	trapezoid body
tzd	trapezoid decussation
VCA	ventral cochlear nu., ant.
VeCb	vestibulocerebellar nu.
veme	vestibulomesencephalic tract
vsc	ventral spinocerebellar tract

W41

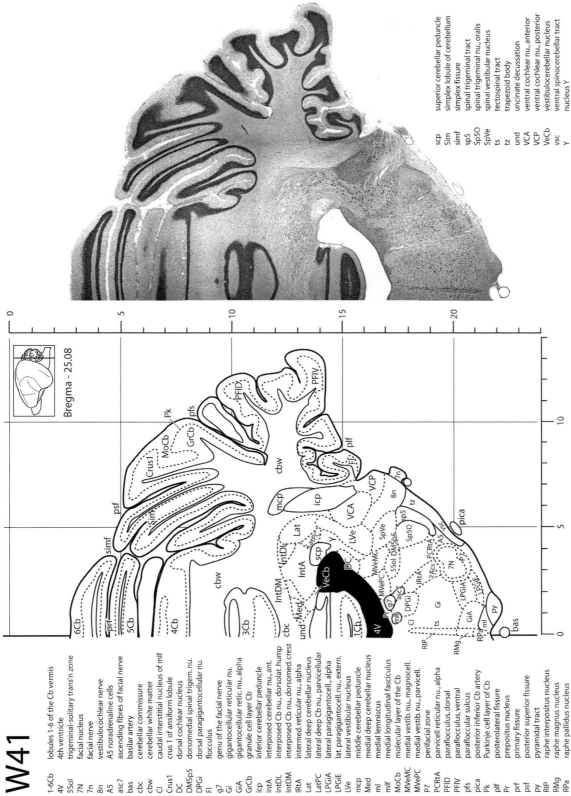

Bregma − 25.08

1-6Cb	lobules 1-6 of the Cb vermis
4V	4th ventricle
5Sol	trigeminal-solitary trans'n zone
7N	facial nucleus
7n	facial nerve
8n	vestibulocochlear nerve
A5	A5 noradrenaline cells
asc7	ascending fibres of facial nerve
bas	basilar artery
cbc	cerebellar commissure
cbw	cerebellar white matter
CI	caudal interstitial nucleus of mlf
Crus1	crus 1 of ansiform lobule
DC	dorsal cochlear nucleus
DMSp5	dorsomedial spinal trigem. nu.
DPGi	dorsal paragigantocellular nu.
Fl	flocculus
g7	genu of the facial nerve
Gi	gigantocellular reticular nu.
GiA	gigantocellular retic. nu., alpha
GrCb	granule cell layer Cb
icp	inferior cerebellar peduncle
IntA	interposed cerebellar nu., ant.
IntDL	interposed Cb nu., dorsolat. hump
IntDM	interposed Cb nu., dorsomed. crest
IRtA	intermed. reticular nu., alpha
Lat	lateral deep cerebellar nucleus
LatPC	lateral deep Cb nu., parvicellular
LPGiA	lateral paragigantocell., alpha
LPGiE	lat. paragigantocell. nu., extern.
LVe	lateral vestibular nucleus
mcp	middle cerebellar peduncle
Med	medial deep cerebellar nucleus
ml	medial lemniscus
mlf	medial longitudinal fasciculus
MoCb	molecular layer of the Cb
MVeMC	medial vestib. nu., magnocell.
MVePC	medial vestib. nu., parvicell.
P7	perifacial zone
PCRtA	parvicell reticular nu., alpha
PFlD	paraflocculus, dorsal
PFlV	paraflocculus, ventral
pfs	paraflocccular sulcus
pica	posterior inferior Cb artery
Pk	Purkinje cell layer of Cb
plf	posterolateral fissure
Pr	prepositus nucleus
prf	primary fissure
psf	posterior superior fissure
py	pyramidal tract
RIP	raphe interpositus nucleus
RMg	raphe magnus nucleus
RPa	raphe pallidus nucleus

scp	superior cerebellar peduncle
Sim	simplex lobule of cerebellum
simf	simplex fissure
sp5	spinal trigeminal tract
Sp5O	spinal trigeminal nu., oralis
SpVe	spinal vestibular nucleus
ts	tectospinal tract
tz	trapezoid body
und	uncinate decussation
VCA	ventral cochlear nu., anterior
VCP	ventral cochlear nu., posterior
VeCb	vestibulocerebellar nucleus
vsc	ventral spinocerebellar tract
Y	nucleus Y

W42

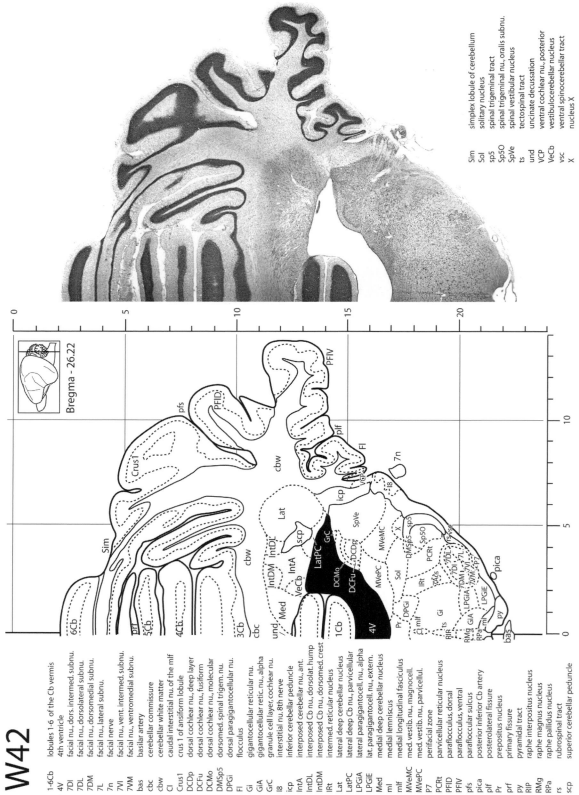

Bregma – 26.22

1-6Cb	lobules 1-6 of the Cb vermis
4V	4th ventricle
7DI	facial nu., dors.intermed.subnu.
7DL	facial nu., dorsolateral subnu.
7DM	facial nu., dorsomedial subnu.
7L	facial nu., lateral subnu.
7n	facial nerve
7VI	facial nu., vent.intermed.subnu.
7VM	facial nu., ventromedial subnu.
bas	basilar artery
cbc	cerebellar commissure
cbw	cerebellar white matter
CI	caudal interstitial nu. of the mlf
Crus1	crus 1 of ansiform lobule
DCDp	dorsal cochlear nu., deep layer
DCFu	dorsal cochlear nu., fusiform
DCMo	dorsal cochlear nu., molecular
DMSp5	dorsomed. spinal trigem. nu.
DPGi	dorsal paragigantocellular nu.
Fl	flocculus
Gi	gigantocellular reticular nu.
GiA	gigantocellular retic. nu., alpha
GrC	granule cell layer, cochlear nu.
I8	interstitial nu. of the mlf
icp	inferior cerebellar peduncle
IntA	interposed cerebellar nu., ant.
IntDL	interposed Cb nu., dorsolat.hump
IntDM	interposed Cb nu., dorsomed. crest
IRt	intermed.reticular nucleus
Lat	lateral deep cerebellar nucleus
LatPC	lateral deep Cb nu., parvicellular
LPGiA	lateral paragigantocell. nu., alpha
LPGiE	lat.paragigantocell.nu., extern.
Med	medial deep cerebellar nucleus
ml	medial lemniscus
mlf	medial longitudinal fasciculus
MVeMC	med.vestib.nu., magnocell.
MVePC	med.vestib.nu., parvicellul.
P7	perifacial zone
PCRt	parvicellular reticular nucleus
PFlD	paraflocculus, dorsal
PFlV	paraflocculus, ventral
pfs	paraflocular sulcus
pica	posterior inferior Cb artery
plf	posterolateral fissure
Pr	prepositus nucleus
prf	primary fissure
py	pyramidal tract
RIP	raphe interpositus nucleus
RMg	raphe magnus nucleus
RPa	raphe pallidus nucleus
rs	rubrospinal tract
scp	superior cerebellar peduncle
Sim	simplex lobule of cerebellum
Sol	solitary nucleus
sp5	spinal trigeminal tract
Sp5O	spinal trigeminal nu., oralis subnu.
SpVe	spinal vestibular nucleus
ts	tectospinal tract
und	uncinate decussation
VCP	ventral cochlear nu., posterior
VeCb	vestibulocerebellar nucleus
vsc	ventral spinocerebellar tract
X	nucleus X

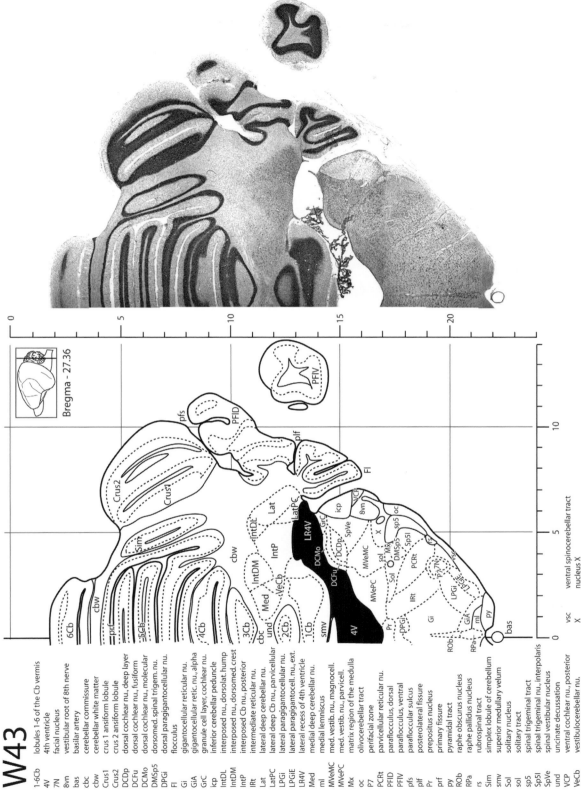

W43

Bregma – 27.36

1-6Cb	lobules 1-6 of the Cb vermis
4V	4th ventricle
7N	facial nucleus
8vn	vestibular root of 8th nerve
bas	basilar artery
cbc	cerebellar commissure
cbw	cerebellar white matter
Crus1	crus 1 ansiform lobule
Crus2	crus 2 ansiform lobule
DCDp	dorsal cochlear nu., deep layer
DCFu	dorsal cochlear nu., fusiform
DCMo	dorsal cochlear nu., molecular
DMSp5	dorsomed. spinal trigem. nu.
DPGi	dorsal paragigantocellular nu.
Fl	flocculus
Gi	gigantocellular reticular nu.
GiA	gigantocellular retic. nu., alpha
GrC	granule cell layer, cochlear nu.
icp	inferior cerebellar peduncle
IntDL	interposed nu., dorsolat. hump
IntDM	interposed nu., dorsomed. crest
IntP	interposed Cb nu., posterior
IRt	intermediate reticular nu.
Lat	lateral deep cerebellar nu.
LatPC	lateral deep Cb nu., parvicellular
LPGi	lateral paragigantocellular nu.
LPGiE	lateral paragigantocell. nu., ext.
LR4V	lateral recess of 4th ventricle
Med	medial deep cerebellar nu.
ml	medial lemniscus
MVeMC	med. vestib. nu., magnocell.
MVePC	med. vestib. nu., parvicell.
Mx	matrix region of the medulla
oc	olivocerebellar tract
P7	perifacial zone
PCRt	parvicellular reticular nu.
PFID	paraflocculus, dorsal
PFIV	paraflocculus, ventral
pfs	parafloccular sulcus
plf	posterolateral fissure
Pr	prepositus nucleus
prf	primary fissure
py	pyramidal tract
ROb	raphe obscurus nucleus
RPa	raphe pallidus nucleus
rs	rubrospinal tract
Sim	simplex lobule of cerebellum
smv	superior medullary velum
Sol	solitary nucleus
sol	solitary tract
sp5	spinal trigeminal tract
Sp5I	spinal trigeminal nu., interpolaris
SpVe	spinal vestibular nucleus
und	uncinate decussation
VCP	ventral cochlear nu., posterior
VeCb	vestibulocerebellar nu.

vsc	ventral spinocerebellar tract	
X	nucleus X	

W44

Bregma - 28.50

4V	4th ventricle
5Cb	lobule 5 of the Cb vermis
5Sol	trigeminal-solitary trans'n zone
6Cb	lobule 6 of the Cb vermis
9Cb	lobule 9 of the Cb vermis
10Cb	lobule 10 of the Cb vermis
10N	vagus nerve nucleus
AmbC	ambiguus nu., compact part
bas	basilar artery
cbw	cerebellar white matter
chp	choroid plexus
Crus1	crus 1 ansiform lobule
Crus2	crus 2 ansiform lobule
DMSp5	dorsomed.spinal trigem. nu.
ECu	external cuneate nucleus
Fl	flocculus
Gi	gigantocellular reticular nu.
GiV	gigantocellular retic. nu., vent.
icp	inferior cerebellar peduncle
IOD	inferior olive, dorsal nu.
IOV	inferior olive, ventral nu.
IRt	intermediate reticular nu.
Li	linear nu. of the medulla
LPGi	lateral paragigantocellular nu.
LPGiE	lat. paragigantocell. nu., ext.
LR4V	lateral recess of 4th ventricle
mlf	medial longitudinal fasciculus
MVeMC	med. vestib. nu., magnocell.
MVePC	med. vestib. nu., parvicell.
oc	olivocerebellar tract
PCRt	parvicellular reticular nu.
PFID	paraflocculus, dorsal
PFIV	paraflocculus, ventral
pfs	parafloccular sulcus
plf	posterolateral fissure
Pr	prepositus nucleus
py	pyramidal tract
ROb	raphe obscurus nucleus
RPa	raphe pallidus nucleus
rs	rubrospinal tract
RVL	rostroventrolat.reticular nu.
sol	solitary tract
SolL	sol nu., lateral part
SolM	sol nu., medial part
SolVL	sol nu., ventrolateral part
sp5	spinal trigeminal tract
Sp5I	spinal trigeminal nu., interpol.
SpVe	spinal vestibular nucleus
ts	tectospinal tract
vsc	ventral spinocerebellar tract
X	nucleus X

W45

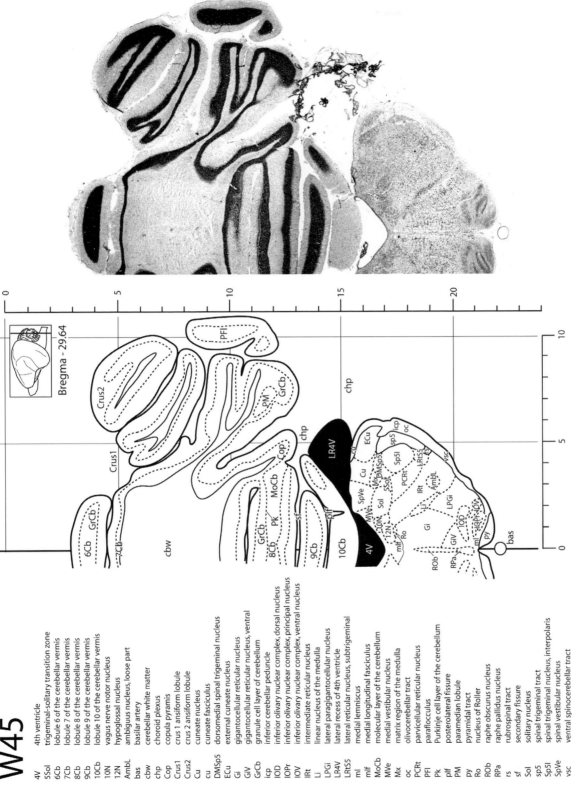

Bregma - 29.64

4V	4th ventricle
5Sol	trigeminal-solitary transition zone
6Cb	lobule 6 of the cerebellar vermis
7Cb	lobule 7 of the cerebellar vermis
8Cb	lobule 8 of the cerebellar vermis
9Cb	lobule 9 of the cerebellar vermis
10Cb	lobule 10 of the cerebellar vermis
10N	vagus nerve notor nucleus
12N	hypoglossal nucleus
AmbL	ambiguus nucleus, loose part
bas	basilar artery
cbw	cerebellar white matter
chp	choroid plexus
Cop	copula pyramis
Crus1	crus 1 ansiform lobule
Crus2	crus 2 ansiform lobule
Cu	cuneate nucleus
cu	cuneate fasciculus
DMSp5	dorsomedial spinal trigeminal nucleus
ECu	external cuneate nucleus
Gi	gigantocellular reticular nucleus
GiV	gigantocellular reticular nucleus, ventral
GrCb	granule cell layer of cerebellum
icp	inferior cerebellar peduncle
IOD	inferior olivary nuclear complex, dorsal nucleus
IOPr	inferior olivary nuclear complex, principal nucleus
IOV	inferior olivary nuclear complex, ventral nucleus
IRt	intermediate reticular nucleus
Li	linear nucleus of the medulla
LPGi	lateral paragigantocellular nucleus
LR4V	lateral recess of 4th ventricle
LRtS5	lateral reticular nucleus, subtrigeminal
ml	medial lemniscus
mlf	medial longitudinal fasciculus
MoCb	molecular layer of the cerebellum
MVe	medial vestibular nucleus
Mx	matrix region of the medulla
oc	olivocerebellar tract
PCRt	parvicellular reticular nucleus
PFl	paraflocculus
Pk	Purkinje cell layer of the cerebellum
plf	posterolateral fissure
PM	paramedian lobule
py	pyramidal tract
Ro	nucleus of Roller
ROb	raphe obscurus nucleus
RPa	raphe pallidus nucleus
rs	rubrospinal tract
sf	secondary fissure
Sol	solitary nucleus
sp5	spinal trigeminal tract
Sp5I	spinal trigeminal nucleus, interpolaris
SpVe	spinal vestibular nucleus
vsc	ventral spinocerebellar tract

W46

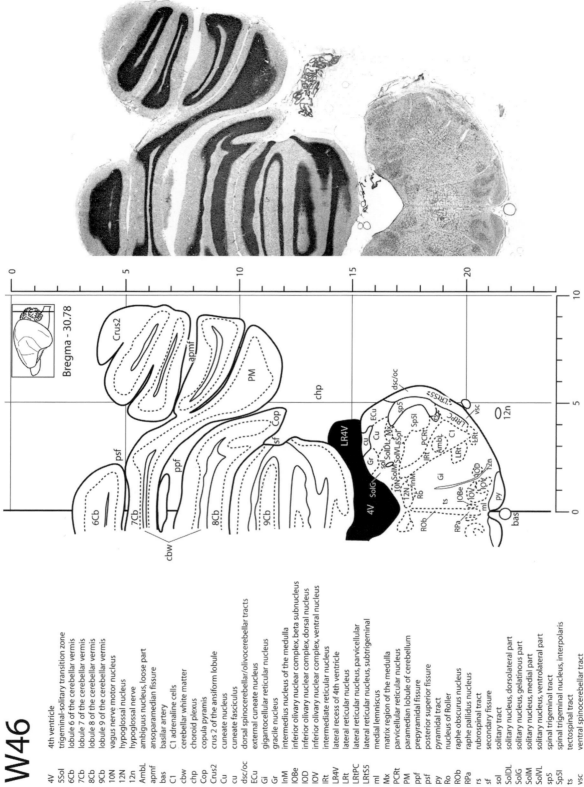

Bregma - 30.78

4V	4th ventricle
5Sol	trigeminal-solitary transition zone
6Cb	lobule 6 of the cerebellar vermis
7Cb	lobule 7 of the cerebellar vermis
8Cb	lobule 8 of the cerebellar vermis
9Cb	lobule 9 of the cerebellar vermis
10N	vagus nerve motor nucleus
12N	hypoglossal nucleus
12n	hypoglossal nerve
AmbL	ambiguus nucleus, loose part
apmf	ansoparamedian fissure
bas	basilar artery
C1	C1 adrenaline cells
cbw	cerebellar white matter
chp	choroid plexus
Cop	copula pyramis
Crus2	crus 2 of the ansiform lobule
Cu	cuneate nucleus
cu	cuneate fasciculus
dsc/oc	dorsal spinocerebellar/olivocerebellar tracts
ECu	external cuneate nucleus
Gi	gigantocellular reticular nucleus
Gr	gracile nucleus
InM	intermedius nucleus of the medulla
IOBe	inferior olivary nuclear complex, beta subnucleus
IOD	inferior olivary nuclear complex, dorsal nucleus
IOV	inferior olivary nuclear complex, ventral nucleus
IRt	intermediate reticular nucleus
LR4V	lateral recess of 4th ventricle
LRt	lateral reticular nucleus
LRtPC	lateral reticular nucleus, parvicellular
LRtSS	lateral reticular nucleus, subtrigeminal
ml	medial lemniscus
Mx	matrix region of the medulla
PCRt	parvicellular reticular nucleus
PM	paramedian lobule of cerebellum
ppf	prepyramidal fissure
psf	posterior superior fissure
py	pyramidal tract
Ro	nucleus of Roller
ROb	raphe obscurus nucleus
RPa	raphe pallidus nucleus
rs	rubrospinal tract
sf	secondary fissure
sol	solitary tract
SolDL	solitary nucleus, dorsolateral part
SolG	solitary nucleus, gelatinous part
SolM	solitary nucleus, medial part
SolVL	solitary nucleus, ventrolateral part
sp5	spinal trigeminal tract
Sp5I	spinal trigeminal nucleus, interpolaris
ts	tectospinal tract
vsc	ventral spinocerebellar tract

W47

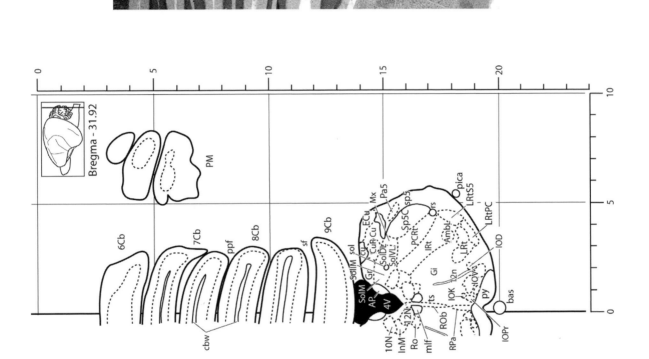

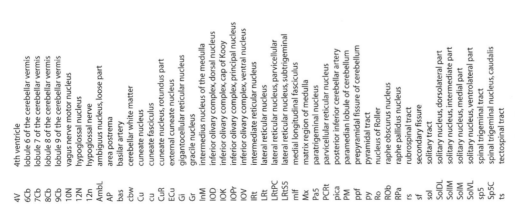

Bregma – 31.92

4V	4th ventricle
6Cb	lobule 6 of the cerebellar vermis
7Cb	lobule 7 of the cerebellar vermis
8Cb	lobule 8 of the cerebellar vermis
9Cb	lobule 9 of the cerebellar vermis
10N	vagus nerve motor nucleus
12N	hypoglossal nucleus
12n	hypoglossal nerve
AmbL	ambiguus nucleus, loose part
AP	area postrema
bas	basilar artery
cbw	cerebellar white matter
Cu	cuneate nucleus
cu	cuneate fasciculus
CuR	cuneate nucleus, rotundus part
ECu	external cuneate nucleus
Gi	gigantocellular reticular nucleus
Gr	gracile nucleus
InM	intermedius nucleus of the medulla
IOD	inferior olivary complex, dorsal nucleus
IOK	inferior olivary complex, cap of Kooy
IOPr	inferior olivary complex, principal nucleus
IOV	inferior olivary complex, ventral nucleus
IRt	intermediate reticular nucleus
LRt	lateral reticular nucleus
LRtPC	lateral reticular nucleus, parvicellular
LRtS5	lateral reticular nucleus, subtrigeminal
mlf	medial longitudinal fasciculus
Mx	matrix region of medulla
Pa5	paratrigeminal nucleus
PCRt	parvicellular reticular nucleus
pica	posterior inferior cerebellar artery
PM	paramedian lobule of cerebellum
ppf	prepyramidal fissure of cerebellum
py	pyramidal tract
Ro	nucleus of Roller
ROb	raphe obscurus nucleus
RPa	raphe pallidus nucleus
rs	rubrospinal tract
sf	secondary fissure
sol	solitary tract
SolDL	solitary nucleus, dorsolateral part
SolIM	solitary nucleus, intermediate part
SolM	solitary nucleus, medial part
SolVL	solitary nucleus, ventrolateral part
sp5	spinal trigeminal tract
Sp5C	spinal trigeminal nucleus, caudalis
ts	tectospinal tract

W48

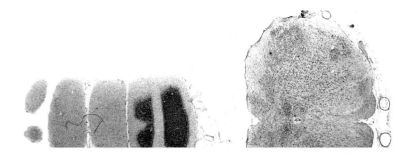

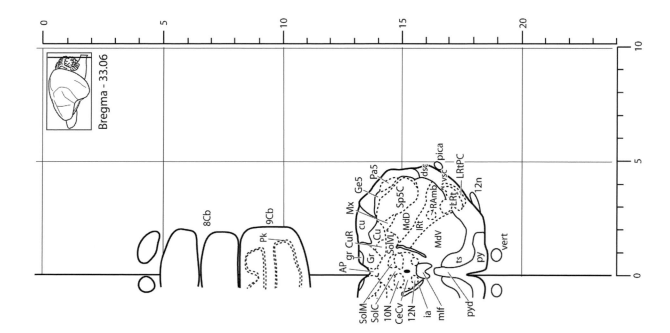

Bregma - 33.06

8Cb	lobule 8 of the cerebellar vermis
9Cb	lobule 9 of the cerebellar vermis
10N	vagus nerve motor nucleus
12N	hypoglossal nucleus
12n	hypoglossal nerve
AP	area postrema
CeCv	central cervical nucleus
Cu	cuneate nucleus
cu	cuneate fasciculus
CuR	cuneate nucleus, rotundus part
dsc	dorsal spinocerebellar tract
Ge5	gelatinous layer of caudal Sp5
Gr	gracile nucleus
gr	gracile fasciculus
ia	internal arcuate fibres
IRt	intermediate reticular nucleus
LRt	lateral reticular nucleus
LRtPC	lateral reticular nucleus, parvicellular part
MdD	medullary reticular nucleus, dorsal
MdV	medullary reticular nucleus, ventral
mlf	medial longitudinal fasciculus
Mx	matrix region of the medulla
Pa5	paratrigeminal nucleus
pica	posterior inferior cerebellar artery
Pk	Purkinje cell layer of the cerebellum
py	pyramidal tract
RAmb	retroambiguus nucleus
SolC	solitary nucleus, commissural part
SolM	solitary nucleus, medial part
SolVL	solitary nucleus, ventrolateral part
Sp5C	spinal trigeminal nucleus, caudalis
ts	tectospinal tract
vert	vertebral artery
vsc	ventral spinocerebellar tract

W49, W50 & W51

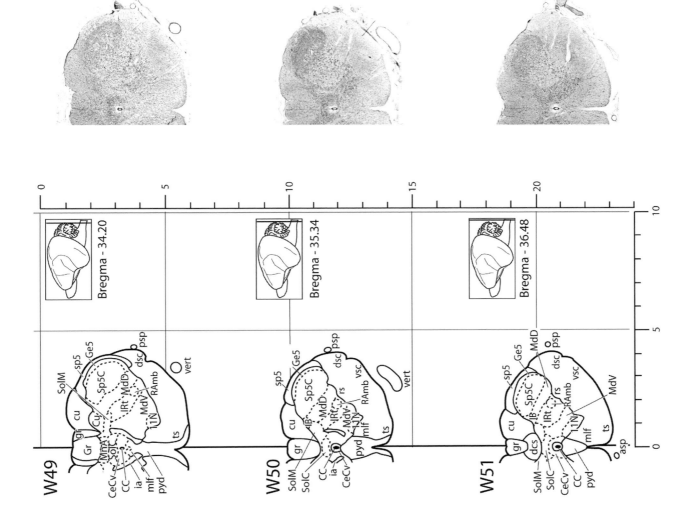

11N	accessory nerve nucleus
asp	anterior spinal artery
CC	central canal of spinal cord
CeCv	central cervical nucleus
Cu	cuneate nucleus
cu	cuneate fasciculus
dcs	dorsal corticospinal tract
dsc	dorsal spinocerebellar tract
Ge5	gelatinous layer of caudal Sp5
Gr	gracile nucleus
gr	gracile fasciculus
ia	internal arcuate fibres
IB	interstitial nucleus of the medulla
IRt	intermediate reticular nucleus
MdD	medullary reticular nucleus, dorsal
MdV	medullary reticular nucleus, ventral
mlf	medial longitudinal fasciculus
MnA	median accessory nucleus of medulla
psp	posterior spinal artery
pyd	pyramidal tract decussation
RAmb	retroambiguus nucleus
rs	rubrospinal tract
SolC	solitary nucleus, commissural part
SolM	solitary nucleus, medial part
sp5	spinal trigeminal tract
Sp5C	spinal trigeminal nucleus, caudalis
ts	tectospinal tract
vert	vertebral artery
vsc	ventral spinocerebellar tract

18

Atlas of the brain of the developing tammar wallaby (*Macropus eugenii*)

K. W. S. Ashwell, L. R. Marotte and J. K. Mai

Introduction

Reproduction of the tammar wallaby has been studied extensively (Tyndale-Biscoe, 2005) and this marsupial is easily bred in captivity, making it an ideal choice as an experimental animal in neurodevelopmental studies. The slightly smaller quokka (*Setonix brachyurus*; adult body weight of 2.7 to 4.2 kg) has also been used extensively in studies of this kind in Western Australia. Although belonging to different genuses, both these macropods have morphologically similar pouch young and an atlas of the brains of early tammar young should also be applicable to early pouch-young quokkas.

In the wild, most tammar young are born in late January. Females mate again within a few hours of birth and the resulting embryo remains quiescent during lactation. In the natural environment, the pouch young is suckled for 8 to 9 months and leaves the pouch in September or October. The quiescent embryo can be stimulated to reactivate by removal of the current pouch young, allowing the sequential births of two or more pouch young within the one year in wallabies maintained in a colony. If the current year's pouch young are not removed, the quiescent embryos naturally reactivate within a few days of the summer solstice (i.e. after 22 December).

The range of ages depicted covers the period from birth, when the rostral brain is 'embryonic', through to P25, when most major subcortical nuclei have begun to emerge. Significant cortical development occurs after P25, particularly in the occipital region, but the major developmental regions of the cerebral cortex are nevertheless present by that age (see text of Chapters 3 and 8, and Figure 3.6).

Other sections from some of the animals depicted in this series have been used for previously published studies of neurodevelopment in this species (Hassiotis *et al.*, 2002; Ashwell *et al.*, 2004, 2008a).

Methods

Ethics, anaesthesia and perfusion

All wallabies used in this study were obtained from a breeding colony. All experimental procedures were approved by the Animal Ethics Experimentation Committee of the ANU, conform to NIH principles of laboratory animal care and were carried out according to the ethical guidelines of the National Health and Medical Research Council (Australia). The ages of animals were determined either directly by noting the elapsed time from the date of birth, which was designated P0, or from measurements of head length and reference to a chart of head lengths of animals of known age. This is accurate to within ±2 days. Gestation length in this species is on average 28.3 days. The steady and measured pace of wallaby postnatal development means that there is little inter-animal variation in developmental stage.

Pouch young were anaesthetised by hypothermia and perfused with normal saline followed by Bouin's fixative. Pouch-young material was stored in 70% ethanol prior to embedding.

Histology, photomicrography and delineation

The heads of pouch young at ages P0, P5 and P12 and the brains of P19 and P25 pouch young were embedded in paraffin and sectioned coronally at a thickness of 10 μm. The sections depicted in the atlas have been stained with haematoxylin and eosin and coverslipped with *DePeX*.

The right side of each section was photographed with the aid of a *.slide* photomicrographic system in the Department of Anatomy at the Heinrich Heine University in Düsseldorf, Germany, as described for the dunnart atlas. Images were placed in *Adobe Illustrator CS2* (as described previously for

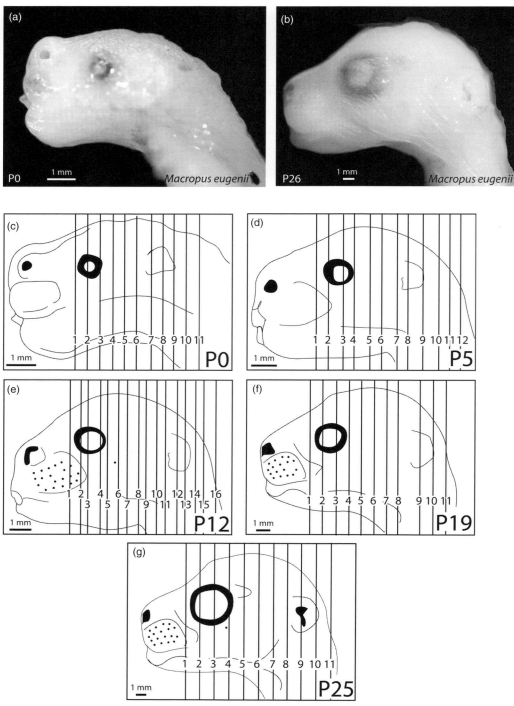

Figure 18.1 Left lateral view photographs of perfused, post-mortem heads of pouch-young wallabies at P0 (a) and P26 (b), showing the profound change in structure of the head during the first postnatal month. Line diagrams (c) to (g) illustrate the rostrocaudal position of coronal sections depicted in the following pages.

the dunnart brain) and delineated. Nomenclature applied to the pouch-young nervous system is adapted from that used for the third edition of the *Atlas of the Developing Rat Nervous System* (Ashwell and Paxinos, 2008). Developmental regions (i.e. neuroepithelium) destined to give rise to adult structures have been denoted by the adult structure's name with an asterisk (e.g. Cx* denotes cerebral cortical neuroepithelium).

Each plate depicts half of a coronal section, because the head is bilaterally symmetrical, with a scale indicating the size in mm of structures in the dehydrated tissue. A small finder diagram has been provided in the top right-hand corner of each line diagram with the distance from the rostral tip of the olfactory bulb/telencephalic vesicle indicated in mm. Atlas plate files are available at www.cambridge.org/9780521519458.

WP0-1 & WP0-2

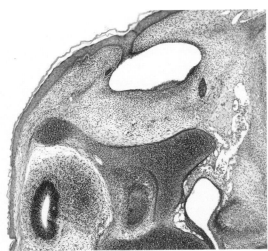

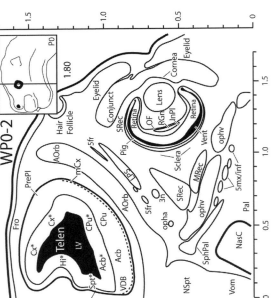

*	denotes precursor of structure
3n	oculomotor nerve
5fr	frontal branch of ophthalmic 5n
5mx/inf	infraorbital nerve
Acb	accumbens nucleus
AOrb	alar orbital bone
Conjunct	conjunctival sac
Cornea	cornea
CPu	caudate putamen (striatum)
CrP	cribriform plate of ethmoid bone
Cx	cerebral cortex
EthB	ethmoid bone
Eyelid	eyelid
Fro	frontal bone
HardG	Harderian gland
Hi	hippocampus
InPl	inner plexiform layer
Lens	lens
LPS	levator palpebrae superioris muscle
LV	lateral ventricle
mCx	marginal zone of developing cortex
MRec	medial rectus muscle
NasC	nasal cavity
ne	neuroepithelium
NSpt	nasal septum
OB	olfactory bulb
OF	optic fibre layer of the retina
olf	olfactory nerve
olfa	olfactory artery
olfepith	olfactory epithelium
ON	olfactory nerve layer of telencephalon
opha	ophthalmic artery
ophv	ophthalmic vein
Pal	palatine bone
Pig	pigment layer of the eye
PrePl	preplate of cortex
Retina	retina
RGn	ganglion cell layer of retina
spa	sphenopalatine artery
SphPal	sphenopalatine ganglion
Spt	septal region of brain
SRec	superior rectus muscle
Telen	telencephalon
term	terminal nerve
VDB	nuclei of vertical limb of diagonal band
Vent	ventricular space of the eye
Vom	vomer (bone)

WP0-3 & WP0-4

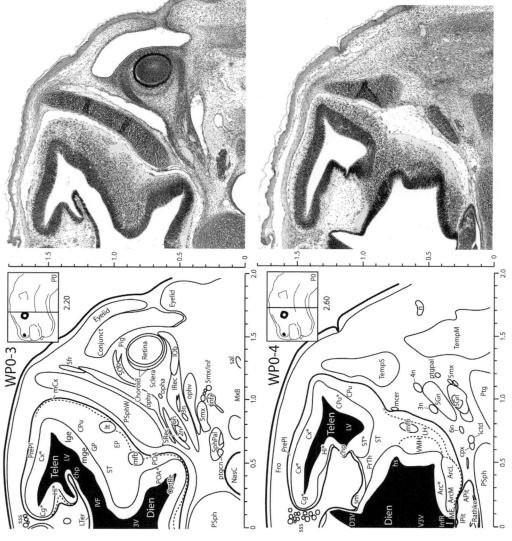

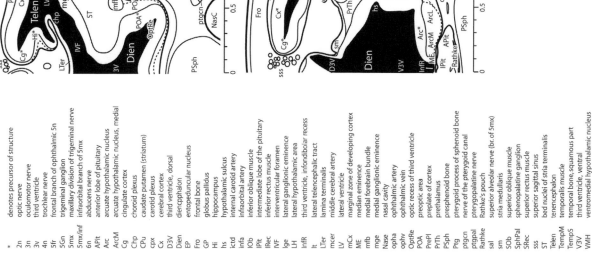

*	denotes precursor of structure
2n	optic nerve
3n	oculomotor nerve
3v	third ventricle
4n	trochlear nerve
5fr	frontal branch of ophthalmic 5n
5Gn	trigeminal ganglion
5mx	maxillary division of trigeminal nerve
5mx/inf	infraorbital branch of 5mx
6n	abducens nerve
APit	anterior lobe of pituitary
Arc	arcuate hypothalamic nucleus
ArcM	arcuate hypothalamic nucleus, medial
Cg	cingulate cortex
Chp	choroid plexus
CPu	caudate putamen (striatum)
cpx	caroid plexus
Cx	cerebral cortex
D3V	third ventricle, dorsal
Dien	diencephalon
EP	entopeduncular nucleus
Fro	frontal bone
GP	globus pallidus
Hi	hippocampus
hs	hypothalamic sulcus
ictd	internal caroid artery
infa	infraorbital artery
IOb	inferior oblique muscle
IPit	intermediate lobe of the pituitary
IRec	inferior rectus muscle
IVF	interventricular foramen
lge	lateral ganglionic eminence
LH	lateral hypothalamic area
InfR	third ventricle, infondiboiar recess
lt	lateral telencephalic tract
LTer	lamina terminalis
mcer	middle cerebral artery
LV	lateral ventricle
mCx	marginal zone of developing cortex
ME	median eminence
mfb	medial forebrain bundle
mge	medial ganglionic eminence
Nase	nasal cavity
opha	ophthalmic artery
ophv	ophthalmic vein
OptRe	optic recess of third ventricle
POA	preoptic area
PrePl	preplate of cortex
PrTh	prethalamus
PSph	presphenoid bone
Ptg	pterygoid process of sphenoid bone
ptgcn	nerve of the pterygoid canal
ptgpal	pterygopalatine nerve
Rathke	Rathke's pouch
sal	superior alveolar nerve (br. of 5mx)
sm	stria medullaris
SOb	superior oblique muscle
SphPal	sphenopalatine ganglion
SRec	superior rectus muscle
ST	superior sagittal sinus
ST	bed nuclei of stria terminalis
Telen	tenencephalon
TempM	temporalis muscle
TempS	temporal bone, squamous part
V3V	third ventricle, ventral
VMH	ventromedial hypothalamic nucleus

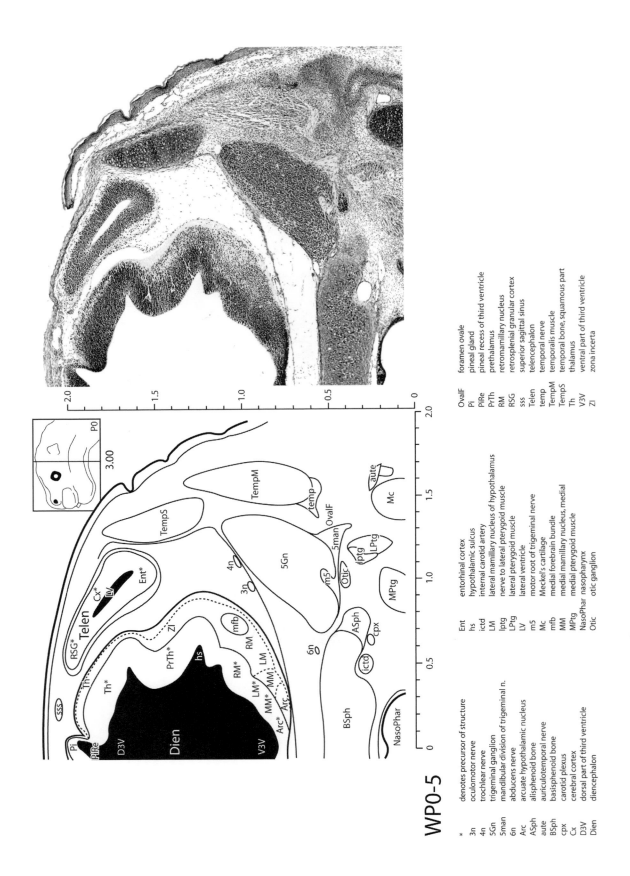

WP0-5

*	denotes precursor of structure	Ent	entorhinal cortex
3n	oculomotor nerve	hs	hypothalamic sulcus
4n	trochlear nerve	ictd	internal carotid artery
5Gn	trigeminal ganglion	LM	lateral mamillary nucleus of hypothalamus
5man	mandibular division of trigeminal n.	lptg	nerve to lateral pterygoid muscle
6n	abducens nerve	LPtg	lateral pterygoid muscle
Arc	arcuate hypothalamic nucleus	LV	lateral ventricle
ASph	alisphenoid bone	m5	motor root of trigeminal nerve
aute	auriculotemporal nerve	Mc	Meckel's cartilage
BSph	basisphenoid bone	mfb	medial forebrain bundle
cpx	carotid plexus	MM	medial mamillary nucleus, medial
Cx	cerebral cortex	MPtg	medial pterygoid muscle
D3V	dorsal part of third ventricle	NasoPhar	nasopharynx
Dien	diencephalon	Otic	otic ganglion

OvalF	foramen ovale
Pi	pineal gland
PiRe	pineal recess of third ventricle
PrTh	prethalamus
RM	retromamillary nucleus
RSG	retrosplenial granular cortex
sss	superior sagittal sinus
Telen	telencephalon
temp	temporal nerve
TempM	temporalis muscle
TempS	temporal bone, squamous part
Th	thalamus
V3V	ventral part of third ventricle
ZI	zona incerta

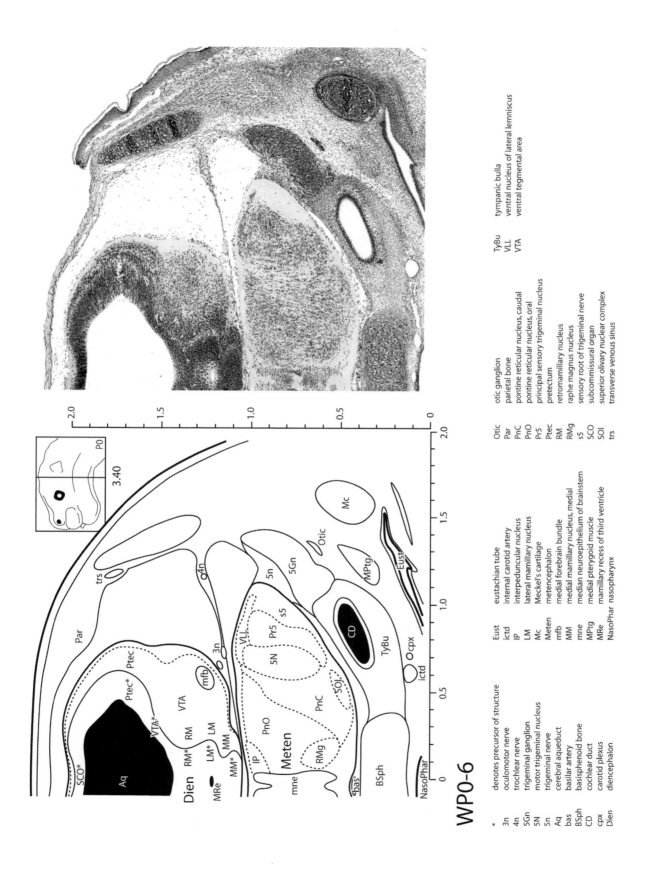

WP0-6

*	denotes precursor of structure	Eust	eustachian tube	Otic	otic ganglion	TyBu	tympanic bulla
3n	oculomotor nerve	ictd	internal carotid artery	Par	parietal bone	VLL	ventral nucleus of lateral lemniscus
4n	trochlear nerve	IP	interpeduncular nucleus	PnC	pontine reticular nucleus, caudal	VTA	ventral tegmental area
5Gn	trigeminal ganglion	LM	lateral mamillary nucleus	PnO	pontine reticular nucleus, oral		
5N	motor trigeminal nucleus	Mc	Meckel's cartilage	Pr5	principal sensory trigeminal nucleus		
5n	trigeminal nerve	Meten	metencephalon	Ptec	pretectum		
Aq	cerebral aqueduct	mfb	medial forebrain bundle	RM	retromamillary nucleus		
bas	basilar artery	MM	medial mamillary nucleus, medial	RMg	raphe magnus nucleus		
BSph	basisphenoid bone	mne	median neuroepithelium of brainstem	s5	sensory root of trigeminal nerve		
CD	cochlear duct	MPtg	medial pterygoid muscle	SCO	subcommissural organ		
cpx	carotid plexus	MRe	mamillary recess of third ventricle	SOl	superior olivary nuclear complex		
Dien	diencephalon	NasoPhar	nasopharynx	trs	transverse venous sinus		

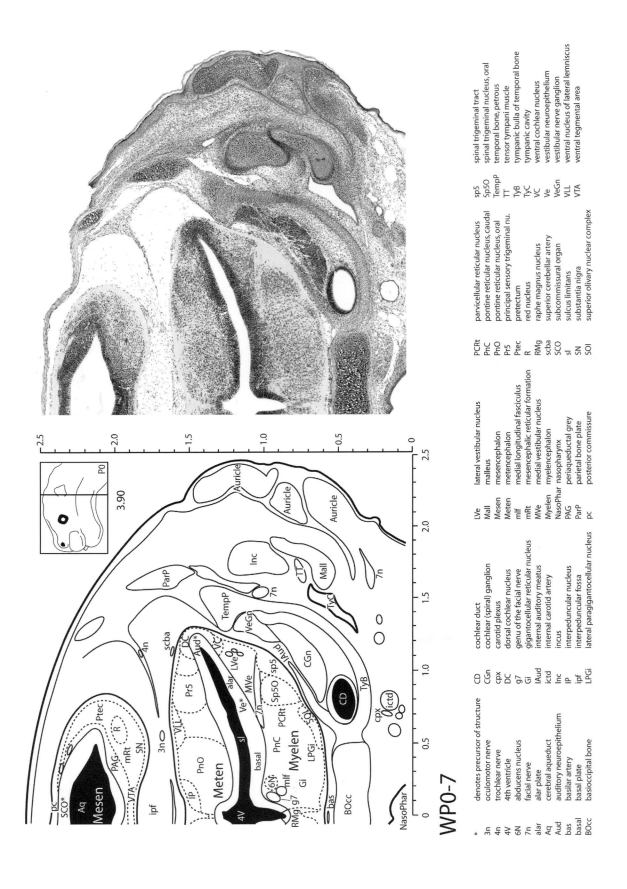

WP0-7

3.90

*		denotes precursor of structure
3n		oculomotor nerve
4n		trochlear nerve
4V		4th ventricle
6N		abducens nucleus
7n		facial nerve
alar		alar plate
Aq		cerebral aqueduct
Aud		auditory neuroepithelium
bas		basilar artery
basal		basal plate
BOcc		basioccipital bone

CD	cochlear duct
CGn	cochlear (spiral) ganglion
cpx	carotid plexus
DC	dorsal cochlear nucleus
g7	genu of the facial nerve
Gi	gigantocellular reticular nucleus
IAud	internal auditory meatus
ictd	internal carotid artery
Inc	incus
IP	interpeduncular nucleus
ipf	interpeduncular fossa
LPGi	lateral paragigantocellular nucleus

LVe	lateral vestibular nucleus
Mall	malleus
Mesen	mesencephalon
Meten	metencephalon
mlf	medial longitudinal fasciculus
mRt	mesencephalic reticular formation
MVe	medial vestibular nucleus
Myelen	myelencephalon
NasoPhar	nasopharynx
PAG	periaqueductal grey
ParP	parietal bone plate
pc	posterior commissure

PCRt	parvicellular reticular nucleus
PnC	pontine reticular nucleus, caudal
PnO	pontine reticular nucleus, oral
Pr5	principal sensory trigeminal nu.
Ptec	pretectum
R	red nucleus
RMg	raphe magnus nucleus
scba	superior cerebellar artery
SCO	subcommissural organ
sl	sulcus limitans
SN	substantia nigra
SOl	superior olivary nuclear complex

sp5	spinal trigeminal tract
SpSO	spinal trigeminal nucleus, oral
TempP	temporal bone, petrous
TT	tensor tympani muscle
TyB	tympanic bulla of temporal bone
TyC	tympanic cavity
VC	ventral cochlear nucleus
Ve	vestibular neuroepithelium
VeGn	vestibular nerve ganglion
VLL	ventral nucleus of lateral lemniscus
VTA	ventral tegmental area

WP0-8

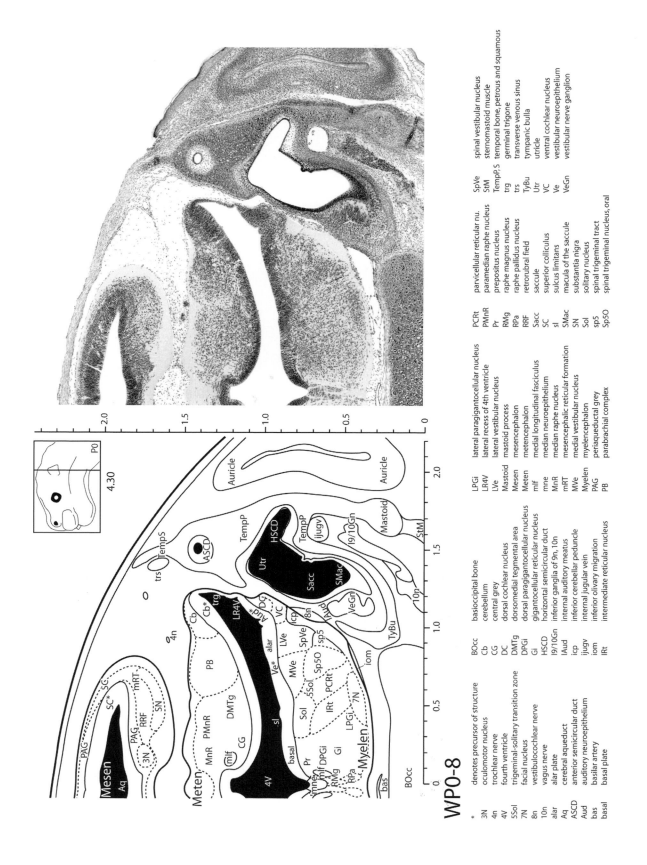

4.30

P0

*	denotes precursor of structure	
3N	oculomotor nucleus	
4n	trochlear nerve	
4V	fourth ventricle	
5Sol	trigeminal-solitary transition zone	
7N	facial nucleus	
8n	vestibulocochlear nerve	
10n	vagus nerve	
alar	alar plate	
Aq	cerebral aqueduct	
ASCD	anterior semicircular duct	
Aud	auditory neuroepithelium	
bas	basilar artery	
basal	basal plate	

BOcc	basioccipital bone	
Cb	cerebellum	
CG	central grey	
DC	dorsal cochlear nucleus	
DMTg	dorsomedial tegmental area	
DPGi	dorsal paragigantocellular nucleus	
Gi	gigantocellular reticular nucleus	
HSCD	horizontal semicircular duct	
I9/10Gn	inferior ganglia of 9n, 10n	
IAud	internal auditory meatus	
icp	inferior cerebellar peduncle	
jjugv	internal jugular vein	
iom	inferior olivary migration	
IRt	intermediate reticular nucleus	

LPGi	lateral paragigantocellular nucleus	
LR4V	lateral recess of 4th ventricle	
LVe	lateral vestibular nucleus	
Mastoid	mastoid process	
Mesen	mesencephalon	
Meten	metencephalon	
mlf	medial longitudinal fasciculus	
mne	median neuroepithelium	
MnR	median raphe nucleus	
mRT	mesencephalic reticular formation	
MVe	medial vestibular nucleus	
Myelen	myelencephalon	
PAG	periaqueductal grey	
PB	parabrachial complex	

PCRt	parvicellular reticular nu.	
PMnR	paramedian raphe nucleus	
Pr	prepositus nucleus	
RMg	raphe magnus nucleus	
RPa	raphe pallidus nucleus	
RRF	retrorubral field	
Sacc	saccule	
SC	superior colliculus	
sl	sulcus limitans	
SMac	macula of the saccule	
SN	substantia nigra	
Sol	solitary nucleus	
sp5	spinal trigeminal tract	
Sp5O	spinal trigeminal nucleus, oral	

SpVe	spinal vestibular nucleus	
StM	sternomastoid muscle	
TempP,S	temporal bone, petrous and squamous	
trg	germinal trigone	
trs	transverse venous sinus	
TyBu	tympanic bulla	
Utr	utricle	
VC	ventral cochlear nucleus	
Ve	vestibular neuroepithelium	
VeGn	vestibular nerve ganglion	

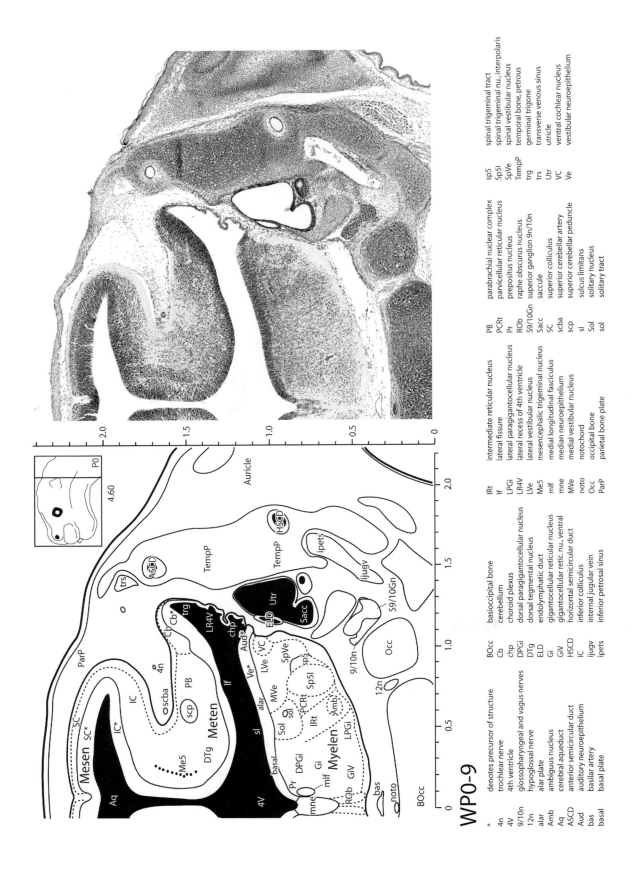

WP0-9

*	denotes precursor of structure	BOcc	basioccipital bone	IRt	intermediate reticular nucleus	PB	parabrachial nuclear complex	
4n	trochlear nerve	Cb	cerebellum	lf	lateral fissure	PCRt	parvicellular reticular nucleus	
4V	4th ventricle	chp	choroid plexus	LPGi	lateral paragigantocellular nucleus	Pr	prepositus nucleus	
9/10n	glossopharyngeal and vagus nerves	DPGi	dorsal paragigantocellular nucleus	LR4V	lateral recess of 4th ventricle	ROb	raphe obscurus nucleus	
12n	hypoglossal nerve	DTg	dorsal tegmental nucleus	LVe	lateral vestibular nucleus	S9/10Gn	superior ganglion 9n/10n	
alar	alar plate	ELD	endolymphatic duct	Me5	mesencephalic trigeminal nucleus	Sacc	saccule	
Amb	ambiguus nucleus	Gi	gigantocellular reticular nucleus	mlf	medial longitudinal fasciculus	SC	superior colliculus	
Aq	cerebral aqueduct	GiV	gigantocellular retic. nu., ventral	mne	median neuroepithelium	scba	superior cerebellar artery	
ASCD	anterior semicircular duct	HSCD	horizontal semicircular duct	MVe	medial vestibular nucleus	scp	superior cerebellar peduncle	
Aud	auditory neuroepithelium	IC	inferior colliculus	noto	notochord	sl	sulcus limitans	
bas	basilar artery	jjugv	internal jugular vein	Occ	occipital bone	Sol	solitary nucleus	
basal	basal plate	ipets	inferior petrosal sinus	ParP	parietal bone plate	sol	solitary tract	

sp5	spinal trigeminal tract
Sp5I	spinal trigeminal nu., interpolaris
SpVe	spinal vestibular nucleus
TempP	temporal bone, petrous
trg	germinal trigone
trs	transverse venous sinus
Utr	utricle
VC	ventral cochlear nucleus
Ve	vestibular neuroepithelium

WPO-10

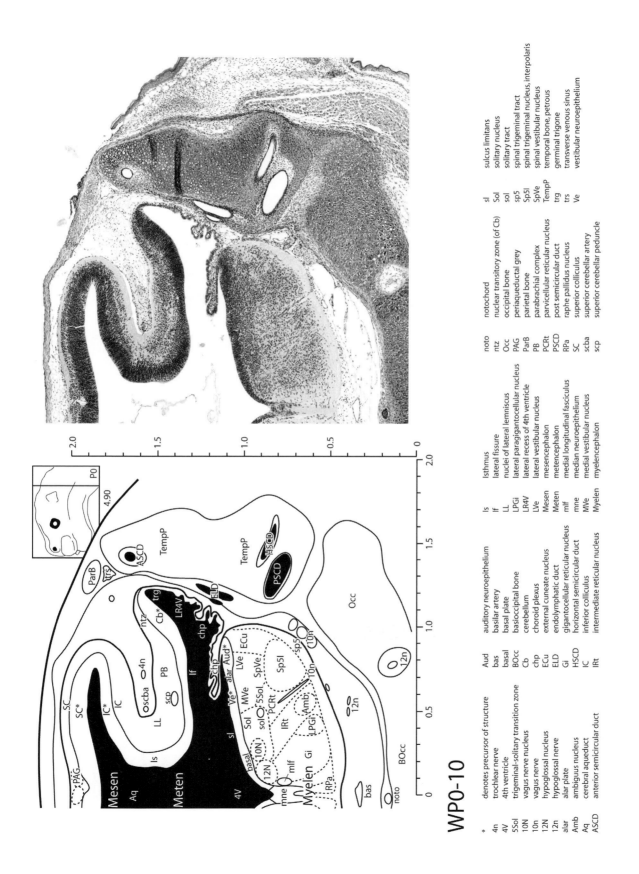

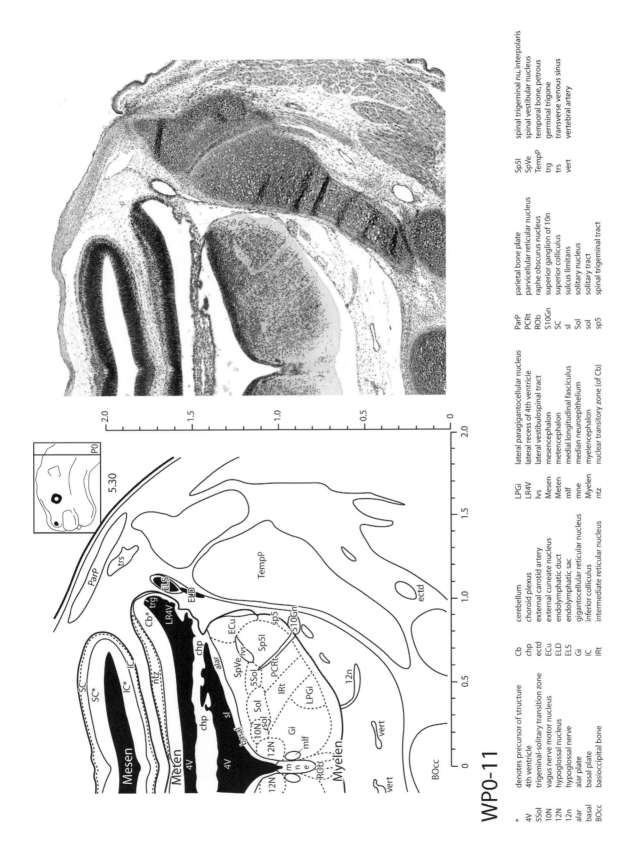

WPO-11

*	denotes precursor of structure	Cb	cerebellum
4V	4th ventricle	chp	choroid plexus
5Sol	trigeminal-solitary transition zone	ectd	external carotid artery
10N	vagus nerve motor nucleus	ECu	external cuneate nucleus
12N	hypoglossal nucleus	ELD	endolymphatic duct
12n	hypoglossal nerve	ELS	endolymphatic sac
alar	alar plate	Gi	gigantocellular reticular nucleus
basal	basal plate	IC	inferior colliculus
BOcc	basioccipital bone	IRt	intermediate reticular nucleus

LPGi	lateral paragigantocellular nucleus	ParP	parietal bone plate
LR4V	lateral recess of 4th ventricle	PCRt	parvicellular reticular nucleus
lvs	lateral vestibulospinal tract	ROb	raphe obscurus nucleus
Mesen	mesencephalon	S10Gn	superior ganglion of 10n
Meten	metencephalon	SC	superior colliculus
mlf	medial longitudinal fasciculus	sl	sulcus limitans
mne	median neuroepithelium	Sol	solitary nucleus
Myelen	myelencephalon	sol	solitary tract
ntz	nuclear transitory zone (of Cb)	sp5	spinal trigeminal tract

Sp5I	spinal trigeminal nu, interpolaris		
SpVe	spinal vestibular nucleus		
TempP	temporal bone, petrous		
trg	germinal trigone		
trs	transverse venous sinus		
vert	vertebral artery		

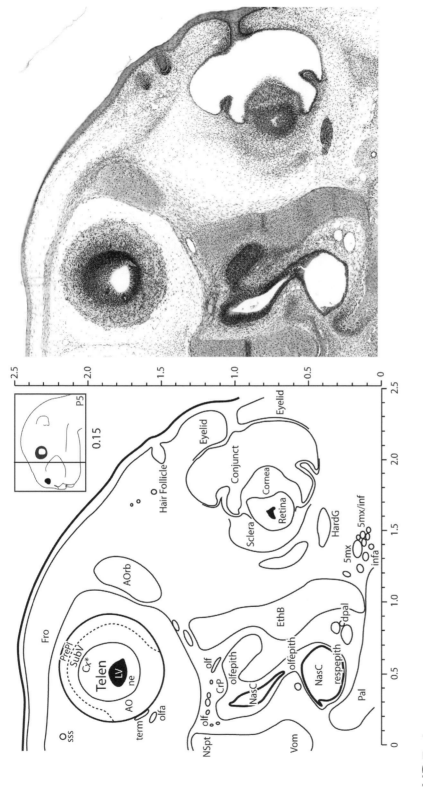

WP5-1

*	denotes precursor of structure
5mx	maxillary division of trigeminal nerve
5mx/inf	infraorbital nerve (branch of 5mx)
AO	anterior olfactory nucleus
AOrb	alar orbital bone
Conjunct	conjunctival sac
Cornea	cornea (limbus)
CrP	cribriform plate of ethmoid bone
Cx	cerebral cortex
dpal	descending palatine artery
EthB	ethmoid bone
Eyelid	eyelid
Fro	frontal bone
HardG	Harderian gland
infa	infraorbital artery
LV	lateral ventricle
NasC	nasal cavity
ne	neuroepithelium
NSpt	nasal septum
olf	olfactory nerve
olfa	olfactory artery
olfepith	olfactory epithelium (sensory)
Pal	palatine bone
PrePl	preplate of cortex
respepith	respiratory epithelium
Retina	retina (developing pigment epithelium)
sss	superior sagittal sinus
SubV	subventricular layer of cortex
Telen	telencephalon
term	terminal nerve
Vom	vomer (bone)

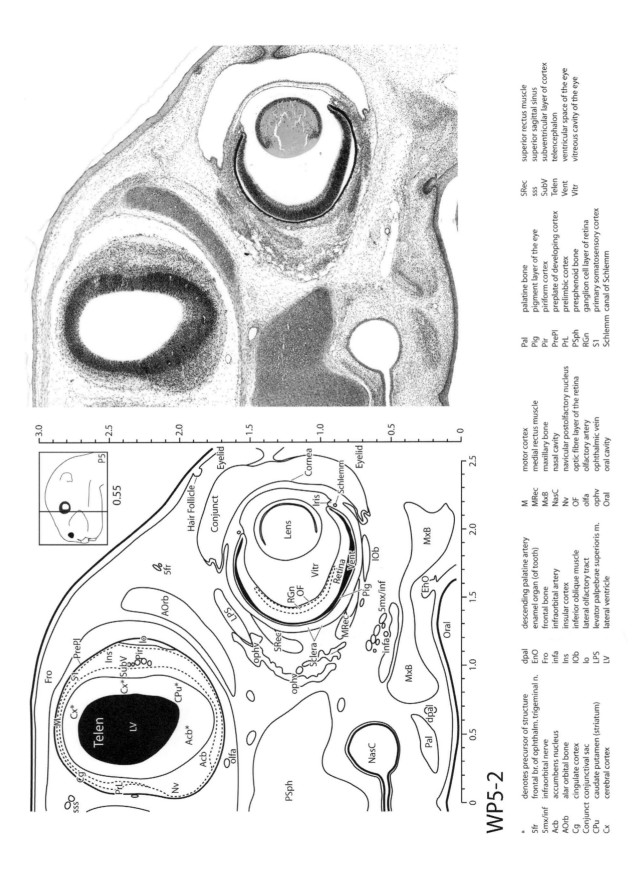

WP5-2

*	denotes precursor of structure	dpal	descending palatine artery	M	motor cortex	Pal	palatine bone	SRec	superior rectus muscle
5fr	frontal br. of ophthalm. trigeminal n.	EnO	enamel organ (of tooth)	MRec	medial rectus muscle	Pig	pigment layer of the eye	sss	superior sagittal sinus
5mx/inf	infraorbital nerve	Fro	frontal bone	MxB	maxillary bone	Pir	piriform cortex	SubV	subventricular layer of cortex
Acb	accumbens nucleus	infa	infraorbital artery	NasC	nasal cavity	PrePl	preplate of developing cortex	Telen	telencephalon
AOrb	alar orbital bone	Ins	insular cortex	Nv	navicular postolfactory nucleus	PrL	prelimbic cortex	Vent	ventricular space of the eye
Cg	cingulate cortex	IOb	inferior oblique muscle	OF	optic fibre layer of the retina	PSph	presphenoid bone	Vitr	vitreous cavity of the eye
Conjunct	conjunctival sac	Io	lateral olfactory tract	olfa	olfactory artery	RGn	ganglion cell layer of retina	
CPu	caudate putamen (striatum)	LPS	levator palpebrae superioris m.	ophv	ophthalmic vein	S1	primary somatosensory cortex	
Cx	cerebral cortex	LV	lateral ventricle	Oral	oral cavity	Schlemm	canal of Schlemm	

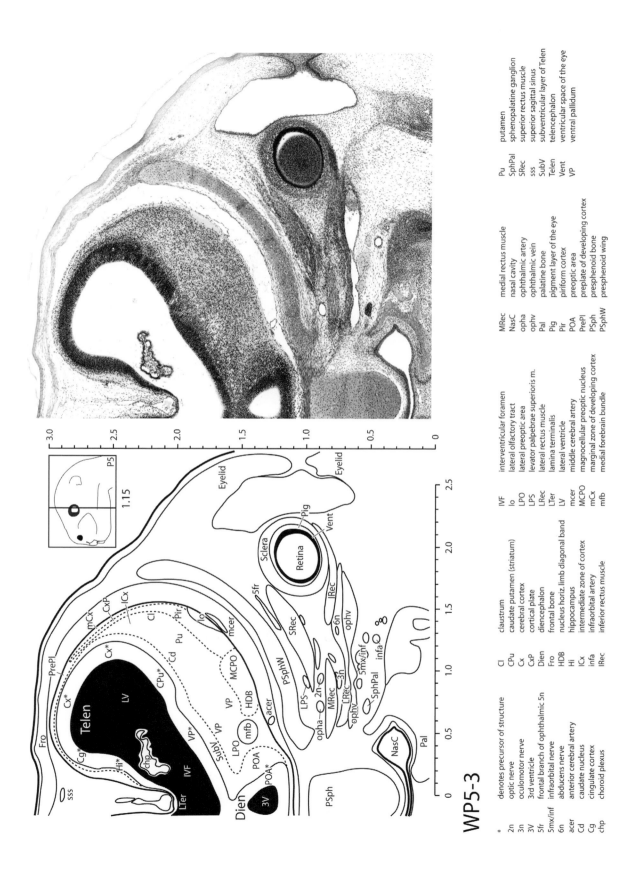

WP5-3

*	denotes precursor of structure	Cl	claustrum	IVF
2n	optic nerve	CPu	caudate putamen (striatum)	lo
3n	oculomotor nerve	Cx	cerebral cortex	LPO
3V	3rd ventricle	CxP	cortical plate	LPS
5fr	frontal branch of ophthalmic 5n	Dien	diencephalon	LRec
5mx/inf	infraorbital nerve	Fro	frontal bone	LTer
6n	abducens nerve	HDB	nucleus horiz. limb diagonal band	LV
acer	anterior cerebral artery	Hi	hippocampus	mcer
Cd	caudate nucleus	ICx	intermediate zone of cortex	MCPO
Cg	cingulate cortex	infa	infraorbital artery	mCx
chp	choroid plexus	IRec	inferior rectus muscle	mfb

IVF	interventricular foramen	MRec	medial rectus muscle
lo	lateral olfactory tract	NasC	nasal cavity
LPO	lateral preoptic area	opha	ophthalmic artery
LPS	levator palpebrae superioris m.	ophv	ophthalmic vein
LRec	lateral rectus muscle	Pal	palatine bone
LTer	lamina terminalis	Pig	pigment layer of the eye
LV	lateral ventricle	Pir	piriform cortex
mcer	middle cerebral artery	POA	preoptic area
MCPO	magnocellular preoptic nucleus	PrePl	preplate of developing cortex
mCx	marginal zone of developing cortex	PSph	presphenoid bone
mfb	medial forebrain bundle	PSphW	presphenoid wing

Pu	putamen
SphPal	sphenopalatine ganglion
SRec	superior rectus muscle
sss	superior sagittal sinus
SubV	subventricular layer of Telen
Telen	telencephalon
Vent	ventricular space of the eye
VP	ventral pallidum

WP5-4

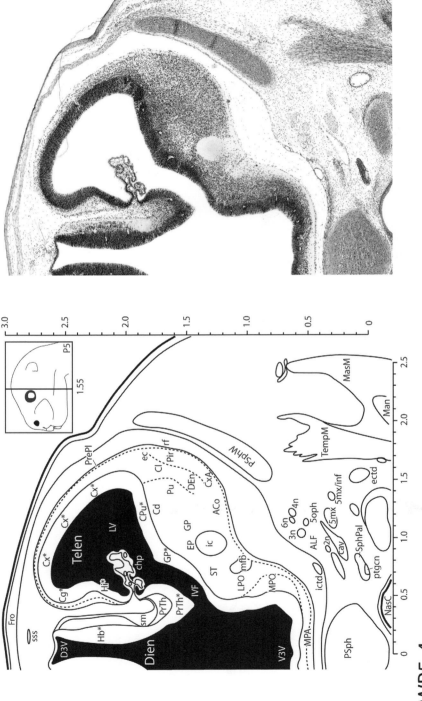

*	denotes precursor of structure			
2n	optic nerve			
3n	oculomotor nerve			
4n	trochlear nerve			
5mx	maxillary division of trigeminal n.			
5mx/inf	infraorbital nerve (branch of 5mx)			
5oph	ophthalmic division of trigeminal n.			
6n	abducens nerve			
ACo	anterior cortical amygdaloid area			
ALF	anterior lacerated foramen			
cav	cavernous sinus			

Cd	caudate nucleus	ectd	external carotid artery	Man	mandible	ptgcn	nerve of the pterygoid canal
Cg	cingulate cortex	EP	entopeduncular nucleus (of pallidum)	MasM	masseter muscle	Pu	putamen
chp	choroid plexus	Fro	frontal bone	mfb	medial forebrain bundle	rf	rhinal fissure
Cl	claustrum	GP	globus pallidus (of dorsal pallidum)	MPA	medial preoptic area	sm	stria medullaris
CPu	caudate putamen (striatum)	Hb	habenular nuclei	MPO	medial preoptic nucleus	SphPal	sphenopalatine ganglion
Cx	cerebral cortex	Hi	hippocampus	NasC	nasal cavity	sss	superior sagittal sinus
CxA	cortex-amygdala transition zone	ic	internal capsule fibres	Pir	piriform cortex	ST	bed nuclei of stria terminalis
D3V	dorsal 3rd ventricle	ictd	internal carotid artery	PrePl	preplate of developing cortex	Telen	telencephalon
DEn	dorsal endopiriform nucleus	IVF	interventricular foramen	PrTh	prethalamus (p3 derivative)	TempM	temporalis muscle
Dien	diencephalon	LPO	lateral preoptic area	PSph	presphenoid bone	V3V	ventral third ventricle
ec	external capsule	LV	lateral ventricle	PSphW	presphenoid wing		

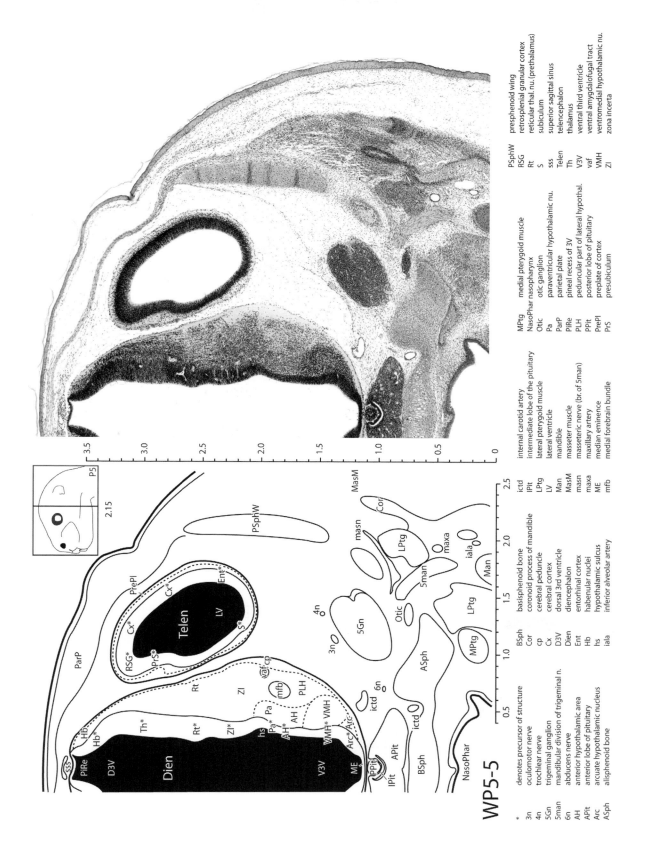

WP5-5

* denotes precursor of structure
3n oculomotor nerve
4n trochlear nerve
5Gn trigeminal ganglion
5man mandibular division of trigeminal n.
6n abducens nerve
AH anterior hypothalamic area
APit anterior lobe of pituitary
Arc arcuate hypothalamic nucleus
ASph alisphenoid bone

BSph basisphenoid bone
Cor coronoid process of mandible
cp cerebral peduncle
Cx cerebral cortex
D3V dorsal 3rd ventricle
Dien diencephalon
Ent entorhinal cortex
Hb habenular nuclei
hs hypothalamic sulcus
iala inferior alveolar artery

ictd internal carotid artery
IPit intermediate lobe of the pituitary
LPtg lateral pterygoid muscle
LV lateral ventricle
Man mandible
MasM masseter muscle
masn masseteric nerve (br. of 5man)
maxa maxillary artery
ME median eminence
mfb medial forebrain bundle

MPtg medial pterygoid muscle
NasoPhar nasopharynx
Otic otic ganglion
Pa paraventricular hypothalamic nu.
ParP parietal plate
PiRe pineal recess of 3V
PLH peduncular part of lateral hypothal.
PPit posterior lobe of pituitary
PrePl preplate of cortex
PrS presubiculum

PSphW presphenoid wing
RSG retrosplenial granular cortex
Rt reticular thal. nu. (prethalamus)
S subiculum
sss superior sagittal sinus
Telen telencephalon
Th thalamus
V3V ventral third ventricle
vaf ventral amygdalofugal tract
VMH ventromedial hypothalamic nu.
ZI zona incerta

2.15

P5

WP5-6

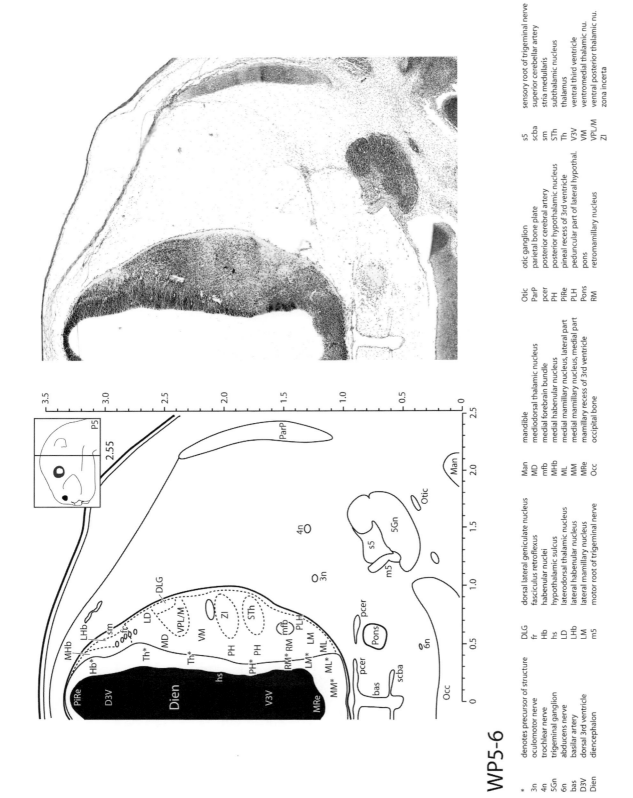

*	denotes precursor of structure	DLG	dorsal lateral geniculate nucleus	Man	mandible	Otic	otic ganglion	s5	sensory root of trigeminal nerve
3n	oculomotor nerve	fr	fasciculus retroflexus	MD	mediodorsal thalamic nucleus	ParP	parietal bone plate	scba	superior cerebellar artery
4n	trochlear nerve	Hb	habenular nuclei	mfb	medial forebrain bundle	pcer	posterior cerebral artery	sm	stria medullaris
5Gn	trigeminal ganglion	hs	hypothalamic sulcus	MHb	medial habenular nucleus	PH	posterior hypothalamic nucleus	STh	subthalamic nucleus
6n	abducens nerve	LD	laterodorsal thalamic nucleus	ML	medial mamillary nucleus, lateral part	PiRe	pineal recess of 3rd ventricle	Th	thalamus
bas	basilar artery	LHb	lateral habenular nucleus	MM	medial mamillary nucleus, medial part	PLH	peduncular part of lateral hypothal.	V3V	ventral third ventricle
D3V	dorsal 3rd ventricle	LM	lateral mamillary nucleus	MRe	mamillary recess of 3rd ventricle	Pons	pons	VM	ventromedial thalamic nu.
Dien	diencephalon	m5	motor root of trigeminal nerve	Occ	occipital bone	RM	retromamillary nucleus	VPL/M	ventral posterior thalamic nu.
								ZI	zona incerta

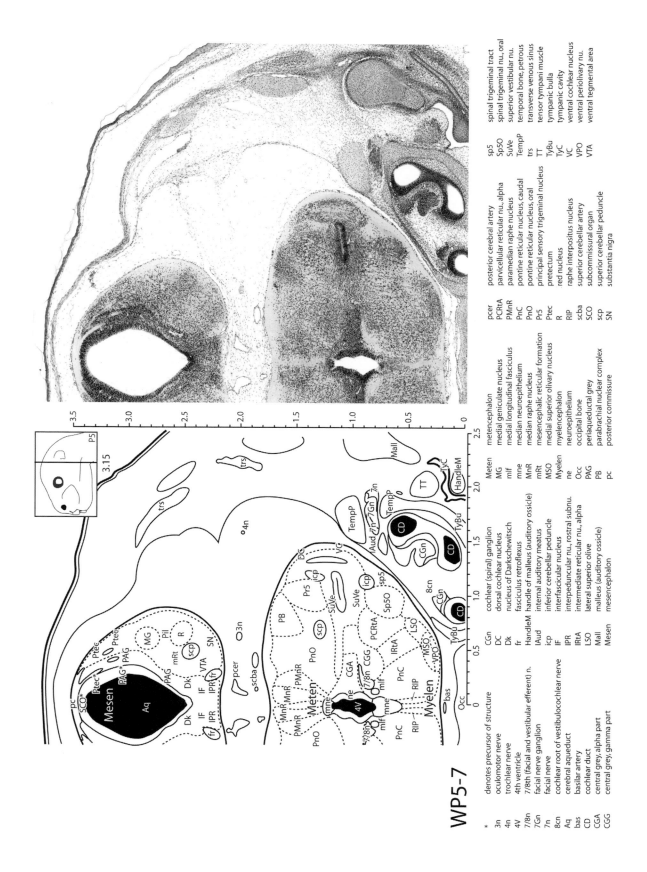

WP5-7

*	denotes precursor of structure	CGn	cochlear (spiral) ganglion
3n	oculomotor nerve	DC	dorsal cochlear nucleus
4n	trochlear nerve	Dk	nucleus of Darkschewitsch
4V	4th ventricle	fr	fasciculus retroflexus
7/8n	7/8th (facial and vestibular efferent) n.	HandleM	handle of malleus (auditory ossicle)
7Gn	facial nerve ganglion	IAud	internal auditory meatus
7n	facial nerve	icp	inferior cerebellar peduncle
8cn	cochlear root of vestibulocochlear nerve	IF	interfascicular nucleus
Aq	cerebral aqueduct	IPR	interpeduncular nu., rostral subnu.
bas	basilar artery	IRtA	intermediate reticular nu., alpha
CD	cochlear duct	LSO	lateral superior olive
CGA	central grey, alpha part	Mall	malleus (auditory ossicle)
CGG	central grey, gamma part	Mesen	mesencephalon

Meten	metencephalon	pcer	posterior cerebral artery
MG	medial geniculate nucleus	PCRtA	parvicellular reticular nu., alpha
mlf	medial longitudinal fasciculus	PMnR	paramedian raphe nucleus
mne	median neuroepithelium	PnC	pontine reticular nucleus, caudal
MnR	median raphe nucleus	PnO	pontine reticular nucleus, oral
mRt	mesencephalic reticular formation	Pr5	principal sensory trigeminal nucleus
MSO	medial superior olivary nucleus	Ptec	pretectum
Myelen	myelencephalon	R	red nucleus
ne	neuroepithelium	RIP	raphe interpositus nucleus
Occ	occipital bone	scba	superior cerebellar artery
PAG	periaqueductal grey	SCO	subcommissural organ
PB	parabrachial nuclear complex	scp	superior cerebellar peduncle
pc	posterior commissure	SN	substantia nigra

sp5	spinal trigeminal tract		
Sp5O	spinal trigeminal nu., oral		
SuVe	superior vestibular nu.		
TempP	temporal bone, petrous		
trs	transverse venous sinus		
TT	tensor tympani muscle		
TyBu	tympanic bulla		
TyC	tympanic cavity		
VC	ventral cochlear nucleus		
VPO	ventral periolivary nu.		
VTA	ventral tegmental area		

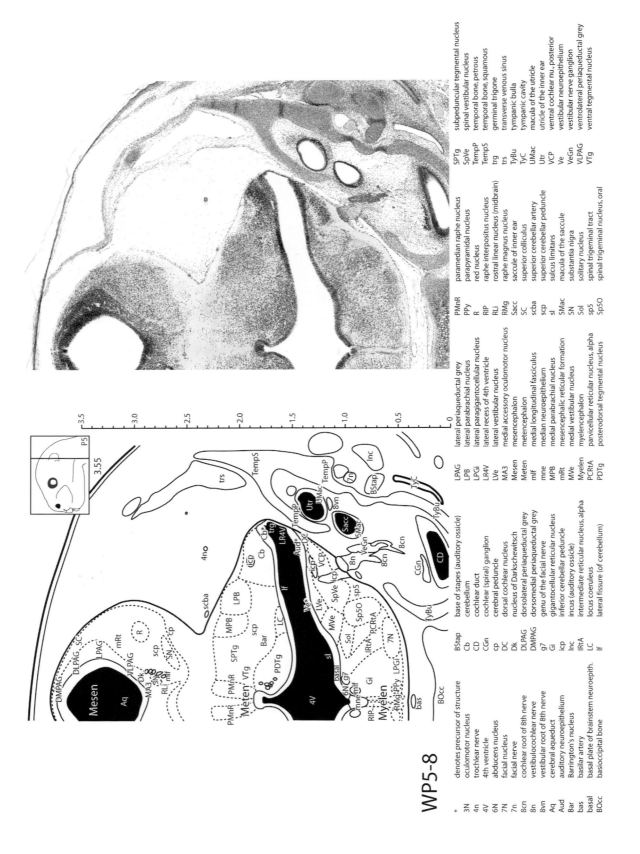

WP5-8

*	denotes precursor of structure
3N	oculomotor nucleus
4n	trochlear nerve
4V	4th ventricle
6N	abducens nucleus
7N	facial nucleus
7n	facial nerve
8cn	cochlear root of 8th nerve
8n	vestibulocochlear nerve
8vn	vestibular root of 8th nerve
Aq	cerebral aqueduct
Aud	auditory neuroepithelium
Bar	Barrington's nucleus
bas	basilar artery
basal	basal plate of brainstem neuroepith.
BOcc	basioccipital bone

BStap	base of stapes (auditory ossicle)
Cb	cerebellum
CD	cochlear duct
CGn	cochlear (spiral) ganglion
cp	cerebral peduncle
DC	dorsal cochlear nucleus
Dk	nucleus of Darkschewitsch
DLPAG	dorsolateral periaqueductal grey
DMPAG	dorsomedial periaqueductal grey
g7	genu of the facial nerve
Gi	gigantocellular reticular nucleus
icp	inferior cerebellar peduncle
Inc	incus (auditory ossicle)
IRtA	intermediate reticular nucleus, alpha
LC	locus coeruleus
If	lateral fissure (of cerebellum)

LPAG	lateral periaqueductal grey
LPB	lateral parabrachial nucleus
LPGi	lateral paragigantocellular nucleus
LR4V	lateral recess of 4th ventricle
LVe	lateral vestibular nucleus
MA3	medial accessory oculomotor nucleus
Mesen	mesencephalon
Meten	metencephalon
mlf	medial longitudinal fasciculus
mne	median neuroepithelium
MPB	medial parabrachial nucleus
mRt	mesencephalic reticular formation
MVe	medial vestibular nucleus
Myelen	myelencephalon
PCRtA	parvicellular reticular nucleus, alpha
PDTg	posterodorsal tegmental nucleus

PMnR	paramedian raphe nucleus
PPy	parapyramidal nucleus
R	red nucleus
RIP	raphe interpositus nucleus
RLi	rostral linear nucleus (midbrain)
RMg	raphe magnus nucleus
Sacc	saccule of inner ear
SC	superior colliculus
scba	superior cerebellar artery
scp	superior cerebellar peduncle
sl	sulcus limitans
SMac	macula of the saccule
SN	substantia nigra
Sol	solitary nucleus
sp5	spinal trigeminal tract
Sp5O	spinal trigeminal nucleus, oral

SPTg	subpeduncular tegmental nucleus
SpVe	spinal vestibular nucleus
TempP	temporal bone, petrous
TempS	temporal bone, squamous
trg	germinal trigone
trs	transverse venous sinus
TyBu	tympanic bulla
TyC	tympanic cavity
UMac	macula of the utricle
Utr	utricle of the inner ear
VCP	ventral cochlear nu., posterior
Ve	vestibular neuroepithelium
VeGn	vestibular nerve ganglion
VLPAG	ventrolateral periaqueductal grey
VTg	ventral tegmental nucleus

WP5-9

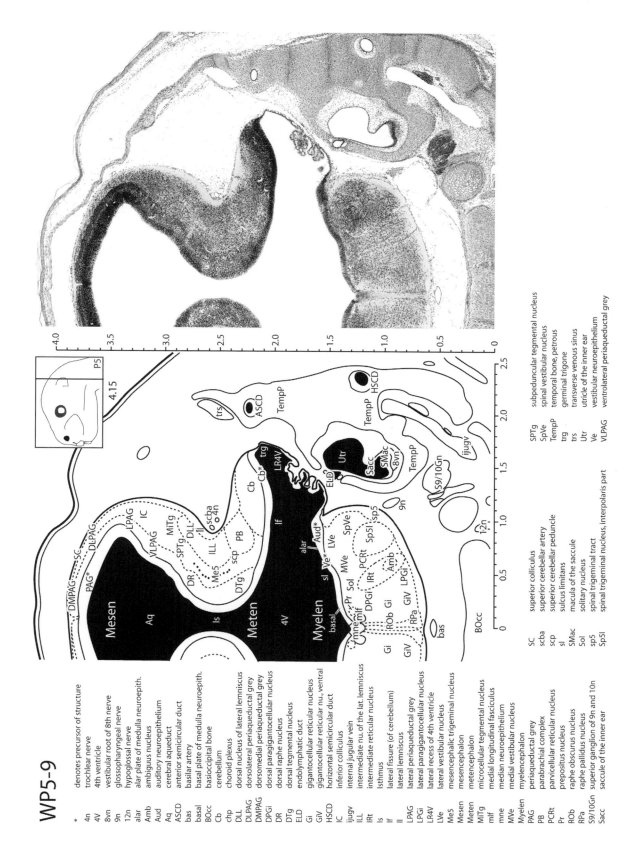

* denotes precursor of structure
4n trochlear nerve
4V 4th ventricle
8vn vestibular root of 8th nerve
9n glossopharyngeal nerve
12n hypoglossal nerve
alar alar plate of medulla neuroepith.
Amb ambiguus nucleus
Aud auditory neuroepithelium
Aq cerebral aqueduct
ASCD anterior semicircular duct
bas basilar artery
basal basal plate of medulla neuroepith.
BOcc basioccipital bone
Cb cerebellum
chp choroid plexus
DLL dorsal nucleus of lateral lemniscus
DLPAG dorsolateral periaqueductal grey
DMPAG dorsomedial periaqueductal grey
DPGi dorsal paragigantocellular nucleus
DR dorsal raphe nucleus
DTg dorsal tegmental nucleus
ELD endolymphatic duct
Gi gigantocellular reticular nucleus
GiV gigantocellular reticular nu., ventral
HSCD horizontal semicircular duct
IC inferior colliculus
jjugv internal jugular vein
ILL intermediate nu. of the lat. lemniscus
IRt intermediate reticular nucleus
Is isthmus
If lateral fissure (of cerebellum)
ll lateral lemniscus
LPAG lateral periaqueductal grey
LPGi lateral paragigantocellular nucleus
LR4V lateral recess of 4th ventricle
LVe lateral vestibular nucleus
Me5 mesencephalic trigeminal nucleus
Mesen mesencephalon
Meten metencephalon
MiTg microcellular tegmental nucleus
mlf medial longitudinal fasciculus
mne median neuroepithelium
MVe medial vestibular nucleus
Myelen myelencephalon
PAG periaqueductal grey
PB parabrachial complex
PCRt parvicellular reticular nucleus
Pr prepositus nucleus
ROb raphe obscurus nucleus
RPa raphe pallidus nucleus
S9/10Gn superior ganglion of 9n and 10n
Sacc saccule of the inner ear

SC superior colliculus
scba superior cerebellar artery
scp superior cerebellar peduncle
sl sulcus limitans
SMac macula of the saccule
Sol solitary nucleus
sp5 spinal trigeminal tract
Sp5I spinal trigeminal nucleus, interpolaris part

SPTg subpeduncular tegmental nucleus
SpVe spinal vestibular nucleus
TempP temporal bone, petrous
trg germinal trigone
trs transverse venous sinus
Utr utricle of the inner ear
Ve vestibular neuroepithelium
VLPAG ventrolateral periaqueductal grey

WP5-10

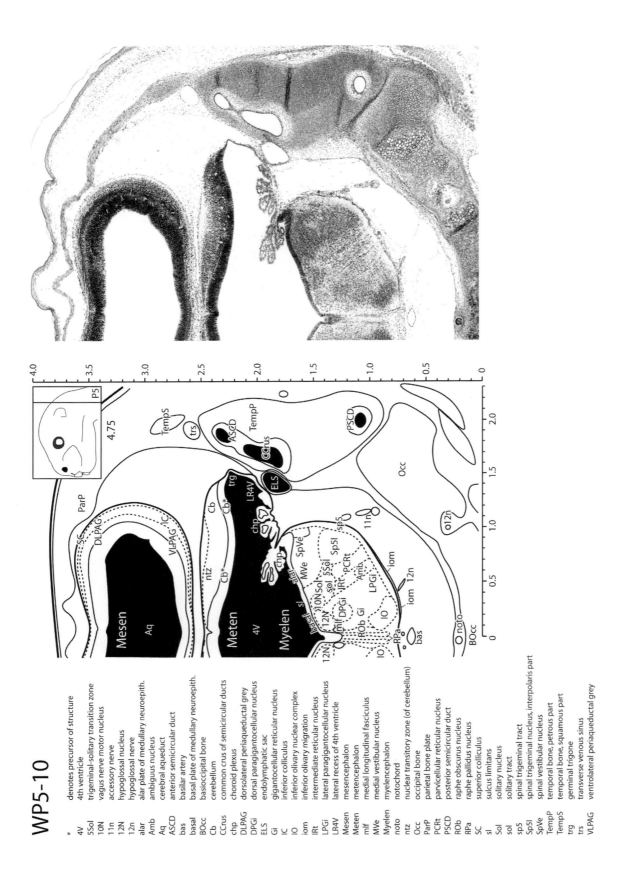

*	denotes precursor of structure
4v	4th ventricle
5Sol	trigeminal-solitary transition zone
10N	vagus nerve motor nucleus
11n	accessory nerve
12N	hypoglossal nucleus
12n	hypoglossal nerve
alar	alar plate of medullary neuroepith.
Amb	ambiguus nucleus
Aq	cerebral aqueduct
ASCD	anterior semicircular duct
bas	basilar artery
basal	basal plate of medullary neuroepith.
BOcc	basioccipital bone
Cb	cerebellum
CCrus	common crus of semicircular ducts
chp	choroid plexus
DLPAG	dorsolateral periaqueductal grey
DPGi	dorsal paragigantocellular nucleus
ELS	endolymphatic sac
Gi	gigantocellular reticular nucleus
IC	inferior colliculus
IO	inferior olivary nuclear complex
iom	inferior olivary migration
IRt	intermediate reticular nucleus
LPGi	lateral paragigantocellular nucleus
LR4V	lateral recess of 4th ventricle
Mesen	mesencephalon
Meten	metencephalon
mlf	medial longitudinal fasciculus
MVe	medial vestibular nucleus
Myelen	myelencephalon
noto	notochord
ntz	nuclear transitory zone (of cerebellum)
Occ	occipital bone
ParP	parietal bone plate
PCRt	parvicellular reticular nucleus
PSCD	posterior semicircular duct
ROb	raphe obscurus nucleus
RPa	raphe pallidus nucleus
SC	superior colliculus
sl	sulcus limitans
Sol	solitary nucleus
sol	solitary tract
sp5	spinal trigeminal tract
Sp5I	spinal trigeminal nucleus, interpolaris part
SpVe	spinal vestibular nucleus
TempP	temporal bone, petrous part
TempS	temporal bone, squamous part
trg	germinal trigone
trs	transverse venous sinus
VLPAG	ventrolateral periaqueductal grey

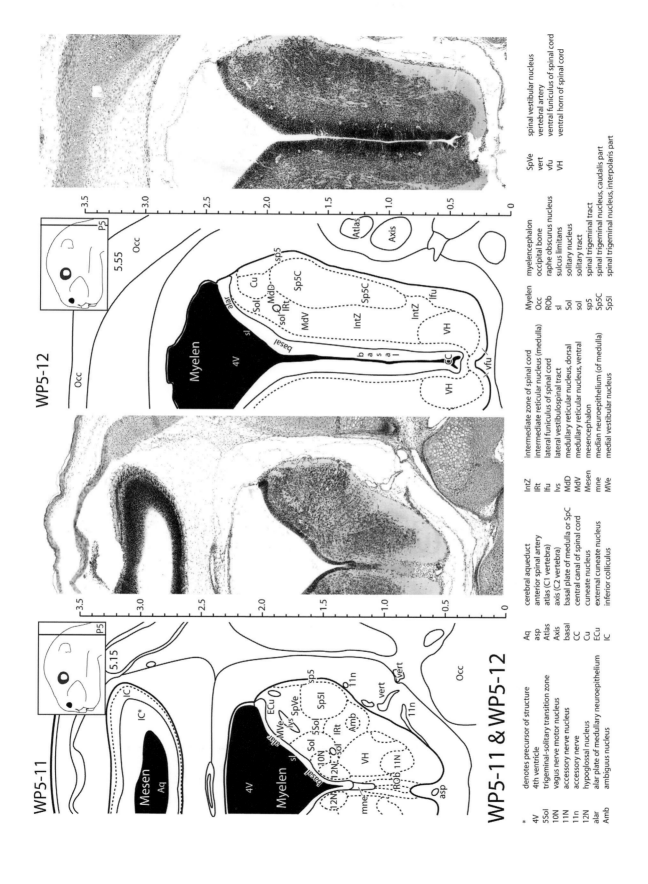

WP5-11

WP5-12

WP5-11 & WP5-12

*   denotes precursor of structure
4V   4th ventricle
5Sol   trigeminal-solitary transition zone
10N   vagus nerve motor nucleus
11N   accessory nerve motor nucleus
11n   accessory nerve
12N   hypoglossal nucleus
alar   alar plate of medullary neuroepithelium
Amb   ambiguus nucleus

Aq   cerebral aqueduct
asp   anterior spinal artery
Atlas   atlas (C1 vertebra)
Axis   axis (C2 vertebra)
basal   basal plate of medulla or SpC
CC   central canal of spinal cord
Cu   cuneate nucleus
ECu   external cuneate nucleus
IC   inferior colliculus

IntZ   intermediate zone of spinal cord
IRt   intermediate reticular nucleus (medulla)
lfu   lateral funiculus of spinal cord
lvs   lateral vestibulospinal tract
MdD   medullary reticular nucleus, dorsal
MdV   medullary reticular nucleus, ventral
Mesen   mesencephalon
mne   median neuroepithelium (of medulla)
MVe   medial vestibular nucleus

Myelen   myelencephalon
Occ   occipital bone
ROb   raphe obscurus nucleus
sl   sulcus limitans
Sol   solitary nucleus
sol   solitary tract
sp5   spinal trigeminal tract
Sp5C   spinal trigeminal nucleus, caudalis part
Sp5I   spinal trigeminal nucleus, interpolaris part

SpVe   spinal vestibular nucleus
vert   vertebral artery
vfu   ventral funiculus of spinal cord
VH   ventral horn of spinal cord

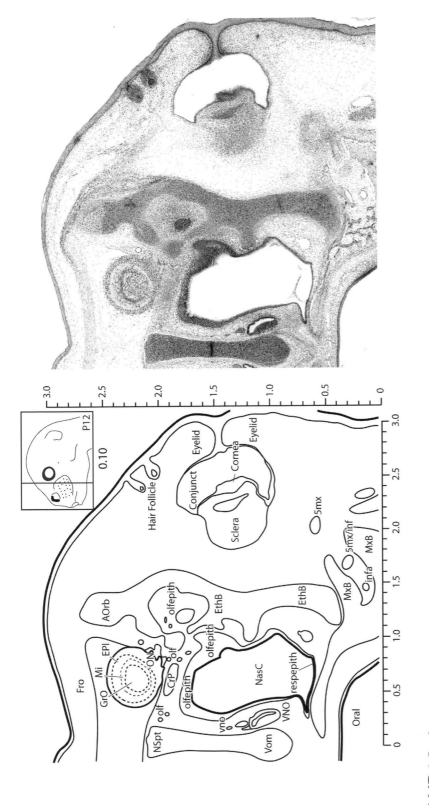

WP12-1

5mx	maxillary division of trigeminal nerve		MxB	maxillary bone
5mx/inf	infraorbital nerve (branch of 5mx)		NasC	nasal cavity
AOrb	alar orbital bone		NSpt	nasal septum
Conjunct	conjunctival sac		olf	olfactory nerve
Cornea	cornea		olfepith	olfactory epithelium
CrP	cribriform plate of ethmoid bone		ON	olfactory nerve fibre layer of olfactory bulb
EPl	external plexiform layer of olfactory bulb		Oral	oral cavity
EthB	ethmoid bone		respepith	respiratory epithelium of nasal cavity
Fro	frontal bone		VNO	vomeronasal organ
GrO	granular cell layer of olfactory bulb		vno	vomeronasal nerve
infa	infraorbital artery		Vom	vomer (bone)
Mi	mitral cell layer of olfactory bulb			

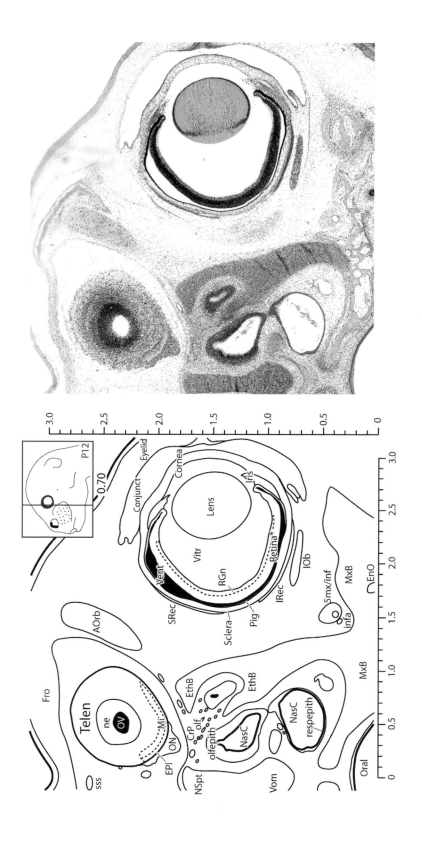

WP12-2

3.0
2.5
2.0
1.5
1.0
0.5
0

P12
0.70

Eyelid
Conjunct
Cornea
Iris
Lens
Vent
Vitr
Retina*
RGn
IRec
SRec
Pig
Sclera
IOb
5mx/inf
infa
MxB
EnO

AOrb

Fro
Telen
ne
OV
Mi
ON
EPI
olf
olfepith
CrP
NSpt
EthB
NasC
Vom
EthB
NasC
respepith
MxB
Oral
SSS

3.0
2.5
2.0
1.5
1.0
0.5
0

5mx/inf	infraorbital nerve (branch of maxillary division)	IRec	inferior rectus muscle
AOrb	alar orbital bone	Iris	iris
Conjunct	conjunctival sac	Lens	lens
Cornea	cornea	Mi	mitral cell layer of olfactory bulb
CrP	cribriform plate of ethmoid bone	MxB	maxillary bone
EnO	enamel organ of tooth	NasC	nasal cavity
EPI	external plexiform layer of olfactory bulb	ne	neuroepithelium (of telencephalon)
EthB	ethmoid bone	NSpt	nasal septum
Eyelid	eyelid	olf	olfactory nerve
Fro	frontal bone	olfepith	olfactory epithelium
infa	infraorbital artery	ON	olfactory nerve layer (of olfactory bulb)
IOb	inferior oblique muscle	Oral	oral cavity

OV	olfactory ventricle (of telencephalon)
Pig	pigment layer of the eye
respepith	respiratory epithelium
Retina	retina
RGn	ganglion cell layer of retina
SRec	superior rectus muscle
sss	superior sagittal sinus
Telen	telencephalon
Vent	ventricular space of the eye
Vitr	vitreous cavity of eye
Vom	vomer (bone)

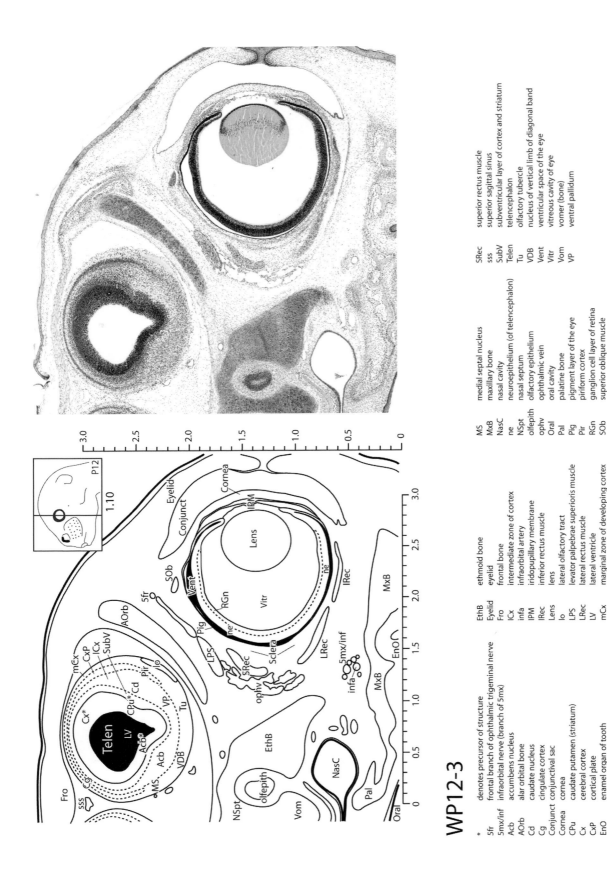

WP12-3

*	denotes precursor of structure
5fr	frontal branch of ophthalmic trigeminal nerve
5mx/inf	infraorbital nerve (branch of 5mx)
Acb	accumbens nucleus
AOrb	alar orbital bone
Cd	caudate nucleus
Cg	cingulate cortex
Conjunct	conjunctival sac
Cornea	cornea
CPu	caudate putamen (striatum)
Cx	cerebral cortex
CxP	cortical plate
EnO	enamel organ of tooth
EthB	ethmoid bone
Eyelid	eyelid
Fro	frontal bone
ICx	intermediate zone of cortex
infa	infraorbital artery
IPM	iridopupillary membrane
IRec	inferior rectus muscle
Lens	lens
lo	lateral olfactory tract
LPS	levator palpebrae superioris muscle
LRec	lateral rectus muscle
LV	lateral ventricle
mCx	marginal zone of developing cortex

MS	medial septal nucleus
MxB	maxillary bone
NasC	nasal cavity
ne	neuroepithelium (of telencephalon)
NSpt	nasal septum
olfepith	olfactory epithelium
ophv	ophthalmic vein
Oral	oral cavity
Pal	palatine bone
Pig	pigment layer of the eye
Pir	piriform cortex
RGn	ganglion cell layer of retina
SOb	superior oblique muscle

SRec	superior rectus muscle
sss	superior sagittal sinus
SubV	subventricular layer of cortex and striatum
Telen	telencephalon
Tu	olfactory tubercle
VDB	nucleus of vertical limb of diagonal band
Vent	ventricular space of the eye
Vitr	vitreous cavity of eye
Vom	vomer (bone)
VP	ventral pallidum

WP12-4

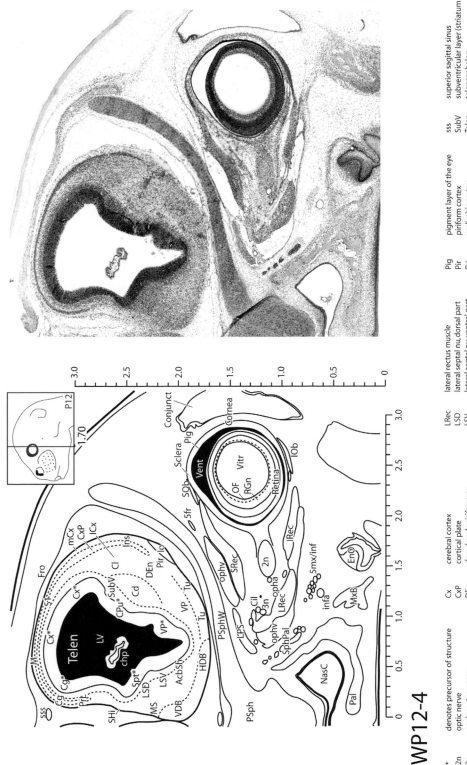

*	denotes precursor of structure	Cx	cerebral cortex
2n	optic nerve	CxP	cortical plate
3n	oculomotor nerve	DEn	dorsal endopiriform nu.
5fr	frontal br of ophthalmic 5n	EnO	enamel organ
5mx/inf	infraorbital n. (br. of 5mx)	Fro	frontal bone
AcbSh	accumbens nu, shell	HDB	nu. horiz limb diagonal band
Cd	caudate nucleus	ICx	intermediate zone of cortex
Cg	cingulate cortex	infa	infraorbital artery
chp	choroid plexus	Ins	insular cortex
Cil	ciliary ganglion	IOb	inferior oblique muscle
Cl	claustrum	IRec	inferior rectus muscle
Conjunct	conjunctival sac	lo	lateral olfactory tract
CPu	caudate putamen (striatum)	LPS	levator palpebrae superioris muscle

LRec	lateral rectus muscle	Pig	pigment layer of the eye	sss	superior sagittal sinus
LSD	lateral septal nu, dorsal part	Pir	piriform cortex	SubV	subventricular layer (striatum and cx)
LSV	lateral septal nu, ventral part	PrL	prelimbic cortex	Telen	telencephalon
LV	lateral ventricle	PSph	presphenoid bone	Tu	olfactory tubercle
M	motor cortex	PSphW	presphenoid wing	VDB	nu. vertical limb diagonal band
mCx	marginal zone of developing cx	Retina	retina	Vent	ventricular space of the eye
MS	medial septal nu.	RGn	retinal ganglion cell layer	Vitr	vitreous cavity of eye
MxB	maxillary bone	S1	primary somatosensory cx	VP	ventral pallidum
NasC	nasal cavity	SHi	septohippocampal nu.		
OF	optic fibre layer of the retina	SOb	superior oblique muscle		
opha	ophthalmic artery	SphPal	sphenopalatine ganglion		
ophv	ophthalmic vein	Spt	septum		
Pal	palatine bone	SRec	superior rectus muscle		

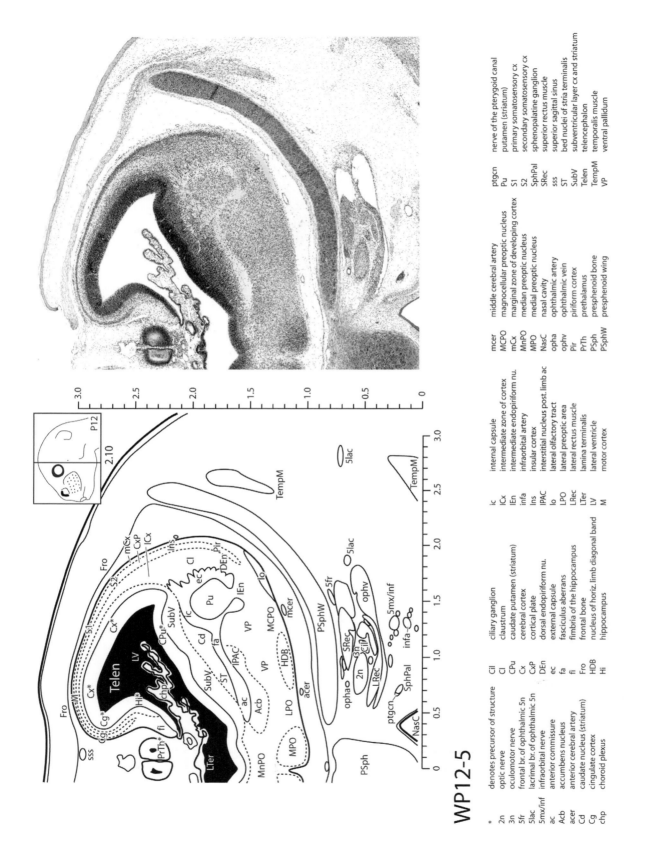

WP12-5

*	denotes precursor of structure
2n	optic nerve
3n	oculomotor nerve
5fr	frontal br. of ophthalmic 5n
5lac	lacrimal br. of ophthalmic 5n
5mx/inf	infraorbital nerve
ac	anterior commissure
Acb	accumbens nucleus
acer	anterior cerebral artery
Cd	caudate nucleus (striatum)
Cg	cingulate cortex
chp	choroid plexus

Cil	ciliary ganglion
Cl	claustrum
CPu	caudate putamen (striatum)
Cx	cerebral cortex
CxP	cortical plate
DEn	dorsal endopiriform nu.
ec	external capsule
fa	fasciculus aberrans
fi	fimbria of the hippocampus
Fro	frontal bone
HDB	nucleus of horiz. limb diagonal band
Hi	hippocampus

ic	internal capsule
ICx	intermediate zone of cortex
IEn	intermediate endopiriform nu.
infa	infraorbital artery
Ins	insular cortex
IPAC	interstitial nucleus post. limb ac
lo	lateral olfactory tract
LPO	lateral preoptic area
LRec	lateral rectus muscle
LTer	lamina terminalis
LV	lateral ventricle
M	motor cortex

mcer	middle cerebral artery
MCPO	magnocellular preoptic nucleus
mCx	marginal zone of developing cortex
MnPO	median preoptic nucleus
MPO	medial preoptic nucleus
NasC	nasal cavity
opha	ophthalmic artery
ophv	ophthalmic vein
Pir	piriform cortex
PrTh	prethalamus
PSph	presphenoid bone
PSphW	presphenoid wing

ptgcn	nerve of the pterygoid canal
Pu	putamen (striatum)
S1	primary somatosensory cx
S2	secondary somatosensory cx
SphPal	sphenopalatine ganglion
SRec	superior rectus muscle
sss	superior sagittal sinus
ST	bed nuclei of stria terminalis
SubV	subventricular layer cx and striatum
Telen	telencephalon
TempM	temporalis muscle
VP	ventral pallidum

WP12-6

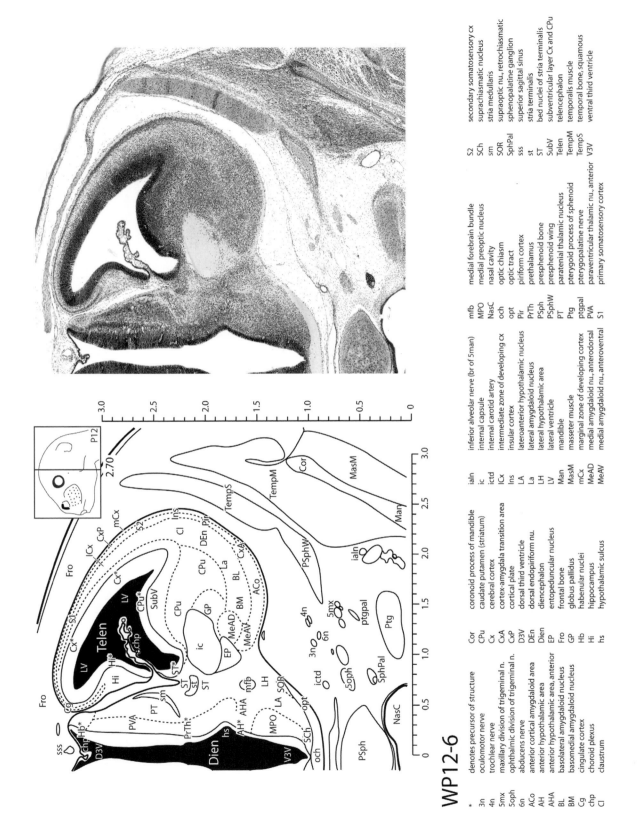

*	denotes precursor of structure	Cor	coronoid process of mandible	ialn	inferior alveolar nerve (br of 5man)	mfb	medial forebrain bundle	S2	secondary somatosensory cx

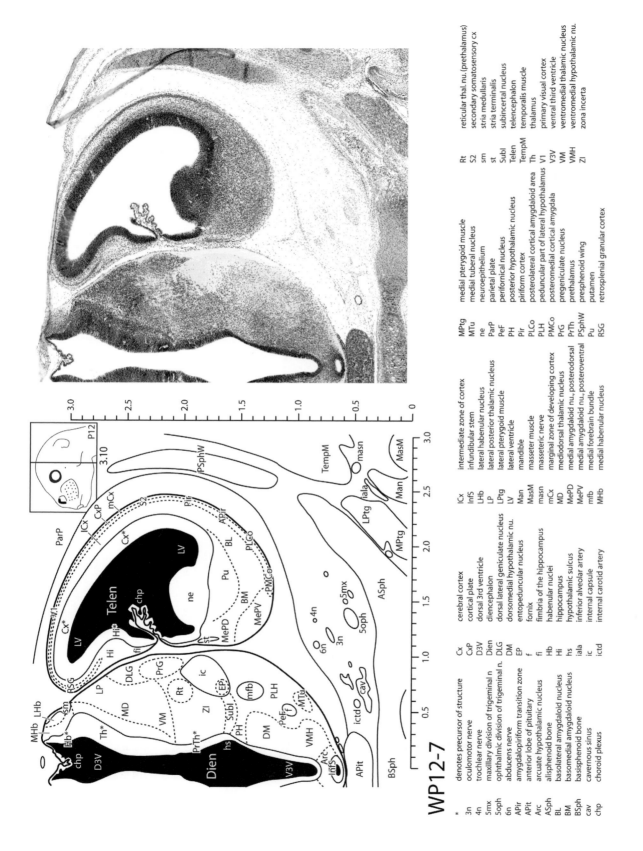

WP12-7

* denotes precursor of structure	Cx cerebral cortex	ICx intermediate zone of cortex	MPtg medial pterygoid muscle	Rt reticular thal. nu. (prethalamus)
3n oculomotor nerve	CxP cortical plate	InfS infundibular stem	MTu medial tuberal nucleus	S2 secondary somatosensory cx
4n trochlear nerve	D3V dorsal 3rd ventricle	LHb lateral habenular nucleus	ne neuroepithelium	sm stria medullaris
5mx maxillary division of trigeminal n	Dien diencephalon	LP lateral posterior thalamic nucleus	ParP parietal plate	st stria terminalis
5oph ophthalmic division of trigeminal n.	DLG dorsal lateral geniculate nucleus	LPtg lateral pterygoid muscle	PeF perifornical nucleus	Subl subincertal nucleus
6n abducens nerve	DM dorsomedial hypothalamic nu.	LV lateral ventricle	PH posterior hypothalamic nucleus	Telen telencephalon
APir amygdalopiriform transition zone	EP entopeduncular nucleus	Man mandible	Pir piriform cortex	TempM temporalis muscle
APit anterior lobe of pituitary	f fornix	MasM masseter muscle	PLCo posterolateral cortical amygdaloid area	Th thalamus
Arc arcuate hypothalamic nucleus	fi fimbria of the hippocampus	masn masseteric nerve	PLH peduncular part of lateral hypothalamus	V1 primary visual cortex
ASph alisphenoid bone	Hb habenular nuclei	mCx marginal zone of developing cortex	PMCo posteromedial cortical amygdala	V3V ventral third ventricle
BL basolateral amygdaloid nucleus	Hi hippocampus	MD mediodorsal thalamic nucleus	PrG pregeniculate nucleus	VM ventromedial thalamic nucleus
BM basomedial amygdaloid nucleus	hs hypothalamic sulcus	MePD medial amygdaloid nu., posterodorsal	PrTh prethalamus	VMH ventromedial hypothalamic nu.
BSph basisphenoid bone	iala inferior alveolar artery	MePV medial amygdaloid nu., posteroventral	PSphW presphenoid wing	ZI zona incerta
cav cavernous sinus	ic internal capsule	mfb medial forebrain bundle	Pu putamen	
chp choroid plexus	ictd internal carotid artery	MHb medial habenular nucleus	RSG retrosplenial granular cortex	

WP12-8

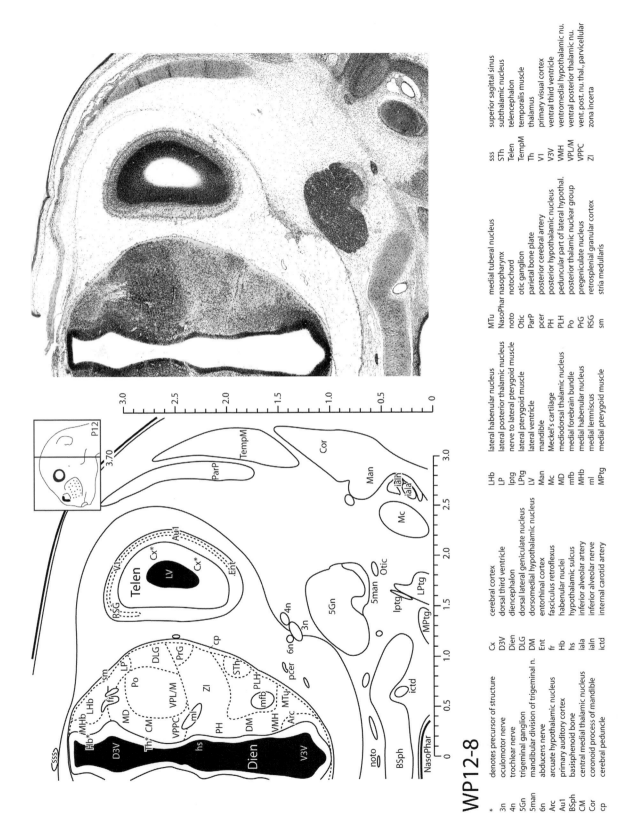

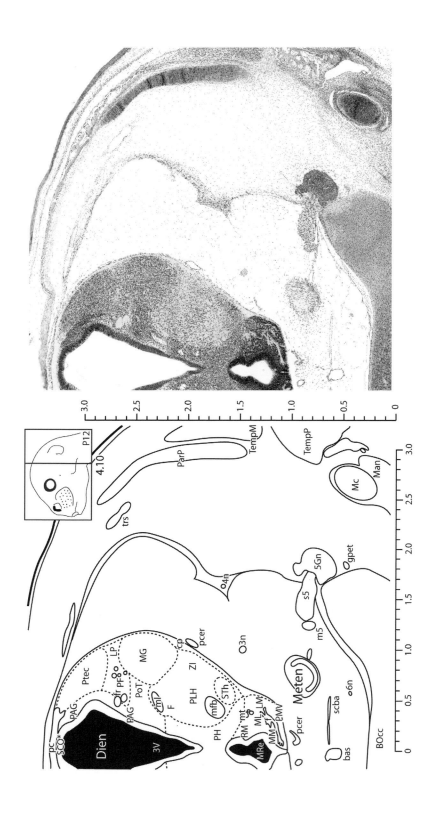

WP12-9

*	denotes precursor of structure	F	nuclei of the fields of Forel	mfb	medial forebrain bundle
3n	oculomotor nerve	f	fornix	MG	medial geniculate nucleus
3V	3rd ventricle	fr	fasciculus retroflexus	ML	medial mamillary nucleus, lateral
4n	trochlear nerve	gpet	greater petrosal nerve	ml	medial lemniscus
5Gn	trigeminal ganglion	LM	lateral mamillary nucleus	MM	medial mamillary nucleus, medial
6n	abducens nerve	LP	lateral posterior thalamic nucleus	MRe	mamillary recess of 3V
bas	basilar artery	m5	motor root of trigeminal nerve	mt	mamillothalamic tract
BOcc	basioccipital bone	Man	mandible	PAG	periaqueductal grey
cp	cerebral peduncle	Mc	Meckel's cartilage	ParP	parietal bone plate
Dien	diencephalon	Meten	metencephalon	pc	posterior commissure

pcer	posterior cerebral artery	SCO	subcommissural organ
PF	parafascicular thalamic nucleus	STh	subthalamic nucleus
PH	posterior hypothalamic nucleus	TempM	temporalis muscle
PLH	peduncular part of lateral hypothalamus	TempP	temporal bone, petrous
PMV	premamillary nucleus, ventral	trs	transverse venous sinus
PoT	posterior thalamic nucleus, triangular	ZI	zona incerta
Ptec	pretectum		
RM	retromamillary nucleus		
s5	sensory root of trigeminal nerve		
scba	superior cerebellar artery		

WP12-10

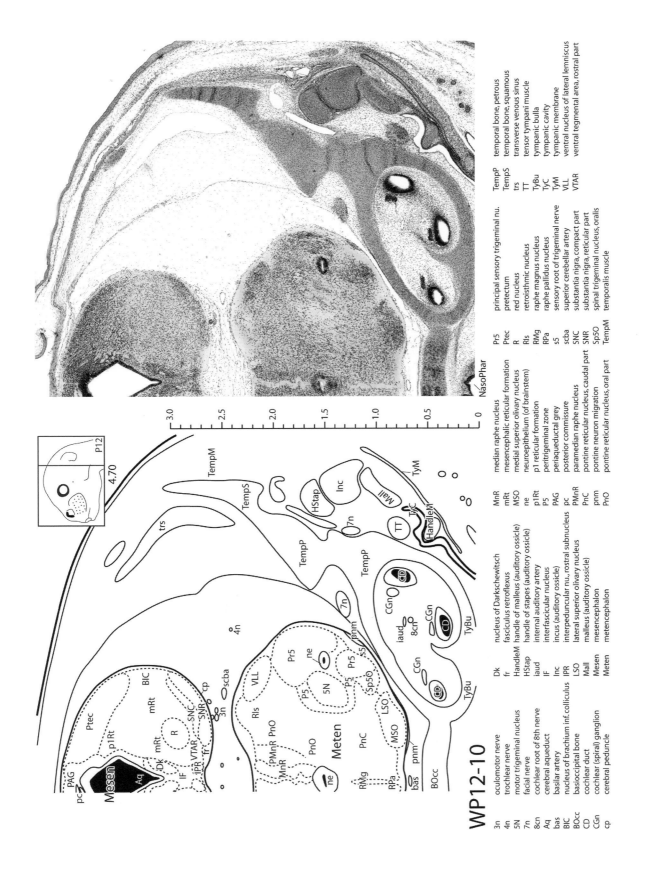

P12   4.70

NasoPhar

3n	oculomotor nerve	
4n	trochlear nerve	
5N	motor trigeminal nucleus	
7n	facial nerve	
8cn	cochlear root of 8th nerve	
Aq	cerebral aqueduct	
bas	basilar artery	
BIC	nucleus of brachium inf. colliculus	
BOcc	basiocciptal bone	
CD	cochlear duct	
CGn	cochlear (spiral) ganglion	
cp	cerebral peduncle	

Dk	nucleus of Darkschewitsch
fr	fasciculus retroflexus
HandleM	handle of malleus (auditory ossicle)
HStap	handle of stapes (auditory ossicle)
iaud	internal auditory artery
IF	interfascicular nucleus
Inc	incus (auditory ossicle)
IPR	interpeduncular nu., rostral subnucleus
LSO	lateral superior olivary nucleus
Mall	malleus (auditory ossicle)
Mesen	mesencephalon
Meten	metencephalon

MnR	median raphe nucleus
mRt	mesencephalic reticular formation
MSO	medial superior olivary nucleus
ne	neuroepithelium (of brainstem)
p1Rt	p1 reticular formation
P5	peritrigeminal zone
PAG	periaqueductal grey
pc	posterior commissure
PMnR	paramedian raphe nucleus
PnC	pontine reticular nucleus, caudal part
pnm	pontine neuron migration
PnO	pontine reticular nucleus, oral part

Pr5	principal sensory trigeminal nu.
Ptec	pretectum
R	red nucleus
Rls	retroisthmic nucleus
RMg	raphe magnus nucleus
RPa	raphe pallidus nucleus
s5	sensory root of trigeminal nerve
scba	superior cerebellar artery
SNC	substantia nigra, compact part
SNR	substantia nigra, reticular part
SpSO	spinal trigeminal nucleus, oralis
TempM	temporalis muscle

TempP	temporal bone, petrous
TempS	temporal bone, squamous
trs	transverse venous sinus
TT	tensor tympani muscle
TyBu	tympanic bulla
TyC	tympanic cavity
TyM	tympanic membrane
VLL	ventral nucleus of lateral lemniscus
VTAR	ventral tegmental area, rostral part

WP12-11

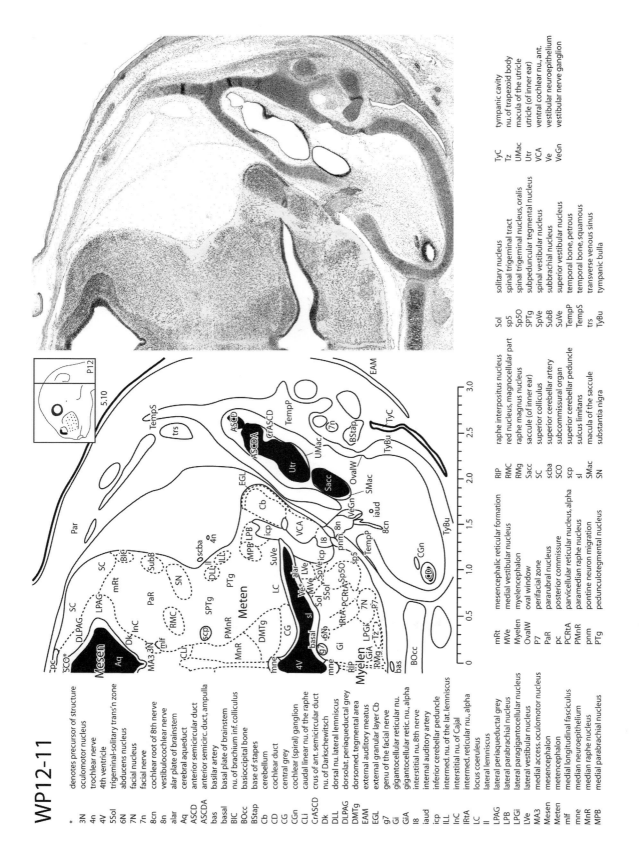

*	denotes precursor of structure		
3N	oculomotor nucleus		
4n	trochlear nerve		
4V	4th ventricle		
5Sol	trigeminal-solitary trans'n zone		
6N	abducens nucleus		
7N	facial nucleus		
7n	facial nerve		
8cn	cochlear root of 8th nerve		
8n	vestibulocochlear nerve		
alar	alar plate of brainstem		
Aq	cerebral aqueduct		
ASCD	anterior semicircular duct		
ASCDA	anterior semicirc. duct, ampulla		
bas	basilar artery		
basal	basal plate of brainstem		
BIC	nu. of brachium inf. colliculus		
BOcc	basioccipital bone		
BStap	base of stapes		
Cb	cerebellum		
CD	cochlear duct		
CG	central grey		
CGn	cochlear (spiral) ganglion		
CLi	caudal linear nu. of the raphe		
CrASCD	crus of ant. semicircular duct		
Dk	nu. of Darkschewitsch		
DLL	dorsal nu. lateral lemniscus		
DLPAG	dorsolat. periaqueductal grey		
DMTg	dorsomed. tegmental area		
EAM	external auditory meatus		
EGL	external granular layer Cb		
g7	genu of the facial nerve		
Gi	gigantocellular reticular nu.		
GiA	gigantocellular retic. nu., alpha		
I8	interstitial nu. 8th nerve		
iaud	internal auditory artery		
icp	inferior cerebellar peduncle		
ILL	intermed. nu. of the lat. lemniscus		
InC	interstitial nu. of Cajal		
IRtA	intermed. reticular nu., alpha		
LC	locus coeruleus		
ll	lateral lemniscus		
LPAG	lateral periaqueductal grey	RIP	raphe interpositus nucleus	Sol	solitary nucleus	TyC	tympanic cavity
LPB	lateral parabrachial nucleus	RMC	red nucleus, magnocellular part	sp5	spinal trigeminal tract	Tz	nu. of trapezoid body
LPGi	lateral paragigantocellular nucleus	RMg	raphe magnus nucleus	SPTg	subpeduncular tegmental nucleus	UMac	macula of the utricle
LVe	lateral vestibular nucleus	Sacc	saccule (of inner ear)	SpVe	spinal vestibular nucleus	Utr	utricle (of inner ear)
MA3	medial access. oculomotor nucleus	SC	superior colliculus	SubB	subbrachial nucleus	VCA	ventral cochlear nu., ant.
Mesen	mesencephalon	scba	superior cerebellar artery	SuVe	superior vestibular nucleus	Ve	vestibular neuroepithelium
Meten	metencephalon	SCO	subcommissural organ	TempP	temporal bone, petrous	VeGn	vestibular nerve ganglion
mlf	medial longitudinal fasciculus	scp	superior cerebellar peduncle	TempS	temporal bone, squamous		
mne	median neuroepithelium	sl	sulcus limitans	trs	transverse venous sinus		
MnR	median raphe nucleus	SMac	macula of the saccule	TyBu	tympanic bulla		
MPB	medial parabrachial nucleus	SN	substantia nigra				
mRt	mesencephalic reticular formation						
MVe	medial vestibular nucleus						
Myelen	myelencephalon						
OvalW	oval window						
P7	perifacial zone						
PaR	pararubral nucleus						
pc	posterior commissure						
PCRtA	parvicellular reticular nucleus, alpha						
PMnR	paramedian raphe nucleus						
pnm	pontine neuron migration						
PTg	pedunculotegmental nucleus						

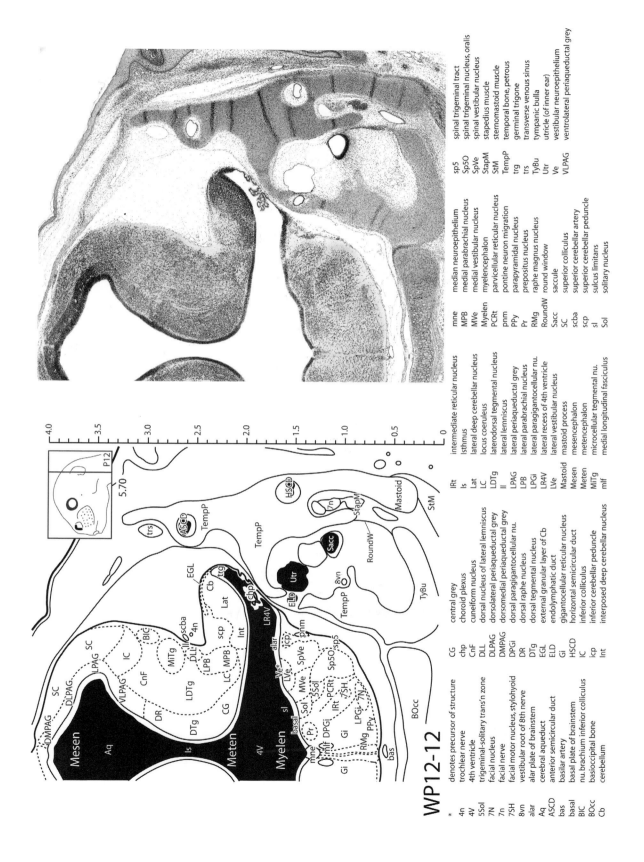

WP12-12

*	denotes precursor of structure		
4n	trochlear nerve	CG	central grey
4V	4th ventricle	chp	choroid plexus
5Sol	trigeminal-solitary trans'n zone	CnF	cuneiform nucleus
7N	facial nucleus	DLL	dorsal nucleus of lateral lemniscus
7n	facial nerve	DLPAG	dorsolateral periaqueductal grey
7SH	facial motor nucleus, stylohyoid	DMPAG	dorsomedial periaqueductal grey
8vn	vestibular root of 8th nerve	DPGi	dorsal paragigantocellular nu.
alar	alar plate of brainstem	DR	dorsal raphe nucleus
Aq	cerebral aqueduct	DTg	dorsal tegmental nucleus
ASCD	anterior semicircular duct	EGL	external granular layer of Cb
bas	basilar artery	ELD	endolymphatic duct
basal	basal plate of brainstem	Gi	gigantocellular reticular nucleus
BIC	nu. brachium inferior colliculus	HSCD	horizontal semicircular duct
BOcc	basioccipital bone	IC	inferior colliculus
Cb	cerebellum	icp	inferior cerebellar peduncle
		Int	interposed deep cerebellar nucleus

IRt	intermediate reticular nucleus	mne	median neuroepithelium
Is	isthmus	MPB	medial parabrachial nucleus
Lat	lateral deep cerebellar nucleus	MVe	medial vestibular nucleus
LC	locus coeruleus	Myelen	myelencephalon
LDTg	laterodorsal tegmental nucleus	PCRt	parvicellular reticular nucleus
ll	lateral lemniscus	pnm	pontine neuron migration
LPAG	lateral periaqueductal grey	PPy	parapyramidal nucleus
LPB	lateral parabrachial nucleus	Pr	prepositus nucleus
LPGi	lateral paragigantocellular nu.	RMg	raphe magnus nucleus
LR4V	lateral recess of 4th ventricle	RoundW	round window
LVe	lateral vestibular nucleus	Sacc	saccule
Mastoid	mastoid process	SC	superior colliculus
Mesen	mesencephalon	scba	superior cerebellar artery
Meten	metencephalon	scp	superior cerebellar peduncle
MiTg	microcellular tegmental nu.	sl	sulcus limitans
mlf	medial longitudinal fasciculus	Sol	solitary nucleus

sp5	spinal trigeminal tract
SpSO	spinal trigeminal nucleus, oralis
SpVe	spinal vestibular nucleus
StapM	stapedius muscle
StM	sternomastoid muscle
TempP	temporal bone, petrous
trg	germinal trigone
trs	transverse venous sinus
TyBu	tympanic bulla
Utr	utricle (of inner ear)
Ve	vestibular neuroepithelium
VLPAG	ventrolateral periaqueductal grey

WP12-13

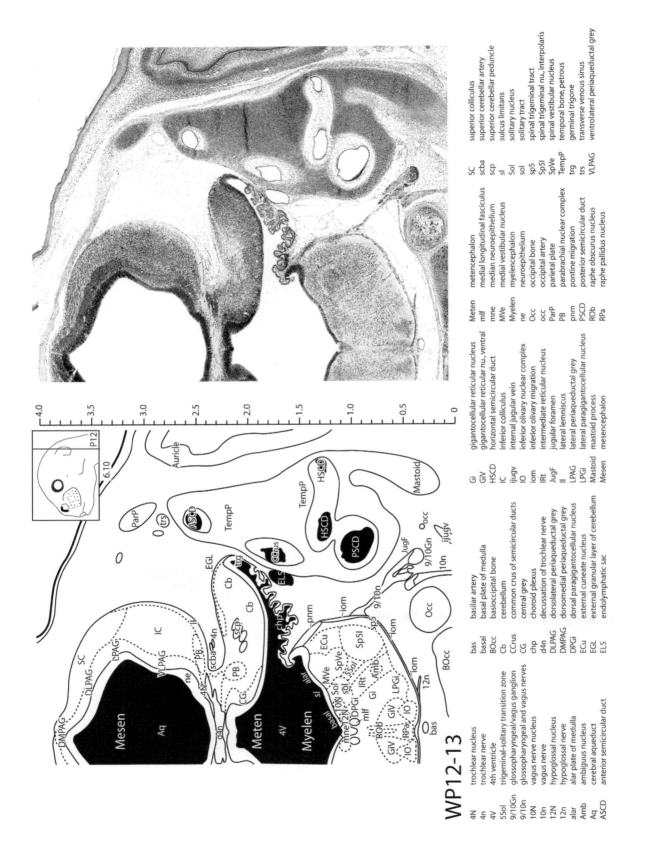

4N	trochlear nucleus	
4n	trochlear nerve	
4V	4th ventricle	
5Sol	trigeminal-solitary transition zone	
9/10Gn	glossopharyngeal/vagus ganglion	
9/10n	glossopharyngeal and vagus nerves	
10N	vagus nerve nucleus	
10n	vagus nerve	
12N	hypoglossal nucleus	
12n	hypoglossal nerve	
alar	alar plate of medulla	
Amb	ambiguus nucleus	
Aq	cerebral aqueduct	
ASCD	anterior semicircular duct	

bas	basilar artery	
basal	basal plate of medulla	
BOcc	basioccipital bone	
Cb	cerebellum	
CCrus	common crus of semicircular ducts	
CG	central grey	
chp	choroid plexus	
d4n	decussation of trochlear nerve	
DLPAG	dorsolateral periaqueductal grey	
DMPAG	dorsomedial periaqueductal grey	
DPGi	dorsal paragigantocellular nucleus	
ECu	external cuneate nucleus	
EGL	external granular layer of cerebellum	
ELS	endolymphatic sac	

Gi	gigantocellular reticular nucleus	
GiV	gigantocellular reticular nu., ventral	
HSCD	horizontal semicircular duct	
IC	inferior colliculus	
jjugv	internal jugular vein	
IO	inferior olivary nuclear complex	
iom	inferior olivary migration	
IRt	intermediate reticular nucleus	
JugF	jugular foramen	
ll	lateral lemniscus	
LPAG	lateral periaqueductal grey	
LPGi	lateral paragigantocellular nucleus	
Mastoid	mastoid process	
Mesen	mesencephalon	

Meten	metencephalon	
mlf	medial longitudinal fasciculus	
mme	median neuroepithelium	
MVe	medial vestibular nucleus	
Myelen	myelencephalon	
ne	neuroepithelium	
Occ	occipital bone	
occ	occipital artery	
ParP	parietal plate	
PB	parabrachial nuclear complex	
pnm	pontine migration	
PSCD	posterior semicircular duct	
ROb	raphe obscurus nucleus	
RPa	raphe pallidus nucleus	

SC	superior colliculus	
scba	superior cerebellar artery	
scp	superior cerebellar peduncle	
sl	sulcus limitans	
Sol	solitary nucleus	
sol	solitary tract	
sp5	spinal trigeminal tract	
Sp5I	spinal trigeminal nu., interpolaris	
SpVe	spinal vestibular nucleus	
TempP	temporal bone, petrous	
trg	germinal trigone	
trs	transverse venous sinus	
VLPAG	ventrolateral periaqueductal grey	

WP12-14

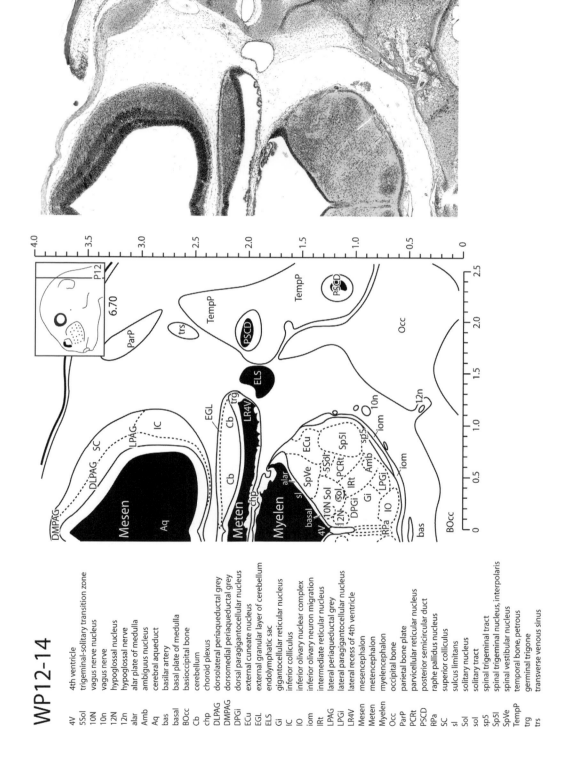

4V	4th ventricle
5Sol	trigeminal-solitary transition zone
10N	vagus nerve nucleus
10n	vagus nerve
12N	hypoglossal nucleus
12n	hypoglossal nerve
alar	alar plate of medulla
Amb	ambiguus nucleus
Aq	cerebral aqueduct
bas	basilar artery
basal	basal plate of medulla
BOcc	basioccipital bone
Cb	cerebellum
chp	choroid plexus
DLPAG	dorsolateral periaqueductal grey
DMPAG	dorsomedial periaqueductal grey
DPGi	dorsal paragigantocellular nucleus
ECu	external cuneate nucleus
EGL	external granular layer of cerebellum
ELS	endolymphatic sac
Gi	gigantocellular reticular nucleus
IC	inferior colliculus
IO	inferior olivary nuclear complex
iom	inferior olivary neuron migration
IRt	intermediate reticular nucleus
LPAG	lateral periaqueductal grey
LPGi	lateral paragigantocellular nucleus
LR4V	lateral recess of 4th ventricle
Mesen	mesencephalon
Meten	metencephalon
Myelen	myelencephalon
Occ	occipital bone
ParP	parietal bone plate
PCRt	parvicellular reticular nucleus
PSCD	posterior semicircular duct
RPa	raphe pallidus nucleus
SC	superior colliculus
sl	sulcus limitans
Sol	solitary nucleus
sol	solitary tract
sp5	spinal trigeminal tract
Sp5I	spinal trigeminal nucleus, interpolaris
SpVe	spinal vestibular nucleus
TempP	temporal bone, petrous
trg	germinal trigone
trs	transverse venous sinus

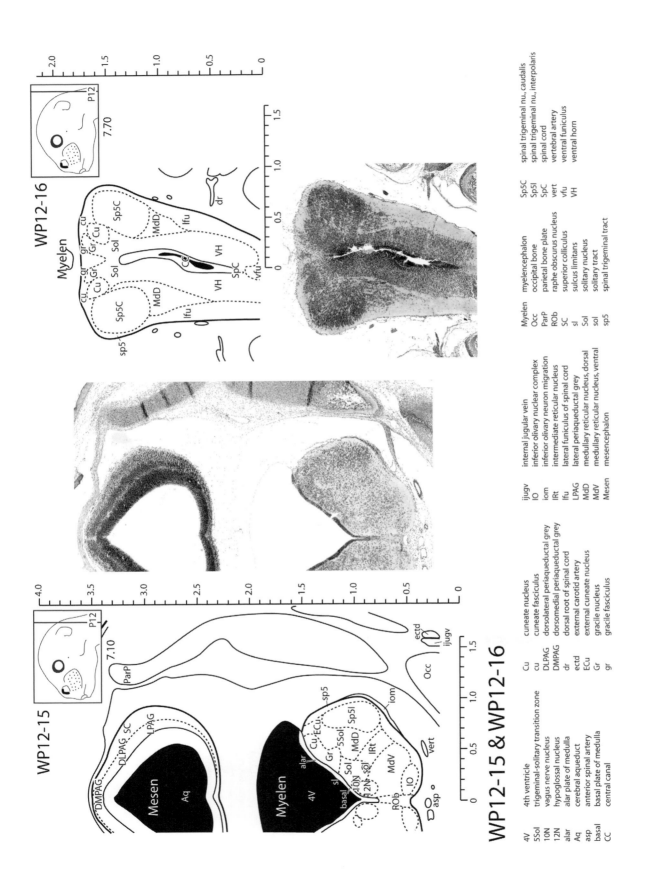

WP12-15 & WP12-16

WP12-15 — P12 — 7.10

Mesen, Aq, DMPAG, DLPAG, SC, LPAG, Myelen, 4V, ParP, Occ, ectd, ijugv, vert, iom, sp5, Sp5I, MdD, Cu, ECu, Gr, Sol, 5Sol, IRt, MdV, IO, ROb, asp, basal, alar, sl, 10N, 12N, CC

WP12-16 — Myelen — P12 — 7.70

cu, gr, Gr, Cu, Sp5C, Sol, MdD, lfu, VH, SpC, vfu, CC, dr, sp5

4V	4th ventricle	Cu	cuneate nucleus	ijugv	internal jugular vein	Myelen	myelencephalon	Sp5C	spinal trigeminal nu, caudalis

4V   4th ventricle
5Sol trigeminal-solitary transition zone
10N   vagus nerve nucleus
12N   hypoglossal nucleus
alar alar plate of medulla
Aq   cerebral aqueduct
asp   anterior spinal artery
basal basal plate of medulla
CC   central canal

Cu   cuneate nucleus
cu   cuneate fasciculus
DLPAG dorsolateral periaqueductal grey
DMPAG dorsomedial periaqueductal grey
dr   dorsal root of spinal cord
ectd external carotid artery
ECu   external cuneate nucleus
Gr   gracile nucleus
gr   gracile fasciculus

ijugv internal jugular vein
IO   inferior olivary nuclear complex
iom   inferior olivary neuron migration
IRt   intermediate reticular nucleus
lfu   lateral funiculus of spinal cord
LPAG lateral periaqueductal grey
MdD   medullary reticular nucleus, dorsal
MdV   medullary reticular nucleus, ventral
Mesen mesencephalon

Myelen myelencephalon
Occ   occipital bone
ParP parietal bone plate
ROb   raphe obscurus nucleus
SC   superior colliculus
sl   sulcus limitans
Sol   solitary nucleus
sol   solitary tract
sp5   spinal trigeminal tract

Sp5C spinal trigeminal nu, caudalis
Sp5I spinal trigeminal nu, interpolaris
SpC   spinal cord
vert vertebral artery
vfu   ventral funiculus
VH   ventral horn

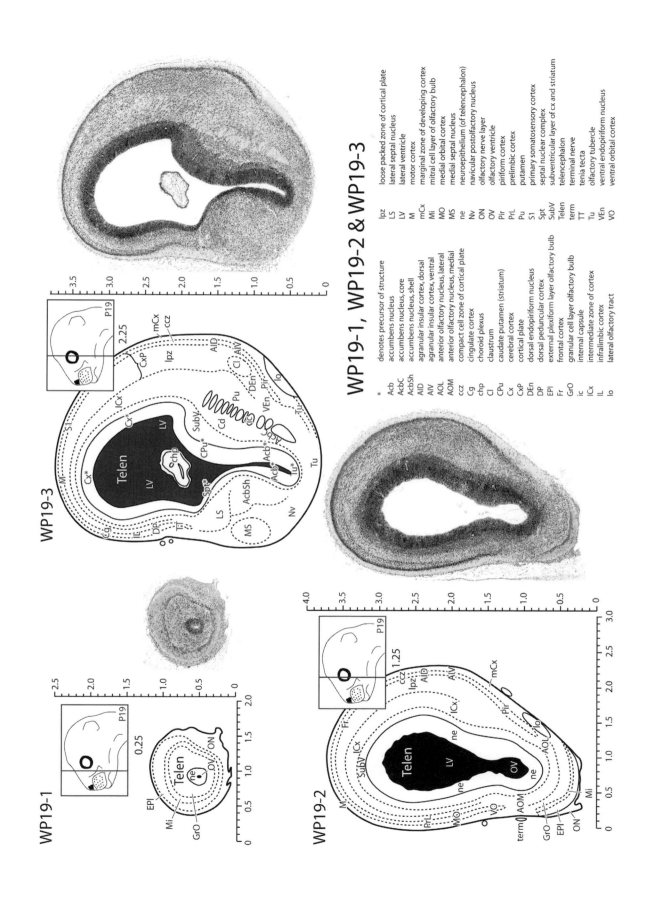

WP19-1, WP19-2 & WP19-3

*	denotes precursor of structure
Acb	accumbens nucleus
AcbC	accumbens nucleus, core
AcbSh	accumbens nucleus, shell
AID	agranular insular cortex, dorsal
AIV	agranular insular cortex, ventral
AOL	anterior olfactory nucleus, lateral
AOM	anterior olfactory nucleus, medial
ccz	compact cell zone of cortical plate
Cg	cingulate cortex
chp	choroid plexus
Cl	claustrum
CPu	caudate putamen (striatum)
Cx	cerebral cortex
CxP	cortical plate
DEn	dorsal endopiriform nucleus
DP	dorsal peduncular cortex
EPl	external plexiform layer olfactory bulb
Fr	frontal cortex
GrO	granular cell layer olfactory bulb
ic	internal capsule
ICx	intermediate zone of cortex
IL	infralimbic cortex
lo	lateral olfactory tract

lpz	loose packed zone of cortical plate
LS	lateral septal nucleus
LV	lateral ventricle
M	motor cortex
mCx	marginal zone of developing cortex
Mi	mitral cell layer of olfactory bulb
MO	medial orbital cortex
MS	medial septal nucleus
ne	neuroepithelium (of telencephalon)
Nv	navicular postolfactory nucleus
ON	olfactory nerve layer
OV	olfactory ventricle
Pir	piriform cortex
PrL	prelimbic cortex
Pu	putamen
S1	primary somatosensory cortex
Spt	septal nuclear complex
SubV	subventricular layer of cx and striatum
Telen	telencephalon
term	terminal nerve
TT	tenia tecta
Tu	olfactory tubercle
VEn	ventral endopiriform nucleus
VO	ventral orbital cortex

WP19-1

WP19-2

WP19-3

WP19-4

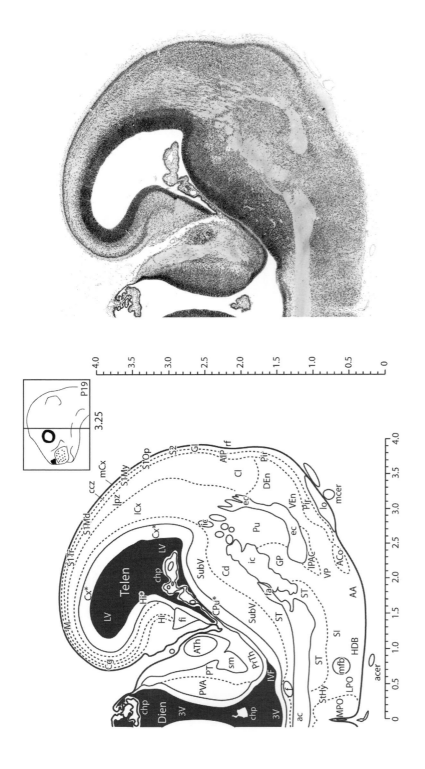

P19 3.25

*	denotes precursor of structure	Cl	claustrum	Hi	hippocampus	mfb	medial forebrain bundle	S2	secondary somatosensory cx
3V	3rd ventricle	CPu	caudate putamen (striatum)	ic	internal capsule	MPO	medial preoptic nucleus	SI	substantia innominata
AA	anterior amygdaloid area	Cx	cerebral cortex	ICx	intermediate zone of cortex	Pir	piriform cortex	sm	stria medullaris
ac	anterior commissure	DEn	dorsal endopiriform nucleus	IPAC	interstitial nu. post. limb anterior commissure	PrTh	prethalamus	ST	bed nuclei of stria terminalis
acer	anterior cerebral artery	Dien	diencephalon	IVF	interventricular foramen	PT	paratenial thalamic nucleus	StHy	striohypothalamic nucleus
ACo	anterior cortical amygdaloid area	ec	external capsule	lo	lateral olfactory tract	Pu	putamen	SubV	subventricular layer of cortex
AIP	agranular insular cortex, post.	f	fornix	LPO	lateral preoptic area	PVA	paraventricular thal. nucleus, ant.	Telen	telencephalon
ATh	anterior thalamic region	fa	fasciculus aberrans	lpz	loose packed zone of cortical plate	rf	rhinal fissure	VEn	ventral endopiriform nu.
ccz	compact cell zone of cortical plate	fi	fimbria of the hippocampus	LV	lateral ventricle	S1Md	prim. somatosens. cx, mandible	VP	ventral pallidum
Cd	caudate nucleus	Gl	granular insular cortex	M	motor cortex	S1My	prim. somatosens. cx, mystacial		
Cg	cingulate cortex	GP	globus pallidus	mcer	middle cerebral artery	S1Op	prim. somatosens. cx, ophthalm.		
chp	choroid plexus	HDB	nucleus horiz. limb diagonal band	mCx	marginal zone of developing cortex	S1Tr	prim. somatosens. cx, trunk		

WP19-5

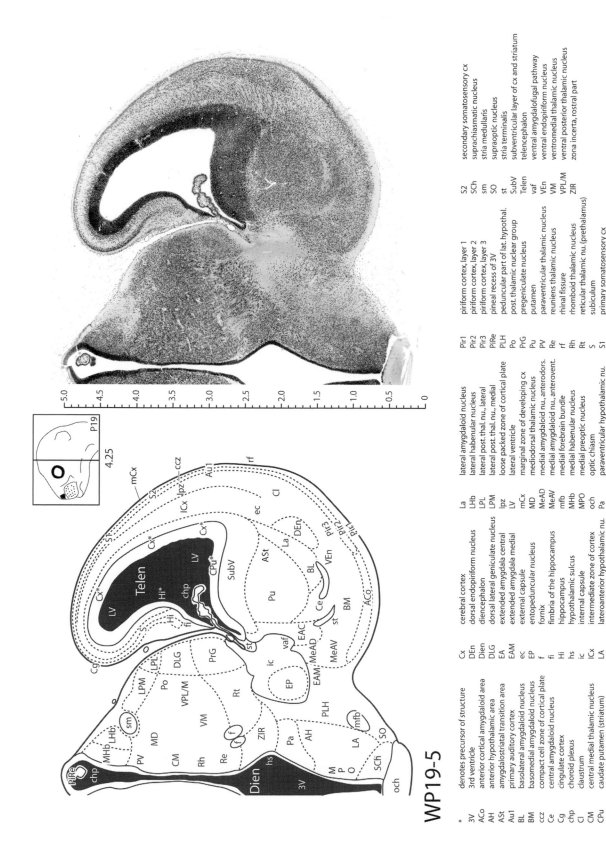

4.25

*	denotes precursor of structure	
3V	3rd ventricle	
ACo	anterior cortical amygdaloid area	
AH	anterior hypothalamic area	
ASt	amygdalostriatal transition area	
Au1	primary auditory cortex	
BL	basolateral amygdaloid nucleus	
BM	basomedial amygdaloid nucleus	
ccz	compact cell zone of cortical plate	
Ce	central amygdaloid nucleus	
Cg	cingulate cortex	
chp	choroid plexus	
Cl	claustrum	
CM	central medial thalamic nucleus	
CPu	caudate putamen (striatum)	

Cx	cerebral cortex	
DEn	dorsal endopiriform nucleus	
Dien	diencephalon	
DLG	dorsal lateral geniculate nucleus	
EA	extended amygdala central	
EAM	extended amygdala medial	
ec	external capsule	
EP	entopeduncular nucleus	
f	fornix	
fi	fimbria of the hippocampus	
Hi	hippocampus	
hs	hypothalamic sulcus	
ic	internal capsule	
ICx	intermediate zone of cortex	
LA	lateroanterior hypothalamic nu.	

La	lateral amygdaloid nucleus	
LHb	lateral habenular nucleus	
LPL	lateral post. thal. nu., lateral	
LPM	lateral post. thal. nu., medial	
lpz	loose packed zone of cortical plate	
LV	lateral ventricle	
mCx	marginal zone of developing cx	
MD	mediodorsal thalamic nucleus	
MeAD	medial amygdaloid nu., anterodors.	
MeAV	medial amygdaloid nu., anterovent.	
mfb	medial forebrain bundle	
MHb	medial habenular nucleus	
MPO	medial preoptic nucleus	
och	optic chiasm	
Pa	paraventricular hypothalamic nu.	

Pir1	piriform cortex, layer 1	
Pir2	piriform cortex, layer 2	
Pir3	piriform cortex, layer 3	
PiRe	pineal recess of 3V	
PLH	peduncular part of lat. hypothal.	
Po	post. thalamic nuclear group	
PrG	pregeniculate nucleus	
Pu	putamen	
PV	paraventricular thalamic nucleus	
Re	reuniens thalamic nucleus	
rf	rhinal fissure	
Rt	reticular thalamic nu. (prethalamus)	
S	subiculum	
S1	primary somatosensory cx	

S2	secondary somatosensory cx	
SCh	suprachiasmatic nucleus	
sm	stria medullaris	
SO	supraoptic nucleus	
st	stria terminalis	
SubV	subventricular layer of cx and striatum	
Telen	telencephalon	
vaf	ventral amygdalofugal pathway	
VEn	ventral endopiriform nucleus	
VM	ventromedial thalamic nucleus	
VPL/M	ventral posterior thalamic nucleus	
ZIR	zona incerta, rostral part	

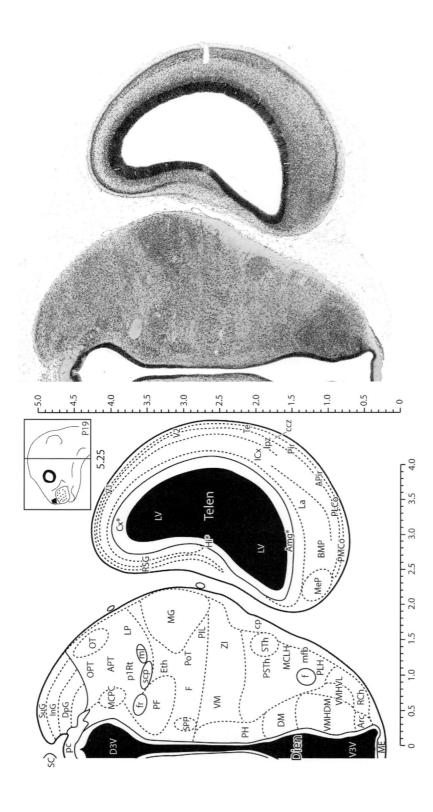

WP19-6

*	denotes precursor of structure	DpG	deep grey of superior colliculus	
Amg	amygdala	Eth	ethmoid thalamic nucleus	
APir	amygdalopiriform transition zone	F	nu. of the fields of Forel	
APT	anterior pretectal nucleus	f	fornix	
Arc	arcuate hypothalamic nucleus	fr	fasciculus retroflexus	
BMP	basomedial amygdaloid nu., post.	Hi	hippocampus	
ccz	compact cell zone of cortical plate	ICx	intermediate zone of cortex	
cp	cerebral peduncle	InG	intermediate grey layer of SC	
Cx	cerebral cortex	La	lateral amygdaloid nucleus	
D3V	dorsal 3rd ventricle	LP	lateral posterior thalamic nucleus	
Dien	diencephalon	lpz	loose packed zone of cortical plate	
DM	dorsomedial hypothalamic nucleus	LV	lateral ventricle	

MCLH	magnocell.nu. lat. hypothalamus	PH	posterior hypothalamic nucleus
MCPC	magnocell. nu. post. commissure	PIL	post. intralaminar thalamic nucleus
ME	median eminence	Pir	piriform cortex
MeP	medial amygdaloid nu., post.	PLCo	posterolateral cortical amyg. area
mfb	medial forebrain bundle	PLH	peduncular part of lateral hypothal.
MG	medial geniculate nucleus	PMCo	posteromed. cortical amygdala
ml	medial lemniscus	PoT	posterior thal. nu., triangular
OPT	olivary pretectal nucleus	PSTh	parasubthalamic nucleus
OT	nucleus of the optic tract	RCh	retrochiasmatic area
p1Rt	p1 reticular formation	RSG	retrosplenial granular cx
pc	posterior commissure	SC	superior colliculus
PF	parafascicular thalamic nucleus	scp	superior cerebellar peduncle

SPF	subparafascicular thal. nucleus
STh	subthalamic nucleus
SuG	superficial grey superior colliculus
Te	temporal cortex
Telen	telencephalon
V1	primary visual cortex
V2	secondary visual cortex
V3V	ventral third ventricle
VM	ventromedial thalamic nucleus
VMHDM	ventromed. hypothal. nu., dorsom.
VMHVL	ventromed. hypothal. nu., ventrolat.
ZI	zona incerta

WP19-7

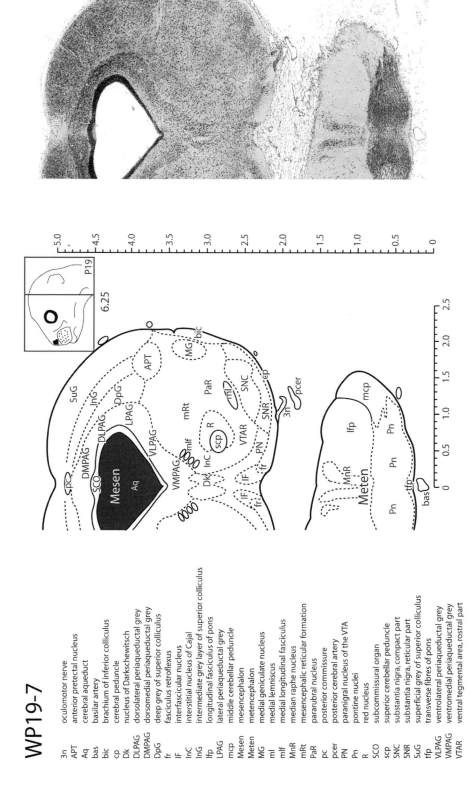

3n	oculomotor nerve
APT	anterior pretectal nucleus
Aq	cerebral aqueduct
bas	basilar artery
bic	brachium of inferior colliculus
cp	cerebral peduncle
Dk	nucleus of Darkschewitsch
DLPAG	dorsolateral periaqueductal grey
DMPAG	dorsomedial periaqueductal grey
DpG	deep grey of superior colliculus
fr	fasciculus retroflexus
IF	interfascicular nucleus
InC	interstitial nucleus of Cajal
InG	intermediate grey layer of superior colliculus
lfp	longitudinal fasciculus of pons
LPAG	lateral periaqueductal grey
mcp	middle cerebellar peduncle
Mesen	mesencephalon
Meten	metencephalon
MG	medial geniculate nucleus
ml	medial lemniscus
mlf	medial longitudinal fasciculus
MnR	median raphe nucleus
mRt	mesencephalic reticular formation
PaR	pararubral nucleus
pc	posterior commissure
pcer	posterior cerebral artery
PN	paranigral nucleus of the VTA
Pn	pontine nuclei
R	red nucleus
SCO	subcommissural organ
scp	superior cerebellar peduncle
SNC	substantia nigra, compact part
SNR	substantia nigra, reticular part
SuG	superficial grey of superior colliculus
tfp	transverse fibres of pons
VLPAG	ventrolateral periaqueductal grey
VMPAG	ventromedial periaqueductal grey
VTAR	ventral tegmental area, rostral part

WP19-8

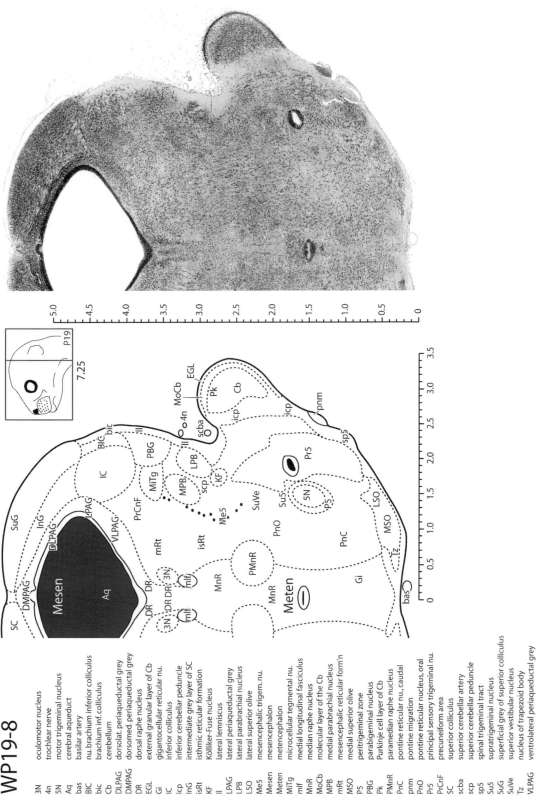

3N	oculomotor nucleus
4n	trochlear nerve
5N	motor trigeminal nucleus
Aq	cerebral aqueduct
bas	basilar artery
BIC	nu. brachium inferior colliculus
bic	brachium inf. colliculus
Cb	cerebellum
DLPAG	dorsolat. periaqueductal grey
DMPAG	dorsomed. periaqueductal grey
DR	dorsal raphe nucleus
EGL	external granular layer of Cb
Gi	gigantocellular reticular nu.
IC	inferior colliculus
icp	inferior cerebellar peduncle
InG	intermediate grey layer of SC
isRt	isthmic reticular formation
KF	Kölliker-Fuse nucleus
ll	lateral lemniscus
LPAG	lateral periaqueductal grey
LPB	lateral parabrachial nucleus
LSO	lateral superior olive
Me5	mesencephalic trigem. nu.
Mesen	mesencephalon
Meten	metencephalon
MiTg	microcellular tegmental nu.
mlf	medial longitudinal fasciculus
MnR	median raphe nucleus
MoCb	molecular layer of the Cb
MPB	medial parabrachial nucleus
mRt	mesencephalic reticular form'n
MSO	medial superior olive
P5	peritrigeminal zone
PBG	parabigeminal nucleus
Pk	Purkinje cell layer of Cb
PMnR	paramedian raphe nucleus
PnC	pontine reticular nu., caudal
pnm	pontine migration
PnO	pontine reticular nucleus, oral
Pr5	principal sensory trigeminal nu.
PrCnF	precuneiform area
SC	superior colliculus
scba	superior cerebellar artery
scp	superior cerebellar peduncle
sp5	spinal trigeminal tract
Su5	supratrigeminal nucleus
SuG	superficial grey of superior colliculus
SuVe	superior vestibular nucleus
Tz	nucleus of trapezoid body
VLPAG	ventrolateral periaqueductal grey

WP19-9

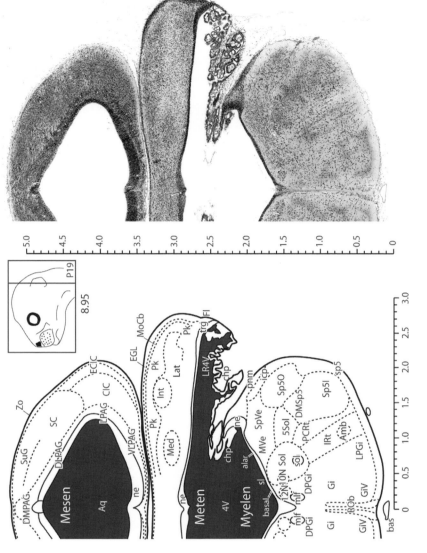

8.95 P19

WP19-10 & WP19-11

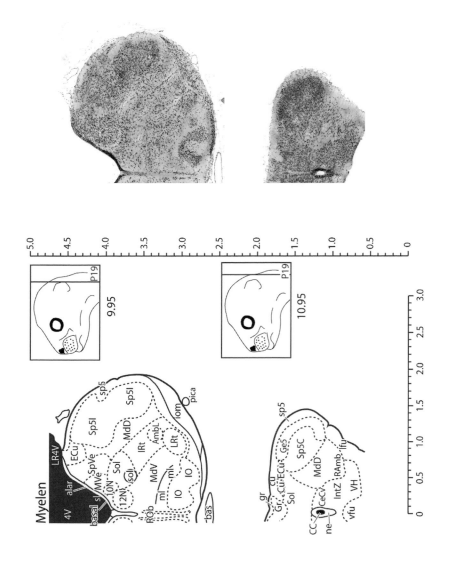

4V	4th ventricle
10N	vagus nerve motor nucleus
12N	hypoglossal nucleus
alar	alar plate of medulla
AmbL	ambiguus nucleus, loose part
bas	basilar artery
basal	basal plate of medulla
CC	central canal of spinal cord
CeCv	central cervical nucleus
Cu	cuneate nucleus
cu	cuneate fasciculus
ECu	external cuneate nucleus
Ge5	gelatinous layer of caudal spinal trigeminal nucleus
Gr	gracile nucleus
gr	gracile fasciculus
IntZ	intermediate zone of spinal cord
IO	inferior olivary nuclear complex
iom	inferior olivary migration
IRt	intermediate reticular nucleus
lfu	lateral funiculus of spinal cord white matter
LR4V	lateral recess of 4th ventricle
LRt	lateral reticular nucleus
MdD	medullary reticular nucleus, dorsal
MdV	medullary reticular nucleus, ventral
ml	medial lemniscus
MVe	medial vestibular nucleus
Myelen	myelencephalon
ne	neuroepithelium
pica	posterior inferior cerebellar artery
RAmb	retroambiguus nucleus
ROb	raphe obscurus nucleus
sl	sulcus limitans
Sol	solitary nucleus
sol	solitary tract
sp5	spinal trigeminal tract
Sp5C	spinal trigeminal nucleus, caudalis
Sp5I	spinal trigeminal nucleus, interpolaris
SpVe	spinal vestibular nucleus
vfu	ventral funiculus of spinal cord white matter
VH	ventral horn of spinal cord

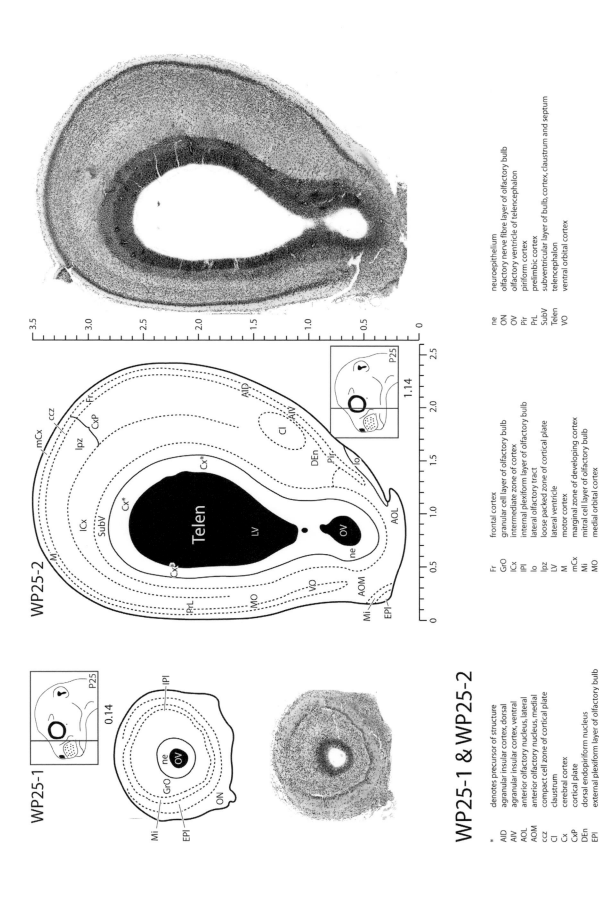

WP25-1 & WP25-2

WP25-1 0.14 P25

WP25-2 1.14 P25

Scale bars: 0 0.5 1.0 1.5 2.0 2.5 ; 0 0.5 1.0 1.5 2.0 2.5 3.0 3.5

Labels (WP25-1): Mi, EPI, GrO, ne, OV, ON, IPl

Labels (WP25-2): mCx, ccz, Fr, CxP, lpz, AID, AIV, CI, DEn, Pir, ICx, Cx*, SubV, Io, M, Cx*, Telen, LV, ne, OV, AOL, PrL, MO, VO, AOM, Mi, EPI

* denotes precursor of structure
AID — agranular insular cortex, dorsal
AIV — agranular insular cortex, ventral
AOL — anterior olfactory nucleus, lateral
AOM — anterior olfactory nucleus, medial
ccz — compact cell zone of cortical plate
CI — claustrum
Cx — cerebral cortex
CxP — cortical plate
DEn — dorsal endopiriform nucleus
EPI — external plexiform layer of olfactory bulb

Fr — frontal cortex
GrO — granular cell layer of olfactory bulb
ICx — intermediate zone of cortex
IPl — internal plexiform layer of olfactory bulb
Io — lateral olfactory tract
lpz — loose packed zone of cortical plate
LV — lateral ventricle
M — motor cortex
mCx — marginal zone of developing cortex
Mi — mitral cell layer of olfactory bulb
MO — medial orbital cortex

ne — neuroepithelium
ON — olfactory nerve fibre layer of olfactory bulb
OV — olfactory ventricle of telencephalon
Pir — piriform cortex
PrL — prelimbic cortex
SubV — subventricular layer of bulb, cortex, claustrum and septum
Telen — telencephalon
VO — ventral orbital cortex

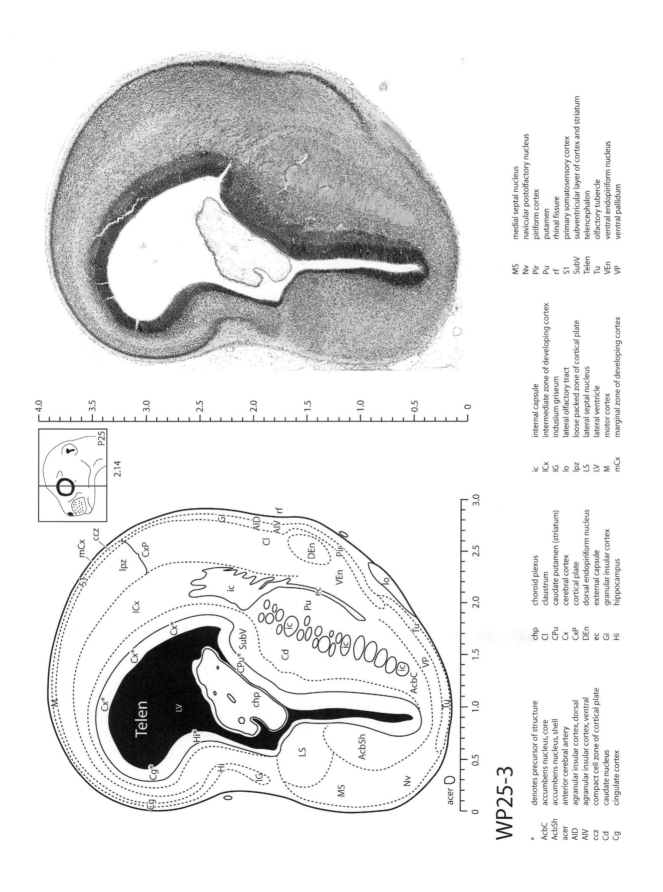

WP25-3

P25

2.14

acer

*	denotes precursor of structure	chp	choroid plexus
AcbC	accumbens nucleus, core	Cl	claustrum
AcbSh	accumbens nucleus, shell	CPu	caudate putamen (striatum)
acer	anterior cerebral artery	Cx	cerebral cortex
AID	agranular insular cortex, dorsal	CxP	cortical plate
AIV	agranular insular cortex, ventral	DEn	dorsal endopiriform nucleus
ccz	compact cell zone of cortical plate	ec	external capsule
Cd	caudate nucleus	GI	granular insular cortex
Cg	cingulate cortex	Hi	hippocampus

ic	internal capsule	MS	medial septal nucleus
ICx	intermediate zone of developing cortex	Nv	navicular postolfactory nucleus
IG	indusium griseum	Pir	piriform cortex
lo	lateral olfactory tract	Pu	putamen
lpz	loose packed zone of cortical plate	rf	rhinal fissure
LS	lateral septal nucleus	S1	primary somatosensory cortex
LV	lateral ventricle	SubV	subventricular layer of cortex and striatum
M	motor cortex	Telen	telencephalon
mCx	marginal zone of developing cortex	Tu	olfactory tubercle
		VEn	ventral endopiriform nucleus
		VP	ventral pallidum

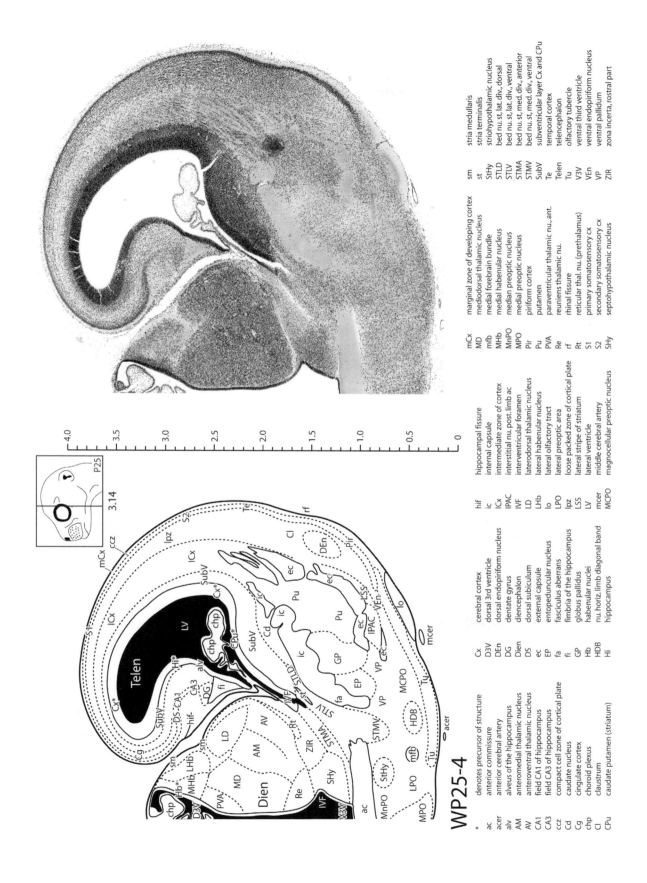

WP25-4

3.14

P25

*	denotes precursor of structure	Cx	cerebral cortex
ac	anterior commissure	D3V	dorsal 3rd ventricle
acer	anterior cerebral artery	DEn	dorsal endopiriform nucleus
alv	alveus of the hippocampus	DG	dentate gyrus
AM	anteromedial thalamic nucleus	Dien	diencephalon
AV	anteroventral thalamic nucleus	DS	dorsal subiculum
CA1	field CA1 of hippocampus	ec	external capsule
CA3	field CA3 of hippocampus	EP	entopeduncular nucleus
ccz	compact cell zone of cortical plate	fa	fasciculus aberrans
Cd	caudate nucleus	fi	fimbria of the hippocampus
Cg	cingulate cortex	GP	globus pallidus
chp	choroid plexus	Hb	habenular nuclei
Cl	claustrum	HDB	nu. horiz. limb diagonal band
CPu	caudate putamen (striatum)	Hi	hippocampus

hif	hippocampal fissure	mCx	marginal zone of developing cortex	sm	stria medullaris
ic	internal capsule	MD	mediodorsal thalamic nucleus	st	stria terminalis
ICx	intermediate zone of cortex	mfb	medial forebrain bundle	SthY	striohypothalamic nucleus
IPAC	interstitial nu. post. limb ac	MHb	medial habenular nucleus	STLD	bed nu. st. lat. div, dorsal
IVF	interventricular foramen	MnPO	median preoptic nucleus	STLV	bed nu. st, lat. div, ventral
LD	laterodorsal thalamic nucleus	MPO	medial preoptic nucleus	STMA	bed nu. st. med. div, anterior
LHb	lateral habenular nucleus	Pir	piriform cortex	STMV	bed nu. st, med. div, ventral
lo	lateral olfactory tract	Pu	putamen	SubV	subventricular layer Cx and CPu
LPO	lateral preoptic area	PVA	paraventricular thalamic nu., ant.	Te	temporal cortex
lpz	loose packed zone of cortical plate	Re	reuniens thalamic nu.	Telen	telencephalon
LSS	lateral stripe of striatum	rf	rhinal fissure	Tu	olfactory tubercle
LV	lateral ventricle	Rt	reticular thal. nu. (prethalamus)	V3V	ventral third ventricle
mcer	middle cerebral artery	S1	primary somatosensory cx	VEn	ventral endopiriform nucleus
MCPO	magnocellular preoptic nucleus	S2	secondary somatosensory cx	VP	ventral pallidum
		SHy	septohypothalamic nucleus	ZIR	zona incerta, rostral part

4.0 3.5 3.0 2.5 2.0 1.5 1.0 0.5 0

WP25-5

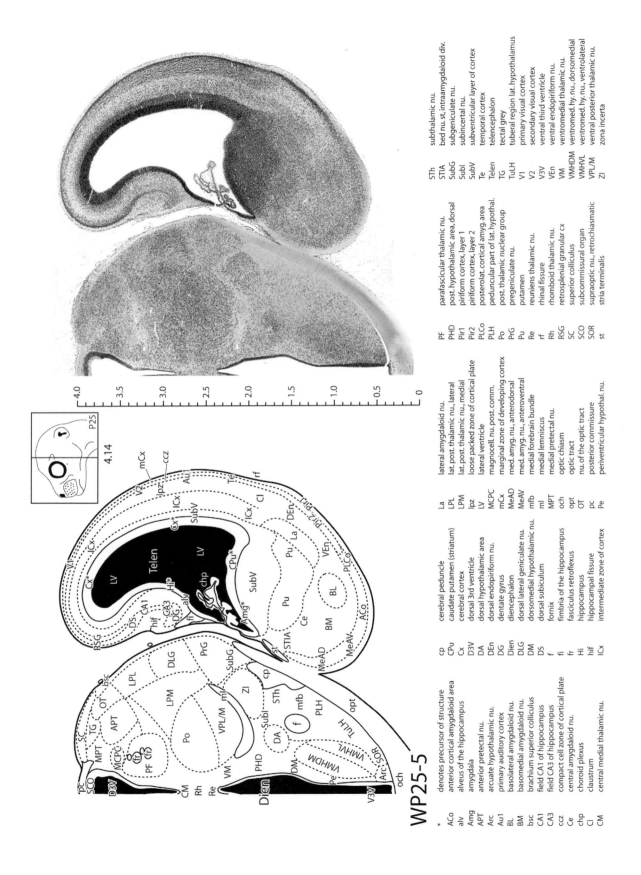

P25

4.14

4.0
3.5
3.0
2.5
2.0
1.5
1.0
0.5
0

*		denotes precursor of structure
ACo		anterior cortical amygdaloid area
alv		alveus of the hippocampus
Amg		amygdala
APT		anterior pretectal nu.
Arc		arcuate hypothalamic nu.
Au1		primary auditory cortex
BL		basolateral amygdaloid nu.
BM		basomedial amygdaloid nu.
bsc		brachium superior colliculus
CA1		field CA1 of hippocampus
CA3		field CA3 of hippocampus
ccz		compact cell zone of cortical plate
Ce		central amygdaloid nu.
chp		choroid plexus
Cl		claustrum
CM		central medial thalamic nu.
cp		cerebral peduncle
CPu		caudate putamen (striatum)
Cx		cerebral cortex
D3V		dorsal 3rd ventricle
DA		dorsal hypothalamic area
DEn		dorsal endopiriform nu.
DG		dentate gyrus
Dien		diencephalon
DLG		dorsal lateral geniculate nu.
DM		dorsomedial hypothalamic nu.
DS		dorsal subiculum
f		fornix
fi		fimbria of the hippocampus
fr		fasciculus retroflexus
Hi		hippocampus
hif		hippocampal fissure
ICx		intermediate zone of cortex
La		lateral amygdaloid nu.
LPL		lat. post. thalamic nu, lateral
LPM		lat. post. thalamic nu, medial
lpz		loose packed zone of cortical plate
LV		lateral ventricle
MCPC		magnocell. nu. post. comm.
mCx		marginal zone of developing cortex
MeAD		med. amyg. nu., anterodorsal
MeAV		med. amyg. nu., anteroventral
mfb		medial forebrain bundle
ml		medial lemniscus
MPT		medial pretectal nu.
och		optic chiasm
opt		optic tract
OT		nu. of the optic tract
pc		posterior commissure
Pe		periventricular hypothal. nu.
PF		parafascicular thalamic nu.
PHD		post. hypothalamic area, dorsal
Pir1		piriform cortex, layer 1
Pir2		piriform cortex, layer 2
PLCo		posterolat. cortical amyg. area
PLH		peduncular part of lat. hypothal.
Po		post. thalamic nuclear group
PrG		pregeniculate nu.
Pu		putamen
Re		reuniens thalamic nu.
rf		rhinal fissure
Rh		rhomboid thalamic nu.
RSG		retrosplenial granular cx
SC		superior colliculus
SCO		subcommissural organ
SOR		supraoptic nu., retrochiasmatic
st		stria terminalis
STh		subthalamic nu.
STIA		bed nu. st, intraamygdaloid div.
SubG		subgeniculate nu.
SubI		subincertal nu.
SubV		subventricular layer of cortex
Te		temporal cortex
Telen		telencephalon
TG		tectal grey
TuLH		tuberal region lat. hypothalamus
V1		primary visual cortex
V2		secondary visual cortex
V3V		ventral third ventricle
VEn		ventral endopiriform nu.
VM		ventromedial thalamic nu.
VMHDM		ventromed.hy. nu., dorsomedial
VMHVL		ventromed.hy.nu., ventrolateral
VPL/M		ventral posterior thalamic nu.
ZI		zona incerta

WP25-6

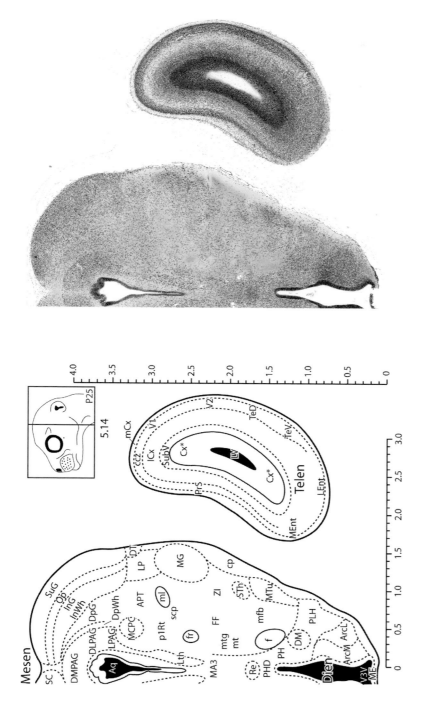

5.14   P25

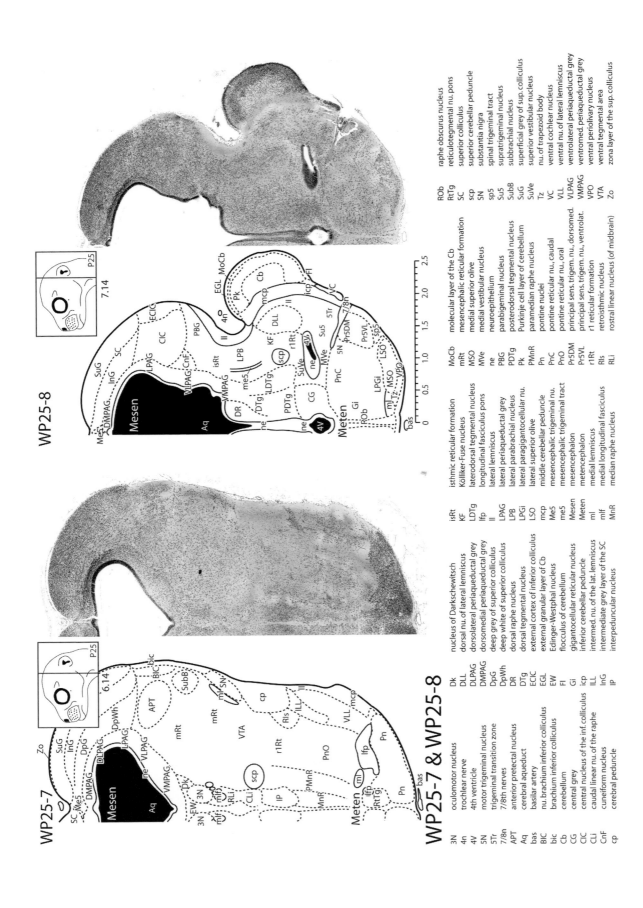

WP25-7

WP25-8

7.14

6.14

P25

WP25-7 & WP25-8

3N	oculomotor nucleus	Dk	nucleus of Darkschewitsch	isRt	isthmic reticular formation	MoCb	molecular layer of the Cb
4n	trochlear nerve	DLL	dorsal nu. of lateral lemniscus	KF	Kölliker-Fuse nucleus	mRt	mesencephalic reticular formation
4V	4th ventricle	DLPAG	dorsolateral periaqueductal grey	LDTg	laterodorsal tegmental nucleus	MSO	medial superior olive
5N	motor trigeminal nucleus	DMPAG	dorsomedial periaqueductal grey	lfp	longitudinal fasciculus pons	MVe	medial vestibular nucleus
5Tr	trigeminal transition zone	DpG	deep grey of superior colliculus	ll	lateral lemniscus	ne	neuroepithelium
7/8n	7/8th nerves	DpWh	deep white of superior colliculus	LPAG	lateral periaqueductal grey	PBG	parabigeminal nucleus
APT	anterior pretectal nucleus	DR	dorsal raphe nucleus	LPB	lateral parabrachial nucleus	PDTg	posterodorsal tegmental nucleus
Aq	cerebral aqueduct	DTg	dorsal tegmental nucleus	LPGi	lateral paragigantocellular nu.	Pk	Purkinje cell layer of cerebellum
bas	basilar artery	ECIC	external cortex of inferior colliculus	LSO	lateral superior olive	PMnR	paramedian raphe nucleus
BIC	nu. brachium inferior colliculus	EGL	external granular layer of Cb	mcp	middle cerebellar peduncle	Pn	pontine nuclei
bic	brachium inferior colliculus	EW	Edinger-Westphal nucleus	Me5	mesencephalic trigeminal nucleus	PnC	pontine reticular nu., caudal
Cb	cerebellum	Fl	flocculus of cerebellum	me5	mesencephalic trigeminal tract	PnO	pontine reticular nu., oral
CG	central grey	Gi	gigantocellular reticular nucleus	Mesen	mesencephalon	Pr5DM	principal sens. trigem. nu., dorsomed.
CIC	central nucleus of the inf. colliculus	icp	inferior cerebellar peduncle	Meten	metencephalon	Pr5VL	principal sens. trigem. nu., ventrolat.
CLi	caudal linear nu. of the raphe	ILL	intermed. nu. of the lat. lemniscus	ml	medial lemniscus	r1Rt	r1 reticular formation
CnF	cuneiform nucleus	InG	intermediate grey layer of the SC	mlf	medial longitudinal fasciculus	RIs	retroisthmic nucleus
cp	cerebral peduncle	IP	interpeduncular nucleus	MnR	median raphe nucleus	RLi	rostral linear nucleus (of midbrain)

ROb	raphe obscurus nucleus
RtTg	reticulotegmental nu. pons
SC	superior colliculus
scp	superior cerebellar peduncle
SN	substantia nigra
sp5	spinal trigeminal tract
Su5	supratrigeminal nucleus
SubB	subbrachial nucleus
SuG	superficial grey of sup. colliculus
SuVe	superior vestibular nucleus
Tz	nu. of trapezoid body
VC	ventral cochlear nucleus
VLL	ventral nu. of lateral lemniscus
VLPAG	ventrolateral periaqueductal grey
VMPAG	ventromed. periaqueductal grey
VPO	ventral periolivary nucleus
VTA	ventral tegmental area
Zo	zona layer of the sup. colliculus

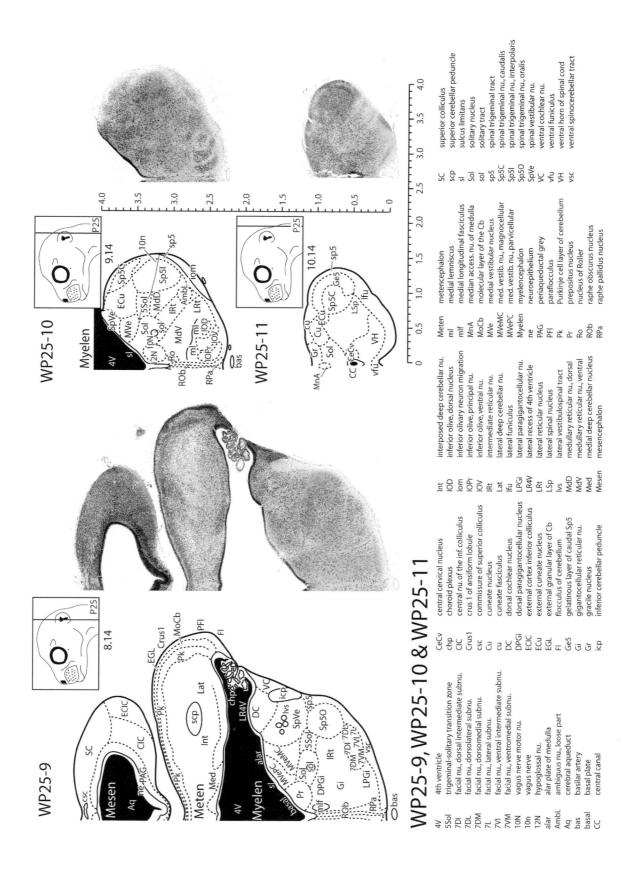

WP25-9, WP25-10 & WP25-11

4V	4th ventricle	CeCv	central cervical nucleus	Int	interposed deep cerebellar nu.	Meten	metencephalon	SC	superior colliculus
5Sol	trigeminal-solitary transition zone	chp	choroid plexus	IOD	inferior olive, dorsal nucleus	ml	medial lemniscus	scp	superior cerebellar peduncle
7DI	facial nu., dorsal intermediate subnu.	CIC	central nu. of the inf. colliculus	iom	inferior olivary neuron migration	mlf	medial longitudinal fasciculus	sl	sulcus limitans
7DL	facial nu., dorsolateral subnu.	Crus1	crus 1 of ansiform lobule	IOPr	inferior olive, principal nu.	MnA	median access. nu. of medulla	Sol	solitary nucleus
7DM	facial nu., dorsomedial subnu.	csc	commissure of superior colliculus	IOV	inferior olive, ventral nu.	MoCb	molecular layer of the Cb	sol	solitary tract
7L	facial nu., lateral subnu.	Cu	cuneate nucleus	IRt	intermediate reticular nu.	MVe	medial vestibular nucleus	sp5	spinal trigeminal tract
7VI	facial nu., ventral intermediate subnu.	cu	cuneate fasciculus	Lat	lateral deep cerebellar nu.	MVeMC	med. vestib. nu., magnocellular	Sp5C	spinal trigeminal nu., caudalis
7VM	facial nu., ventromedial subnu.	DC	dorsal cochlear nucleus	lfu	lateral funiculus	MVePC	med. vestib. nu., parvicellular	Sp5I	spinal trigeminal nu., interpolaris
10N	vagus nerve motor nu.	DPGi	dorsal paragigantocellular nucleus	LPGi	lateral paragigantocellular nu.	Myelen	myelencephalon	Sp5O	spinal trigeminal nu., oralis
10n	vagus nerve	ECIC	external cortex inferior colliculus	LR4V	lateral recess of 4th ventricle	ne	neuroepithelium	SpVe	spinal vestibular nu.
12N	hypoglossal nu.	ECu	external cuneate nucleus	LRt	lateral reticular nucleus	PAG	periaqueductal grey	VC	ventral cochlear nu.
alar	alar plate of medulla	EGL	external granular layer of Cb	LSp	lateral spinal nucleus	PFI	paraflocculus	vfu	ventral funiculus
AmbL	ambiguus nu., loose part	Fl	flocculus of cerebellum	lvs	lateral vestibulospinal tract	Pk	Purkinje cell layer of cerebellum	VH	ventral horn of spinal cord
Aq	cerebral aqueduct	Ge5	gelatinous layer of caudal Sp5	MdD	medullary reticular nu., dorsal	Pr	prepositus nucleus	vsc	ventral spinocerebellar tract
bas	basilar artery	Gi	gigantocellular reticular nu.	MdV	medullary reticular nu., ventral	Ro	nucleus of Roller		
basal	basal plate	Gr	gracile nucleus	Med	medial deep cerebellar nucleus	ROb	raphe obscurus nucleus		
CC	central canal	icp	inferior cerebellar peduncle	Mesen	mesencephalon	RPa	raphe pallidus nucleus		

List of structures and abbreviations

A

A1 noradrenaline cells **A1**
A2 noradrenaline cells **A2**
A5 noradrenaline cells **A5**
A6 noradrenaline cells **A6**
A7 noradrenaline cells **A7**
A11 dopamine cells **A11**
A13 dopamine cells **A13**
abducens nerve **6n**
abducens nucleus **6N**
accessory nerve **11n**
accessory nerve nucleus **11N**
accessory neurosecretory nucleus **ANS**
accessory olfactory bulb **AOB**
accessory trigeminal motor nucleus **ASCS**
accumbens nucleus **Acb**
  core **AcbC**
  shell **AcbSh**
agranular insular cortex **AI**
  dorsal part **AID**
  posterior part **AIP**
  ventral part **AIV**
alar part of orbital bone **AOrb**
alar plate of brainstem neuroepithelium **alar**
alisphenoid bone **ASph**
alveus of the hippocampus **alv**
ambiguus nucleus **Amb**
  compact part **AmbC**
  loose part **AmbL**
amygdala **Amg**
amygdalohippocampal area **AHi**
  anterior part **AHiA**
  anterolateral part **AHiAL**
  posterior part **AHiP**
amygdaloid fissure **af**

amygdalopiriform transition area **APir**
amygdalostriatal transition area **ASt**
ansoparamedian fissure of the cerebellum **apmf**
anterior amygdaloid area **AA**
  dorsal part **AAD**
  ventral part **AAV**
anterior cerebral artery **acer**
anterior commissure **ac**
  anterior limb **aca**
  intrabulbar limb **aci**
  posterior limb **acp**
anterior cortical amygdaloid area **ACo**
anterior hypothalamic area **AH**
  anterior part **AHA**
  central part **AHC**
  dorsal part **AHD**
  posterior part **AHP**
anterior lacerated foramen **ALF**
anterior lobe of pituitary gland **APit**
anterior lobule of cerebellum **Ant**
anterior olfactory area (nucleus) **AO**
  dorsal part **AOD**
  external part **AOE**
  lateral part **AOL**
  medial part **AOM**
  posterior part **AOP**
  ventral part **AOV**
  ventroposterior part **AOVP**
anterior pretectal nucleus **APT**
  dorsal part **APTD**
  ventral part **APTV**
anterior semicircular duct **ASCD**
anterior semicircular duct, ampulla **ASCDA**
anterior spinal artery **asp**
anterior thalamic region **ATh**
anterior ventral periventricular nucleus **AVPe**

anterior ventral thalamic nucleus **AV**
   dorsomedial part **AVDM**
   ventrolateral part **AVVL**
anterodorsal thalamic nucleus **AD**
anteromedial thalamic nucleus **AM**
   ventral part **AMV**
arcuate hypothalamic nucleus **Arc**
   dorsal part **ArcD**
   lateral part **ArcL**
   lateroposterior part **ArcLP**
   medial **ArcM**
   medioposterior part **ArcMP**
area postrema **AP**
ascending fibres of facial nerve **asc7**
atlas (first cervical vertebra) **Atlas**
auditory cortex **Au**
auditory neuroepithelium **Aud**
auditory ossicles (malleus, incus, stapes) **audos**
auriculotemporal nerve **aute**
axis (second cervical vertebra) **Axis**

B

B9 serotonin cells **B9**
Barrington's nucleus **Bar**
basal nucleus (of Meynert) **B**
basal plate of brainstem neuroepith. **basal**
base of stapes **BStap**
basilar artery **bas**
basioccipital bone **BOcc**
basisphenoid bone **BSph**
basolateral amygdaloid nucleus **BL**
   anterior part **BLA**
   posterior part **BLP**
   ventral part **BLV**
basomedial amygdaloid nucleus **BM**
   anterior part **BMA**
   posterior part **BMP**
   ventral part **BMV**
bed nucleus of accessory olfactory tract **BAOT**
bed nucleus of the stria terminalis **ST**
   fusiform part **Fu**
   intra-amygdaloid part **STIA**
   lateral division, dorsal part **STLD**
   lateral division, juxtacapsular part **STLJ**
   lateral division, posterior part **STLP**
   lateral division, ventral part **STLV**
   medial division, anterior part **STMA**
   medial division, dorsal part **STMD**
   medial division, posterior intermediate part **STMPI**

   medial division, posterior part **STMP**
   medial division, posterolateral part **STMPL**
   medial division, posteromedial part **STMPM**
   medial division, ventral part **STMV**
blood vessel (unspecified) **bv**
brachium of the inferior colliculus **bic**
brachium of the superior colliculus **bsc**
branchiomeric column of brainstem and
   spinal cord **bm**

C

C1 adrenaline cells **C1**
calcarine (v) sulcus of cerebral cortex **ca (v)**
canal of Schlemm **Schlemm**
carotid nerve plexus **cpx**
caudal entorhinal cortex **CEnt**
caudal interstitial nucleus of medial longit. fasc. **CI**
caudal linear nucleus of the raphe system **CLi**
caudate nucleus **Cd**
caudate and putamen (dorsal striatum) **CPu**
caudoventrolateral reticular nucleus **CVL**
cavernous sinus **cav**
cell bridges of the ventral striatum **CB**
central amygdaloid nucleus **Ce**
   capsular part **CeC**
   lateral division **CeL**
   medial division **CeM**
central canal of medulla and spinal cord **CC**
central cervical nucleus **CeCv**
central grey **CG**
   alpha part **CGA**
   beta part **CGB**
   gamma part **CGG**
central medial thalamic nucleus **CM**
central nucleus of the inferior colliculus **CIC**
centrolateral thalamic nucleus **CL**
cephalic flexure of brainstem **ceph**
cerebellar commissure **cbc**
cerebellar white matter **cbw**
cerebellomesencephalic fissure **ceme**
cerebellum **Cb**
cerebral aqueduct **Aq**
cerebral cortex **Cx**
cerebral peduncle **cp**
cervical flexure **cef**
choroid plexus **chp**
ciliary ganglion **Cil**
cingulate cortex **Cg**
   area 1 **Cg1**
   area 2 **Cg2**

cingulate (λ) sulcus of cerebral cortex **cg (λ)**

cingulum **cg**

claustrum **Cl**

cochlea **Cochlea**

cochlear duct **CD**

cochlear (spiral) ganglion **CGn**

cochlear root of eighth nerve **8cn**

commissural nucleus of inferior colliculus **Com**

commissure of the inferior colliculus **cic**

commissure of the superior colliculus **csc**

common crus of semicircular ducts **CCrus**

compact cell zone of the cortical plate **ccz**

conjunctival sac **conjunt**

copula pyramis of cerebellum **Cop**

cornea **Cornea**

cornu Ammonis **CA**

coronoid process of the mandible **Cor**

corpus callosum **cc**

cortex-amygdala transition area **CxA**

    layer 1 **CxA1**

    layer 2 **CxA2**

    layer 3 **CxA3**

cortical plate (developing cortex) **CxP**

cortical sulcus η **η**

cribriform plate of the ethmoid bone **CrP**

crus 1 of the ansiform lobule of Cb **Crus1**

crus 2 of the ansiform lobule of Cb **Crus2**

crus of anterior semicircular duct **CrASCD**

cuneate fasciculus **cu**

cuneate nucleus **Cu**

cuneate nucleus, rotundus part **CuR**

cuneiform nucleus **CnF**

    dorsal part **CnFD**

    intermediate part **CnFI**

    magnocellular part **CnFMC**

    parvicellular part **CnFPC**

    ventral part **CnFV**

D

deep cerebral white matter **dcw**

deep grey layer of the superior colliculus **DpG**

deep white layer of the superior colliculus **DpWh**

dentate gyrus **DG**

descending palatine artery **dpal**

diagonal band nucleus **DB**

diencephalon **Dien**

dorsal acoustic stria **das**

dorsal cochlear nucleus **DC**

    deep layer **DCDp**

    fusiform layer **DCFu**

    granular layer **DCGr**

    molecular layer **DCMo**

dorsal cortex of the inferior colliculus **DCIC**

dorsal corticospinal tract **dcs**

dorsal endopiriform nucleus **DEn**

dorsal funiculus of spinal cord white matter **dfu**

dorsal horn of spinal cord **DH**

dorsal hypothalamic area **DA**

dorsal lamina of principal inferior olivary nucleus **dl**

dorsal lateral geniculate nucleus **DLG**

    alpha segment **DLG**α

    beta segment **DLG**β

dorsal lateral olfactory tract **dlo**

dorsal nucleus of the lateral lemniscus **DLL**

dorsal nucleus of the spinal cord **D**

dorsal paragigantocellular nucleus **DPGi**

dorsal part of the claustrum **DCl**

dorsal part of third ventricle **D3V**

dorsal peduncular cortex **DP**

dorsal periaqueductal grey **DPAG**

dorsal ramus of spinal nerve **dra**

dorsal raphe nucleus **DR**

    caudal part **DRC**

    dorsal part **DRD**

    interfascicular part **DRI**

    lateral part **DRL**

    ventral part **DRV**

dorsal root of spinal cord **dr**

dorsal root (spinal) ganglion **DRG**

dorsal spinocerebellar/olivocerebellar

    tracts **dsc/oc**

dorsal spinocerebellar tract **dsc**

dorsal subiculum **DS**

dorsal tegmental nucleus **DTg**

    central part **DTgC**

    pericentral part **DTgP**

dorsal tegmental tract **dtg**

dorsal temporal cortex **TeD**

dorsal tenia tecta **DTT**

    layer 1 **DTT1**

    layer 2 **DTT2**

    layer 3 **DTT3**

dorsal terminal nucleus of the optic tract **DT**

dorsal transition zone of the cortex **DTr**

dorsal tuberomamillary nucleus **DTM**

dorsolateral orbital cortex **DLO**

dorsolateral periaqueductal grey **DLPAG**

dorsolateral tract of Lissauer **dlL**

dorsomedial hypothalamic nucleus **DM**

dorsal part **DMD**
ventral part **DMV**
dorsomedial periaqueductal grey **DMPAG**
dorsomedial spinal trigeminal nucleus **DMSp5**
dorsomedial tegmental area **DMTg**
dysgranular insular cortex **DI**

E

ectorhinal cortex **Ect**
Edinger–Westphal nucleus **EW**
enamel organ of the tooth **EnO**
endolymphatic duct **ELD**
endolymphatic sac **ELS**
entopeduncular nucleus **EP**
entorhinal cortex **Ent**
ependymal and subependymal layers **E**
episupraoptic nucleus **ESO**
ethmoid bone **EthB**
ethmoid thalamic nucleus **Eth**
ethmoid turbinate **EthT**
extended amygdala **EA**
central part **EAC**
medial part **EAM**
external capsule **ec**
external carotid artery **ectd**
external cortex of the inferior colliculus **ECIC**
layer 1 **ECIC1**
layer 2 **ECIC2**
layer 3 **ECIC3**
external cuneate nucleus **ECu**
external granular layer of developing Cb **EGL**
external medullary lamina of the diencephalon **eml**
external plexiform layer of acc. olfactory bulb **EPlA**
external plexiform layer of main olfactory bulb **EPl**
eyelid **Eyelid**

F

facial and vestibulocochlear nerves **7/8n**
facial nerve **7n**
facial nerve (geniculate) ganglion **7Gn**
facial (motor) nucleus **7N**
dorsal intermediate subnucleus **7DI**
dorsolateral subnucleus **7DL**
dorsomedial subnucleus **7DM**
lateral subnucleus **7L**
stylohyoid part **7SH**

ventral intermediate subnucleus **7VI**
ventromedial subnucleus **7VM**
fasciculus retroflexus **fr**
field CA1 of the hippocampus **CA1**
field CA2 of the hippocampus **CA2**
field CA3 of the hippocampus **CA3**
fields of Forel **FF**
fimbria of the hippocampus **fi**
fimbriodentate sulcus of the hippocampus **fds**
flocculus of the cerebellum **Fl**
floor plate of the developing spinal cord **fp**
foramen ovale **OvalF**
forelimb region of primary somatosensory cortex **S1Fl**
fornix **f**
fourth ventricle **4V**
frontal (α) sulcus of cerebral cortex **fr (α)**
frontal association cortex **FrA**
frontal bone **Fro**
frontal branch of ophthalmic division, trigeminal n. **5fr**
frontal cortex (unspecified) **Fr**

G

ganglion cell layer of retina **RGn**
gelatinous layer of caudal spinal trigeminal nucleus **Ge5**
gemini hypothalamic nucleus **Gem**
general somatic afferent nuclei **gsa**
genu of the facial nerve **g7**
germinal trigone of the developing fourth ventricle **trg**
gigantocellular reticular nucleus **Gi**
alpha part **GiA**
ventral part **GiV**
globus pallidus **GP**
glomerular layer of the accessory olfactory bulb **GlA**
glomerular layer of the olfactory bulb **Gl**
glossopharyngeal and vagus nerves **9/10n**
glossopharyngeal nerve **9n**
glossopharyngeal, vagus and accessory nerves **9/10/11n**
glossopharyngeal/vagus ganglia **9/10Gn**
gracile fasciculus **gr**
gracile nucleus **Gr**
granular insular cortex **GI**
granule cell cluster **GCC**
granule cell cluster magna **GCCM**
granule cell layer of the accessory olfactory bulb **GrA**
granule cell layer of the cerebellum **GrCb**
granule cell layer of the cochlear nucleus **GrC**

granule cell layer of the dentate gyrus **GrDG**
granule cell layer of the olfactory bulb **GrO**
greater petrosal nerve **gpet**

H

habenular commissure **hbc**
habenular nuclei **Hb**
hair follicle **Hair Follicle**
handle of the malleus **HandleM**
handle of the stapes **HStap**
Harderian gland **HardG**
hippocampal fissure **hif**
hippocampus **Hi**
horizontal semicircular duct **HSCD**
hypoglossal nerve **12n**
hypoglossal nucleus **12N**
hypothalamic sulcus **hs**
hypothalamus **Hy**

I

incus **Inc**
indusium griseum **IG**
inferior alveolar artery **iala**
inferior alveolar nerve **ialn**
inferior cerebellar peduncle **icp**
inferior colliculus **IC**
inferior ganglia of glossopharyngeal/vagus nerves **I9/10Gn**
inferior oblique muscle **IOb**
inferior olivary neuron migratory stream **iom**
inferior olivary nuclear complex **IO**
    cap of Kooy **IOK**
    dorsal nucleus **IOD**
    dorsomedial cell group **IODM**
    principal nucleus **IOPr**
    subnucleus A **IOA**
    subnucleus B **IOB**
    subnucleus C **IOC**
    subnucleus beta **IOBe**
    ventral nucleus **IOV**
inferior petrosal sinus **ipets**
inferior rectus muscle **IRec**
inferior salivatory nucleus **IS**
infralimbic cortex **IL**
infraorbital artery **infa**
infraorbital nerve, branch of maxillary division, 5n **5mx/inf**
infundibular recess of third ventricle **InfR**

infundibular stem **InfS**
inner plexiform layer of the retina **InPl**
insular cortex **Ins**
interanterodorsal thalamic nucleus **IAD**
interanteromedial thalamic nucleus **IAM**
interbrachial sulcus of cerebral cortex **ib**
intercalated amygdaloid nucleus **I**
    main part **IM**
interfascicular nucleus **IF**
intergeniculate leaflet **IGL**
intermediate endopiriform nucleus **IEn**
intermediate grey layer of superior colliculus **InG**
intermediate lobe of the pituitary gland **IPit**
intermediate nucleus of lateral lemniscus **ILL**
intermediate reticular nucleus **IRt**
intermediate reticular nucleus, alpha **IRtA**
intermediate white layer of superior colliculus **InWh**
intermediate zone of developing cortex **ICx**
intermediate zone of spinal cord **IntZ**
intermediodorsal thalamic nucleus **IMD**
intermediolateral cell column of spinal cord **IML**
intermediomedial cell column of spinal cord **IMM**
intermedius nucleus of the medulla **InM**
internal arcuate fibres **ia**
internal auditory artery **iaud**
internal auditory meatus **IAud**
internal capsule **ic**
internal carotid artery **ictd**
internal jugular vein **ijugv**
internal plexiform layer of olfactory bulb **IPl**
interpeduncular fossa **IPF**
interpeduncular nucleus **IP**
    apical subnucleus **IPA**
    caudal subnucleus **IPC**
    dorsal subnucleus **IPD**
    lateral subnucleus **IPL**
    rostral subnucleus **IPR**
    ventral subnucleus **IPV**
interposed deep cerebellar nucleus **Int**
    anterior part **IntA**
    dorsolateral hump **IntDL**
    dorsomedial crest **IntDM**
    posterior part **IntP**
interstitial nucleus of Cajal **InC**
interstitial nucleus of Cajal, shell part **InCSh**
interstitial nucleus of eighth (vestibulocochlear) nerve **I8**
interstitial nucleus of posterior limb, anterior
    commissure **IPAC**
    lateral part **IPACL**
    medial part **IPACM**
interstitial nucleus of the medulla **IB**

interventricular foramen **IVF**
intradecussational nucleus of sup. cerebellar peduncle **ID**
iridopupillary membrane **IPM**
iris **Iris**
isocortex **IsoCx**
isthmic neuromere **is**
isthmic reticular formation **IsRt**
isthmus **Is**

J

jugular foramen **JugF**
jugular sulcus of cerebral cortex **j**
juxtaparaventricular nucleus, lateral hypothalamus **JPLH**

K

Kölliker–Fuse nucleus **KF**

L

lacrimal branch of ophthalmic division of
    trigeminal n. **5lac**
lacunosum moleculare layer of the hippocampus **LMol**
lambdoid sulcus of cerebral cortex (**λ**)
lamina terminalis **LTer**
lateral accumbens shell **LAcbSh**
lateral amygdaloid nucleus **La**
    dorsal part **LaD**
    dorsolateral part **LaDL**
    posterior part **LaP**
    ventral part **LaV**
    ventrolateral part **LaVL**
    ventromedial part **LaVM**
lateral corticospinal tract **lcs**
lateral deep cerebellar nucleus **Lat**
lateral deep cerebellar nucleus, parvicellular part **LatPC**
lateral entorhinal cortex **LEnt**
lateral fissure of the cerebellum **lf**
lateral funiculus of the spinal cord white matter **lfu**
lateral ganglionic eminence, developing Telen **lge**
lateral habenular nucleus **LHb**
    lateral part **LHbL**
    medial part **LHbM**
lateral horn of the spinal cord **LatH**
lateral hypothalamic area **LH**
lateral lemniscus **ll**

lateral mamillary nucleus **LM**
lateral olfactory tract **lo**
lateral orbital cortex **LO**
lateral parabrachial nucleus **LPB**
    central part **LPBC**
    crescent part **LPBCr**
    dorsal part **LPBD**
    external part **LPBE**
    internal part **LPBI**
lateral paragigantocellular nucleus **LPGi**
    alpha part **LPGiA**
    external part **LPGiE**
lateral periaqueductal grey **LPAG**
lateral posterior thalamic nucleus **LP**
    lateral part **LPL**
    laterocaudal part **LPLC**
    laterorostral part **LPLR**
    medial part **LPM**
    mediocaudal part **LPMC**
    mediorostral part **LPMR**
lateral preoptic area **LPO**
lateral pterygoid muscle **LPtg**
lateral recess of fourth ventricle **LR4V**
lateral rectus muscle **LRec**
lateral reticular nucleus **LRt**
    parvicellular part **LRtPC**
    subtrigeminal part **LRtS5**
lateral septal nucleus **LS**
    dorsal part **LSD**
    intermediate part **LSI**
    ventral part **LSV**
lateral spinal nucleus **LSp**
lateral stripe of the striatum **LSS**
lateral superior olivary nucleus **LSO**
lateral telencephalic tract **lt**
lateral terminal nucleus of the optic tract **LT**
lateral ventricle **LV**
lateral vestibular nucleus **LVe**
lateral vestibulospinal tract **lvs**
lateroanterior hypothalamic nucleus **LA**
laterodorsal tegmental nucleus **LDTg**
    dorsal part **LDTgD**
    ventral part **LDTgV**
laterodorsal thalamic nucleus **LD**
    dorsomedial part **LDDM**
    ventrolateral part **LDVL**
lateroventral periolivary nucleus **LVPO**
layer 1 of cerebral cortex **1Cx**
layer 2 of cerebral cortex **2Cx**
layer 3 of cerebral cortex **3Cx**
layer 4 of cerebral cortex **4Cx**

layer 5 of cerebral cortex **5Cx**
layer 6a,b of cerebral cortex **6a,bCx**
lens **Lens**
levator palpebrae superioris muscle **LPS**
linear nucleus of the medulla **Li**
lithoid nucleus **Lth**
lobules 1 to 10 of cerebellar vermis **1Cb** to **10Cb**
locus coeruleus **LC**
longitudinal fasciculus of the pons **lfp**
loose-packed zone of the developing cortical plate **lpz**

M

macula of the saccule **SMac**
macula of the utricle **UMac**
magnocellular nucleus of lateral hypothalamus **MCLH**
magnocellular nucleus of posterior commissure **MCPC**
magnocellular preoptic nucleus **MCPO**
malleus **Mall**
mamillary body **MB**
mamillary peduncle **mp**
mamillary recess of the 3V **MRe**
mamillotegmental tract **mtg**
mamillothalamic tract **mt**
mandible **Man**
mandibular division of the trigeminal nerve **5man**
marginal zone of medial geniculate nucleus **MZMG**
marginal zone of the developing cortex **mCx**
masseter muscle **MasM**
masseteric nerve **masn**
mastoid process **Mastoid**
matrix region of the medulla **Mx**
maxillary artery **maxa**
maxillary bone (maxilla) **MxB**
maxillary division of the trigeminal nerve **5mx**
maxillary sinus **MaxS**
maxilloturbinate bone **MaxT**
Meckel's cartilage **Mc**
medial accessory oculomotor nucleus **MA3**
medial amygdaloid nucleus **Me**
   anterior part **MeA**
   anterodorsal part **MeAD**
   anteroventral part **MeAV**
   posterior part **MeP**
   posterodorsal part **MePD**
   posteroventral part **MePV**
medial deep cerebellar nucleus **Med**
medial entorhinal cortex **MEnt**
medial forebrain bundle **mfb**
medial frontal sulcus **mf**

medial ganglionic eminence, developing Telen **mge**
medial geniculate nucleus **MG**
   dorsal part **MGD**
   medial part **MGM**
   ventral part **MGV**
medial habenular nucleus **MHb**
medial lemniscus **ml**
medial longitudinal fasciculus **mlf**
medial mamillary nucleus, lateral part **ML**
medial mamillary nucleus, medial part **MM**
medial mamillary nucleus, median part **MnM**
medial orbital cortex **MO**
medial parabrachial nucleus **MPB**
medial parabrachial nucleus,
    external part **MPBE**
medial paralemniscal nucleus **MPL**
medial preoptic area **MPA**
medial preoptic nucleus **MPO**
   lateral part **MPOL**
   medial part **MPOM**
medial pretectal nucleus **MPT**
medial pterygoid muscle **MPtg**
medial rectus muscle **MRec**
medial septal nucleus **MS**
medial superior olive **MSO**
medial terminal nucleus of the optic tract **MT**
medial trapezoid nucleus **MTz**
medial tuberal nucleus **MTu**
medial vestibular nucleus **MVe**
   magnocellular part **MVeMC**
   parvicellular part **MVePC**
median accessory nucleus of the medulla **MnA**
median eminence **ME**
   external layer **MEE**
   internal layer **MEI**
median neuroepithelium of the
    brainstem **mne**
median preoptic nucleus **MnPO**
median raphe nucleus **MnR**
mediodorsal thalamic nucleus **MD**
   central part **MDC**
   lateral part **MDL**
   medial part **MDM**
medioventral periolivary nucleus **MVPO**
medulla **Md**
medullary reticular nucleus, dorsal **MdD**
medullary reticular nucleus, ventral **MdV**
mesencephalic neuromere **mes**
mesencephalic reticular formation **mRt**
mesencephalic tegmentum **MeTg**
mesencephalic trigeminal nucleus **Me5**

mesencephalic trigeminal tract **me5**

mesencephalon **Mesen**

microcellular tegmental nucleus **MiTg**

middle cerebellar peduncle **mcp**

middle cerebral artery **mcer**

mitral cell layer of accessory olfactory bulb **MiA**

mitral cell layer of the olfactory bulb **Mi**

molecular layer of the cerebellum **MoCb**

molecular layer of the dentate gyrus **MoLDG**

molecular layer of the subiculum **MoS**

motor cortex **M**

motor root of the trigeminal nerve **m5**

motor trigeminal nucleus **5N**

    masseter part **5Ma**

    parvicellular part **5PC**

    temporalis part **5Te**

    ventromedial part **5VM**

myelomeres of developing neural tube **my**

N

nasal bone **NasB**

nasal cavity **NasC**

nasal septum **NSpt**

nasal turbinate bone **NasT**

nasopharynx **NasoPhar**

navicular postolfactory nucleus **Nv**

nerve (unidentified) **n**

nerve of the pterygoid canal **ptgcn**

nerve to lateral pterygoid muscle **lptg**

neuroepithelium **ne**

nigrostriatal tract **ns**

notochord **noto**

nuclear transitory zone of developing cerebellum **ntz**

nuclei of the lateral lemniscus **LL**

nucleus of Darkschewitsch **Dk**

nucleus of origin for efferents of vestibular

   nerve **EVe**

nucleus of Roller **Ro**

nucleus of the brachium of inferior colliculus **BIC**

nucleus of the central acoustic tract **CAT**

nucleus of the fields of Forel **F**

nucleus of the horizontal limb of diagonal band **HDB**

nucleus of the lateral olfactory tract **LOT**

    layer 1 **LOT1**

    layer 2 **LOT2**

    layer 3 **LOT3**

nucleus of the optic tract **OT**

nucleus of the posterior commissure **PCom**

nucleus of the stria medullaris **SM**

nucleus of the trapezoid body **Tz**

nucleus of the vertical limb diagonal band **VDB**

nucleus X **X**

nucleus Y **Y**

O

occipital artery **occ**

occipital bone **Occ**

oculomotor nerve **3n**

oculomotor nucleus **3N**

oculomotor nucleus, parvicellular part **3PC**

olfactory artery **olfa**

olfactory bulb **OB**

olfactory epithelium **olfepith**

olfactory nerve **olf**

olfactory nerve layer of the olfactory bulb **ON**

olfactory tubercle **Tu**

olfactory ventricle **OV**

olivary pretectal nucleus **OPT**

olivocerebellar tract **oc**

ophthalmic artery **opha**

ophthalmic division of trigeminal nerve **5oph**

ophthalmic vein **ophv**

optic chiasm **och**

optic fibre layer of the retina **OF**

optic nerve **2n**

optic nerve layer of the superior colliculus **Op**

optic recess of the third ventricle **OptRe**

optic stalk **os**

optic tract **opt**

oral cavity **Oral**

organ of Corti **Corti**

oriens layer of the hippocampus **Or**

otic ganglion **Otic**

oval paracentral thalamic nucleus **OPC**

oval window **OvalW**

P

p1 reticular formation **p1Rt**

palate **Palate**

palatine bone **Pal**

paraabducens nucleus **Pa6**

parabigeminal nucleus **PBG**

parabrachial nuclear complex **PB**

parabrachial pigmented nucleus **PBP**

paracentral thalamic nucleus **PC**
paracommissural nucleus of post. commiss. **PaC**
parafascicular thalamic nucleus **PF**
parafloccular sulcus of the cerebellum **pfs**
paraflocculus of the cerebellum **PFl**
   dorsal part **PFlD**
   ventral part **PFlV**
parainterfascicular nu. of vent. tegmental area **PIF**
paralambdoid septal nucleus **PLd**
paralemniscal nucleus **PL**
paramedian lobule of the cerebellum **PM**
paramedian raphe nucleus **PMnR**
paramedian reticular nucleus **PMn**
paranigral nucleus of ventral tegmental area **PN**
parapedunculotegmental nucleus **PPTg**
parapyramidal nucleus **PPy**
pararubral nucleus **PaR**
parasubiculum **PaS**
parasubthalamic nucleus **PSTh**
paratenial thalamic nucleus **PT**
paratrigeminal nucleus **Pa5**
paratrochlear nucleus **Pa4**
paraventricular hypothalamic nucleus **Pa**
   dorsal part **PaD**
   dorsolateral part **PaDL**
   intermediate part **PaI**
   lateral magnocellular part **PaLM**
   lateral part **PaL**
   magnocellular **PaM**
   medial magnocellular part **PaMM**
   medial parvicellular part **PaMP**
   ventral part **PaV**
   ventrolateral part **PaVL**
paraventricular thalamic nucleus **PV**
   anterior part **PVA**
   posterior part **PVP**
parietal bone **ParB**
parietal plate **ParP**
parietal (β) sulcus of cerebral cortex **pa (β)**
parietal (ε) sulcus of cerebral cortex **pa (ε)**
parvicellular reticular nucleus **PCRt**
parvicellular retic. nucleus, alpha part **PCRtA**
peduncular part of lateral hypothalamus **PLH**
pedunculotegmental nucleus **PTg**
periaqueductal grey **PAG**
perifacial zone **P7**
perifornical nucleus **PeF**
perifornical part of lateral hypothalamus **PeFLH**
perilemniscal nucleus, ventral part **PLV**
peripeduncular nucleus **PP**
perirhinal cortex **PRh**

peritrigeminal zone **P5**
periventricular fibres **pv**
periventricular grey **PVG**
periventricular hypothalamic nucleus **Pe**
periventricular preoptic nucleus **PePO**
pigment layer of the eye **Pig**
pineal gland **Pi**
pineal recess of the third ventricle **PiRe**
pinna of the ear **Pinna**
piriform cortex **Pir**
   layer 1 **Pir1**
   layer 1a **Pir1a**
   layer 1b **Pir1b**
   layer 2 **Pir2**
   layer 3 **Pir3**
polymorphic layer of dentate gyrus **PoDG**
pontine neuron migratory stream **pnm**
pontine nuclei **Pn**
pontine raphe nucleus **PnR**
pontine reticular formation **PnRt**
pontine reticular nucleus, caudal part **PnC**
pontine reticular nucleus, oral part **PnO**
pontine reticular nucleus, ventral part **PnV**
posterior cerebral artery **pcer**
posterior commissure **pc**
posterior communicating artery **pcoma**
posterior cortical amygdaloid area **PCo**
posterior hypothalamic area **PHA**
posterior hypothalamic nucleus **PH**
   dorsal part **PHD**
   ventral part **PHV**
posterior inferior cerebellar artery **pica**
posterior intralaminar thalamic nucleus **PIL**
posterior limitans thalamic nucleus **PLi**
posterior lobe of pituitary gland **PPit**
posterior mamillary nucleus **MP**
posterior semicircular duct **PSCD**
posterior spinal artery **psp**
posterior superior fissure of the
     cerebellum **psf**
posterior thalamic nuclear group **Po**
posterior thalamic nucleus,
     triangular part **PoT**
posterodorsal preoptic nucleus **PDPO**
posterodorsal raphe nucleus **PDR**
posterodorsal tegmental nucleus **PDTg**
posterolateral cortical amygdaloid area **PLCo**
   layers 1 to 3 **PLCo1 to 3**
posterolateral fissure of the cerebellum **plf**
posteromedial cortical amygdala **PMCo**
postsubiculum **Post**

precentral fissure of the cerebellum **pcn**
precommissural nucleus **PrC**
preculminate fissure of the cerebellum **pcuf**
precuneiform area **PrCnF**
prefrontal (τ) sulcus of cerebral cortex **pf (τ)**
pregeniculate nucleus, maguocellular **PrGMC**
pregeniculate nucleus, parvicellular **PrGPC**
pregeniculate nucleus (ventral lat. genic. nu) **PrG**
prelimbic cortex **PrL**
premamillary nucleus, dorsal part **PMD**
premamillary nucleus, ventral part **PMV**
preoptic area **POA**
preparietal (L) sulcus **pp (L)**
preplate of developing cortex **PrePl**
prepositus nucleus **Pr**
prepyramidal fissure of the cerebellum **ppf**
prerubral field **PR**
presphenoid bone **PSph**
presphenoid bone wing **PSphW**
presubiculum **PrS**
pretectum **Ptec**
prethalamus **PrTh**
primary and secondary motor cortex **M1/M2**
primary auditory cortex **Au1**
primary fissure of the cerebellum **prf**
primary motor cortex **M1**
primary somatosensory cortex **S1**
  forelimb area **S1FL**
  hindlimb area **S1HL**
  mandible area **S1Md**
  mystacial area **S1My**
  ophthalmic area **S1Op**
  shoulder area **S1Sh**
  tongue area **S1To**
  trunk area **S1Tr**
primary visual cortex **V1**
primordial plexiform layer **ppl**
principal sensory trigeminal nucleus **Pr5**
  dorsomedial part **Pr5DM**
  ventrolateral part **Pr5VL**
prosencephalon **Pros**
prosomere 1 **p1**
prosomere 2 **p2**
prosomere 3 **p3**
pterygoid process of the sphenoid bone **Ptg**
pterygopalatine nerve **ptgpal**
Purkinje cell layer of the cerebellum **Pk**
putamen **Pu**
pyramidal cell layer of the hippocampus **Py**
pyramidal decussation **Pyd**
pyramidal tract **py**

R

radiatum layer of the hippocampus **Rad**
raphe interpositus nucleus **RIP**
raphe magnus nucleus **RMg**
raphe obscurus nucleus **ROb**
raphe pallidus nucleus **RPa**
Rathke's pouch **Rathke**
recess of the inferior colliculus **ReIC**
red nucleus **R**
  magnocellular part **RMC**
  parvicellular part **RPC**
respiratory epithelium of nasal cavity **respepith**
reticular thalamic nucleus (prethalamus) **Rt**
reticulostrial nucleus **RtSt**
reticulotegmental nucleus of the pons **RtTg**
reticulotegmental nucleus, pericentral
    part **RtTgP**
retina **Retina**
retroambiguus nucleus **RAmb**
retrochiasmatic area **RCh**
retrochiasmatic area, lateral part **RChL**
retroethmoid nucleus **REth**
retroisthmic nucleus **RIs**
retromamillary decussation **rmd**
retromamillary nucleus **RM**
  lateral part **RML**
  medial part **RMM**
retroparafascicular nucleus **RPF**
retroreuniens area **RRe**
retrorubral field **RRF**
retrorubral nucleus **RR**
retrosplenial dysgranular cortex **RSD**
retrosplenial granular cortex **RSG**
  part a **RSGa**
  part b **RSGb**
reuniens thalamic nucleus **Re**
Rexed's laminae 1–10 of spinal cord **R1–10**
rhinal fissure **rf**
rhinal incisure **ri**
rhombencephalon **Rhomb**
rhomboid thalamic nucleus **Rh**
rhombomere 1 reticular formation **r1Rt**
rhombomeres 1–8 of developing brainstem **r1–8**
roof plate of the developing spinal cord **rfp**
rostral interstitial nucleus **RI**
rostral linear nucleus (of midbrain) **RLi**
rostroventrolateral reticular nucleus **RVL**
round window **RoundW**
rubrospinal tract **rs**

S

saccule of the inner ear **Sacc**
sagulum nucleus **Sag**
scala tympani **ScTy**
scala vestibuli **ScVe**
secondary fissure of the cerebellum **sf**
secondary motor cortex **M2**
secondary somatosensory cortex **S2**
secondary visual cortex **V2**
    lateral area **V2L**
    medial area **V2M**
sensory root of the trigeminal nerve **s5**
septal olfactory organ **SOO**
septofimbrial nucleus **SFi**
septohippocampal nucleus **SHi**
septohypothalamic nucleus **SHy**
septum (of developing brain) **Spt**
simplex fissure **simf**
simplex lobule of the cerebellum **Sim**
simplex lobule of the cerebellum A **SimA**
simplex lobule of the cerebellum B **SimB**
solitary nucleus **Sol**
    commissural part **SolC**
    dorsolateral part **SolDL**
    dorsomedial part **SolDM**
    gelatinous part **SolG**
    intermediate part **SolIM**
    lateral part **SolL**
    medial part **SolM**
    parvicellular **SolPC**
    ventral part **SolV**
    ventrolateral part **SolVL**
solitary tract **sol**
somatic efferent nuclear column **se**
special somatic afferent nuclear column **ssa**
sphenopalatine artery **spa**
sphenopalatine ganglion **SphPal**
spinal cord **SpC**
spinal nerve **spn**
spinal trigeminal nucleus **Sp5**
    caudalis part **Sp5C**
    interpolaris part **Sp5I**
    oralis part **Sp5O**
spinal trigeminal tract **sp5**
spinal vestibular nucleus **SpVe**
stapedius muscle **StapM**
sterno(cleido)mastoid muscle **StM**
stratum lucidum of the hippocampus **SLu**
stria medullaris **sm**

stria terminalis **st**
striohypothalamic nucleus **StHy**
subbrachial nucleus **SubB**
subcoeruleus nucleus, alpha part **SubCA**
subcoeruleus nucleus, dorsal part **SubCD**
subcoeruleus nucleus, ventral part **SubCV**
subcommissural organ **SCO**
subfornical organ **SFO**
subgeniculate nucleus **SubG**
subiculum **S**
subiculum, transition area **STr**
subincertal nucleus **SubI**
submedius thalamic nucleus **Sub**
subparafascicular thalamic nucleus **SPF**
    parvicellular part **SPFPC**
subparaventricular zone of the
       hypothalamus **SPa**
subpeduncular tegmental nucleus **SPTg**
substantia innominata **SI**
    basal part **SIB**
substantia nigra **SN**
    compact part **SNC**
    compact part, dorsal tier **SNCD**
    compact part, medial tier **SNCM**
    lateral part **SNL**
    reticular part **SNR**
subthalamic nucleus **STh**
subventricular layer of the developing
    cortex **SubV**
sulcus limitans **sl**
sulcus µ of cortex **µ**
superficial grey layer of the superior
      colliculus **SuG**
superior alveolar nerve **sal**
superior cerebellar artery **scba**
superior cerebellar peduncle **scp**
    decussation **scpd**
superior cervical ganglion **SCGn**
superior colliculus **SC**
superior gang. of glossopharyng.,
      vagus nerves **S9/10Gn**
superior ganglion of the vagus nerve **S10Gn**
superior medullary velum **smv**
superior oblique muscle **SOb**
superior olive **SOl**
superior paraolivary nucleus **SPO**
superior rectus muscle **SRec**
superior sagittal sinus **sss**
superior salivatory nucleus **SuS**
superior thalamic radiation **str**
superior vestibular nucleus **SuVe**

suprachiasmatic nucleus **SCh**
suprageniculate thalamic nucleus **SG**
supragenual nucleus **SGe**
supraoculomotor cap **Su3C**
supraoculomotor periaqueductal grey **Su3**
supraoptic decussation **sod**
supraoptic nucleus **SO**
   anterior part **SOA**
   retrochiasmatic part **SOR**
supratrigeminal nucleus **Su5**
sympathetic trunk **Symp**

T

tectal grey **TG**
tectospinal tract **ts**
telencephalon **Telen**
temporal association cortex **TeA**
temporal bone, petrous part **TempP**
temporal bone, squamous part **TempS**
temporal (δ) sulcus of cerebral cortex **te (δ)**
temporal nerve **temp**
temporalis muscle **TempM**
tensor tympani muscle **TT**
terete hypothalamus nucleus/ temporal
   cx (unspecified) **Te**
terminal nerve **term**
thalamus **Th**
third ventricle **3V**
transverse fibres of the pons **tfp**
transverse venous sinus **trs**
trapezoid body **tz**
triangular nucleus of lateral lemniscus **TrLL**
triangular septal nucleus **TS**
trigeminal ganglion **5Gn**
trigeminal nerve **5n**
trigeminal transition zone **5Tr**
trigeminal-solitary transition zone **5Sol**
trigeminothalamic tract **tth**
trochlear nerve **4n**
trochlear nerve decussation **d4n**
trochlear nucleus **4N**
tuberal part of the hypothalamus **Tub**
tuberal pituitary **TubPit**
tuberal region of the lateral
    hypothalamus **TuLH**
tympanic bulla **TyBu**
tympanic cavity **TyC**
tympanic membrane **TyM**

U

uncinate fasciculus of the cerebellum **un**
uncinate fasciculus of the cerebellum, decussation **und**
utricle of the inner ear **Utr**

V

vagus nerve **10n**
vagus nerve (autonomic motor) nucleus **10N**
vascular organ of the lamina terminalis **VOLT**
vein (unidentified) **v**
ventral amygdalofugal pathway **vaf**
ventral anterior thalamic nucleus **VA**
ventral cochlear nucleus **VC**
   anterior part **VCA**
   capsular part **VCCap**
   granule cell layer **VCAGr**
   posterior part **VCP**
ventral corticospinal tract **vcs**
ventral endopiriform nucleus **VEn**
ventral funiculus of the spinal cord **vfu**
ventral hippocampal commissure **vhc**
ventral horn of the spinal cord **VH**
ventral lamina of the principal inferior olivary nucleus **vl**
ventral lateral geniculate nu. see pregeniculate nu. **PrG**
ventral median fissure **vmnf**
ventral nucleus of the lateral lemniscus **VLL**
ventral orbital cortex **VO**
ventral pallidum **VP**
ventral part of claustrum **VCl**
ventral periolivary nucleus **VPO**
ventral posterior thalamic (lateral and medial)
    nucleus **VPL/M**
ventral posterior thalamus nucleus, parvicellular
    part **VPPC**
ventral posterolateral thalamic nucleus **VPL**
ventral posteromedial thalamic nucleus **VPM**
ventral ramus **vra**
ventral reuniens thalamic nucleus **VRe**
ventral roots of spinal cord **vr**
ventral spinocerebellar tract **vsc**
ventral subiculum **VS**
ventral tegmental area **VTA**
ventral tegmental area, rostral area **VTAR**
ventral tegmental decussation **vtgd**
ventral tegmental nucleus **VTg**

ventral temporal cortex **TeV**
ventral tenia tecta **VTT**
ventral third ventricle **V3V**
ventral tuberomamillary nucleus **VTM**
ventricular space of the eye **Vent**
ventrolateral hypothalamic nucleus **VLH**
ventrolateral orbital cortex **VLO**
ventrolateral periaqueductal grey **VLPAG**
ventrolateral preoptic nucleus **VLPO**
ventrolateral thalamic nucleus **VL**
ventromedial hypothalamic nucleus **VMH**
   central part **VMHC**
   dorsomedial part **VMHDM**
   ventrolateral part **VMHVL**
ventromedial periaqueductal grey **VMPAG**
ventromedial preoptic nucleus **VMPO**
ventromedial thalamic nucleus **VM**
vertebral artery **vert**
vestibular nerve ganglion **VeGn**
vestibular neuroepithelium **Ve**
vestibular root of the eighth cranial nerve **8vn**
vestibule of the inner ear **Vest**
vestibulocerebellar nucleus **VeCb**
vestibulocochlear nerve **8n**
vestibulomesencephalic tract **veme**

visceral afferent nuclear column **va**
visceral efferent nuclear column **ve**
vitreous cavity of the eye **Vitr**
vomer (bone) **Vom**
vomeronasal nerve **vno**
vomeronasal nerve layer, access. olfactory
   bulb **vn**
vomeronasal organ **VNO**

X

xiphoid thalamic nucleus **Xi**

Z

zona incerta **ZI**
   caudal part **ZIC**
   dorsal part **ZID**
   rostral part **ZIR**
   ventral part **ZIV**
zona layer of the sup. colliculus **Zo**
zona limitans **ZL**

Index to figures and atlas plates

Abb'n	Short Title	Chapter Figure Numbers; Atlas Plate Numbers
5mx/inf	infraorbital n., br. of 5mx	WP0–1 to 3, WP5–1 to 4, WP12–1 to 5
5N	motor trigeminal nu.	Figures 3.2, 4.1, 4.5; D24, D25, W39, WP0–6, WP12–10, WP19–8, WP25–8
5n	trigeminal nerve	Figures 4.1 and 12.2; D24, WP0–6
5oph	ophthalmic division of 5n	Figure 4.3; WP5–4, WP12–6 and 7
5PC	motor trigeminal nu., parvicellular	W38
5Sol	trigeminal-solitary trans. zone	Figure 4.3; D27 to 32, W41, W44 to W46, WP0–8, WP0–10 and 11, WP5–10 to 12, WP12–11 to 16, WP19–9, WP25–9 and 10
5Te	motor trigeminal nu., temporalis	W38
5Tr	trigeminal transition zone	D25, W39, WP25–8
5VM	motor trigeminal nu., ventromed.	W38
6N	abducens nu.	Figures 3.2, 3.3, 4.1; D26, W40, WP0–7, WP5–8, WP12–11
6n	abducens nerve	Figure 4.1; W36, W37, W39, WP0–3 to 5, WP12–6 to 9, WP5–3 to 6
7/8n	7/8th nerves	Figure 2.4; WP5–7, WP25–8
7DI	facial nu., dorsal intermed. subnu.	D26, D27, W42, WP25–9
7DL	facial nu., dorsolateral subnu.	D26, D27, W42, WP25–9
7DM	facial nu., dorsomedial subnu.	D26, D27, W42, WP25–9
7Gn	facial nerve ganglion	WP5–7
7L	facial nu., lateral subnu.	D26, D27, W42, WP25–9
7N	facial nu.	Figures 3.2, 3.3, 4.1, 4.4; W41, W43, WP0–8, WP5–8, WP12–11 and 12
7n	facial nerve	Figures 4.1 and 12.2; D25, D26 to D42, WP0–7, WP5–7 and 8, WP12–10 to 12
7SH	facial motor nu., stylohyoid	WP12–12
7VI	facial nu., ventral intermed. subnu.	D26, D27, W42, WP25–9
7VM	facial nu., ventromedial subnu.	D26, D27, W42, WP25–9
8cn	cochlear root of eighth nerve	Figures 4.4 and 11.1; D25 to D27, WP5–7 and 8, WP12–10 and 11
8n	vestibulocochlear nerve	Figures 4.1, 4.4, 11.1, 12.2; D24, D25, W41, WP0–8, WP5–8, WP12–11
8vn	vestibular root of eighth nerve	Figure 4.4; D25 to D27, W40, W43, WP5–8 and 9, WP12–12
9/10/11n	glossopharyn., vagus, access. n.	Figure 2.4
9/10Gn	glossopharyngeal/vagus ganglion	WP12–13
9/10n	glossopharyngeal and vagus n.	WP0–9, WP12–13
9n	glossopharyngeal nerve	Figure 4.1; WP5–9
10N	vagus nerve motor nu.	Figures 3.3, 4.1 to 4.3; D28 to D32, W44 to W48, WP0–10 and 11, WP5–10 to 12, WP12–13 to 16, WP19–9 to 11, WP25–10
10n	vagus nerve	Figures 4.1 and 4.2; D30, WP0–8, WP0–10, WP12–13 and 14, WP25–10
11N	accessory nerve nu.	Figure 4.1; D32, W49 to W51, WP5–11
11n	accessory nerve	Figure 4.1; WP5–10 and 11
12N	hypoglossal nu.	Figures 3.2, 3.3, Figure 4.1 to 4.3; D30 to D32, W45 to W48, WP0–10 and 11, WP5–10 to 12, WP12–13 to 15, WP19–9 to 11, WP25–9 and 10
12n	hypoglossal nerve	Figure 4.1; W46 to W48, WP0–9 to 11, WP5–9 and 10, WP12–13 and 14
A		
α	sulcus α of cerebral cortex	Figures 2.4, 2.5, 2.8, 4.4, 8.1
α'	sulcus α' of cerebral cortex	Figures 2.4, 2.5, 2.8, 8.1
A1	A1 noradrenaline cells	D32
A2	A2 noradrenaline cells	D32
A5	A5 noradrenaline cells	D24, D25, D26, W41
A6	A6 noradrenaline cells	Figure 4.5
A7	A7 noradrenaline cells	D23, D24, W36, W37
A11	A11 dopamine cells	D14, W27, W29
A13	A13 dopamine cells	D13
AA	anterior amygdaloid area	D10, D11, W22, W23, WP19–4
AAD	anterior amygdaloid area, dorsal	D8, D9
AAV	anterior amygdaloid area, ventral	D8, D9
ac	anterior commissure	Figures 2.2, 3.2, 3.4, 3.5, 6.3, 7.3, 7.4, 8.5, 11.1; D7 to D9, W20 to W22, WP12–5, WP19–4, WP25–4

Abb'n	Short Title	Chapter Figure Numbers; Atlas Plate Numbers
aca	anterior commissure, anterior limb	Figures 7.4 and 12.4; D4 to D6, W13 to W19
Acb	accumbens nu.	D3, D6, WP0–2, WP5–2, WP12–3, WP12–5, WP19–3
AcbC	accumbens nu., core	Figures 7.3, 7.4, 12.4; D4, D5, W15 to W19, WP19–3, WP25–3
AcbSh	accumbens nu., shell	Figures 7.3, 7.4, 12.4; D4, D5, W14 to W19, WP12–4, WP19–3, WP25–3
acer	anterior cerebral artery	D5, D6, W17 to W19, W21, WP5–3, WP12–5, WP19–4, WP25–3 and 4
aci	anterior commiss., intrabulbar limb	Figure 12.4; D1 to D3, W1 to W12
ACo	anterior cortical amygdaloid area	Figures 7.1, 7.2, 12.1; D8, D10, D11, W23 to W25, WP5–4, WP12–6, WP19–4 and 5, WP25–5
acp	anterior commissure, posterior limb	D9 to D10, W21
Acs5	accessory trigeminal nu.	D23 to D25
Acs7	accessory facial nucleus	D26, W41, W42
AD	anterodorsal thalamic nu.	Figures 6.1, 6.2, 6.3; D11, W24 to W27
af	amygdaloidal fissure	W22, W23
AH	anterior hypothalamic area	Figure 6.4; WP5–5, WP12–6, WP19–5
AHA	anterior hypothalamic area, ant	Figure 6.3; D10, D11, WP12–6
AHC	anterior hypothalamic area, central	Figures 6.3 and 6.4; W24
AHD	anterior hypothalamic area, dorsal	W24
AHi	amygdalohippocampal area	D15 to D18
AHiAL	amygdalohippo. area, anterolat.	D14, W26
AHiA	amygdalohippo. area, anterior	Figure 7.2
AHiP	amygdalohippo. area, posterior	W27 to W29
AHP	anterior hypothalamic area, post.	Figures 6.3 and 6.4; D12, W25
AI	agranular insular cortex	D6
AID	agranular insular cx, dorsal	Figures 8.2 and 8.5; D5, W9 to W23, WP19–2 and 3, WP25–2 and 3
AIP	agranular insular cx, post.	Figure 8.2; D7 to D11, W24, WP19–4
AIV	agranular insular cx, vent.	Figure 8.2; D5, W10 to W24, WP19–2 and 3, WP25–2 and 3
alar	alar plate of developing brainstem	Figures 3.1 to 3.3; WP0–7 to 11, WP5–9 to 12, WP12–11 to 16, WP19–9 to 11, WP25–9
ALF	anterior lacerated foramen	WP5–4
alv	alveus of the hippocampus	D9 to D15, D17, W23 to W29, W31 to W33, WP25–4 and 5
AM	anteromedial thalamic nu.	Figures 6.2 and 6.3; D11, D12, W24, W25, WP25–4
Amb	ambiguus nucleus	Figures 3.2, 4.1, 4.3; WP0–9 and 10, WP5–9 to 12, WP12–13 and 14, WP19–9
AmbC	ambiguus nu., compact part	Figure 3.3; D28, W44
AmbL	ambiguus nu., loose part	D29 to D32, W45 to W47, WP19–10, WP25–10
Amg	amygdala	Figures 2.2, 3.3, 6.3, 8.3, 12.2; WP19–6, WP25–5
AMV	anteromedial thal. nu., ventral	D11
ANS	accessory neurosecretory nu.	W24
Ant	anterior lobule of Cb	D25
AO	anterior olfactory nu.	Figures 2.2, 2.8, 12.1, 12.2; WP5–1
AOB	accessory olfactory bulb	Figures 2.7 and 12.1; D2, W10, W11
AOD	anterior olfactory nu., dorsal	Figures 12.3 and 12.4; D2 and D3, W8 to W12
AOE	anterior olfactory nu., external	Figure 12.4; D2, W8 to W11
AOL	anterior olfactory nu., lateral	Figures 12.3 and 12.4; D1 and D2, W6 to W12, WP19–2, WP25–2
AOM	anterior olfactory nu., medial	Figures 12.1, 12.3, 12.4; D2 and D3, WP19–2, WP25–2
AOP	anterior olfactory nu., posterior	W12, W13
AOrb	alar orbital bone	WP0–1 and 2, WP5–1 and 2, WP12–1 to 3
AOV	ant. olfactory nu., ventral	Figures 12.3 and 12.4; D1 to D3, W11
AOVP	ant. olfactory area, ventroposterior	W12, W13
AP	area postrema	D31, W47, W48
APir	amygdalopiriform transition	Figure 7.2; D15 to D18, W26, W27, WP12–7, WP19–6
APit	anterior lobe pituitary	Figure 6.6; WP0–4, WP5–5, WP12–7
apmf	ansoparamedian fissure	W46
APT	anterior pretectal nu.	Figures 6.1 and 6.2; D17, W31, WP19–6 and 7, WP25–5 to WP25–7
APTD	anterior pretectal nu., dorsal	Figures 6.1 and 6.5; D14 to D16, W29, W30

Abb'n	Short Title	Chapter Figure Numbers; Atlas Plate Numbers
APTV	anterior pretectal nu., ventral	Figures 6.1 and 6.5; D14 to D16, W29, W30
Aq	cerebral aqueduct	Figures 2.2, 3.2, 3.3, 4.6, 4.7, 6.2, 6.5; D17 to D23, W30 to W36, WP0–6 to 10, WP5–7 to 11, WP12–10 to 16, WP19–7 to 9, WP25–6 to 9
Arc	arcuate hypothalamic nu.	Figures 6.3 and 6.6; D13, D14, W25, WP0–4 and 5, WP5–5, WP12–7, WP12–8, WP19–6, WP25–5
ArcD	arcuate hypothal. nu., dorsal	W26
ArcL	arcuate hypothal. nu., lateral	W26, WP0–4, WP25–6
ArcLP	arcuate hypothal. nu., lateropost.	D15, W27, W28
ArcM	arcuate hypothal. nu., medial	W26, WP0–4, WP25–6
ArcMP	arcuate hypothal. nu., mediopost.	D15, D16, W27, W28
asc7	ascending fibres of facial nerve	D27, W41
ASCD	anterior semicircular duct	WP0–8 to 10, WP5–9 and 10, WP12–11 to 13
ASCDA	anterior semicircular duct, ampulla	WP12–11
asp	anterior spinal artery	W51, WP5–11, WP12–15
ASph	alisphenoid bone	WP0–5, WP5–5, WP12–7
ASt	amygdalostriatal transition area	Figure 7.1; D11 to D13, W23 to W25, WP19–5
ATh	anterior thalamic region	WP19–4
Atlas	atlas (1st cervical vertebra)	WP5–12
Au1	primary auditory cortex	Figures 2.8, 8.2, 8.4, 8.5; D14 to D16, W23 to W30, WP12–8, WP19–5, WP25–5
Aud	auditory neuroepithelium	WP0–7 to 10
audos	auditory ossicles	Figure 11.1
aute	auriculotemporal nerve	WP0–5
AV	anteroventral thalamic nu.	Figures 6.1, 6.2, 6.3; D11, D12, W26, W27, WP25–4
AVDM	ant. vent. thalamic nu., dorsomed.	Figure 6.2; W24, W25
AVPe	ant. ventral periventric. nu.	D9, W22
AVVL	ant. vent. thalamic nu., ventrolat.	Figure 6.2; W24, W25
Axis	axis (second cervical vertebra)	WP5–12
B		
B	basal nu. (of Meynert)	W23 to W25
β	sulcus β of cerebral cortex	Figures 2.5, 2.8, 8.1
B9	B9 serotonin cells	D21, W34
BAOT	bed nu. access. olfactory tract	Figure 12.1; W22
Bar	Barrington's nu.	Figure 4.5; D24, D25, W37, WP5–8
bas	basilar artery	Figure 5.3; W36, W40 to W47, WP0–6 to 10, WP5–6 to 10, WP12–9 to 14, WP19–7 to WP19–10, WP25–7 and 8, WP25–9 and 10
basal	basal plate of developing brainstem	Figures 3.1 to 3.3; WP0–7 to 11, WP5–8 to 12, WP12–11 to 15, WP19–9 and 10, WP25–9
BIC	nu. brachium inf. colliculus	D19 to D21, W33 to W35, WP12–10 to 12, WP19–8, WP25–7 and 8
bic	brachium of inf. colliculus	D19 to D21, W33 to W36, WP19–7 and 8, WP25–7
BL	basolateral amygdaloid nu.	WP12–6 and 7, WP19–5, WP25–5
BLA	basolat. amygdaloid nu., ant.	Figures 7.1 and 7.2; D10 to D13, W22 to W24
BLP	basolat. amygdaloid nu., post.	Figure 7.2; D14, D15, W25, W26
BLV	basolat. amygdaloid nu., ventral	Figure 7.2; D12 to D14, W25
BM	basomedial amygdaloid nu.	WP12–6 and 7, WP19–5, WP25–5
bm	branchiomeric column	Figure 4.1
BMA	basomed. amygdaloid nu., ant.	Figures 7.1 and 7.2; D10 to D13, W22 to W24
BMP	basomed. amygdaloid nu., post.	Figure 7.2; D12 to D17, W25, W26, WP19–6
BMV	basomed. amygdaloid nu., vent.	D14
BOcc	basioccipital bone	Figure 2.1; WP0–7 to 11, WP12–9 to 14, WP5–8 to 11
bp	brachium pontis fibres	D21
bsc	brachium of superior colliculus	Figure 6.5; D13 to D16, D19, D21, W29 to W32, WP25–5
BSph	basisphenoid bone	WP0–5 and 6, WP5–5, WP12–7 and 8

Abb'n	Short Title	Chapter Figure Numbers; Atlas Plate Numbers
Cl	claustrum	Figures 2.2 and 7.1; D5, D10, D11, W8 to W17, WP5–3 and 4, WP12–4 to 6, WP19–3 to 5, WP25–2 to 5
CLi	caudal linear nu. of the raphe	Figure 4.6; D20, D21, W32 to W34, WP12–11, WP25–7
CM	central medial thalamic nu.	Figures 6.1 and 6.2; D10 to D14, W25 to W28, WP12–8, WP19–5, WP25–5
CnF	cuneiform nu.	D22, D23, WP12–12, WP25–8
CnFD	cuneiform nu., dorsal part	W35, W36
CnFI	cuneiform nu., intermediate	W35, W36
CnFMC	cuneiform nu., magnocellular	Figure 4.7
CnFPC	cuneiform nu., parvicellular	Figure 4.7
CnFV	cuneiform nu., ventral part	W35, W36
Cochlea	cochlea	Figure 11.1
Com	commissural nu. inf. colliculus	Figure 4.7
Conjunct	conjunctival sac	WP0–1 to 4, WP5–1 to 3, WP12–1 to 4
Cop	copula pyramis	Figure 5.1; W45, W46
Cor	coronoid process of mandible	WP5–5, WP12–6, WP12–8
Cornea	cornea	WP0–2, WP5–1 and 2, WP12–1 to 4
Corti	organ of Corti	Figure 11.1
cp	cerebral peduncle	Figure 4.6; D14 to D20, W26 to W34, WP5–5 and 8, WP12–8 to 10, WP19–6 and 7, WP25–5 to 8
CPu	caudate putamen (striatum)	WP0–2 to 4, WP5–2 to 4, WP12–3 to 6, WP19–3 to 5, WP25–3 to 5
cpx	carotid plexus	WP0–4 to 7
CrASCD	crus of ant. semicircular duct	WP12–11
CrP	cribriform plate ethmoid	WP0–1, WP5–1, WP12–1 and 2
Crus1	crus 1 ansiform lobule of Cb	Figures 2.4, 5.1, 5.4; W38 to W45, WP25–9
Crus2	crus 2 ansiform lobule of Cb	Figures 2.4, 2.5, 5.1; W43 to W46
csc	commissure of superior colliculus	Figure 6.5; D16, D17, W31, W32, WP25–9
Cu	cuneate nu.	Figures 4.2 and 4.3; D30 to D32, W45 to W49, WP5–12, WP12–15 and 16, WP19–11, WP25–11
cu	cuneate fasciculus	Figures 3.5, 4.3, 13.2, 13.3; D31 and D32, W45 to W51, WP12–16, WP19–11, WP25–11
CuR	cuneate nu., rotundus part	Figure 4.3; W47, W48
CVL	caudoventrolat. reticular nu.	D30
Cx	cerebral cortex	Figures 2.4, 2.5, 2.7; D2, D21, W4, WP0–1 to 5, WP5–1 to 5, WP12–3 to 8, WP19–2 to 6, WP25–2 to 6
CxA	cortex-amygdala transition	Figures 7.1, 7.2; D8, D9, D11 to D13, W21, W24, WP5–4, WP12–6
CxA1	cortex-amyg. trans'n, layer 1	W22, W23
CxA2	cortex-amyg. trans'n, layer 2	W22, W23
CxA3	cortex-amyg. trans'n, layer 3	W22, W23
CxP	cortical plate	WP5–3, WP12–3 to 7, WP19–3, WP25–2 and 3
D		
D	dorsal nu. of the spinal cord	Figure 13.2
δ	sulcus δ of cerebral cortex	Figure 8.1
D3V	dorsal third ventricle	D10 to D14, W24, W26 to W28, WP0–4 and 5, WP5–4 to 6, WP12–7 and 8, WP19–6, WP25–4 and 5
d4n	decussation of 4n	Figure 3.2; WP12–13
DA	dorsal hypothalamic area	D13, W25 to W27, WP25–5
das	dorsal acoustic stria	Figure 4.4; D27
DB	diagonal band nu.	Figures 3.2 and 12.1
DC	dorsal cochlear nu.	Figures 4.1, 5.2, 11.1; D29, W41, WP0–7 and 8, WP5–7 and 8, WP25–9
DCDp	dorsal cochlear nu., deep layer	Figure 4.4; D27, D28, W42, W43
DCFu	dorsal cochlear nu., fusiform	Figure 4.4; D28, W42, W43
DCGr	dorsal cochlear nu., granular	Figure 4.4
DCIC	dorsal cortex of inferior colliculus	Figures 4.7 and 11.1; D22 to D25, W37

Abb'n	Short Title	Chapter Figure Numbers; Atlas Plate Numbers
DCl	dorsal part of claustrum	D6 to D9, W18 to W23
DCMo	dorsal cochlear nu., molecular	Figure 4.4; D27, D28, W42, W43
dcs	dorsal corticospinal tract	Figures 13.2 and 13.3; W51
dcw	deep cerebral white matter	D5 to D15, D18, D19, W6 to W36
DEn	dorsal endopiriform nu.	Figures 7.1 and 7.2; D3 to D12, W12 to W26, WP5–4, WP12–4 to 6, WP19–3 to 5, WP25–2 to 5
dfu	dorsal funiculus	Figures 2.1 and 13.3
DG	dentate gyrus	Figures 2.2, 7.3, 8.2, 8.3, 8.4; D8 to D10, D15, W20 to W32, WP25–4 and 5
DH	dorsal horn of spinal cord	Figure 2.1
DI	dysgranular insular cortex	W20 to W23
Dien	diencephalon	Figure 3.3; WP0–3 to 6, WP5–3 to 6, WP12–6 to 9, WP19–4 to 6, WP25–4 to 6
Dk	nu. of Darkschewitsch	D15 to D18, W30, W31, WP5–7 and 8, WP12–10 and 11, WP19–7, WP25–7 and 8
dl	dorsal lamina of IOPr	Figure 5.4
DLG	dorsal lateral geniculate nu.	Figure 6.1; W27, W28, WP5–6, WP12–7, WP12–8, WP19–5, WP25–5
DLGα	dorsal lateral geniculate nu., alpha	Figure 6.2; D13 to D15, W29 to W31
DLGβ	dorsal lateral geniculate nu., beta	Figure 6.2; D13 to D15, W29 to W31
DLL	dorsal nu. of lateral lemniscus	Figure 4.7; D22, D23, W36, W37, WP5–9, WP12–11, WP12–12, WP25–7
dlL	dorsolateral tract of Lissauer	Figure 13.2
DLO	dorsolateral orbital cortex	W5 to W7
dlo	dorsal lateral olfactory tract	D2
DLPAG	dorsolateral periaqueductal grey	Figures 4.6 and 4.7; D18 to D24, W32 to W35, WP5–8 to 10, WP12–11 to 15, WP19–7 to 9, WP25–6 to 7
DM	dorsomedial hypothalamic nu.	Figure 6.3; W26, WP12–7 and 8, WP19–6, WP25–5 and 6
DMD	dorsomedial hypothal. nu., dorsal	Figure 6.3; D13, D14, W25, W27
DMPAG	dorsomedial periaqueductal grey	Figures 4.6 and 4.7; D17 to D24, W32 to W36, WP5–8 and 9, WP12–12 to 15, WP19–7 to 9, WP25–6 to 8
DMSp5	dorsomedial spinal trigem. nu.	Figures 4.2 and 4.3; D26 to D30, W41 to W45, WP19–9
DMTg	dorsomedial tegmental area	D23, W36 to W38, WP0–8, WP12–11
DMV	dorsomedial hypothal. nu., ventral	D14, W27
DP	dorsal peduncular cortex	Figure 12.1; WP19–3
DPAG	dorsal periaqueductal grey	W31
dpal	descending palatine artery	WP5–1 and 2
DpG	deep grey of superior colliculus	Figure 4.6; D16 to D22, W31 to W35, WP19–6 and 7, WP25–6 and 7
DPGi	dorsal paragigantocellular nu.	D26 to D29, W40 to W43, WP0–8 and 9, WP5–9 and 10, WP12–12 to 14, WP19–9, WP25–9
DpWh	deep white of superior colliculus	Figure 4.6; D18 to D21, W31 to W35, WP25–6 and 7
DR	dorsal raphe nu.	Figures 4.6 and 12.1; D19, D20, W33, WP5–9, WP12–12, WP19–8, WP25–8
dr	dorsal root of spinal cord	Figures 2.1, 3.1, 13.2; WP12–16
dra	dorsal ramus of spinal nerve	Figure 2.1
DRC	dorsal raphe nu., caudal	D23, D24, W37, W38
DRD	dorsal raphe nu., dorsal	Figure 4.7; D21, D22, W34 to W36
DRG	dorsal root (spinal) ganglion	Figures 2.1 and 3.1
DRI	dorsal raphe nu., interfascicular	D21, D23
DRL	dorsal raphe nu., lateral	D21 to D23, W34, W35
DRV	dorsal raphe nu., ventral	Figure 4.7; D21, D22, W34 to W36
DS	dorsal subiculum	D8, D10 to D18, W18 to W33, WP25–4 and 5
dsc	dorsal spinocerebellar tract	Figure 4.3; D31 and D32, W48 to W51
dsc/oc	dorsal spinocereb./olivocereb. tr.	W46
DT	dorsal terminal nu.	D17, D18, W32, WP25–6
DTg	dorsal tegmental nu.	W37, W38, WP0–9, WP5–9, WP12–12, WP25–8
DTgC	dorsal tegmental nu., central	D23 to D25
DTgP	dorsal tegmental nu., pericent.	D23, D25
dtg	dorsal tegmental tract	Figure 4.6
DTM	dorsal tuberomamillary nu.	D15

Abb'n	Short Title	Chapter Figure Numbers; Atlas Plate Numbers
Fr	frontal cortex	WP19–2, WP25–2
FrA	frontal association cortex	Figure 8.5; D3 and D4, W5 to W12
Fro	frontal bone	WP0–2 and 4, WP5–1 to 4, WP12–1 to 6
Fu	bed nu. st, fusiform part	W21
G		
g7	genu of the facial nerve	Figure 3.3; D26, W39 to W41, WP0–7, WP5–8, WP12–11
GCC	granule cell cluster	Figures 7.3, 12.3 and 12.4; D3 to D7, W14 to 20
GCCM	granule cell cluster magna	Figures 7.3, 7.4, 12.4; D5, W19
Ge5	gelatinous layer caudal Sp5	Figure 4.3; D32, W48 to W51, WP19–11, WP25–11
Gem	gemini hypothalamic nu.	W29
GI	granular insular cortex	Figures 3.4, 8.2, 8.3, 8.5; D6 to D11, W13 to W20, W22, WP19–4, WP25–3
Gi	gigantocellular reticular nu.	Figures 3.3, 4.1, 4.2; D26 to D30, W40 to W47, WP0–7 to 11, WP5–8 to 10, WP12–11 to 14, WP19–8 and 9, WP25–8 and 9
GiA	gigantocellular retic. nu., alpha	D26, D27, W40 to W43, WP12–11
GiV	gigantocellular retic. nu., ventral	D28, D29, W44, W45, WP0–9, WP5–9, WP12–13, WP19–9
Gl	glomerular layer olfactory bulb	Figures 12.3 and12.4; D1 to D3, W1 to W12
GlA	glomerular layer acc. olfact. bulb	Figure 12.3; D3, W7 to W9
GP	globus pallidus	Figures 6.3 and 7.1; D10, D11, W22 to W24, WP0–3, WP5–4, WP12–6, WP19–4, WP25–4
gpet	greater petrosal nerve	WP12–9
Gr	gracile nu.	Figures 4.2 and 4.3; D31 and D32, W46 to W51, WP12–15 and 16, WP19–11, WP25–11
gr	gracile fasciculus	Figures 3.5, 4.3, 13.2, 13.3; W48 to W51, WP12–16, WP19–11
GrA	granule cell layer AOB	Figure 12.3; W7 to W9
GrC	granule cell, cochlear nu.	Figure 4.4; D28, W42, W43
GrCb	granule cell layer of cerebellum	D24, W41, W45
GrDG	granular dentate gyrus	D10 to D18, W26, W27, W29, W30
GrO	granular cell layer olf. bulb	Figures 12.3 and 12.4; D1 to D3, W1 to W12, WP12–1, WP19–1 and 2, WP25–1
gsa	general somatic afferent nuclei	Figure 4.1
H		
Hair Follicle	hair follicle	WP5–1 and 2, WP12–1
HandleM	handle malleus (auditory ossicle)	WP5–7, WP12–10
HardG	Harderian gland	WP0–1, WP5–1
Hb	habenular nuclei	Figures 2.2 and 3.2; WP5–4 to 6, WP12–6 to 8, WP25–4
hbc	habenular commissure	D14, W29
HDB	nu. horiz. limb diagonal band	Figures 6.3, 7.3, 7.4; D6 to D9, W20 to W22, WP5–3, WP12–4 and 5, WP19–4, WP25–4
Hi	hippocampus	Figures 2.2, 3.2, 3.4, 12.2; WP0–2 to 4, WP5–3 and 4, WP12–5 to 7, WP19–4 to 6, WP25–3 to 5
hif	hippocampal fissure	D9 to D11, W26 to W30, WP25–4 and 5
hs	hypothalamic sulcus	Figure 6.4; WP0–4 and 5, WP5–5 and 6, WP12–6 to 8, WP19–5
HSCD	horizontal semicircular duct	WP0–8 to 10, WP5–9, WP12–12 and 13
HStap	handle of stapes (auditory ossicle)	WP12–10
Hy	hypothalamus	Figures 2.2, 3.2, 3.3, 6.1,12.2
I		
I	intercalated nuclei of amygdala	Figure 7.2; D10, D11, W22 to W25
I8	interstitial nu. of 8th nerve	Figure 4.4; D26, D27, W40, W42, WP12–11
I9/10Gn	inferior ganglia of 9n and 10n	WP0–8
ia	internal arcuate fibres	Figure 4.3; W48 to W50
IAD	interanterodorsal thalamic nu.	D11, W24
iala	inferior alveolar artery	WP5–5, WP12–7 and 8

Abb'n	Short Title	Chapter Figure Numbers; Atlas Plate Numbers
ialn	inferior alveolar nerve	WP12-6, WP12-8
IAM	interanteromedial thalamic nu.	Figure 6.1; D11, W25
IAud	internal auditory meatus	WP0-7 and 8, WP5-7
iaud	internal auditory artery	WP12-10 and 11
IB	interstitial nu. of the medulla	W50 and W51
ib (ζ)	interbrachial (ζ) sulcus of cortex	Figures 2.4, 2.8, 8.1
IC	inferior colliculus	Figures 2.2, 3.2, 3.3, 5.1; W38, WP0-9 to 11, WP5-9 to 11, WP12-12 to 14, WP19-8
ic	internal capsule	Figures 2.2, 3.4, 3.5, 6.1 to 6.4, 7.1, 7.2, 7.4, 8.3, 12.2; D5 to D13, W16 to W26, W28, WP5-4, WP12-5 to 7, WP19-3 to 5, WP25-3 and 4
icp	inferior cerebellar peduncle	Figures 4.1, 4.3, 4.4, 5.2; D25 to D30, W40 to W45, WP0-8, WP5-7 and 8, WP12-11 and 12, WP19-8 and 9, WP25-8 and 9
ictd	internal carotid artery	WP0-4 to 7, WP5-4 and 5, WP12-6 to 8
ICx	intermediate zone of cortex	WP5-3, WP12-3 to 7, WP19-2 to 6, WP25-2 to 6
ID	intradecussational nu. sup. Cb ped.	D20
IEn	intermediate endopiriform nu.	W12 to W15, W17, W18, WP12-5
IF	interfascicular nu.	D17, W30, W31, WP5-7, WP12-10, WP19-7
IG	indusium griseum	Figures 7.3 and 8.2; D4 to D7, W15 to W19, WP25-3
IGL	intergeniculate leaflet	D13, D14
ijugv	internal jugular vein	WP0-8 and 9, WP5-9, WP12-13, WP12-15
IL	infralimbic cortex	W13, W14, WP19-3
ILL	intermed nu. of the lat. lemniscus	D21 to D23, W36, W37, WP5-9, WP12-11, WP25-7
IM	intercalated amygdaloid nu., main	Figure 7.1
IMD	intermediodorsal thalamic nu.	D12 to D14, W26 to W28
IML	intermediolateral cell column	Figures 6.6 and 13.2
IMM	intermediomedial nu. of spinal cord	Figure 13.2
InC	interstitial nu. of Cajal	D16 to D20, W30, W31, WP12-11, WP19-7
Inc	incus (auditory ossicle)	WP0-7, WP5-8, WP12-10
InCSh	interstitial nu. of Cajal, shell	D19, W30, W31
infa	infraorbital artery	WP0-3, WP5-1 to 3, WP12-1 to 5
InfR	infundibular recess of 3V	W25, WP0-4
InfS	infundibular stem	Figures 3.2 and 6.3; W27, WP12-7
InG	intermediate grey layer SC	D16 to D21, W31 to W36, WP19-6 to 8, WP25-6 to 8
InM	intermedius nu. of the medulla	D30, W46, W47
InPl	inner plexiform layer of retina	WP0-2
Ins	insular cortex	WP5-2, WP12-4 to 6
Int	interposed deep cerebellar nu.	Figure 5.2; WP12-12, WP19-9, WP25-9
IntA	interposed Cb nu., ant.	Figure 5.2; D26, D27, W40 to W42
IntDL	interposed Cb nu., dorsolat. hump	D26, D27, W41 to W43
IntDM	interposed Cb nu., dorsomed. crest	Figure 5.2; W41 to W43
IntP	interposed Cb nu., posterior	D28, W43
IntZ	intermediate zone of SpC	Figure 2.1, WP5-12, WP19-11
InWh	intermediate white layer SC	D16 to D21, W31 to W35, WP25-6
IO	inferior olivary nu.	Figures 3.3, 4.2, 5.3; WP5-10, WP12-13 to 15, WP19-10
IOA	inferior olivary nu., subnu. A	Figures 5.3 and 5.4; D29, D30
IOB	inferior olivary nu., subnu. B	Figures 5.3 and 5.4; D29
IOb	inferior oblique muscle	WP0-3, WP5-2, WP12-2, WP12-4
IOBe	inferior olivary nu., subnu. beta	Figures 5.3 and 5.4; D29, W46
IOC	inferior olivary nu., subnu. C	Figures 5.3 and 5.4; D30
IOD	inferior olivary nu., dorsal nu.	Figures 5.3 and 5.4; D28 to D30, W44 to W47, WP25-10
IODM	inferior olivary nu., dorsomed.	Figures 5.3 and 5.4
IOK	inferior olivary nu., cap of Kooy	Figures 5.3 and 5.4; D30, W47
iom	inferior olivary neuron migration	WP0-8, WP5-10, WP12-13 to 15, WP19-10, WP25-10
IOPr	inferior olive, principal nu.	Figures 5.3 and 5.4; D28, D29, W45, W47, WP25-10

Abb'n	Short Title	Chapter Figure Numbers; Atlas Plate Numbers
IOV	inferior olive, ventral nu.	Figure 5.4; W44 to W47, WP25–10
IP	interpeduncular nu.	Figures 4.6 and 6.1; WP0–6 and 7, WP25–7
IPA	interpeduncular nu., apical subnu.	D20, W33
IPAC	interstitial nu. post. limb ac	D6 to D10, W19, W21, W22, WP12–5, WP19–4, WP25–4
IPACL	interstit. nu. post. limb ac, lateral	W20
IPACM	interstit. nu. post. limb ac, medial	W20
IPC	interpeduncular nu., caudal	D20, W32, W33
IPD	interpeduncular nu., dorsal	D18, D19
ipets	inferior petrosal sinus	WP0–9
ipf	interpeduncular fossa	D18, W33, WP0–7
IPit	intermediate lobe of the pituitary	WP0–4, WP5–5
IPL	interpeduncular nu., lateral	D18, D20, W31 to W33
IPl	internal plexiform layer olfact. bulb	Figures 12.3 and 12.4; D1 and D2, W1 to W5, W11, WP25–1
IPM	iridopupillary membrane	WP12–3
IPR	interpeduncular nu., rostral	W31, WP5–7, WP12–10
IPV	interpeduncular nu., ventral	D18, D19
IRec	inferior rectus muscle	WP0–3, WP5–3, WP12–2 to 4
Iris	iris (of eye)	WP5–2, WP12–2
IRt	intermediate reticular nu.	Figures 4.1, 4.2, 4.3, 5.3; D28 to D32, W42 to W51, WP0–8 to 10, WP5–9 to 12, WP12–12 to 15, WP19–9 and 10, WP25–9 and 10
IRtA	intermediate reticular nu., alpha	D25 to D27, W40, W41, WP5–7 and 8, WP12–11
IS	inferior salivatory nu.	Figure 4.1
Is	Isthmus	Figure 3.2; WP0–10, WP5–9, WP12–12
is	isthmic neuromere	Figure 3.2
IsoCx	isocortex	Figures 2.2, 3.2, 12.2
isRt	isthmic reticular formation	D21, WP19–8, WP25–8
IVF	interventricular foramen	W23, WP0–3, WP5–3 and 4, WP19–4, WP25–4
J		
j	jugular sulcus	Figures 2.4, 2.5, 2.8
JPLH	juxtaparaventricular lateral hypoth.	D11
JugF	jugular foramen	WP12–13
K		
KF	Kölliker-Fuse nu.	Figure 4.5; D22 to D24, W37 to W39, WP19–8, WP25–8
L		
λ	lambdoid sulcus of cerebral cx	Figures 2.4, 2.5, 2.8, 8.1
LA	lateroanterior hypothal. nu.	Figure 6.4; D10, D11, W23, WP12–6, WP19–5
La	lateral amygdaloid nu.	Figure 7.2; W24, WP12–6, WP19–5 and 6, WP25–5
LAcbSh	lateral accumbens shell	D5, W16 to W19
LaD	lateral amygdaloid nu., dorsal	W25, W26
LaDL	lateral amygdaloid nu., dorsolateral	Figures 7.1 and 7.2; D12, D13
LaP	lateral amygdaloid nu., posterior	Figure 7.2
Lat	lateral deep cerebellar nu.	Figure 5.2; D27, D28, W40 to W43, WP12–12, WP19–9, WP25–9
LatH	lateral horn of spinal cord	Figure 2.1
LatPC	lateral deep Cb nu., parvicellular	D27, D28, W41 to W43
LaV	lateral amygdaloid nu., ventral	W25, W26
LaVL	lateral amygdaloid nu., ventrolateral	Figure 7.1; D12 to D14
LaVM	lateral amygdaloid nu., ventromed.	Figure 7.1
LC	locus coeruleus	Figures 4.5 and 12.1; D24, D25, W38, W39, WP5–8, WP12–11 and 12
lcs	lateral corticospinal tract	Figure 13.3

Abb'n	Short Title	Chapter Figure Numbers; Atlas Plate Numbers
LD	laterodorsal thal. nu.	Figures 6.1 and 6.2; D12, WP5–6; WP25–4
LDDM	laterodorsal thal. nu., dorsomed.	W25 to W27
LDTg	laterodorsal tegmental nu.	Figure 4.5; D24, D25, W36 to W38, WP12–12, WP25–8
LDTgD	laterodorsal tegmental nu., dorsal	D23
LDTgV	laterodorsal tegmental nu., ventral	D23, D24, W36, W37
LDVL	laterodorsal thalamic nu., ventrolat.	W25 to W27
Lens	lens	Figure 3.3; WP0–2, WP5–2, WP12–2 and 3
LEnt	lateral entorhinal cortex	Figure 8.2; D16 to D20, W27 to W34, WP25–6
lf	lateral fissure of cerebellum	WP0–9 and 10, WP5–8 and 9
lfp	longitudinal fasciculus pons	D22, D23, W34 to W37, WP19–7, WP25–7
lfu	lateral funiculus of spinal cord	Figures 2.1 and 13.3; WP5–12, WP12–16, WP19–11, WP25–11
lge	lateral ganglionic eminence	Figure 3.2; WP0–3
LH	lateral hypothalamic area	Figures 6.3 and 6.4; D10, WP0–4, WP12–6
LHb	lateral habenular nu.	Figures 6.1 and 6.2; D12, D14, W27 to W29, WP5–6, WP12–7 and 8, WP19–5, WP25–4
LHbL	lateral habenular nu., lateral	D13
LHbM	lateral habenular nu., medial	D13
Li	linear nu. of the medulla	W44, W45
LL	nuclei of lateral lemniscus	Fig11.1; WP0–10
ll	lateral lemniscus	Figures 4.4 and 4.7; D21 to D23, W34 to W38, WP5–9, WP12–11 to 13, WP19–8, WP25–7 and 8
LM	lateral mamillary nu.	Figure 6.3; D16, W29, WP0–5 and 6, WP5–6, WP12–9
LMol	lacunosum moleculare layer	D14, D16 to D19, W27, W31, W32
LO	lateral orbital cortex	Figures 8.2 and 12.3; D3 and D4, W5 to W9
lo	lateral olfactory tract	Figures 12.1, 12.3, 12.4; D1 to D9, W8 to W21, WP5–2 and 3, WP12–3 to 5, WP19–1 to 4, WP25–1 to 4
LOT	nu. lateral olfactory tract	Figures 3.4 and 12.1; D10, D11, W21
LOT1	nu. lateral olfactory tract, layer 1	D10, D11, W21
LOT2	nu. lateral olfactory tract, layer 2	D10, D11, W21
LOT3	nu. lateral olfactory tract, layer 3	D10, D11, W21
LP	lateral posterior thalamic nu.	Figures 6.1, 6.2, 6.5; D13, D14, WP12–7 to 9, WP19–6, WP25–6
LPAG	lateral periaqueductal grey	Figures 4.6 and 4.7; D18 to D23, W31 to W36, WP5–8 and 9, WP12–11 to 15, WP19–7 to 9, WP25–6 to 8
LPB	lateral parabrachial nu.	Figures 4.5 and 4.7; D25, W37 to W39, WP5–8, WP12–11 and 12, WP19–8, WP25–8
LPBC	lateral parabrachial nu., central	D24
LPBCr	lateral parabrachial nu., crescent	D24
LPBD	lateral parabrachial nu., dorsal	D24
LPBE	lateral parabrachial nu., external	D24, W38
LPBI	lateral parabrachial nu., internal	D24
LPGi	lateral paragigantocellular nu.	D26, D28, D29, W43 to W45, WP0–7 to 11, WP5–9 to 10, WP12–11 to 14, WP19–9, WP25–8 to 9
LPGiA	lateral paragigantocell. nu., alpha	D27, W40 to W42
LPGiE	lateral paragigantocell. nu., extern.	W41 to W44
LPL	lateral post thal. nu., lateral	D15, D16, WP19–5, WP25–5
LPLC	lateral post thal. nu., laterocaud.	W30
LPLR	lateral post thal, nu., laterorost.	W28, W29
LPM	lateral post thal. nu., medial	D15, D16, WP19–5, WP25–5
LPMC	lateral post thal. nu., mediocaud.	D17, W30, W31
LPMR	lateral post thal. nu., mediorost.	W28, W29
LPO	lateral preoptic area	Figures 6.3, 7.3, 12.1; D7 to D9, W21 to W23, WP5–3 and 4, WP12–5, WP19–4, WP25–4
LPS	levator palpebrae super. muscle	WP0–2 to 4, WP5–2 and 3
lptg	nerve to lat. pterygoid muscle	WP0–5, WP12–8

Abb'n	Short Title	Chapter Figure Numbers; Atlas Plate Numbers
MdD	medullary reticular nu., dorsal	Figures 4.2, 4.3, 5.3; D31 and D32, W48 to W51, WP5–12, WP12–15, WP19–10 and 11, WP25–10
MDL	mediodorsal thalamic nu., lat	D12, D13, W25 to W28
MDM	mediodorsal thalamic nu., med	D12, D13, W26 to W28
MdV	medullary reticular nu., ventral	Figures 4.2, 4.3, 5.3; D31 and D32, W48 to W51, WP5–12, WP12–15, WP19–10, WP25–10
ME	median eminence	Figure 6.6; D14, W27, WP0–4, WP5–5, WP19–6, WP25–6
Me	medial amygdaloid nu.	Figures 7.1 and 12.1
Me5	mesencephalic trigeminal nu.	Figure 4.1; D17 to D22, W31 to W37, WP0–9, WP5–9, WP19–8, WP25–7 and 8
me5	mesencephalic trigeminal tract	D18, D19, D22, W34, W35, W38, WP25–7 and 8
MeA	medial amygdaloid nu., anterior	Figure 7.2; D10, D11, W22 to W24
MeAD	medial amyg. nu., anterodorsal	D12, D13, WP12–6, WP19–5, WP25–5
MeAV	medial amyg. nu., anteroventral	D12, D13, WP12–6, WP19–5, WP25–5
Med	medial deep cerebellar nu.	Figure 5.2; D27, D28, W40 to W43, WP19–9, WP25–9
MEE	median eminence, external layer	W25, W26
MEI	median eminence, internal layer	W25, W26
MEnt	medial entorhinal cortex	Figure 8.2; D19, D20, W29 to W34, WP25–6
MeP	medial amyg. nu., posterior	Figure 7.2; W25, WP19–6
MePD	medial amyg. nu., posterodorsal	D14, WP12–7
MePV	medial amyg. nu., posteroventral	D14, WP12–7
mes	mesencephalic neuromere	Figure 3.2
Mesen	mesencephalon	Figures 3.2 and 3.3; WP0–7 to 11, WP5–7 to 11, WP12–10 to 15, WP19–7 to 9, WP25–6 to 9
Meten	metencephalon	Figure 3.3; WP0–6 to 8, 10 and 11, WP5–7 to 9, WP12–9 to 14, WP19–7 to 9, WP25–7 to 10
MeTg	mesencephalic tegmentum	Figure 2.2
mf	medial frontal sulcus	Figure 8.5; W6 to W17
mfb	medial forebrain bundle	Figures 6.3 and 6.6; D5 to D14, W23 to W25, W27, WP0–4 to 6, WP5–3 to 6, WP12–6 to 9, WP19–4 to 6, WP25–4 to 6
MG	medial geniculate nu.	Figures 4.6, 6.1, 6.2, 6.5, 11.1; WP5–7, WP12–9, WP19–6 and 7, WP25–6
MGD	medial geniculate nu., dorsal	Figure 6.2; D16 to D18, W30 to W32
mge	medial ganglionic eminence	Figure 3.2; WP0–3
MGM	medial geniculate nu., medial	Figure 6.2; D16 to D18, W32
MGV	medial geniculate nu., ventral	Figure 6.2; D16 to D18, W30 to W32
MHb	medial habenular nu.	Figures 6.1 and 6.2; D12 to D14, W28, W29, WP5–6, WP12–7 and 8, WP19–5, WP25–4
Mi	mitral cell layer of olfactory bulb	Figures 12.3 and 12.4; D1 to D4, W1 to W11, WP12–1 and 2, WP19–1 and 2, WP25–1 and 2
MiA	mitral cell layer accessory olf. bulb	D3, W7 to W9
MiTg	microcellular tegmental nu.	Figure 4.7; D20 to D23, W34, W35, WP5–9, WP12–12, WP19–8
ML	medial mamillary nu., lateral	Figure 6.3; D16, W29, WP5–6, WP12–9
ml	medial lemniscus	Figures 4.1, 4.4, 4.6; D13 to D29, W27 to W46, WP12–8 and 9, WP19–6 and 7, WP19–10, WP25–5 to 10
mlf	medial longitudinal fasciculus	Figures 4.1, 4.3, 4.6, 4.7; D17 to D32, W32 to W51, WP0–7 to 11, WP5–7 to 10, WP12–11 to 13, WP19–7 to 9, WP25–7 to 11
MM	medial mamillary nu., medial	Figure 6.3; D16, W29, WP0–5, WP0–6, WP5–6, WP12–9
MMn	medial mamillary nu., median	Figure 6.3; D16
MnA	median accessory nu. medulla	Figure 4.3; D32, W49, WP25–11
mne	median neuroepithelium	WP0–6 to 10, WP5–7 to 11, WP12–11 to 13
MnPO	median preoptic nu.	Figure 7.3; D7, W21, W22, WP12–5, WP25–4
MnR	median raphe nu.	Figures 4.7 and 12.1; D21 to D23, W34 to W37, WP0–8, WP5–7, WP12–10 and 11, WP19–7 and 8, WP25–7
MO	medial orbital cortex	Figures 8.2, 8.5, 12.4; D3 and D4, W5 to W12, WP19–2, WP25–2

Abb'n	Short Title	Chapter Figure Numbers; Atlas Plate Numbers
MoCb	molecular layer of the Cb	D26 to D31, W41, W45, WP19–8 and 9, WP25–7 to 9
MoDG	molecular dentate gyrus	D10 to D15, D17 to 19, W29 to W32
MoS	molecular layer of subiculum	W33, W34
mp	mamillary peduncle	D17, W29 to W31
mp (σ)	medial parietal (σ) sulcus	Figure 8.5; W18 to W21
MP	posterior mamillary nu.	Figure 6.3; D17, W30
MPA	medial preoptic area	Figures 6.3, 7.3, 7.4; D7, D8, D10, D11, W21 to W23, WP5–4
MPB	medial parabrachial nu.	Figures 4.5 and 4.7; D24, D25, W36 to W39, WP5–8, WP12–11 and 12, WP19–8
MPBE	medial parabrachial nu. external	Figure 4.5; D24, W38
MPL	medial paralemniscal nu.	W36
MPO	medial preoptic nu.	Figures 6.3 and 7.4; W20, WP5–4, WP12–5 and 6, WP19–4 and 5, WP25–4
MPOL	medial preoptic nu., lateral	D8, D9, W22, W23
MPOM	medial preoptic nu., medial	D8, D9, W22, W23
MPT	medial pretectal nu.	Figure 6.5; D14, D15, W29, W30, WP25–5
MPtg	medial pterygoid muscle	WP0–5 and 6, WP5–5, WP12–7 and 8
MRe	mamillary recess of 3V	W28, W29, WP0–6, WP5–6, WP12–9
MRec	medial rectus muscle	WP0–2, WP5–2 and 3
mRt	mesencephalic reticular formation	D17 to D20, W31 to W33, WP0–7 and 8, WP5–7 and 8, WP12–10 and 11, WP19–7 and 8, WP25–7
MS	medial septal nu.	Figures 7.3 and 7.4; D5, D6, W19, W20, WP12–3 and 4, WP19–3, WP25–3
MSO	medial superior olive	Figures 4.4 and 11.1; D24, D25, W39, W40, WP5–7, WP12–10, WP19–8, WP25–8
MT	medial terminal nu. of optic tract	Figure 4.6; D17, D18, W30
mt	mamillothalamic tract	Figures 6.1, 6.2, 6.3; D12 to D16, W25 to W28, WP12–9, WP25–6
mtg	mamillotegmental tract	W29, WP25–6
MTu	medial tuberal nu.	Figure 6.3; D14, WP12–7 and 8, WP25–6
MTz	medial trapezoid nucleus	Figures 4.4, 11.1; D22, W38 to W40
MVe	medial vestibular nu.	Figures 4.2 and 5.2; W45, WP0–7 to 10, WP5–8 to 11, WP12–11 to 13, WP19–9 and 10, WP25–8 to 10
MVeMC	medial vestibular nu., magnocell.	Figure 4.4; D25 to D29, W40 to W44, WP25–9
MVePC	medial vestibular nu., parvicell.	Figure 4.4; D26 to D29, W39 to W44, WP25–9
MVPO	medioventral periolivary nu.	Figure 4.4; D22, D23, W38 to W40
Mx	matrix region of the medulla	Figure 4.3; D28 to D32, W43, W45 to W48
MxB	maxillary bone	WP0–3 and 4, WP5–2, WP12–1 to 4
my	myelomeres	Figure 3.2
Myelen	myelencephalon	Figure 3.3; WP0–7, 8, 10, 11, WP5–7 to 12, WP12–11 to 16, WP19–9 to 11, WP25–9 to 11
MZMG	marg. zone med. geniculate nu.	D16, W30 to W32
N		
NasB	nasal bone	Figure 12.1
NasC	nasal cavity	Figures 2.4, 2.5, 3.3, 12.1; WP0–1 to 3, WP5–1 to 4, WP12–1 to 6
NasoPhar	nasopharynx	WP0–5 to 7, WP5–5, WP12–8, WP12–10
NasT	nasal turbinate	Figure 12.1
noto	notochord	WP0–9, WP0–10, WP5–10, WP12–8
ns	nigrostriatal tract	D13 to D15, W25 to W27
NSpt	nasal septum	WP0–1 and 2, WP5–1, WP12–1 to 3
ntz	nuclear transitory zone of Cb	WP0–10 and 11, WP5–10
Nv	navicular postolfactory nu.	D4, W14 to W16, WP5–2, WP19–3, WP25–3
O		
OB	olfactory bulb	Figures 2.2, 2.4, 2.5, 2.7, 2.8, 3.2, 8.5, 12.1, 12.2; WP0–1
oc	olivocerebellar tract	Figure 4.2; W43 to W45
Occ	occipital bone	WP0–9 and 10, WP5–6 and 7, WP5–10 and 11, WP12–13 to 15

Abb'n	Short Title	Chapter Figure Numbers; Atlas Plate Numbers
occ	occipital artery	WP12–13
och	optic chiasm	Figures 2.2, 3.2, 6.1, 6.3, 6.6; D8 to D11, W22, W23, WP12–6, WP19–5, WP25–5
OF	optic fibre layer of the retina	WP0–2, WP5–2, WP12–4
olf	olfactory nerve	WP0–1, WP5–1, WP12–1 and 2
olfa	olfactory artery	WP0–1, WP5–1 and 2
olfepith	olfactory epithelium	WP0–1, WP5–1, WP12–1 to 3
ON	olfactory nerve layer olfactory bulb	Figures 12.3 and 12.4; D1 and D2, W1 to W11, WP0–1, WP12–1 and 2, WP19–1 and 2, WP25–1
Op	optic nerve layer superior coll.	Figures 4.7 and 6.5; D16 to D21, W31 to W35, WP25–6
OPC	oval paracentral thalamic nu.	D14, W27, W28
opha	ophthalmic artery	WP0–2 and 3, WP5–3, WP12–4 and 5
ophv	ophthalmic vein	WP0–2 and 3, WP5–2 and 3, WP12–3 to 5
OPT	olivary pretectal nu.	Figures 6.1, 6.2, 6.5; D14, D15, W29, W30, WP19–6
opt	optic tract	Figures 3.5, 6.1, 6.3, 6.4, 7.1, 7.2, 8.3, 8.4; D10 to D15, W24 to W31, WP12–6, WP25–5
OptRe	optic recess of 3V	WP0–3
Or	oriens layer of the hippocampus	D10 to D17, W26, W27, W31
Oral	oral cavity	Figure 12.1; WP5–2, WP12–1 to 3
os	optic stalk	Figure 3.2
OT	nu. of the optic tract	Figure 6.5; D15, D16, W30, W31, WP19–6, WP25–5 and 6
Otic	otic ganglion	WP0–5 and 6, WP5–5 and 6, WP12–8
OV	olfactory ventricle	Figures 12.2 to 12.4; W3 to W13, WP12–2, WP19–1, WP25–1
OvalF	foramen ovale (of skull)	WP0–5
OvalW	oval window	WP12–11
P		
p0	prosomere 0	Figure 3.2
p1	prosomere 1 of diencephalon	Figure 3.2
p1Rt	prosomere 1 reticular formation	Figure 6.2; D16, D17, W29, W30, WP12–10, WP19–6, WP25–6
p2	prosomere 2 of diencephalon	Figure 3.2
p3	prosomere 3 of diencephalon	Figure 3.2
P5	peritrigeminal zone	D22 to D25, W39, WP12–10, WP19–8
P7	perifacial zone	D26, D27, W41 to W43, WP12–11
Pa	paraventricular hypothalamic nu.	Figures 6.2, 6.3, 6.6; D10, WP5–5, WP19–5
pa (β)	parietal (β) sulcus of cerebral cx	Figure 8.5; W19 to W26
pa (ε)	parietal (ε) sulcus of cerebral cx	W20 to W25
Pa4	paratrochlear nu.	D21, W34, W35
Pa5	paratrigeminal nu.	Figure 4.3; D31, W47 and W48
Pa6	paraabducens nu.	W40
PaC	paracommissural nu. post. comm..	D16
PaD	paravent. hypothal. nu., dorsal	Figures 6.3 and 6.4; W25
PaDL	paravent. hypothal. nu., dorsolat.	Figure 6.3
PAG	periaqueductal grey	Figures 2.2, 4.7, 6.2, 6.5, 11.1; D16, D17, W30, WP0–7 to 10, WP5–7 and 9, WP12–9 and 10, WP25–9
PaI	paravent. hypothal. nu., intermed.	Figures 6.3 and 6.4; W25
Pal	palatine bone	WP0–2, WP5–1 to 3, WP12–3 and 4
Palate	palate	Figure 12.1
PaL	paravent. hypothal. nu., lateral	Figures 6.3 and 6.4; W24, W25
PaLM	paravent. hy. nu., lat. magnocell.	D11, D12
PaM	paravent. hy. nu., magnocell.	Figures 6.3 and 6.4; W24, W25
PaMM	paravent. hy. nu., med. magnocell.	D11, D12
PaMP	paravent. hy. nu., med. parvicell.	D11, D12
PaR	pararubral nu.	D19, D20, W31, W32, WP12–11, WP19–7

Abb'n	Short Title	Chapter Figure Numbers; Atlas Plate Numbers
ParP	parietal plate	WP0–5 to 11, WP5–6, WP12–7 to 14
PaS	parasubiculum	Figure 8.2; D19, D20, W31 to W35
PaV	paravent. hypothal. nu., ventral	Figures 6.3 and 6.4
PaVL	paravent. hypothal. nu., ventrolat.	Figure 6.3
PB	parabrachial nuclear complex	WP0–8 to 10, WP5–7, WP5–9, WP12–13
PBG	parabigeminal nu.	D19 to D21, W34, W35, WP19–8, WP25–8
PBP	parabrachial pigmented nu.	D16, D17, D20, W30 to W32
PC	paracentral thalamic nu.	Figures 6.1 and 6.2; D12, D13, W24 to W28
pc	posterior commissure	Figures 3.2, 6.1, 6.2, 6.5; D15 to D17, W30, WP0–7, WP5–7, WP12–9 to 11, WP19–6 and 7, WP25–5
pcer	posterior cerebral artery	WP5–6 and 7, WP12–8 and 9, WP19–7
pcn	precentral fissure of Cb	Figure 5.1
PCo	posterior cortical amygdaloid area	Figure 12.1
PCom	nu. of posterior commissure	D15, W30
pcoma	posterior communicating artery	W23
PCRt	parvicellular reticular nu.	Figures 4.1, 4.2, 4.3; D28 to D30, W42 to W47, WP0–7 to 11, WP5–9 and 10, WP12–12 and 14, WP19–9
PCRtA	parvicellular reticular nu., alpha	D24 to D27, W39 to W41, WP5–7 to 9, WP12–11
pcuf	preculminate fissure	Figure 5.1; D26, W38 to W40
PDPO	posterodorsal preoptic nu.	D9
PDR	posterodorsal raphe nu.	D21, D22
PDTg	posterodorsal tegmental nu.	Figure 4.5; WP5–8, WP25–8
Pe	periventricular hypothalamic nu.	Figures 6.3 and 6.4; D8, D10 to D14, W24 to W27, WP25–5
PeF	perifornical nu.	Figures 6.3 and 6.4; D14, W25 to W27, WP12–7
PeFLH	perifornical part of lat. hypothal.	W26
PePO	periventricular preoptic nu.	Figure 6.3
pf (τ)	prefrontal (τ) sulcus	Figure 8.5; W6 to W12
PF	parafascicular thalamic nu.	D14, D15, W29, WP12–9, WP19–6, WP25–5
PFl	paraflocculus of cerebellum	Figures 2.4, 2.5, 2.7, 2.8, 5.4; D27 to D29, W39, W45, WP25–9
PFlD	paraflocculus, dorsal	Figure 5.1; W40 to W44
PFlV	paraflocculus, ventral	Figure 5.1; W40 to W44
pfs	parafloccular sulcus	W39 to W44
PH	posterior hypothalamic nu.	Figure 6.3; D14, D15, WP5–6, WP12–7 to 9, WP19–6, WP25–6
PHA	posterior hypothalamic area	W29
PHD	posterior hypothalamic area, dorsal	D13, W27, W28, WP25–5 and 6
PHV	posterior hypothalamic area, vent.	W27
Pi	pineal gland	Figures 3.2, 3.3, 6.2, 6.6, 8.5; D15, W30, WP0–5
pica	posterior inferior Cb artery	W40 to W42, W47, W48, WP19–10
PIF	parainterfascicular nu. of VTA	D17, W32
Pig	pigment layer of the eye	WP0–2 to 3, WP5–2 and 3, WP12–2 to 4
PIL	posterior intralaminar thalamic nu.	Figure 6.2; D16 to D18, W30, W31, WP5–7, WP19–6
Pinna	pinna of the ear	Figure 11.1
Pir	piriform cortex	Figures 2.2, 2.8, 3.4, 7.2, 8.1, 8.2, 8.5, 12.1, 12.2, 12.3, 12.4; D4, D16, W12, W13, WP5–2 to 4, WP12–3 to 7, WP19–3 to 6, WP25–2 to 4
Pir1	piriform cortex, layer 1	D4, D5, D11 to D15, W14 to W26, WP19–5, WP25–5
Pir1a	piriform cortex, layer 1a	D6 to D10
Pir1b	piriform cortex, layer 1b	D6 to D10
Pir2	piriform cortex, layer 2	D4 to D15, W14 to W26, WP19–5, WP25–5
Pir3	piriform cortex, layer 3	D4 to D15, W14 to W26, WP19–5
PiRe	pineal recess of 3V	Figures 6.5 and 6.6; D15, W30, WP0–5, WP5–5 and 6, WP19–5
Pk	Purkinje cell layer of Cb	D24 to D31, W41, W45, W48, WP19–8 and 9, WP25–7 to 9
PL	paralemniscal nu.	Figure 4.7; D21 to D23, W35
PLCo	posterolateral cortical amygdala	Figure 7.2; D12 to D15, W26, WP12–7, WP19–6, WP25–5

Abb'n	Short Title	Chapter Figure Numbers; Atlas Plate Numbers
psf	posterior superior fissure of Cb	Figure 5.1; W40, W41, W46
psp	posterior spinal artery	W51
PSph	presphenoid bone	WP0–3 and 4, WP5–2 to 4, WP12–4 to 6
PSphW	presphenoid wing	WP0–3, WP5–3 to 5, WP12–4 to 7
PSTh	parasubthalamic nu.	D15, W27, W28, WP19–6
PT	paratenial thalamic nu.	Figure 6.2; D10, D11, W24, W25, WP12–6, WP19–4
Ptec	pretectum	Figures 2.2 and 3.2; WP0–6, WP0–7, WP5–7, WP12–9 and 10
PTg	pedunculotegmental nu.	D21 to D23, W34 to W36, WP12–11
Ptg	pterygoid process of sphenoid	WP0–4, WP12–6
ptgcn	nerve of the pterygoid canal	WP0–3, WP5–4, WP12–5
ptgpal	pterygopalatine nerve	WP0–4, WP12–6
Pu	putamen	Figures 2.2, 3.4, 6.3, 7.1, 7.2, 8.3, 12.2; D5 to D12, W16 to W24, WP5–3 and 4, WP12–5, WP12–7, WP19–1 to 5, WP25–3 to 5
pv	periventricular fibres	D15
PV	paraventricular thalamic nu.	Figure 3.2; D12, D13, WP19–5
PVA	paraventricular thal. nu., anterior	Figure 6.1, 6.2, 6.3; D10, D11, W24, W25, WP12–6, WP19–4, WP25–4
PVG	periventricular grey	Figure 6.5; D15, W29
PVP	paraventricular thal. nu., posterior	Figure 6.1, 6.2; D14, W26 to W28
Py	pyramidal cell layer hippocampus	D10 to D17, W27, W31
py	pyramidal tract	Figures 3.5, 4.1, 4.2, 4.4, 5.3, 5.4; D21 to D32, W38 to W48
pyd	pyramidal decussation	D32, W48 to W51

R		
R	red nu.	WP0–7, WP5–7 and 8, WP12–10, WP19–7
r0-r8	rhombomeres 0 to 8	Figure 3.2
R1-R10	Rexed's laminae 1 to 10 of SpC	Figures 13.2 to 13.4
r1Rt	r1 reticular formation	WP25–7 and 8
Rad	radiatum layer of hippocampus	D10 to D17, W27, W31, W32
RAmb	retroambiguus nu.	D32, W48 to W51, WP19–11
Rathke	Rathke's pouch	WP0–4
RCh	retrochiasmatic area	Figures 3.2, 6.3, 6.4; D12, W24, WP19–6
RChL	retrochiasmatic area, lateral	Figure 6.3; D12, W24, W25
Re	reuniens thalamic nu.	Figures 6.1, 6.2, 6.4; D11 to D13, W25 to W27, WP19–5, WP25–4 to 6
ReIC	recess of the inferior colliculus	W37
Respepith	respiratory epithelium	WP5–1, WP12–1 and 2
REth	retroethmoid nu.	D16, W30
Retina	retina	Figures 3.3, 6.6; WP0–2 and 3, WP5–1 and 2, WP12–2, WP12–4
rf	rhinal fissure	Figures 2.2, 2.4, 2.5, 2.7, 2.8, 3.4, 7.2, 8.1, 8.3 to 8.5, 12.2; D2 to D18, W4 to W37, WP5–4, WP19–4 and 5, WP25–3 to 5
rfp	roof plate of spinal cord	Figure 3.1
RGn	ganglion cell layer of retina	WP0–2, WP5–2, WP12–2 to 4
Rh	rhomboid thalamic nu.	D13, W25, W26, WP19–5, WP25–5
Rhomb	rhombencephalon	Figure 3.2
RI	rostral interstitial nu.	W29
ri	rhinal incisure	D3, W4 to W13
RIP	raphe interpositus nu.	W38 to W42, WP5–7 and 8, WP12–11
RIs	retroisthmic nu.	WP12–10, WP25–7
RLi	rostral linear nu. (midbrain)	Figure 4.6; D18, D19, W30, WP5–8, WP25–7
RM	retromamillary nu.	Figures 3.2, 6.3, 8.2; D16, W28, WP0–5 and 6, WP5–6, WP12–9
RMC	red nucleus, magnocellular part	Figure 4.6; D17 to D20, W30 to W32, WP12–11
rmd	retromamillary decussation	D16
RMg	raphe magnus nu.	Figure 4.4; D22 to D29, W39, W41, W42, WP0–6 to 8, WP5–8, WP12–10 to 12
RML	retromamillary nu., lateral part	Figure 6.3; W29
RMM	retromamillary nu., medial	Figure 6.3; W29

Abb'n	Short Title	Chapter Figure Numbers; Atlas Plate Numbers
Ro	nu. of Roller	D30, D31, W45 to W47, WP25–10
ROb	raphe obscurus nu.	Figure 4.2; D27 to D31, W43 to W47, WP0–9, WP0–11, WP5–9 to 11, WP12–13 and 15, WP19–9 and 10, WP25–8 to 10
RoundW	round window	WP12–12
RPa	raphe pallidus nu.	Figure 4.2; D25 to D30, W39 to W47, WP0–8 and 10, WP5–9 and 10, WP12–10, WP12–13 and 14, WP25–9 and 10
RPC	red nucleus, parvicellular part	Figures 4.6, 6.2; D16 to D19, W30 to W32
RPF	retroparafascicular nu.	Figure 6.5; D15, D16
RR	retrorubral nu.	W34, W35
RRe	retroreuniens area	W28
RRF	retrorubral field	Figure 4.6; W33, W34, WP0–8
rs	rubrospinal tract	Figure 13.4; D23 to D32, W33 to W47, W50 and W51
RSD	retrosplenial dysgranular cx	Figure 8.2; D12 to D17
RSG	retrosplenial granular cx	D12, WP0–5, WP5–5, WP12–7 and 8, WP19–6, WP25–5
RSGa	retrosplenial granular cx, part a	Figures 8.2, 8.5; D13 to D18, W31 to W38
RSGb	retrosplenial granular cx, part b	Figures 8.2, 8.5; D13 to D18, W32 to W38
Rt	reticular thalamic nu. (prethalamus)	Figures 6.1, 6.2, 6.3, 7.1; D11 to D13, W24 to W29, WP5–5, WP12–7, WP19–5, WP25–4
RtSt	reticulostrial nu.	Figure 7.1; D11, D12, W24 to W26
RtTg	reticulotegmental nu. pons	D20 to D22, W34 to W37, WP25–7
RtTgP	reticulotegmental nu., pericentral	W37
RVL	rostroventrolateral reticular nu.	D28, D29, W44
S		
S	subiculum	Figure 8.2; D9, W18, WP5–5, WP19–5
S1	primary somatosensory cx	Figures 3.4, 8.2; D5 to D9, W13, W14, WP5–2, WP12–4 to 6, WP19–3 and 5, WP25–3 and 4
S1OGn	superior ganglion of 10n	WP0–11
S1FL	prim. somato. cx, forelimb area	Figures 2.8, 8.5; D10, W15 to W19
S1HL	prim. somato. cx, hindlimb area	Figures 2.8, 8.5; D10 to D12, W19 to W25
S1Md	prim. somato. cx, mandible area	Figure 8.5; D10 to D12, W15 to W25, WP19–4
S1My	prim. somato. cx, mystacial area	Figures 2.8, 8.3, 8.5; D10 to D13, W15 to W26, WP19–4
S1Op	prim. somato. cx, ophthalmic area	Figure 8.5; D10, D11, W19 to W25, WP19–4
S1Sh	prim. somato. cx, shoulder area	D11
S1To	prim. somato. cx, tongue area	Figure 8.5
S1Tr	prim. somato. cx, trunk area	Figures 2.8, 8.5; W20 to W25, WP19–4
S2	secondary somatosensory cx	Figures 8.2, 8.3, 8.5; D7 to D13, W18 to W26, WP12–5 to 7, WP19–4 and 5, WP25–4
s5	sensory root of trigeminal nerve	D23 to D25, W37 to W39, WP0–6, WP5–6, WP12–9 and 10
S9/10Gn	superior ganglion of 9n/10n	WP0–9, WP5–9
Sacc	saccule	WP0–8 and 9, WP5–8 and 9, WP12–11 and 12
Sag	sagulum nu.	Figure 4.7; D22, D23, W36, W37
sal	superior alveolar nerve	WP0–3
SC	superior colliculus	Figures 2.2, 2.7, 3.2, 3.3, 5.1, 6.1, 6.2, 8.5; D22, W34, W35, WP0–8 to 11, WP5–8 to 10, WP12–11 to 15, WP19–6, WP19–8 and 9, WP25–5 to 9
scba	superior cerebellar artery	WP0–7, 9 and 10, WP5–6 to 9, WP12–9 to 13, WP19–8
SCGn	superior cervical ganglion	Figure 6.6
SCh	suprachiasmatic nu.	Figures 6.3, 6.6; D9, W23, WP12–6, WP19–5
SChDM	SCh, dorsomedial part	D10 to 11
Schlemm	canal of Schlemm	WP5–2
SChVL	SCh, ventrolateral part	D10 to 11
Sclera	sclera	WP0–1 to 4, WP5–1 to 3, WP12–1 to 4

Abb'n	Short Title	Chapter Figure Numbers; Atlas Plate Numbers
SCO	subcommissural organ	Figure 6.5; D15, D16, W30, WP0–6 and 7, WP5–7, WP12–9 and 11, WP19–7, WP25–5
scp	superior cerebellar peduncle	Figures 4.5, 4.7; D22 to D26, W27 to W42, WP0–9 and 10, WP5–7 to 9, WP12–11 to 13, WP19–6 to 8, WP25–6 to 9
scpd	superior Cb peduncle, decussation	Figure 4.6; D21, W33 to W35
ScTy	scala tympani	Figure 11.1
ScVe	scala vestibuli	Figure 11.1
se	somatic efferent nuclear column	Figure 4.1
sf	secondary fissure of cerebellum	Figure 5.1; W45 to W47
SFi	septofimbrial nu.	Figures 7.3, 7.4; D7 to D9, W21 to W23
SFO	subfornical organ	Figures 7.3, 7.4; D7, W21 to W23
SG	suprageniculate thalamic nu.	Figure 6.2; D16 to D18, W31
SGe	supragenual nu.	W39
SHi	septohippocampal nu.	Figure 7.3; D4 to D7, W16, W17, WP12–4
SHy	septohypothalamic nu.	Figure 7.3; D6 to D9, W19, W20, WP25–4
SI	substantia innominata	WP19–4
SIB	substantia innominata, basal	D6 to D10, W20 to W23
Sim	simplex lobule of cerebellum	Figures 2.4, 5.1; D24, D25, W38 to W43
SimA	simplex lobule of cerebellum, A	W40
SimB	simplex lobule of cerebellum, B	W40
simf	simplex fissure	W39 to W41
sl	sulcus limitans	Figures 3.1 to 3.3, 4.1, 4.2; WP0–7 to 11, WP5–8 to 12, WP12–11 to 15, WP19–9 and 10, WP25–9 and 10
SLu	stratum lucidum, hippocampus	D10 to D13
SM	nucleus of the stria medullaris	D11
sm	stria medullaris	Figures 3.5, 6.2, 6.3; D10 to D13, W23 to W28, WP0–4, WP5–4 and 6, WP12–6 to 8, WP19–4 and 5, WP25–4
SMac	macula of the saccule	WP0–8, WP5–8 and 9, WP12–11
smv	superior medullary velum	W43
SN	substantia nigra	Figure 6.1; WP0–7 and 8, WP5–7 and 8, WP12–11, WP25–7
SNC	substantia nigra, compact part	Figure 4.6; D16, D17, D19, WP12–10, WP19–7
SNCD	subst. nigra, compact, dors tier	D18, W30 to W33
SNCM	subst. nigra, compact, med tier	Figure 4.6; W32
SNL	substantia nigra, lateral part	Figure 4.6; D17 to D19, W30 to W32
SNR	substantia nigra, reticular part	Figure 4.6; D16 to D20, W29 to W33, WP12–10, WP19–7
SO	supraoptic nu.	WP19–5
SOA	supraoptic nu., anterior	Figure 6.3; D10, D11, W23, W24
SOb	superior oblique muscle	WP0–3, WP12–3 and 4
sod	supraoptic decussation	W24 to W26
SOl	superior olive	WP0–6 and 7
Sol	solitary nucleus	Figures 4.1, 4.2, 4.3, 6.6; D27, W42 to W45, WP0–8 to 11, WP5–8 to 12, WP12–11 to 16, WP19–9 to 11, WP25–9 to 11
sol	solitary tract	Figures 3.3, 4.2, 4.3; D28 to D32, W43, W44, W46, W47, WP0–9 to 11, WP5–10 to 12, WP12–13 to 16, WP19–9 to 11, WP25–9 and 10
SolC	solitary nu., commissural part	Figure 4.3; D32, W48 to W51
SolDL	solitary nu., dorsolateral part	Figures 4.2, 4.3; D28, D30 to D32, W46, W47
SolDM	solitary nu., dorsomedial part	D28, D29
SolG	solitary nu., gelatinous part	Figures 4.2, 4.3; D30, D31, W46
SolIM	solitary nu., intermediate part	Figure 4.2; W47
SolL	solitary nu., lateral part	Figures 4.2, 4.3; D29, D30, W44
SolM	solitary nu., medial part	Figures 4.2, 4.3; D28 to D32, W44 to 51
SolV	solitary nu., ventral part	Figures 4.2, 4.3; D28, D31
SolVL	solitary nu., ventrolateral part	Figures 4.2, 4.3; D28 to D32, W44, W46 to W48
SOO	septal olfactory organ	Figure 12.1

Abb'n	Short Title	Chapter Figure Numbers; Atlas Plate Numbers
SOR	supraoptic nu., retrochiasmatic	Figures 6.3, 6.4; D12, W25, W26, WP12-6, WP25-5
Sp5	spinal trigeminal nu.	Figure 4.1
sp5	spinal trigeminal tract	Figures 3.5, 4.1, 4.3, 4.4; D26 to D32, W40 to W51, WP0-7 to 11, WP5-7 to 12, WP12-11 to 16, WP19-8 to 11, WP25-7 to 11
Sp5C	spinal trigeminal nu., caudalis	Figures 4.1, 4.2, 4.3; D32, W47 to W51, WP5-12, WP12-16, WP19-11, WP25-11
Sp5I	spinal trigeminal nu., interpolaris	Figures 3.3, 4.1, 4.2, 4.3, 5.3; D28 to D32, W43 to W46, WP0-9 to 11, WP5-9 to 12, WP12-13 to 16, WP19-9 to 11, WP25-10
Sp5O	spinal trigeminal nu., oralis	Figures 3.3, 4.1, 4.4; D26, D27, W41, W42, WP0-7 and 8, WP5-7 and 8, WP12-10 to 12, WP19-9, WP25-9
SPa	subparaventricular zone hypothal.	D10
spa	sphenopalatine artery	WP0-1
SpC	spinal cord	Figures 2.2, 3.2; WP12-16
SPF	subparafascicular thalamic nu.	D15, WP19-6
SPFPC	subparafascic. thal. nu., parvicell.	W29
SphPal	sphenopalatine ganglion	Figure 6.6; WP0-2 and 3, WP5-3 and 4, WP12-4 to 6
spn	spinal nerve	Figures 2.1, 3.1
SPO	superior paraolivary nu.	Figure 4.4; D23 to D25, W39, W40
Spt	septum	Figures 2.2, 3.2, 6.3, 8.5, 12.2; WP0-2, WP12-4, WP19-3
SPTg	subpeduncular tegmental nu.	D22, W34, W35, WP5-8 and 9, WP12-11
SpVe	spinal vestibular nu.	Figures 4.4, 5.2; D27 to D29, W41 to W45, WP0-8 to 11, WP5-8 to 11, WP12-11 to 14, WP19-9 and 10, WP25-9 and 10
SRec	superior rectus muscle	WP0-2 and 3, WP5-2 and 3, WP12-2 to 5
ssa	special somatic afferent nu. column	Figure 4.1
sss	superior sagittal sinus	WP0-3 to 5, WP5-1 to 5, WP12-2 to 6, WP12-8
st	stria terminalis	Figures 3.5, 7.1, 7.2; D10 to D13, W22 to W26, W28, WP12-6 and 7, WP19-5, WP25-4 and 5
StapM	stapedius muscle	WP12-12
STh	subthalamic nu.	D14, D15, W27, W28, WP5-6, WP12-8 and 9, WP19-6, WP25-5 and 6
StHy	striohypothalamic nu.	D8 to D10, W23, W24, WP19-4, WP25-4
ST	bed nu. stria terminalis	Figures 3.3, 12.1; WP0-3 and 4, WP5-4, WP12-5 and 6, WP19-4
STIA	bed nu. st, intraamygdaloid div.	D12 to D14, W24 to W26, WP25-5
STLD	bed nu. st, lat. div., dorsal part	D6, W19, W20, WP25-4
STLJ	bed nu. st, lat. div., juxtacapsular	D7, D8
STLP	bed nu. st, lat. div., posterior part	D7 to D9, W21, W22
STLV	bed nu. st, lat. div., ventral part	D7 to D9, W20 to W22, WP25-4
StM	sterno(cleido)mastoid muscle	WP0-8, WP12-12
STMA	bed nu. st, med. div., ant. part	D6, W20, W21, WP25-4
STMD	bed nu. st, med. div., dorsal part	D7
STMP	bed nu. st, med. div., post. part	D8, D9
STMPI	bed nu. st, med. div., posteroint.	W23
STMPL	bed nu. st, med. div., posterolat.	D10, W23
STMPM	bed nu. st, med. div., posteromed.	D10, W23
STMV	bed nu. st, med. div., ventral part	Figure 7.3; D6 to D9, W20, W22, WP25-4
STr	subiculum, transition area	W34
str	superior thalamic radiation	D13 to D15
Su3	supraoculomotor PAG	Figure 4.6; D19, D20, W32, W33
Su3C	supraoculomotor cap	Figure 4.6; D19, D20, W32, W33
Su5	supratrigeminal nu.	D23, D24, W38, W39, WP19-8, WP25-7 and 8
Sub	submedius thalamic nu.	Figure 6.2; D12, W25, W26
SubB	subbrachial nu.	Figure 4.7; D19, W32 to W34, WP12-11, WP25-8
SubCA	subcoeruleus nu., alpha part	Figure 4.5
SubCD	subcoeruleus nu., dorsal part	Figure 4.5; D23, D24, W38
SubCV	subcoeruleus nu., ventral part	D23, D24, W38

Abb'n	Short Title	Chapter Figure Numbers; Atlas Plate Numbers
SubG	subgeniculate nu.	D14 to D16, W27 to W29, WP25–5
SubI	subincertal nu.	D12, D13, W26, WP12–7, WP25–5
SubV	subventricular layer of cortex etc	WP5–1 to 3, WP12–3 to 6, WP19–2 to 5, WP25–2 to 6
SuG	superficial grey layer sup. coll.	Figures 4.7, 6.5; D16 to D21, W31 to W36, WP19–6 to 9, WP25–6 and 7
SuS	superior salivatory nu.	Figures 4.1, 6.6; D27
SuVe	superior vestibular nu.	Figure 5.2; D24 to D26, W39, WP5–7, WP12–11, WP19–8, WP25–8
Symp	sympathetic trunk	Figure 2.1
T		
Te	terete hypothalamic nu.	D14, W28
te (δ)	temporal (δ) sulcus of cerebral cx	Figures 8.4, 8.5; W26 to W35
TeA	temporal association cortex	Figure 8.2; D14 to D19
TeD	dorsal temporal cortex	Figures 8.4, 8.5; W26 to W38, WP25–6
Telen	telencephalon	Figure 3.3; WP0–1 to 5, WP5–1 to 5, WP12–2 to 8, WP19–1 to 6, WP25–1 to 6
temp	temporal nerve	WP0–5
TempM	temporalis muscle	WP0–4 and 5, WP5–4, WP12–5 to 10
TempP	temporal bone, petrous part	WP0–7 to 11, WP5–7 to 10, WP12–9 to 14
TempS	temporal bone, squamous part	WP0–4 and 5, WP0–8, WP5–8 and 10, WP12–6, WP12–10 and 11
term	terminal nerve	WP0–1, WP5–1, WP19–2
TeV	ventral temporal cortex	Figure 8.5; W23 to W37, WP25–6
tfp	transverse fibres of pons	D19, D21, W33 to W37, WP19–7
TG	tectal grey	Figure 6.5; D15, W30, W31, WP25–5
Th	thalamus	Figures 2.2, 3.2, 3.3, 6.1, 6.3, 12.2; WP0–5, WP5–5 and 6, WP12–7 and 8
trg	germinal trigone of fourth ventricle	WP0–8 to 11, WP5–8 to 10, WP12–12 to 14, WP19–9
TrLL	triangular nu., lateral lemniscus	W36
trs	transverse venous sinus	Figures 2.4, 2.5, 2.7, 2.8; WP0–6 to 11, WP5–7 to 10, WP12–9 to 14
TS	triangular septal nu.	Figure 7.4; D8, D9, W21 to W23
ts	tectospinal tract	D29, W32, W33, W36, W39 to W42, W44, W46 to W51
TT	tensor tympani muscle	WP0–7, WP5–7, WP12–10
tth	trigeminothalamic tract	D26, W33, W37
Tu	olfactory tubercle	Figures 2.2, 6.3, 7.3, 12.1 to 12.4; D3 to D8, W14 to W20, WP12–3 and 4, WP19–3, WP25–3 and 4
Tub	tuberal part of hypothalamus	Figure 3.2
TubPit	tuberal part of pituitary gland	Figure 6.6
TuLH	tuberal region of lateral hypothal.	Figure 6.3; D12, D13, W25 to W27, WP25–5
TyBu	tympanic bulla	WP0–6 to 8, WP5–7 and 8, WP12–10 to 12
TyC	tympanic cavity	WP0–7, WP5–7 and 8, WP12–10 and 11
TyM	tympanic membrane	Figure 11.1; WP12–10
Tz	nucleus of trapezoid body	D22 to D24, WP12–11, WP19–8, WP25–8
tz	trapezoid body	Figure 4.4; D23 to D28, W38 to W41
tzd	trapezoid body decussation	W38 to W40
U		
UMac	macula of the utricle	WP5–8, WP12–11
un	uncinate fasciculus of the Cb	D25, W39
und	uncinate decussation	W41 to W43
Utr	utricle	WP0–8 and 9, WP5–8 and 9, WP12–11 and 12
V		
V1	primary visual cortex	Figures 2.8, 8.2, 8.4, 8.5; D13 to D20, W26 to W38, WP12–7 and 8, WP19–6, WP25–5 and 6
V2L	secondary visual cx, lateral area	Figures 8.2, 8.4, 8.5; D13 to D20, W26 to W38, WP19–6, WP25–5 and 6
V2M	secondary visual cx, medial area	Figure 8.2; D13 to D19

Abb'n	Short Title	Chapter Figure Numbers; Atlas Plate Numbers
V3V	ventral third ventricle	Figure 6.6; D10, W26 to W28, WP0–4 and 5, WP5–4 to 6, WP12–6 to 8, WP19–6, WP25–4 to 6
VA	ventral anterior thalamic nu.	Figure 6.2; D12, W25, W26
va	visceral afferent nuclear column	Figure 4.1
vaf	ventral amygdalofugal pathway	WP5–5, WP19–5
VC	ventral cochlear nu.	Figure 4.1; WP0–7 to 9, WP5–7, WP25–8 and 9
VCA	ventral cochlear nu., anterior part	Figures 4.4, 5.2, 11.1; D25 to D27, W40, W41, WP12–11
VCAGr	ventral cochlear nu., granule cell	D25, D26
VCCap	ventral cochlear nu., capsular part	D28
VCl	ventral part of claustrum	D6 to D9, W18 to W22
VCP	ventral cochlear nu., posterior part	Figures 4.4, 11.1; D26 to D28, W41 to W43, WP5–8
vcs	ventral corticospinal tract	Figure 13.3
VDB	nu. vertical limb diagonal band	Figures 2.2, 7.3, 7.4; D5, D6, W18 to W20, WP0–2, WP12–3 and 4
Ve	vestibular nuclei or neuroepithel.	Figure 4.1; WP0–7 to 10, WP5–8 and 9, WP12–11 and 12
ve	visceral efferent nuclear column	Figure 4.1
VeCb	vestibulocerebellar nu.	Figure 5.2; D27, W40 to W43
VeGn	vestibular nerve ganglion	WP0–7 and 8, WP5–8, WP12–11
veme	vestibulomesencephalic tract	D25, D27, W40
VEn	ventral endopiriform nucleus	Figure 7.2; D5 to D11, W19 to W26, WP19–3 to 5, WP25–3 to 5
Vent	ventricular space of the eye	WP0–2, WP5–2 and 3, WP12–2 to 4
vert	vertebral artery	W48 to W50, WP0–11, WP5–11, WP12–15
Vest	vestibule of the inner ear	Figure 11.1
vfu	ventral funiculus of spinal cord	Figures 2.1, 13.3; WP5–12, WP12–16, WP19–11, WP25–11
VH	ventral horn of spinal cord	Figure 2.1; WP5–12, WP12–16, WP19–11, WP25–11
vhc	ventral hippocampal commissure	Figures 2.2, 3.4, 7.3, 7.4; D8, D9, W22 to W25
Vitr	vitreous cavity of eye	WP5–2, WP12–2 to 4
VL	ventrolateral thalamic nu.	Figures 6.1, 6.2; D12, D13, W25, W26
vl	ventral lamina of IOPr	Figure 5.4
VLH	ventrolateral hypothalamic nu.	Figure 6.3; D11, W23
VLL	ventral nu. of lateral lemniscus	D21 to D23, W35 to W38, WP0–6 and 7, WP12–10, WP25–7
VLO	ventrolateral orbital cortex	W8, W9
VLPAG	ventrolateral periaqueductal grey	Figures 4.6, 4.7; D19 to D23, W32 to W36, WP5–8 to 10, WP12–12 and 13, WP19–7 to 9, WP25–7 and 8
VLPO	ventrolateral preoptic nu.	Figure 6.3; D8, D9, W22
VM	ventromedial thalamic nu.	Figures 6.1, 6.2; D12 to D14, W25 to W28, WP5–6, WP12–7, WP19–5 and 6, WP25–5
VMH	ventromedial hypothalamic nu.	Figures 6.3, 6.6; D13, WP0–4, WP5–5, WP12–7 and 8
VMHC	ventromedial hypothal nu., central	D13
VMHDM	ventromedial hy. nu., dorsomed.	Figure 6.3; D13, W26, WP19–6, WP25–5
VMHVL	ventromedial hy. nu., ventrolat.	Figure 6.3; D13, D14, W26, WP19–6, WP25–5
vmnf	ventral median fissure	Figure 2.1
VMPAG	ventromedial periaqueductal grey	WP19–7, WP25–7 and 8
VMPO	ventromedial preoptic nu.	Figure 6.3; D7, D8, W22
vn	vomeronasal nerve layer of AOB	W6 to W9
VNO	vomeronasal organ	Figure 12.1; WP12–1
vno	vomeronasal nerve	WP12–1
VO	ventral orbital cortex	Figures 8.2, 8.5; D3 and D4, W5 to W9, WP19–2, WP25–2
VOLT	vascular organ of lamina terminalis	D7, W21
Vom	vomer (bone)	Figure 12.1; WP0–1, WP5–1, WP12–1 to 3
VP	ventral pallidum	Figures 7.3, 7.4, 12.4; D3 to D9, W14 to W22, WP5–3, WP12–3 to 5, WP19–4, WP25–3 and 4
VPL	ventral posterolateral thalamic nu.	Figures 6.1, 6.2; D13 to D15, W25 to W29
VPL/M	ventral posterior thalamic nu.	WP5–6, WP12–8, WP19–5, WP25–5

Abb'n	Short Title	Chapter Figure Numbers; Atlas Plate Numbers
VPM	ventral posteromedial thalamic nu.	Figures 6.1, 6.2; D13 to D15, W26 to W29
VPO	ventral periolivary nu.	D24, D25, WP5–7, WP25–8
VPPC	vent. post. nu. thalamus, parvicell.	D14, D15, W27, W28, WP12–8
vr	ventral roots of spinal cord	Figures 2.1 and 3.1
vra	ventral ramus of spinal nerve	Figure 2.1
VRe	ventral reuniens thalamic nu.	Figures 6.1 and 6.4; D11 to D13, W26
VS	ventral subiculum	D16, D19, D20, W27 to W33
vsc	ventral spinocerebellar tract	D22, D23, D25 to D32, W38 to W48, W50, W51, WP25–9
VTA	ventral tegmental area	Figure 4.6; D18, D19, W33, WP0–6 and 7, WP5–7, WP25–7
VTAR	ventral tegmental area, rostral part	D16, W30, WP12–10, WP19–7
VTg	ventral tegmental nu.	D23, W36, W37, WP5–8
vtgd	ventral tegmental decussation	D19, D20
VTM	ventral tuberomamillary nu.	W28
VTT	ventral tenia tecta	W13
X		
X	nucleus X	D28, D29, W42 to W44
Xi	xiphoid thalamic nu.	D12
Y		
Y	nucleus Y	W41
Z		
ZI	zona incerta	Figures 6.1, 6.2 to 6.4, 7.1; D12, WP0–5, WP5–5 and 6, WP12–7 to 9, WP19–6, WP25–5 and 6
ZIC	zona incerta, caudal part	D16, W29, W30
ZID	zona incerta, dorsal part	D13 to D15, W26 to W28
ZIR	zona incerta, rostral part	W25, WP19–5, WP25–4
ZIV	zona incerta, ventral part	D13 to D15, W26 to W28
ZL	zona limitans	Figure 3.2
Zo	zona layer of the superior colliculus	D16, D18 to D21, W31 to W36, WP19–9, WP25–7

Glossary

Acetylcholine A neurotransmitter (often abbreviated as ACh) formed from the esterification of choline and acetyl CoA. ACh is used by cholinergic pathways in the brain and at the neuromuscular junction.

Acetylcholinesterase (AChE) The enzyme that catalyses the hydrolysis of the neurotransmitter acetylcholine. The enzyme is present in high concentrations in parts of the brain where acetylcholine is used as a neurotransmitter and analysis of its histochemical distribution is useful for distinguishing functionally significant brain regions.

Action potential A moving zone of altered membrane potential that travels actively and without decrement down an axon, carrying information to the axon terminal.

Afferent An axon that projects into the central nervous system or a part of the central nervous system, implying a sensory or input function.

Allocortex The part of the cerebral cortex or pallium that has three to five layers at some stage during development. Allocortex includes the hippocampal formation (also known as archicortex) and olfactory cortex (also known as paleocortex).

Ameridelphian A marsupial from North or South America.

Amygdala An almond-shaped group of neurons (from the Latin for *almond*) located in the temporal region of the brain. The amygdala is part of the limbic system and is concerned with emotional responses.

Anterior commissure One of the fibre bundles connecting the two cerebral hemispheres: it is the major commissure in marsupials and monotremes. The anterior commissure develops in the rostral wall of the diencephalic vesicle.

Antidiuretic hormone (ADH) A hormone released from the posterior pituitary (neurohypophysis) to control reabsorption of water by the kidneys.

Arachnoid Literally meaning *spider-like* (from a reference to Arachne, a skilled weaver of Colophon, who challenged Athene to a weaving contest and was subsequently turned into a spider). The middle of the three layers of the meninges, the arachnoid is the delicate, spider's web-like membrane attached to the inside of the dura mater. Immediately inside the arachnoid is the subarachnoid space, which is filled with cerebrospinal fluid.

Area centralis The central area of the retina in mammalian predators and arboreal species. It is characterised by a high spatial density of retinal ganglion cells, minimal convergence in the pathway from photoreceptors to ganglion cells, and is used for high-resolution vision in assessing depth.

Australidelphian A marsupial from Greater Australia, i.e. mainland Australia, New Guinea and associated islands, or Tasmania.

Axon The long output process of a nerve cell. It usually carries the action potential away from the cell body and may be coated with myelin to speed up transmission velocity of the action potential.

Barrel field A specialised region of primary somatosensory cortex associated with vibrissal inputs and characterised by clustering of layer 4Cx granule cells to form discrete structures known as barrels. Barrels are found in the cortex of most rodents and some phalangerids.

Barrington's nucleus Otherwise known as the pontine micturition (urination) centre, this region has direct excitatory connections with parasympathetic neurons in the sacral spinal cord, which in turn control urinary bladder smooth muscle.

Basal metabolic rate (BMR) is the minimal rate of metabolism compatible with endothermy and represents the minimal energy expenditure of a mammal at rest under controlled conditions. It is usually expressed as O_2 consumption in ml per g body weight per hour.

Binocular vision The ability to use both eyes synchronously without double vision. It is used by predators and arboreal species to judge depth.

Blood–brain barrier   A barrier between the circulation and the tissue of the brain. The barrier is created by bands of tight junctions between the endothelial cells of the central nervous system capillaries.

Blood–CSF barrier   A barrier between the cerebral circulation and the cerebrospinal fluid. The barrier is manifest at the tight junctions of the choroid plexus and arachnoid membrane. Unlike the blood–brain barrier, the blood–CSF barrier does not depend on endothelial tight junctions.

Branchiomeric   Skeletal muscle derived from the mesoderm (primitive connective tissue and muscle precursors) of the embryonic pharyngeal arches 1 to 4 and also used to refer to the motorneuron groups in the brainstem that supply those muscles.

Brodmann's areas   A series of almost 50 areas which make up the iso- (or neo-) cortex. These were originally described and numbered by Korbinian Brodmann based on structural differences between regions.

Cajal–Retzius cells   Early generated stellate (star-shaped) neurons of the superficial cerebral cortex. They have cell bodies in layer 1Cx and project axons into layer 2Cx.

Calbindin   A calcium-binding protein found in specific classes of neurons and used as a chemoarchitectonic marker. It was originally identified in the gut as a calcium-binding and vitamin-D-dependent protein. In the central nervous system, calbindin acts as a calcium-stabilising protein and intracellular messenger in many neurons (e.g. Purkinje cells of the cerebellum and dopaminergic neurons of the substantia nigra).

Calretinin   A vitamin-D-dependent calcium-binding protein found in local circuit neurons of the cerebral cortex and thalamus.

Caudal   Literally meaning *towards the tail*. The term refers to a direction towards the end of the spinal cord.

Caudatoputamen   Combined caudate and putamen nuclei of the dorsal striatum (part of the basal ganglia).

Cenozoic   Geological time period from 65 mya to present, divided into the Tertiary (65 mya to 2.5 mya) and the Quaternary (2.5 mya to present).

Cerebellar peduncles   Bundles of axons connecting the cerebellum with the mid- and hindbrain. The three peduncles (superior, middle and inferior cerebellar peduncles) contain cerebellar inputs and outputs.

Cerebellum   Literally meaning the *little brain* (Latin). The cerebellum is derived from the rostral embryonic rhombencephalon and develops dorsal to the fourth ventricle. The cerebellum uses information from the vestibular apparatus, proprioceptive receptors of the spinal cord and other senses, to co-ordinate motor activity.

Cerebrospinal fluid (CSF)   A clear, colourless fluid that fills the ventricular system within the brain and the subarachnoid space around the brain. The CSF is mainly produced by the choroid plexus of the lateral, third and fourth ventricles and is reabsorbed into the systemic veins by arachnoid granulations associated with the major dural venous sinuses.

CGRP   Calcitonin-gene-related peptide. It is a calcitonin-related peptide involved in the perception of pain.

Chemoarchitect/ure, -onics   The use of chemical markers with functional significance (e.g. acetylcholinesterase, NADPH diaphorase, immunoreactivity to calcium-binding proteins or neurofilament proteins) to differentiate between functional areas of the brain.

Choroid plexus   Highly vascular structures that produce most of the CSF within the lateral, third and fourth ventricles. The choroid plexus consists of cuboidal choroidal epithelium surrounding scattered pial cells and collagen, which in turn surround fenestrated choroidal capillaries.

Chromophore   Cells which produce and retain pigment; *or* an atom or group of atoms, which are responsible for absorption of specific wavelengths of light.

Cingulum   From the Latin for *girdle* or *belt*. This axon bundle runs in a sagittal plane in the depths of the cingulate cortex. The cingulum is part of the Papez circuit at the core of the limbic system, and carries axons connecting the cingulate cortex with the entorhinal cortex.

Circumventricular organs   Specialised neurohumeral structures found associated with the ventricular system of the brain. Examples include the subfornical organ, the subcommissural organ and the vascular organ of the lamina terminalis.

Claustrum   From the Latin for *shutter* or *lock*. The claustrum is a thin layer of grey matter lying deep to the insular cortex. The claustrum and associated endopiriform nucleus are part of the pallium.

Cloaca   From the Latin for *sewer* or *drain*. The cloaca is the common caudal opening for the gastrointestinal, urinary and reproductive tracts seen in monotremes, many birds, reptiles and amphibians.

Commissural   Connections between opposite sides of the brain. Many of these pass by the anterior, hippocampal or posterior commissures in marsupials.

Cone photoreceptors   One of two types of photoreceptor in the retina (see rods). Cones are less sensitive to light than rods, but are connected in pathways with little convergence so they serve high acuity vision, usually in the central retina. The presence of different cone types with dissimilar spectral sensitivities makes colour vision possible.

Conspecific   Belonging to the same species.

Contralateral   On the opposite side of the body. Used with reference to representations of the body surface or musculature in, or on, the brain, or the course of sensory or motor pathways in the central nervous system.

Corpus callosum   One of the fibre bundles connecting the two cerebral hemispheres, but present only in placental mammals. From the Latin and Old French for *hard body*.

Corpus striatum   From the Latin for *striped body*. Corpus striatum refers to the caudate and putamen (together called the dorsal striatum) and the globus pallidus (part of the dorsal pallidum).

Cortex   The laminated surface of the cerebral or cerebellar hemispheres. The term cortex is derived from the Latin for the *rind* (of a fruit). The term pallium is also used in comparative neuroanatomy for the dorsal part of the telencephalon, which includes cortical and associated nuclear structures.

Corticobulbar tract   Axons projecting from neuronal cell bodies in the cerebral cortex (isocortex) to the brainstem.

Corticocortical   Connections between different parts of the cerebral cortex. These axons are also called associational connections.

Corticopontine tract   Axons projecting from neuronal cell bodies in the cerebral cortex (isocortex) to the pontine nuclei of the brainstem.

Corticoreticular tract   Axons projecting from neuronal cell bodies in the cerebral cortex to the reticular formation of the brainstem.

Corticospinal tract   Axons projecting from neuronal cell bodies in the cerebral cortex (isocortex) to the spinal cord.

Crepuscular   Being active around dawn or dusk.

Cytoarchitect/ure, -onics   The use of patterns of neurons and glia as revealed in Nissl stains (e.g. large vs. small neurons, dense vs. loosely packed) to differentiate between brain regions.

Cytochrome oxidase   A key enzyme in oxidative cellular metabolism, usually abbreviated as CO. Histochemical demonstration of the distribution of this enzyme reveals areas of high oxidative metabolic rate (e.g. clustered axon terminals of barrel fields or visual cortex). It is therefore a useful marker in differentiating sensory and non-sensory cortical regions.

Dasyurid   A member of the marsupial family Dasyuridae, including dunnarts, antechinus, quolls, phascogales and planigales.

Dichromatic vision   Dichromatic animals must match any colour they see with a mixture of no more than two pure wavelengths (i.e. two cone photoreceptor types).

Didelphids   Members of the marsupial order Didelphimorphia, from North and South America. The common name opossum is usually applied to these mammals.

Diencephalon   The caudal part of the forebrain or prosencephalon. The diencephalon is usually considered to include the dorsal and ventral thalamus (prethalamus), the hypothalamus, the pretectum and the epithalamus.

Diprotodont   A member of the marsupial order Diprotodontia. The term refers to the presence of only a single pair of incisor teeth in the lower jaw. Living diprotodont marsupials include possums, kangaroos, wallabies, wombats and koalas.

Diurnal   Being active during the day (e.g. macropods).

Dopamine   A catecholamine neurotransmitter used by neurons of the substantia nigra pars compacta (projecting to the striatum) and ventral tegmental area of the midbrain (projecting to the frontal cortex and limbic structures).

Dorsal   Towards the back of the animal (from the Latin for *back*).

Dorsal columns   Bundles of somatosensory axons running longitudinally in the dorsal white matter of the spinal cord. These ascending pathways (fasciculus gracilis and cuneatus) are concerned with conscious proprioception, vibration and discriminative (i.e. high spatial resolution) aspects of touch.

Dura mater   From the Latin for *hard* or *tough mother*. The dura mater is the tough external layer of the meninges covering the brain and spinal cord.

Ectoturbinal   An elevation of the ethmoid bone that projects laterally in the nasal cavity. It is covered with mucosa during life.

Efferent   An axon projecting out of the central nervous system or part of the central nervous system. In the case of spinal cord efferents, the term implies motor function in its broadest sense (i.e. direct or indirect control of smooth or skeletal muscles or glands).

Eimer's organ   Sensory organs in the epidermis of the skin, originally described in placental moles and consisting of stacked epidermal cells innervated by myelinated nerve fibres which form terminal swellings immediately beneath the external keratinised layer of the epidermis. The complex consists of a Merkel-cell–neurite complex in the epidermis and a lamellated corpuscle in the subjacent dermis.

Embryonic diapause   A state of dormancy of the blastocyst (early embryo) seen in some marsupials (all kangaroos, wallabies and rat kangaroos). During the diapause, cell division and growth of the embryo either cease or continue at a very slow rate until an appropriate signal is received from the mother.

Encephalisation   The measure of relative brain size, excluding the effects of body size. The encephalisation quotient is the ratio of actual brain volume or mass over the expected brain volume or mass for a reference animal of that body size.

Endocast   A casting of the interior of a skeletal cavity, usually the cranial cavity, used as an indicator of the shape and size of the contained internal organ(s) (e.g. the brain and meninges in the case of a cranial endocast).

Endoturbinal   An elevation of the ethmoid bone that projects medially in the nasal cavity. It is covered with mucosa during life.

Entopeduncular nucleus   Equivalent to the medial globus pallidus. The entopeduncular nucleus consists of large neurons embedded in the lateral internal capsule. The entopeduncular nucleus receives afferents from the lateral globus pallidus and substantia nigra pars compacta and projects to the ventral anterior, ventral lateral and intralaminar nuclei of the thalamus.

Entorhinal cortex   Literally meaning *the cortex within the rhinal fissure*, but some entorhinal cortex actually extends onto the dorsal bank of the rhinal fissure. The entorhinal cortex is characterised by an acellular layer 4Cx (the lamina dissecans), so that neurons are grouped into bands above and below. The entorhinal cortex provides the major input to the dentate gyrus and significant input to the hippocampus and subiculum.

Eucalypt(us)   Any tree of the genus *Eucalyptus*, native to Australia. Members of this genus are adapted for survival in the dry conditions of Australia and produce eucalyptus oil in their leaves. Eucalyptus pollen first appears in the fossil record in the early Oligocene (about 34 mya).

Eutherian   See placental mammal.

Exon   A nucleic-acid sequence in a mature RNA molecule after removal of introns. The mature RNA molecule may be messenger RNA (mRNA), transfer RNA (tRNA) or ribosomal RNA (rRNA). The term exon may also be used for the matching sequence in DNA or its RNA transcript.

External capsule   A shell of white matter external (lateral) to the putamen.

External granular layer   A proliferative cell layer that develops over the cerebellum during pouch life in marsupials (and pre- or perinatally in placentals) and gives rise to the granule cells of the cerebellum.

Fasciculus   A bundle of axons serving a particular function. The term comes from the Latin for *little bundle*.

Fasciculus aberrans   An axonal bundle joining the internal capsule to the anterior commissure. It is a distinctive feature of the forebrains of diprotodonts and allows commissural axons from the dorsal isocortex to cross the midline without taking the long path through the external capsule.

Fasciculus retroflexus   An axonal bundle from neurons of the habenular nuclei to the region of the interpeduncular nucleus and tegmentum of the midbrain. An alternative name is the habenulo-interpeduncular tract.

Field metabolic rate (FMR)   The energy cost for an animal of free existence and includes energy costs of tissue maintenance and thermoregulation as well as daily activity in the wild. It is usually expressed as CO_2 generation in ml per g of body weight per hour.

Foliation   Folding of the surface of the cerebellar cortex.

Fornix   From the Latin for *arch*. An arching axon bundle arising from the hippocampal formation and adjacent cortex and projecting to the mamillary bodies of the hypothalamus, septal nuclei and preoptic area. Part of the Papez circuit at the core of the limbic system.

Funiculus   From the Latin for *thin rope* or *cord*. The term refers to a collection of axons in the white matter of the spinal cord (i.e. dorsal, ventral or lateral funiculi).

GABA   Gamma-aminobutyric acid. An inhibitory neurotransmitter derived from glutamate and used for brief synaptic events by local circuit neurons in the cortex and subcortical structures.

Glabrous   Skin that is normally devoid of hair (e.g. palms and digit pads).

Glia   Supporting cells of the central nervous system. They may be responsible for making myelin (oligodendrocytes), maintaining the blood–brain barrier and extracellular fluid (astrocytes), or immune responses in the brain and spinal cord (microglia).

Globus pallidus   From the Latin for *pale globe*. Part of the basal ganglia, the lateral globus pallidus receives input from the caudate and putamen and projects to the subthalamic nucleus.

Glutamate   An amino acid which is also the most prevalent excitatory neurotransmitter in the vertebrate nervous system.

Golgi tendon organ   A type of sensory receptor found in tendons, usually at the junction with the skeletal muscle. They serve proprioception by detecting the tension in the tendon.

Grey matter   Those parts of the central nervous system containing neuronal cell bodies, axon terminals and dendritic trees.

Griffonia simplicifolia isolectin B4   A molecule derived from *Griffonia* (or *Bandeiraea*) *simplicifolia* which, when bonded with a marker enzyme like horseradish peroxidase, can be used to label small-calibre nociceptive axons in the dorsal horn of the spinal cord and the trigeminal sensory nuclei.

Gyrus   An elevation of the cerebral cortical surface, defined by surrounding grooves or sulci.

Habenula   Nuclei forming part of the epithalamus. Medial and lateral habenular nuclei are usually identified.

Hippocampal commissure   An axon bundle that joins the hippocampal formations of each side. Placental mammals are said to have both dorsal and ventral hippocampal commissures, whereas marsupials are said to have only a ventral hippocampal commissure.

Hippocampus   (from the Latin for *seahorse*) A specialised region of allocortex, with three layers. The hippocampal formation has a rolled shape (like a seahorse tail) and consists of the dentate gyrus, hippocampus proper (also called *cornu Ammonis* – the horn of Ammon) and the subiculum (from the Latin for *support*), a transitional area between the hippocampus and entorhinal cortex.

Homology   Usually meaning phylogenetic homology – the fundamental similarity of structures (e.g. brain nuclei) in different organisms thought to be the result of their evolution from a precursor structure in a common ancestor.

Hypothalamus   A part of the diencephalon as classically defined. The hypothalamus develops mainly from the proneuromere rostral to prosomere 3 and includes neuronal groups involved with homeostasis (maintaining a constant internal body environment), sexual behaviour and motivation.

Immunoreactivity   The labelling of naturally occurring brain chemicals (e.g. neurotransmitters, neurofilament proteins, enzymes) by the specific binding of antibodies with sections of brain tissue.

Indel   A mutation involving the insertion or deletion (or both) of nucleotides in DNA. Indels are uncommon in mRNA coding regions of DNA because they produce frame shifts that are likely to be lethal, but are common in non-coding regions.

Inferior colliculus   An auditory structure in the roof of the midbrain.

Inferior olivary nuclear complex   A group of neurons situated in the ventral medulla oblongata, most of which project axons to the cerebellum (olivocerebellar tract). Dorsal, principal and ventral groups are identified. The term 'olivary' refers to the olive-like bulge made by this neuronal complex on the ventral surface of the medulla of some mammals.

Inferior olive   See inferior olivary nuclear complex.

Insula   A region of the cerebral cortex immediately dorsal to the rhinal fissure and external to the putamen.

Internal capsule   A large sheet of axons that passes between the thalamus (medially) and dorsal striatopallidal structures (putamen and globus pallidus, laterally). In mammals, the internal capsule may either separate the caudate and putamen or pass as discrete fascicles through the substance of a combined caudatoputamen. The internal capsule carries ascending and descending connections between the isocortex, on the one hand, and the thalamus, striatum or brainstem, on the other.

Intron   A region of DNA within a gene that is not translated into protein. The intron may be translated into precursor mRNA, but is subsequently removed by splicing during the formation of mature mRNA.

Ipsilateral   On the same side of the body.

Isocortex   The part of the cerebral cortex or pallium which has six layers at some stage during development. The term neocortex is also used for this region, but is not the preferred term because it implies an invalid phylogeny of the cortex.

Isthmus   The narrowing or constriction at the caudal end of the midbrain or mesencephalon. This region develops from the isthmic neuromere of the embryonic brain.

Lagomorphs   Rabbits and hares.

Lamellated receptor   A sensory receptor in the skin which has a layered structure (e.g. Pacinian corpuscle) and is supplied by myelinated nerve fibres.

Lanceolate receptor   A mechanoreceptor found in sinus hairs and characterised by a flattened axon terminal sandwiched between two Schwann cells, giving it an elongated lance-like structure.

Lateral   Towards the side.

Limbic system   From the Latin for *fringe* or *hem*, this is a group of structures lying around the rim or edge of the telencephalon. The key elements are the hippocampal formation, cingulate gyrus, parahippocampal gyrus, amygdala and septal nuclei. The limbic system is important in emotion and memory.

Locus coeruleus   From the Latin for *sky blue place*. A column of blue-black pigmented neurons located in the lateral floor of the rostral fourth ventricle. The axons from these neurons provide most of the noradrenergic innervation of the brain.

Macroneuron   A neuron with a large cell body (e.g. Purkinje cells of the cerebellum, large pyramidal neurons of the isocortex), which are usually generated early in development (perinatally in marsupials).

Macropod   A member of the marsupial superfamily Macropodoidea, including kangaroos, wallabies, potoroos, bettongs, the quokka and pademelons.

Male semelparity   Multiple copulation during a single short breeding season in the male's lifetime. In other words, the males live to only one breeding season and subsequently die *en masse*.

Mamillary body   From the Latin for *little breast*. These are small, paired elevations that project ventrally from the caudal hypothalamus. The mamillary bodies are part of the Papez circuit involved in memory and emotion.

Mamillothalamic tract   A bundle of axons projecting from neurons in the mamillary bodies of the hypothalamus to the anterior nucleus of the thalamus.

Marsupial   A type of mammal in which females possess a pouch (Latin: *marsupium*), where young complete post-embryonic development.

Maxilloturbinal   A bony elevation extending from the maxillary bone into the nasal cavity.

Medial   Towards the midline.

Medial forebrain bundle A diffuse bundle of axons running longitudinally throughout the length of the forebrain. The medial forebrain bundle reciprocally connects ventral telencephalic structures with the hypothalamus.

Medulla oblongata The caudal part of the brainstem between the pons and the spinal cord. From the Latin for *elongated marrow* or *pith*. The medulla contains ascending and descending axonal pathways, cranial nerve nuclei, as well as reflex centres controlling vital automatic functions.

Meissner corpuscle A mechanoreceptor consisting of encapsulated nerve endings in the upper dermis of the skin. They are rapidly adapting receptors responsible for the perception of light touch and are concentrated in the digit pads, soles and palms, face and tongue.

Melanopsin A photopigment found in specialised photosensitive retinal ganglion cells that are concerned with the control of circadian (day/night) rhythms and other non-visual responses to light.

Meninges Membranes covering the brain. From outside to inside these are: dura mater, arachnoid and pia mater.

Merkel receptor A slowly adapting mechanoreceptor supplied by myelinated nerve fibres that convey information about sustained pressure. Merkel receptors are clustered into structures known as touch domes and are found in glabrous and hairy skin, in hair follicles and in the mucosa in the oral and anal regions of the gastrointestinal tract.

Mesencephalon The midbrain. A region derived from the walls of the mesencephalic vesicle of the embryonic brain. The midbrain includes the tectum, periaqueductal grey, tegmentum, substantia nigra and assorted fibre tracts of passage. The term may also be used to include the more caudal isthmus.

Mesopic A form of vision used in low light level (but not dark) conditions. It may depend on a combination of photopic and scotopic vision.

Mesotocin A peptide hormone of nine amino acids, which differs by only one amino acid from the hormone oxytocin secreted by the posterior pituitary of placental mammals. Mesotocin serves the same function in kangaroos as oxytocin in placental mammals, i.e. stimulating contraction of the uterus during parturition as well as contraction of myoepithelial cells in the mammary glands during milk ejection.

Metatherian See marsupial.

Microneuron A small neuron, particularly used in reference to the small neurons of the cerebral and cerebellar cortex. Examples include the granule cells of the cerebellum and the stellate cells of the isocortex. Microneurons tend to be generated later in development than macroneurons.

Midbrain See mesencephalon.

Monocular Vision from one eye/ vision using each eye separately.

Monotreme An egg-laying mammal. The name refers to the prominent single opening (cloaca) for gastrointestinal, urinary and genital systems. Living monotremes include the short- and long-beaked echidnas and the platypus.

Müller cells Elongated glial cells of the retina, extending from the inner to the outer retinal surface. Müller cells may dedifferentiate into multipotent progenitor cells to subsequently produce a variety of retinal cells.

Multituberculate A type of early mammal (Upper Jurassic to Upper Eocene) characterised by having molars with multiple low cusps arranged in paired longitudinal rows.

Muscle spindle A sensory receptor embedded within the belly of a skeletal muscle. Each muscle spindle includes 3 to 12 intrafusal muscle fibres and terminals of sensory axons spiralling around the central and pericentral regions of the spindle. The muscle spindle detects changes in the length of the muscle belly during contraction. Axons of γ motorneurons act on the intrafusal muscle fibres to regulate the sensitivity of the muscle spindle apparatus.

Myoblasts A type of progenitor cell that gives rise to muscle cells.

Myofibres Multinucleated, single muscle cells with peripherally located nuclei and centrally located actin and myosin contractile proteins.

Myotopic organisation The organisation of a group of motorneurons in a motor nucleus or column, whereby motoneurons supplying individual muscles are grouped together in a particular part of the nucleus or column and separated from those supplying other muscles.

Myotubes Developing muscle fibres derived from the fusion of myoblasts. They are elongated multinucleated cells that have some peripherally located myofibrils and will develop into mature muscle fibres that have peripherally located nuclei and centrally placed contractile protein.

Myrtaceae The myrtle family of vegetation, including myrtles, eucalypts, melaleucas and leptospermums.

Mystacial Pertaining to the vibrissal (whisker) pad of the snout.

NADPH diaphorase Literally nicotinamide adenine dinucleotide phosphate diaphorase, an enzyme found in neurons using NO (nitric oxide) as a neurotransmitter.

Nasoturbinal A bony elevation of the nasal bone extending into the nasal cavity.

Neurofilament Structural proteins in the axon and cell body of a neuron.

Neurogenesis The production of immature neurons by cell division in the proliferative zones of the developing brain and spinal cord (e.g. ventricular and subventricular zones).

Nissl stain   A type of histological stain that reveals the rough endoplasmic reticulum (protein-manufacturing organelle) of the neuron cell body.

NMDA   *N*-methyl-D-aspartic acid is a synthetic amino-acid derivative that mimics the action of the neurotransmitter glutamate at the NMDA receptor, but has no effect on other glutamate receptors.

Nociception   The perception of painful stimuli, usually the result of the action of damaging agents on the skin or internal body structures.

Nocturnal   Being active at night. See also diurnal and crepuscular.

Noradrenaline   A neurotransmitter used by neurons of the locus coeruleus. The locus coeruleus projects to diverse regions within the brain (olfactory bulb and cortex, isocortex, hippocampus, thalamus, hypothalamus, striatum, cerebellum, brainstem and spinal cord).

Nothofagus   The Antarctic beeches. A type of once-widespread Gondwanan rainforest vegetation appearing in the fossil record during the Cretaceous, but now confined to pockets in New Guinea, New Caledonia, New Zealand, southern South America and restricted areas of rainforests in south-eastern Australia.

Nucleus   When used at the cellular level, may refer to the organelle containing the DNA and chromosomes; *or* may be used at the tissue level for a group of nerve cells in the central nervous system that serve a similar function.

Obex   The region where the fourth ventricle meets the central canal of the spinal cord.

Olfactory bulb   Paired, laminated, rostral protrusions of the telencephalon, concerned with the processing of sensory input from the olfactory epithelium of the nasal cavity and projecting to more caudal parts of the olfactory system (e.g. anterior olfactory nucleus/area, olfactory tubercle and piriform cortex).

Olfactory tubercle   A laminated region of the ventral telencephalon, receiving olfactory input and considered part of the ventral striatum.

Opossum   Common name for members of the American marsupial orders Didelphimorphia (didelphids) and Paucituberculata (shrew opossums). The term comes from the Algonquian word *wapathemwa*.

Orthologue   Homologous gene sequences found in different species due to speciation events rather than gene duplication.

Oxytocin   A hormone secreted by the posterior pituitary of placental mammals. It stimulates uterine contraction during parturition (birth) and contraction of myoepithelial cells during the milk-ejection reflex.

Pacinian corpuscle   A vibration-sensitive, rapidly adapting mechanoreceptor in the dermis of the skin, mesenteries of the gut, periosteum and around muscle fibres. They are about 1 mm in length and have many concentric lamellae like an onion. Each is supplied by a single large myelinated nerve fibre that loses its myelin at its termination in the capsule.

Pallidum   Neuronal groups deep within the telencephalon, so named because of the pallor of the region in humans. The dorsal pallidum includes the globus pallidus and entopeduncular nucleus; the ventral pallidum is within the substantia innominata ventral to the anterior commissure.

Pallium   The dorsal part of the telencephalon. It includes laminated parts (the cortex) and assorted corticonuclear components (e.g. claustrum and endopiriform nuclei, parts of the amygdala).

Papez circuit   A central circuit of the limbic system. Key elements of the Papez circuit include (nuclei in roman script; intervening fibre bundles in italics) the hippocampal formation, *fornix*, mamillary bodies, *mamillothalamic tract*, anterior nucleus of the thalamus, *thalamocortical fibres*, cingulate gyrus, *cingulum*, entorhinal cortex, and *alvear* and *perforant pathways*.

Parabrachial nuclei   Nuclei within the dorsal tegmentum of the midbrain/pons junction, which act as the functional interface between medullary autonomic reflex centres and forebrain behavioural centres.

Parvalbumin   A calcium-binding protein found in select neurons in sensory pathways (e.g. auditory and visual system components) and often used as a chemoarchitectonic marker. Parvalbumin acts to buffer calcium in highly active neurons that use GABA as a neurotransmitter.

Pelage hairs   Hairs of a mammal's coat, as opposed to vibrissae on the head.

Peramelid   A member of the marsupial order Peramelemorphia, including bilbies and bandicoots.

Periallocortex   Transitional cortex between the six-layered isocortex and the three- to five-layered allocortex. It is found around the rhinal fissure and the medial isocortical edge.

Periaqueductal grey (PAG)   Tissue surrounding the cerebral aqueduct of the midbrain. The PAG is organised into longitudinal functional columns concerned with responses to stressful or threatening situations.

Petaurid   A member of the marsupial family Petauridae, including wrist-winged gliders, Leadbeater's possum and the striped possum.

Phalanger(id)   A member of the marsupial family Phalangeridae, including brush-tailed possums and cuscuses.

Photopic vision   Vision under strong light conditions. In many primates it allows colour and high acuity vision and is dependant on cone photoreceptors.

Photoreceptor   A light-sensitive cell in the outer part of the retina. The two kinds of photoreceptors are the rods and cones.

Pia mater   From the Latin for *tender mother*, this is the innermost of the meningeal coverings of the brain. The delicate pia mater directly covers the brain tissue itself.

Pineal gland   A gland attached to the dorsum of the diencephalon. The pineal gland secretes melatonin and plays a central role in seasonal control of reproduction.

Piriform cortex   From the Latin for *pear-shaped*. The piriform cortex is the major cortical component of the olfactory system.

Placental   A mammal whose *in utero* development is supported for a prolonged period after organogenesis by an invasive chorio-allantoic placenta.

Pleistocene   The geological period from about 2.5 mya to 10 kya. It is the first part of the Quaternary period.

Polyprotodont   A member of a group of marsupials with four or five incisors on each side of the upper jaw and three incisors on each side of the lower jaw. All American marsupials and living Australian marsupial carnivores fall into this broad group.

Pons   The rostroventral part of the hindbrain and the site of settling of the pontine nuclear groups. Pontine nuclei are part of a circuit that is critically important for the co-ordination of fine motor activity and involves the motor cortex, cerebellar hemispheres, the dentate deep cerebellar nucleus and motor thalamus.

Pouch young   Marsupial young developing within the maternal pouch (marsupium).

Precerebellar nuclei   Those nuclei projecting axons into the cerebellum. The group includes the inferior olivary complex, pontine nuclei and the lateral reticular nucleus among others.

Prehensile   Referring to the ability to grip, i.e. by the manus (hand) or pedis (foot) of the distal limbs, or by the tail.

Pretectum   That part of the diencephalon developing from embryonic prosomere 1, between the thalamus and the superior colliculus of the midbrain. The pretectum has five nuclei in all mammals and serves in a variety of visual reflexes.

Prethalamus   That part of the diencephalon developing from prosomere 3. The prethalamus (also known as the ventral thalamus) includes the reticular nucleus of the thalamus, pregeniculate nucleus (ventral lateral geniculate nucleus), zona incerta and subthalamus.

Proprioception   The sense of muscle tension and length, and position of parts of the body in space. Proprioception may be conscious (i.e. perceived at the cortical level) or nonconscious (e.g. that used by the cerebellum) and depends on information from joints, muscles, tendons and skin receptors.

Prosencephalon   The forebrain, which includes the telencephalon and diencephalon. The forebrain develops from three prosomeres and the proneuromere region.

Prosimian   Lemurs, tarsiers and galagos.

Prosomere   One of the rostrocaudally sequential segments of the forebrain vesicle during embryonic development.

Prototherian   See monotreme.

Purkinje cell   Output neuron of the cerebellar cortex.

Pyramidal neuron   Output neuron of the cerebral cortex, so named because of the pyramid-shaped cell body.

Raphe nucleus   One of the nuclei located along the midline of the midbrain and hindbrain. Many (but not all) of the raphe nuclei neurons are serotonergic, projecting either rostrally into the forebrain or caudally to the spinal cord.

Receptive field   The region of the body, body surface or visual field wherein a stimulus (e.g. touch or light source) elicits a response in a receptor or sensory nerve cell.

Red nucleus   A collection of neurons in the rostral midbrain. The magnocellular (large-celled) part receives input from the interposed deep nucleus of the cerebellum and gives rise to the rubrospinal tract. The parvicellular (small-celled) part receives input from the dentate nucleus of the cerebellum and projects to the inferior olivary nuclear complex (rubro-olivary tract).

Restiform body   The larger, lateral part of the inferior cerebellar peduncle, including olivocerebellar, cuneocerebellar, reticulocerebellar and dorsal spinocerebellar axons.

Reticular formation   From the Latin for *small net*. A network of neurons extending throughout the rostrocaudal extent of the brainstem. The network is concerned with the generation of patterned motor activity, automatic reflexes, sleep, arousal and focussed sensory attention.

Reticulospinal tracts   Tracts projecting from neurons of the brainstem reticular formation to the spinal cord. The reticulospinal tracts are concerned with control of patterned motor activity (e.g. walking or climbing), autonomic function and lung ventilation.

Rexed's laminae   A system of 10 layers with different cyto- and myeloarchitectural features in the grey matter of the spinal cord.

Rhombencephalon   Broadly equivalent to the hindbrain, consisting of the pons, medulla oblongata and cerebellum.

Rhombomere   One of a series of rostrocaudally arranged segments of the developing hindbrain.

Ringwulst   A dense annulus or ring of connective tissue surrounding the mesenchymal sheath of vibrissal follicles in some mammals. The ringwulst projects into the blood sinus within the follicle.

Rod photoreceptors   A type of photoreceptor in the retina that is particularly important for vision in low light conditions, but does not serve colour vision.

Rostral Literally meaning *towards the beak or prow* (from Latin). The term refers to a direction towards the front end of the brain or spinal cord.

Rubrospinal tract A bundle of axons projecting from large neurons of the red nucleus of the midbrain to the spinal cord. The pathway crosses the midline in the ventral tegmental decussation.

Ruffini ending A type of slowly adapting mechanoreceptor that is sensitive to skin stretch and is found in the deep layers of glabrous skin.

Saccule From the Latin for a *small sac*. The saccule is part of the vestibular apparatus within the inner ear and detects gravity as well as linear acceleration in the sagittal (paramedian) plane.

Sagittal In a plane parallel to the midline or median plane.

Sauropsid Reptiles and birds.

Sclerophyll Literally meaning vegetation with leathery leaves to conserve water.

Scotopic vision Vision under very low light levels, usually dependant on rod photoreceptors.

Semicircular canal One of a system of three canals in the inner ear, arranged at right angles to each other. The canals detect angular acceleration (rotation) of the head by sensing the movement of fluid around their interior.

Septum From the Latin for *wall* or *partition*. The septum of the brain lies medial to the lateral ventricle and consists of medial and lateral nuclear groups. The septum is often considered part of the limbic system.

Serotonin 5-hydroxytryptamine (5HT), a monoamine neurotransmitter derived from tryptophan and involved in diverse functions (autonomic regulation, emotional behaviour, nociception to name a few). It is used by neurons in the raphe nuclei and PAG of the brainstem.

SMI-32 antibody A commercial antibody that has been raised to bind the non-phosphorylated neurofilament proteins present in cell bodies and axons of large neurons. It is often used to identify distinct cortical areas on the basis of different patterns of labelling.

Somatosensory The perception of sensory information from the body surface (pain, touch, vibration, temperature) or from deeper musculoskeletal structures (joints, muscle spindles).

Somatotopic organisation The representation of different parts of the body in an often continuous (but not always), topographically corresponding map in, or on the surface of, a central nervous system structure e.g. somatosensory cortex, VPL, VPM.

Spinal cord The caudal end of the central nervous system. The spinal cord is segmentally organised into cervical, thoracic, lumbar, sacral and coccygeal regions and contains motor and somatosensory systems concerned with the trunk, limbs and internal organs.

Spinocerebellar tracts Dorsal and ventral spinocerebellar tracts convey somatosensory information to the cerebellum for use in co-ordination of limb and trunk musculature.

Spinothalamic tract A bundle of axons arising from neurons in the spinal cord and projecting to the VPL, intralaminar nuclei (CL) and posterior group nuclei of the thalamus. The spinothalamic tract conveys simple touch, pain and temperature information.

Striatum Nuclei deep within the telencephalon, so named because of the striated appearance of the dorsal part of the region in humans. The dorsal striatum includes the caudate and putamen and the ventral striatum is the olfactory tubercle and nucleus accumbens.

Substantia gelatinosa From the Latin for *jelly-like substance*. This superficial part of the dorsal horn of the spinal cord is involved in pain perception. A similar layer is present in the superficial parts of the caudal spinal trigeminal nucleus.

Substantia nigra From the Latin for *black substance*. A nucleus of the midbrain between the red nucleus and the axons of the cerebral peduncles. The compact part of the substantia nigra has closely packed, pigmented neurons that project dopaminergic axons to the caudate and putamen. The reticular part of the substantia nigra receives input from the striatum and projects to the thalamus.

Subthalamus A motor nucleus within the prethalamus that engages in reciprocal connections with the dorsal pallidum (globus pallidus and entopeduncular nucleus).

Subventricular zone (SVZ) A proliferative zone of the developing telencephalon that emerges around the ventricular germinal zone during fetal life in placentals, and during postnatal life in marsupials. The SVZ is believed to give rise to microneurons.

Succinic dehydrogenase (SDH) An enzyme that catalyses the conversion of succinic acid to fumaric acid in the Krebs cycle. It is used as a marker of metabolic activity and is often elevated in sensory nerve terminals (e.g. trigeminal nerve afferents and trigeminal pathway axons).

Sulcus A groove on the brain surface (e.g. the cerebral hemisphere). Sulci often separate functionally important regions of the cerebral cortex.

Superior colliculus Paired elevations (Latin: colliculus – *little hill*) on the roof of the midbrain. The superficial layers of the superior colliculi receive visual input, whereas the deeper layers receive somatosensory and auditory projections. The superior colliculus as a whole appears to integrate sensory information from multiple sources to build up a sensory map of surrounding space at the midbrain level.

Synapse   A specialised region, usually at the end of an axon, where information is transmitted from one axon (or dendrite) to the dendrite or axon of another neuron by release of neurotransmitter chemical or electrotonic coupling.

Tapetum   A reflective layer at the back of the retina that maximises the available light for photoreceptors in low light conditions.

Tectum   From the Latin for *roof*. The tectum is the most superficial or dorsal part of the midbrain and consists of a rostral optic tectum (the superior colliculus of mammals) and a more caudal auditory region (the inferior colliculus of mammals).

Tegmentum   From the Latin for *covering* or *protection*. The tegmentum lies ventral to the tectum of the midbrain and the fourth ventricle floor of the hindbrain.

Telencephalon   The most rostral part of the brain, derived from the walls of the telencephalic vesicle of embryonic life. The telencephalon consists of a pallium (laminated cortex and olfactory structures), as well as deeper grey matter structures such as the striatum, pallidum and septum.

Thalamus   From the Greek for *bedroom* or *chamber*. The thalamus develops from prosomere 2 of the embryonic diencephalic vesicle. It acts as a relay station for ascending sensory projections to the cortex (somatosensory, visual and auditory) as well as motor feedback loops to the cortex. The thalamus also has component nuclei involved in projections to association areas of the cerebral cortex.

Thylacinid   A member of the Thylacinidae, a now extinct family of large polyprotodont marsupial carnivores.

Thylacoleonid   A member of the Thylacoleonidae, an extinct group of large, diprotodont marsupial carnivores.

Tonotopic organisation   The representation of different frequencies of sound in a continuous map in, or on the surface of, central nervous system auditory structures (e.g. auditory cortex, MG, IC).

Tract   A bundle of axons (from the Latin for *drawn out*). An alternative term is fasciculus (from the Latin for little *bundle*).

Trichromatic vision   Trichromatic animals can match any colour they see with a mixture of three pure wavelengths (i.e. three types of cone photoreceptor).

Trigeminal sensory nuclei   A complex of neuronal groups concerned with processing somatosensory information from the head and upper neck. Elements include the mesencephalic nucleus (concerned with pressure on teeth and masticatory proprioception), the principal nucleus (concerned with discriminative touch from the face) and the spinal trigeminal nucleus (concerned with simple touch, pain and temperature from the head and neck).

Urogenital sinus   Common opening for the urinary and reproductive tracts in female marsupials.

Utricle   Part of the vestibular apparatus within the inner ear. The utricle detects linear acceleration in the horizontal plane.

Ventral   Towards the underside of the body. From the Latin for *belonging to the belly*.

Ventral thalamus   Also called the prethalamus. This region of the diencephalon develops from prosomere 3 rostral to the (dorsal) thalamus. In the adult brain, the ventral or prethalamus includes the zona incerta, subthalamic nucleus, reticular thalamic nucleus and the pregeniculate thalamic nucleus (ventral lateral geniculate nucleus of the thalamus).

Ventricular (germinal) zone   The initial proliferative zone of the developing brain. The ventricular zone emerges during embryonic life in both placentals and marsupials and gives rise to macroneurons.

Ventricular system   Cerebrospinal-fluid-filled spaces within the brain. The ventricular system consists of paired lateral, and midline third and fourth, ventricles. All are derived from the vesicles of the embryonic brain.

Vestibular system   The structures concerned with detecting and processing information about orientation, linear and angular acceleration of the head. The system includes the vestibular apparatus of the inner ear (utricle, saccule, semicircular canals), the vestibular division of the vestibulo-cochlear nerve and the vestibular nuclei of the brainstem.

Vibrissae   Whiskers, usually concentrated on the snout, but also present on the chin and around the eye.

Visuotopic organisation   The representation of the visual field of an animal in an ordered topographic map in, or on the surface of, a central nervous system structure (e.g. visual cortex, DLG, SC). A related term is **retinotopic organisation**, which is the mapping of retinal topography to a brain structure, but note that because of the optics of the eye, the retina 'sees' an inverted representation of the visual world.

Vomeronasal organ   Also known as Jacobson's organ, this structure in the vomer of the nasal septum is concerned with detection of pheromones.

White matter   That part of the central nervous system containing almost exclusively myelinated axons.

Zona incerta   A part of the ventral or prethalamus, situated between the thalamus and subthalamus. The zona incerta (from the Latin for *uncertain region*) appears to function as a rostral extension of the reticular formation of the brainstem and has projections to diverse areas of the cerebral cortex.

References

Abbie, A. A. (1939). The origin of the corpus callosum and the fate of the structures related to it. *Journal of Comparative Neurology*, **70**, 9–44.

Abbie, A. A. (1940). The excitable cortex in Perameles, Sarcophilus, Dasyurus, Trichosurus, and Wallabia (Macropus). *Journal of Comparative Neurology*, **72**, 469–487.

Abdel-Mannan, O., Cheung, A. F. P. and Molnár, Z. (2008). Evolution of cortical neurogenesis. *Brain Research Bulletin*, **75**, 398–404.

Adey, W. R. (1953). An experimental study of the central olfactory connections in a marsupial (Trichosurus vulpecula). *Brain*, **76**, 311–330.

Adey, W. R. and Kerr, D. I. (1954). The cerebral representation of deep somatic sensibility in the marsupial phalanger and the rabbit: an evoked potential and histological study. *Journal of Comparative Neurology*, **100**, 597–624.

Ahnelt, P. K., Hokoc, J. N. and Rohlich, P. (1995). Photoreceptors in a primitive mammal, the South American opossum, Didelphis marsupialis aurita: characterization with anti-opsin immunolabeling. *Visual Neuroscience*, **12**, 793–804.

Ahnelt, P. K., Hokoc, J. N. and Rohlich, P. (1996). The opossum photoreceptors – a model for evolutionary trends in early mammalian retina. *Revista Brasileira de Biologia*, **56**, 199–207.

Aitkin, L. M. (1979). The auditory midbrain. *Trends in Neurosciences*, **1**, 308–310.

Aitkin, L. M. (1986). *The Auditory Midbrain. Structure and Function in the Central Auditory Pathway*. New Jersey: Humana Press.

Aitkin, L. M. (1995). The auditory neurobiology of marsupials: a review. *Hearing Research*, **82**, 257–266.

Aitkin, L. M. (1996). The anatomy of the cochlear nuclei and superior olivary complex of arboreal Australian marsupials. *Brain Behavior and Evolution*, **48**, 103–114.

Aitkin, L. M. (1998). *Hearing: The Brain and Auditory Communication in Marsupials*. Berlin: Springer.

Aitkin, L. M. and Gates, G. R. (1983). Connections of the auditory cortex of the brush-tailed possum, (*Trichosurus vulpecula*). *Brain Behavior and Evolution*, **22**, 75–88.

Aitkin, L. M. and Kenyon, C. E. (1981). The auditory brainstem of a marsupial. *Brain Behavior and Evolution*, **19**, 126–143.

Aitkin, L. M. and Nelson, J. E. (1989). Peripheral and central auditory specialization in a gliding marsupial, the feathertail glider, *Acrobates pygmaeus*. *Brain Behavior and Evolution*, **33**, 325–333.

Aitkin, L. M., Bush, B. M. and Gates, G. R. (1978). The auditory midbrain of a marsupial: the brush-tailed possum (*Trichosurus vulpecula*). *Brain Research*, **150**, 29–44.

Aitkin, L. M., Gates, G. R. and Phillips, S. C. (1984). Responses of neurons in inferior colliculus to variation in sound-source azimuth. *Journal of Neurophysiology*, **52**, 1–17.

Aitkin, L. M., Byers, M. and Nelson, J. E. (1986a). Brainstem auditory nuclei and their connections in a carnivorous marsupial, the northern native cat (*Dasyurus hallucatus*). *Brain Behavior and Evolution*, **29**, 1–16.

Aitkin, L. M., Irvine, D. R., Nelson, J. E., Merzenich, M.M. and Clarey, J. C. (1986b). Frequency representation in the auditory midbrain and forebrain of a marsupial, the northern native cat (*Dasyurus hallucatus*). *Brain Behavior and Evolution*, **29**, 17–28.

Aitkin, L. M., Nelson, J., Farrington, M. and Swann, S. (1991). Neurogenesis in the brain auditory pathway of a marsupial, the northern native cat (*Dasyurus hallucatus*). *Journal of Comparative Neurology*, **309**, 250–260.

Aitkin, L. M., Nelson, J., Farrington, M. and Swann, S. (1994a). The morphological development of the inferior colliculus in a marsupial, the northern quoll (*Dasyurus hallucatus*). *Journal of Comparative Neurology*, **343**, 532–541.

Aitkin, L. M., Nelson J. E. and Shepherd, R. K. (1994b). Hearing, vocalization and the external ear of a marsupial, the northern quoll, *Dasyurus hallucatus*. *Journal of Comparative Neurology*, **349**, 377–388.

Aitkin, L. M., Nelson, J. E., Martsi-McClintock, A. and Swann, S. (1996a). Features of the structural development of the inferior colliculus in relation to the onset of hearing in a marsupial: the northern quoll, *Dasyurus hallucatus*. *Journal of Comparative Neurology*, **375**, 77–88.

Aitkin, L. M., Nelson, J. and Shepherd, R. (1996b). Development of hearing and vocalization in a marsupial, the northern quoll, *Dasyurus hallucatus*. *Journal of Experimental Zoology*, **276**, 394–402.

Aitkin, L. M., Cochran, S., Frost, S., Martsi-McClintock, A. and Masterton, B. (1997). Features of the auditory development of the short-tailed Brazilian opossum, *Monodelphis domestica*: evoked responses, neonatal vocalization and synapses in the inferior colliculus. *Hearing Research*, **113**, 69–75.

Allen, L. S. and Gorski, R. A. (1990). Sex differences in the bed nucleus of the stria terminalis of the human brain. *Journal of Neuroscience* **9**, 497–506.

Almeida, M. R., Hesse, A., Steinmetz, A. *et al.* (1991). Transthyretin Leu 68 in a form of cardiac amyloidosis. *Basic Research in Cardiology*, **86**, 567–571.

Altland, K. and Richardson, S. J. (2009). Histidine 31: the Achilles heel of human transthyretin. Microheterogeneity is not enough to understand the molecular causes of amyloidogenicity. In *Recent Advances in Transthyretin Evolution, Structure and Biological Functions*. (ed. S. J. Richardson and V. Cody). Berlin: Springer Verlag. pp. 210–214.

Altman, J. and Bayer, S. A. (1979a). Development of the diencephalon in the rat. IV Quantitative study of the time of origin of neurons and the internuclear gradients in the thalamus. *Journal of Comparative Neurology*, **188**, 455–472.

Altman, J. and Bayer, S. A. (1979b). Development of the diencephalon in the rat. V Thymidine-radiographic observations on internuclear and intranuclear gradients in the thalamus. *Journal of Comparative Neurology*, **188**, 473–500.

Altman, J. and Bayer, S. A. (1990). Mosaic organization of the hippocampal neuroepithelium and the multiple germinal sources of dentate granule cells. *Journal of Comparative Neurology*, **301**, 325–342.

Altman, J. and Bayer, S. A. (1995). *Atlas of Prenatal Rat Brain Development*. Boca Raton: CRC Press.

Amadeo, A., Ortino, B. and Frassoni, C. (2001). Parvalbumin and GABA in the developing somatosensory thalamus of the rat: an immunohistochemical ultrastructural correlation. *Anatomy and Embryology*, **203**, 109–119.

Angevine, J. B. (1970). Time of neuron origin in the diencephalon of the mouse. An autoradiographic study. *Journal of Comparative Neurology*, **139**, 129–187.

Anker, R. L. (1977). The prenatal development of some of the visual pathways in the cat. *Journal of Comparative Neurology*, **173**, 185–204.

Aoki, M. and McIntyre, A. K. (1973). Pyramidal effects on some forelimb motoneurone populations of the arboreal brush-tailed possum (*Trichosurus vulpecula*). *Brain Research*, **60**, 485–488.

Aoki, M. and McIntyre, A. K. (1976). Long spinal and pyramidal actions on hindlimb motoneurons of the marsupial brush-tailed possum, *Trichosurus vulpecula*. *Journal of Neurophysiology*, **39**, 331–339.

Apelt, J., Bigi, M., Wunderlich, P. and Schliebs, R. (2004). Ageing-related increases in oxidative stress correlates with developmental pattern of beta-secretase activity and beta-amyloid plaque formation in transgenic Tg2576 mice with Alzheimer's-like pathology. *International Journal of Developmental Neuroscience*, **22**, 475–484.

Arai, R., Jacobowitz, D. M. and Deura, S. (1994). Distribution of calretinin, calbindin-D28k, and parvalbumin in the rat thalamus. *Brain Research Bulletin*, **32**, 595–614.

Archer, M. (1984). The Australian marsupial radiation. In *Vertebrate Zoogeography and Evolution in Australasia* (ed. M. Archer and G. Clayton). Carlisle: Hesperian Press. pp. 633–808.

Archer, M. and Kirsch, J. (2006). The evolution and classification of marsupials. In *Marsupials* (ed. P. Armati, C. Dickman and I. Hume). Cambridge: Cambridge University Press. pp. 1–21.

Archer, M., Arena, R., Bassarova, M. *et al.* (1999). The evolutionary history and diversity of Australian mammals. *Australian Mammalogy*, **21**, 1–45.

Ariëns-Kappers, C. U. A., Huber, G. C. and Crosby, E. C. (1960). *The Comparative Anatomy of the Nervous System, Including Man*. New York: Hafner.

Armati, P., Dickman, C. and Hume, I. (2006). *Marsupials*. Cambridge: Cambridge University Press.

Arrese, C. (2002). Pupillary mobility in four species of marsupials with differing lifestyles. *Journal of Zoology, London*, **256**, 191–197.

Arrese, C. and Runham, P. (2002). Radio-tracking the honey possum (*Tarsipes rostratus*). *Australian Mammalogy*, **23**, 169–172.

Arrese, C., Dunlop, S. A., Harman, A. M. *et al.* (1999). Retinal structure and visual acuity in a polyprotodont marsupial, the fat-tailed dunnart (*Sminthopsis crassicaudata*). *Brain Behavior and Evolution*, **53**, 111–126.

Arrese, C., Archer, M., Runham, P., Dunlop, S. A. and Beazley, L. D. (2000). Visual system in a diurnal marsupial, the numbat (*Myrmecobius fasciatus*): retinal organization, visual acuity and visual fields. *Brain Behavior and Evolution*, **55**, 163–175.

Arrese, C., Archer, M. and Beazley, L. D. (2002a). Visual capabilities in a crepuscular marsupial, the honey possum (*Tarsipes rostratus*): a visual approach to ecology. *Journal of Zoology, London*, **256**, 151–158.

Arrese, C. A., Hart, N. S., Thomas, N., Beazley, L. D. and Shand, J. (2002b). Trichromacy in Australian marsupials. *Current Biology*, **12**, 657–660.

Arrese, C. A., Rodger, J., Beazley, L. D. and Shand, J. (2003). Topographies of retinal cone photoreceptors in two Australian marsupials. *Visual Neuroscience*, **20**, 307–311.

Arrese, C. A., Oddy, A. Y., Runham, P. B. *et al.* (2005). Cone topography and spectral sensitivity in two potentially trichromatic marsupials, the quokka (*Setonix brachyurus*) and quenda (*Isoodon obesulus*). *Proceedings of The Royal Society. B. Biological Sciences*, **272**, 791–796.

Arrese, C. A., Beazley, L. D., Ferguson, M. C., Oddy, A. and Hunt, D. M. (2006a). Spectral tuning of the long wavelength-sensitive cone pigment in four Australian marsupials. *Gene*, **381**, 13–17.

Arrese, C. A., Beazley, L. D. and Neumeyer, C. (2006b). Behavioural evidence for marsupial trichromacy. *Current Biology*, **16**, R193-R194.

Asenjo, A. B., Rim, J. and Oprian, D. D. (1994). Molecular determinants of human red/green color discrimination. *Neuron*, **12**, 1131–1138.

Ashwell, K. W. S. (2008a). Encephalization of Australian and New Guinean marsupials. *Brain Behavior and Evolution*, **71**, 181–199.

Ashwell, K. W. S. (2008b). Topography and chemoarchitecture of the striatum and pallidum in a monotreme, the short-beaked echidna (*Tachyglossus aculeatus*). *Somatosensory and Motor Research*, **25**, 171–187.

Ashwell, K. W. S. and Mai, J. K. (1997). Developmental expression of the CD15 epitope in the brainstem and spinal cord of the mouse. *Anatomy and Embryology*, **196**, 13–25.

Ashwell, K. W. S. and Paxinos, G. (2005). Cyto- and chemoarchitecture of the dorsal thalamus of the monotreme *Tachyglossus aculeatus*, the short beaked echidna. *Journal of Chemical Neuroanatomy*, **30**, 161–183.

Ashwell, K. W. S. and Paxinos, G. (2007). The pretectal nuclei in two monotremes: the short-beaked echidna (*Tachyglossus aculeatus*) and the platypus (*Ornithorhynchus anatinus*). *Brain Structure and Function*, **212**, 359–369.

Ashwell, K. W. S. and Paxinos, G. (2008). *Atlas of the Developing Rat Nervous System*. 3rd edn. London: Elsevier Academic.

Ashwell, K. W. S. and Zhang, L-L. (1997). Cyto- and myeloarchitectonic organisation of the spinal cord of an echidna (Tachyglossus aculeatus). *Brain Behavior and Evolution* **49**, 276–294.

Ashwell, K. W. S., Waite, P. M. E. and Marotte, L. R. (1996a). Ontogeny of the projection tracts and commissural fibres in the forebrain of the tammar wallaby (*Macropus eugenii*): timing in comparison with other mammals. *Brain Behavior and Evolution*, **47**, 8–22.

Ashwell, K. W. S., Marotte, L. R., Lixin, L. and Waite, P. M. E. (1996b). Anterior commissure of the wallaby (*Macropus eugenii*): adult morphology and development. *Journal of Comparative Neurology*, **366**, 478–494.

Ashwell, K. W. S., Mai, J. K. and Andressen, C. (2004). CD15 immunoreactivity in the developing brain of a marsupial, the tammar wallaby (*Macropus eugenii*). *Anatomy and Embryology*, **209**, 157–168.

Ashwell, K. W. S., Zhang, L.-L. and Marotte, L. R. (2005a). Cyto- and chemoarchitecture of the cortex of the tammar wallaby (*Macropus eugenii*): areal organization. *Brain Behavior and Evolution*, **66**, 114–136.

Ashwell, K. W. S., Hardman, C. D. and Paxinos, G. (2005b). Cyto- and chemoarchitecture of the amygdala of a monotreme, *Tachyglossus aculeatus* (the short beaked echidna). *Journal of Chemical Neuroanatomy* **30**, 82–104.

Ashwell, K. W. S., Paxinos, G. and Watson, C. R. R. (2007a). Cyto- and chemoarchitecture of the cerebellum of the short-beaked echidna (*Tachyglossus aculeatus*). *Brain Behavior and Evolution*, **70**, 71–89.

Ashwell, K. W. S., Paxinos, G. and Watson, C. R. R. (2007b). Precerebellar and vestibular nuclei of the short-beaked echidna (*Tachyglossus aculeatus*). *Brain Structure and Function*, **212**, 209–221.

Ashwell, K. W. S., Marotte, L. R. and Cheng, G. (2008a). Development of the olfactory system in a wallaby (*Macropus eugenii*). *Brain Behavior and Evolution*, **71**, 216–230.

Ashwell, K. W. S., McAllan, B. M., Mai, J. K. and Paxinos, G. (2008b). Cortical cyto- and chemoarchitecture in three small Australian marsupial carnivores: *Sminthopsis macroura*, *Antechinus stuartii* and *Phascogale calura*. *Brain Behavior and Evolution*, **72**, 215–232.

Auclair, F., Bélanger, M-C. and Marchand, R. (1993). Ontogenetic study of early brain stem projections to the spinal cord in the rat. *Brain Research Bulletin*, **30**, 281–289.

Augee, M. L., Carrick, F. N., Grant, T. R. and Temple-Smith P. D. (2008). Order Monotremata: platypus and echidnas. In *The Mammals of Australia* (ed. S. Van Dyck and R. Strahan). 3rd edn. Sydney: Reed New Holland, pp. 30–31.

Auladell, C., Perez-Sust, P., Super, H. and Soriano, E. (2000). The early development of thalamocortical and corticothalamic projections in the mouse. *Anatomy and Embryology*, **201**, 169–179.

Austad, S. N. (1997). Comparative aging and life histories in mammals. *Experimental Gerontology*, **32**, 23–38.

Austad, S. N. (2005). Diverse aging rates in metazoans: targets for functional genomics. *Mechanisms of Ageing and Development*, **126**, 43–49.

Austad, S. N. and Fischer, K. E. (1991). Mammalian aging, metabolism and ecology: evidence from the bats and marsupials. *Journal of Gerontology*, **46**, B47–B53.

Baker, G. E. (1990). Prechiasmatic reordering of fibre diameter classes in the retinofugal pathway of ferrets. *European Journal of Neuroscience*, **2**, 24–33.

Baker, G. E. and Jeffery, G. (1989). Distribution of uncrossed axons along the course of the optic nerve and chiasm of rodents. *Journal of Comparative Neurology*, **289**, 455–461.

Ballantyne, J. (1984) Anatomy of the ear. In *Ultrastructural Atlas of the Inner Ear*. (ed. I. Friedmann and J. Ballantyne). London: Butterworths, pp. 1–7.

Barbour, H. R., Archer, M. A., Hart, N. S. *et al.* (2002). Retinal characteristics of the ornate dragon lizard, *Ctenophorus ornatus*. *Journal of Comparative Neurology*, **450**, 334–344.

Barker, I. K., Beveridge, I., Bradley, A. and Lee, A. K. (1978). Observations on spontaneous stress-related mortality among males of the dasyurid marsupial Antechinus stuartii Macleay. *Australian Journal of Zoology*, **26**, 435–447.

Barnett, J. L. (1973). A stress response in *Antechinus stuartii* (Macleay). *Australian Journal of Zoology*, **21**, 501–513.

Barton, R. A. and Dean, P. (1993). Comparative evidence indicating neural specialization for predatory behaviour in mammals. *Proceedings of The Royal Society. B. Biological Sciences*, **254**, 63–68.

Bates, C. A. and Killackey, H. P. (1985). The organization of the neonatal rat's brainstem trigeminal complex and its role in the formation of central trigeminal patterns. *Journal of Comparative Neurology*, **240**, 265–287.

Bathgate, R. A., Sernia, C. and Gemmell, R. T. (1992a). Mesotocin and oxytocin in the brain and plasma of an Australian marsupial, the northern brown bandicoot, *Isoodon macrourus*. *Comparative Biochemistry and Physiology. Comparative Physiology*, **102**, 43–48.

Bathgate, R. A., Sernia, C. and Gemmell, R. T. (1992b). Brain content and plasma concentrations of arginine vasopressin in an Australian marsupial, the brushtail possum, *Trichosurus vulpecula*. *General and Comparative Endocrinology*, **88**, 217–223.

Bathgate, R. A., Parry, L. J., Fletcher, T. P. *et al.* (1995). Comparative aspects of oxytocin-like hormones in marsupials. *Advances in Experimental Medicine and Biology*, **395**, 639–655.

Baudinette, R. V., Runciman, S. I. C., Frappell, P. F. and Gannon, B. J. (1988). Development of the marsupial cardiorespiratory system. In *The Developing Marsupial: Models for Biomedical Research* (ed. C. H. Tyndale-Biscoe and P. A. Janssens). Berlin: Springer-Verlag, pp. 132–147.

Baudry, M., Dubrin, R., Beasley, L., Leon, M. and Lynch, G. (1986). Low-levels of calpain activity in chiroptera brain – implications for mechanisms of aging. *Neurobiology of Aging*, **7**, 255–258.

Bauer-Moffett, C. and King, J. S. (1981). The development of the inferior olivary complex in preweanling opossums. Identification of midbrain, cerebellar and spinal terminals. *Anatomy and Embryology*, **162**, 249–280.

Bayer, S. A. and Altman, J. (1991). Development of the endopiriform nucleus and the claustrum in the rat brain. *Neuroscience*, **45**, 391–412.

Bayer, S. A. and Altman, J. (2004). Development of the telencephalon: neural stem cells, neurogenesis and neuronal migration. In *The Rat Nervous System* (ed. G Paxinos). 3rd edn. San Diego, London: Elsevier Academic. pp. 27–73.

Beard, L. A. and Grigg, G. C. (2000). Reproduction in the short-beaked echidna, *Tachyglossus aculeatus*: field observations at an elevated site in south-east Queensland. *Proceedings of the Linnean Society of New South Wales*, **122**, 89–99.

Beazley, L. D. and Dunlop, S. A. (1983). The evolution of an area centralis and visual streak in the marsupial *Setonix brachyurus*. *Journal of Comparative Neurology*, **216**, 211–231.

Beck, P. D., Popischal, M. W. and Kaas, J. H. (1996). Topography, architecture and connections of somatosensory cortex in opossums: evidence for five somatosensory areas. *Journal of Comparative Neurology*, **366**, 109–133.

Belford, G. R. and Killackey, H. P. (1978). Anatomical correlates of the forelimb in the ventrobasal complex and the cuneate nucleus of the neonatal rat. *Brain Research,* **158**, 450–455.

Belford, G. R. and Killackey, H. P. (1979a). The development of vibrissae representation in subcortical trigeminal centers of the neonatal rat. *Journal of Comparative Neurology*, **188**, 63–74.

Belford, G. R., and Killackey, H. P. (1979b). Vibrissae representation in subcortical trigeminal centers of the neonatal rat. *Journal of Comparative Neurology*, **183**, 305–321.

Bellingham, J., Chaurasia, S. S., Melyan, Z. *et al.* (2006). Evolution of melanopsin photoreceptors: discovery and characterization of a new melanopsin in nonmammalian vertebrates. *PLoS Biology*, **4**, e254.

Benson, M. D. (2009). Clinical implications of TTR in amyloidosis. Microheterogeneity is not enough to understand the molecular causes of amyloidogenicity. In *Recent Advances in Transthyretin Evolution, Structure and Biological Functions* (ed. S.J. Richardson and V. Cody). Berlin: Springer Verlag. pp. 173–190.

Berman, N. (1977). Connections of the pretectum in the cat. *Journal of Comparative Neurology*, **174**, 227–254.

Berson, D. M., Dunn, F. A. and Takao, M. (2002). Phototransduction by retinal ganglion cells that set the circadian clock. *Science*, **295**, 1070–1073.

Blackburn, D. G. (1993). Lactation – historical patterns and potential for manipulation. *Journal of Dairy Science*, **76**, 3195–3212.

Blumer, W. (1963). Ascending and descending spinal tracts of the quokka (*Setonix brachyurus*). *Journal of Anatomy*, **97**, 490.

Bonini, N. M, and Fortini, M. E. (2003). Human neurodegenerative disease modeling using *Drosophila*. *Annual Reviews in Neuroscience*, **26**, 627–656.

Bonney, K. R. and Wynne, C. D. L. (2002). Visual discrimination learning and strategy behaviour in the fat-tailed dunnart (*Sminthopsis crassicaudata*). *Journal of Comparative Psychology*, **116**, 55–62.

Bonney K. R. and Wynne C. D. L. (2003). Configural learning in two species of marsupial (*Setonix brachyurus* and *Sminthopsis crassicaudata*). *Journal of Comparative Psychology*, **117**, 188–199.

Bonney, K. R. and Wynne, C. D. L. (2004). Studies of learning and problem solving in two species of Australian marsupials. *Neuroscience and Biobehavioural Reviews*, **28**, 583–594.

Bortoff, G. A. and Strick, P. L. (1993). Corticospinal terminations in two new-world primates: further evidence that corticomotoneuronal connections provide part of the neural substrate for manual dexterity. *Journal of Neuroscience*, **13**, 5105–5118.

Bowmaker, J. K. (2008). Evolution of vertebrate visual pigments. *Vision Research*, **48**, 2022–2041.

Bowmaker, J. K. and Hunt, D. M. (2006). Evolution of vertebrate visual pigments. *Current Biology*, **16**, R484–489.

Braak, H. and Braak, E. (1991). Neuropathological stageing of Alzheimer-related changes. *Acta Neuropathologica*, **82**, 239–259.

Bradley, A. J, McDonald, I. R. and Lee, A. K. (1980). Stress and mortality in a small marsupial (*Antechinus stuartii*, Macleay). *General and Comparative Endocrinology*, **40**, 188–200.

Braekevelt, C. R. (1983). Retinal photoreceptor fine structure in the domestic sheep. *Acta Anatomica*, **116**, 265–275.

Bravo, H., Olavarría, J. and Martinich, S. (1990). Patterns of interhemispheric and striate-peristriate connections in visual cortex of the South American marsupial *Marmosa elegans* (mouse opossum). *Anatomy and Embryology* **182**, 583–589.

Bredt, D. S., Glatt, C. E., Hwang, P. M. *et al.* (1991). Nitric oxide synthase protein and mRNA are discretely localised in neuronal populations of the CNS together with NADPH diaphorase. *Neuron*, **7**, 615–624.

Brenowitz, G. L., Tweedle, C. D. and Johnson, J. I. (1980). The development of receptors in the glabrous forepaw skin of pouch young opossums. *Neuroscience*, **5**, 1303–1310.

Broome, L. S. (2001). Intersite differences in population demography of Mountain Pygmy-possums *Burramys parvus* Broom (1986-1998): implications for metapopulation conservation and ski resorts in Kosciusko National Park, Australia. *Biological Conservation*, **102**, 309–323.

Broomhead, A. (1974). The mediodorsal thalamic nucleus of the brush-tailed possum, *Trichosurus vulpecula*. *Journal of Anatomy*, **118**, 392.

Brunjes, P. C., Jazaeri, A. and Sutherland, M. J. (1992). Olfactory bulb organization and development in *Monodelphis domestica* (grey short-tailed opossum). *Journal of Comparative Neurology*, **320**, 544–554.

Bulfone, A., Puelles, L., Porteus, M. H. *et al.* (1993). Spatially restricted expression of Dlx-1, Dlx-2 (Tes-1), Gbx-2 and Wnt-3 in the embryonic day 12.5 mouse forebrain defines potential transverse and longitudinal boundaries. *Journal of Neuroscience*, **13**, 3155–3172.

Butler, A. B. and Hodos, W. (2005). *Comparative Vertebrate Neuroanatomy. Evolution and Adaptation.* 2nd edn. New Jersey: Wiley, Hoboken.

Butler, C. M., Harry, J. L., Deakin, J. E., Cooper, D. W. and Renfree, M. B. (1998). Developmental expression of the androgen receptor during virilization of the urogenital system of a marsupial. *Biology of Reproduction,* **59**, 725–732.

Buttery, R. G., Haight J. R. and Bell, K. (1990). Vascular and avascular retinae in mammals. A funduscopic and fluorescein angiographic study. *Brain Behavior and Evolution*, **35**, 156–175.

Buttery, R. G., Hinrichsen, C. F., Weller, W. L. and Haight, J. R. (1991). How thick should a retina be? A comparative study of mammalian species with and without intraretinal vasculature. *Vision Research*, **31**, 169–187.

Byers, J. A. (1999). The distribution of play behaviour among Australian marsupials. *Journal of Zoology*, **247**, 349–356.

Caballero-Bleda, M., Fernandez, B. and Puelles, L. (1992). The pretectal complex of the rabbit: distribution of acetylcholinesterase and reduced nicotinamide adenine dinucleotide diaphorase activities. *Acta Anatomica*, **144**, 7–16.

Cabana T. and Martin, G. F. (1984). Developmental sequence of descending spinal pathways. Studies using retrograde transport techniques in the North American opossum (*Didelphis virginiana*). *Brain Research*, **317**, 247–263.

Cailliet , G. M., Andrews, A. H., Burton, E. J. *et al.* (2001). Age determination and validation studies of marine fishes: do deep-dwellers live longer? *Experimental Gerontology*, **36**, 739–764.

Calaby, J. H. (1960). Observations on the banded ant-eater *Myrmecobius fasciatus* Waterhouse (Marsupialia), with particular reference to its food habits. *Proceedings of the Zoological Society*, **135**, 183–207.

Calder, W. A. III. (1984). *Size, Function and Life History.* Cambridge, MA: Harvard University Press.

Calderone, J. B. and Jacobs, G. H. (1999). Cone receptor variations and their functional consequences in two species of hamster. *Visual Neuroscience*, **16**, 53–63.

Callard, G. V., Petro, Z. and Tyndale-Biscoe, C. H. (1982). Aromatase activity in marsupial brain, ovaries, and adrenals. *General and Comparative Endocrinology*, **46**, 541–546.

Campbell, M. J. and Morrison, J. H. (1989). Monoclonal antibody to neurofilament protein (SMI-32) labels a subpopulation of pyramidal neurons in the human and monkey neocortex. *Journal of Comparative Neurology*, **282**, 191–205.

Campenhausen, M. V. and Kirschfeld, K. (1998). Spectral sensitivity of the accessory optic system of the pigeon. *Journal of Comparative Physiology A*, **183**, 1–6.

Cannon, J. R., Bakker, H. R., Bradshaw, S. D. and McDonald, I. R. (1976). Gravity as the sole navigational aid to the newborn quokka. *Nature*, **259**, 42.

Carmichael, S. T. and Price, J. L. (1994). Architectonic subdivision of the orbital and medial prefrontal cortex in the macaque monkey. *Journal of Comparative Neurology*, **346**, 366–402.

Carrodeguas, J. A., Rodolosse, A., Garza, M. V. *et al.* (2005). The chick embryo appears a model for research into β-amyloid precursor protein. *Neuroscience*, **134**, 1285–1300.

Carter, C. S., Ramsey, M. M., Ingram, R. L. *et al.* (2002). Models of growth hormone and IGF-1 deficiency: applications to studies of aging processes and life-span determination. *Journals of Gerontology. A. Biological Sciences and Medical Sciences*, **57**, B177–B188.

Cartmill, M. (1972). Arboreal adaptations and the origin of the order Primates. In *The Functional and Evolutionary Biology of Primates.* (ed. R. Tuttle). Chicago: Aldine-Alherton. pp.97–122.

Cartmill, M. (1974). Rethinking primate origins. *Science*, **184**, 436–443.

Caruso, C., Candore, G., Romano, G. C. *et al.* (2000). HLA, aging, and longevity: a critical reappraisal. *Human Immunology*, **61**, 942–949.

Carvalho, L. D. S., Cowing, J. A., Wilkie, S. E., Bowmaker, J. K. and Hunt, D. M. (2006). Shortwave visual sensitivity in tree and flying squirrels reflects changes in lifestyle. *Current Biology*, **16**, R81–R83.

Catalano, S. M., Robertson, R. T. and Killackey. H. P. (1991). Early ingrowth of thalamocortical afferents to the neocortex of the prenatal rat. *Proceedings of the National Academy of Science USA*, **88**, 2999–3003.

Catalano, S. M., Robertson, R. T. and Killackey, H. P. (1996). Individual axon morphology and thalamocortical topography in developing rat somatosensory cortex. *Journal of Comparative Neurology*, **367**, 36–53.

Catania, K. C., Jain, N., Franca, J. G., Volchan, E. and Kaas, J. H. (2000). The organisation of somatosensory cortex in the short-tailed opossum (*Monodelphis domestica*). *Somatosensory and Motor Research*, **17**, 39–51.

Caverson, M. M., Ciriello, J., Calaresu, F.R. and Krukoff, T. L. (1987). Distribution and morphology of vasopressin-, neurophysin II, and oxytocin- immunoreactive cell bodies in the forebrain of the cat. *Journal of Comparative Neurology*, **259**, 211–236.

Celio, M. R. (1990). Calbindin D-28k and parvalbumin in the rat nervous system. *Neuroscience*, **35**, 375–475.

Chalupa, L. M. and Snider, C. J. (1998). Topographic specificity in the retinocollicular projection of the developing ferret: an anterograde tracing study. *Journal of Comparative Neurology*, **392**, 35–47.

Chalupa, L. M., Snider, C. J. and Kirby, M. A. (1996). Topographic organisation in the retinocollicular pathway of the fetal cat demonstrated by retrograde labeling of ganglion cells. *Journal of Comparative Neurology*, **368**, 295–303.

Charvet, C. J., Owerkowicz, T. and Striedter, G. F. (2009). Phylogeny of the telencephalic subventricular zone in sauropsids: evidence for the sequential evolution of pallial and subpallial subventricular zones. *Brain Behavior and Evolution*, **73**, 285–294.

Chase, J. (1982). The evolution of retinal vascularisation in mammals. A comparison of vascular and avascular retinae. *Ophthalmology*, **89**, 1518–1525.

Chase, J. and Graydon, M. L. (1990). The eye of the northern brown bandicoot, *Isoodon macrourus*. In *Bandicoots and Bilbies*, (ed. J. H. Seebeck, P. R. Brown, R. L. Wallis and C. M. Kemper). Sydney, Australia: Surrey Beatty and Sons. pp. 117–122.

Chauvet, M. T., Hurpet, D., Chauvet, J. and Acher, R. (1980). Phenypressin (Phe2-Arg8-vasopressin), a new neurohypophysial peptide found in marsupials. *Nature*, **287**, 640–642.

Chauvet, M. T., Hurpet, D., Chauvet, J. and Acher, R. (1981). A reptilian neurohypophysial hormone, mesotocin (Ile8-oxytocin), in Australian marsupials. *FEBS Letters*, **129**, 120–122.

Chauvet, M. T., Hurpet, D., Chauvet, J. and Acher, R. (1983a). Identification of mesotocin, lysine vasopressin, and phenypressin in the eastern grey kangaroo (*Macropus giganteus*). *General and Comparative Endocrinology*, **49**, 63–72.

Chauvet, M. T., Colne, T., Hurpet, D., Chauvet, J. and Acher, R. (1983b). Marsupial neurohypophysial hormones: identification of mesotocin, lysine vasopressin, and phenypressin in the quokka wallaby (*Setonix brachyurus*). *General and Comparative Endocrinology*, **51**, 309–315.

Chauvet, M. T., Colne, T., Hurpet, D., Chauvet, J. and Acher, R. (1983c). A multigene family for the vasopressin-like hormones? Identification of mesotocin, lysipressin and phenypressin in Australian macropods. *Biochemical and Biophysical Research Communications*, **116**, 258–263.

Chauvet, J., Hurpet, D., Chauvet, M. T. and Acher, R. (1984). Divergent neuropeptide evolutionary drifts between American and Australian marsupials. *Bioscience Reports*, **4**, 245–252.

Chauvet, J., Rouille, Y., Chauvet, M. T. and Acher, R. (1987). Evolution of marsupials traced by their neurohypophyseal hormones: micro-identification of mesotocin and arginine vasopressin in two Australian families, Dasyuridae and Phascolarctidae. *General and Comparative Endocrinology*, **67**, 399–408.

Chelvanayagam, D. K., Dunlop, S. A. and Beazley, L. D. (1998). Axon order in the visual pathway of the quokka wallaby. *Journal of Comparative Neurology*, **390**, 333–341.

Chen, X. M., Milne, N. and O'Higgins, P. (2005). Morphological variation of the thoracolumbar vertebrae in Macropodidae and its functional relevance. *Journal of Morphology*, **266**, 167–181.

Cheng, G., Marotte, L. R., Mai, J. K. and Ashwell, K. W. S. (2002). Early development of the hypothalamus of a wallaby (*Macropus eugenii*). *Journal of Comparative Neurology*, **453**, 199–215.

Cheng, G., Marotte, L. R. and Ashwell, K. W. (2003). Cyto- and chemoarchitecture of the hypothalamus of a wallaby (*Macropus eugenii*) with special emphasis on oxytocin- and vasopressinergic neurons. *Anatomy and Embryology*, **207**, 233–253.

Cheung, T. C. and Hearn, J. P. (2002). Molecular cloning and tissue expression of the gonadotrophin-releasing hormone receptor in the tammar wallaby (*Macropus eugenii*). *Reproduction Fertility and Development*, **14**, 157–164.

Cheung, A. F., Kondo, S., Abdel-Mannan, O. *et al.* (2009). The subventricular zone is the developmental milestone of a 6-layered neocortex: comparisons in metatherian and eutherian mammals. *Cerebral Cortex*, **20**, 1071–1081.

Chiaia, N. L., Rhoades, R. W., Bennett-Clarke, C. A., Fish, S. E. and Killackey, H. P. (1991). Thalamic processing of vibrissal information in the rat. I. Afferent input to the medial ventral posterior and posterior nuclei. *Journal of Comparative Neurology*, **314**, 201–216.

Christensen, J. L. and Hill, R. M. (1970a). Receptive fields of single cells of a marsupial visual cortex of *Didelphis virginiana*. *Experientia*, **26**, 43–44.

Christensen, J. L. and Hill, R. M. (1970b). Response properties of single cells of a marsupial visual cortex. *American Journal of Optometry, Archives of the American Academy of Optometry*, **47**, 547–556.

Christensen, P., Maisey, K. and Perry, D. H. (1984). Radio-tracking the numbat, *Myrmecobius fasciatus*, in the Perup Forest of Western Australia. *Australian Wildlife Research*, **11**, 275–288.

Chuah, M. I., Tennent, R. and Teague, R. (1997). Developmental anatomy of the primary olfactory pathway in the opossum *Monodelphis domestica*. *Histology and Histopathology*, **12**, 799–806.

Chung, W. C., Swaab, D. F. and De Vries, G. J. (2000). Apoptosis during sexual differentiation of the bed nucleus of the stria terminalis in the rat brain. *Journal of Neurobiology*, **43**, 234–243.

Cifelli, R. L. and Davis, B. M. (2003). Paleontology. Marsupial origins. *Science*, **302**, 1899–1900.

Clarke, A. and Rothery, P. (2008). Scaling of body temperature in mammals and birds. *Functional Ecology*, **22**, 58–67.

Clemens, W. A., Richardson, B. J. and Baverstock, P. R. (1989). Biogeography and phylogeny of the Metatheria. In *Fauna of Australia*, vol. B1 (ed. D. W. Walton and B. J. Richardson), Canberra: Australian Government Publishing Service. pp. 527–548.

Clezy, J. K., Dennis, B. J. and Kerr, D. I. (1961). A degeneration study of the somaesthetic afferent systems in the marsupial phalanger, *Trichosurus vulpecula*. *Australian Journal of Experimental Biology and Medical Science*, **39**, 19–27.

Coleman, G. T., Zhang, H. Q., Murray, G. M., Zazhariah, M. K. and Rowe, M. J. (1999). Organisation of somatosensory areas I and II in marsupial cerebral cortex: parallel processing in the possum sensory cortex. *Journal of Neurophysiology*, **81**, 2316–2324.

Coleman, L. A. and Beazley, L. D. (1988). The accessory optic system of the wallaby, *Setonix brachyurus*: anatomy in normal animals and after early unilateral eye removal. *Journal of Comparative Neurology*, **273**, 359–376.

Coles, R.B. and Guppy, A. (1986). Biophysical aspects of hearing in the tammar wallaby, *Macropus eugenii*. *Journal of Experimental Biology*, **121**, 371–394.

Coles, R.B. and Hill, K. G. (1981). Ear movements in the tammar wallaby, *Macropus eugenii*. *Proceedings of the Australian Physiological and Pharmacological Society*, **12**, 169.

Collins, P. (1983). Morphological features of the surface of the sub-commissural organ and aqueduct in the red-necked wallaby (*Wallabia rufogriseus*). *Journal of Anatomy*, **137**, 665–673.

Collin, S. P. (1997). Specialisations of the teleost visual system: adaptive diversity from shallow-water to deep-sea. *Acta Physiologica Scandinavica. Supplementum*, **638**, 5–24.

Collin, S. P. and Pettigrew, J. D. (1988a). Retinal topography in reef teleosts. I. Some species with well-developed areae but poorly-developed streaks. *Brain Behavior and Evolution*, **31**, 269–282.

Collin, S. P. and Pettigrew, J. D. (1988b). Retinal topography in reef teleosts. II. Some species with prominent horizontal streaks and high-density areae. *Brain Behavior and Evolution*, **31**, 283–295.

Comans, P. E., McLennan, I. S. and Mark, R. F. (1987). Mammalian motoneuron cell death: development of the lateral motor column of a wallaby (*Macropus eugenii*). *Journal of Comparative Neurology*, **260**, 627–634.

Comans, P. E., McLennan, I. S., Mark, R. F. and Hendry, I. A. (1988). Mammalian motoneuron development: effect of peripheral deprivation on motoneuron numbers in a marsupial. *Journal of Comparative Neurology*, **270**, 111–120.

Commins, D. and Yahr, P. (1984). Acetylcholinesterase activity in the sexually dimorphic area of the gerbil brain: sex differences and the influences of adult gonadal steroids. *Journal of Comparative Neurology*, **224**, 123–131.

Cone-Wesson, B. K., Hill, K. G. and Liu, G. B. (1997). Auditory brainstem response in tammar wallaby (*Macropus eugenii*). *Hearing Research*, **105**, 119–129.

Contreras-Rodriguez, J., Gonvalez-Soriano, J., Martinez-Sainz, P. and Rodriguez-Veiga, E. (2002). The thalamic reticular and perireticular nuclei in developing rabbits: patterns of parvalbumin expression. *Developmental Brain Research*, **136**, 123–133.

Corbett, L. K. (1995) *The Dingo in Australia and Asia*. Sydney: The University of New South Wales Press.

Cosenza, R. M., Sousa Neto, J. A. and Machado, A. B. (1990) The pineal recess of the opossum: a scanning and transmission electron microscope study. *Microscopía Electrónica y Biología Celular*, **14**, 101–113.

Covenas, R., De Leon, M., Alonso, J. R. *et al.* (1991). Distribution of parvalbumin-immunoreactivity in the rat thalamus using a monoclonal antibody. *Archives Italiennes de Biologie*, **129**, 199–210.

Cowing, J. A., Poopalasundaram, S., Wilkie, S. E. *et al.* (2002). The molecular mechanism for the spectral shifts between vertebrate ultraviolet- and violet-sensitive cone visual pigments. *Biochemical Journal*, **367**, 129–135.

Cowing, J. A., Arrese, C. A., Davies, W. L., Beazley, L. D. and Hunt, D. M. (2008). Cone visual pigments in two marsupial species: the fat-tailed dunnart (*Sminthopsis crassicaudata*) and the honey possum (Tarsipes rostratus). *Proceedings of The Royal Society. B. Biological Sciences*, **275**, 1491–1499.

Cowley, A. R. (1973). The nuclei of the cochlear nerve of the red kangaroo, *Megaleia rufus*. *Journal für Hirnforschung*, **14**, 287–301.

Cowley, A. R. (1976). The nuclei of the vestibular nerve of the red kangaroo, *Megaleia rufus*. *Journal für Hirnforschung*, **17**, 61–72.

Craigie, E. H. (1945). The architecture of the cerebral capillary bed. *Biological Reviews*, **20**, 133–146.

Crewther, D. P., Crewther, S. G. and Sanderson, K. J. (1984). Primary visual cortex in the brush-tailed possum: receptive field properties and corticocortical connections. *Brain Behavior and Evolution*, **24**, 184–197.

Crewther, D. P., Nelson, J. E. and Crewther, S. G. (1988). Afferent input for target survival in marsupial visual development. *Neuroscience Letters*, **86**, 147–154.

Croft, D. B. and Eisenberg, J. F. (2006). Behaviour. In *Marsupials* (ed. P. J. Armati, C. R. Dickman and I. D. Hume). Cambridge: Cambridge University Press. pp. 229–298.

Culberson, J. L. (1987). Projection of cervical dorsal root fibres to the medulla oblongata in the brush tailed possum, *Trichosurus vulpecula*. *American Journal of Anatomy*, **179**, 232–242.

Culberson, J. L. and Albright, B. C. (1984). Morphologic evidence for fiber sorting in the fasciculus cuneatus. *Experimental Neurology*, **85**, 358–370.

Cummings, D. M. and Brunjes, P. C. (1995). Migrating luteinizing hormone-releasing hormone (LHRH) neurons and processes are associated with a substrate that expresses S100. *Developmental Brain Research*, **88**, 148–157.

Cummings, D. M., Malun, D. and Brunjes, P. C. (1997). Development of the anterior commissure in the opossum: midline extracellular space and glia coincide with early axon decussation. *Journal of Neurobiology*, **32**, 403–414.

Curlewis, J. D., Renfree, M. B., Sheldrick, E. L. and Flint, A. P. (1988). Mesotocin and luteal function in macropodid marsupials. *Journal of Endocrinology*, **117**, 367–372.

Curlewis, J. D., Saunders, M. C., Kuang, J., Harrison, G. A. and Cooper, D. W. (1998). Cloning and sequence analysis of a pituitary prolactin cDNA from the brushtail possum (*Trichosurus vulpecula*). *General and Comparative Endocrinology*, **111**, 61–67.

Dahlström, A. and Fuxe, K. (1964a). Localisation of monoamines in the lower brainstem. *Experientia*, **20**, 398–399.

Dahlström, A. and Fuxe, K. (1964b). Evidence for the existence of monoamine-containing neurons in the central nervous system. I. Demonstration of monoamines in the cell bodies of brain stem neurons. *Acta Physiologica Scandinavica*, **62**, 1–155.

Dallos, P., Harris, D., Ozdamar, O. and Ryan, A. (1978). Behavioural, compound action potential, and single unit thresholds: relationship in normal and abnormal ears. *Journal of Acoustic Society of America*, **64**, 151–157.

Darian-Smith, I. (1973). The trigeminal system. In *Handbook of Sensory Physiology*. (ed. A. Iggo). New York: Springer-Verlag. pp. 271–314.

Darlington, R. B., Dunlop, S. A. and Finlay, B. L. (1999). Neural development in metatherian and eutherian mammals: variation and constraint. *Journal of Comparative Neurology*, **411**, 359–368.

Davies, W. L., Carvalho, L. S., Cowing, J. A. *et al.* (2007). Visual pigments of the platypus: a novel route to mammalian colour vision. *Current Biology*, **17**, R161–163.

Dawson, D. R. and Killackey, H. P. (1987). The organisation and mutability of the forepaw and hindpaw representations in the somatosensory cortex of the neonatal rat. *Journal of Comparative Neurology*, **256**, 246–256.

Dawson, T. J. and Hulbert, A. J. (1970). Standard metabolism, body temperature, and surface areas of Australian marsupials. *American Journal of Physiology*, **218**, 1233–1238.

Dawson, T. J. and Taylor, C. R. (1973). Energetic cost of locomotion in kangaroos. *Nature*, **246**, 313–314.

De Biasi, S., Arcelli, P. and Spreafico, R. (1994). Parvalbumin immunoreactivity in the thalamus of guinea pig: light and electron microscopic correlation with gamma-aminobutyric acid immunoreactivity. *Journal of Comparative Neurology*, **348**, 556–569.

De Carlos, J. A., Lopez-Mascaraque, L. and Valverde, F. (1996). Early olfactory fiber projections and cell migration into the rat telencephalon. *International Journal of Developmental Neuroscience*, **14**, 853–866.

De Leon, M., McRae, T., Rusinek, H. *et al.* (1997). Cortisol reduces hippocampal glucose metabolism in normal elderly but not in Alzheimer's disease. *Journal of Clinical Endocrinology*, **82**, 3251–3259.

De Muizon, C., Cifelli, R. L. and Paz, R. C. (1997). The origin of the dog-like borhyaenoid marsupials of South America. *Nature*, **389**, 486–489.

Deaner, R. O., Isler, K., Burkart, J. and van Schaik, C. (2007). Overall brain size, and not encephalisation quotient, best predicts cognitive ability across non-human primates. *Brain Behavior and Evolution*, **70**, 115–124.

Deeb, S. S., Wakefield, M. J., Tada, T. *et al.* (2003). The cone visual pigments of an Australian marsupial, the tammar wallaby (*Macropus eugenii*): sequence, spectral tuning, and evolution. *Molecular Biology and Evolution*, **20**, 1642–1649.

Dempster, E. (1994). Vocalisations of adult northern quolls, *Dasyurus hallucatus*. *Australian Mammalogy*, **17**, 43–49.

Dennis, B. J. and Kerr, D. I. (1961). Somaesthetic pathways in the marsupial phalanger, *Trichosurus vulpecula*. *Australian Journal of Experimental Biology and Medical Science*, **39**, 29–42.

Diamond, I. T., Fitzpatrick, D. and Schmechel, D. (1993). Calcium-binding proteins distinguish large and small cells of the ventral posterior and lateral geniculate nuclei of the prosimian galago and the tree shrew (*Tupaia belangeri*). *Proceedings of the National Academy of Science USA*, **90**, 1425–1429.

Dickman C. R. (1986). An experimental study of competition between two species of dasyurid marsupials. *Ecological Monographs*, **56**, 221–241.

Dickman, C. R. and Woodford Ganf, R. (2007). *A Fragile Balance. The Extraordinary Story of Australian Marsupials.* Melbourne: Craftsman House.

Dillon, L. S. (1963). Comparative studies of the brain in the Macropodidae. Contribution to the phylogeny of the mammalian brain. II. *Journal of Comparative Neurology*, **120**, 43–51.

Ding, Y. and Marotte, L. R. (1996). The initial stages of development of the retinocollicular projection in the wallaby (*Macropus eugenii*): distribution of ganglion cells in the retina and their axons in the superior colliculus. *Anatomy and Embryology*, **194**, 301–317.

Ding, Y. and Marotte, L. R. (1997). Retinotopic order in the optic nerve and superior colliculus during development of the retinocollicular projection in the wallaby (*Macropus eugenii*). *Anatomy and Embryology*, **196**, 141–158.

Dinopoulos, A., Papadopoulos, G. C., Michaloudi, H. *et al.* (1992). Claustrum in the hedgehog (*Erinaceus europaeus*) brain: cytoarchitecture and connections with cortical and subcortical structures. *Journal of Comparative Neurology*, **316**, 187–205.

Dkhissi-Benyahya, O., Szel, A., Degrip, W. J. and Cooper, H. M. (2001). Short and mid-wavelength cone distribution in a nocturnal Strepsirrhine primate (*Microcebus murinus*). *Journal of Comparative Neurology*, **438**, 490–504.

Doetsch, F. (2003a). The glial identity of neural stem cells. *Nature Neuroscience*, **11**, 1127–1134.

Doetsch, F. (2003b), A niche for adult neural stem cells. *Current Opinion in Genetics and Development*, **13**, 543–550.

Doetsch, F. and Hen, R. (2005). Young and excitable: the function of new neurons in the adult mammalian brain. *Current Opinion in Neurobiology*, **15**, 121–128.

Doetsch, F. and Scharff, C. (2001). Challenges for brain repair: insights from adult neurogenesis in birds and mammals. *Brain Behavior and Evolution*, **58**, 306–322.

Dollery, C. T., Bulpitt, C. J. and Kohner, E. M. (1969). Oxygen supply to the retina from the retinal and choroidal circulations at normal and increased arterial oxygen tensions. *Investigative Ophthalmology*, **8**, 588–594.

Doran, A. and Garson J. G. (1879). Morphology of the mammalian ossicula auditus. *Journal of Anatomy and Physiology*, **13**, 401–406.

Doré. L., Jacobson, C. D. and Hawkes, R. (1990). Organisation and postnatal development of zebrin II antigenic compartmentation in the cerebellar vermis of the grey opossum, *Monodelphis domestica*. *Journal of Comparative Neurology*, **291**, 431–449.

Douglas, R. H., Harper, R. D. and Case, J. F. (1998). The pupil response of a teleost fish, *Porichthys notatus*: description and comparison to other species. *Vision Research*, **38**, 2697–2710.

Dreher, B. and Robinson, S. R. (1988). Development of the retinofugal pathway in birds and mammals: evidence for a common 'timetable'. *Brain Behavior and Evolution*, **31**, 369–390.

Dunlop, S. A. and Beazley, L. D. (1985). Changing distribution of retinal ganglion cells during area centralis and visual streak formation in the marsupial *Setonix brachyurus*. *Brain Research*, **355**, 81–90.

Dunlop, S. A., Longley, W. A. and Beazley, L. D. (1987). Development of the area centralis and visual streak in the grey kangaroo *Macropus fuliginosus*. *Vision Research*, **27**, 151–164.

Dunlop, S. A., Ross, W. M. and Beazley, L. D. (1994). The retinal ganglion cell layer and optic nerve in a marsupial, the honey possum (*Tarsipes rostratus*). *Brain Behavior and Evolution*, **44**, 307–323.

Dunlop, S. A., Tee, L. B., Lund, R. D. and Beazley, L. D. (1997). Development of primary visual projections occurs entirely postnatally in the fat-tailed dunnart, a marsupial mouse, *Sminthopsis crassicaudata*. *Journal of Comparative Neurology*, **384**, 26–40.

Dunlop, S. A., Tee, L. B. and Beazley, L. D. (2000). Topographic order of retinofugal axons in a marsupial: implications for map formation in visual nuclei. *Journal of Comparative Neurology*, **428**, 33–44.

Dunlop, S. A., Rodger, J. and Beazley, L. D. (2007). Compensatory and transneuronal plasticity after early collicular ablation. *Journal of Comparative Neurology*, **500**, 1117–1126.

Dziegielewska, K. M., Hinds, L. A., Møllgård, K., Reynolds, M. L. and Saunders, N. R. (1988). Blood–brain, blood–cerebrospinal fluid and cerebrospinal fluid–brain barriers in a marsupial (*Macropus eugenii*) during development. *Journal of Physiology*, **403**, 367–388.

Ebner, F. F. (1969). A comparison of primitive forebrain organisation in metatherian and eutherian mammals. *Annals of the New York Academy of Science*, **167**, 241–257.

Ek, C. J., Habgood, M. D., Dziegielewska, K. M. and Saunders, N. R. (2003). Structural characteristics and barrier properties of the choroid plexuses in developing brain of the opossum (*Monodelphis domestica*). *Journal of Comparative Neurology*, **460**, 451–464.

Ek, C. J., Dziegielewska, K. M., Stolp, H. and Saunders, N. R. (2006). Functional effectiveness of the blood brain barrier to small water-soluble molecules in developing and adult opossum (*Monodelphis domestica*). *Journal of Comparative Neurology*, **496**, 13–26.

Ellendorff, F., Tyndale-Biscoe, C. H. and Mark, R. F. (1988). Ontogenetic changes in olfactory bulb neural activities of pouched young tammar wallabies. In *The Endocrine Control of the Fetus* (ed. W. Künzel and A. Jensen), Berlin: Springer-Verlag. pp. 193–200.

Elston, G. and Manger, P. (1999). The organisation and connections of somatosensory cortex in the brush-tailed possum (*Trichosurus vulpecula*): evidence for multiple, topographically organised and interconnected representations in an Australian marsupial. *Somatosensory and Motor Research*, **16**, 312–337.

Endo, T., Yoshino, J., Kado, K. and Tochinai, S. (2007). Brain regeneration in anuran amphibians. *Development, Growth and Differentiation*, **49**, 121–129.

Eshet, R., Gil-Ad, I., Apelboym, O. *et al.* (2004). Modulation of brain insulin-like growth factor I (IGF-I) binding sites and hypothalamic GHRH and somatostatin levels by exogenous growth hormone and IGF-I in juvenile rats. *Journal of Molecular Neuroscience*, **22**, 179–188.

Ernest, S. K. M. (2003). Life history characteristics of placental non-volant mammals. *Ecology*, **84**, 3402.

Etgen, A. M. and Fadem, B. H. (1989). Ontogeny of estrogen binding sites in the brain of grey short-tailed opossums (*Monodelphis domestica*). *Developmental Brain Research*, **49**, 131–133.

Eugenín, J. and Nicholls, J. G. (2000). Control of respiration in the isolated central nervous system of the neonatal opossum, Monodelphis domestica. *Brain Research Bulletin*, **53**, 605–613.

Evans, E.F. (1972). The frequency response and other properties of single fibres in the guinea-pig cochlear nerve. *Journal of Physiology*, **226**, 263–287.

Fadem, B. H., Walters, M. and MacLusky, N. J. (1993). Neural aromatase activity in a marsupial, the grey short-tailed opossum (*Monodelphis domestica*): ontogeny during postnatal development and androgen regulation in adulthood. *Developmental Brain Research*, **74**, 199–205.

Fan, Z. W., Eng, J., Shaw, G. and Yalow, R. S. (1988). Cholecystokinin octapeptide purified from brains of Australian marsupials. *Peptides*, **9**, 429–431.

Farber, J. P. (1983). Expiratory motor responses in the suckling opossum. *Journal of Applied Physiology*, **54**, 919–925.

Farber, J. P. (1985). Motor responses to positive-pressure breathing in the developing opossum. *Journal of Applied Physiology*, **58**, 1489–1495.

Farber, J. P. (1988). Medullary inspiratory activity during opossum development. *American Journal of Physiology*, **254**, R578–584.

Farber, J. P. (1989). Medullary expiratory activity during opossum development. *Journal of Applied Physiology*, **66**, 1606–1612.

Farber, J. P. (1993a). GABAergic effects on respiratory neuronal discharge during opossum development. *American Journal of Physiology*, **264**, R331–336.

Farber, J. P. (1993b). Maximum discharge rates of respiratory neurons during opossum development. *Journal of Applied Physiology*, **75**, 2040–2044.

Farber, J. P. (1995). Effects on breathing of medullary bicuculline microinjections in immature opossums. *American Journal of Physiology*, **269**, R1295–R1300.

Farber, J. P., Fisher, J. T. and Sant' Ambrogio, G. (1984). Airway receptor activity in the developing opossum. *American Journal of Physiology*, **246**, R753–R758.

Farmer, S. W., Licht, P., Gallo, A. B. *et al.* (1981). Studies on several marsupial anterior pituitary hormones. *General and Comparative Endocrinology*, **43**, 336–345.

Fasick, J. I., Applebury, M. L. and Oprian, D. D. (2002). Spectral tuning in the mammalian short-wavelength sensitive cone pigments. *Biochemistry*, **41**, 6860–6865.

Faulks, I. and Mark, R. (1982). The somatotopic organisation of the thalamic ventrobasal complex in a wallaby, the tammar (*Macropus eugenii*). *Proceedings of the Australian Physiological and Pharmacological Society*, **13**, 179P.

Faulstich, M., Kossl, M. and Reimer, K. (1996). Analysis of non-linear cochlear mechanics in the marsupial *Monodelphis domestica*: ancestral and modern mammalian features. *Hearing Research*, **94**, 47–53.

Ferguson, I. A., Hardman, C. D., Marotte, L. R. *et al.* (1999). Serotonergic neurons in the brainstem of the wallaby, *Macropus eugenii*. *Journal of Comparative Neurology*, **411**, 535–549.

Fernandez, C. and Schmidt, R. S. (1963). The opossum ear and evolution of the coiled cochlea. *Journal of Comparative Neurology*, **121**, 151–159.

Fiala, J. C. (2007). Mechanisms of amyloid plaque pathogenesis. *Acta Neuropathologica*, **114**, 551–571.

Fidler, A. E., Lawrence, S. B., Vanmontfort, D. M., Tisdall, D. J. and McNatty, K. P. (1998). The Australian brushtail possum (*Trichosurus vulpecula*) gonadotrophin alpha-subunit: analysis of cDNA sequence and pattern of expression. *Journal of Molecular Endocrinology*, **20**, 345–353.

Finlay, B. L., Hersman, M. N. and Darlington, R. B. (1998). Patterns of vertebrate neurogenesis and the paths of vertebrate evolution. *Brain Behavior and Evolution*, **52**, 232–242.

Fisher, D. O., Double, M. C., Blomberg, S. P., Jennions, M. D. and Cockburn, A. (2006). Post-mating sexual selection increases lifetime fitness of polyandrous females in the wild. *Nature*, **444**, 89–92.

Fite, K. V. (1973). The visual fields of the frog and toad: a comparative study. *Behavioural Biology*, **9**, 707–718.

Fitzgerald, M. (1987). Spontaneous and evoked activity of fetal primary afferents in vivo. *Nature*, **326**, 603–605.

Flannery, T. F. (1994). *The Future Eaters*. Sydney: Reed Books.

Flannery, T. F. (1995). *Mammals of New Guinea*. Sydney: Reed Books.

Flett, D. L., Marotte, L. R. and Mark, R.F. (1988). Retinal projections to the superior colliculus and dorsal lateral geniculate nucleus in the tammar wallaby (*Macropus eugenii*): I. Normal topography. *Journal of Comparative Neurology*, **271**, 257–273.

Flett, D. L., Lim, C. H., Ho, S. M., Mark, R. F. and Marotte, L. R. (2006). Retinocollicular synaptogenesis and synaptic transmission during formation of the visual map in the superior colliculus of the wallaby (*Macropus eugenii*). *European Journal of Neuroscience*, **23**, 3043–3050.

Flurkey, K., Brandvain, Y., Klebanov, S. *et al.* (2007). PohnB6F1: a cross of wild and domestic mice that is a new model of extended female reproductive life-span. *The Journals of Gerontology A. Biological Sciences and Medical Sciences*, **62**, 1187–1198.

Forrester, J. V., Dick, A. D., McMenamin, P. G. and Lee, W. R. (2002). *The Eye: Basic Sciences in Practice*. London: WB Saunders.

Fox, C. A., Adam, D. E., Watson, R. E. Jr, Hoffman, G. E. and Jacobson, C. D. (1990). Immunohistochemical localisation of cholecystokinin in the medial preoptic area and anterior hypothalamus of the Brazilian grey short-tailed opossum: a sex difference. *Journal of Neurobiology*, **21**, 705–718.

Fox, C. A., Ross, L. R., Handa, R. J. and Jacobson, C. D. (1991a). Localisation of cells containing estrogen receptor-like immunoreactivity in the Brazilian opossum brain. *Brain Research*, **546**, 96–105.

Fox, C. A., Ross, L. R. and Jacobson, C. D. (1991b). Ontogeny of cells containing estrogen receptor-like immunoreactivity in the Brazilian opossum brain. *Developmental Brain Research*, **63**, 209–219.

Fox, C. A., Jeyapalan, M., Ross, L. R. and Jacobson, C. D. (1991c). Ontogeny of cholecystokinin-like immunoreactivity in the Brazilian opossum brain. *Developmental Brain Research*, **64**, 1–18.

Frame, T. (2009) *Evolution in the Antipodes. Charles Darwin and Australia*. Sydney: UNSW Press.

Franca, J. G., Volchan, E., Jain, N. *et al.* (2000). Distribution of NADPH-diaphorase cells in visual and somatosensory cortex in four mammalian species. *Brain Research*, **864**, 163–175.

Frankenberg, E. (1979). Pupillary response to light in gekkonid lizards having various times of daily activity. *Vision Research*, **19**, 235–245.

Franklin, K. B. J. and Paxinos, G. (2007) *The Mouse Brain in Stereotaxic Co-ordinates*. 3rd edn. San Diego, London: Elsevier Academic.

Frappell, P. B. (2008). Ontogeny and allometry of metabolic rate and ventilation in the marsupial: matching supply and demand from ectothermy to endothermy. *Comparative Biochemistry and Physiology A – Molecular and Integrative Physiology*, **150**, 181–188.

Frappell P. B., and McFarlane, P. M. (2006). Development of the respiratory system in marsupials. *Respiratory Physiology and Neurobiology* **154**, 252–267.

Frassoni, C., Bentivoglio, M., Spreafico R. *et al.* (1991). Postnatal development of calbindin and parvalbumin immunoreactivity in the thalamus of the rat. *Developmental Brain Research*, **58**, 243–249.

Freedman, M. S., Lucas, R. J., Soni, B. *et al.* (1999). Regulation of mammalian circadian behaviour by non-rod, non-cone, ocular photoreceptors. *Science*, **284**, 502–504.

Freeman, B. and Tancred, E. (1978). The number and distribution of ganglion cells in the retina of the brush-tailed possum, *Trichosurus vulpecula*. *Journal of Comparative Neurology*, **177**, 557–567.

Freyer, C., Zeller, U. and Renfree, M. B. (2003). The marsupial placenta: a phylogenetic analysis. *Journal of Experimental Zoology A*, **299**, 59–77.

Freyer, C., Zeller, U. and Renfree, M. B. (2007). Placental function in two distantly related marsupials. *Placenta*, **28**, 249–257.

Friant, M. (1961). Apropos of the fasciculus aberrans of the ventral commissure of the brain of diprotodont marsupials. *Acta Neurologica (Belgium)*, **61**, 174–176.

Friauf, E. and Lohmann, C. (1999). Development of auditory brainstem circuitry. Activity-dependent and activity-independent processes. *Cell and Tissue Research*, **297**, 187–195.

Friauf, E. and Shatz, C. J. (1991). Changing patterns of synaptic input to subplate and cortical plate during development of visual cortex. *Journal of Neurophysiology*, **66**, 2059–2071.

Friend, J. A. and Burrows, R. G. (1983). Bringing up young numbats. *South West Australian Nature Society*, **12**, 3–9.

Frost, S. B. and Masterton, R. B. (1994). Hearing in primitive mammals: *Monodelphis domestica* and *Marmosa elegans*. *Hearing Research*, **76**, 67–72.

Frost, S. B., Milliken, G. W., Plautz, E. J., Masterton, R. B. and Nudo, R. J. (2000). Somatosensory and motor representations in cerebral cortex of a primitive mammal (*Monodelphis domestica*): a window into the early evolution of sensorimotor cortex. *Journal of Comparative Neurology*, **421**, 29–51.

Furber, S. E. and Watson, C. R. (1979). Ratio of inferior olivary cells to Purkinje cells in a marsupial (*Trichosurus vulpecula*). *Acta Anatomica*, **104**, 363–367.

Gabbott, P. L. A. and Bacon, S. J. (1996). Local circuit neurons in the medial prefrontal cortex (areas 24a, b, c, 25 and 32) in the monkey: II. Quantitative areal and laminar distributions. *Journal of Comparative Neurology*, **364**, 609–636.

Gabernet, L., Meskenaïte, V. and Hepp-Reymond, M-C. (1999). Parcellation of the lateral premotor cortex of the macaque monkey based on staining with neurofilament antibody SMI-32. *Experimental Brain Research*, **128**, 188–193.

Garde, E., Heide-Jørgensen, M. P., Hansen, S. H., Nachman, G. S. and Forchhammer, M. C. (2007). Age-specific growth and remarkable longevity in narwhals (*Monodon monoceros*) from west Greenland as estimated by aspartic acid racemisation. *Journal of Mammalogy*, **88**, 49–58.

Garrett, B., Østerballe, R., Stomianka, L. and Geneser, F. A. (1994). Cytoarchitecture and staining for acetylcholinesterase and zinc in the visual cortex of the parma wallaby (*Macropus parma*). *Brain Behavior and Evolution*, **43**, 162–172.

Gasse, H. and Meyer, W. (1995a). Neuron-specific enolase as a marker of hypothalamo-neurohypophyseal development in postnatal *Monodelphis domestica* (Marsupialia). *Neuroscience Letters*, **189**, 54–56.

Gasse, H. and Meyer, W. (1995b). Immunohistochemical demonstration of adenohypophyseal hormones during postnatal ontogenesis in the grey short-tailed opossum, *Monodelphis domestica* (Marsupialia). *European Journal of Morphology*, **33**, 373–380.

Gates, G. R. and Aitken, L. M. (1982). Auditory cortex in the marsupial possum *Trichosurus vulpecula*. *Hearing Research*, **7**, 1–11.

Gates, T. S., Weedman, D. L., Pongstaporn, T. and Ryugo, D. K. (1996). Immunocytochemical localisation of glycine in a subset of cartwheel cells of the dorsal cochlear nucleus in rats. *Hearing Research*, **96**, 157–166.

Gaughwin, M. D. (1979). The occurrence of flehmen in a marsupial – the hairy-nosed wombat (*Lasiorhinus latifrons*). *Animal Behaviour*, **27**, 1063–1065.

Geiser, F. (2004). Metabolic rate and body temperature reduction during hibernation and daily torpor. *Annual Review of Physiology*, **66**, 239–274.

Geiser, F., Westman W., McAllan, B. and Brigham, R. M. (2006). Development of thermoregulation and torpor in a marsupial: energetic and evolutionary implications. *Journal of Comparative Physiology B*, **176**, 107–116.

Gemmell, R. T. and Nelson, J. (1988a). Ultrastructure of the olfactory system of three newborn marsupial species. *Anatomical Record*, **221**, 655–662.

Gemmell, R. T. and Nelson, J. (1988b). The ultrastructure of the pituitary and the adrenal gland of three newborn marsupials (*Dasyurus hallucatus, Trichosurus vulpecula* and *Isoodon macrourus*). *Anatomy and Embryology*, **177**, 395–402.

Gemmell, R. T. and Nelson, J. (1989). Vestibular system of the newborn marsupial cat, *Dasyurus hallucatus*. *Anatomical Record*, **225**, 203–208.

Gemmell, R. T. and Nelson, J. (1992). Development of the vestibular and auditory system of the northern native cat, *Dasyurus hallucatus*. *Anatomical Record*, **234**, 136–143.

Gemmel, R. T. and Rose, R. W. (1989). The senses involved in movement of some newborn Macropodoidea and other marsupials from cloaca to pouch. In *Kangaroos, Wallabies and Rat-kangaroos* (ed. G. Grigg, P. Jarman, J. Hume). Sydney: Surrey Beatty and Sons. pp. 339–247.

Gemmell, R. T. and Selwood, L. (1994). Structural development in the newborn marsupial, the stripe-faced dunnart, *Sminthopsis macroura*. *Acta Anatomica*, **149**, 1–12.

Gemmell, R. T. and Sernia, C. (1989). Immunocytochemical location of oxytocin and mesotocin within the hypothalamus of two Australian marsupials, the bandicoot *Isoodon macrourus* and the brushtail possum *Trichosurus vulpecula*. *General and Comparative Endocrinology*, **75**, 96–102.

Gemmell, R. T., Peters, B. and Nelson, J. (1988). Ultrastructural identification of Merkel cells around the mouth of the newborn marsupial. *Anatomy and Embryology*, **177**, 403–408.

Gemmell, R. T., Chua, T., Bathgate, R. A. and Sernia, C. (1993). Posterior pituitary of the newborn marsupial possum, *Trichosurus vulpecula*. *Anatomical Record*, **237**, 228–235.

Gemmell, R. T., Veitch, C. and Nelson, J. (1999). Birth in the northern brown bandicoot, *Isoodon macrourus* (Marsupialia: Peramelidae). *Australian Journal of Zoology*, **47**, 517–528.

Gerfen, C. R. (2004). Basal ganglia. In *The Rat Nervous System* (ed. G. Paxinos). London: Elsevier. pp. 455–508.

Geyer, S., Zilles, K., Luppino, G. and Matelli, M. (2000). Neurofilament protein distribution in the macaque monkey dorsolateral premotor cortex. *European Journal of Neuroscience*, **12**, 1554–1566.

Ghosh, A. and Shatz, C. J. (1992). Pathfinding and target selection by developing geniculocortical axons. *Journal of Neuroscience*, **12**, 39–55.

Gill, M. S. (2006). Endocrine targets for pharmacological intervention in aging in *Caenorhabditis elegans*. *Aging Cell*, **5**, 23–30.

Gilland, E. and Baker, R. (2005). Evolutionary patterns of cranial nerve efferent nuclei in vertebrates. *Brain Behavior and Evolution*, **66**, 234–254.

Gillilan, L. A. (1972). Blood supply to primitive mammalian brains. *Journal of Comparative Neurology*, **145**, 209–221.

Gilmore, D. P. (2002). Sexual dimorphism in the central nervous system of marsupials. *International Review of Cytology*, **214**, 193–224.

Glendenning, K. K. (2006). Thalamic development of the grey short-tailed opossum. *Journal of Mammalogy*, **87**, 554–562.

Glendenning, K. K. and Masterton, R. B. (1998). Comparative morphometry of mammalian central auditory systems: variation in nuclei and form of the ascending system. *Brain Behavior and Evolution*, **51**, 59–89.

Glezer, I. I., Hof, P. R., Leranth, C. and Morgane, P. J. (1993). Calcium-binding protein-containing neuronal populations in mammalian visual cortex: a comparative study in whales, insectivores, bats, rodents and primates. *Cerebral Cortex*, **3**, 249–272.

Glezer, I. I., Hof, P. R. and Morgane, P. J. (1998). Comparative analysis of calcium-binding protein-immunoreactive neuronal populations in the auditory and visual systems of the bottlenose dolphin (*Tursiops truncatus*) and the macaque monkey (*Macaca fascicularis*). *Journal of Chemical Neuroanatomy*, **15**, 203–237.

Godement, P., Salaun, J. and Imbert, M. (1984). Prenatal and postnatal development of retinogeniculate and retinocollicular projections in the mouse. *Journal of Comparative Neurology*, **230**, 552–575.

Godthelp, H., Archer, M., Cifelli, R., Hand, S. J. and Gilkeson, C. F. (1992). Earliest known Australian Tertiary mammal fauna. *Nature*, **356**, 514–516.

Goldby, F. (1939). An experimental investigation of the motor cortex and its connections in the phalanger, *Trichosurus vulpecula*. *Journal of Anatomy*, **74**, 12–33.

Goldby, F. (1941). The normal histology of the thalamus in the phalanger, *Trichosurus vulpecula*. *Journal of Anatomy*, **75**, 197–U10.

Goldman, B. D. (2001). Mammalian photoperiod system: formal properties and neuroendocrine mechanisms of photoperiodic time measurement. *Journal of Biological Rhythms*, **16**, 283–301.

Goldsmith, T. C. (2004). Aging as an evolved characteristic – Weismann's theory reconsidered. *Medical Hypotheses*, **62**, 304–308.

Gorbanova, V. and Seluanov, A. (2009). Coevolution of telomerase activity and body mass in mammals: from mice to beavers. *Mechanisms of Ageing and Development*, **130**, 3–9.

Gorbunova, V., Bozzella, M. J. and Seluanov, A. (2008). Rodents for comparative aging studies: from mice to beavers. *Age*, **30**, 111–119.

Gorski, R. A., Gordon, J. H., Shryne, J. E. and Southam, A. M. (1978). Evidence for a morphological sex difference within the medial preoptic area of the rat brain. *Brain Research*, **148**, 333–346.

Gorski, R. A., Harlan, R. E., Jacobson, C. D., Shryne, J. E. and Southam, A. M. (1980). Evidence for the existence of a sexually dimorphic nucleus in the preoptic area of the rat. *Journal of Comparative Neurology*, **193**, 529–539.

Götz, J., Schild, A., Hoerndli, F. and Pennanen, L. (2004). Amyloid-induced neurofibrillary tangle formation in Alzheimer's disease: insights from transgenic mouse and tissue culture models. *International Journal of Developmental Neuroscience*, **22**, 453–465.

Götz, J., Deters, N., Doldissen, A. *et al.* (2007). A decade of tau transgenic animal models and beyond. *Brain Pathology*, **17**, 91–103.

Gould E., Reeves A. J., Graziano M. S. A. and Gross C. G. (1999a). Neurogenesis in the neocortex of adult primates. *Science*, **286**, 548–552.

Gould, E., Reeves A. J., Fallah M. *et al.* (1999b). Hippocampal neurogenesis in adult Old World primates. *Proceedings of the National Academy of Sciences USA*, **96**, 5263–5267.

Goyal, R., Rasey, S. K. and Wall, J. T. (1992). Current hypotheses of structural pattern formation in the somatosensory system and their potential relevance to humans. *Brain Research*, **583**, 316–319.

Grabiec, M., Turlejski, K. and Djavadian, R. L. (2009). The partial 5-HT1A receptor agonist buspirone enhances neurogenesis in the opossum *(Monodelphis domestica) European Neuropsychopharmacology*, **19**, 431–439.

Grant, G. and Koerber, H. R. (2004). Spinal cord cytoarchitecture. In *The Rat Nervous System* (ed. G. Paxinos). London: Elsevier. pp. 121–128.

Graves, J. A. and Watson, J. M. (1991). Mammalian sex chromosomes: evolution of organisation and function. *Chromosoma*, **101**, 63–68.

Gregory, J. E., McIntyre, A. K. and Proske, U. (1986). Vibration-evoked responses from lamellated corpuscles in the legs of kangaroos. *Experimental Brain Research*, **62**, 648–653.

Groenewegen, H. J. and Witter, M. P. (2004). Thalamus. In *The Rat Nervous System* (ed. G. Paxinos). London: Elsevier. pp. 407–453.

Gruart, A., Lopez-Ramos, J. C., Munoz, M. D., and Delgado-Garcia, J. M. (2008). Aged wild-type and APP, PS1, and APP+PS1 mice present similar deficits in associative learning and synaptic plasticity independent of amyloid load. *Neurobiology of Disease*, **30**, 439–450.

Grubb, M. S., Rossi, F. M., Changeux, J. P. and Thompson, I. D. (2003). Abnormal functional organisation in the dorsal lateral geniculate nucleus of mice lacking the beta 2 subunit of the nicotinic acetylcholine receptor. *Neuron*, **40**, 1161–1172.

Guillery, R. W. and Taylor, J. S. (1993). Different rates of axonal degeneration in the crossed and uncrossed retinofugal pathways of *Monodelphis domestica*. *Journal of Neurocytology*, **22**, 707–716.

Guillery, R. W., Jeffery, G. and Saunders, N. (1999). Visual abnormalities in albino wallabies: a brief note. *Journal of Comparative Neurology*, **403**, 33–38.

Gummer, A. W. and Mark, R. F. (1994). Patterned neural activity in brain stem auditory areas of a prehearing mammal, the tammar wallaby (*Macropus eugenii*). *NeuroReport*, **5**, 685–688.

Haberly, L. B. (1983). Structure of the piriform cortex of the opossum. I. Description of neuron types with Golgi methods. *Journal of Comparative Neurology*, **213**, 163–187.

Haberly, L. B. and Price, J. L. (1978). Association and commissural fiber systems of the olfactory cortex of the rat. I. Systems originating in the piriform cortex and adjacent areas. *Journal of Comparative Neurology*, **178**, 711–740.

Haight, J. R. and Murray. P. F. (1981). The cranial endocast of the early Miocene marsupial, *Wynyardia bassiana*: an assessment of taxonomic relationships based upon comparisons with recent forms. *Brain Behavior and Evolution*, **19**, 17–36.

Haight, J. R. and Nelson, J. E. (1987). A brain that doesn't fit its skull: a comparative study of the brain and endocranium of the koala *Phascolarctos cinereus* (Marsupialia: Phascolarctidae). In *Possums and Opossums: Studies in Evolution* (ed. M. Archer). Sydney: Surrey Beatty and Sons. pp. 331–352.

Haight, J. R. and Neylon. L. (1978a). An atlas of the dorsal thalamus of the marsupial brush-tailed possum, *Trichosurus vulpecula*. *Journal of Anatomy*, **126**, 225–245.

Haight, J. R. and Neylon, L. (1978b). The organisation of neocortical projections from the ventroposterior thalamic complex in the marsupial brush-tailed possum, *Trichosurus vulpecula*: a horseradish peroxidase study. *Journal of Anatomy*, **126**, 459–485.

Haight, J. R. and Neylon, L. (1978c). Morphological variation in the brain of the marsupial brush-tailed possum, *Trichosurus vulpecula*. *Brain Behavior and Evolution*, **15**, 415–445.

Haight, J. R. and Neylon, L. (1979). The organisation of neocortical projections from the ventrolateral thalamic nucleus in the brush-tailed possum, *Trichosurus vulpecula*, and the problem of motor and somatic sensory convergence within the mammalian brain. *Journal of Anatomy*, **129**, 673–694.

Haight, J. R. and Neylon, L. (1981a). A description of the dorsal thalamus of the marsupial native cat, *Dasyurus viverrinus* (Dasyuridae). *Brain Behavior and Evolution*, **19**, 155–179.

Haight, J. R. and Neylon, L. (1981b). An analysis of some thalamic projections to parietofrontal neocortex in the marsupial native cat *Dasyurus viverrinus* (Dasyuridae). *Brain Behavior and Evolution*, **19**, 193–204.

Haight, J. R. and Sanderson, K. J. (1988). Retinal projections in two Australian polyprotodont marsupials: kowari, *Dasyuroides byrnei*, and fat-tailed dunnart, *Sminthopsis crassicaudata* (Dasyuridae). *Brain Behavior and Evolution*, **31**, 96–110.

Haight, J. R. and Weller, W. L. (1973). Neocortical topography in the brush-tailed possum: variability and functional significance of sulci. *Journal of Anatomy*, **116**, 473–474.

Haight, J. R., Sanderson, K. J., Neylon, L. and Patten, G. S. (1980). Relationships of the visual cortex in the marsupial brush-tailed possum, *Trichosurus vulpecula*, a horseradish peroxidase and autoradiographic study. *Journal of Anatomy*, **131**, 387–413.

Hajieva, P., Kuhlmann, C., Luhmann, H. J., and Behl, C. (2009). Impaired calcium homeostasis in aged hippocampal neurons. *Neuroscience Letters*, **451**, 119–123.

Halata, Z. and Munger, B. L. (1980). Sensory nerve endings in rhesus monkey sinus hairs. *Journal of Comparative Neurology*, **192**, 645–663.

Hall, L. S. and Hughes, R. L. (1985). The embryological and cytodifferentiation of the anterior pituitary in the marsupial, *Isoodon macrourus*. *Anatomy and Embryology*, **172**, 353–363.

Hamel, E. G. (1967). The striatum of the kangaroo, *Macropus*. *Alabama Journal of Medical Science*, **4**, 390–398.

Hamilton, T. C. and Johnson, J. I. (1973). Somatotopic organisation related to nuclear morphology in the cuneate-gracile complex of opossums *Didelphis marsupialis virginiana*. *Brain Research*, **51**, 125–140.

Hardie, N.A., Martsi-McClintock, A., Aitkin, L. and Shepherd, R.K. (1998). Neonatal sensorineural hearing loss affects synaptic density in the auditory midbrain. *NeuroReport*, **9**, 2019–2022.

Harman, A. M. (1991). Generation and death of cells in the dorsal lateral geniculate nucleus and superior colliculus of the wallaby, *Setonix brachyurus* (quokka). *Journal of Comparative Neurology*, **313**, 469–478.

Harman, A. M. (1997). Development and cell generation in the hippocampus of a marsupial, the quokka wallaby (*Setonix brachyurus*). *Developmental Brain Research*, **104**, 41–54.

Harman, A. M. and Beazley, L. D. (1986). Development of visual projections in the marsupial, *Setonix brachyurus*. *Anatomy and Embryology*, **175**, 181–188.

Harman, A. M. and Beazley, L. D. (1987). Patterns of cytogenesis in the developing retina of the wallaby *Setonix brachyurus*. *Anatomy and Embryology*, **177**, 123–130.

Harman, A. M. and Beazley, L. D. (1989). Generation of retinal cells in the wallaby, *Setonix brachyurus* (quokka). *Neuroscience*, **28**, 219–232.

Harman, A. M. and Jeffery, G. (1995). Development of the chiasm of a marsupial, the quokka wallaby. *Journal of Comparative Neurology*, **359**, 507–521.

Harman, A. M., and Moore S. (1999). Number of neurons in the retinal ganglion cell layer of the quokka wallaby do not change throughout life. *Anatomical Record*, **256**, 78–83.

Harman, A. M., Coleman, L. A. and Beazley, L. D. (1990). Retinofugal projections in a marsupial, *Tarsipes rostratus* (honey possum). *Brain Behavior and Evolution*, **36**, 30–38.

Harman, A. M., Sanderson, K. J. and Beazley, L. D. (1992). Biphasic retinal neurogenesis in the brush-tailed possum, *Trichosurus vulpecula*: further evidence for the mechanics involved in formation of ganglion cell density gradients. *Journal of Comparative Neurology*, **325**, 595–606.

Harman, A.M., Eastough, N. J. and Beazley, L. J. (1995). Development of the visual cortex in the wallaby – phylogenetic implications. *Brain Behavior and Evolution*, **45**, 138–152.

Harman, A. M., Moore, S., Hoskins, R. and Keller, P. (1999). Horse vision and an explanation for the visual behaviour originally explained by the 'ramp retina'. *Equine Veterinary Journal*, **31**, 384–390.

Harman, A.M., Meyer, P. and Ahmat, A. (2003a). Neurogenesis in the hippocampus of an adult marsupial. *Brain Behavior and Evolution*, **62**, 1–12.

Harman, A.M., MacDonald A, Meyer, P. and Ahmat, A. (2003b). Numbers of neurons in the retinal ganglion cell layer of the rat do not change throughout life. *Gerontology*, **49**, 350–355.

Harper, J. M. (2008). Wild-derived mouse stocks: an underappreciated tool for aging research. *Age*, **30**, 135–145.

Harrison, P. H. (1991). Development of hindlimb muscle spindles in the marsupial *Macropus eugenii* (tammar wallaby). *Developmental Brain Research*, **62**, 277–280.

Harrison, P. H. and Porter, M. (1992). Development of the brachial spinal cord in the marsupial *Macropus eugenii* (tammar wallaby). *Developmental Brain Research*, **70**, 139–144.

Härtig, W., Stieler, J., Boerema, A. S. *et al.* (2007). Hibernation model of tau phosphorylation in hamsters: selective vulnerability of cholinergic basal forebrain neurons – implications for Alzheimer's disease. *European Journal of Neuroscience*, **25**, 69–80.

Hartman, C. G. (1920). Studies in the development of the opossum *Didelphis virginiana*. V. The phenomena of parturition. *Anatomical Record*, **19**, 251–261.

Harvey, P. H., Pagel M. D. and Rees, J. A. (1991). Mammalian metabolism and life histories. *American Naturalist*, **137**, 556–566.

Hassiotis, M., Ashwell, K. W. S., Marotte, L. R., Lensing-Höhn, S. and Mai, J. K. (2002). GAP-43 immunoreactivity in the brain of the developing and adult wallaby (*Macropus eugenii*). *Anatomy and Embryology*, **206**, 97–118.

Hassiotis, M., Paxinos, G. and Ashwell, K. W. S. (2005). Cyto- and chemoarchitecture of the cerebral cortex of the Australian echidna (*Tachyglossus aculeatus*). II Laminar organisation and synaptic density. *Journal of Comparative Neurology*, **482**, 94–122.

Hattar, S., Liao, H. W., Takao, M., Berson, D. M. and Yau, K. W. (2002). Melanopsin-containing retinal ganglion cells: architecture, projections, and intrinsic photosensitivity. *Science*, **295**, 1065–1070.

Haug, H. (1987). Brain sizes, surfaces and neuronal sizes of the cortex cerebri: a stereological investigation of man and his variability and a comparison with some mammals (primates, whales, marsupials, insectivores, and one elephant). *American Journal of Anatomy*, **180**, 126–142.

Hawkes, R., Colonnier, M. and Leclerc, N. (1985). Monoclonal antibodies reveal sagittal banding in the rodent cerebellar cortex. *Brain Research*, **333**, 359–365.

Hayhow, W. R. (1966). The accessory optic system in the marsupial phalanger, *Trichosurus vulpecula*. An experimental degeneration study. *Journal of Comparative Neurology*, **126**, 653–672

Hayhow, W. R. (1967). The lateral geniculate nucleus of the marsupial phalanger, *Trichosurus vulpecula*. An experimental study of cytoarchitecture in relation to the intranuclear optic nerve projection fields. *Journal of Comparative Neurology*, **131**, 571–604.

Hazlerigg, D. and Loudon, A. (2008). New insights into ancestral seasonal life timers. *Current Biology*, **18**, R795–804.

Hazlett, J. C., Dom, R. and Martin, G. F. (1972). Spino-bulbar, spino-thalamic and medical lemniscal connections in the American opossum, *Didelphis marsupialis virginiana*. *Journal of Comparative Neurology*, **146**, 95–118.

Heath, C. J. and Jones, E. G. (1971). Interhemispheric pathways in the absence of a corpus callosum. An experimental study of commissural connections in the marsupial phalanger. *Journal of Anatomy*, **109**, 253–270.

Hebel, R. (1976). Distribution of retinal ganglion cells in five mammalian species (pig, sheep, ox, horse, dog). *Anatomy and Embryology*, **150**, 45–51.

Heffner, R. and Masterton, B. (1975). Variation in form of the pyramidal tract and its relationship to digital dexterity. *Brain Behaviour and Evolution*, **12**, 161–200.

Helgen, K. M., Wells, R. T., Kear, B. P., Gerdtz, W. R. and Flannery, T. F. (2006). Ecological and evolutionary significance of sizes of giant extinct kangaroos. *Australian Journal of Zoology*, **54**, 293–303.

Hemmi, J. M. (1999). Dichromatic colour vision in an Australian marsupial, the tammar wallaby. *Journal of Comparative Physiology A*, **185**, 509–515.

Hemmi, J. M. and Grunert, U. (1999). Distribution of photoreceptor types in the retina of a marsupial, the tammar wallaby (*Macropus eugenii*). *Visual Neuroscience*, **16**, 291–302.

Hemmi, J. M. and Mark, R. F. (1998). Visual acuity, contrast sensitivity and retinal magnification in a marsupial, the tammar wallaby (*Macropus eugenii*). *Journal of Comparative Physiology A*, **183**, 379–387.

Hemmi, J. M., Maddess, T. and Mark, R. F. (2000). Spectral sensitivity of photoreceptors in an Australian marsupial, the tammar wallaby (*Macropus eugenii*). *Vision Research*, **40**, 591–599.

Hendrickson, A. and Provis, J. (2006). Comparison of development of the primate fovea centralis with peripheral retina. In *Retinal Development* (ed. E. Sernagor, S. Eglen, W. Harris, R. Wong). Cambridge: Cambridge University Press. pp. 126–149.

Hendrickson, A., Djajadi, H. R., Nakamura, L., Possin, D. E. and Sajuthi, D. (2000). Nocturnal tarsier retina has both short and long/medium-wavelength cones in an unusual topography. *Journal of Comparative Neurology*, **424**, 718–730.

Heng, M. Y., Detlof, P. J. and Albin, R. L. (2008). Rodent genetic models for Huntington's disease. *Neurobiology of Disease*, **32**, 1–9.

Henkel, C. K. and Martin, G. F. (1977). The vestibular complex of the American opossum *Didelphis virginiana*, I. Conformation, cytoarchitecture and primary vestibular input. *Journal of Comparative Neurology*, **172**, 299–320.

Henry, G. H. and Mark, R. F. (1992). Partition of function in the morphological subdivisions of the lateral geniculate nucleus of the tammar wallaby (*Macropus eugenii*). *Brain Behavior and Evolution*, **39**, 358–370.

Herculano-Houzel, S., Collins, C. E., Wong, P., Kaas, J. H. and Lent, R. (2008). The basic non-uniformity of the cerebral cortex. *Proceedings of the National Academy of Science USA*, **105**, 12593–12598.

Herrmann, K., Antonini, A. and Shatz, C. J. (1994). Ultrastructural evidence for synaptic interactions between thalamocortical axons and subplate neurons. *European Journal of Neuroscience*, **6**, 1729–1742.

Herron, P., Baskerville, K. A., Chang, H. T. and Doetsch, G. S. (1997). Distribution of neurons immunoreactive for parvalbumin and calbindin in the somatosensory thalamus of the racoon. *Journal of Comparative Neurology*, **388**, 120–129.

Hickey, T. L. and Guillery, R.W. (1974). An autoradiographic study of retinogeniculate pathways in the cat and the fox. *Journal of Comparative Neurology*, **156**, 239–253.

Hill, J. P. and Hill, W. C. O. (1955). The growth stages of the pouch young of the native cat (*Dasyurus viverrinus*) together with observations on the anatomy of the new-born young. *Transactions of the Zoological Society of London*, **28**, 349–425.

Hill, K. G., Cone-Wesson, B. and Liu, G. B. (1998). Development of auditory function in the tammar wallaby *Macropus eugenii*. *Hearing Research*, **117**, 97–106.

Hinman, J. D. and Abraham, C. R. (2007). What's behind the decline? The role of white matter in brain aging. *Neurochemical Research*, **32**, 2023–2031.

Ho, S. (1997). Rhythmic motor activity and interlimb co-ordination in the developing pouch young of a wallaby (*Macropus eugenii*). *Journal of Physiology*, **501**, 623–636.

Ho, S. (1998). Strychnine- and bicuculline-induced changes in the firing pattern of motoneurones during in vitro fictive locomotion reveal a possible N-methyl-D-aspartic acid (NMDA)-mediated suppression of motor discharge in wallaby (*Macropus eugenii*) pouch young. *Somatosensory and Motor Research*, **15**, 325–332.

Ho, S. M. and Stirling, R. V. (1998). Development of muscle afferents in the spinal cord of the tammar wallaby. *Developmental Brain Research*, **106**, 79–91.

Hof, P. R., Nimchinsky, E. A. and Morrison, J. H. (1995). Neurochemical phenotype of corticocortical connections in the macaque monkey: quantitative analysis of a subset of neurofilament protein-immunoreactive projection neurons in frontal, parietal, temporal, and cingulate cortices. *Journal of Comparative Neurology*, **362**, 109–133.

Hof, P. R., Ungerleider, L. G., Webster, M. J. *et al.* (1996). Neurofilament protein is differentially distributed in subpopulations of corticocortical projection neurons in the macaque monkey visual pathways. *Journal of Comparative Neurology*, **376**, 112–127.

Hof, P. R., Glezer, I. I. , Condé, F. *et al.* (1999). Cellular distribution of the calcium-binding proteins parvalbumin, calbindin, and calretinin in the neocortex of mammals: phylogenetic and developmental patterns. *Journal of Chemical Neuroanatomy*, **16**, 77–116.

Hoffman, P. N., Cleveland, D. W., Griffin, J. W. *et al.* (1987). Neurofilament gene expression: a major determinant of axonal caliber. *Proceedings of the National Academy of Science USA*, **84**, 3472–3476.

Hoffmann, K. P., Distler, C., Mark, R. F. *et al.* (1995). Neural and behavioural effects of early eye rotation on the optokinetic system in the wallaby, *Macropus eugenii*. *Journal of Neurophysiology*, **73**, 727–735.

Hokoc, J. N. and Oswaldo-Cruz, E. (1979). A regional special-isation in the opossum's retina: quantitative analysis of the ganglion cell layer. *Journal of Comparative Neurology*, **183**, 385–395.

Hollis, D. and Lyne, A. (1974). Innervation of vibrissa follicles in the marsupial *Trichosurus vulpecula*. *Australian Journal of Zoology*, **22**, 263–276.

Holst, M. C. (1986). *The Olivocerebellar Projection in a Marsupial and a Monotreme*. PhD thesis. Sydney: The University of New South Wales.

Hongo, T., Jankowska, E. and Lundberg, A. (1969). The rubrospinal tract I. Effects on alpha motoneurons innervating hindlimb muscles in the cat. *Experimental Brain Research*, **7**, 344–364.

Hope, P. J., Turnbull, H., Farr, S. *et al.* (2000). Peripheral administration of CRF and urocortin: effects on food intake and the HPA axis in the marsupial *Sminthopsis crassicaudata*. *Peptides*, **21**, 669–677.

Hore, J. and Porter, R. (1971). The role of the pyramidal tract in the production of cortically evoked movements in the brush-tailed possum *(Trichosurus vulpecula)*. *Brain Research*, **30**, 232–234.

Hore, J. and Porter, R. (1972). Pyramidal and extrapyramidal influences on some hindlimb motoneuron populations of the arboreal brush-tailed possum *Trichosurus vulpecula*. *Journal of Neurophysiology*, **35**, 112–121.

Hoyt, W. F. and Luis, O. (1963). The primate chiasm. Details of visual fiber organisation studied by silver impregnation techniques. *Archives of Ophthalmology*, **70**, 69–85.

Huberman, A. D., Wang, G. Y., Liets, L. C. *et al.* (2003). Eye-specific retinogeniculate segregation independent of normal neuronal activity. *Science*, **300**, 994–998.

Huberman, A. D., Dehay, C., Berland, M., Chalupa, L. M. and Kennedy, H. (2005). Early and rapid targeting of eye-specific axonal projections to the dorsal lateral geniculate nucleus in the fetal macaque. *Journal of Neuroscience*, **25**, 4014–4023.

Huberman, A. D., Feller, M. B. and Chapman, B. (2008). Mechanisms underlying development of visual maps and receptive fields. *Annual Review of Neuroscience*, **31**, 479–509.

Huffman, K. J., Nelson, J., Clarey, J. and Krubitzer, L. (1999a). Organisation of somatosensory cortex in three species of marsupials, *Dasyurus hallucatus*, *Dactylopsila trivirgata* and *Monodelphis domestica*: neural correlates of morphological specialisations. *Journal of Comparative Neurology*, **403**, 5–32.

Huffman, K. J., Molnár, Z., van Dellen, A. *et al.* (1999b). Formation of cortical fields on a reduced cortical sheet. *Journal of Neuroscience*, **19**, 9939–9952.

Hughes, A. (1973). The development of dorsal root ganglia and ventral horns in the opossum. A quantitative study. *Journal of Embryology and Experimental Morphology*, **30**, 359–376.

Hughes, A. (1975). A comparison of retinal ganglion cell topography in the plains and tree kangaroo. *Journal of Physiology*, **244**, 61P–63P.

Hughes, A. (1977). The topography of vision in mammals of contrasting lifestyle: comparative optics and retinal organisation. In *Handbook of Sensory Ecology*, vol. VII/5 (ed. F. Cresitelli). Berlin, Heidelberg, New York: Springer. pp. 613–756.

Hughes, A. (1985). New perspectives in retinal organisation. *Progress in Retinal Research*, **4**, 243–311.

Hughes, R. L., Hall, L. S., Tyndale-Biscoe, C. H. and Hinds, L. A. (1989). Evolutionary implications of macropod organogenesis. In *Kangaroos, Wallabies and Rat-kangaroos* (ed. G. Grigg, P. Jarman, I. Hume). Sydney: Surrey Beatty & Sons. pp 377–405.

Hulbert, A. J. (1988). Metabolism and the development of endo-thermy. In *The Developing Marsupial: Models for Biomedical Research*. (ed. C. H. Tyndale-Biscoe and P. A. Janssens). Berlin, New York: Springer-Verlag. pp. 148–161.

Hume, I. D. (1999). *Marsupial Nutrition*. Cambridge: Cambridge University Press.

Hunt, D. M., Carvalho, L. S., Cowing, J. A. *et al.* (2007). Spectral tuning of shortwave-sensitive visual pigments in vertebrates. *Photochemistry and Photobiology*, **83**, 303–310.

Hunt, D. M., Carvalho, L. S., Davies, W. L. and Cowing, J. A. (2009a). Opsin evolution in birds and mammals. *Philosophical Transactions of the Royal Society. Series B.* **364**, 2941–2955.

Hunt, D. M., Chan, J., Carvalho, L. S. *et al.* (2009b). Cone visual pigments in two species of South American marsupials. *Gene*, **433**, 50–55.

Hunt, M., Slotnick, B. and Croft, D. (1999). Olfactory function in the red kangaroo (*Macropus rufus*) assessed using odor-cued taste avoidance. *Physiology of Behaviour*, **67**, 365–368.

Hurpet, D., Chauvet, M. T., Chauvet, J. and Acher, R. (1980). Identification of lysine vasopressin in two Australian marsupials, the red kangaroo and the tammar. *Biochemical and Biophysical Research Communications*, **95**, 1585–1590.

Hurpet, D., Chauvet, M. T., Chauvet, J. and Acher, R. (1982). Marsupial hypothalamo-neurohypophyseal hormones. The brush-tailed possum (*Trichosurus vulpecula*) active peptides. *International Journal of Peptide and Protein Research*, **19**, 366–371.

Huxley, T. H. (1880). On the application of the laws of evolution to the arrangement of the Vertebrata, and more particularly of the Mammalia. *Proceedings of the Zoological Society of London*, **43**, 649–662.

Ibbotson, M. R. and Mark, R. F. (1994). Wide-field nondirectional visual units in the pretectum: do they suppress ocular following of saccade-induced visual stimulation? *Journal of Neurophysiology*, **72**, 1448–1450.

Ibbotson, M. R. and Mark, R. F. (2003). Orientation and spatio-temporal tuning of cells in the primary visual cortex of an Australian marsupial, the wallaby *Macropus eugenii*. *Journal of Comparative Physiology. A. Sensory, Neural and Behavioural Physiology*, **189**, 115–123.

Ibbotson, M. R., Mark, R. F. and Maddess, T. L. (1994). Spatiotemporal response properties of direction-selective neurons in the nucleus of the optic tract and dorsal terminal nucleus of the wallaby, *Macropus eugenii*. *Journal of Neurophysiology*, **72**, 2927–2943.

Ibbotson, M. R., Marotte, L. R. and Mark, R. F. (2002). Investigations into the source of binocular input to the nucleus of the optic tract in an Australian marsupial, the wallaby *Macropus eugenii*. *Experimental Brain Research*, **147**, 80–88.

Ibbotson, M. R., Price, N. S. and Crowder, N. A. (2005). On the division of cortical cells into simple and complex types: a comparative viewpoint. *Journal of Neurophysiology*, **93**, 3699–3702.

Iggo, A. (1974). Cutaneous receptors. In *The Peripheral Nervous System*. (ed. H. Hubbard). New York: Plenum. pp. 347–404.

Iqbal, J. and Jacobson, C. D. (1995a). Ontogeny of arginine vasopressin-like immunoreactivity in the Brazilian opossum brain. *Developmental Brain Research*, **89**, 11–32.

Iqbal, J. and Jacobson, C. D. (1995b). Ontogeny of oxytocin-like immunoreactivity in the Brazilian opossum brain. *Developmental Brain Research*, **90**, 1–16.

Iqbal, J., Elmquist, J. K., Ross, L. R., Ackermann, M. R. and Jacobson, C. D. (1995a). Postnatal neurogenesis of the hypothalamic paraventricular and supraoptic nuclei in the Brazilian opossum brain. *Developmental Brain Research*, **85**, 151–160.

Iqbal, J., Swanson, J. J., Prins, G. S. and Jacobson, C. D. (1995b). Androgen receptor-like immunoreactivity in the Brazilian opossum brain and pituitary: distribution and effects of castration and testosterone replacement in the adult male. *Brain Research*, **703**, 1–18.

Iqbal, K., Alonso, A. D. C., Chen, S., *et al.* (2005). Tau pathology in Alzheimer disease and other tauopathies. *Biochimica et Biophysica Acta*, **1739**, 198–210.

Irvine, D. R. F. (1986). The auditory brainstem. In *Progress in Sensory Physiology* (ed. D. Ottoson). New York: Springer. pp. 1–279.

Irvine, D. R. F. (1992). Physiology of auditory brainstem. In *The Mammalian Auditory Pathway: Neurophysiology*. (ed. A. N. Popper and R. R. Fay). New York: Springer. pp. 153–231.

Ito, H. and Murakami, T. (1984). Retinal ganglion cells in two teleost species, *Sebastiscus marmoratus* and *Navodon modestus*. *Journal of Comparative Neurology*, **229**, 80–96.

Ito, M. (1984). *The Cerebellum and Neural Control*. New York: Raven Press.

Ivanco, T. L., Pellis, S. M. and Whishaw, I. Q. (1996). Skilled forelimb movements in prey catching and in reaching by rats (*Rattus norvegicus*) and opossums (*Monodelphis domestica*): relations to anatomical differences in motor systems. *Behavioural Brain Research*, **79**, 163–181.

Iwaniuk, A. N. and Whishaw, I. Q. (2000). On the origin of skilled forelimb movements. *Trends in Neuroscience*, **23**, 372–376.

Iwaniuk, A. N., Pellis, S. M. and Whishaw, I. Q. (1999). Is digital dexterity related to corticospinal projections? A re-analysis of the Heffner and Masterton data set using modern comparative statistics. *Behavioural Brain Research*, **101**, 173–187.

Iwaniuk, A. N., Nelson, J. E. and Pellis, J. M. (2001). Do big-brained animals play more? Comparative analysis of play and relative brain size in mammals. *Journal of Comparative Psychology*, **115**, 29–41.

Jacobs, G. H. (1993). The distribution and nature of colour vision among the mammals. *Biological Reviews*, **68**, 413–471.

Jacobs, G. H., Calderone, J. B., Fenwick, J. A., Krogh, K. and Williams, G. A. (2003). Visual adaptations in a diurnal rodent, *Octodon degus*. *Journal of Comparative Physiology. A. Sensory, Neural and Behavioural Physiology*, **189**, 347–361.

Jacobson, C. D., Shryne, J. E., Shapiro, F. and Gorski, R. A. (1980). Ontogeny of the sexually dimorphic nucleus of the preoptic area. *Journal of Comparative Neurology*, **193**, 541–548.

Jacobson, C. D., Davis, F. C. and Gorski, R. A. (1985). Formation of the sexually dimorphic nucleus of the preoptic area: neuronal growth, migration and changes in cell number. *Developmental Brain Research*, **21**, 7–18.

Jacobson, R. D., Virag, I. and Skene, J. H. P. (1986). A protein associated with axon growth, GAP-43, is widely distributed and developmentally regulated in rat CNS. *Journal of Neuroscience*, **6**, 1843–1855.

Jacquin, M. F., Renehan, W. E., Klein, B. G., Mooney, R. D. and Rhoades, R. W. (1986). Functional consequences of neonatal infraorbital nerve section in rat trigeminal ganglion. *Journal of Neuroscience*, **6**, 3706–3720.

James, A. C., Mark, R. F. and Sheng, X-M. (1993). Geometry of the projection of the visual field onto the superior colliculus of the wallaby (*Macropus eugenii*). II. Stability of the projection after prolonged rearing with rotational squint. *Journal of Comparative Neurology*, **330**, 315–323.

Janssens, P. A. and Messer, M. (1988). Changes in nutritional metabolism during weaning. In *The Developing Marsupial. Models for Biomedical Research*. (ed. H. Tyndale-Biscoe and P. A. Janssens). Berlin: Springer-Verlag. pp. 162–175.

Jaubert-Miazza, L., Green, E., Lo, F. S. *et al.* (2005). Structural and functional composition of the developing retinogeniculate pathway in the mouse. *Visual Neuroscience*, **22**, 661–676.

Jeffery, G. (2001). Architecture of the optic chiasm and the mechanisms that sculpt its development. *Physiological Reviews*, **81**, 1393–1414.

Jeffery, G. and Harman, A. M. (1992). Distinctive pattern of organisation in the retinofugal pathway of a marsupial: II. Optic chiasm. *Journal of Comparative Neurology*, **325**, 57–67.

Jeffery, G., Harman, A. and Flügge, G. (1998). First evidence of diversity in eutherian chiasmatic architecture: tree shrews, like marsupials, have spatially segregated crossed and uncrossed chiasmatic pathways. *Journal of Comparative Neurology*, **390**, 183–193.

Jerison, H. J. (1973) *Evolution of the Brain and Intelligence*. New York: Academic Press.

Jerison, H. J. (1976). Paleoneurology and the evolution of the mind. *Scientific American*, **234**: 94–101.

Johnsen, A. H. and Shulkes, A. (1993). Gastrin and cholecystokinin in the eastern grey kangaroo, *Macropus giganteus giganteus*. *Peptides*, **14**, 1133–1139.

Johnson, C. (2006). *Australia's Mammal Extinctions. A 50,000 Year History*. Cambridge: Cambridge University Press.

Johnson, G. L. (1968). Ophthalmoscopic studies on the eyes of mammals. *Philosophical Transactions of the Royal Society of London. Series B*, **254**, 207–220.

Johnson, J. I. (1977). Central nervous system in marsupials. In *The Biology of Marsupials*. (ed. D. Hunsaker). New York: Academic. pp. 157–278.

Johnson, J. I. and Marsh, M. P. (1969) Laminated lateral geniculate in the nocturnal marsupial *Petaurus breviceps* (sugar glider). *Brain Research*, **15**, 250–254.

Johnson, J. I., Haight, J. and Megirian, D. (1973). Convolutions related to sensory projections in cerebral neocortex of marsupial wombats. *Journal of Anatomy*, **114**, 153.

Johnson, J. I., Kirsch, J. A. and Switzer, R. C. (1982a). Phylogeny through brain traits: fifteen characters which adumbrate mammalian genealogy. *Brain Behavior and Evolution*, **20**, 72–83.

Johnson, J. I., Switzer, R. C. and Kirsch, J. A. (1982b). Phylogeny through brain traits: the distribution of categorizing characters in contemporary mammals. *Brain Behavior and Evolution*, **20**, 97–117.

Johnson, J. I., Kirsch, J. A. W., Reep, R. L. and Switzer, R. C. (1994). Phylogeny through brain traits: more characters for the analysis of mammalian evolution. *Brain Behavior and Evolution*, **43**, 319–347.

Johnson, K. A. (1989). Thylacomyidae. In *Fauna of Australia, Mammalia 1B.* (ed. D. W. Walton and B. J. Richardson). Canberra: Australian Government Publishing Service. pp. 625–635.

Johnson, P. M. and Strahan, R. (1982). A further description of the musky rat-kangaroo, *Hypsiprymnodon moschatus*, 1876. *Australian Zoologist*, **21**, 27–46.

Jones, A. S., Lamont, B. B., Fairbanks, M. M. and Rafferty, C. M. (2003). Kangaroos avoid eating seedlings with or near others with volatile essential oils. *Journal of Chemical Ecology*, **29**, 2621–2635.

Jones, E. G. (1966a). Structure and distribution of muscle spindles in the forepaw lumbricals of the phalanger, *Trichosurus vulpecula*. *Anatomical Record*, **155**, 287–304.

Jones, E. G. (1966b). The innervation of muscle spindles in the Australian opossum, *Trichosurus vulpecula*, with special reference to the motor nerve endings. *Journal of Anatomy*, **100**, 733–759.

Jones, E. G. (2007). *The Thalamus.* Cambridge: Cambridge University Press.

Jones, S. E., Dziegielewska, K. M., Saunders, N. R. *et al.* (1988). Early cortical plate specific glycoprotein in a marsupial species belongs to the same family as fetuin and alpha 2HS glycoprotein. *FEBS Letters*, **236**, 411–414.

Jones, S. E., Christie, D. L., Dziegielewska, K. M., Hinds, L. A. and Saunders. N. R. (1991). Developmental profile of a fetuin-like glycoprotein in neocortex, cerebrospinal fluid and plasma of postnatal tammar wallaby (*Macropus eugenii*). *Anatomy and Embryology*, **183**, 313–320.

Jones, T. E. and Munger, B. L. (1985). Early differentiation of the afferent nervous system in glabrous snout skin of the opossum (*Monodelphis domestica*). *Somatosensory Research*, **3**, 169–184.

Joschko, M. A. and Sanderson, K. J. (1987). Cortico-cortical connections of the motor cortex in the brush-tailed possum (*Trichosurus vulpecula*). *Journal of Anatomy*, **150**, 31–42.

Kageyama, G. H. and Robertson, R. T. (1993). Development of geniculocortical projections to visual cortex in rat: evidence of early ingrowth and synaptogenesis. *Journal of Comparative Neurology*, **335**, 123–148.

Kahn, D. M. and Krubitzer, L. (2002). Retinofugal projections in the short-tailed opossum (*Monodelphis domestica*). *Journal of Comparative Neurology*, **447**, 114–127.

Kahn, D. M., Huffman, K. J. and Krubitzer, L. (2000). Organisation and connections of V1 in *Monodelphis domestica*. *Journal of Comparative Neurology*, **428**, 337–353.

Karlen, S. J. and Krubitzer, L. (2006a). Phenotypic diversity is the cornerstone of evolution: variation in cortical field size within short-tailed opossums. *Journal of Comparative Neurology*, **499**, 990–999.

Karlen, S. J. and Krubitzer, L. (2006b). The evolution of the neocortex in mammals: intrinsic and extrinsic contributions to the cortical phenotype. *Novartis Foundation Symposium*, **270**, 146–169.

Karlen, S. J. and Krubitzer, L. (2007). The functional and anatomical organisation of marsupial neocortex: evidence for parallel evolution across mammals. *Progress in Neurobiology*, **82**, 122–141.

Karlen, S. J. and Krubitzer, L. (2009). Effects of bilateral enucleation on the size of visual and nonvisual areas of the brain. *Cerebral Cortex*, **19**, 1360–1371.

Karns, L. R., Ng, S-C., Freeman, J. A. and Fishman, M. C. (1987). Cloning of complementary DNA for GAP-43, a neuronal growth related protein. *Science*, **236**, 597–600.

Kear, B. P. (2003). Macropodoid endocranial casts from the early Miocene of Riversleigh, northwestern Queensland. *Alcheringa*, **27**, 295–302.

Keay, K. A. and Bandler, R. (2004). Periaqueductal grey. In *The Rat Nervous System* (ed. G. Paxinos), 3rd edn. San Diego, London: Elsevier Academic. pp. 243–257.

Kenny, G. C. and Scheelings, F. T. (1979). Observations of the pineal region of non-eutherian mammals. *Cell and Tissue Research*, **198**, 309–324.

Kent, A. R. and Harman, A. M. (1998). The effects of a transient increase in temperature on cell generation and cell death in the hippocampus and amygdala of the wallaby, *Setonix brachyurus* (quokka). *Experimental Brain Research*, **122**, 301–308.

Kielan-Jaworowska, Z. (1984). Evolution of the therian mammals in the Late Cretaceous of Asia. Part VI. Endocranial casts of Eutherian mammals. *Palaeontologia Polonica*, **46**, 157–171.

Kielan-Jaworowska, Z. and Lancaster, T. E. (2004). A new reconstruction of multituberculate endocranial casts and encephalisation quotient of *Kryptobaatar*. *Acta Paleontologica Polonica*, **49**, 177–188.

Kielan-Jaworowska, Z. and Trofimov, B. A. (1986). Endocranial cast of the Cretaceous Eutherian mammal *Barunlestes*. *Acta Palaeontologica Polonica*, **31**, 137–144.

Killackey, H. P. (1985). Intrinsic order in the developing rat trigeminal system. In *Development, Organisation and Processing in Somatosensory Pathways* (ed. M. Rowe and W. Willis). New York: Alan R Liss. pp. 43–51.

Killackey, H. P. and Belford, G. R. (1979). The formation of afferent patterns in the somatosensory cortex of the neonatal rat. *Journal of Comparative Neurology*, **183**, 285–303.

Killackey, H. P. and Ebner, F. (1973). Convergent projection of three separate thalamic nuclei onto a single cortical area. *Science*, **179**, 283–285.

King, J. A., Mehl, A. E., Tyndale-Biscoe, C. H., Hinds, L. and Millar, R. P. (1989). A second form of gonadotropin-releasing hormone (GnRH), with chicken GnRH II-like properties, occurs together with mammalian GnRH in marsupial brains. *Endocrinology*, **125**, 2244–2252.

King, J. A., Fidler, A., Lawrence, S. *et al.* (2000). Cloning and expression, pharmacological characterisation, and internalisation kinetics of the pituitary GnRH receptor in a metatherian species of mammal. *General and Comparative Endocrinology*, **117**, 439–448.

King, J. S., Martin, G. F. and Biggert, T. P. (1968). The basilar pontine grey of the opossum (*Didelphis virginiana*). I. Morphology. *Journal of Comparative Neurology*, **133**, 439–445.

King, J. S., Morgan, J. K., Bishop, G. A., Hazlett, J. C. and Martin, G. F. (1987). Development of the basilar pons in the North American opossum: dendrogenesis and maturation of afferent and efferent connections. *Anatomy and Embryology*, **176**, 191–202.

Kirov, S. A., Sorra, K. E. and Harris, K. M. (1999). Slices have more synapses than perfusion-fixed hippocampus from both young and mature rats. *Journal of Neuroscience*, **19**, 2876–2886.

Kirsch, J. A., Johnson, J. I. and Switzer, R. C. (1983). Phylogeny through brain traits: the mammalian family tree. *Brain Behavior and Evolution*, **22**, 70–74.

Kitchener, P. D., Hutton, E. J. and Knott, G. W. (2006). Primary sensory afferent innervation of the developing superficial dorsal horn in the South American opossum *Monodelphis domestica*. *Journal of Comparative Neurology*, **495**, 37–52.

Kleemann, G. A. and Murphy, C.T. (2009). The endocrine regulation of aging in *Caenorhabditis elegans*. *Molecular and Cellular Endocrinology*, **299**, 51–57.

Kokjohn, T. A. and Roher, A. E. (2009). Amyloid precursor protein transgenic mouse models and Alzheimer's disease: understanding the paradigms, limitations, and contributions. *Alzheimer's and Dementia*, **5**, 340–347.

Kolb, H. and Wang, H. H. (1985). The distribution of photoreceptors, dopaminergic amacrine cells and ganglion cells in the retina of the North American opossum (*Didelphis virginiana*). *Vision Research*, **25**, 1207–1221.

Konishi, M. (1970). Comparative neurophysiological study of hearing and vocalisation in songbirds. *Zeitschrift für Vogel Physiologie*, **66**, 257–272.

Kornack, D. R. (2000). Neurogenesis and the evolution of cortical diversity: mode, tempo, and portioning during development and persistence into adulthood. *Brain, Behavior and Evolution*, **55**, 336–344.

Kostovic, I. and Rakic, P. (1990). Developmental history of the transient subplate zone in the visual and somatosensory cortex of the macaque monkey and human brain. *Journal of Comparative Neurology*, **297**, 441–470.

Koteja, P. (1991). On the relation between basal and field metabolic rates in birds and mammals. *Functional Ecology*, **5**, 56–64.

Kratzing, J. E. (1978). The olfactory apparatus of the bandicoot (*Isoodon macrourus*): fine structure and presence of a septal olfactory organ. *Journal of Anatomy*, **125**, 601–613.

Kratzing, J. E. (1984a). The anatomy and histology of the nasal cavity of the koala (*Phascolarctos cinereus*). *Journal of Anatomy*, **138**, 55–65.

Kratzing, J. E. (1984b). The structure and distribution of nasal glands in four marsupial species. *Journal of Anatomy*, **139**, 553–564.

Kratzing, J. E. (1986). Morphological maturation of the olfactory epithelium of Australian marsupials. In *Ontogeny of Olfaction* (ed. W. Breipohl). Berlin: Springer-Verlag. pp. 57–70.

Krause, W. J. (1991). The vestibular apparatus of the opossum (*Didelphis virginiana*) prior to and immediately after birth. *Acta Anatomica*, **142**, 57–59.

Krause, W. J. and Saunders, N. R. (1994). Brain growth and neocortical development in the opossum. *Annals of Anatomy*, **176**, 395–407.

Krous, H. F., Jordan, J., Wen, J. and Farber, J. P. (1985). Development morphometry of the vagus nerve in the opossum. *Developmental Brain Research*, **20**, 155–159.

Krubitzer, L. (1995). The organisation of neocortex in mammals: are species differences really so different? *Trends in Neuroscience*, **18**, 408–417.

Krubitzer, L and Hunt, D. (2006). Captured in the net of space and time: understanding cortical field evolution. In *The Evolution of Nervous Systems in Mammals* (ed. J. Kaas and L. Krubitzer). Oxford: Academic Press. pp. 49–72.

Krubitzer, L., Manger, P., Pettigrew, J. and Calford, M. (1995). Organisation of somatosensory cortex in monotremes: in search of the prototypical plan. *Journal of Comparative Neurology*, **351**, 261–306.

Kryger, Z., Galli-Resta, L., Jacobs, G. H. and Reese, B. E. (1998). The topography of rod and cone photoreceptors in the retina of the ground squirrel. *Visual Neuroscience*, **15**, 685–691.

Kubota, K., Kubota, J., Fukuda, N. *et al.* (1963). Comparative anatomical and neurohistological observations on the tongue of the marsupials. *Anatomical Record*, **147**, 337–353.

Kubota, K., Shimizu, T., Shibanai, S., Nagae, K. and Nagata, S. (1989). Histological properties and biological significance of pouch in red kangaroo, *Macropus rufus*. *Anatomischer Anzeiger*, **168**, 169–179.

Kudielka, B. M. and Kirschbaum, C. (2005). Sex differences in HPA axis responses to stress: a review. *Biological Psychology*, **69**, 113–132

Kudo, M., Glendenning, K.K., Frost, S.B. and Masterton, R.B. (1986). Origin of mammalian thalamocortical projections. I. Telencephalic projections of the medial geniculate body in the opossum (*Didelphis virginiana*). *Journal of Comparative Neurology*, **245**, 176–197.

Kudo, M., Aitken, L. M. and Nelson, J. E. (1989). Auditory forebrain organisation of an Australian marsupial, the northern native cat (*Dasyurus hallucatus*). *Journal of Comparative Neurology*, **279**, 28–42.

Kuehl-Kovarik, M. C., Ross, L. R., Elmquist, J. K. and Jacobson, C. D. (1993). Localisation of cholecystokinin binding sites in the adult and developing Brazilian opossum brain. *Journal of Comparative Neurology*, **336**, 40–52.

Kuehl-Kovarik, M. C., Iqbal, J. and Jacobson, C. D. (1997). Autoradiographic localisation of arginine vasopressin binding sites in the brain of adult and developing Brazilian opossums. *Brain Behavior and Evolution*, **49**, 261–275.

Kullander, K., Carlson, B. and Hallböök, F. (1997). Molecular phylogeny and evolution of the neurotrophins from monotremes and marsupials. *Journal of Molecular Evolution*, **45**, 311–321.

La Vail, M. M., Rapaport, D. H. and Rakic, P. (1991). Cytogenesis in the monkey retina. *Journal of Comparative Neurology*, **309**, 86–114.

Ladhams, A. and Pickles, J. O. (1996). Morphology of the monotreme organ of Corti and macula lagena. *Journal of Comparative Neurology*, **366**, 335–347.

Lane, M. A, Truettner, J. S., Brunschwig, J- P. *et al.* (2007). Age-related differences in the local cellular and molecular responses to injury in developing spinal cord of the opossum, *Monodelphis domestica*. *European Journal of Neuroscience*, **25**, 1725–1742.

Langworthy, O. R. (1928). The behavior of pouch-young opossums correlated with the myelinisation of tracts in the nervous system. *Journal of Comparative Neurology*, **46**, 201–247.

Larsell, O. (1970). *The Comparative Anatomy and Histology of the Cerebellum from Monotremes through Apes.* (ed. J. Jansen) Minneapolis: University of Minnesota Press.

Larsell, O. E., McCrady, E. and Zimmerman, A. A. (1935). Morphological and functional development of the membranous labyrinth in the opossum. *Journal of Comparative Neurology*, **63**, 95–118.

Larsen, D. D. and Krubitzer, L. (2008). Genetic and epigenetic contributions to the cortical phenotype in mammals. *Brain Research Bulletin*, **75**, 391–397.

Lavallée, A. and Pflieger, J-F. (2009). Developmental expression of spontaneous activity in the spinal cord of postnatal opossums, *Monodelphis domestica*: an anatomical study. *Brain Research*, **1282**, 1–9.

Lavery, W. L. (2000). How relevant are animal models to human aging? *Journal of the Royal Society of Medicine*, **93**, 296–298.

Lawson, S. N. and Waddell, P. J. (1991). Soma neurofilament immunoreactivity is related to cell size and fibre conduction velocity in rat primary sensory neurons. *Journal of Physiology*, **435**, 41–63.

Laxson, L. C. and King, J. S. (1983). The formation and growth of the cortical layers in the cerebellum of the opossum. *Anatomy and Embryology*, **167**, 391–409.

Leamey, C. A., Marotte, L. R. and Waite, P. M. (1996). Timecourse of development of the wallaby trigeminal pathway. II. Brainstem to thalamus and the emergence of cellular aggregations. *Journal of Comparative Neurology*, **364**, 494–514.

Leamey, C. A., Ho, S. M. and Marotte, L. R. (1998). Morphological development of afferent segregation and onset of synaptic transmission in the trigeminothalamic pathway of the wallaby (*Macropus eugenii*). *Journal of Comparative Neurology*, **399**, 47–60.

Leamey, C. A., Flett, D. L., Ho, S. M. and Marotte, L. R. (2007). Development of structural and functional connectivity in the thalamocortical somatosensory pathway in the wallaby. *European Journal of Neuroscience*, **25**, 3058–3070.

Leao, R. N., Sun. H., Svahn, K., *et al.* (2006). Topographic organisation in the auditory brainstem of juvenile mice is disrupted in congenital deafness. *Journal of Physiology*, **571**, 563–578.

Leatherland, J. F. and Renfree, M. B. (1983a). Structure of the pars distalis in the adult tammar wallaby (*Macropus eugenii*). *Cell and Tissue Research*, **229**, 155–174.

Leatherland, J. F. and Renfree, M. B. (1983b). Structure of the pars distalis in pouch-young tammar wallabies (*Macropus eugenii*). *Cell and Tissue Research*, **230**, 587–603.

Lee, A. K. and Cockburn, A. (1985). *Evolutionary Ecology of Marsupials.* Cambridge: Cambridge University Press.

Lee, V. M-Y., Kenyon, T. K. and Trojanowski, J. Q. (2005). Transgenic animal models of tauopathies. *Biochimica et Biophysica Acta*, **1739**, 251–259.

Lende, R. A. (1963). Cerebral cortex: a sensorimotor amalgam in the marsupialia. *Science*, **141**, 730–732.

Lende, R. A. (1969). Comparative approach to the neocortex: localisation in monotremes, marsupials and insectivores. *Annals of the New York Academy of Science*, **167**, 262–275.

Lentle, R. G., Kruger, M. C., Mellor, D. J., Birtles, M. and Moughan, P. J. (2006). Limb development in pouch young of the brushtail possum (*Trichosurus vulpecula*) and tammar wallaby (Macropus eugenii). *Journal of Zoology*, **270**, 122–131.

Letinic, K. and Kostovic, I. (1998). Postnatal development of calcium-binding proteins calbindin and parvalbumin in human visual cortex. *Cerebral Cortex*, **8**, 660–669.

Leuba, G. and Saini, K. (1997). Colocalisation of parvalbumin, calretinin and calbindin D-28k in human cortical and subcortical visual structures. *Journal of Chemical Neuroanatomy*, **13**, 41–52.

Li, Y., Erzurumlu, R. S., Chen, C., Jhaveri, S. and Tonegawa, S. (1994). Whisker-related neuronal patterns fail to develop in the trigeminal brainstem nuclei of NMDAR1 knockout mice. *Cell*, **76**, 427–437.

Liberman, M.C. (1982). The cochlear frequency map for the cat: labeling auditory-nerve fibers of known characteristic frequency. *Journal of the Acoustical Society of America*, **72**, 1441–1449.

Lim, C. H. and Ho, S. M. (1997). Early detection of optic nerve-evoked response in the superior colliculus of the neonatal rat. *Neuroscience Letters*, **235**, 141–144.

Lin, P. J., Phelix, C. and Krause, W. J. (1988). An immunohistochemical study of olfactory epithelium in the opossum before and after birth. *Zeitschrift für Mikroskopisch-anatomische Forschung*, **102**, 272–282.

Linauts, M. and Martin, G. F. (1978). The organisation of olivocerebellar projections in the opossum, *Didelphis virginiana*, as revealed by the retrograde transport of horseradish peroxidase. *Journal of Comparative Neurology*, **179**, 355–381.

Lincoln, D. W. and Renfree, M. B. (1981). Milk ejection in a marsupial, *Macropus agilis*. *Nature*, **289**, 504–506.

Lippolis, G., Westman, W., McAllan, B. M. and Rogers, L. J. (2005). Lateralisation of escape responses in the stripe-faced dunnart, *Sminthopsis macroura* (Dasyuridae: Marsupialia). *Laterality*, **10**, 457–470.

Liu, C. N. (1956). Afferent nerves to Clarke's and the lateral cuneate nuclei in the cat. *AMA Archives of Neurology and Psychiatry*, **75**, 67–77.

Liu, G. B. (2003). Functional development of the auditory brainstem in the tammar wallaby (*Macropus eugenii*): the superior olivary complex and its relationship with the auditory brainstem response (ABR). *Hearing Research*, **175**, 152–164.

Liu, G. B. and Mark, R. F. (2001). Functional development of the inferior colliculus (IC) and its relationship with the auditory brainstem response (ABR) in the tammar wallaby (*Macropus eugenii*). *Hearing Research*, **157**, 112–123.

Liu, G. B., Hill, K. B. and Mark, R. F. (2001). Temporal relationship between the auditory brainstem responses of auditory nerve root and cochlear nucleus during development in the tammar wallaby (*Macropus eugenii*). *Audiology and Neurootology*, **6**, 140–153.

Liu, L., Orozco, I. J., Planel, E. *et al.* (2008). A transgenic rat that develops Alzheimer's disease-like amyloid pathology, deficits in synaptic plasticity and cognitive impairment. *Neurobiology of Disease*, **31**, 46–57.

Ljubojevic M., Herak-Kramberger C. M., Hagos Y. *et al.* (2004). Rat renal cortical OAT1 and OAT3 exhibit gender differences determined by both androgen stimulation and estrogen inhibition. *American Journal of Physiology-Renal Physiology*, **287**, F124–F138.

Locket, N. A. (1999). Vertebrate photoreceptors. In *Adaptive Mechanisms in the Ecology of Vision*. (ed. S. N. Archer, M. B. A. Djamgoz, E. R. Loew, J. C. Partridge and S. Vallerga), Dordrecht: Kluwer Academic Publishers. pp. 163–196.

Long, J., Archer, M., Flanner, T. and Hand, S. (2002). *Prehistoric Mammals of Australia and New Guinea*. Sydney: UNSW Press.

Long, K. O. and Fisher, S. K. (1983). The distributions of photoreceptors and ganglion cells in the California ground squirrel, *Spermophilus beecheyi*. *Journal of Comparative Neurology*, **221**, 329–340.

Lonsbury-Martin, B.L., Martin, G.K., Probst, R. and Coats, A.C. (1987). Acoustic distortion products in rabbit ear canal. I. Basic features and physiological vulnerability. *Hearing Research*, **28**, 173–189.

Loo, S. K. and Halata, Z. (1985). The sensory innervation of the nasal glabrous skin in the short-nosed bandicoot (*Isoodon macrourus*) and the opossum (*Didelphis virginiana*). *Journal of Anatomy*, **143**, 167–180.

Loo, S. K. and Halata, Z. (1991). Innervation of hairs in the facial skin of marsupial mammals. *Journal of Anatomy*, **174**, 207–219.

Lossi, L., Cantile, C., Tamagno, I. and Merighi, A. (2005). Apoptosis in the mammalian CNS: Lessons from animal models. *Veterinary Journal*, **170**, 52–66.

Lovegrove, B. G. (2003). The influence of climate on the basal metabolic rate of small mammals: a slow-fast metabolic continuum. *Journal of Comparative Physiology B. Biochemical Systemic and Environmental Physiology*, **173**, 87–112.

Lovegrove, B. G. (2009). Age at first reproduction and growth rate are independent of basal metabolic rate in mammals. *Journal of Comparative Physiology B. Biochemical Systemic and Environmental Physiology*, **179**, 391–401.

Lovegrove, B. G., and Haines, L. (2004). The evolution of placental mammal body sizes: evolutionary history, form and function. *Oecologia*, **138**, 13–27.

Lucas, R. J., Freedman, M. S., Munoz, M., Garcia-Fernandez, J. M. and Foster, R. G. (1999). Regulation of the mammalian pineal by non-rod, non-cone, ocular photoreceptors. *Science*, **284**, 505–507.

Luo, Z-X. and Wibble, J. R. (2005). A late Jurassic digging mammal and early mammalian diversity. *Science*, **308**, 103–107.

Luskin, M. B. and Shatz, C. J. (1985). Neurogenesis of the cat's primary visual cortex. *Journal of Comparative Neurology*, **242**, 611–631.

Lyne, A. G. (1970). The development of hair follicles in the marsupial *Trichosurus vulpecula*. *Australian Journal of Biological Science*, **23**, 1241–1253.

Lyne, A. G., Henrikson, R. C. and Hollis, D. E. (1970). Development of the epidermis of the marsupial *Trichosurus vulpecula*. *Australian Journal of Biological Science*, **23**, 1067–1075.

Lyne, D. (1957). The develoment and replacement of pelage hairs in the bandicoot *Perameles nasuta* Geoffroy (Marsupialia: Peramelidae). *Australian Journal of Biological Science*, **10**, 197–216.

Lysakowski, A. and Goldberg, J. M. (2004). Morphophysiology of the vestibular periphery. In: *The Vestibular System* (ed. S. M. Highstein, R. R. Fay and A. N. Popper), New York: Springer-Verlag. pp. 57–152.

Lythgoe, J. N. (1979). *The Ecology of Vision*. Oxford: Clarendon Press.

Ma, J., Znoiko, S., Othersen, K. L. *et al.* (2001). A visual pigment expressed in both rod and cone photoreceptors. *Neuron*, **32**, 451–461.

Ma, P. M. (1991). The barrelettes – architectonic vibrissal representations in the brainstem trigeminal complex of the mouse. I. Normal structural organisation. *Journal of Comparative Neurology*, **309**, 161–199.

Macphail, E. M. (1982). *Brain and Intelligence in Vertebrates*. Oxford: Clarendon Press.

Macrini, T. E. (2007). Observations on the turbinal elements of the ethmoid bone of marsupials. *Journal of Morphology*, **268**, 1101.

Macrini, T. E., Rougier, G. W. and Rowe, T. (2007a). Description of a cranial endocast from the fossil mammal *Vincelestes neuquenianus* (Theriiformes) and its relevance to the evolution of endocranial characters in therians. *Anatomical Record*, **290**, 875–892.

Macrini, T. E., Rowe, T. and Van de Berg, J. L. (2007b). Cranial endocasts from a growth series of *Monodelphis domestica* (Didelphidae, Marsupialia): a study of individual and ontogenetic variation. *Journal of Morphology*, **268**, 844–865.

Macrini, T. E., de Muizon, C., Cifelli, R. L. and Rowe, T. (2007c). Digital cranial endocast of *Pucadelphys andinus*, a Paleocene metatherian. *Journal of Vertebrate Paleontology*, **27**, 99–107.

Magalhaes-Castro, B. and Saraiva, P. E. (1971). Sensory and motor representation in the cerebral cortex of the marsupial *Didelphis azarae azarae*. *Brain Research*, **34**, 291–299.

Magalhães, J. P., Costa J., and Church G. M. (2007). An analysis of the relationship between metabolism, developmental schedules, and longevity using phylogenetic independent contrasts. *The Journals of Gerontology: Biological Sciences*, **62A**, 149–160.

Mai, J. K., Andressen, C. and Ashwell, K. W. S. (1998). Demarcation of prosencephalic regions by CD15-positive radial glia. *European Journal of Neuroscience*, **10**, 746–751.

Maldonado, T. A., Jones, R. E. and Norris, D. O. (2002). Timing of neurodegeneration and beta amyloid deposition in the brain of aging kokanee salmon. *Journal of Neurobiology*, **53**, 21–35.

Maley, B. E. and King, J. S. (1980). Early development of the inferior olivary complex in pouch young opossums. I. A light microscopic study. *Journal of Comparative Neurology*, **194**, 721–739.

Maloney, S. K., Fuller, A., Meyer, L. C. R. *et al.* (2008). Brain thermal inertia, but no evidence for selective brain cooling, in free ranging western grey kangaroos (*Macropus fuliginosus*). *Journal of Comparative Physiology B. Biochemical Systemic and Environmental Physiology*, **179**, 241–251.

Malun, D. and Brunjes, P. C. (1996). Development of olfactory glomeruli: temporal and spatial interactions between olfactory receptor axons and mitral cells in opossums and rats. *Journal of Comparative Neurology*, **368**, 1–16.

Malz, C. R. and Kuhn, H-J. (2002). Calretinin and FMRFamide immunoreactivity in the nervus terminalis of prenatal tree shrews (*Tupaia belangeri*). *Developmental Brain Research*, **135**, 39–44.

Manger, P. R., Fahringer, H. M., Pettigrew, J. D. and Siegel, J. M. (2002). The distribution and morphological characteristics of serotonergic cells in the brain of monotremes. *Brain Behavior and Evolution*, **60**, 315–332.

Manley, G.A. (2001). Evidence for an active process and a cochlear amplifier in nonmammals. *Journal of Neurophysiology*, **86**, 541–549.

Marchand, R. and Bélanger, M. C. (1991). Ontogenesis of the axonal circuitry associated with the olfactory system of the rat embryo. *Neuroscience Letters*, **129**, 285–290.

Marino, L. (2006). Absolute brain size: did we throw the baby out with the bathwater? *Proceedings of the National Academy of Science USA*, **103**, 13606–13611.

Mark, R. F. and Marotte, L. R. (1992). Australian marsupials as models for the developing mammalian visual system. *Trends in Neuroscience*, **15**, 51–57.

Mark, R. F., James, A. C. and Sheng, X-M. (1993). Geometry of the representation of the visual field on the superior colliculus of the wallaby (*Macropus eugenii*). I. Normal projection. *Journal of Comparative Neurology*, **330**, 303–314.

Mark, R. F., Flett, D. L., Marotte, L. R. and Waite, P. M. (2002). Developmental onset of functional activity in the wallaby whisker cortex in response to stimulation of the infraorbital nerve. *Somatosensory and Motor Research*, **19**, 198–206.

Marlow, B. J. (1961). Reproductive behaviour of the marsupial mouse, *Antechinus flavipes* (Waterhouse) (Marsupialia) and the development of the pouch young. *Australian Journal of Zoology*, **9**, 203–218.

Marotte, L. R. (1990). Development of retinotopy in projections from the eye to the dorsal lateral geniculate nucleus and superior colliculus of the wallaby (*Macropus eugenii*). *Journal of Comparative Neurology*, **293**, 524–539.

Marotte, L. R. and Mark, R. F. (1988). Retinal projections to the superior colliculus and dorsal lateral geniculate nucleus in the tammar wallaby (*Macropus eugenii*): II. Topography after rotation of an eye prior to retinal innervation of the brain. *Journal of Comparative Neurology*, **271**, 274–92.

Marotte, L. R. and Sheng, X-M. (2000). Neurogenesis and the identification of developing layers in the visual cortex of the wallaby (*Macropus eugenii*). *Journal of Comparative Neurology*, **416**, 131–142.

Marotte, L. R., Rice, F. L. and Waite, P. M. E. (1992). The morphology and innervation of facial vibrissae in the tammar wallaby, *Macropus eugenii*. *Journal of Anatomy*, **180**, 401–417.

Marotte, L. R., Leamey, C. A. and Waite, P. M. E. (1997). Timecourse of development of the wallaby trigeminal pathway: III. Thalamocortical and corticothalamic projections. *Journal of Comparative Neurology*, **387**, 194–214.

Marotte, L. R., Vidovic, M., Wheeler, E. and Jhaveri, S. (2004). Brain-derived neurotrophic factor is expressed in a gradient in the superior colliculus during development of the retinocollicular projection. *European Journal of Neuroscience*, **20**, 843–847.

Martin, G. (1999). Optical structure and visual fields in birds: their relationship with foraging behaviour and ecology. In *Adaptive Mechanisms in the Ecology of Vision*, (ed. S. N. Archer, M. B. A. Djamgoz, E. R. Loew, J. C. Partridge and S. Vallerga), Dordrecht, Boston, London: Kluwer Academic Publishers. pp. 485–508.

Martin, G. F. (1967). Interneocortical connections in the opossum, *Didelphis virginiana*. *Anatomical Record*, **157**, 607–615.

Martin, G. F. (1968). The pattern of neocortical projections to the mesencephalon of the opossum, Didelphis virginiana. *Brain Research*, **11**, 593–610.

Martin, G. F. and Dom, R. (1970). The rubrospinal tract of the opossum (*Didelphis virginiana*). *Journal of Comparative Neurology*, **138**, 19–30.

Martin, G. F. and Megirian, D. (1972). Corticobulbar projections of the marsupial phalanger (*Trichosurus vulpecula*). II. Projections to the mesencephalon. *Journal of Comparative Neurology*, **144**, 165–192.

Martin, G. F., Megirian, D. and Roebuck, A. (1970). The corticospinal tract of the marsupial phalanger (*Trichosurus vulpecula*). *Journal of Comparative Neurology*, **139**, 245–258.

Martin, G. F., Megirian, D. and Roebuck, A. (1971). Corticobulbar projections of the marsupial phalanger (*Trichosurus vulpecula*). I. Projections to the pons and medulla oblongata. *Journal of Comparative Neurology*, **142**, 275–295.

Martin, G. F., Megirian, D. and Conner, J. B. (1972). The origin, course and termination of the corticospinal tracts of the Tasmanian potoroo (*Potorous apicalis*). *Journal of Anatomy*, **111**, 263–281.

Martin. G. F., Dom, R., King, J. S., Robards, M. and Watson, C. R. R. (1975a). The inferior olivary nucleus of the opossum (*Didelphis marsupialis virginiana*), its organisation and connections. *Journal of Comparative Neurology*, **160**, 507–534.

Martin, G. F., Bresnahan, J. C., Henkel, C. K. and Megirian, D. (1975b). Corticobulbar fibres in the North American opossum (*Didelphis marsupialis virginiana*) with notes on the Tasmanian brush-tailed possum (*Trichosurus vulpecula*) and other marsupials. *Journal of Anatomy*, **120**, 439–484.

Martin, G. F., Culberson, J. L. and Hazlett, J. C. (1983). Observations on the early development of ascending spinal pathways. Studies using the North American opossum. *Anatomy and Embryology*, **166**, 191–207.

Martin, G. F., De Lorenzo, G., Ho, R. H., Humberstron, A. O. and Waltzer, R. (1985). Serotonergic innervation of the forebrain in the North American opossum. *Brain Behavior and Evolution*, **26**, 196–228.

Martin, G. F., Cabana, T., Hazlett, J. C., Ho, R. and Waltzer, R. (1987). Development of brainstem and cerebellar projections to the diencephalon with notes on thalamocortical projections: studies in the North American opossum. *Journal of Comparative Neurology*, **260**, 186–200.

Martin, G. F., Cabana, T. and Waltzer, R. (1988). The origin of projections from the medullary reticular formation to the spinal cord, the diencephalon and the cerebellum at different stages of development in the North American opossum: studies using single and double labeling techniques. *Neuroscience*, **25**, 87–96.

Martin, G. F., Ho, R. H. and Hazlett, J. C. (1989). The early development of major projections to the dorsal striatum in the North American opossum. *Developmental Brain Research*, **47**, 161–170.

Martin, G. F., Ghooray, G., Ho, R. H., Pindzola, R. R. and Xu, X-M. (1991). The origin of serotoninergic projections to the lumbosacral spinal cord at different stages of development in the North American opossum. *Developmental Brain Research*, **58**, 203–213.

Martin, G. F., Pindzola, R. R. and Xu, X. M. (1993). The origin of descending projections to the lumbar spinal cord at different stages of development in the North American opossum. *Brain Research Bulletin*, **30**, 303–317.

Martin, G. F., Terman, J. R., and Wang, X. M. (2000). Regeneration of descending spinal axons after transection of the thoracic spinal cord during early development in the North American opossum, *Didelphis virginiana*. *Brain Research Bulletin*, **53**, 677–687.

Martinez, M. C., Blanco, J., Bullón, M. M. and Agudo, F. J. (1987). Structure of the piriform cortex of the adult rat. A Golgi study. *Journal für Hirnforschung*, **28**, 341–348.

Martinez-Marcos, A. and Halpern, M. (2006). Efferent connections of the main olfactory bulb in the opossum (*Monodelphis domestica*): a characterisation of the olfactory entorhinal cortex in a marsupial. *Neuroscience Letters*, **395**, 51–56.

Martinich, S., Rosa, M. G. and Rocha-Miranda, C. E. (1990). Patterns of cytochrome oxidase activity in the visual cortex of a South American opossum (*Didelphis marsupialis aurita*). *Brazilian Journal of Medical and Biological Research*, **23**, 883–887.

Martinich, S., Pontes, M. N. and Rocha-Miranda, C. E. (2000). Patterns of corticocortical, corticotectal and commissural connections in the opossum visual cortex. *Journal of Comparative Neurology*, **416**, 224–244.

Masterton, R.B., Heffner, H. and Ravizza, R. (1969). The evolution of human hearing. *Journal of the Acoustic Society of America*, **45**, 966–985.

Math, F. and Davrainville, J. L. (1980). Electrophysiological study on the postnatal development of mitral cell activity in the rat olfactory bulb. *Brain Research*, **190**, 243–247.

Mayner, L. (1985). The anatomy of the sensorimotor cortex and thalamus in the wallaby (*Macropus eugenii*). PhD thesis. Canberra: Australian National University.

Mayner, L. (1989a). A nomenclature for the cortical sulcal features of the tammar wallaby, *Macropus eugenii*. *Brain Behavior and Evolution*, **33**, 293–302.

Mayner, L. (1989b). A cyto-architectonic study of the cortex of the tammar wallaby, *Macropus eugenii*. *Brain Behavior and Evolution*, **33**, 303–316.

Mayner, L. (1989c). A cyto-architectonic description of the thalamus of the tammar wallaby, *Macropus eugenii*. *Brain Behavior and Evolution*, **33**, 342–355.

McAllan, B. M. (2003). Timing of reproduction in carnivorous marsupials In *Predators with Pouches. The Biology of Carnivorous Marsupials*. (ed. M. Jones, C. Dickman and M. Archer). Collingwood: CSIRO Publishing. pp. 147–168.

McAllan, B. M. (2009). Reproductive parameters of post "die-off" male *Antechinus flavipes* and *Antechinus stuartii* (Dasyuridae: Marsupialia). *Australian Mammalogy*, **31**, 17–23.

McAllan, B. M., Joss, J. M. P. and Firth, B. T. (1991). Phase delay of the natural photoperiod alters reproductive timing in the marsupial *Antechinus stuartii*. *Journal of Zoology*, **225**, 633–646.

McAllan, B. M., Roberts, J. R. and O'Shea, T. (1996). Seasonal changes in the renal morphometry of *Antechinus stuartii* (Marsupialia: Dasyuridae). *Australian Journal of Zoology*, **44**, 337–354.

McAllan, B. M., O'Shea, T. and Roberts, J. R. (1997). Seasonal changes in the reproductive anatomy of male *Antechinus stuartii* (Marsupialia: Dasyuridae). *Journal of Morphology*, **231**, 261–275.

McAllan, B. M., Roberts, J. R. and O'Shea, T. (1998). Seasonal changes in glomerular filtration rate in *Antechinus stuartii* (Marsupialia: Dasyuridae). *Journal of Comparative Physiology B*, **168**, 41–49.

McAllan, B. M., Westman, W. and Joss, J. M. (2002). The seasonal reproductive cycle of a marsupial, *Antechinus stuartii*: effects of oral administration of melatonin. *General and Comparative Endocrinology*, **128**, 82–90.

McAllan, B. M., Hobbs S. and Norris D. O. (2006). Effects of stress on the neuroanatomy of a marsupial. *Journal of Experimental Zoology A. Comparative Experimental Biology*, **305A**, 154.

McAllan, B. M., Westman, W., Koertner, G. and Cairns, S. C. (2008a). Sex, season and melatonin administration affects daily activity rhythms in a marsupial, the brown antechinus, *Antechinus stuartii*. *Physiology and Behavior*, **93**, 130–138.

McAllan, B. M., Westman, W., Crowther, M. S. and Dickman, C. R. (2008b). Morphology, growth, and reproduction in the Australian house mouse: differential effects of moderate temperatures. *Biological Journal of the Linnean Society*, **94**, 21–30

McCluskey, S. U., Marotte, L. R. and Ashwell, K. W. S. (2008). Development of the vestibular apparatus and central vestibular connections in a wallaby (*Macropus eugenii*). *Brain Behavior and Evolution*, **71**, 271–286.

McConnell, S. J. (1986). Seasonal changes in the circadian plasma melatonin profile of the tammar wallaby, *Macropus eugenii*. *Journal of Pineal Research*, **3**, 119–125.

McConnell, S. J. and Hinds, L. A. (1985). Effects of pinealectomy on plasma melatonin, prolactin and progesterone concentrations during seasonal reproductive quiescence in the tammar, *Macropus eugenii*. *Journal of Reproduction and Fertility*, **75**, 433–440.

McConnell, S. J., Tyndale-Biscoe, C. H. and Hinds, L. A. (1986). Change in duration of elevated concentrations of melatonin is the major factor in photoperiod response of the tammar, *Macropus eugenii. Journal of Reproduction and Fertility*, **77**, 623–632.

McCrady, E. J. (1938). The embryology of the opossum. *American Anatomical Memoirs*, **16**, 1–233.

McCrady, E.J., Wever, E.G. and Bray, C.W. (1937). The development of hearing in the opossum. *Journal of Experimental Zoology*, **75**, 503–517.

McCrady, E.J., Wever, E.G. and Bray, C.W. (1940). A further investigation of the development of hearing in the opossum. *Journal of Comparative Psychology*, **30**, 17–21.

McDonald, I. R., Lee A. K., Bradley A. J. and Than K. A. (1981). Endocrine changes in dasyurid marsupials with differing mortality patterns. *General and Comparative Endocrinology*, **44**, 292–301.

McEwen, B. (1999). Stress and hippocampal plasticity. *Annual Reviews in Neuroscience*, **22**, 105–122.

McFarlane, J. R., Rudd, C. D., Foulds, L. M., Fletcher, T. P. and Renfree, M. B. (1997). Isolation and partial characterisation of tammar wallaby luteinizing hormone and development of a radioimmunoassay. *Reproduction Fertility and Development*, **9**, 475–480.

McIlwain, J. T. (1996). *An Introduction to the Biology of Vision*. New York: Cambridge University Press.

McMenamin, P. G. (2007). The unique paired retinal vascular pattern in marsupials: structural, functional and evolutionary perspectives based on observations in a range of species. *British Journal of Ophthalmology*, **91**, 1399–1405.

McMenamin, P. G. and Krause, W. J. (1993). Morphological observations on the unique paired capillaries of the opossum retina. *Cell and Tissue Research*, **271**, 461–468.

McNab, B. K. (2006). The energetics of reproduction in endotherms and its implication for their conservation. *Integrative and Comparative Biology*, **46**, 1159–1168.

Megirian, D., Johnson, J. I. and Haight, J. R. (1972). Relation between cerebral cortex convolutions and sensory projections in the wombat *Vombatus ursinus. Journal of Physiology (Paris)*, **65**, 448A.

Megirian, D., Weller, L., Martin, G. F. and Watson, C. R. (1977). Aspects of laterality in the marsupial *Trichosurus vulpecula* (brush-tailed possum). *Annals of the New York Academy of Sciences*, **299**, 197–212.

Meiri, K. F., Pfenninger, K. H. and Willard, M. B. (1986). Growth associated protein, GAP-43, a polypeptide that is induced when neurons extend axons, is a component of growth cones and corresponds to pp46, a major polypeptide of a subcellular fraction enriched in growth cones. *Proceedings of the National Academy of Science USA*, **83**, 3537–3541.

Meissirel, C., Wikler, K. C., Chalupa, L. M. and Rakic, P. (1997). Early divergence of magnocellular and parvocellular functional subsystems in the embryonic primate visual system. *Proceedings of the National Academy of Science USA*, **94**, 5900–5905.

Meister, M., Wong, R. O., Baylor, D. A. and Shatz, C. J. (1991). Synchronous bursts of action potentials in ganglion cells of the developing mammalian retina. *Science*, **252**, 939–943.

Melrose H. L., Lincoln, S. J., Tyndall, G. M. and Farrer, M. J. (2006). Parkinson's disease: a rethink of rodent models. *Experimental Brain Research*, **173**, 196–204.

Melyan, Z., Tarttelin, E. E., Bellingham, J., Lucas, R. J. and Hankins, M. W. (2005). Addition of human melanopsin renders mammalian cells photoresponsive. *Nature*, **433**, 741–745.

Merbs, S. L. and Nathans, J. (1992). Absorption spectra of human cone pigments [see comments]. *Nature*, **356**, 433–435.

Meredith, R. W., Westerman, M. and Springer, M. S. (2008a). A timescale and phylogeny for 'Bandicoots' (Peramelemorphia: Marsupialia) based on sequences for five nuclear genes. *Molecular Phylogenetics and Evolution*, **47**, 1–20.

Meredith, R. W., Westerman, M., Case, J. A. and Springer, M. S. (2008b). A phylogeny and timescale for marsupial evolution based on sequences for five nuclear genes. *Journal of Mammalian Evolution*, **15**, 1–36.

Merzenich, M. M. and Brugge, J. F. (1973). Representation of the cochlea on the superior temporal plane of the macaque monkey. *Brain Research*, **50**, 271–296.

Merzenich, M. M. and Schreiner, C. E. (1992). Mammalian auditory cortex – some comparative observations. In *The Evolutionary Biology of Hearing*. (ed. D. B. Webster, R. R. Fay and A. N. Popper). New York: Springer. pp. 673–689.

Merzenich, M. M., Kitzes, L. and Aitkin, L. (1973). Anatomical and physiological evidence for auditory specialisation in the mountain beaver (*Aplodontia rufa*). *Brain Research*, **58**, 331–344.

Merzenich, M. M., Knight, P. L. and Roth, G. L. (1975). Representation of cochlea within primary auditory cortex in the cat. *Journal of Neurophysiology*, **38**, 231–249.

Meskenaite, V. (1997). Calretinin-immunoreactive local circuit neurons in area 17 of the cynomolgus monkey, *Macaca fascicularis. Journal of Comparative Neurology*, **379**, 113–132.

Meyer, J. (1981). A quantitative comparison of the parts of the brains of two Australian marsupials and some eutherian mammals. *Brain Behavior and Evolution*, **18**, 60–71.

Mihailoff, G. A. and King, J. S. (1975). The basilar pontine grey of the opossum: a correlated light and electron microscopic study. *Journal of Comparative Neurology*, **159**, 521–552.

Mihailoff, G. A., Martin, G. F. and Linauts, M. (1980). The pontocerebellar system in the opossum, *Didelphis virginiana*. A horseradish peroxidase study. *Brain Behavior and Evolution*, **17**, 179–208.

Miller, R. A., Austad, S., Burke, D. *et al.* (1999). Exotic mice as models for aging research: polemic and prospectus. *Neurobiology of Aging*, **20**, 217–231.

Miller, R. A., Harper, J. M., Dysko, R. C., Durkee, S. J. and Austad, S. N. (2002). Longer life spans and delayed maturation in wild-derived mice. *Experimental Biology and Medicine*, **227**, 500–508.

Mills, D. M. and Rubel, E. W. (1996). Development of the cochlear amplifier. *Journal of the Acoustical Society of America*, **100**, 428–441.

Mills, D. M. and Shepherd, R. K. (2001). Distortion product otoacoustic emission and auditory brainstem responses in the echidna

(*Tachyglossus aculeatus*). *Journal of the Association for Research in Otolaryngology*, **2**, 130–146.

Mlonyeni, M. (1973). The number of Purkinje cells and inferior olivary neurons in the cat. *Journal of Comparative Neurology*, **147**, 1–10.

Molnár, Z., Knott, G. W., Blakemore, C. and Saunders, N. R. (1998). Development of thalamocortical projections in the South American grey short-tailed opossum (*Monodelphis domestica*). *Journal of Comparative Neurology*, **398**, 491–514.

Moore, D. R. and Irvine, D. R. (1979). The development of some peripheral and central auditory responses in the neonatal cat. *Brain Research*, **163**, 49–59.

Moreno, N. and González, A. (2007). Evolution of the amygdaloid complex in vertebrates, with special reference to the anamnio-amniotic transition. *Journal of Anatomy*, **211**, 151–163.

Morest, D. K. and Winer, J. A. (1986). The comparative anatomy of neurons: homologous neurons in the medial geniculate body of the opossum and cat. *Advances in Anatomy Embryology and Cell Biology*, **97**, 1–96.

Morgan, P. J. and Hazlerigg, D. G. (2008). Photoperiodic signaling through the melatonin receptor turns full circle. *Journal of Neuroendocrinology*, **20**, 820–826.

Morris, J. R. and Lasek, R. J. (1982). Stable polymers of the axonal cytoskeleton: the axoplasmic ghost. *Journal of Cell Biology*, **92**, 192–198.

Morrisson, P. R. (1962). Body temperatures in some Australian mammals. II Paramelidae. *Australian Journal of Biological Science*, **15**, 386–394.

Mosconi, T. M. and Rice, F. L. (1993). Sequential differentiation of sensory innervation in the mystacial pad of the ferret. *Journal of Comparative Neurology*, **333**, 309–325.

Mosconi, T. M., Rice, F. L. and Song, M. J. (1993). Sensory innervation in the inner conical body of the vibrissal follicle-sinus complex of the rat. *Journal of Comparative Neurology*, **328**, 232–251.

Mountz, J. D., Van Zant, G., Allison, D. B., Zhang, H.-G. and Hsu, H.-C. (2002). Beneficial influences of systemic cooperation and sociological behaviour on longevity. *Mechanisms of Ageing and Development*, **123**, 963–973.

Munger, B. L. and Rice, F. L. (1986). Successive waves of differentiation of cutaneous afferents in rat mystacial skin. *Journal of Comparative Neurology*, **252**, 404–414.

Murialdo, G., Barreca, A., Nobili, F. *et al.* (2001). Relationships between cortisol, dehydroepiandrosterone sulphate and insulin-like growth factor-I system in dementia. *Journal of Endocrinological Investigation*, **24**, 139–146.

Murphy, W. J., Pevzner, P. A. and O'Brien, S. J. (2004). Mammalian phylogenetics comes of age. *Trends in Genetics*, **20**, 631–639.

Nagy, K. A. (2005). Field metabolic rate and body size. *Journal of Experimental Biology*, **208**, 1621–1625.

Nagy, K. A. and Bradshaw, S. D. (2000). Scaling of energy and water fluxes in free-living arid-zone Australian marsupials. *Journal of Mammalogy*, **81**, 962–970.

Nambu, T., Sakurai, T., Mizukami, K. *et al.* (1999). Distribution of orexin neurons in adult rat brain. *Brain Research*, **827**, 243–260.

Narkiewicz, O. and Mamos, L. (1990). Relation of the insular claustrum to the neocortex in Insectivora. *Journal für Hirnforschung*, **31**, 623–633.

Nathans, J., Thomas, D. and Hogness, D. S. (1986). Molecular genetics of human colour vision: the genes encoding blue, green, and red pigments. *Science*, **232**, 193–202.

Naylor, R. J. (2005). Investigating the suitability of two marsupial species as models for Alzheimer's disease. Honours thesis. Melbourne: The University of Melbourne.

Naylor, R., Richardson, S. J. and McAllan, B. M. (2008). Boom and bust: a review of the physiological life history pattern in the marsupial genus *Antechinus*. *Journal of Comparative Physiology B. Biochemical Systemic and Environmental Physiology*, **178**, 545–562.

Nelson, J. E. (1992). Developmental staging in a marsupial *Dasyurus hallucatus*. *Anatomy and Embryology*, **185**, 335–354.

Nelson, J., Knight, R. M. and Kingham, C. (2003). Perinatal sensory and motor development in marsupials with special reference to the northern quoll, *Dasyurus hallucatus*. In *Predators with Pouches. The Biology of Carnivorous Marsupials*. (ed. M. Jones, C. Dickman and M. Archer). Collingwood: CSIRO Publishing. pp. 205–217.

Neve, R. L., Finch, E. A., Bird, E. D. and Benowitz, L. I. (1988). Growth-associated protein GAP-43 is expressed selectively in associative regions of the adult human brain. *Proceedings of the National Academy of Science USA*, **85**, 3638–3642.

Neveu, M. M. and Jeffery, G. (2007). Chiasm formation in man is fundamentally different from that in the mouse. *Eye*, **21**, 1264–1270.

Neveu, M. M., Holder, G. E., Ragge, N. K. *et al.* (2006). Early midline interactions are important in mouse optic chiasm formation, but are not critical in man: a significant distinction between man and mouse. *European Journal of Neuroscience*, **23**, 3034–3042.

Neylon, L. and Haight, J. R. (1983). Neocortical projections of the suprageniculate and posterior thalamic nuclei in the marsupial brush-tailed possum, *Trichosurus vulpecula* (Phalangeridae), with a comparative commentary on the organisation of the posterior thalamus in marsupial and placental mammals. *Journal of Comparative Neurology*, **217**, 357–375.

Nilsson, M. A., Arnason, U., Spencer, P. B. S. and Janke, A, (2004). Marsupial relationships and a timeline for marsupial radiation in South Gondwana. *Gene*, **340**, 189–196.

Nilsson, O. and Baron, J. (2004). Fundamental limits on longitudinal bone growth: growth plate senescence and epiphyseal fusion. *Trends in Endocrinology and Metabolism*, **15**, 370–374.

Nimchinsky, E. A., Hof, P. R., Young, W. G. and Morrison, J. H. (1996). Neurochemical, morphologic, and laminar characterisation of cortical projection neurons in the cingulate motor areas of the macaque monkey. *Journal of Comparative Neurology*, **374**, 136–160.

Nixon, R. A. (2003). The calpains in aging and aging-related diseases. *Ageing Research Reviews*, **2**, 407–418.

Nixon, R. A., Paskevich, P. A., Sihag, R. K. and Thayer, C. Y. (1994). Phosphorylation on carboxyl terminus domains of neurofilament proteins in retinal ganglion cell neurons in vivo: influences on regional neurofilament accumulation, interneurofilament spacing, and axon caliber. *Journal of Cell Biology*, **126**, 1031–1046.

Nomura, S., Itoh, K., Sugimoto, T. *et al.* (1986). Mystacial vibrissae representation within the trigeminal sensory nuclei of the cat. *Journal of Comparative Neurology*, **253**, 121–133.

Noriega, A. L. and Wall, J. T. (1991). Parcellated organisation in the trigeminal and dorsal column nuclei of primates. *Brain Research*, **565**, 188–194.

Nudo, R. J. and Masterton, R. B. (1990). Descending pathways to the spinal cord, III: Sites of origin of the corticospinal tract. *Journal of Comparative Neurology*, **296**, 559–583.

Nyberg-Hansen, R. and Brodal, A. (1963) Sites of termination of corticospinal fibers in the cat. An experimental study with silver impregnation methods. *Journal of Comparative Neurology*, **120**, 369–391.

Nyberg-Hansen, R. and Brodal, A. (1964). Sites and mode of termination of rubrospinal fibres in the cat. *Journal of Anatomy*, **98**, 235–253.

O'Day, K. (1935). A preliminary note on the presence of double cones and oil droplets in the retina of marsupials. *Journal of Anatomy*, **70**, 465–467.

O'Day, K. (1938a). The retina of the Australian marsupial. *Medical Journal of Australia*, **1**, 326–328.

O'Day, K. (1938b). The visual cells of the platypus (*Ornithorhynchus*). *British Journal of Ophthalmology*, **22**, 321–328.

O'Donoghue, D. L., Martin, G. F. and King, J. S. (1987). The timing of granule cell differentiation and mossy fibre morphogenesis in the opossum. *Anatomy and Embryology*, **175**, 341–354.

Ohtsuka, T. (1985). Relation of spectral types to oil droplets in cones of turtle retina. *Science*, **229**, 874–877.

Okano, T., Kojima, D., Fukada, Y., Shichida, Y. and Yoshizawa, T. (1992). Primary structures of chicken cone visual pigments: vertebrate rhodopsins have evolved out of cone visual pigments. *Proceedings of the National Academy of Science USA*, **89**, 5932–5936.

Olkowicz, S., Turlejski, K., Bartkowska, K., Wielkopolska, E. and Djavadian, R. L. (2008). Thalamic nuclei in the opossum *Monodelphis domestica*. *Journal of Chemical Neuroanatomy*, **36**, 85–97.

Olmos, J. S. E, Beltramino, C. A. and Alheid, G. (2004). Amygdala and extended amygdala of the rat: a cytoarchitectonical, fibroarchitectonical and chemoarchitectonical survey. In *The Rat Nervous System*. (ed. G. Paxinos). 3rd edn. San Diego, London: Elsevier Academic. pp. 509–603.

Olsson, I. A. S., Hansen, A. I. and Sandøe, P. (2007). Ethics and refinement in animal research. *Science*, **317**, 1680.

Olszewski, J. (1950). On the anatomical and functional organisation of the spinal trigeminal nucleus. *Journal of Comparative Neurology*, **92**, 401–413.

Oswaldo-Cruz, E. and Rocha-Miranda, C. E. (1967). The diencephalon of the opossum in stereotaxic coordinates. I. Epithalamus and dorsal thalamus. *Journal of Comparative Neurology*, **129**, 1–37.

Oswaldo-Cruz, E. and Rocha-Miranda, C.E. (1968). *The Brain of the Opossum (Didelphis marsupialis)*. Rio de Janiero: Instituto de Biofisica, Universidada do Rio de Janiero.

Oswaldo-Cruz, E., Hokoc, J. N. and Sousa, A. P. (1979). A schematic eye for the opossum. *Vision Research*, **19**, 263–278.

Owen, R. (1837). On the structure of the brain in marsupial animals. *Philosophical Transactions of the Royal Society of London B*, **127**, 87–96.

Packer, A. D. (1941). An experimental investigation of the visual system in the phalanger, *Trichosurus vulpecula*. *Journal of Anatomy*, **75**, 309–329.

Paddle, R. (2000). *The Last Tasmanian Tiger. The History and Extinction of the Thylacine*. Cambridge: Cambridge University Press.

Panda, S., Nayak, S. K., Campo, B. *et al.* (2005). Illumination of the melanopsin-signaling pathway. *Science*, **307**, 600–604.

Pardon, M-C. and Rattray, I. (2008). What do we know about the long-term consequences of stress on ageing and the progression of age-related neurodegenerative disorders? *Neuroscience and Biobehavioural Reviews*, **32**, 1103–1120.

Park, K. M., Kim J. I., Ahn Y., Bonventre A. J. and Bonventre J. V. (2004). Testosterone is responsible for enhanced susceptibility of males to ischemic renal injury. *Journal of Biological Chemistry*, **279**, 52282–52292.

Parry, J. W., Poopalasundaram, S., Bowmaker, J. K. and Hunt, D. M. (2004). A novel amino acid substitution is responsible for spectral tuning in a rodent violet-sensitive visual pigment. *Biochemistry*, **43**, 8014–8020.

Paterson, A., Chong, N. W., Brinklow, B. R., Loudon, A. S. and Sugden, D. (1992). Characterisation of 2-[125I]iodomelatonin binding sites in the brain of a marsupial, Bennett's wallaby *(Macropus rufogriseus rufogriseus)*. *Comparative Biochemistry and Physiology. A. Physiology*, **102**, 55–58.

Pawlik, M., Fuchs, E., Walker, L. C. and Levy, E. (1999). Primate-like amyloid-β sequence but no cerebral amyloidosis in aged tree shrews. *Neurobiology of Aging*, **20**, 47–51.

Paxinos, G. and Watson, C. R. R. (2007). *The Rat Brain in Stereotaxic Co-ordinates*. 6th ed. San Diego, London: Elsevier Academic.

Paxinos, G., Kus, L., Ashwell, K. W. S. and Watson, C. R. R. (1999a). *Chemoarchitectonic Atlas of the Rat Forebrain*. San Diego: Academic.

Paxinos, G., Carrive, P., Wang, H-Q. and Wang, P-Y. (1999b). *Chemoarchitectonic Atlas of the Rat Brainstem*. San Diego: Academic.

Peake, W. T. and Rosowski, J. J. (1991). Impedance matching, optimum velocity, and ideal middle ears. *Hearing Research*, **53**, 1–6.

Pearce, A. R., James, A. C. and Mark, R. F. (2000). Development of functional connections between thalamic fibres and the visual cortex of the wallaby revealed by current source density analysis in vivo. *Journal of Comparative Neurology*, **418**, 441–456.

Pearce, A. R. and Marotte, L. R. (2003). The first thalamocortical synapses are made in the cortical plate in the developing visual cortex of the wallaby *(Macropus eugenii)*. *Journal of Comparative Neurology*, **461**, 205–216.

Pearson, L. J., Sanderson, K. J. and Wells, R. T. (1976). Retinal projections in the ringtailed possum, *Pseudocheirus peregrinus*. *Journal of Comparative Neurology*, **170**, 227–240.

Peichl, L. (1992a). Morphological types of ganglion cells in the dog and wolf retina. *Journal of Comparative Neurology*, **324**, 590–602.

Peichl, L. (1992b). Topography of ganglion cells in the dog and wolf retina. *Journal of Comparative Neurology*, **324**, 603–620.

Peichl, L., Kunzle, H. and Vogel, P. (2000). Photoreceptor types and distributions in the retinae of insectivores. *Visual Neuroscience*, **17**, 937–948.

Pellis, S. M., Pellis, V. C. and Nelson, J. E. (1992). The development of righting reflexes in the pouch young of the marsupial, *Dasyurus hallucatus*. *Developmental Psychobiology*, **25**, 105–125.

Petersen, C. C. (2003). The barrel cortex - integrating molecular, cellular and systems physiology. Pflügers Archives: *European Journal of Physiology*, **447**, 126–134.

Pflieger, J. F. and Cabana, T. (1996). The vestibular primary afferents and the vestibulospinal projections in the developing and adult opossum, *Monodelphis domestica*. *Anatomy and Embryology*, **194**, 75–88.

Phelan, J. P. and Austad S. N. (1994). Selecting animal models of human aging: inbred strains often exhibit less biological uniformity than F1 hybrids. *Journal of Gerontology*, **49**, B1–B11.

Phillips, D. P., Calford, M. B., Pettigrew, J. D., Aitkin, L. M. and Semple, M. N. (1982). Directionality of sound pressure transformation in the cat's pinna. *Hearing Research*, **8**, 13–28.

Pindzola, R. R., Ho, R. H. and Martin, G. F. (1990). Development of catecholaminergic projections to the spinal cord in the North American opossum, *Didelphis virginiana*. *Journal of Comparative Neurology*, **294**, 399–417.

Pires, S. S., Shand, J., Bellingham, J. *et al.* (2007). Isolation and characterisation of melanopsin (Opn4) from the Australian marsupial *Sminthopsis crassicaudata* (fat-tailed dunnart). *Proceedings of The Royal Society. B. Biological Sciences*, **274**, 2791–2799.

Polyak, S. (1957). *The Vertebrate Visual System*. Chicago: University of Chicago Press.

Pompeiano, O. and Brodal, A. (1957). Experimental demonstration of a somatopical origin of rubrospinal fibers in the cat. *Journal of Comparative Neurology*, **108**, 225–251.

Pratico, D., Clark, C. M., Lee, V. M. Y. *et al.* (2000). Increased 8,12-iso-iPF(2 alpha)-VI in Alzheimer's disease: correlation of a non-invasive index of lipid peroxidation with disease severity. *Annals of Neurology*, **48**, 809–812.

Preuss, T. M., Stepniewska, I., Jain, N. and Kaas, J. H. (1997). Multiple divisions of macaque precentral motor cortex identified with neurofilament antibody SMI-32. *Brain Research*, **767**, 148–153.

Price, G. J. (2008). Taxonomy and paleobiology of the largest-ever marsupial, *Diprotodon*, Owen, 1838 (Diprotodontidae, Marsupialia). *Zoological Journal of the Linnean Society*, **153**, 369–397.

Price, J. L. (1973). An autoradiographic study of complementary laminar patterns of termination of afferent fibers to the olfactory cortex. *Journal of Comparative Neurology*, **150**, 87–108.

Price, S. R. and Briscoe, J. (2004). The generation and diversification of spinal motor neurons: signals and responses. *Mechanisms of Development*, **121**, 1103–1115.

Primmer, S. R. (2002). In search of a model species for aging research: a study of the life span of tree shrews. *Journal of Anti-Aging Medicine*, **5**, 179–201.

Promislow, D. E. L. and Harvey P. H. (1990). Living fast and dying young: a comparative analysis of life-history variation among mammals. *Journal of Zoology*, **220**, 417–437.

Provis, J. (1977). The organisation of the facial nucleus of the brush-tailed possum *(Trichosurus vulpecula)*. *Journal of Comparative Neurology*, **172**, 177–188.

Pubols, B. H. Jr. and Pubols, L. M. (1966). Somatic sensory representation in the thalamic ventrobasal complex of the Virginia opossum. *Journal of Comparative Neurology*, **127**, 19–34.

Puelles, L. and Rubenstein, J. L. (1993). Expression patterns of homeobox and other putative regulatory genes in the embryonic mouse forebrain suggest a neuromeric organisation. *Trends in Neuroscience*, **16**, 472–479.

Puelles, L., Kuwana, E., Puelles, E. *et al.* (2000). Pallial and subpallial derivatives in the embryonic chick and mouse telencephalon, traced by the expression of the genes Dlx-2, Emx-1, Nkx-2.1, Pax-6, and Tbr-1. *Journal of Comparative Neurology*, **424**, 409–438.

Puelles, L., Martínez, S., Martínez-de-la-Torre, M. and Rubenstein, J. L. R. (2004). Gene maps and related histogenetic domains in the forebrain and midbrain. In *The Rat Nervous System*. (ed. G. Paxinos). 3rd edn. San Diego, London: Elsevier Academic. pp. 3–25.

Puelles, L., Martinez-de-la-Torre, S., Paxinos, G., Watson, C. R. R. and Martinez, M. (2007). *The Chick Brain in Stereotaxic Co-ordinates: An Atlas Correlating Avian and Mammalian Neuroanatomy*. San Diego, London: Elsevier Academic.

Putnam, S. P., Megirian, D. and Manning, J. W. (1968). Marsupial interhemispheric relation. *Journal of Comparative Neurology*, **132**, 227–234.

Qin, Y. Q., Wang, X. M. and Martin, G. F. (1993). The early development of major projections from caudal levels of the spinal cord to the brainstem and cerebellum in the grey short-tailed Brazilian opossum, *Monodelphis domestica*. *Developmental Brain Research*, **75**, 75–90.

Qiu, X., Kumbalasiri, T., Carlson, S. M. *et al.* (2005). Induction of photosensitivity by heterologous expression of melanopsin. *Nature*, **433**, 745–749.

Quiroga, J. C. and Dozo, M. T. (1988). The brain of *Thylacosmilus atrox*. Extinct South American saber-tooth carnivore marsupial. *Journal für Hirnforschung*, **29**, 573–586.

Rakic, P. (1977). Prenatal development of the visual system in rhesus monkey. *Philosophical Transactions of the Royal Society of London. B. Biological Science*, **278**, 245–260.

Rakic, P. (2008). Confusing cortical columns. *Proceedings of the National Academy of Science USA*, **105**, 12099–12100.

Rapaport, D. (2006). Retinal Neurogenesis. In *Retinal Development* (ed. E. Sernagor, S. Eglen, W. Harris and R. O. Wong). Cambridge: Cambridge University Press. pp. 30–58.

Rapaport, D. H., Wilson, P. D. and Rowe, M. H. (1981). The distribution of ganglion cells in the retina of the North American opossum (*Didelphis virginiana*). *Journal of Comparative Neurology*, **199**, 465–480.

Rapaport, D. H., Rakic, P. and LaVail, M. M. (1996). Spatiotemporal gradients of cell genesis in the primate retina. *Perspectives in Developmental Neurobiology*, **3**, 147–159.

Rapaport, D. H., Wong, L. L., Wood, E. D., Yasumura, D. and LaVail, M. M. (2004). Timing and topography of cell genesis in the rat retina. *Journal of Comparative Neurology*, **474**, 304–324.

Rausell, E., Bae, C. S., Viñuela, A., Huntley, G. W. and Jones, E. G. (1992). Calbindin and parvalbumin cells in monkey VPL thalamic nucleus: distribution, laminar cortical projections, and relations to spinothalamic terminations. *Journal of Neuroscience*, **12**, 4088–4111.

Ravizza, R.J., Heffner, H.E. and Masterton, R.B. (1969). Hearing in primitive mammals. I. Opossum (*Didelphis virginiana*). *Journal of Auditory Research*, **9**, 1–7.

Raynaud, F. and Marcilhac, A. (2007). Implication of calpain in neuronal apoptosis: a possible regulation of Alzheimer's disease. *FEBS Journal*, **273**, 3437–3443.

Read, A. F. and Harvey, P. H. (1989). Life history differences among the eutherian radiations. *Journal of Zoology*, **219**, 329–353.

Reece, L. J. and Lim, C. H. (1998). Onset of optic nerve conduction and synaptic potentials in superior colliculus of fetal rats studied in vitro. *Developmental Brain Research*, **106**, 25–38.

Rees, S. and Hore, J. (1970). The motor cortex of the brush-tailed possum (*Trichosurus vulpecula*): motor representation, motor function and the pyramidal tract. *Brain Research*, **20**, 439–451.

Reese, B. E. and Baker, G. E. (1990). The course of fibre diameter classes through the chiasmatic region in the ferret. *European Journal of Neuroscience*, **2**, 34–49.

Reimer, K. (1993). Tonotopic organisation of the inferior colliculus of the grey short-tailed opossum *Monodelphis domestica*, as revealed by the 2-DG method. In *Gene-brain-behaviour*. (ed. N. Elsner and M. Heisenberg). Stuttgart: Thieme. pp. 266–279.

Reimer, K. (1995). Hearing in the marsupial *Monodelphis domestica* as determined by auditory-evoked brainstem responses. *Audiology*, **34**, 334–342.

Reimer, K. (1996). Ontogeny of hearing in the marsupial, *Monodelphis domestica*, as revealed by brainstem auditory evoked potentials. *Hearing Research*, **92**, 143–150.

Reimer, K. and Baumann, S. (1995). Behavioural audiogram of the Brazilian grey short-tailed opossum, *Monodelphis domestica* (Metatheria, Didelphidae). *Zoology*, **99**, 121–127.

Renfree, M. B. (1979). Initiation of development of diapausing embryo by mammary denervation during lactation in a marsupial. *Nature*, **278**, 549–551.

Renfree, M. B. (1983). Marsupial reproduction: the choice between placentation and lactation. In *Oxford Reviews of Reproductive Biology 5* (ed. C. A. Finn). Oxford: Clarendon Press. pp. 1–29.

Renfree, M. B. (2006). Life in the pouch: womb with a view. *Reproduction Fertility and Development*, **18**, 721–734.

Renfree, M. B., Lincoln, D. W., Almeida, O. F. and Short, R. V. (1981). Abolition of seasonal embryonic diapause in a wallaby by pineal denervation. *Nature*, **293**, 138–139.

Renfree, M. B., Holt, A. B., Green, S. W., Carr, J. P. and Cheek, D. B. (1982). Ontogeny of the brain in a marsupial (*Macropus eugenii*) throughout pouch life. I. Brain growth. *Brain Behavior and Evolution*, **20**, 57–71.

Renfree, M. B., Fletcher, T. P., Blanden, D. R. *et al.* (1989). Physiological and behavioural events around the time of birth in macropodid marsupials. In *Kangaroos, Wallabies, and Rat Kangaroos* (ed. G. Grigg, P. Jarman, I. Hume). Sydney: Surrey Beatty and Sons. pp. 323–337.

Renfree, M. B., Wilson, J. D., Short, R. V., Shaw, G. and George, F. W. (1992). Steroid hormone content of the gonads of the tammar wallaby during sexual differentiation. *Biology of Reproduction*, **47**, 644–647.

Renfree, M. B., O, W. S., Short, R. V. and Shaw, G. (1996). Sexual differentiation of the urogenital system of the fetal and neonatal tammar wallaby, *Macropus eugenii*. *Anatomy and Embryology*, **194**, 111–134.

Résibois, A. and Rogers, J. H. (1992). Calretinin in rat brain: an immunohistochemical study. *Neuroscience*, **46**, 101–134.

Revel, F. G., Masson-Pevet, M., Pevet, P., Mikkelsen, J. D. and Simonneaux, V. (2009). Melatonin controls seasonal breeding by a network of hypothalamic targets. *Neuroendocrinology*, **90**, 1–14.

Rexed, B. (1952). The cytoarchitectonic organisation of the spinal cord in the cat. *Journal of Comparative Neurology*, **96**, 415–495.

Rexed, B. (1954). A cytoarchitectonic atlas of the spinal cord in the cat. *Journal of Comparative Neurology*, **100**, 297–379.

Reynolds, M. L., Cavanagh, M. E., Dziegielewska, K. M. *et al.* (1985). Postnatal development of the telencephalon of the tammar wallaby (*Macropus eugenii*). An accessible model of neocortical differentiation. *Anatomy and Embryology*, **173**, 81–94.

Reynolds, M. L. and Saunders, N. R. (1988). Differentiation of the neocortex. In *The Developing Marsupial* (ed. C. H. Tyndale-Biscoe and P. A. Janssens). Berlin: Springer Verlag. pp. 101–116.

Rice, F. L., Gomez, C., Barstow, C., Burnet, A. and Sands, P. (1985). A comparative analysis of the development of the primary somatosensory cortex: interspecies similarities during barrel and laminar development. *Journal of Comparative Neurology*, **236**, 477–495.

Rice, F. L., Mance, A. and Munger, B. L. (1986). A comparative light microscopic analysis of the sensory innervation of the mystacial pad. I. Innervation of vibrissal follicle-sinus complexes. *Journal of Comparative Neurology*, **252**, 154–174.

Richardson, S. J. (2007). Cell and molecular biology of transthyretin and thyroid hormones. *International Reviews of Cytology*, **258**, 137–193.

Richardson, S. J., Lemkine, G. F., Alfama, G., Hassani, Z. and Demeneix, B. A. (2007). Cell division and apoptosis in the adult neural stem cell niche are differentially affected in transthyretin null mice. *Neuroscience Letters*, **421**, 234–238.

Ricklefs, R. E. (2008) The evolution of senescence from a comparative perspective. *Functional Ecology*, **22**, 379–392.

Riley, H. A. (1929). The mammalian cerebellum. A comparative study of the arbor vitae and folial patterns. *Research Publications of the Association of Nervous and Mental Disorders*, **6**, 37–192.

Risold, P. Y. (2004). The septal region. In *The Rat Nervous System* (ed. G. Paxinos) 3rd edn. San Diego, London: Elsevier Academic. pp. 605–632.

Robinson, S. R. (1982). Interocular transfer in a marsupial: the brush-tailed possum (*Trichosurus vulpecula*). *Brain Behavior and Evolution*, **21**, 114–124.

Robinson, S. R. and Dreher, B. (1990). The visual pathways of eutherian mammals and marsupials develop according to a common timetable. *Brain Behavior and Evolution*, **36**, 177–195.

Robinson, S. R. and Webster, M. J. (1985). The morphology of relay neurons in the dorsal lateral geniculate nucleus of the marsupial brush-tailed possum (*Trichosurus vulpecula*). *Journal of Comparative Neurology*, **235**, 196–206.

Rocha-Miranda, C. E., Bombardieri, R. A. Jr, de Monasterio, F. M. and Linden, R. (1973). Receptive fields in the visual cortex of the opossum. *Brain Research*, **63**, 362–367.

Rockel, A. J., Heath, C. J. and Jones, E. G. (1972). Afferent connections to the diencephalon in the marsupial phalanger and the question of sensory convergence in the "posterior group" of the thalamus. *Journal of Comparative Neurology*, **145**, 105–129.

Rockel, A. J., Hiorns, R. W. and Powell, T. P. (1980), The basic uniformity in structure of the neocortex. *Brain*, **103**, 221–244.

Rodger, J., Dunlop, S. A. and Beazley, L. D. (1998). The ipsilateral retinal projection in the fat-tailed dunnart, *Sminthopsis crassicaudata*. *Visual Neuroscience*, **15**, 677–684.

Rodger, J., Dunlop, S. A., Beaver, R. and Beazley, L. D. (2001). The development and mature organisation of the end-artery retinal vasculature in a marsupial, the dunnart *Sminthopsis crassicaudata*. *Vision Research*, **41**, 13–21.

Rodieck, R. W. (1973). *The Vertebrate Retina: Principles of Structure and Function*. San Francisco: W. H. Freeman.

Rosa, M. G., Krubitzer, L. A., Molnár, Z. and Nelson, J. E. (1999). Organisation of visual cortex in the northern quoll, *Dasyurus hallucatus*: evidence for a homologue of the second visual area in marsupials. *European Journal of Neuroscience*, **11**, 907–915.

Rose, K. D. (2006). *The Beginning of the Age of Mammals*. Baltimore: Johns Hopkins.

Rouillé, Y., Chauvet, M. T., Chauvet, J. and Acher, R. (1988). Dual duplication of neurohypophysial hormones in an Australian marsupial: mesotocin, oxytocin, lysine vasopressin and arginine vasopressin in a single gland of the northern bandicoot (*Isoodon macrourus*). *Biochemical and Biophysical Research Communications*, **154**, 346–350.

Rowe, M. J. (1990). Organisation of the cerebral cortex in monotremes and marsupials. In *Cerebral Cortex*. (ed. E. G. Jones and A. Peters) Vol. 8B. New York: Plenum. pp. 263–333.

Rudd, C. D., Short, R. V., Shaw, G. and Renfree, M. B. (1996). Testosterone control of male-type sexual behaviour in the tammar wallaby (*Macropus eugenii*). *Hormones and Behavior*, **30**, 446–454.

Ruigrok, T. J. H. (2004). Precerebellar nuclei and red nucleus. In *The Rat Nervous System*. (ed. G. Paxinos). 3rd edn. London: Elsevier. pp. 167–204.

Russell, E. M. (1985). The metatherians: order Marsupialia. In *Social Odours in Mammals* (ed. R. E. Brown and D. W. MacDonald). Oxford: Clarendon. pp. 45–104.

Russell, E. M. (1986). Observations on the behaviour of the honey possum, *Tarsipes rostratus* (Marsupiala: Tarsipedidae) in captivity. *Australian Journal of Zoology*, **121**: 1–63.

Russell, A., Gilmore, D. P., Mackay, S. *et al.* (2003). The role of androgens in development of the scrotum of the grey short-tailed Brazilian opossum (*Monodelphis domestica*). *Anatomy and Embryology*, **206**, 381–389.

Sakaguchi, D. S., Van Hoffelen, S. J., Grozdanic, S. D. *et al.* (2005). Neural progenitor cell transplants into the developing and mature central nervous system. *Annals of the New York Academy of Science*, **1049**, 118–134.

Samarasinghe, D. and Delahunt, B. (1980). The ependyma of the saccular pineal gland in the non-eutherian mammal *Trichosurus vulpecula*. A scanning electron microscopic study. *Cell and Tissue Research*, **213**, 417–432.

Samollow, P. B. (2008). The opossum genome: insights and opportunities from an alternative mammal. *Genome Research*, **18**, 1199–1215.

Sánchez-Villagra, M. R. and Sultan, F. (2002). The cerebellum at birth in therian mammals, with special reference to rodents. *Brain Behavior and Evolution*, **59**, 101–113.

Sánchez-Villagra, M. R., Gemballa, S., Nummela, S., Smith, K. K. and Maier, W. (2002). Ontogenetic and phylogenetic transformations of the ear ossicles in marsupial mammals. *Journal of Morphology*, **251**, 219–238.

Sánchez-Villagra, M. R., Ladeveze, S., Horovitz, I. *et al.* (2007). Exceptionally well-preserved North American Paleogene metatherians: adaptations and discovery of a major gap in the opossum fossil record. *Biological Letters*, **3**, 318–322.

Sanderson, K. J. and Aitkin, L. M. (1990). Neurogenesis in a marsupial: the brush-tailed possum (*Trichosurus vulpecula*). I. Visual and auditory pathways. *Brain Behavior and Evolution*, **35**, 325–338.

Sanderson, K. J. and Pearson, L. J. (1977). Retinal projections in the native cat, *Dasyurus viverrinus*. *Journal of Comparative Neurology*, **174**, 347–357.

Sanderson, K. J. and Weller, W. L. (1990a). Gradients of neurogenesis in possum neocortex. *Developmental Brain Research*, **55**, 269–274.

Sanderson, K. J. and Weller, W. L. (1990b). Neurogenesis in a marsupial: the brush-tailed possum (*Trichosurus vulpecula*). II. Sensorimotor pathways. *Brain Behavior and Evolution*, **35**, 339–349.

Sanderson, K. J. and Wilson, P. M. (1997). Neurogenesis in septum, amygdala and hippocampus in the marsupial brushtailed possum (*Trichosurus vulpecula*). *Revista Brasileira de Biologia*, **57**, 323–335.

Sanderson, K. J., Pearson, L. J. and Haight, J. R. (1979). Retinal projections in the Tasmanian devil, *Sarcophilus harrisii*. *Journal of Comparative Neurology*, **188**, 335–345.

Sanderson, K. J., Haight, J. R. and Pearson, L. J. (1980). Transneuronal transport of tritiated fucose and proline in the visual pathways of the brush-tailed possum. *Trichosurus vulpecula*. *Neuroscience Letters*, **20**, 243–248.

Sanderson, K. J., Dixon, P. G. and Pearson, L. J. (1982). Postnatal development of retinal projections in the brush-tailed possum, Trichosurus vulpecula. *Brain Research*, **281**, 161–180.

Sanderson, K. J., Haight, J. R. and Pettigrew, J. D. (1984). The dorsal lateral geniculate of macropodid marsupials: cytoarchitecture and retinal projections. *Journal of Comparative Neurology*, **224**, 85–106.

Sanderson, K. J., Nelson, J. E., Crewther, D. P., Crewther, S. G. and Hammond, V. E. (1987). Retinogeniculate projections in diprotodont marsupials. *Brain Behavior and Evolution*, **30**, 22–42.

Santacana, M., Heredia, M. and Valverde, F. (1992a). Development of the main efferent cells of the anterior olfactory bulb and of the bulbar component of the anterior commissure. *Developmental Brain Research*, **65**, 75–83.

Santacana, M., Heredia, M. and Valverde, F. (1992b). Transient pattern of exuberant projections of olfactory axons during development in the rat. *Developmental Brain Research*, **70**, 213–222.

Santangelo A. M., de Souza, F. S. J., Franchini, L. F. *et al.* (2007). Ancient exaptation of a CORE-SINE retroposon into a highly conserved mammalian neuronal enhancer of the proopiomelanocortin gene. *PLoS Genetics*, **3**, 1813–1826.

Saper, C. B. (2004a). Hypothalamus. In *The Human Nervous System*. (ed. G. Paxinos). 2nd edn. San Diego, London: Elsevier Academic. pp 513–550.

Saper, C. B. (2004b). Central autonomic system. In *The Rat Nervous System*. (ed. G. Paxinos). 3rd edn. San Diego, London: Elsevier Academic. pp. 761–796.

Saunders, M. C., Deakin, J., Harrison, G. A. and Curlewis, J. D. (1998). cDNA cloning of growth hormone from the brushtail possum *(Trichosurus vulpecula)*. *General and Comparative Endocrinology*, **111**, 68–75.

Schatz, C. J. and Luskin, M. B. (1986). Relationship between the geniculocortical afferents and their cortical target cells during development of the cat's primary visual cortex. *Journal of Neuroscience*, **6**, 3655–3668.

Schneider, C. (1968). Contribution to the knowledge on the brain of *Notoryctes typhlops*. *Anatomischer Anzeiger*, **123**, 1–24.

Schneider, N. Y., Fletcher, T. P., Shaw, G. and Renfree, M. B. (2008). The vomeronasal organ of the tammar wallaby. *Journal of Anatomy*, **213**, 93–105.

Schroeder, B. E. and Koo, E. H. (2005). To think or not to think: synaptic activity and A beta release. *Neuron*, **48**, 873–875.

Schulze, C., Spaethe, A. and Halata, Z. (1993). The sensory innervation of the periodontium of the third premolar in *Monodelphis domestica*. *Acta Anatomica*, **146**, 42–45.

Schwab, C., Hosokawa, M. and McGeer, P. (2004). Transgenic mice over-expressing amyloid beta protein are an incomplete model of Alzheimer disease. *Experimental Neurology*, **188**, 52–64.

Schwanzel-Fukuda, M., Fadem, B. H., Garcia, M. S. and Pfaff, D. W. (1988). Immunocytochemical localisation of luteinizing hormone-releasing hormone (LHRH) in the brain and nervus terminalis of the adult and early neonatal grey short-tailed opossum (*Monodelphis domestica*). *Journal of Comparative Neurology*, **276**, 44–60.

Schwob, J. E. and Price, J. L. (1984). The development of axonal connections in the central olfactory system of rats. *Journal of Comparative Neurology*, **223**, 177–202.

Sciote, J. J. and Rowlerson, A. (1998). Skeletal fiber types and spindle distribution in limb and jaw muscles of the adult and neonatal opossum, *Monodelphis domestica*. *Anatomical Record*, **251**, 548–562.

Sefton, A. J., Dreher, B. and Harvey, A. (2004). Visual system. In *The Rat Nervous System*. (ed. G. Paxinos). 3rd edn. San Diego, London: Elsevier Academic. pp. 1083–1165.

Segall, W. (1969). The auditory ossicles (malleus, incus) and their relationship to the tympanic membrane in marsupials. *Acta Anatomica*, **73**, 176–191.

Segall, W. (1971). The auditory region (ossicles, sinuses) in gliding mammals and selected representatives of non-gliding genera. *Fieldiana, Zoology*, **58**, 27–59.

Sekaran, S., Foster, R. G., Lucas, R. J. and Hankins, M. W. (2003). Calcium imaging reveals a network of intrinsically light-sensitive inner-retinal neurons. *Current Biology*, **13**, 1290–1298.

Selman, C., McLaren, J. S., Collins, A. R., Duthie, G. G. and Speakman, J. R. (2008). The impact of experimentally elevated energy expenditure on oxidative stress and lifespan in the short-tailed field vole *Microtus agrestis*. *Proceedings of the Royal Society. B. Biological Sciences*, **275**, 1907–1916.

Selwood, L. and Woolley, P. A. (1991). A timetable of embryonic development and ovarian and uterine development in the stripe-faced dunnart *Sminthopsis macroura* (Marsupialia: Dasyuridae). *Journal of Reproduction and Fertility*, **91**, 213–227.

Semple, M. N., Aitkin, L. M., Calford, M. B., Pettigrew, J. D. and Phillips, D. P. (1983). Spatial receptive fields in the cat inferior colliculus. *Hearing Research*, **10**, 203–215.

Sernia, C., Lello, P. and Thomas, W. G. (1990). Angiotensin receptors in an Australian marsupial, the brushtail possum *Trichosurus vulpecula*. *General and Comparative Endocrinology*, **77**, 116–126.

Setchell, P. J. (1974). The development of thermoregulation and thyroid function in the marsupial *Macropus eugenii* (Desmarest). *Comparative Biochemistry and Physiology A*, **47**, 1115–1121.

Shah, N. M., Pisapia, D. J., Maniatis, S. *et al.* (2004). Visualizing sexual dimorphism in the brain. *Neuron*, **43**, 313–319.

Shand, J., Chin, S. M., Harman, A. M. and Collin, S. P. (2000a). The relationship between the position of the retinal area centralis and feeding behaviour in juvenile black bream *Acanthopagrus butcheri* (Sparidae: Teleostei). *Philosophical Transactions of the Royal Society of London. B. Biological Science*, **355**, 1183–1186.

Shand, J., Chin, S. M., Harman, A. M., Moore, S. and Collin, S. P. (2000b). Variability in the location of the retinal ganglion cell area centralis is correlated with ontogenetic changes in feeding behaviour in the black bream, *Acanthopagrus butcheri* (Sparidae, Teleostei). *Brain Behavior and Evolution*, **55**, 176–190.

Shang, F., Ashwell, K. W. S., Marotte, L. R. and Waite, P. M. E. (1997). Development of commissural neurons in the wallaby *(Macropus eugenii)*. *Journal of Comparative Neurology*, **387**, 507–523.

Shapiro, L. E., Leonard, C. M., Sessions, C. E., Dewsbury, D. A. and Insel, T. R. (1991). Comparative neuroanatomy of the sexually dimorphic hypothalamus in monogamous and polygamous voles. *Brain Research*, **541**, 232–240.

Shapiro, L. S., Roland, R. M., Li, C. S. and Halpern, M. (1996). Vomeronasal system involvement in response to conspecific odors in adult male opossums, *Monodelphis domestica*. *Behavioral Brain Research*, **77**, 101–113.

Shatz, C. J. (1983). The prenatal development of the cat's retino-geniculate pathway. *Journal of Neuroscience*, **3**, 482–499.

Shatz, C. J. (1990). Competitive interactions between retinal ganglion cells during prenatal development. *Journal of Neurobiology*, **21**, 197–211.

Shaw, G. and Renfree, M. B. (2006). Parturition and perfect prematurity: birth in marsupials. *Australian Journal of Zoology*, **54**, 139–149.

Shaw, G., Renfree, M. B. and Short, R. V. (1990). Primary genetic control of sexual differentiation in marsupials. *Australian Journal of Zoology*, **37**, 443–450.

Shaw, G., Renfree, M. B., Leihy, M. W. *et al.* (2000). Prostate formation in a marsupial is mediated by the testicular androgen 5 alpha-androstane-3 alpha, 17 beta-diol. *Proceedings of the National Academy of Science USA*, **97**, 12256–12259.

Sheng, X-M. (1989). An anatomical study of the development of cortical visual pathways in the wallaby (*Macropus eugenii*). PhD thesis. Canberra: Australian National University.

Sheng, X-M., Marotte, L. R. and Mark, R. F. (1990). Development of connections to and from the visual cortex in the wallaby (*Macropus eugenii*). *Journal of Comparative Neurology*, **300**, 196–210.

Sheng X-M., Marotte, L. R. and Mark, R. F. (1991). Development of the laminar distribution of thalamocortical axons and cortico-thalamic cell bodies in the visual cortex of the wallaby. *Journal of Comparative Neurology*, **307**, 17–38.

Sherman, D. M. and Krause, W. J. (1990). Morphological, developmental and immunohistochemical observations on the opossum pituitary with emphasis on the pars intermedia. *Acta Histochemica*, **89**, 37–56.

Shipley, M. T., Ennis, M. and Puche, A. C. (2004). Olfactory system. In *The Rat Nervous System* (ed. G. Paxinos). 3rd edn. London: Elsevier. pp. 923–964.

Short, J. and Turner, B. (1999). Ecology of burrowing bettongs (*Bettongia lesueur*) (Marsupiala: Potoroidae), on Dorre and Bernier Islands, Western Australia. *Wildlife Research* **26**, 651–669.

Silver, J., Lorenz, S. E., Wahlsten, D. and Coughlin, J. (1982). Axonal guidance during development of the great cerebral commissures: descriptive and experimental studies, in vivo, on the role of preformed glial pathways. *Journal of Comparative Neurology*, **210**, 10–29.

Simerly, R. B. (2004). Anatomical substrates of hypothalamic integration. In *The Rat Nervous System*. (ed. G. Paxinos) 3rd edn. San Diego, London: Elsevier Academic. pp. 335–368.

Simon, D. K. and O'Leary, D. D. (1992). Development of topographic order in the mammalian retinocollicular projection. *Journal of Neuroscience*, **12**, 1212–1232.

Simpson, J. I., Giolli, R. A. and Blanks, R. H. I. (1988). The pretectal nuclear complex and the accessory optic system. *Reviews of Oculomotor Research*, **2**, 335–364.

Singleton G. R., Krebs C. J., Davis S., Chambers L. and Brown P. R. (2001). Reproductive changes in fluctuating house mouse populations in southeastern Australia. *Proceedings of The Royal Society. B. Biological Sciences*, **268**, 1741–1748.

Skene, J. H. P., Jacobson, R. D., Snipes, G. J. *et al.* (1986). A protein induced during nerve growth (GAP-43) is a major component of growth cone membranes. *Science*, **233**, 783–786.

Smith, G. E. (1902). On a peculiarity of the cerebral commissure in certain marsupials, not hitherto recognised as a distinctive feature of the diprotodontia. *Proceedings of the Royal Society. B. Biological Sciences*, **70**, 226–231.

Smith, G. E. (1903a). Notes on the morphology of the mammalian cerebellum. *Journal of Anatomy*, **37**, 329–332.

Smith, G. E. (1903b). Further observations of the natural mode of subdivision of the mammalian cerebellum. *Anatomischer Anzeiger*, **23**, 368–384.

Smith, K. K. (2001). Early development of the neural plate, neural crest and facial region of marsupials. *Journal of Anatomy*, **199**, 121–131.

Snipes, G., Chan, S., McGuire, C. *et al.* (1987). Evidence for the coidentification of GAP-43, a growth associated protein, and F1, a plasticity associated protein. *Journal of Neuroscience*, **7**, 4066–4075.

Snyder, G. K., Gannon, B., Baudinette, R. V. and Nelson, J. (1989). Functional organisation of cerebral microvasculature in a marsupial, the northern quoll (*Dasyurus hallucatus*). *Journal of Experimental Zoology*, **251**, 349–354.

Sousa, A. P., Oswaldo-Cruz, E. and Gattass, R. (1971). Somatotopic organisation and response properties of neurons of the ventrobasal complex in the opossum. *Journal of Comparative Neurology*, **142**, 231–247.

Sousa, A. B., Gattass, R. and Oswaldo-Cruz, E. (1978). The projection of the opossum's visual field on the cerebral cortex. *Journal of Comparative Neurology*, **177**, 569–587.

Speakman, J. R. (2005a). Body size, energy metabolism and lifespan. *Journal of Experimental Biology*, **208**, 1717–1730.

Speakman, J. R. (2005b). Correlations between physiology and lifespan – two widely ignored problems with comparative studies. *Aging Cell*, **4**, 167–175.

Speakman, J. R. (2008). The physiological costs of reproduction in small mammals. *Philosophical Transactions of the Royal Society. B. Biological Sciences*, **363**, 375–398.

Spires, T. L. and Hyman, B. T. (2005). Transgenic models of Alzheimer's disease: learning from animals. *Journal of the American Society for Experimental Neurotherapeutics*, **2**, 423–437.

Stokes, J. H. (1912). The acoustic complex and its relations in the brain of the opossum (*Didelphys virginiana*). *American Journal of Anatomy*, **12**, 401–445.

Stone, J. (1965). A quantitative analysis of the distribution of ganglion cells in the cat's retina. *Journal of Comparative Neurology*, **124**, 337–352.

Stoothoff, W. H. and Johnson, G. V. W. (2005). Tau phosphorylation: physiological and pathological consequences. *Biochimica et Biophysica Acta*, **1739**, 280–297.

Strachan, J., Chang, L. Y., Wakefield, M. J., Graves, J. A. and Deeb, S. S. (2004). Cone visual pigments of the Australian marsupials, the stripe-faced and fat-tailed dunnarts: sequence and inferred spectral properties. *Visual Neuroscience*, **21**, 223–229.

Strasmann, T. J., Halata, Z. and Loo, S. K. (1987). Topography and ultrastructure of sensory nerve endings in the joint capsules of the Kowari (*Dasyuroides byrnei*), an Australian marsupial. *Anatomy and Embryology*, **176**, 1–12.

Strasmann, T. J., Feilscher, T. H., Baumann, K. I. and Halata, Z. (1999). Distribution of sensory receptors in joints of the upper cervical column in the laboratory marsupial *Monodelphis domestica*. *Annals of Anatomy*, **181**, 199–206.

Striedter, G. F. (2005). *Principles of Brain Evolution*. Sunderland: Sinauer Associates.

Stubbs, J., Palmer, A., Vidovic, M. and Marotte, L. R. (2000). Graded expression of EphA3 in the retina and ephrin-A2 in the superior colliculus during initial development of coarse topography in the wallaby retinocollicular projection. *European Journal of Neuroscience*, **12**, 3626–3636.

Sumner, P., Arrese, C. A. and Partridge, J. C. (2005). The ecology of visual pigment tuning in an Australian marsupial: the honey possum *Tarsipes rostratus*. *Journal of Experimental Biology*, **208**, 1803–1815.

Sun, C., Warland, D. K., Ballesteros, J. M., van der List, D. and Chalupa, L. M. (2008). Retinal waves in mice lacking the beta2 subunit of the nicotinic acetylcholine receptor. *Proceedings of the National Academy of Science USA*, **105**, 13 638–13 643.

Sutherland, D. R., Banks, P. B., Jacob, J. and Singleton G. R. (2004). Shifting age structure of house mice during a population outbreak. *Wildlife Research*, **31**, 613–618.

Swanson, L. W. and Petrovich, G. D. (1998). What is the amygdala? *Trends in Neuroscience*, **21**, 323–331.

Swanson, J. J., Kuehl-Kovarik, M. C., Elmquist J. K., Sakaguchi, D. S. and Jacobson, C. D. (1999). Development of the facial and hypoglossal motor nuclei in the neonatal Brazilian opossum brain. *Developmental Brain Research*, **112**, 159–172.

Sweet, G. (1906). Contribution to our knowledge of the anatomy of *Notoryctes typhlops*, Stirling. Part III. The eye. *Quarterly Journal of Microscopic Science*, **50**, 547–572.

Switzer, R. C. and Johnson, J. I. (1977). Absence of mitral cells in monolayer in monotremes. Variations in vertebrate olfactory bulbs. *Acta Anatomica*, **99**, 36–42.

Szalay, F. S. (1982). A new appraisal of marsupial phylogeny and classification. In *Carnivorous Marsupials*, (ed. M. Archer). Sydney: Royal Zoological Society of New South Wales. pp. 621–640.

Szalay, F. S. (1994). *Evolutionary History of the Marsupials and an Analysis of Osteological Characters*. Cambridge: Cambridge University Press.

Szel, A., Rohlich, P., Mieziewska, K., Aguirre, G. and van Veen, T. (1993). Spatial and temporal differences between the expression of short- and middle-wave sensitive cone pigments in the mouse retina: a developmental study. *Journal of Comparative Neurology*, **331**, 564–577.

Tancred, E. (1981). The distribution and sizes of ganglion cells in the retinas of five Australian marsupials. *Journal of Comparative Neurology*, **196**, 585–603.

Tansley, K. D. (1965). *Vision in Vertebrates*. London: Chapman and Hall.

Tanzi, R. E., Moir, R. D. and Wagner, S. L. (2004). Clearance of Alzheimer's A beta peptide: the many roads to perdition. *Neuron*, **43**, 605–608.

Tarozzo, G., Peretto, P., Perroteau, I. *et al.* (1994). GnRH neurons and other cell populations migrating from the olfactory neuroepithelium. *Annals of Endocrinology*, **55**, 249–54.

Tayebati, S. K. (2005). Animal models of cognitive dysfunction. *Mechanisms of Ageing and Development*, **127**, 100–108.

Taylor J. S. and Guillery, R. W. (1995) Does early monocular enucleation in a marsupial affect the surviving uncrossed retinofugal pathway? *Journal of Anatomy*, **186**, 335–342.

Temple-Smith, P. D. and Grant, T. R. (2001). Uncertain breeding: a short history of reproduction in Monotremes. *Reproduction Fertility and Development*, **13**, 487–497.

Terman, J. R., Wang, X-M. and Martin, G. F. (1998). Origin, course, and laterality of spinocerebellar axons in the North American opossum, *Didelphis virginiana*. *Anatomical Record*, **251**, 528–547.

Torvik, A. (1956). Afferent connections to the sensory trigeminal nuclei, the nucleus of the solitary tract and adjacent structures; an experimental study in the rat. *Journal of Comparative Neurology*, **106**, 51–141.

Towe, A. L. (1973). Relative numbers of pyramidal tract neurons in mammals of different sizes. *Brain Behavior and Evolution*, **7**, 1–17.

Tracey, D. (2004a). Ascending and descending pathways in the spinal cord. In *The Rat Nervous System*. (ed. G. Paxinos). 3rd edn. London: Elsevier. pp. 149–164.

Tracey, D. (2004b). Somatosensory system. In *The Rat Nervous System*. (ed. G. Paxinos). 3rd edn. London: Elsevier. pp. 797–815.

Treloar, H. B., Purcell, A. L. and Greer, C. A. (1999). Glomerular formation in the developing rat olfactory bulb. *Journal of Comparative Neurology*, **413**, 289–304.

Tulsi, R. S. (1979a). A scanning electron microscopic study of the pineal recess of the adult brush-tailed possum (*Trichosurus vulpecula*). *Journal of Anatomy*, **129**, 521–530.

Tulsi, R. S. (1979b). The subcommissural organ of the adult and pouch young brush-tailed possum (*Trichosurus vulpecula*). A scanning electron microscopic study. *Cell and Tissue Research*, **203**, 107–114.

Tulsi, R. S. (1982). Reissner's fibre in the sacral cord and filum terminale of the possum *Trichosurus vulpecula*: a light, scanning, and electron microscopic study. *Journal of Comparative Neurology*, **211**, 11–20.

Tulsi, R. S. and Kennaway, D. J. (1979). Observations on the secretions of the subcommissural organ and the pineal in the adult brush-tailed possum (*Trichosurus vulpecula*). *Neuroendocrinology*, **28**, 264–272.

Tyler, C. J., Dunlop, S. A., Lund, R. D. *et al.* (1998). Anatomical comparison of the macaque and marsupial visual cortex: common features that may reflect retention of essential cortical elements. *Journal of Comparative Neurology*, **400**, 449–468.

Tyndale-Biscoe, C. H. and Hinds, L. A. (1984). Seasonal patterns of circulating progesterone and prolactin and response to bromocriptine in the female tammar *Macropus eugenii*. *General and Comparative Endocrinology*, **53**, 58–68.

Tyndale-Biscoe C. and Janssens P. (1988). Introduction. In *The Developing Marsupial. Models for Biomedical Research*. (ed. C. Tyndale-Biscoe and P. Janssens). Berlin, Heidelberg, New York, London, Paris, Tokyo: Springer-Verlag. pp. 1–7.

Tyndale-Biscoe, C. H., Hearn, J. P. and Renfree, M. B. (1974). Control of reproduction in macropodid marsupials. *Journal of Endocrinology*, **63**, 589–614.

Tyndale-Biscoe, H. (2005). *Life of Marsupials*. Collingwood: CSIRO publishing.

Tyndale-Biscoe, H., and Renfree, M. B. (1987). *Reproductive Physiology of Marsupials*. Cambridge: Cambridge University Press.

Valentino, K. L. and Jones, E. G. (1982). The early formation of the corpus callosum: a light and electron microscopic study in foetal and neonatal rats. *Journal of Neurocytology*, **11**, 583–609.

Valtschanoff, J. G., Weinberg, R. J., Kharazia, V. N. *et al.* (1993). Neurons in rat cerebral cortex that synthesise nitric oxide: NADPH diaphorase histochemistry, NOS immunocytochemistry, and colocalisation with GABA. *Neuroscience Letters*, **157**, 157–161.

Van Brederode, J. F. , Helliesen, M. K. and Hendrickson, A. E. (1991). Distribution of the calcium-binding proteins parvalbumin and calbindin-D28k in the sensorimotor cortex of the rat. *Neuroscience*, **44**, 157–171.

Van der Gucht, E., Hof, P. R., van Brussel, L., Burnat, K. and Arckens, L. (2007). Neurofilament protein and neuronal activity markers define regional architectonic parcellation in the mouse visual cortex. *Cerebral Cortex*, **17**, 2805–2819.

Van der Loos, H. (1976). Barreloids in mouse somatosensory thalamus. *Neuroscience Letters*, **2**, 1–6.

Van der Loos, H. and Woolsey, T. A. (1973). Somatosensory cortex: structural alterations following early injury to sense organs. *Science*, **179**, 395–398.

Van Dongen, P. A. M. (1998). Brain size in vertebrates. In *The Central Nervous System of Vertebrates* (ed. R. Nieuwenhuys, H. J. Ten Donkelaar and C. Nicholson). Berlin: Springer. pp. 2099–2134.

Van Dyck, S. and Strahan, R. (2008). *The Mammals of Australia*. 3rd edn. Sydney: New Holland Publishers.

Van Horn, R. N. (1970). Vibrissae structure in the rhesus monkey. *Folia Primatologica*, **13**, 241–285.

Vaney, D. I., Young, H. M. and Gynther, I. C. (1991). The rod circuit in the rabbit retina. *Visual Neuroscience*, **7**, 141–154.

Vargas, C. D., Volchan, E., Hokoç, J. N. *et al.* (1997). On the functional anatomy of the nucleus of the optic tract-dorsal terminal nucleus commissural connection in the opossum (*Didelphis marsupialis aurita*). *Neuroscience*, **76**, 313–321.

Veitch, C. E., Nelson, J. and Gemmell, R. T. (2000). Birth in the brush-tail possum, *Trichosurus vulpecula* (Marsupialia: Phalangeridae). *Australian Journal of Zoology*. **48**, 691–700.

Verley, R. and Axelrad, H. (1975). Postnatal ontogenesis of potentials elicited in the cerebral cortex by afferent stimulation. *Neuroscience Letters*, **1**, 99–104.

Verzola, D., Gandolfo, M. T., Salvatore, F. *et al.* (2004) Testosterone promotes apoptotic damage in human renal tubular cells. *Kidney International*, **65**, 1252–1261.

Vidal, P.-P. and Sans, A. (2004). Vestibular system. In: *The Rat Nervous System* (ed. G. Paxinos). 3rd edn. San Diego, London: Elsevier Academic. pp. 965–996.

Vidovic, M., Marotte, L. R. and Mark, R. F. (1999). Marsupial retino-collicular system shows differential expression of messenger RNA encoding EphA receptors and their ligands during development. *Journal of Neuroscience Research*, **57**, 244–254.

Vidovic, M. and Marotte, L. R. (2003). Analysis of EphB receptors and their ligands in the developing retinocollicular system of the wallaby reveals dynamic patterns of expression in the retina. *European Journal of Neuroscience*, **18**, 1549–1558.

Vidyasagar, T. R., Wye-Dvorak, J., Henry, G. H. and Mark, R. F. (1992). Cytoarchitecture and visual field representation in area 17 of the tammar wallaby (*Macropus eugenii*). *Journal of Comparative Neurology*, **325**, 291–300.

Volchan, E., Bernardes, R. F., Rocha-Miranda, C. E., Gleiser, L. and Gawryszewski, L. G. (1988). The ipsilateral field representation in the striate cortex of the opossum. *Experimental Brain Research*, **73**, 297–304.

Voogd, J. (2004). Cerebellum. In *The Rat Nervous System* (ed. G. Paxinos G). 3rd edn. San Diego, London: Elsevier Academic. pp. 205–242.

Voss, H. (1963). Besitzen die Monotremen (*Echidna* und *Ornithorhynchus*) ein- oder mehrfaserige Muskelspindeln? *Anatomischer Anzeiger*, **113**, 255–258.

Vozzo, R., Wittert, G. A., Chapman, I. M. *et al.* (1999). Evidence that nitric oxide stimulates feeding in the marsupial *Sminthopsis crassicaudata*. *Comparative Biochemistry and Physiology. C. Toxicology and Pharmacology*, **123**, 145–151.

Waite, P. M. (2004). Trigeminal sensory system. In *The Rat Nervous System*. (ed. G. Paxinos). 3rd edn. San Diego, London: Elsevier Academic. pp 817–851.

Waite, P. and Weller, W. (1997). Development of somatosensory pathways from the whiskers. In *Marsupial Biology. Recent Research, New Perspectives*. (ed. N. R. Saunders and L. A. Hinds). Sydney: UNSW Press. pp. 327–344.

Waite, P. M., Marotte, L. R. and Mark, R. F. (1991). Development of whisker representation in the cortex of the tammar wallaby *Macropus eugenii*. *Developmental Brain Research*, **58**, 35–41.

Waite, P. M. E., Marotte, L. R. and Leamey, C. A. (1994). Timecourse of development of the wallaby trigeminal pathway. I. Periphery to brainstem. *Journal of Comparative Neurology*, **350**, 75–95.

Waite, P. M., Marotte, L. R., Leamey, C. A. and Mark, R. F. (1998). Development of whisker-related patterns in marsupials: factors controlling timing. *Trends in Neuroscience*, **21**, 265–269.

Waite, P. M., Gorrie, C. A., Herath, N. P. and Marotte, L. R. (2006). Whisker maps in marsupials: nerve lesions and critical periods. *Anatomical Record. A. Discoveries in Molecular Cellular and Evolutionary Biology*, **288**, 174–181.

Wakefield, M. J., Anderson, M., Chang, E. *et al.* (2008). Cone visual pigments of monotremes: filling the phylogenetic gap. *Visual Neuroscience*, **25**, 257–264.

Walls, G. L. (1939). Notes on the retinae of two opossum genera. *Journal of Morphology*, **64**, 67–87.

Walls, G. (1942). *The Vertebrate Eye and Its Adaptive Radiation.* Bloomfield Hills: Cranbrooke.

Walsh, T. M. and Ebner, F. F. (1973). Distribution of cerebellar and somatic lemniscal projections in the ventral nuclear complex of the Virginia opossum. *Journal of Comparative Neurology*, **147**, 427–446.

Wang, X-M., Xu, X-M., Qin, Y-Q. and Martin, G. F. (1992). The origins of supraspinal projections to the cervical and lumbar spinal cord at different stages of development in the grey short-tailed Brazilian opossum, *Monodelphis domestica. Developmental Brain Research*, **68**, 203–216.

Wang, X-M., Qin, Y-Q., Xu, X-M. and Martin, G. F. (1994). Developmental plasticity of reticulospinal and vestibulospinal axons in the north American opossum, *Didelphis virginiana. Journal of Comparative Neurology*, **349**, 288–302.

Ward, L. and Watson, C. (1973). An experimental study of the ventrolateral thalamic nucleus of the brush-tailed possum. *Journal of Anatomy*, **116**, 116.

Warner, F. J. (1969). The development of the diencephalon in *Trichosurus vulpecula. Okajimas Folia Anatomica Japonica*, **46**, 265–295.

Warner, F. J. (1970). The development of the pretectal nuclei in *Trichosurus vulpecula. Okajimas Folia Anatomica Japonica*, **47**, 73–100.

Warner, F. J. (1980). The development of the forebrain in *Trichosurus vulpecula. Okajimas Folia Anatomica Japonica*, **57**, 265–320.

Warner, G. and Watson, C. R. (1972), The rubrospinal tract in a diprotodont marsupial (*Trichosurus vulpecula*). *Brain Research*, **41**, 180–183.

Warrant, E. J. (1999). Seeing better at night: life style, eye design and the optimum strategy of spatial and temporal summation. *Vision Research*, **39**, 1611–1630.

Watson, C. R. (1971). The corticospinal tract of the quokka wallaby (*Setonix brachyurus*). *Journal of Anatomy*, **109**, 127–133.

Watson, C. R. and Freeman, B. W. (1977). The corticospinal tract in the kangaroo. *Brain Behavior and Evolution*, **14**, 341–351.

Watson, C. R. and Herron, P. (1977). The inferior olivary complex of marsupials. *Journal of Comparative Neurology*, **176**, 527–537.

Watson, C. R., Broomhead, A. and Holst, M. C. (1976). Spinocerebellar tracts in the brush-tailed possum, *Trichosurus vulpecula. Brain Behavior and Evolution*, **13**, 142–153.

Watson, G. S. and Craft, S. (2004). Modulation of memory by insulin and glucose: neuropsychological observations in Alzheimer's disease. *European Journal of Pharmacology*, **490**, 97–113.

Way, J. B. (1970). Bilateral corticopontine projections in two diprotodonts – a degeneration study. *Journal of Anatomy*, **106**, 204.

Way, J. S. (1975). A degeneration study of some habenular efferents to the midbrain in a wallaby. *American Journal of Anatomy*, **142**, 1–13.

Way, J. S. and Kaelber, W. W. (1969). A degeneration study of efferent connections of the habenular complex in the opossum. *American Journal of Anatomy*, **124**, 31–46.

Weinert, B. T. and Timiras, P. S. (2003). Physiology of aging. Invited review: theories of aging. *Journal of Applied Physiology*, **95**, 1706–1716.

Welker, C. and Woolsey, T. A. (1974). Structure of layer IV in the somatosensory neocortex of the rat: description and comparison with the mouse. *Journal of Comparative Neurology*, **158**, 437–453.

Weller, W. L. (1972). Barrels in somatic sensory neocortex of the marsupial *Trichosurus vulpecula* (brush-tailed possum). *Brain Research*, **43**, 11–24.

Weller, W. L. (1979). Experimental alteration of neocortex by removal of vibrissal follicles from pouch young brush-tailed possums, *Trichosurus vulpecula. Journal of Anatomy*, **128**, 659.

Weller, W. L. (1980). 'Barreloids' in a subdivision of the ventral posterior thalamic nucleus in the brush-tailed possum *Trichosurus vulpecula. Journal of Anatomy*, **130**, 214.

Weller, W. L. (1983). Vibrissa-related spatial patterns of succinic dehydrogenase activity in the somatosensory system of brush-tailed possums *Trichosurus vulpecula. Journal of Anatomy*, **136**, 663.

Weller, W. L. (1993). SmI cortical barrels in an Australian marsupial, *Trichosurus vulpecula* (brush-tailed possum): structural organisation, patterned distribution and somatotopic relationships. *Journal of Comparative Neurology*, **337**, 471–492.

Weller, W. L. and Haight, J. R. (1973). Barrels and somatotopy in S1 neocortex of the brush-tailed possum. *Journal of Anatomy*, **116**, 474.

Weller, W. L. and Johnson, J. I. (1975). Barrels in cerebral cortex altered by receptor disruption in newborn, but not in five-day-old mice (Cricetidoe and Muridae). *Brain Research*, **83**, 504–508.

Westman, W., Körtner, G. and Geiser, F. (2002). Developmental thermoenergetics of the dasyurid marsupial, *Antechinus stuartii. Journal of Mammalogy*, **83**, 81–90.

White, C. R. and Seymour, R. S. (2004). Does basal metabolic rate contain a useful signal? Mammalian BMR allometry and correlations with a selection of physiological, ecological, and life-history variables. *Physiological and Biochemical Zoology.* **77**, 929–941.

White, C. R., Blackburn, T. M. and Seymour, R. S. (2009). Phylogenetically informed analysis of the allometry of mammalian basal metabolic rate supports neither geometric nor quarter-power scaling. *Evolution*, **63**, 2658–2667.

Wiesel, T. N., Hubel, D. H. and Lam, D. M. (1974). Autoradiographic demonstration of ocular-dominance columns in the monkey striate cortex by means of transneuronal transport. *Brain Research*, **79**, 273–279.

Wilkes, G. E. and Janssens, P. A. (1988). The development of renal function. In *The Developing Marsupial: Models for Biomedical Research.* (ed. C. H. Tyndale-Biscoe and P. A. Janssens). Berlin, New York: Springer Verlag. pp. 176–189.

Wilkinson, G. S. and South, J. M. (2002). Life history, ecology and longevity in bats. *Aging Cell*, **1**, 124–131.

Willard, F.H. (1993). Postnatal development of auditory nerve projections to the cochlear nucleus in *Monodelphis domestica.* In

The Mammalian Cochlear Nuclei: Organisation and Function. (ed. M. A. Merchán, J. M. Juiz, D. A. Geoffrey and E. Mugnaini). New York: Plenum Press. pp. 29–42.

Willard, F. H. and Martin, G. F. (1983). The auditory brainstem nuclei and some of their projections to the inferior colliculus in the North American opossum. *Neuroscience*, **10**, 1203–1232.

Willard, F.H. and Martin, G.F. (1984). Collateral innervation of the inferior colliculus in the North American opossum: a study using fluorescent markers in a double-labeling paradigm. *Brain Research*, **303**, 171–182.

Willard, F.H. and Martin, G.F. (1986). The development and migration of large multipolar neurons into the cochlear nuclei of the North American opossum. *Journal of Comparative Neurology*, **248**, 119–132.

Willard, F.H. and Munger, B.L. (1988). Sequential waves of differentiation in the cochlea of Monodelphis domestica. *Society for Neuroscience Abstracts*, **14**, 425.

Williamson, P., Fletcher, T. P. and Renfree, M. B. (1990). Testicular development and maturation of the hypothalamic-pituitary-testicular axis in the male tammar, *Macropus eugenii*. *Journal of Reproduction and Fertility*, **88**, 549–557.

Wilson, P. M. and Astheimer, L. B. (1989). Laminar and non-laminar patterns of acetylcholinesterase activity in the marsupial lateral geniculate nucleus. *Brain Research*, **486**, 236–260.

Wilson, P. M. and Watson, C. (1980). Acetylthiocholinesterase staining in an interpedunculotegmental pathway in four species: *Procavia capensis* (dassie), *Cavia procellus* (guinea-pig), *Trichosurus vulpecula* (brush-tail possum) and *Rattus rattus* (hooded rat). *Brain Research*, **201**, 418–422.

Wimborne, B. M., Mark, R. F. and Ibbotson, M. R. (1999). Distribution of retinogeniculate cells in the tammar wallaby in relation to decussation at the optic chiasm. *Journal of Comparative Neurology*, **405**, 128–140.

Winter, J. W. (1976). The behaviour and social organisation of the brushtailed possum (*Trichosurus vulpecula*: Kerr). PhD thesis. Brisbane: University of Queensland.

Withers, P. C., Richardson, K. C. and Wooller, R. D. (1990). Metabolic physiology of euthermic and torpid honey possums, *Tarsipes rostratus*. *Australian Journal of Zoology*, **37**, 685–693.

Withers, P. C., Cooper, C. E. and Larcombe, A. N. (2006). Environmental correlates of physiological variables in marsupials. *Physiological and Biochemical Zoology*, **79**, 437–453.

Withington, D. J., Mark, R. F., Thornton, S. K., Liu, G. B. and Hill, K. G. (1995). Neural responses to free-field auditory stimulation in the superior colliculus of the wallaby (*Macropus eugenii*). *Experimental Brain Research*, **105**, 233–240.

Witter, M. P. and Amaral, D. G. (2004). Hippocampal formation. In *The Rat Nervous System*. (ed. G. Paxinos) 3rd edn. San Diego, London: Elsevier Academic. pp. 635–704.

Wong, P. and Kaas, J. K. (2009). An architectonic study of the neocortex of the short-tailed opossum (*Monodelphis domestica*). *Brain Behavior and Evolution*, **73**, 206–228.

Wong, P., Cai, H. B., Borchelt, D. R. and Price, D. L. (2002). Genetically engineered mouse models of neurodegenerative diseases. *Nature Neuroscience*, **5**, 633–639.

Wong, R. O. and Hughes, A. (1987). The morphology, number, and distribution of a large population of confirmed displaced amacrine cells in the adult cat retina. *Journal of Comparative Neurology*, **255**, 159–177.

Wong, R. O., Wye-Dvorak, J. and Henry, G. H. (1986). Morphology and distribution of neurons in the retinal ganglion cell layer of the adult tammar wallaby – *Macropus eugenii*. *Journal of Comparative Neurology*, **253**, 1–12.

Wong, R. O., Meister, M. and Shatz, C. J. (1993). Transient period of correlated bursting activity during development of the mammalian retina. *Neuron*, **11**, 923–938.

Wong-Riley, M. T. (1989). Cytochrome oxidase: an endogenous metabolic marker for neuronal activity. *Trends in Neuroscience*, **12**, 94–101.

Woodruff-Pak, D. S. (2008). Animal models of Alzheimer's disease: therapeutic implications. *Journal of Alzheimer's Disease*, **15**, 507–521.

Woolley, P. (1990a). Reproduction in *Sminthopsis macroura* (Marsupialia: Dasyuridae) I. The female. *Australian Journal of Zoology*, **38**, 187–205.

Woolley, P. (1990b). Reproduction in *Sminthopsis macroura* (Marsupialia: Dasyuridae) II. The male. *Australian Journal of Zoology*, **38**, 207–217.

Woolsey, T. and van der Loos, H. (1970). The structural organisation of layer IV in the somatosensory region (SI) of mouse cerebral cortex. The description of a cortical field composed of discrete cytoarchitectonic units. *Brain Research*, **17**, 205–242.

Woolsey, T. A., Welker, C. and Schwartz, R. H. (1975). Comparative anatomical studies of the SmI face cortex with special reference to the occurrence of "barrels" in layer IV. *Journal of Comparative Neurology*, **164**, 79–94.

Wye-Dvorak, J. (1984). Postnatal development of primary visual projections in the tammar wallaby (*Macropus eugenii*). *Journal of Comparative Neurology*, **228**, 491–508.

Wye-Dvorak, J., Levick, W. R. and Mark, R. F. (1987). Retinotopic organisation in the dorsal lateral geniculate nucleus of the tammar wallaby (*Macropus eugenii*). *Journal of Comparative Neurology*, **263**, 198–213.

Xie, Q., Mackay, S., Ullmann, S. L. et al. (1998). Postnatal development of Leydig cells in the opossum (*Monodelphis domestica*): an immunocytochemical endocrinological study. *Biology of Reproduction*, **58**, 664–669.

Xu, H.-P., Chen, M., Manivannan, A., Lois, N. and Forrester, J. V. (2008). Age-dependent accumulation of lipofuscin in perivascular and subretinal microglia in experimental mice *Aging Cell*, **7**, 58–68.

Xu, Z., Marszalek, J. R., Lee, M. K. et al. (1996). Subunit composition of neurofilaments specifies axonal diameter. *Journal of Cell Biology*, **133**, 1061–9.

Yamamoto, Y., Ueta, Y., Hara, Y. et al. (2000). Postnatal development of orexin/hypocretin in rats. *Molecular Brain Research*, **78**, 108–119.

Yamamoto, Y., McKinley, M. J., Nakazato, M. et al. (2006). Postnatal development of orexin-A and orexin-B like immunoreactivities in the eastern grey kangaroo (*Macropus giganteus*) hypothalamus. *Neuroscience Letters*, **392**, 124–128.

Yokoyama, S. and Radlwimmer, F. B. (2001). The molecular genetics and evolution of red and green colour vision in vertebrates. *Genetics*, **158**, 1697–1710.

Yokoyama, S., Radlwimmer, F. B. and Kawamura, S. (1998). Regeneration of ultraviolet pigments of vertebrates. *FEBS Letters*, **423**, 155–158.

Yokoyama, S., Takenaka, N., Agnew, D. W. and Shoshani, J. (2005). Elephants and human colour-blind deuteranopes have identical sets of visual pigments. *Genetics*, **170**, 335–344.

Young, H. M. and Pettigrew, J. D. (1991). Cone photoreceptors lacking oil droplets in the retina of the Echidna, *Tachyglossus aculeatus* (Monotremata). *Visual Neuroscience*, **6**, 409–420.

Zeller, U. (1999). Mammalian reproduction: origin and evolutionary transformations. *Zoologischer Anzeiger*, **238**, 117–131.

Zhang Q. H. and Kelly J. W. C. (2003). ys10 mixed disulfides make transthyretin more amyloidogenic under mildly acidic conditions. *Biochemistry*, **42**, 8756–8761.

Zheng, L-M., Pfaff, D. W. and Schwanzel-Fukuda, M. (1990). Synaptology of luteinizing hormone-releasing hormone (LHRH)-immunoreactive cells in the nervus terminalis of the grey short-tailed opossum (*Monodelphis domestica*). *Journal of Comparative Neurology*, **295**, 327–337.

Zhu, Q. and Julien, J. P. (1999). A key role for GAP-43 in the retinotectal topographic organisation. *Experimental Neurology*, **155**, 228–242.

Zhuang, B-Q. and Sockanathan, S. (2006). Dorsal-ventral patterning: a view from the top. *Current Opinion in Neurobiology*, **16**, 20–24.

Ziehen, T. (1897). Das Centralnervensystem der Monotremen und Marsupialier. Teil I. Makroskopische Anatomie. *Jena Denkschriften*, **6**, 168–187.

Zoghbi, H. Y. and Botas J. (2002). Mouse and fly models of neurodegeneration. *Trends in Genetics*, **18**, 463–471.

Zou, D. J. (1994). Respiratory rhythm in the isolated central nervous system of newborn opossum. *Journal of Experimental Biology*, **197**, 201–213.

Zou, X-C. and Martin, G. F. (1995). The distribution of GAP-43 immunoreactivity in the central nervous system of adult opossums (*Didelphis virginiana*) with notes on their development. *Brain Behavior and Evolution*, **45**, 63–83.

Index to text